Neuromechanical Basis of KINESIOLOGY

SECOND EDITION

Roger M. Enoka, PhD
The Cleveland Clinic Foundation

Human Kinetics

Library of Congress Cataloging-in-Publication Data

Enoka, Roger M., 1949-
Neuromechanical basis of kinesiology / Roger M. Enoka. -- 2nd ed.
p. cm.
Includes bibliographical references and index.
ISBN 0-87322-665-8
1. Kinesiology. 2. Human mechanics. I. Title.
QP303.E56 1994
612.7'6--dc20 93-50146
CIP

ISBN: 0-87322-665-8

Acquisitions Editor: Richard Frey, PhD
Developmental Editor: Larret Galasyn-Wright
Assistant Editors: Sally Bayless, Jacqueline Blakley, Anna Curry, Ed Giles, Julie Lancaster, Dawn Roselund, and Lisa Sotirelis
Copyeditor: Jane Bowers
Proofreaders: Pam Johnson and Pat Wilson
Indexer: Schroeder Indexing Service
Typesetter: Julie Overholt
Text Designer: Keith Blomberg
Layout Artist: Denise Lowry, Denise Peters, Tara Welsch
Cover Designer: Jack Davis
Printer: Braun-Brumfield

Printed in the United States of America 10 9 8 7 6 5 4 3 2

Human Kinetics
P.O. Box 5076, Champaign, IL 61825-5076
1-800-747-4457

Canada: Human Kinetics, Box 24040, Windsor, ON N8Y 4Y9
1-800-465-7301 (in Canada only)

Europe: Human Kinetics, P.O. Box IW14, Leeds LS16 6TR, United Kingdom
(44) 1132 781708

Australia: Human Kinetics, 2 Ingrid Street, Clapham 5062, South Australia
(08) 371 3755

New Zealand: Human Kinetics, P.O. Box 105-231, Auckland 1
(09) 523 3462

To Bonny, Maro, Joel, and Seth

Giovanni Alfonso Borelli (1608-1679)

The unique illustrations that introduce each part and chapter are the work of Giovanni Alfonso Borelli (1608-1679), an Italian physiologist and physicist who was first to explain muscular movement and other body functions according to the laws of mechanics (statics and dynamics). He was appointed professor of mathematics at Messina in 1649 and at Pisa in 1656. In 1674 Borelli moved to Rome where he lived under the protection of Queen Christina of Sweden. His best known work is *De Motu Animalium* (On the Movement of Animals), published in 1680, in which he sought to explain the movement of the animal body on mechanical principles. Based on this work, he is regarded as the founder of the iatrophysical school (iatrophysics is the combination of physics and medicine). Borelli also wrote many astronomical works, including a treatise in 1666 that considered the influence of attraction on the satellites of Jupiter; he was also the first to suggest that comets travel in a parabolic path.

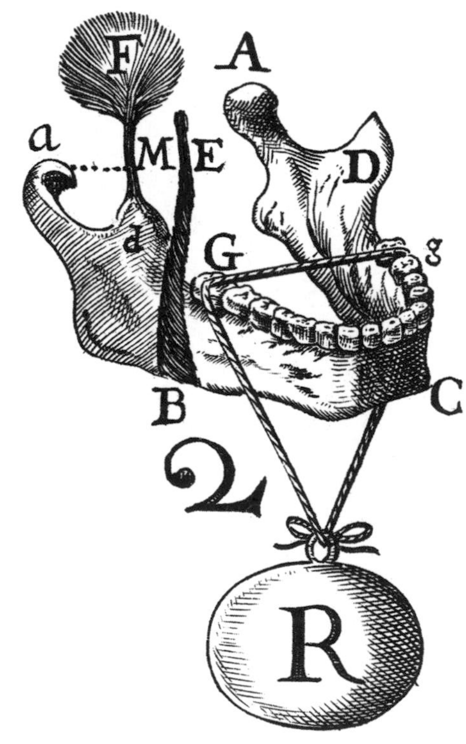

Borelli's work appears on this page and pages 1, 3, 35, 39, 81, 121, 123, 151, 193, 229, 231, 271, and 303 courtesy of The Huntington Library in San Marino, California. The illustrations are from *De Motu Animalium* by Giovanni Alfonso Borelli, 1680, RB 4500, volume 1. We thank The Huntington Library for their generosity in sharing the work of this pioneer in biomechanics.

Contents

Preface to the Second Edition

As a graduate student (1974-1981) in the Department of Kinesiology at the University of Washington, I was greatly influenced by the perspectives of Robert S. Hutton in motor control and Doris I. Miller in biomechanics. One of their innovative ventures was the development of an undergraduate course in kineoenergetics that combined the area of motor control, as it evolved from neurophysiology, with the field of biomechanics. Although this was a rigorous course that demanded the full attention of the students, the focus on human movement generated substantial enthusiasm among the students. I had the opportunity to serve as a teaching assistant for the course and, during the latter part of my graduate training, to serve as the instructor. One difficulty we encountered with this course was finding an adequate textbook. We used *Understanding the Scientific Bases of Human Movement* by Alice L. O'Connell and Elizabeth B. Gardner, but this text was supplemented extensively with material from the research literature and material created by Drs. Hutton and Miller.

In August 1981, I began as an assistant professor in the Department of Physical Education at the University of Arizona. One of my duties was to offer an undergraduate course in kinesiology. In this endeavor, I relied substantially on my experiences at the University of Washington, but I was again confronted with the difficulty of selecting an appropriate textbook. Rather than assign a single textbook, I made several textbooks available to the students and used my lecture notes as the principal resource. Each year, the lecture notes became more extensive. In 1988, Human Kinetics Publishers published this material as the first edition of *Neuromechanical Basis of Kinesiology*.

From 1988 to 1993, I used the first edition as the required textbook for my undergraduate kinesiology course. Over this period, I have received substantial feedback from students at the University of Arizona and from colleagues throughout North America and abroad on the strengths and weaknesses of the book. Because the first edition was largely written to meet my own teaching needs, a number of aspects of the book could have been improved. In writing the second edition, I have attempted to address these concerns by focusing on five goals: (a) to broaden the scope of the book so that it can meet the needs of a broader audience, including those involved in rehabilitation, those principally interested in biomechanics, and those who wish to study the effects of training on the neuromuscular system; (b) to update the material in the book so that it reflects current concepts and opinions in mid 1994; (c) to expand the reference list so that interested readers can pursue individual topics in more detail; (d) to increase the number and types of questions at the end of each chapter; and (e) to expand the graphic content by increasing the number of figures included in each chapter. I hope you will find these features useful and the second edition of *Neuromechanical Basis of Kinesiology* a valuable resource.

Roger M. Enoka
July 1, 1994

Acknowledgments

In the second edition of this text, I have retained the original conceptual organization but have attempted to expand the contents so that it might be useful to a greater number of students of human movement. I have been aided in this endeavor by constructive comments from several colleagues, especially Diana Glendinning (University of Arizona), Mark Grabiner (The Cleveland Clinic Foundation), Joseph Hamill (University of Massachusetts, Amherst), Carol Putnam (Dalhousie University), and Mark Rogers (Northwestern University). In addition, the questions and curiosity of numerous undergraduate students, graduate students, and postdoctoral fellows at the University of Arizona have provided considerable encouragement in the preparation of the second edition.

I must also acknowledge the capable assistance that I received from Peter B. Worden in the production of this text, including the supervision of Aaron E. James, Jr., and Michael Webster, who drew most of the figures.

Introduction

The term *kinesiology* is derived from two Greek verbs, *kinein* and *logos*, which mean to move and to discourse, respectively. Based on this literal definition, kinesiology refers to discourse on movement, or, as in current usage, the study of motion. It is generally agreed that the Greek philosopher Aristotle (384-322 B.C.) was the founder of kinesiology, because several of his treatises were the first to describe the actions of muscles and to subject them to geometrical analysis. Although the system Aristotle devised for explaining motion contained some contradictions, his pioneering efforts laid the foundation for the subsequent work of Galen (131-201), Galileo (1564-1643), Newton (1642-1727), and Borelli (1608-1679). The work of these philosophers and scientists has led us to view human motion as the consequence of the interaction between muscles and the external forces imposed by the surroundings on the system. For as Aristotle wrote, ''The animal that moves makes its change of position by pressing against that which is beneath it'' (Aristotle, 1968; E. Forster, Trans. p. 489). This statement emphasizes that the study of movement must focus on (a) characterizing the physical interaction between an animal and its surroundings and (b) determining the way in which the animal organizes the physical interaction (pressing). Within this framework, we can regard movements as the consequences of an interaction between a biological system and its surroundings. Several factors, including the following, influence this interaction (see also Higgins, 1985):

1. The structure of the environment—Shape and stability
2. The field of external forces—Orientation relative to gravity, movement speed
3. The structure of the system—Bony arrangement, net muscle activity, segmental organization of the body, scale or size, motor integration (such as the need to provide postural support)
4. The role of the psychological state—Degree of attentiveness, motivation
5. The task to be performed—The framework for the organization of the movement

This perspective is well captured by Higgins (1985): ''Movement is inseparable from the structure supporting it and the environment defining it'' (p. 144).

The goal of this text is to examine the neuromechanical basis of kinesiology, where **kinesiology** is defined as the study of movement. The organization of the text is based on the Aristotelian model that movement involves an interaction between an animal and its surroundings. Accordingly, the text has been organized into three parts that examine the mechanical basis of movement (Part I), develop a biological model with which to emphasize the control of movement (Part II), and integrate the mechanical and biological material (Part III).

Part I, ''The Force-Motion Relationship,'' examines the *mechanical* basis of movement and discusses selected principles of physics as they relate to biomechanics and the study of movement.

Part II, ''The Single Joint System,'' develops a *biological* model with which to emphasize the control of movement as an interaction between the nervous system and skeletal muscle.

Part III, ''Adaptability of the Motor System,'' involves an *integration* of the mechanical and biological material and extends the single joint system model to provide a more complete description of the motor system and its acute and chronic adaptive capabilities.

The intent of this text is to provide a scientific basis for the study of human movement. Therefore, the ideas and principles presented are discussed in scientific terms, and more attention is paid to precise definitions and measurements than is commonly done in everyday conversation. Also, this text uses metric units of measurement. Metric units (Appendix A) are preferred because of their common usage and precise definitions.

If we are to advance the study of human movement, those of us studying movement must have as our basis a set of rigorously defined terms and concepts. I hope that this text provides such a foundation. This goal is well illustrated by the analogy from Sherlock Holmes on the following page: How could Holmes know of Watson's intentions? The answer, of course, is that he used his well-known ability to apply deductive reasoning. In a similar vein, movement can be considered the conclusion of a process, and our task, based on rigorously defined terms and concepts, is to identify the intervening steps between the starting point and the conclusion.

THE STRAND MAGAZINE.

Vol. xxvi. DECEMBER, 1903. No. 156.

THE RETURN OF SHERLOCK HOLMES.

By A. CONAN DOYLE.

III.—*The Adventure of the Dancing Men.*

HOLMES had been seated for some hours in silence with his long, thin back curved over a chemical vessel in which he was brewing a particularly malodorous product. His head was sunk upon his breast, and he looked from my point of view like a strange, lank bird, with dull grey plumage and a black top-knot.

"So, Watson," said he, suddenly, "you do not propose to invest in South African securities?"

I gave a start of astonishment. Accustomed as I was to Holmes's curious faculties, this sudden intrusion into my most intimate thoughts was utterly inexplicable.

"How on earth do you know that?" I asked.

He wheeled round upon his stool, with a steaming test-tube in his hand and a gleam of amusement in his deep-set eyes.

"Now, Watson, confess yourself utterly taken aback," said he.

"I am."

"I ought to make you sign a paper to that effect."

"Why?"

"Because in five minutes you will say that it is all so absurdly simple."

"I am sure that I shall say nothing of the kind."

"You see, my dear Watson"—he propped his test-tube in the rack and began to lecture with the air of a professor addressing his class—"it is not really difficult to construct a series of inferences, each dependent upon its predecessor and each simple in itself. If, after doing so, one simply knocks out all the central inferences and presents one's audience with the starting-point and the conclusion, one may produce a startling, though possibly a meretricious, effect. Now, it was not really difficult, by an inspection of the groove between your left forefinger and thumb, to feel sure that you did *not* propose to invest your small capital in the goldfields."

"I see no connection."

"Very likely not; but I can quickly show you a close connection. Here are the missing links of the very simple chain: 1. You had chalk between your left finger and thumb when you returned from the club last night. 2. You put chalk there when you play billiards to steady the cue. 3. You never play billiards except with Thurston. 4. You told me four weeks ago that Thurston had an option on some South African property which would expire in a month, and which he desired you to share with him. 5. Your cheque-book is locked in my drawer, and you have not asked for the key. 6. You do not propose to invest your money in this manner."

"How absurdly simple!" I cried.

"Quite so!" said he, a little nettled. "Every problem becomes very childish when once it is explained to you. Here is an unexplained one. See what you can make of that, friend Watson." He tossed a sheet of paper upon the table and turned once more to his chemical analysis.

Note. From *The Illustrated Sherlock Holmes* (p. 369) by A.C. Doyle, 1985, New York: Clarkson N. Potter, Inc. Copyright 1985 by Clarkson N. Potter, Inc. Reprinted by permission.

PART I

The Force-Motion Relationship

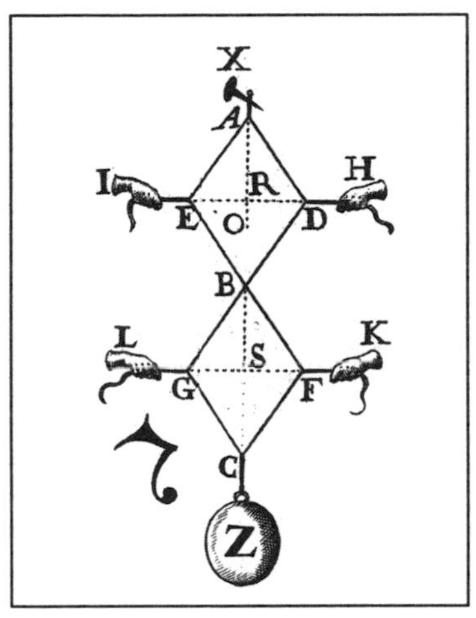

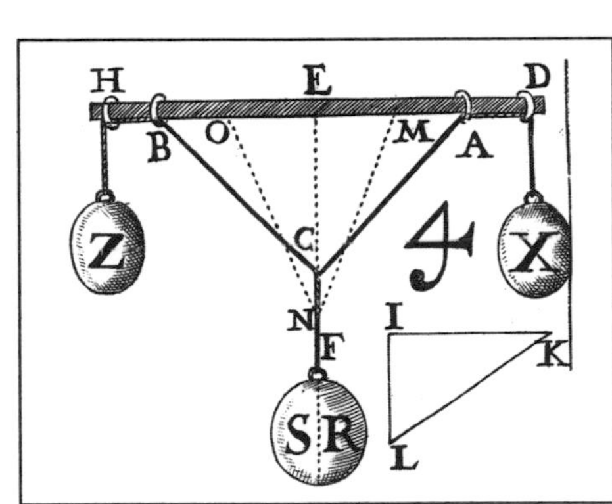

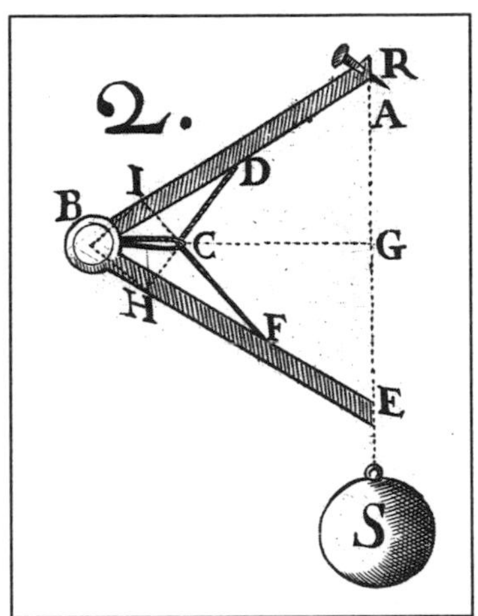

Movement has historically fascinated inquirers from numerous disciplines. But as the frontiers of sciencc have expanded, it has become apparent that movement has as its basis complex biological and mechanical interactions. Many of those who have followed these developments closely have approached the study of movement from neurophysiological and biomechanical perspectives, or a **neuromechanical** focus. Because **biomechanics** is defined as *the application of the principles of mechanics to the study of biological systems*, a neuromechanical focus combines both biology and physics.

Part I provides a mechanical characterization of the interaction between the world in which movement occurs and the body parts that are moved (biomechanics). The discussion includes an introduction of terms and concepts commonly used to describe motion, an examination of the notion of force and its relationship to movement, and the identification of the general techniques used to analyze motion. Although many aspects of the force-motion relationship are illustrated with numerical examples, it is important that students not get lost in the mathematics but rather focus on the concepts. For those students who need assistance with the mathematics, Appendix B provides a brief review of elementary mathematics.

Objectives

The goal of this text is to describe movement as the interaction of a biological model with the physical world in which we live. In Part I, the aim is to define the mechanical bases of movement. The specific objectives are

- to describe movement in precise, well-defined terms;
- to define force and its various effects;
- to consider the role of force in movement; and
- to analyze movement from three different mechanical perspectives.

CHAPTER 1 Motion

Although it is not difficult to appreciate the aesthetic qualities or the difficulty of a movement such as a triple-twisting backward one-and-a-half somersault dive, it is another matter to describe the movement in precise terms. The accurate and precise description of human movement is accomplished by the use of the terms *position*, *velocity*, and *acceleration*. Such a description of motion, one that ignores the causes of motion, is known as a **kinematic** description. These terms are often used in everyday language, but without concern for or even knowledge of their precise meanings. In biomechanics, as in any scientific endeavor, the observations and principles that are elaborated are only as good as the concepts and definitions on which they are based. The complexity of movement makes it important, indeed crucial, that our analyses rely on the rigorous definitions of these three motion descriptors. We must use correctly both the kinematic terms and their units of measurement.

Position, Velocity, and Acceleration

The **position** of an object refers to its location in space relative to some baseline value or axis. For example, the term *3-m diving board* indicates the position of the board above the waterline. Similarly, the height of a high jump bar is specified relative to the ground, the position of the finish line in a race is described with respect to the start, the third and fifth positions in ballet refer to the position of one foot relative to the other, and so on. When an object experiences a change in position, it has been displaced and **motion** has occurred. Motion cannot be detected instantaneously; rather it relies on the comparison of the object's position at one instant in time with its position at another instant. *Motion, therefore, is an event that occurs in* **space** *and* **time**.

When an object is described as experiencing a **displacement**, the reference is to the spatial (space) element of motion, that is, the change in location of the object. Alternatively, an account of both the spatial and temporal (time) elements of motion involves the terms *speed* and *velocity*. The distinction between speed and velocity concerns scalar and vector quantities, respectively (as outlined in Appendix B). Specifically, speed is simply the magnitude of the velocity vector, and as such it has no regard for the change in direction. Speed defines how fast, whereas velocity tells both how fast and in what direction. **Velocity** is defined as the rate of change in position with respect to time. In other words, how rapidly did the change in position occur and in what direction? Because displacement refers to a change in position, velocity can be described as the time rate (derivative) of displacement.

Figure 1.1 represents two observations, separated in time by 3 s, of the vertical position of an object above some baseline value. The change in vertical position over this 3-s period was 2 m; therefore, the rate of change in position was 2 m in 3 s, that is, 2 m/3 s, or 0.67 m/s. Thus the average velocity of the object moving from Position 1 to Position 2 was 0.67 m/s, where m/s refers to meters per second (this unit of measurement can also be expressed as $m \cdot s^{-1}$). Stated more explicitly,

$$\text{velocity} = \frac{\Delta\ \text{position}}{\Delta\ \text{time}} \tag{1.1}$$

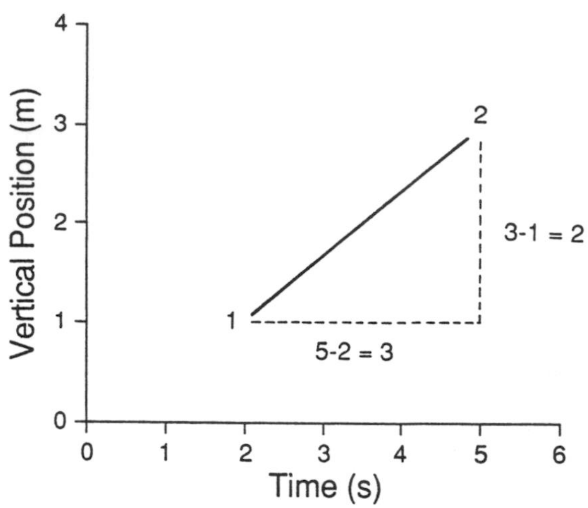

Figure 1.1 A position-time graph.

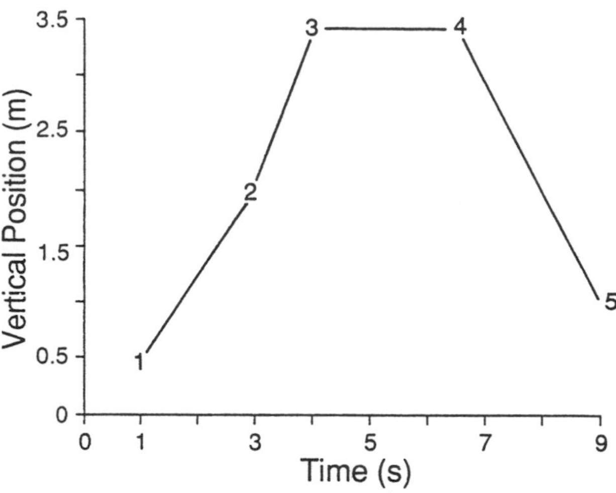

Figure 1.2 The variation in velocity associated with unequal changes in magnitude and direction in a position-time graph.

where Δ (delta) indicates a *change* in some parameter. Graphically, therefore, velocity refers to the slope of the position-time graph. Because a line graph (such as shown in Figure 1.4) depicts the relationship between two (or sometimes more) variables, *a change in the* **slope** *of the line as it becomes more or less steep indicates a change in the relationship between the variables.* The slope of the line, therefore, is determined numerically by subtracting an initial-position value from some final position (Δ position) and dividing the change in position by the amount of time it took for the change to occur (Δ time). Slope, therefore, refers to the rate of change in a variable such that the steeper the slope, the greater the rate of change, and vice versa.

Throughout this text, many concepts are presented in the form of graphs. Typically, a graph shows the relationship between at least two variables. Figure 1.1, for example, shows the relationship between position and time. The relationship can be indicated as the line, or data points, plotted on the graph. The main feature of a graph is to show the trend or pattern of the relationship between the variables; more precise quantitative data are presented as tables or sets of numerical values. In evaluating a graph, first determine the variables involved and then examine the relationship between the variables. The relationship between position and time in Figure 1.1 is relatively straightforward; it can be represented by a single measurement, the slope of the line.

Vertical displacement can vary not only in **magnitude** (i.e., size) but also in **direction** (i.e., up or down). Figure 1.2 illustrates some of these alternatives by noting the position of an object at five instances in time. Use of Equation 1.1 produces velocities of 0.75, 1.50, 0, and −1.00 m/s for movement of the object from Positions 1 to 2, 2 to 3, 3 to 4, and 4 to 5, respectively. The steeper the slope of the position-time graph (e.g., Position 2 to Position 3 = 1.50 m/s vs. Position 1 to Position 2 = 0.75 m/s), the greater the velocity. Conversely, a downward slope (e.g., Position 4 to Position 5) indicates a negative velocity. No change in position (e.g., Position 3 to Position 4) represents a zero velocity (i.e., no change in position). This example illustrates an important point about velocity: When the sign of the velocity value (positive, negative, or zero) changes, the movement has changed direction. Conversely, when the direction of movement changes, the velocity-time graph must pass through zero. Figure 1.2 indicates that an object initially moved in one direction (arbitrarily called the positive direction—note the positive slope), then was stationary, and finally moved in the other direction (negative slope). Because velocity has both magnitude and direction, it is a vector quantity.

It is not sufficient, however, to describe motion only in terms of the occurrence and rate of a displacement. For example, a ball held 1.23 m above the ground and dropped will reach the ground 0.5 s later. The change in position is 1.23 m, and the average velocity is 2.46 m/s (i.e., 1.23 m/0.5 s). But the ball does not travel with a constant velocity; the velocity changes over time. Starting with a zero velocity at release, the speed of the ball increases to 4.91 m/s just prior to contact with the ground. This rate of change in velocity with respect to time is referred to as **acceleration**. The acceleration that the ball experiences while it falls is constant and has a value of 9.81 m/s^2. If the velocity of an object is measured in meters per second (m/s), then acceleration indicates the change in meters per second each second (m/s^2). Consequently,

$$\text{acceleration} = \frac{\Delta\ \text{velocity}}{\Delta\ \text{time}} \tag{1.2}$$

Acceleration, therefore, represents the change in velocity with respect to time, or the change in the rate of change in position with respect to time. A running back in football who is described as having good acceleration can change velocity (speed and direction) quickly.

Because velocity is defined as the rate of change in position with respect to time, velocity can be represented graphically as the slope of the position-time graph. Similarly, acceleration, defined as the rate of change in velocity with respect to time, can be represented as the slope of the velocity-time graph. Thus, Figure 1.2 could be relabeled from a vertical position-time graph to a vertical velocity-time graph, and the relationships identified for the position-time curve and velocity would apply similarly to the velocity-time profile and acceleration. That is, acceleration refers to the slope of the velocity-time graph. For example, suppose Figure 1.2 were relabeled as a vertical velocity-time graph. If point 2 had the coordinates of 2.0 m/s and 3 s, and point 3 had the coordinates of 3.5 m/s and 4 s, the rate of change from point 2 to point 3, the acceleration, could be calculated with Equation 1.2 as follows:

$$\text{acceleration} = \frac{3.5 - 2.0}{4 - 3} \; \frac{\text{m/s}}{\text{s}}$$

$$= 1.5 \text{ m/s/s (or m/s}^2 \text{ or m}\cdot\text{s}^{-2})$$

Similarly, the acceleration from points 1 to 2, 3 to 4, and 4 to 5 would be 0.75, 0, and −1.00 m/s^2, respectively. As with velocity, acceleration can have both magnitude and direction and hence is a vector quantity.

The acceleration experienced by the ball in the previous example was due to the gravitational attraction between two masses, planet Earth and the ball. The force of **gravity** produces a constant acceleration of approximately 9.81 m/s^2 at sea level. In general, an object acted upon by a force experiences an acceleration. A constant force (i.e., gravity) applied to an unsupported object produces a constant acceleration; conversely, the absence of a force means that the object is at rest or is traveling at a constant velocity (i.e., no acceleration). What would the velocity-time graph look like for constant acceleration and for zero acceleration? Because acceleration can be depicted as the slope of the velocity-time graph, it should be possible to visualize the shape of the velocity-time graph when an acceleration is present and when it is absent. When acceleration is present, the slope of the velocity-time graph is nonzero. Conversely, when acceleration is absent, the velocity-time graph has a zero slope.

In this discussion of kinematic variables, careful attention has been paid to the units of measurement. In the study of biomechanics, the units of measurement can be a useful tool in understanding concepts related to the mechanical analysis of movement. Similarly, identifying a criterion measure for each parameter can help us appreciate the magnitude of a variable. For example, if a movement is performed at an average velocity of 2.9 m/s, is this fast or slow? Although the answer depends on the details of the movement, it is useful to know that a 100-m race completed in 10 s has an average velocity of 10 m/s. This reference value can be compared with other velocity values. Similarly, acceleration values can be compared with the acceleration produced by gravity (9.81 m/s^2). In contrast, it is not particularly useful to remember more than a few conversion factors because these can always be looked up (see Appendix C).

Equations of Motion

From the elementary definitions of velocity (Equation 1.1) and acceleration (Equation 1.2), we can derive algebraic expressions involving time (t), position ($\mathbf{r}$), velocity ($\mathbf{v}$), and acceleration ($\mathbf{a}$) that are often used to describe motion. In this section, we derive three equations of motion that are valid for conditions where acceleration is constant. (Throughout the text, symbols representing vector terms—e.g., position, velocity, acceleration—are boldfaced.) In these equations of motion, $\mathbf{v}_i$ and $\mathbf{v}_f$ refer to initial and final velocities, respectively, and similarly $\mathbf{r}_i$ and $\mathbf{r}_f$ refer to initial and final positions. The terms *initial* and *final* refer to the values at the beginning and end of an epoch.

1. *Express final velocity as a function of initial velocity, acceleration, and time* ($t = \Delta$ time).

$$\text{average acceleration} = \frac{\Delta \text{ velocity}}{\Delta \text{ time}}$$

$$\mathbf{a} = \frac{\mathbf{v}_f - \mathbf{v}_i}{t}$$

$$\mathbf{v}_f = \mathbf{v}_i + \mathbf{a}t \qquad (1.3)$$

For example, consider the dropped ball example. The ball was dropped from a height of 1.23 m, and 0.5 s later it reached the ground with a final velocity of 4.91 m/s. According to Equation 1.3, the variables that affect the final velocity are the initial velocity ($\mathbf{v}_i$) of the ball, its acceleration ($\mathbf{a}$), and the duration of the fall (t). In this example, $\mathbf{v}_i$ was zero, $\mathbf{a}$ was that due to gravity (9.81 m/s^2), and t was 0.5 s. These values can be inserted into Equation 1.3 to verify the final velocity. Suppose we wanted to determine the final velocity of a pitched baseball as it crossed the plate. What would we need to know? As before, we would need to know $\mathbf{v}_i$, $\mathbf{a}$, and t. The major difficulty would be in determining $\mathbf{a}$, because other forces, in addition to gravity, would be acting on the ball. Typically, $\mathbf{a}$ is determined by deriving it numerically from film or video images of a movement. Alternatively, $\mathbf{a}$ can be measured directly with an instrument known as an accelerometer.

2. *Derive an expression for position in terms of initial and final velocity, acceleration, and time.*

$$\text{average velocity} = \frac{\Delta \text{ position}}{\Delta \text{ time}}$$

$$\frac{\mathbf{v}_f + \mathbf{v}_i}{2} = \frac{\mathbf{r}_f - \mathbf{r}_i}{t}$$

Substitute Equation 1.3 for $\mathbf{v}_f$:

$$\mathbf{r}_f - \mathbf{r}_i = \frac{\mathbf{v}_i + \mathbf{a}t + \mathbf{v}_i}{2}\, t$$

$$\mathbf{r}_f - \mathbf{r}_i = \frac{2\mathbf{v}_i + \mathbf{a}t}{2}\, t$$

$$\mathbf{r}_f - \mathbf{r}_i = \mathbf{v}_i t + \frac{1}{2}\mathbf{a}t^2 \qquad (1.4)$$

Equation 1.4 indicates that the change in position of an object (or the distance that the object travels from one point in time to another, provided that direction does not change) depends on three variables: its initial velocity ($\mathbf{v}_i$), the acceleration ($\mathbf{a}$) it experiences, and time (t). In other words, the object's position will be different if any of these variables changes. This relationship can be used to determine how the position of an object changes as time varies. For example, consider an individual who is diving off a 10-m tower (Problem 119); by varying the value of t from 0 to 1.5 s in 0.1-s increments, we can determine the trajectory (position-time graph) of the diver during the performance. The initial velocity of the diver is zero, so Equation 1.4 reduces to

$$\mathbf{r}_f - \mathbf{r}_i = \frac{1}{2}\mathbf{a}t^2$$

for this problem. If we assume that the effects of air resistance are so small that we can ignore them, then the acceleration is simply that due to gravity. The set of position-time data (as in Table 1.1) can be determined by doing several calculations using the above equation and incrementing the value of t by 0.1 s each time.

3. *Relate final velocity to initial velocity, acceleration, and position.*

$$\text{average velocity} = \frac{\Delta \text{ position}}{\Delta \text{ time}}$$

$$\frac{\mathbf{v}_f + \mathbf{v}_i}{2} = \frac{\mathbf{r}_f - \mathbf{r}_i}{t}$$

In this condition, t is unknown, so we rearrange Equation 1.3 to express t as the dependent variable [$t = (\mathbf{v}_f - \mathbf{v}_i)/\mathbf{a}$] and substitute for t in the previous expression:

$$\frac{\mathbf{v}_f + \mathbf{v}_i}{2} = \frac{\mathbf{r}_f - \mathbf{r}_i}{(\mathbf{v}_f - \mathbf{v}_i)/\mathbf{a}}$$

$$\frac{\mathbf{v}_f + \mathbf{v}_i}{2} = (\mathbf{r}_f - \mathbf{r}_i)\,\frac{\mathbf{a}}{\mathbf{v}_f - \mathbf{v}_i}$$

$$2\mathbf{a}(\mathbf{r}_f - \mathbf{r}_i) = (\mathbf{v}_f + \mathbf{v}_i)(\mathbf{v}_f - \mathbf{v}_i)$$

$$2\mathbf{a}(\mathbf{r}_f - \mathbf{r}_i) = (\mathbf{v}_f^2 - \mathbf{v}_i^2)$$

$$\mathbf{v}_f^2 = \mathbf{v}_i^2 + 2\mathbf{a}(\mathbf{r}_f - \mathbf{r}_i) \qquad (1.5)$$

As with both Equations 1.3 and 1.4, Equation 1.5 illustrates that a kinematic variable (final velocity in this case) can be determined from three other parameters. For example, the speed of a bullet at some known distance from a gun (e.g., 10 m) depends on the velocity of the bullet as it leaves the gun (initial velocity), the acceleration the bullet experiences (i.e., due to air resistance), and the specified distance (10 m in this example).

When initial velocity is zero (e.g., the object in motion began at rest), the equations are further simplified:

$$\mathbf{v}_f = \mathbf{a}t$$

$$\mathbf{r}_f - \mathbf{r}_i = \frac{1}{2}\mathbf{a}t^2$$

$$\mathbf{v}_f^2 = 2\mathbf{a}(\mathbf{r}_f - \mathbf{r}_i)$$

EXAMPLE

A juggler performs in a room with a ceiling that is 2 m above his hands. He throws a ball vertically so that it just reaches the ceiling. Figure 1.3 shows a sketch of the initial conditions.

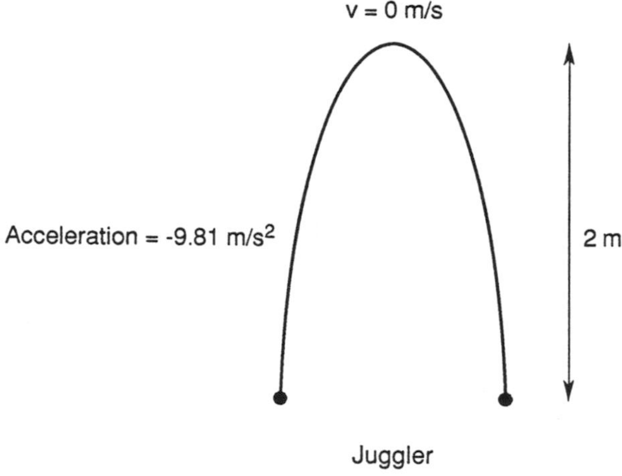

Figure 1.3 Initial conditions for the juggler example.

A. With what initial vertical velocity does the juggler throw the ball ($\mathbf{v}_f = 0$)?

$$\mathbf{v}_f^2 = \mathbf{v}_i^2 + 2\mathbf{a}(\mathbf{r}_f - \mathbf{r}_i)$$

$$\mathbf{v}_i^2 = 0 - [2 \times (-9.81) \times 2]$$

$$\mathbf{v}_i = \sqrt{39.2}$$

$$\mathbf{v}_i = 6.3 \text{ m/s}$$

B. How long does it take for the ball to reach the ceiling?

$$\mathbf{v}_f = \mathbf{v}_i + \mathbf{a}t$$

$$t = \frac{\mathbf{v}_f - \mathbf{v}_i}{\mathbf{a}}$$

$$t = \frac{-6.3}{-9.81}$$

$$= 0.64 \text{ s}$$

C. The juggler throws up a second ball with the same initial velocity at the instant that the first ball is at the ceiling. How long after the second ball is thrown do the balls pass each other? The answer to this question can be determined with careful thought rather than the rampant application of equations. Begin by sketching the initial conditions. The trajectories of the two balls are essentially identical. The first ball has a vertical velocity at the peak of its trajectory of 0 m/s and a final velocity (at the moment it is caught) of −6.3 m/s; the second ball has the converse values of 6.3 m/s at release and 0 m/s at its peak. Because the trajectory is parabolic, it is symmetrical. Therefore, the balls will meet halfway, or at 0.32 s, which is $t/2$. Incidentally, the velocity of the balls as they pass will be about 3.1 m/s.

D. When the balls pass each other, how far are they above the juggler's hands?

$$\mathbf{r}_f - \mathbf{r}_i = \mathbf{v}_i t + \frac{1}{2}\mathbf{a}t^2$$

$$\Delta\mathbf{r} = (6.3 \times 0.32) + 0.5[-9.81 \times (0.32)^2]$$

$$= 2.02 - 0.50$$

$$\Delta\mathbf{r} = 1.5 \text{ m above the hands}$$

Numerical Calculation

Kinematic analyses are usually based on a set of position-time data that are obtained with a recording device such as a movie camera or a video camera. A film or video record of a movement is a set of still images (frames) that are subsequently projected individually onto a measuring device, and the locations of selected landmarks with respect to some reference are determined. The instrument used in this procedure, called a digitizer, is capable of determining the (x,y) coordinates of the selected landmarks. Once we have a set of position-time data, we can use standard numerical analysis techniques to determine the associated velocity- and acceleration-time data. Table 1.1 provides an example of this procedure. Additional sets of position-time data for a variety of movements are provided in Appendix E.

Table 1.1 presents a set of position-time data and the derived velocity- and acceleration-time data for an abstract movement. The 13 position values, each recorded at a different instant in time, represent the vertical path that an object travels over a 1-s epoch. The object first rises above an initial position (0.0 m) to a height of 1.0 m before being displaced by an equal amount (−1.0 m) below the original position and finally returning to 0.0 m. The velocity of the object during this motion is calculated by applying Equation 1.1 to selected intervals of time for which position information is available. For example, from Table 1.1 we could select the intervals of 0.0 to 1.0 s, 0.0 to 0.25 s, or 0.0 to 0.1 s. If we applied Equation 1.1 to each of the intervals, the average velocity for each interval would be

$$0.0 \text{ to } 1.0 = \frac{0.0 - 0.0}{1.0 - 0.0}$$

$$= 0 \text{ m/s}$$

$$0.0 \text{ to } 0.25 = \frac{1.0 - 0.0}{0.25 - 0.0}$$

$$= 4 \text{ m/s}$$

$$0.0 \text{ to } 0.1 = \frac{0.59 - 0.0}{0.1 - 0.0}$$

$$= 5.9 \text{ m/s}$$

Similarly, if we measured the position at only the times of 0.0, 0.5, and 1.0 s, we would get the impression that the object had not moved. Clearly, the smaller the intervals of time we measure, the closer the calculated velocity will match that experienced by the object. However, economy of effort suggests that we do not want to measure too frequently and that there must be some intermediate value. Most human movements can be measured adequately with frame rates that range from 50 to 100 frames per second, which correspond to intervals of 0.1 to 0.2 s between consecutive data points.

In Table 1.1, velocity has been determined for each interval over which position data were recorded. For example, the displacement during the first interval (0.59 − 0.0 = 0.59 m) is divided by the time elapsed during the interval (0.1 − 0.0 = 0.1 s) to yield the velocity for that interval (0.59/0.1 = 5.9 m/s). The calculated value (5.9 m/s) represents the *average* velocity over that interval and consequently is recorded at the midpoint in time of the interval (0.05 s). Similarly, the first acceleration value (−23.0 m/s^2), which is determined with Equation 1.2, is listed at the midpoint in time (0.10 s) of the first velocity interval (0.05 to 0.15). By this procedure, *the average value of velocity is determined for each position interval, and the average acceleration is calculated for each velocity interval.* Thus, from a set of 13 position-time observations, 12 velocity-time and 11 acceleration-time values are calculated.

The graphical relationship between a motion descriptor (e.g., position, velocity) and its rate of change has already been mentioned. The slopes of the position- and

Table 1.1 Calculation of Velocity and Acceleration From a Set of Position-Time Data

Position (m)	Time (s)	Velocity (Δ position/Δ time) (m/s)	Acceleration (Δ velocity/Δ time) (m/s²)
0.00	0.000		
	0.050	(0.59 – 0.00)/(0.100 – 0.000) = 5.9	
0.59	0.100		(3.6 – 5.9)/(0.150 – 0.050) = –23.0
	0.150	(0.95 – 0.59)/(0.200 – 0.100) = 3.6	
0.95	0.200		(1.0 – 3.6)/(0.255 – 0.150) = –34.7
	0.225	(1.00 – 0.95)/(0.250 – 0.200) = 1.0	
1.00	0.250		(–1.0 – 1.0)/(0.275 – 0.225) = –40.0
	0.275	(0.95 – 1.00)/(0.300 – 0.250) = –1.0	
0.95	0.300		(–3.6) – [–1.0])/(0.350 – 0.275) = –34.7
	0.350	(0.59 – 0.95)/(0.400 – 0.300) = –3.6	
0.59	0.400		(–5.9 – [–3.6])/(0.450 – 0.350) = –23.0
	0.450	(0.00 – 0.59)/(0.500 – 0.400) = –5.9	
0.00	0.500		(–5.9 – [–5.9])/(0.550 – 0.450) = 0.0
	0.550	(–5.9 – 0.00)/(0.600 – 0.500) = –5.9	
–0.59	0.600		(–3.6 – [–5.9])/(0.650 – 0.550) = 23.0
	0.650	(–0.95 – [–0.59])/(0.700 – 0.600) = –3.6	
–0.95	0.700		(–1.0 – [–3.6])/(0.725 – 0.650) = 34.7
	0.725	(–1.00 – [–0.95])/(0.750 – 0.700) = – 1.0	
–1.00	0.750		(1.0 – [–1.0])/(0.775 – 0.725) = 40.0
	0.775	(–0.95 – [–1.00])/(0.800 – 0.750) = 1.0	
–0.95	0.800		(3.6 – 1.0)/(0.850 – 0.775) = 34.7
	0.850	(–0.59 – [–0.95])/(0.900 – 0.800) = 3.6	
–0.59	0.900		(5.9 – 3.6)/(0.950 – 0.850) = 23.0
	0.950	(0.00 – [–0.59])/(1.000 – 0.900) = 5.9	
0.00	1.000		

velocity-time graphs represent velocity and acceleration, respectively. Further evidence of these relationships is provided in Table 1.1. When position increases (Positions 1 to 4 and 10 to 13), velocity is positive; when position decreases (Positions 4 to 10), velocity is negative. A similar dependency exists between the slope (increase or decrease) of velocity and the sign of the acceleration values. These relationships are shown in Figure 1.4.

Graphical Relationship

Using these graphical associations, we can estimate the rate of change in a kinematic variable from the shape of the graph for that variable as it changes with time. Figure 1.5 illustrates this process for the changes in thigh angle of a skilled runner for one stride (defined as one complete cycle, from left-foot takeoff to left-foot takeoff in this example). The thigh angle is measured with respect to the right horizontal, and its measurement is indicated in the upper panel of the figure (the measured angle is for the limb with the filled-in shoe). The angle is measured in radians (1 rad = 57.3°), the SI unit for angle (see Appendix A for a list of SI units).

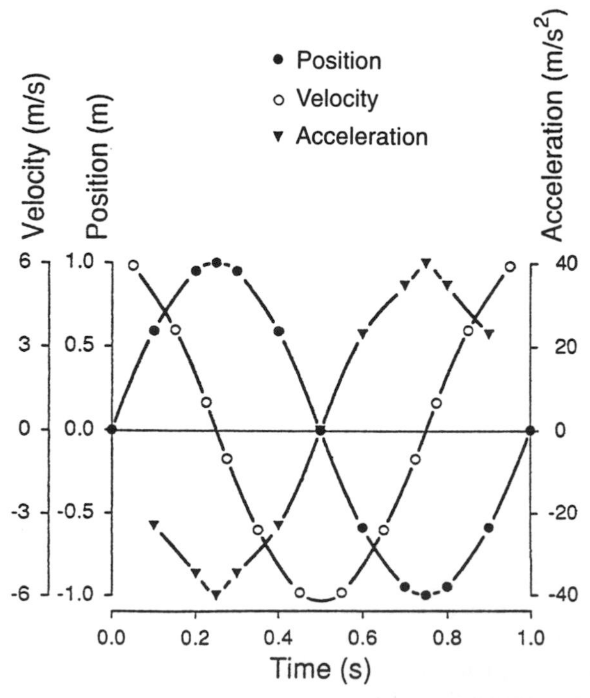

Figure 1.4 Graph of the kinematic data derived in Table 1.1.

The first step in deriving the velocity-time graph from the position-time graph is to identify the relative **minima** and **maxima**. Any—and there may be several—peaks and valleys in the curve should be noted. These points denote instants at which the rate of change has a value of zero. That is, the slope of the graph is neither upward (positive) nor downward (negative) but zero. In Figure 1.5, the thigh angle-time function has one minimum and one maximum. From these points of zero slope, and thus zero velocity, a perpendicular line is extended to the time axis of the velocity-time graph to mark the locations in time of zero velocity. In Figure 1.5 these occur at about 0.03 s and 0.30 s, respectively.

The second step is to determine the slope of the position-time graph between these maxima and minima. The slope of the graph will be the same (i.e., positive or negative) between these points because, as points of zero velocity, they identify the locations in time when the position-time curve changes its slope (i.e., changes direction). In each interval between the minima and maxima, the slope may become more or less steep, but it will remain either upward (positive) or downward (negative). In the thigh angle-time figure, there is one minimum and one maximum and, therefore, three such intervals (i.e., from beginning of the movement to the minimum, from the minimum to the maximum, and from the maximum to the end of the movement). The slopes of the position-time graph associated with these intervals are negative, positive, and negative, respectively. Thus the velocity-time graph has values (positive or negative) similar to the slope of the thigh angle-time function for each interval. For example, for the first interval, from the beginning of the movement to the minimum, both the position-time slope and the velocity values are negative. Because a negative velocity value is associated with a downward position-time slope, the negative velocities in Figure 1.5 indicate a backward rotation of the thigh (i.e., a reduction in the measured angle). In total, the velocity-time graph of Figure 1.5 indicates two intervals of backward thigh rotation separated in time by an interval of forward thigh rotation. The variation in the magnitude of the velocity over time indicates how the speed of this rotation varies, whereas the sign (positive or negative) indicates the direction (forward or backward) of rotation.

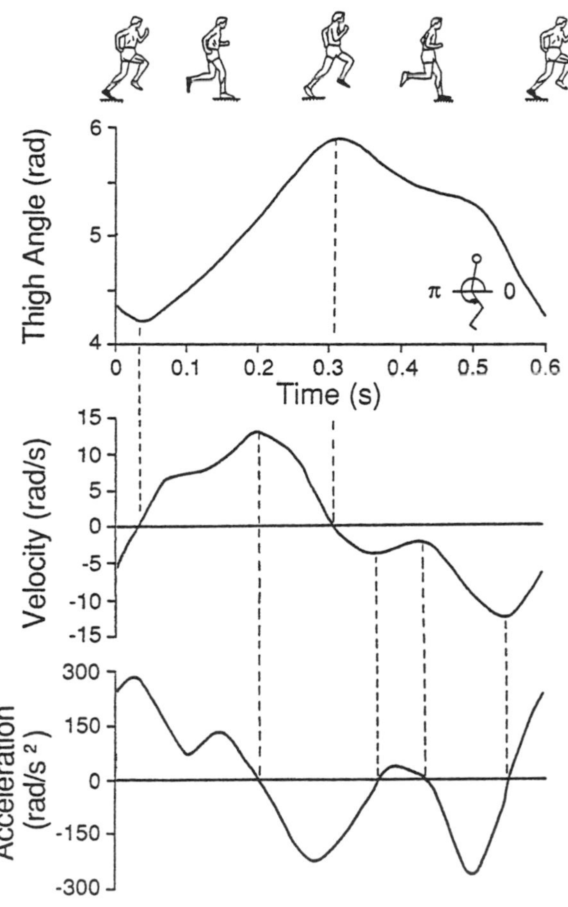

Figure 1.5 The angular velocity-time profile can be graphically derived from the angular position-time graph, and the angular acceleration-time profile can be derived from the angular velocity-time graph.

The derivation of the acceleration-time relationship from the velocity-time graph is accomplished by the same two-stage procedure: (a) identification of the relative minima and maxima and (b) determination of the slope during the identified intervals. From Figure 1.5, the velocity-time curve contained four minima and maxima, and thus there were four instances at which the acceleration-time graph has to cross zero. The resulting acceleration-time relationship was a five-interval alternating positive-negative curve. The interpretation of an acceleration-time graph is generally more complicated than position- and velocity-time graphs. In Figure 1.5, a positive acceleration indicates an acceleration of the thigh in the forward direction; during the first acceleration interval, the thigh rotated first backward and then forward (seen from the velocity-time graph), but throughout the interval the thigh experienced an acceleration in the forward direction (recall the proportionality between acceleration and force). This concept is important, and we will return to it several more times; in general, an object (or body segment) may be moving in one direction (positive or negative velocity) yet experiencing an acceleration in the opposite direction. *It is not possible to tell the direction of acceleration from the direction of movement.*

For example, consider the motion of a ball that a juggler tosses into the air and then catches. The motion of the ball can be shown on a position-time graph in which position is represented as the vertical position above the juggler's hand (Figure 1.6). When drawn this way, the motion of the ball appears as a bell-shaped graph (actually, it is parabolic, as we will discuss later). What is the shape of the acceleration-time graph? The answer is simple. We could determine the answer using the technique shown in Figure 1.5, but an object in free fall, such as a ball tossed into the air, experiences a constant acceleration of 9.81 m/s^2. Although the ball

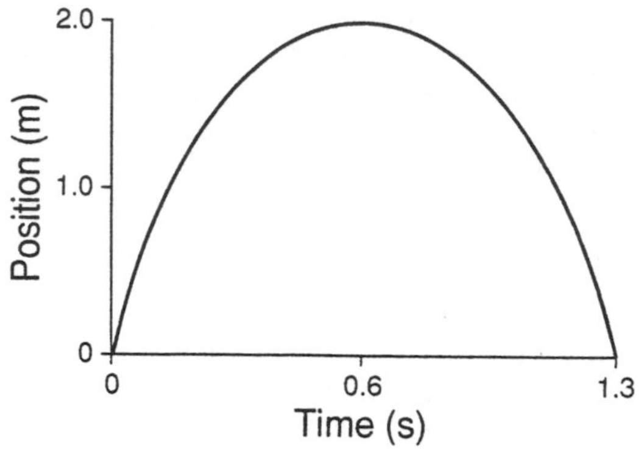

Figure 1.6 The trajectory of a ball tossed by a juggler. Zero position indicates the position of the juggler's hands.

moved up and down (Figure 1.6), this displacement does not provide any intuitive clue about the direction of the acceleration that the ball experienced.

By the graphical technique outlined in this section, we can determine the precise magnitude of the rate of change. This procedure is **qualitative** in nature; it merely gives a positive or negative sign for the rate of change and possibly an approximate value. In contrast, Table 1.1 indicates a **quantitative** approach by which the values of the derivatives can be determined more accurately. A qualitative analysis tells us what type or kind, whereas a quantitative analysis tells us how much.

Linear and Angular Motion

In the preceding discussion, the magnitude of displacement was indicated with either of two units of measurement, meters (m) or radians (rad); the distinction between the two is that of **linear** and **angular** motion, respectively. Linear motion refers to an equivalent displacement in space of all the parts of the object. Conversely, when all the parts of the object do not experience the same displacement, the object has rotated, and hence we describe its motion as angular. A combination of linear (**translation**) and angular (**rotation**) motion in a single plane is called **planar** motion and involves rotation about a point that is itself moving. When such motion occurs in more than one plane, we talk of general motion (three-dimensional). In most human movement, body segments undergo both linear and angular motion.

A **meter**, the unit of measurement for linear motion, is defined as the wavelength of one line in an isotope of the element krypton. A **radian** (Appendix B) is the ratio of a distance on the circumference of the circle to the radius of the circle. When this distance on the circumference equals the radius, the ratio has a value of one and the object has rotated 1 rad (57.3°). For example, consider a discus being thrown; the length of the thrower's arm from the shoulder to the discus is 63 cm. As the arm rotates about the shoulder joint, the discus moves in a circular path. When the discus has moved along its path 63 cm (equal to the length of the arm and thus equal to the radius of the circle), the arm and the discus have been rotated through the angle of 1 rad.

If the intent in measuring is to describe rotary motion, then angular units are appropriate; otherwise linear terms should be used. The commonly used symbols and associated units of measurement for linear and angular position, velocity, and acceleration are outlined in Table 1.2. As indicated, the symbols are usually Latin letters for linear terms and Greek for angular.

Table 1.2 Linear and Angular Symbols and the SI Units of Measurement

	Linear			Angular	
	Symbol		Unit	Symbol	Unit
	Scalar	Vector			
Position	r	$\mathbf{r}$	m	θ (theta)	rad
Displacement	Δr	$\Delta\mathbf{r}$	m	Δθ (theta)	rad
Velocity	v	$\mathbf{v}$	m/s	ω (omega)	rad/s
Acceleration	a	$\mathbf{a}$	m/s^2	α (alpha)	rad/s^2

Angle-Angle Diagrams

In measuring human movement, we usually graph some variable (e.g., thigh angle, ball height) against time. But because human movement is accomplished by the rotation of body segments about one another, it is often more revealing to examine the relationship between two angles during a movement. Such a graph, called an **angle-angle diagram** (Cavanagh & Grieve, 1973), usually plots a relative angle (i.e., the angle between two adjacent body segments) against the absolute angle of a body segment (i.e., the angle relative to a reference in the surroundings).

Figure 1.7 shows two angle-angle diagrams of part of a weight-lifting movement (the clean and jerk) in which the barbell was lifted from Position 1 through Position 10. The knee-trunk diagram shows that the movement comprises three distinct phases: (a) Positions 1 to 5, extension of the knee joint and slight forward rotation of the trunk; (b) Positions 5 to 8, backward rotation of the trunk and flexion of the knees; and (c) Positions 8 to 10, knee-joint extension and some backward-forward trunk rotation. Similarly, the ankle-thigh diagram comprises three phases: (a) forward thigh rotation-ankle plantarflexion; (b) constant thigh angle-ankle

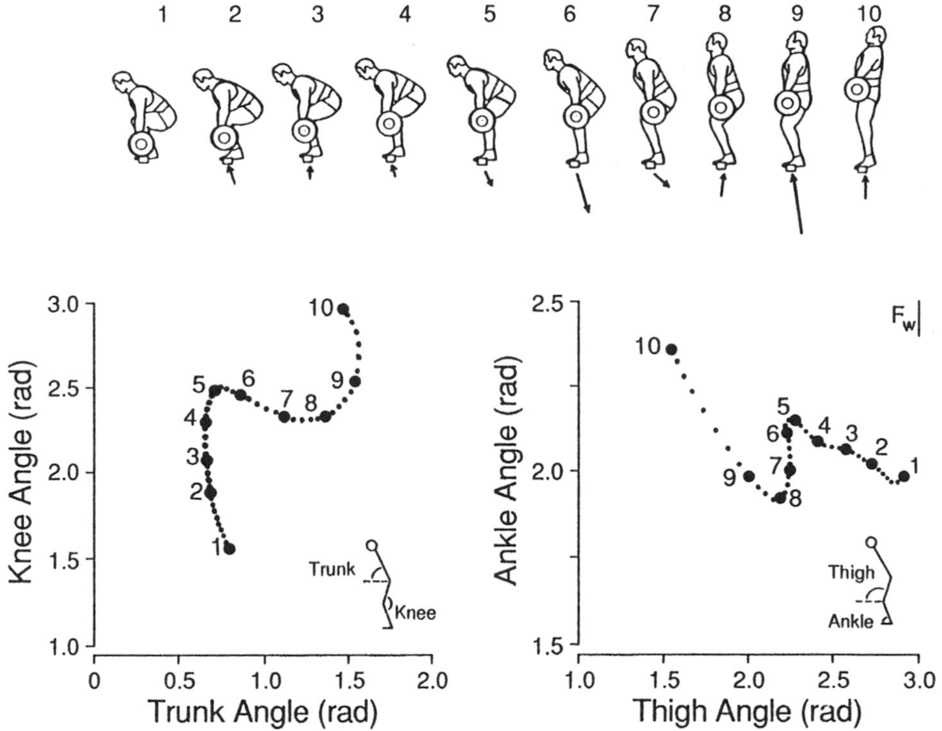

Figure 1.7 Angle-angle relationships during the first part of a weight-lifting event. The numbers in the diagrams correspond to the positions of the lifter shown at the top of the figure. The arrows beneath the figures indicate the net force acting on the lifter ($\mathbf{F}_w$ = body weight).

Note. From "Load- and Skill-Related Changes in Segmental Contributions to a Weightlifting Movement" by R.M. Enoka, 1988, *Medicine and Science in Sports and Exercise*, **20**, pp. 178-187. Copyright 1988 by Williams and Wilkins. Reprinted by permission.

dorsiflexion; and (c) forward thigh rotation-ankle plantarflexion. Figure 1.7 shows the extent to which the three phases for the two angle-angle diagrams coincide; the movement is accomplished by coordinated displacements about the lower extremity joints. In Figure 1.7 there is a constant 10-ms interval between each dot; so the greater the distance between the dots, the greater the velocity of movement.

Angular Kinematics

Because human movement typically involves both translation and rotation, it is necessary to know the relationships between the linear and angular measures of position, velocity, and acceleration. When a rigid body of fixed length (r) rotates about a point from Position 1 to Position 2 (Figure 1.8a), the displacement (s) experienced by the end of the rigid body is given by Equation 1.6:

$$s = r\theta \tag{1.6}$$

The linear velocity (v) of the end of the rigid body is determined as the rate of change in s:

$$\frac{\Delta s}{\Delta t} = \frac{\Delta(r\theta)}{\Delta t}$$

Because r has a fixed magnitude and does not change over time, this expression reduces to

$$\frac{\Delta s}{\Delta t} = \frac{r\Delta\theta}{\Delta t}$$

$$\mathbf{v} = r\omega \tag{1.7}$$

Equation 1.7 indicates that the linear velocity (**v**) of any point on a rigid body is equal to the product of the distance from the axis of rotation to that point (r) and the angular velocity of the rigid body (ω). For different points along a rigid body, therefore, both r and v vary. As anyone who has been part of a rotating human chain on skates knows, the person farthest from the axis of rotation experiences the greatest linear velocity. Furthermore, the direction of the linear velocity vector (**v**) is always tangent to the path of the rigid body (Figure 1.8b).

The variables, **r**, **v**, and ω are vectors and have both magnitude and direction. The magnitude and direction of **r** and **v** are straightforward, as shown in Figure 1.8b.

When a vector is shown as a curved arrow, its direction is actually perpendicular to the page. In Figure 1.8b, ω is directed toward you and away from the page; we will discuss why later. Equation 1.7 is the scalar form of the relationship between linear and angular velocity. The vector relationship is indicated in Equation 1.8:

$$\mathbf{v} = \omega \times \mathbf{r} \quad (1.8)$$

Equation 1.8 states that linear velocity (**v**) is equal to the cross product (×) of angular velocity (ω) and position (**r**). The cross product is a vector operator that is used to multiply vectors, the result (product) of which is a vector that is perpendicular to the plane of the original vectors.

In order to determine the relationship between linear and angular acceleration, we need to use Equation 1.8 and to account for the change in both magnitude and direction of each vector:

$$\frac{\Delta \mathbf{v}}{\Delta t} = \frac{\Delta(\omega \times \mathbf{r})}{\Delta t}$$

$$\mathbf{a} = (\omega \times \frac{\Delta \mathbf{r}}{\Delta t}) + (\frac{\Delta \omega}{\Delta t} \times \mathbf{r})$$

$$= (\omega \times \mathbf{v}) + (\alpha \times \mathbf{r})$$

$$\mathbf{a} = \omega \times (\omega \times \mathbf{r}) + (\alpha \times \mathbf{r})$$

$$\mathbf{a} = \mathbf{r}\omega^2 + \mathbf{r}\alpha \quad (1.9)$$

The term $\mathbf{r}\omega^2$ accounts for the change in direction of **v**, and the term $\mathbf{r}\alpha$ represents the change in magnitude of **v** (Figure 1.8c). Because the direction of **v** changes during angular motion, $\mathbf{r}\omega^2$ is never zero, but $\mathbf{r}\alpha$ may be zero if the magnitude of **v** is constant.

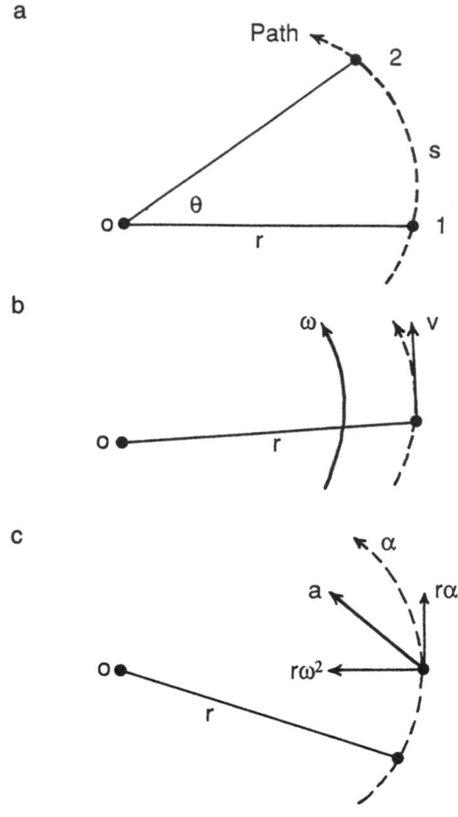

Figure 1.8 Relationships between the linear and angular motion of a rigid body of fixed length: (a) position; (b) velocity; (c) acceleration.

Acceleration and Muscle Activity

To pursue further the relationship between acceleration and force, consider the following example of an elbow extension-flexion movement in a horizontal plane passing through the shoulder joint (Figure 1.9). The movement begins with the upper arm raised to the side so that it is horizontal with an angle of 0.70 rad (40°) between the upper arm and the forearm. In one continuous movement of moderate speed, the upper arm is held stationary while the elbow joint is extended horizontally to 3.14 rad (180°) and flexed back to the starting position (0.70 rad). In general, **flexion** at a joint results in a decrease of the angle between two adjacent body segments that meet at the joint (upper arm and forearm in this example), whereas **extension** refers to an increase in the angle. What would be the shape of the position-time graph associated with this movement? The upper arm remains stationary while the forearm rotates about the elbow joint; a graph of elbow angle over time should adequately describe the event. In this simple movement, the minimum angle (0.70 rad) occurs at the beginning and end of the movement, and the maximum angle is at complete extension (3.14 rad). The position-time graph is shown in Figure 1.9.

Without considering the previously described relationships between position, velocity, and acceleration, answer two questions about the movement (focus on the elbow joint rotation): When is the velocity zero? When is the acceleration zero? First, at which point in the extension-flexion event is the velocity zero? The velocity is zero when there is no displacement—at the beginning and end of the movement and, for an instant, when the direction of the movement changes from extension to flexion. Let us focus on the zero velocity at the direction change. The velocity-time graph in Figure 1.9 indicates that angular velocity is zero and hence that the change in direction occurs at the maximum displacement (3.14 rad). This could be deduced graphically from the previous comments on the equivalency between velocity and the slope of the position-time function. As the graph illustrates, when the velocity is positive (above zero), the elbow is extending; when the velocity is negative, the elbow is flexing. Thus a change in the sign of velocity (e.g., positive to negative) indicates a change in the direction of movement.

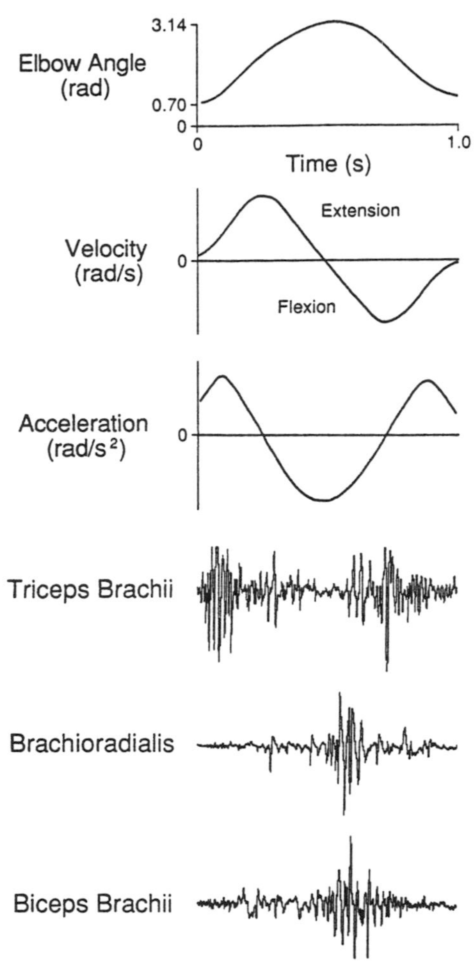

Figure 1.9 Kinematic graphs and EMG patterns for a simple extension-flexion movement of the forearm-hand segment about the elbow joint. The kinematic features of the movement are largely determined by the net muscle activity.

Next consider the acceleration question. Apart from the start and the end of the movement, when is acceleration zero? An intuitive answer to this question is more difficult, as the criteria for declaring a zero acceleration are not obvious. With velocity the decision was more straightforward and merely involved locating instances of no displacement. The parallel approach for acceleration is to identify intervals in which the change in velocity is zero.

The answer to this question is most easily obtained by referring to a velocity-time graph. Given the graphic relationship between a variable and its derivative, the angular acceleration-time curve in Figure 1.9 indicates that acceleration is zero when the value of velocity is at its maximum (0.3 s) and minimum (0.7 s). That is, acceleration is zero when the slope of the velocity-time graph is zero. The acceleration-time graph could be described as triphasic. It comprises three epochs of acceleration: From the beginning until about 0.3 s, the movement consists of an acceleration in the extension direction; subsequently, acceleration experiences a sign change and for most of the movement (0.3 to 0.7 s) the acceleration is in the direction of elbow flexion; and the movement concludes with another epoch of extension-directed acceleration. Note that the acceleration-time profile cannot be predicted directly from the position-time graph.

Muscle Activity

Could this acceleration-time profile have been determined by any means other than the velocity-time graph? In other words, is it possible to identify the instances of zero acceleration without knowledge of the velocity-time function? To answer this question, consider the pattern of muscular activity necessary to produce the movement. Often kinesiologists describe movements in terms of muscle groups rather than a specific muscle. In this elbow extension-flexion example, the movement involves the muscle groups identified as the elbow extensors and elbow flexors. The elbow flexor group includes biceps brachii, brachialis, and brachioradialis as the major muscles (prime movers) contributing to elbow flexion (but this does not mean that these muscles cannot also control elbow extension—more about this later). The elbow extensor group includes the triceps brachii muscle. In terms of the pattern of muscular activity, the zero acceleration question can be restated as follows: What is the pattern of elbow extensor and elbow flexor activity during the movement?

Part II, "The Single Joint System," will examine the manner in which the nervous system activates muscle. Essentially this communication is an electrochemical process in which the final stage is electrical in nature. Therefore, to determine whether a muscle is active, we need only monitor the electrical activity of the muscle. This technique, known as **electromyography** (EMG), involves placing electrodes on the skin over the muscle to monitor the electrical input (excitation) of the muscle. Figure 1.9 includes EMG records of the elbow extensors and flexors during the elbow extension-flexion movement. Although both muscle groups are active throughout portions of the movement (a phenomenon known as **coactivation**), there is a high correlation between the net muscle activity (EMG) and the acceleration-time profile. Later we will consider in more detail the mechanical effect of muscle activity on body segments. One such effect is that it exerts a force and can cause body segments to rotate. In the elbow extension-flexion movement there are no other significant horizontal forces, so the acceleration about the elbow joint is largely determined by the activity of the muscles about the joint. Thus, in response to the zero-acceleration question, it is possible to identify the instances of zero acceleration if the pattern of net muscular activity is known.

Given this relationship between force and acceleration, the acceleration-time graph shown in Figure 1.9 must also represent the pattern of net muscle activity about the elbow joint. But it often seems difficult to understand, at least initially, why a simple elbow extension-flexion movement would involve a triphasic pattern of net muscle activity. To move a limb at moderate speed toward a target involves a burst of muscle activity to accelerate the limb toward the target followed by activity in an opposing muscle, which accelerates the limb in the opposite direction to slow down (brake) the motion of the limb. The elbow extension-flexion movement involves displacement to two different targets, first to complete extension and then back to the initial position. This movement involves two sets of muscle activity, one to each target. Movement to the first target, complete elbow extension, is accomplished by activation of the elbow extensors followed by activation of the elbow flexors to brake the motion of the limb. Movement to the second target, a return to the initial position, is accomplished by activation of the elbow flexors followed by activation of the elbow extensors to brake the movement. When these sequences are superimposed, the net pattern consists of elbow extensor activity, followed by elbow flexor activity, and concluding with elbow extensor activity. This pattern of muscle activity results in the triphasic acceleration-time curve (Figure 1.9).

In general, the relationship exhibited in this elbow extension-flexion example between the activity of the muscles about a joint and the acceleration of the body segment is a special case. The acceleration that a body segment experiences is a reflection of *all* the forces acting on the segment. For example, if the elbow extension-flexion movement were performed in a vertical plane, gravity would exert an influence on the movement and the acceleration-time graph would differ from that shown in Figure 1.9. When the movement is performed in a horizontal plane, the muscles exert the only significant force on the forearm in this plane and thus produce the special relationship between muscle activity and acceleration of the forearm.

Kinematics of Gait

To illustrate the use of motion descriptors (position, velocity, and acceleration) in the analysis of human movement, let us consider some kinematic characteristics of gait. Human gait involves alternating sequences in which the body is supported first by one limb, which is contacting the ground, and then by the other limb. Although this sounds quite straightforward, its control is complex enough that, despite our technological advances, no machine has yet been built that mimics human gait.

Human gait has two modes, walking and running. The distinction between the two lies in the percentage of each cycle that the body is supported by foot contact with the ground. During walking (open symbols in Figure 1.10) at least one foot is always on the ground, and for a brief period of each cycle both feet are on the ground; walking can be characterized as an alternating sequence of single and double support. In contrast, running (solid symbols in Figure 1.10) involves alternating sequences of support and nonsupport, with the proportion of the cycle spent in support varying with speed; as speed increases, the time of support decreases (Figure 1.10). But during a single cycle of either walking or running, each limb experiences a sequence of support and nonsupport. The period of support is referred to as the **stance phase**, and the period of nonsupport is known as the **swing phase**. These intervals are separated by two events, the instant at which the foot contacts the ground, or footstrike (FS), and the instant at which the foot leaves the ground, or takeoff (TO). Gait cycles are usually defined relative to these events. For example, one complete cycle, from left foot takeoff to left foot takeoff, is defined as a **stride**.

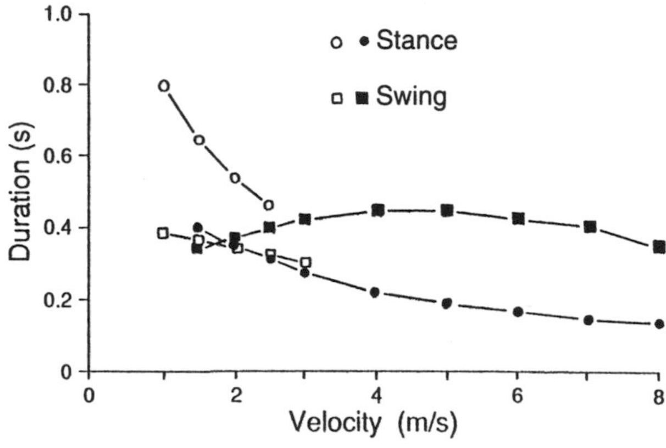

Figure 1.10 Changes in the duration of the stance and swing phases as gait velocity increases.
Note. From "Adaptability in Frequency and Amplitude of Leg Movements During Human Locomotion at Different Speeds" by J. Nilsson and A. Thorstensson, 1987, *Acta Physiologica Scandinavica*, **129**, p. 109. Copyright 1987 by Blackwell Scientific Publications. Adapted by permission.

Figure 1.11 summarizes these relationships. The stride contains two steps. A **step** is defined as the part of the cycle from the takeoff (or footstrike) of one foot to the takeoff (or footstrike) of the other foot. Within a stride, four events of footstrike and takeoff occur, two for each limb. These are right footstrike (rFS), right takeoff (rTO), left footstrike (lFS), and left takeoff (lTO). The swing phase exists between the events of TO and FS, whereas stance occurs from FS to TO. Figure 1.11 shows how the durations of the stride and of the stance and swing phases change with gait speed.

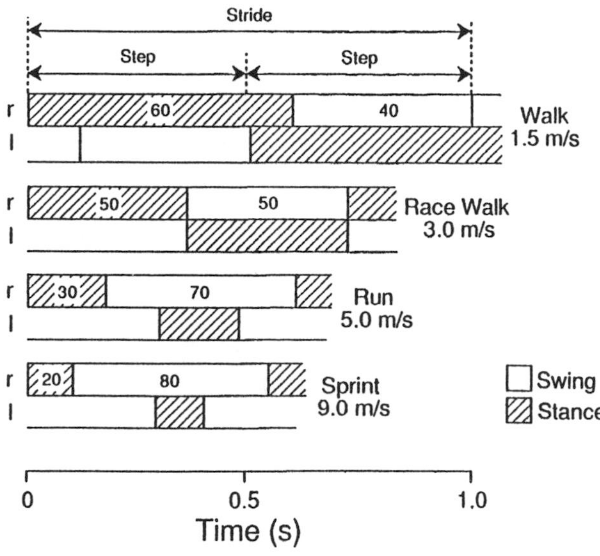

Figure 1.11 The events and phases characterizing walking and running gaits (r = right; l = left).
Note. From "Biomechanics of Running Gait" by C.L. Vaughan, 1984, *CRC Critical Reviews in Biomedical Engineering*, **12**, p. 6. Copyright 1984 by CRC Press, Inc. Adapted by permission.

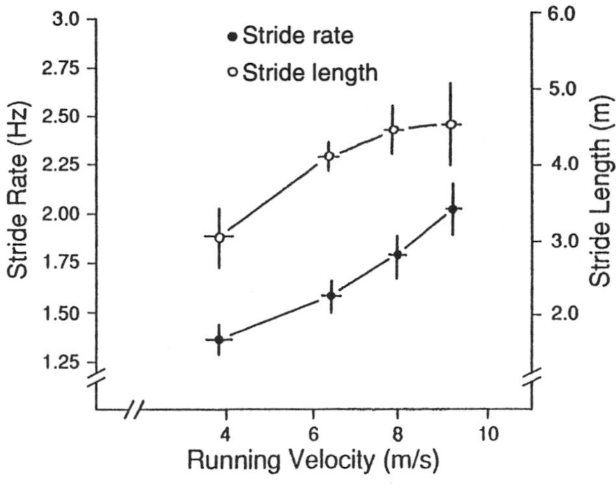

Figure 1.12 Average changes in stride length and stride rate with running velocity.
Note. From "Mechanical Factors Influencing Running Speed" by P. Luhtanen and P.V. Komi. In *Biomechanics VI-B* (p. 25) by E. Asmussen and K. Jorgensen (Eds.), 1978, Baltimore: University Park Press. Copyright 1978 by University Park Press.

Stride Length and Rate

Running speed depends on two variables, stride length and stride rate (Vaughan, 1984). If stride length remains constant, then as stride time decreases (i.e., stride rate increases) running speed increases. If stride rate remains constant, speed increases as stride length increases. These effects of stride rate and stride length on running speed are illustrated in Figure 1.12. Within certain limits a number of length-rate combinations will produce a desired speed. The average combinations are shown in Figure 1.12. For example, an individual running at a speed of 8 m/s will use a stride rate of about 1.75 Hz and a stride length of about 4.6 m. Figure 1.12 illustrates that, on average, a runner increases speed over the range from 4 to 9 m/s by increasing stride rate continually, although more slowly (the slope is not that steep) at lower velocities, but only increases stride length up to about 8 m/s. Notice that the contribution of changes in stride length and rate to running velocity are different at low and high velocities; this is apparent by the differences in the slope of each curve (stride length and rate) at different velocities.

In contrast to the average data shown in Figure 1.12, the strategy adopted by four runners for increasing running speed is depicted in Figure 1.13. The data were obtained by measuring stride length from footprints and stride rate from foot switches that indicated the stance phase. Consider the results obtained for Subject SU, whose change in speed from 4.3 to 8.5 m/s seemed to be accomplished in two phases: Initial

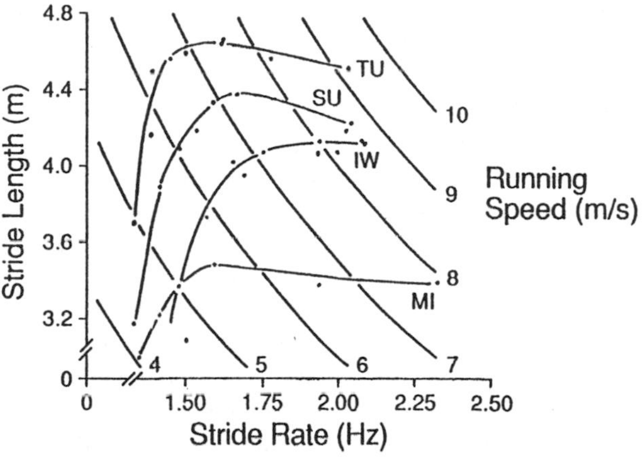

Figure 1.13 The relationships among stride rate, stride length, and running speed.
Note. From "Temporal Patterns in Running" by M. Saito, K. Kobayashi, M. Miyashita, and T. Hoshikawa. In *Biomechanics IV* (p. 107) by R.C. Nelson and C.A. Morehouse (Eds.), 1974, Baltimore: University Park Press. Copyright 1974 by University Park Press.

changes in speed (4.3 to 7.0 m/s) were due to a combined increase in stride length (3.2 to 4.4 m) and stride rate (1.4 to 1.7 Hz); subsequent speed increases (7.0 to 8.5 m/s) were achieved by a slight decrease in stride length (about 20 cm) and a sustained increase in stride rate (1.7 to 2.0 Hz). In general, the trained runners (Subjects TU, SU, and IW) increased stride length up to 7.0 m/s, whereas the untrained runner (MI) did so only up to about 5.5 m/s. All four runners, however, achieved initial increases in speed (up to 6.0 m/s for the trained runners) mainly by increasing stride length. Clearly, the combination of stride length and rate chosen to achieve a desired speed varies among runners. Furthermore, it seems that anthropometric variables (e.g., stature, leg length, limb segment mass) are not the primary determinants of preferred stride frequency and length (Cavanagh & Kram, 1989). The typical explanation given for the strategy of changing stride length rather than rate is that it requires less energy to lengthen the stride within reasonable limits than to increase stride rate.

Kinematic Effects of Speed

Increasing running speed by increasing stride length requires an alteration of the kinematics of the limbs. The changes needed include both the range of motion about a joint (quantity) and the pattern of displacement (quality). For example, Figure 1.14 shows that angular displacement about the knee joint increases as the runner goes from a walk to a run and that the stance phase (indicated by the shaded horizontal bar) includes only knee extension during a sprint but both flexion and extension during walking and running. Similarly, as running speed increases, there is an increase in arm motion, which includes an increase in the range of motion about both the shoulder and elbow joints. As a consequence of these changes, the vertical displacement of the whole-body center of gravity is reduced as speed increases (Cavagna, Saibene, & Margaria, 1964).

Performance variables that do not differ between runners capable of running 9 m/s and those who can achieve 11 m/s include (a) step length, (b) the minimum distance between the heel and the buttocks during the swing phase, (c) vertical velocity at takeoff, and (d) the height of the foot during the swing phase as it passes the support leg. In contrast, those runners who can achieve 11 m/s exhibit (a) a 15% greater stride rate, (b) less time in the stance phase, (c) a shorter horizontal distance between the foot and a vertical projection of the center of gravity at takeoff, (d) a less extended knee joint at takeoff, and (e) a more vertical trunk position. These differences demonstrate that the maximum running speed that an individual can achieve is influenced by the kinematic details of the movement.

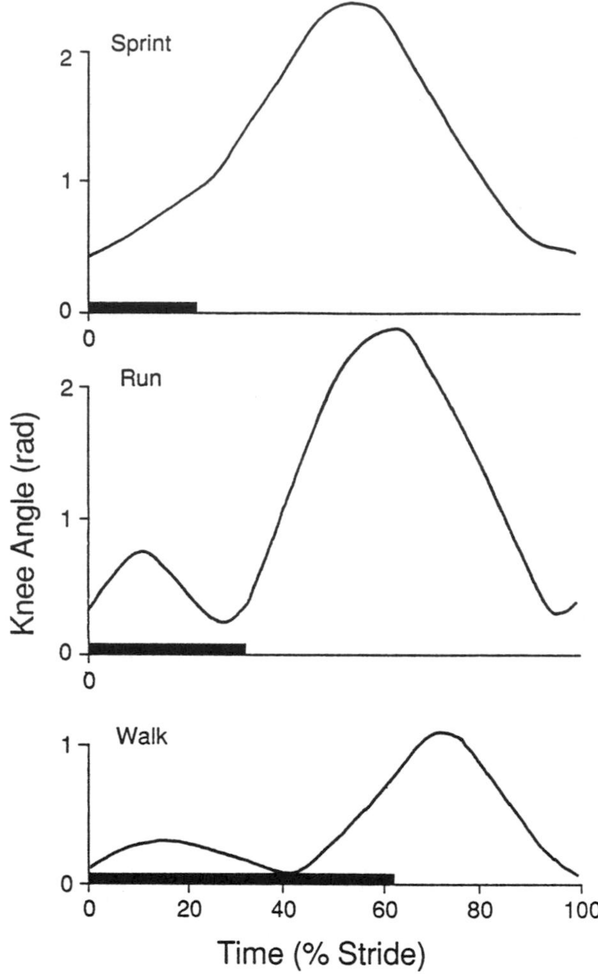

Figure 1.14 Knee angle during a stride for a walk, run, and sprint. A knee angle of 0 rad indicates complete extension.

Note. From ''Biomechanics of Running Gait'' by C.L. Vaughan, 1984, *CRC Critical Reviews in Biomechanical Engineering*, **12**, p. 11. Copyright 1984 by CRC Press, Boca Raton, Florida. Adapted by permission.

Angle-Angle Diagrams

Cyclical activities, such as walking and running, are ideal movements to represent in angle-angle diagrams because the beginning and the end of an event are located at about the same point on the diagram. This type of diagram has proved useful in comparing movement forms (Hershler & Milner, 1980a, 1980b; D.I. Miller, 1978). For example, comparing the knee-thigh diagram of a normal subject during running with that of a subject who has had a lower extremity amputated can be useful in evaluating the effectiveness of prostheses in restoring normal-looking gait. In this type of analysis, emphasis is placed on comparing the shape of the respective angle-angle diagrams (e.g., Figures 1.15 vs. 1.16).

The interpretation of Figure 1.15 involves following the curve around in a counterclockwise direction. This diagram illustrates the angle-angle diagram for one limb during a running stride. As mentioned previously, each limb experiences two events during a stride, footstrike (FS) and takeoff (TO). From footstrike to takeoff (stance phase—dotted line), the foot is in contact with the ground. Conversely, from takeoff to footstrike (swing phase—solid line), the foot of the illustrated limb is not in contact with the ground. In addition, during the swing phase (lTO to lFS), the other foot first contacts and then leaves the ground (rFS to rTO). Accordingly, Figure

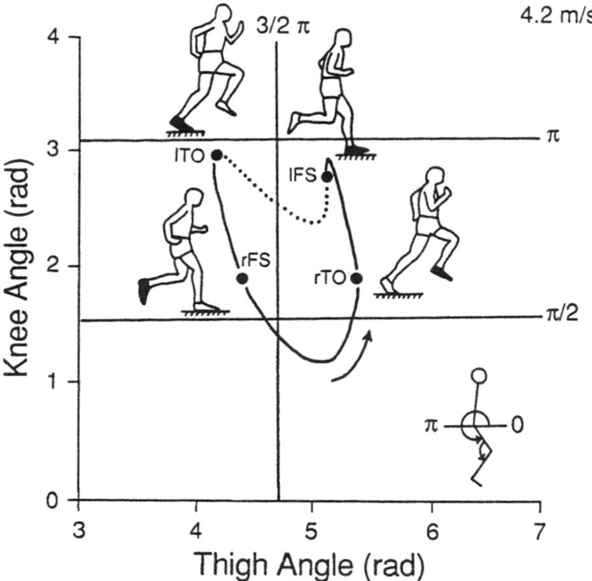

Figure 1.15 Knee-thigh diagram of the left limb of a skilled runner.
Note. From "Below-Knee Amputee Running Gait" by R.M. Enoka, D.I. Miller, and E.M. Burgess, 1982, *American Journal of Physical Medicine*, **61**, p. 70. Copyright 1982 by Williams and Wilkins. Reprinted by permission.

1.15 reveals that (a) after lTO, the thigh rotates forward about the hip joint and the knee flexes to a minimum angle; (b) following the minimum angle, the knee extends until just before lFS while the thigh continues to rotate forward and then begins rotating backward; and (c) during stance (lFS to lTO), the knee first flexes and then extends, while the thigh rotates backward.

Angle-angle diagrams have several important features, and these features are apparent in Figure 1.15. A relative angle (knee) is plotted against an absolute angle (thigh). The graph illustrates the combined actions of the knee flexion-extension and thigh forward-backward rotation. In addition, three reference axes are shown with which to evaluate the range of motion of the movement: (a) The 3/2 π axis indicates a thigh angle at which the thigh would be in a vertical position, (b) the π axis (3.14 rad) represents a knee angle of complete extension, and (c) the π/2 axis shows a right angle (1.57 rad) for the knee joint. According to Figure 1.15, therefore, the thigh passes in front of and behind the 3/2π line, the knee joint is never fully extended, and the smallest knee angle is less than a right angle during a normal running stride at 4.2 m/s.

In contrast to this normal knee-thigh diagram, the three graphs for subjects with below-knee amputations depicted in Figure 1.16 indicate a substantial difference during the stance phase (i.e., the region from lFS to lTO). Specifically, the amputee knee-thigh diagrams reveal a knee-joint pattern of a constant angle followed by flexion rather than the normal flexion-extension sequence. Because Figure 1.16 shows the knee-thigh diagrams for the prosthetic limbs of the below-knee amputees, the pattern of the graphs is perhaps not surprising. The failure to flex the knee during stance, shown in Figure 1.16 by the knee angle not decreasing immediately after FS, means that the amputees just used their limbs as a rigid lever about which to rotate while the prosthetic foot was on the ground. This type of graphic display could be used in a clinical setting to monitor a rehabilitation program aimed at correcting this strategy so that the gait would appear more normal.

Angle-angle diagrams have also been used to represent the kinematics of the arms during running. Because the motion of the arms is frequently not confined to the sagittal plane during running, imaging techniques that can capture three-dimensional motion are necessary. When this is done, the displacement of the upper arm about the shoulder and the relative angle between the upper and lower arms (elbow angle) for an individual running at 11.4 m/s has the form shown in Figure 1.17.

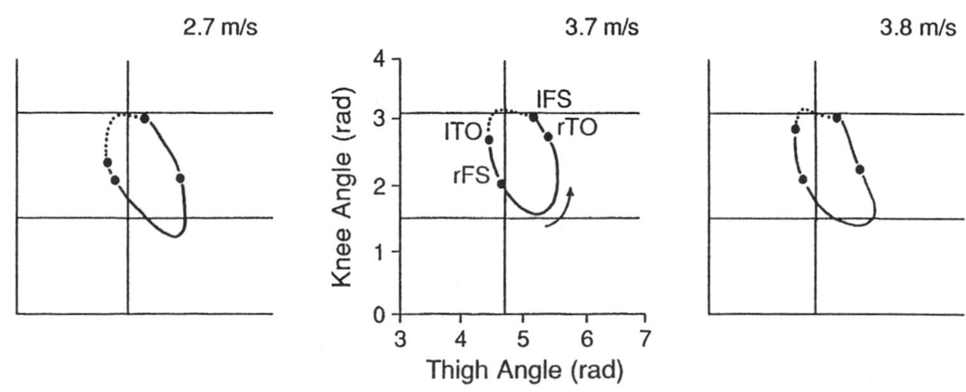

Figure 1.16 Knee-thigh diagrams for three below-knee amputees running at speeds from 2.7 to 3.8 m/s.
Note. From "Below-Knee Amputee Running Gait" by R.M. Enoka, D.I. Miller, and E.M. Burgess, 1982, *American Journal of Physical Medicine*, **61**, p. 78. Copyright 1982 by Williams and Wilkins. Reprinted by permission.

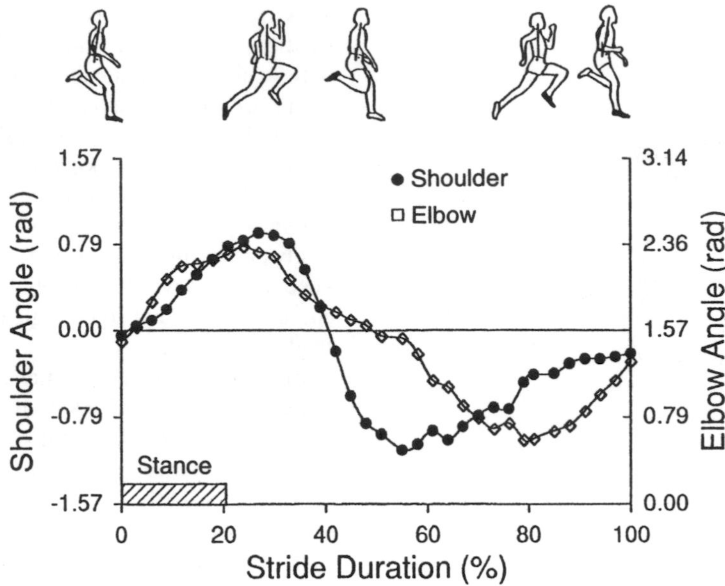

Figure 1.17 Displacement about the right shoulder and elbow joints during a running stride.
Note. From "Temporal and Kinematic Analysis of Arm Motion in Sprinters" by C. Li and A.E. Atwater, 1984, Presented at the Olympic Scientific Congress, Eugene, OR. Reprinted by permission.

A positive shoulder angle indicates flexion (forward of vertical) and a negative one represents extension (backward of vertical). Of course, the amplitude and timing of the displacement vary as a function of running speed (Lusby & Atwater, 1983). Although Figure 1.17 includes information on the timing of the displacement at the two angles, this information can be presented more succinctly in an angle-angle diagram (Figure 1.18). Essentially, the pattern of displacement about the shoulder and elbow joints is confined to the upper right and lower left quadrants of the angle-angle diagram. The upper right quadrant represents concurrent shoulder and elbow flexion, whereas the lower quadrant indicates concurrent shoulder and elbow extension. The stance phase of the ipsilateral leg (rFS to rTO in Figure 1.18) is mainly accompanied by concurrent flexion at the two joints. But the pattern is not one of a tight coupling of the two actions (i.e., flexion or extension), because there are instances when opposing motion occurs at the two joints. Can you see where this occurs in Figure 1.18? One example is in the phase from lFS to lTO when the shoulder extends and the elbow flexes. The elbow-shoulder angle-angle diagrams, like those for the leg, provide a qualitative means to evaluate the pattern (i.e., shape or structure) of the movement.

Another useful feature of these cyclic angle-angle diagrams is that, in addition to shape comparisons, the size of the diagram indicates the range of motion experienced at each joint during the event. For example, we would expect that increases in stride length as a runner increases speed are due to changes in the range of motion (amount of motion) at various lower extremity joints.

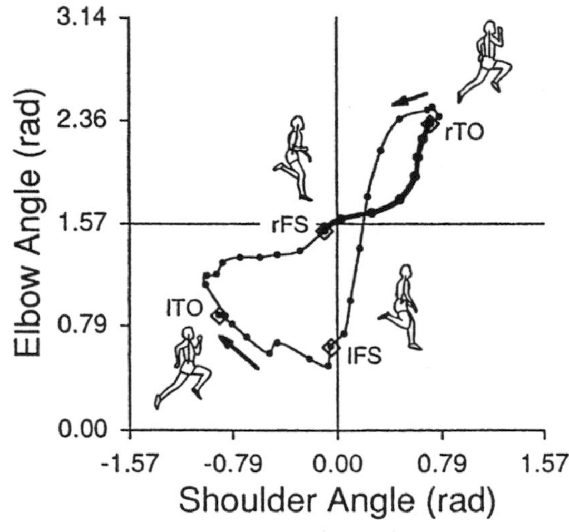

Figure 1.18 Elbow-shoulder angle-angle diagram based on the data shown in Figure 1.17.
Note. From "Temporal and Kinematic Analysis of Arm Motion in Sprinters" by C. Li and A.E. Atwater, 1984, Presented at the Olympic Scientific Congress, Eugene, OR. Reprinted by permission.

Figure 1.19 confirms this expectation by showing that, as speed increases (3.9 vs. 7.6 m/s), the amount of rotation both of the thigh and about the knee joint increases; the larger angle-angle diagram represents the faster speed.

At this point a good exercise to test your grasp of the angle-angle diagram approach is to sketch a knee-thigh diagram as a runner goes uphill and then downhill (Milliron & Cavanagh, 1990). The key to this exercise

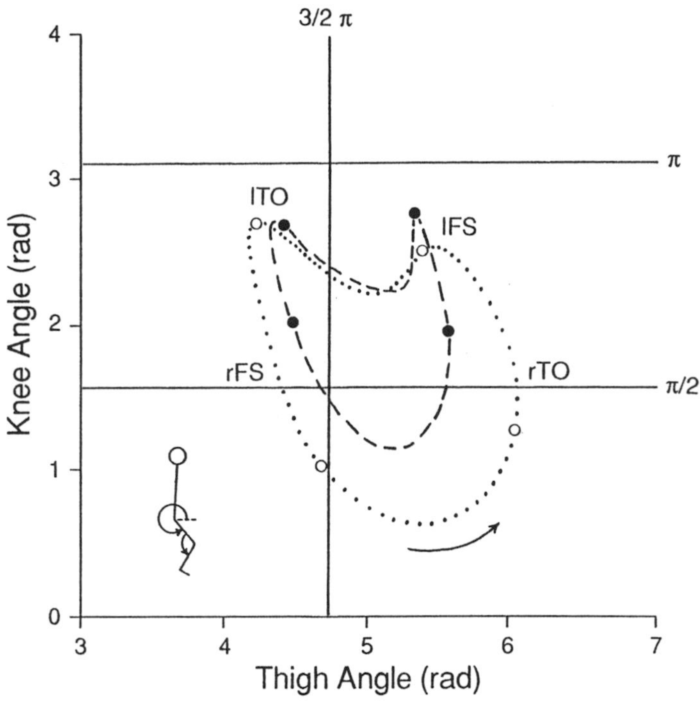

Figure 1.19 Knee-thigh angle-angle diagram as a function of running speed.
Note. Adapted from Miller, Enoka, McCullough, Burgess, Hutton, and Frankel, 1979.

is to think of how the movement would change relative to the three reference axes. For example, it seems reasonable to expect that, compared with a runner on level ground, a runner going uphill would extend the knee less, flex the knee less at the minimum angle, and have the thigh remain in front of vertical (forward rotation) for a greater part of the stride. Try sketching this relationship.

The angle-angle diagram format, first proposed by Cavanagh and Grieve (1973), has largely been confined to representing position information. Some investigators (e.g., D.I. Miller, 1978) have experimented with plots of angular velocity and acceleration, but these attempts have not been readily accepted, probably due to the complexity of the relationships. Similarly, given current computer graphics capabilities, it is surprising that no three-dimensional angle-angle diagrams, with time (e.g., running speed) as the third axis, have yet been published.

Kinematics of Throwing and Kicking

The throw and the kick are two of the basic elements of human movement. Although the purpose of both is *to project an object so that it has a flight phase*, the distinction between the two is the manner in which the body imparts the flight phase to the object. In a throw, the object is supported by a limb, usually the hand, and displaced through a range of motion while the limb increases the *quantity of motion*, or **momentum**, of the object. Typically several body segments, in a proximal-to-distal sequence, contribute to the object's momentum (Putnam, 1993). Although the kick is also characterized by a proximal-to-distal involvement of body segments, it differs from a throw in that the kick is a striking event in which the momentum of the object is increased by a brief impact between a limb and the object. In this section we will consider some basic features of the throwing motion and the kicking motion and then examine the characteristics of the flight phase experienced by the projectile. Irrespective of the manner by which the momentum of the object is increased, once the object loses contact with the limb, it follows a trajectory that is prescribed by the laws of projectile motion.

Throwing Motion

Although the throw can be characterized by the progressive contribution of the body segments to the momentum of the object to be projected (with a constant mass, the change in momentum corresponds to the change in velocity), the task can be accomplished with a variety of motions. These differences in form include the overarm throw (e.g., baseball, cricket, javelin, darts), the underarm throw (e.g., softball pitch), the push throw (e.g., shot put), and the pull throw (e.g., discus, hammer). The kinematics of the throwing motion are typically three-dimensional in nature, especially when the throw is for maximum distance (Feltner, 1989; Feltner & Dapena, 1986, 1989). For example, the contributions of the body segments to the overarm throw (Figure 1.20) include displacements in the vertical, side-to-side, and forward-backward directions (Figure 1.21). In contrast, when the throwing task requires accuracy, such as throwing a dart or shooting a free throw in basketball, the throwing motion is generally planar and the strategy appears to be to use the minimum number of body segments in the movement.

The sequence of a typical baseball pitch is shown in Figure 1.20. A qualitative inspection of this sequence indicates that the movement involves the progressive contribution of the body segments, beginning from the base of support and progressing through to the hand. The baseball pitch consists of two phases (Figure 1.20): (a) Positions a to k—the velocity of the ball is increased mainly by the action of the legs; and (b) Positions l to u—the velocity of the ball is increased by the action of the trunk and arms. The second phase, which produces the greater increase in the velocity of the ball, involves the progressive increase in the angular velocity of the body segments in the following order: pelvis, upper trunk and upper arm, forearm, and hand (Atwater, 1979). This means that the peak angular momentum of the pelvis occurs before that of the upper trunk and upper arm, the peak angular momentum of the upper trunk and upper arm occurs before that of the forearm, and so on. In this progression of segmental activity, which is also seen in striking movements such as the kick, proximal segments begin to rotate before the distal segments, and the proximal segments begin to slow down before the distal segments have reached peak angular velocity. There is an optimal delay in the timing of the rotation of one segment to the next in order to maximize the velocity of the projectile (Alexander, 1991). The result of this proximal-to-distal progression is that the velocity of the ball does not increase substantially until late in the movement. Figure 1.22 shows that the greatest

Figure 1.20 Sequence of a typical baseball pitch.
Note. From "Dynamics of the Shoulder and Elbow Joints of the Throwing Arm During a Baseball Pitch" by M. Feltner and J. Dapena, 1986, *International Journal of Sport Biomechanics*, **2**, p. 236. Copyright 1986 by Human Kinetics. Reproduced with permission.

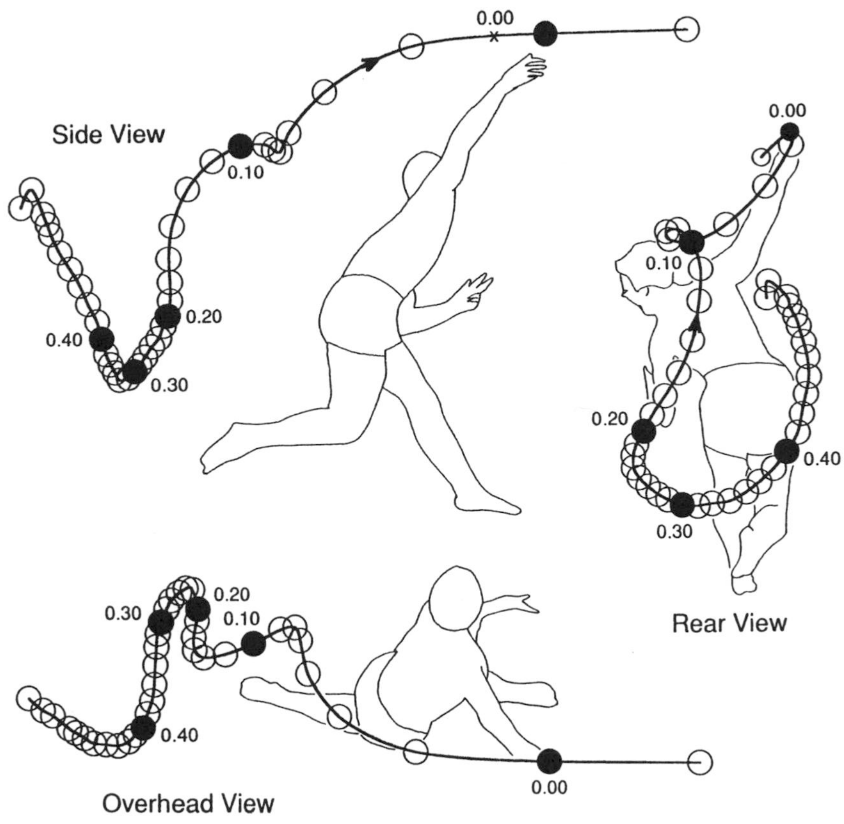

Figure 1.21 Three views of the path of the ball during an overarm throw. The ball was released at 37.3 m/s (83.5 mph), and the instant of release is indicated as the final shaded-ball position (0.00). Prior to release, the shaded-ball positions indicate intervals of 100 ms. These measurements were obtained from a film (64 frames/s) of the movement.
Note. From *Biomechanics of throwing: Correction of common misconceptions* by A.E. Atwater, 1977, a paper presented at the Joint Meeting of the National College Physical Education Association for Men and the National Association for Physical Education of College Women, Orlando, FL. Adapted with permission.

increase in ball velocity for an overarm throw occurred in the 60 ms prior to release, which corresponds to the interval depicted in Figure 1.20 by the last three ball positions prior to release (r-t).

In addition to this proximal-to-distal sequence, the overarm throw is characterized by the extreme range of motion exhibited by the arm. These displacements occur at the shoulder, elbow, and wrist joints but are most extensive about the shoulder joint. Because of its geometry, the shoulder joint is capable of displacement about three separate axes of rotation: rotation about an anterior-posterior axis, which is referred to as abduction-adduction; rotation about a side-to-side axis, which is known as flexion-extension; and rotation about a longitudinal axis, which is described as external-internal rotation. Figure 1.23 shows that during a baseball pitch there is about 1.57 rad of elbow flexion-extension, 1.57 rad of external rotation, and 1.0 rad of internal rotation. Position p in Figure 1.20 shows the arm in the position of maximum external rotation. This extreme position is thought to cause many of the arm injuries sustained by baseball pitchers.

Kicking Motion

Previously we described the kick as a striking skill in which a flight phase is imparted to an object as the result of a brief impact between a limb (or implement) and the object. By this criterion, we include as striking skills such activities as kicking a ball, hitting a volleyball, and striking a projectile in racket and bat sports. As with the throw, the motion underlying a striking skill can vary depending on the objectives of the activity; these goals can be related to horizontal distance, time in the air, accuracy, or the speed of the movement. For many striking skills (e.g., soccer kick, punt, volleyball serve, tennis serve), the motion is similar to that for the overarm throw and involves a proximal-to-distal sequence of body segment contributions to the velocity of the endpoint (hand, foot, implement) that will strike the projectile (Figure 1.24). This is apparent in Figure 1.24: The thigh reaches a peak positive (forward) angular velocity before the shank does and the angular velocity of the thigh decreases while that for the shank continues to increase through to contact with the ball.

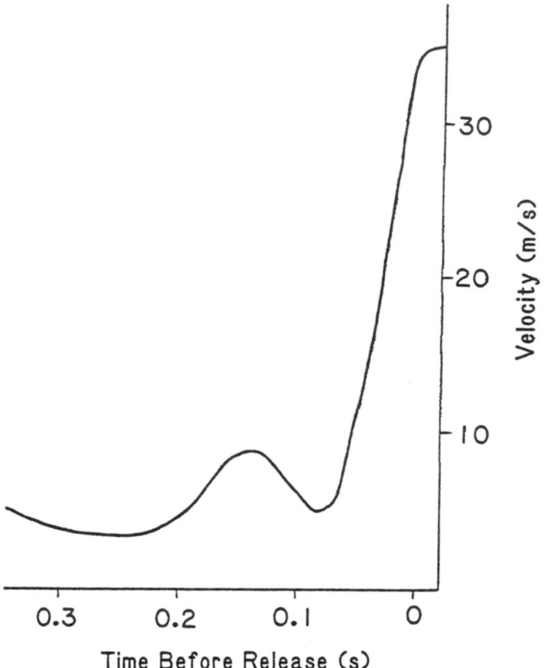

Figure 1.22 Resultant velocity (horizontal + vertical) of the ball during an overarm throw.
Note. From *Overarm Throwing Patterns: A Kinematographic Analysis* by A.E. Atwater, 1970, a paper presented at the National Convention of the American Association for Health, Physical Education and Recreation, Seattle, WA. Adapted with permission.

Luhtanen (1988) describes a similar sequence for the upper arm, forearm, and hand during a volleyball serve.

Because striking skills alter the momentum of the projectile through impact, an important difference between a throw and a kick is the rigidity of the limb during contact with the ball. When performing striking skills, athletes frequently manipulate the rigidity of the limb to influence the impact. For example, the change in ball velocity during a kick is greater when the limb is more rigid at impact. This is accomplished by contracting the muscles in the foot and those that cross the ankle joint so that the foot is more rigidly attached to the shank. One extreme example of variation in rigidity is the underhand pass (bump, or dig) in volleyball. This skill requires that the individual allow the volleyball to bounce off the ventral surface of the forearms with a prescribed trajectory. Skillful players accomplish this task by varying the rigidity of the arms and thereby determining the extent to which the ball bounces off the forearms. Similarly, the strings on rackets can be strung to various levels of tightness, which represents one factor that contributes to rigidity in racket sports.

Projectile Motion

The relationships between position, velocity, and acceleration describe the trajectory that a projected (thrown) object will travel. Two forces act to change the trajectory

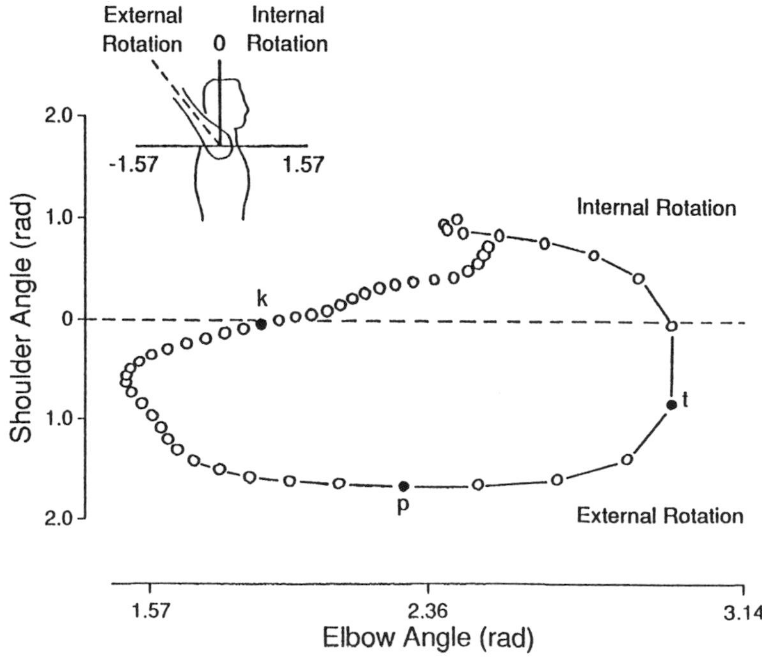

Figure 1.23 Shoulder-elbow angle-angle diagram for a baseball pitch. The interval between data points is 5 ms; the farther apart the data points, the faster the movement. The markers k, p, and t correspond to positions identified in Figure 1.20.
Note. From "Dynamics of the Shoulder and Elbow Joints of the Throwing Arm During a Baseball Pitch" by M. Feltner and J. Dapena, 1986, *International Journal of Sport Biomechanics*, **2**, p. 249. Copyright 1986 by Human Kinetics. Reprinted with permission.

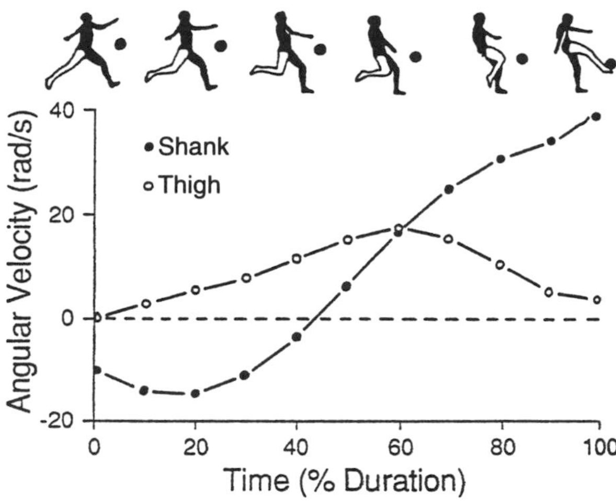

Figure 1.24 Angular velocity of the thigh and shank of the kicking leg during the final step of a kick
Note. From "A Segment Interaction Analysis of Proximal-to-Distal Sequential Segment Motion Patterns" by C.A. Putnam, 1991, *Medicine and Science in Sports and Exercise*, **23**, p. 134. Copyright 1991 by Williams and Wilkins. Adapted by permission.

of a projectile: gravity and air resistance. The simplest case of projectile motion is one in which we assume that the effect of air resistance is so small (e.g., due to the mass of the object or the absence of wind) that we can neglect it. Given this condition, we can use the following facts to help us explore the motion of a projectile:

- The effect of gravity is to cause the trajectory to deviate from a straight line into a curved path that can be described as a parabola.
- When the release height and landing occur at the same level, the time taken for the projectile to reach the peak of its trajectory will be identical to the time taken to go from the peak to landing.
- The vertical velocity of the projectile will change from an upward value (positive) at release, to zero at the peak of the trajectory (when it changes direction), to downward (negative) when it returns to the ground.
- The time that the object spends in the air depends on the magnitude of its vertical velocity at release.
- The only force that the projectile experiences will be that due to gravity, and this will cause a vertical acceleration of -9.81 m/s^2.
- Because there is no force acting in the horizontal direction, the horizontal acceleration of the object will be zero, which means that the object's horizontal velocity will remain constant. Consequently, the horizontal distance traveled by the projectile can be determined as the product of horizontal velocity and the flight time of the projectile.
- The flight time will depend on the vertical velocity at release and the height of release above the landing surface.

Although the trajectory of a projectile is parabolic, the shape of the parabola depends on the velocity—both magnitude and direction—of release. Three different parabolas are shown in Figure 1.25. Based on the laws of projectile motion, we can describe the velocity vector at selected instances throughout the trajectory; this is shown for the javelin in Figure 1.25. The diagram shows that provided no force (or negligible force) acts on the javelin while it is in flight, its trajectory will be characterized by a constant horizontal velocity, a positive vertical velocity on the upward phase, a negative vertical velocity on the downward phase, and zero vertical velocity at the peak of the trajectory. The vertical velocity will have the greatest absolute value at the beginning and at the end of the flight. These features apply to such projectiles as a shot during the shot put, a gymnast performing a vault, a high jumper clearing the bar, and a basketball player performing a jump shot. Although some basketball players appear to hang in the air during a jump shot, the trajectory of their center of gravity remains parabolic. The illusion is created by the relative displacement of the arms and legs (Bishop & Hay, 1979).

Using the definitions of velocity (Equation 1.1) and acceleration (Equation 1.2), let us consider an example of projectile-motion analysis. A ball is thrown at an angle of 1.05 rad with respect to the horizontal and 2.5 m above the ground with a resultant velocity along the line of projection of 6 m/s. For solving problems of projectile motion, it is convenient to begin with a sketch of initial conditions (see Figure 1.26).

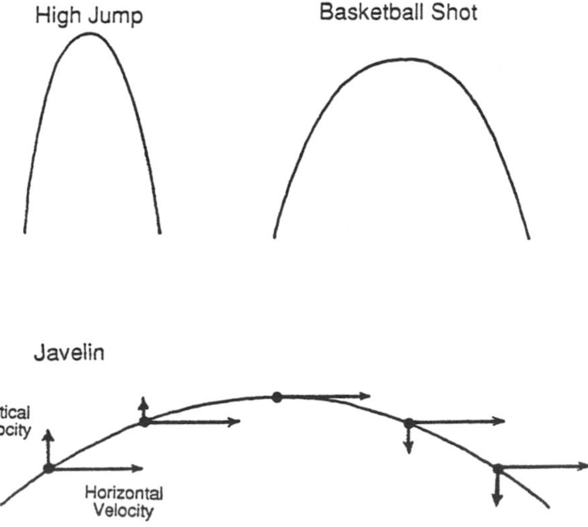

Figure 1.25 The parabolic trajectory experienced by projectiles.

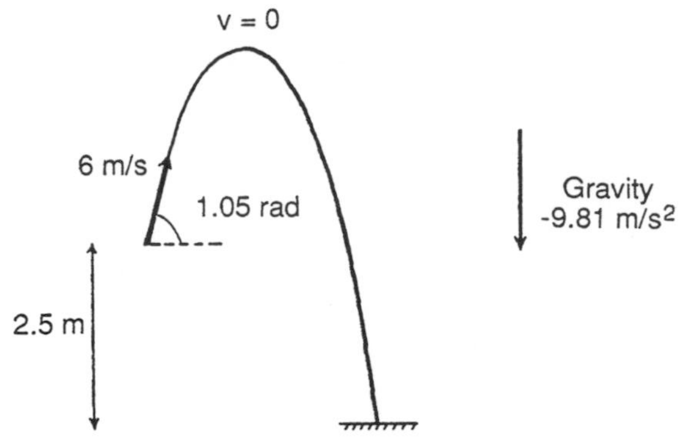

Figure 1.26 Initial conditions for the projectile-motion problem.

1. How long does the ball take to reach its highest point? The trajectory is parabolic, so the ball will continue rising until the vertical velocity is zero (where $\bar{a}$ = average acceleration and $v_i = 6 \sin 1.05 = 5.2$ m/s).

$$\bar{a} = \frac{\Delta v}{\Delta t}$$

$$= \frac{v_f - v_i}{t}$$

$$-9.81 = (0 - 5.2)t$$

$$t = -5.2/-9.81$$

$$t = 0.53 \text{ s}$$

2. How high does the ball get? The maximum height reached depends on the vertical component of the release velocity plus the height above the ground at which the ball was released. We already know the following: vertical velocity at release, vertical velocity at the peak of the trajectory, vertical acceleration, release height, time to reach the peak, and that average velocity = $\bar{v}$.

$$\bar{v} = \frac{\Delta r}{\Delta t}$$

$$\frac{5.2 + 0}{2} = \frac{\Delta r}{0.53}$$

$$\Delta r = \frac{0.53(5.2 + 0)}{2}$$

$$= 1.38 \text{ m}$$

$$\text{Total height} = 1.38 + 2.5$$

$$= 3.88 \text{ m}$$

(Note that 2.5 = release height.)

3. What is the vertical velocity of the ball just before it hits the ground? To answer this question, we must consider the second part of the trajectory, from the peak down to contact with the ground. The initial velocity is 0 m/s, and the ball falls 3.88 m from the peak to the ground and experiences an acceleration of –9.81 m/s² ($\bar{a}$ = average acceleration).

$$\bar{a} = \frac{\Delta v}{\Delta t}$$

$$-9.81 = \frac{(v_f - v_i)}{t}$$

$$-9.81 = \frac{(v_f - 0)}{t}$$

$$v_f = -9.81t$$

Once we know how long it takes the ball to go from its peak to the ground, we can determine its final velocity just prior to impact.

Alternatively, we could use one of the equations of motion that we developed earlier. Equation 1.5 gives us a relationship between final velocity, initial velocity, acceleration, and displacement:

$$v_f^2 = v_i^2 + 2a(\Delta r)$$

$$v_f^2 = 2 \times -9.81 \times 3.88$$

$$v_f^2 = -76.1$$

$$v_f = -8.73 \text{ m/s}$$

(Note that negative = downward.)

4. How long does it take the ball to reach the ground from the peak? The ball took 0.53 s to reach its peak, but to reach the ground the ball must fall an additional 2.5 m (the release height). The total height from the peak to the ground is 3.88 m (–3.88 = downward).

$$\bar{r} = \frac{\Delta r}{\Delta t}$$

$$\frac{v_f - v_i}{2} = \frac{-3.88}{t}$$

$$\frac{v_f + 0}{2} = \frac{-3.88}{t}$$

Substitute in the known relationship (from Step 3) for v_f:

$$\frac{-9.81 \times t}{2} = \frac{-3.88}{t}$$

$$t^2 = \frac{-3.88 \times 2}{-9.81}$$

$$t^2 = 0.79$$

$$t = 0.89 \text{ s}$$

Thus for the unknown expression of v_f in Step 3,

$$v_f = -9.81t$$

$$v_f = -9.81 \times 0.89$$

$$v_f = -8.73 \text{ m/s}$$

(Note that negative = downward.)

The time for the ball to reach the ground also could be determined by using Equation 1.3, one of the equations of motion:

$$v_f = v_i + at$$

$$= 0 + -9.81t$$

$$t = \frac{-8.73}{-9.81}$$

$$t = 0.89 \text{ s}$$

5. How long did the ball spend in flight?

$$t = t_{up} + t_{down}$$

$$= 0.53 + 0.89$$

$$= 1.42 \text{ s}$$

6. What horizontal distance did the ball travel (that is, how far was it thrown)? In contrast to the previous calculations, this one uses the horizontal, as opposed to the vertical, information. Because we were given the velocity at release, we can determine the horizontal velocity at release. We also know that in the absence of significant air resistance the horizontal velocity is constant because it experiences a zero horizontal acceleration. Where $v_i = 6 \cos 1.05 = 3$ m/s:

$$\bar{v} = \frac{\Delta r}{\Delta t}$$

$$\frac{v_i + v_f}{2} = \frac{\Delta r}{t}$$

$$\frac{3 + 3}{2} = \frac{\Delta r}{1.42}$$

$$\Delta r = 1.42 \times 3$$

$$= 4.24 \text{ m}$$

This example demonstrates that the time of flight for a projectile depends on the vertical velocity at release (or takeoff) and the release height. The vertical velocity at release determines the height reached by the projectile and the time taken to reach the maximum height. Because parabolic trajectories are symmetrical, the time it takes the ball to return to the release height is identical to that taken to reach the peak point. If the landing height differs from the release height, then the time to travel this extra distance must also be considered in determining the total flight time.

When the goal of a projectile event is to maximize the horizontal distance covered by the projectile, it is necessary to choose the correct combination of flight time (vertical velocity + release height) and horizontal velocity. This involves selecting the **optimum** angle of release (or takeoff). The necessary angle depends largely on the relative positions of the release and landing heights. When the release and landing heights are identical, then the optimum angle of release is 0.785 rad (45°). But when the landing height is lower than the release height, the optimum angle of projection is less than 0.785 rad. And when the landing height is higher than the release height, the optimum angle of projection is greater than 0.785 rad. Furthermore, as the velocity of release increases at a given height of release, the more closely the optimum angle approaches 0.785 rad. Similarly, as the height of release increases for a constant velocity of release, the lower (below 0.785 rad) the optimum angle of projection (Hay, 1985).

In this example, we determined all the details of the ball's trajectory by simply applying the definitions for velocity and acceleration; we did not need to resort to unwieldy, complicated-looking formulas. Note that this procedure applies only to conditions in which the effects of air resistance are so small that they can be ignored.

Because the trajectory of a projectile (i.e., parabolic) is predictable, a common practice in sports is questionable. Often coaches have told players, ''Keep your eyes on the ball.'' Most athletes interpret this suggestion to mean that they should watch the projectile *all the way* to the bat, racket, foot, and so forth so that they can adjust performance to accommodate changes in the trajectory of the projectile. This is absurd. For example, suppose you are playing defense in a volleyball game, and an opponent hits (spikes) the ball at you at a speed of 100 mph (45 m/s). The shortest time with which a person can react to a stimulus, the reaction time, is about 120 ms. In 120 ms, the volleyball will travel 5.4 m. In other words, no matter what happens to the trajectory of the ball over the last 5.4 m, you will not be able to react fast enough to generate an appropriate change in performance—you may as well close your eyes for the last 5.4 m. Consequently, the key is not to watch the ball all the way to your hands but rather to predict the trajectory of the ball so that you can initiate, in plenty of time, an appropriate response. Is there a correlation between success in a sport (e.g., volleyball, tennis, soccer, football) and how quickly an athlete can predict the trajectory of the projectile? Indeed, a strategy in some sports, such as volleyball, is to deliver the ball (projectile) with no spin, which makes its trajectory less predictable.

Some coaches think that it is necessary for the player to watch the projectile all the way to contact so that the player can learn the magnitude of the error made when the projectile is missed. For example, there is an argument in baseball about whether a batter can ''see'' how much the bat misses the ball in a noncontact swing. But there is no scientific evidence to support the claim that the visual images can be captured and processed under such conditions, so it does not seem appropriate to suggest that it is necessary to watch the projectile all the way through to contact.

Summary

This chapter provides the foundation for the study of human movement. It examines the rigorously defined relationships between position, velocity, and acceleration and establishes the concepts and definitions that are necessary to precisely describe movement. The study of kinematics has its foundation in physics, yet the application to human movement typically tests the limits of understanding of these relationships. Without a clear understanding of these principles, it is difficult to proceed to the next levels in the study of human movement. For this reason, the chapter repeats important concepts and presents many numerical and graphical examples of these relationships. This chapter is not an encyclopedia on human movement, but it provides the tools for the study of movement.

TRUE-FALSE QUESTIONS

1. The unit of measurement for displacement can be m. **TRUE FALSE**
2. The prefix k represents 0.001. **TRUE FALSE**
3. The SI unit of measurement for weight is kg. **TRUE FALSE**
4. Force is a base parameter in the SI system. **TRUE FALSE**
5. Velocity is a derived parameter in the SI system. **TRUE FALSE**
6. The abbreviation for second is s in the SI system. **TRUE FALSE**
7. The symbol Δ indicates a *change* in some parameter. **TRUE FALSE**
8. One km is smaller than 1 mile. **TRUE FALSE**
9. Slope can only have a positive sign. **TRUE FALSE**
10. The prefix m represents 0.001. **TRUE FALSE**
11. A radian is a vector quantity. **TRUE FALSE**
12. One degree is equal to 57.3 rad. **TRUE FALSE**
13. A right angle is equal to 3.14 rad. **TRUE FALSE**
14. Displacement describes the spatial aspect of motion. **TRUE FALSE**
15. Velocity can be expressed in radians per second. **TRUE FALSE**
16. A zero velocity means that the object is, at least momentarily, stationary. **TRUE FALSE**
17. *Velocity* is defined as the rate of change in position with respect to time. **TRUE FALSE**
18. *Speed* refers to the magnitude of the velocity vector. **TRUE FALSE**
19. Acceleration and movement are not always in the same direction. **TRUE FALSE**
20. Acceleration is always zero when velocity is zero. **TRUE FALSE**
21. A qualitative analysis tells us type or kind. **TRUE FALSE**
22. The symbol for angular acceleration is α. **TRUE FALSE**
23. Gravity is an acceleration of -9.81 m/s^2. **TRUE FALSE**
24. A scalar variable is one that possesses both magnitude and direction. **TRUE FALSE**
25. Acceleration can be seen as the slope of a velocity-time graph. **TRUE FALSE**
26. An object acted upon by a force will experience an acceleration. **TRUE FALSE**
27. The symbol μ is used to represent angular position. **TRUE FALSE**
28. The combination of translation and rotation is referred to as angular motion. **TRUE FALSE**
29. Position can be described as the derivative of velocity. **TRUE FALSE**
30. The unit of measurement m/s defines the magnitude of the acceleration vector. **TRUE FALSE**
31. The slope of the position-time graph provides information about acceleration. **TRUE FALSE**
32. When a velocity-time graph is at a minimum, acceleration will be negative. **TRUE FALSE**
33. The symbol θ is used to represent angular position. **TRUE FALSE**
34. A radian is defined as the ratio of two distances. **TRUE FALSE**

35. Zero velocity can represent a change in movement direction. **TRUE FALSE**

36. A description of motion that ignores the causes of motion is known as a kinematic description. **TRUE FALSE**

37. The direction of acceleration is usually the same as the direction of the movement. **TRUE FALSE**

38. One of the equations of motion is $\mathbf{v}_f^2 = \mathbf{v}_i^2 + 2\mathbf{a}t$. **TRUE FALSE**

39. The following expression is one of the equations of motion: $\mathbf{r}_f - \mathbf{r}_i = \mathbf{v}_i + \frac{1}{2}\mathbf{a}t^2$. **TRUE FALSE**

40. One of the equations of motion is $\mathbf{v}_f^2 = \mathbf{v}_i^2 + 2\mathbf{a}(\mathbf{r}_f - \mathbf{r}_i)$. **TRUE FALSE**

41. It is not possible to depict time on an angle-angle diagram. **TRUE FALSE**

42. The linear velocity of different points along a rotating rigid body is equal. **TRUE FALSE**

43. The direction of the linear velocity vector is perpendicular to the path of a rotating rigid body. **TRUE FALSE**

44. A vector indicated as a curved arrow has a direction that is tangential to the page. **TRUE FALSE**

45. Angular velocity is a vector. **TRUE FALSE**

46. The cross product is a mathematical operator for vectors. **TRUE FALSE**

47. The $\mathbf{r}\omega^2$ term accounts for the change in the magnitude of ω. **TRUE FALSE**

48. The direction of **v** changes continuously in angular motion. **TRUE FALSE**

49. At complete extension, elbow joint angle is about 3.14 rad. **TRUE FALSE**

50. Brachioradialis is an elbow extensor muscle. **TRUE FALSE**

51. Triceps brachii is an elbow extensor muscle. **TRUE FALSE**

52. The unit of measurement for angular acceleration is rad/s^2. **TRUE FALSE**

53. Electromyography measures the electrical output of muscle. **TRUE FALSE**

54. Biceps brachii can control elbow extension. **TRUE FALSE**

55. Muscle activity can cause a limb to accelerate. **TRUE FALSE**

56. There is always at least one foot on the ground in walking. **TRUE FALSE**

57. As the speed of locomotion increases, the time of support decreases. **TRUE FALSE**

58. A stride consists of two steps. **TRUE FALSE**

59. If stride rate remains constant, speed increases as stride length increases. **TRUE FALSE**

60. A step can be defined as the interval from right footstrike to right footstrike. **TRUE FALSE**

61. The swing phase is from takeoff to footstrike. **TRUE FALSE**

62. Stride length for an adult during maximum sprinting has an approximate value of 2.5 Hz. **TRUE FALSE**

63. When running at 8 m/s, stride rate will be about 20 Hz. **TRUE FALSE**

64. Knee flexion during the stance phase decreases with running speed. **TRUE FALSE**

65. It costs less energy to increase stride length than stride rate. **TRUE FALSE**

66. Differences in maximal running speeds (9 vs. 11 m/s) do not depend on step length but are influenced by stride rate. **TRUE FALSE**

67. An angle of 1.0 rad is larger than a right angle. **TRUE FALSE**

68. Knee angle never reaches π rad during a normal running stride. **TRUE FALSE**

69. The range of motion on an angle-angle diagram is measured in radians. **TRUE FALSE**

70. A relative angle is defined as one between adjacent body segments. **TRUE FALSE**

71. An angle-angle diagram shows the range of motion for two joints. **TRUE FALSE**

72. Angle-angle diagrams usually plot a relative angle against an absolute angle. **TRUE FALSE**

73. Ankle dorsiflexion refers to extending the ankle and pointing the toes. **TRUE FALSE**

74. Arm motion during running typically consists of concurrent elbow and shoulder flexion and concurrent elbow and shoulder extension. **TRUE FALSE**

75. The kick is a striking event. **TRUE FALSE**

76. The kinematics of the overarm throw typically involve a distal-to-proximal sequence of activation. **TRUE FALSE**

77. In the overarm throw, the velocity of the ball does not increase significantly until late in the movement. **TRUE FALSE**

78. Displacement is greater about the shoulder joint compared to the elbow joint during the overarm throw. **TRUE FALSE**

79. During the kick, the thigh reaches a peak forward angular velocity before the shank. **TRUE FALSE**

80. The trajectory that a projectile follows while in the air is linear. **TRUE FALSE**

81. Air resistance never affects the trajectory of a projectile. **TRUE FALSE**

82. The speed of a projectile is measured in m/s. **TRUE FALSE**

83. The vertical velocity of a projectile at release determines how high the projectile goes. **TRUE FALSE**

84. If air resistance is negligible, the horizontal velocity of a projectile remains constant. **TRUE FALSE**

85. Flight time depends on horizontal velocity at release and on the height of release above the landing surface. **TRUE FALSE**

86. The vertical velocity of a projectile is zero at the peak of its trajectory. **TRUE FALSE**

87. The symbol $\bar{\mathbf{a}}$ represents average acceleration. **TRUE FALSE**

88. The path of a projectile is hyperbolic. **TRUE FALSE**

MULTIPLE-CHOICE QUESTIONS

89. What would be your height and weight in SI units if your height were 5 ft 5 in. and you weighed 150 lb?
    A. 1.65 cm, 668 N
    B. 1.65 m, 669 kg
    C. 1.65 cm, 68 kg
    D. 1.65 m, 669 N
    E. 1.65 m, 68 kg

90. Which is not one of the seven base units of the SI system?
    A. Length
    B. Force
    C. Temperature
    D. Time
    E. Mass

91. What is the unit of measurement for angular displacement?
    A. m
    B. rad/s
    C. rad
    D. m/s

92. Which base unit is Hz derived from?
    A. Frequency
    B. Length
    C. Time
    D. Mass

93. Which base unit is the unit of measurement for velocity derived from?
    A. Time
    B. Length
    C. Angle
    D. Frequency

94. Which is not a unit of measurement for acceleration?
    A. rad/s/s
    B. $m \cdot s^{-2}$
    C. m/s^{-2}
    D. rad/s^2

95. When velocity is at a maximum, what will acceleration be?
    A. Minimum
    B. Positive
    C. Negative
    D. Zero
    E. Maximum

96. Which variable or variables are not a vector quantity?
    A. Displacement
    B. Acceleration
    C. Time
    D. Position

97. What is the symbol for angular velocity?
    A. $\mathbf{v}$
    B. θ
    C. α
    D. ω

98. If an object experienced an acceleration of 10 m/s^2, what would its velocity be after 2 s? Assume that it began with an initial velocity of 3 m/s.
    A. 20 m/s
    B. 13 m/s
    C. 23 m/s
    D. 30 m/s

99. With regard to Figure 1.2, identify the incorrect statements.
    A. The object is going downward from Position 4 to Position 5.
    B. The magnitude of the velocity from Position 2 to Position 3 is less than that for Position 1 to Position 2.
    C. Velocity is zero from Position 3 to Position 4.
    D. The unit of measurement for velocity is m/s.
    E. The graph gives information about the magnitude only and not about the direction of the velocity vector.

100. The slope of which type of graph indicates velocity?
   A. Position-time
   B. Angle-angle
   C. Force-time
   D. Acceleration-time

101. When a velocity-time graph reaches a maximum, which variable has a value of zero?
   A. Displacement
   B. Acceleration
   C. Position
   D. Velocity

102. When a velocity-time graph crosses zero, which variable or variables have a value of zero?
   A. Acceleration
   B. Velocity
   C. Displacement
   D. Position

103. When a velocity-time graph crosses zero, which variable or variables have a maximum or a minimum?
   A. Velocity
   B. Displacement
   C. Position
   D. Acceleration

104. What information is not included on an angle-angle diagram?
   A. Range of motion
   B. Velocity
   C. Direction of displacement
   D. Time

105. Which equation expresses the relationship between linear and angular velocity?
   A. $v = r\omega$
   B. $v = r\omega^2$
   C. $v = r\theta$
   D. $v = r\alpha$

106. What is the unit of measurement for the term $\mathbf{r}\omega^2$?
   A. $m \cdot rad/s^2$
   B. $m \cdot rad/s/s$
   C. m/s^2
   D. m/s^{-2}

107. The swing phase is best defined as the interval between which events?
   A. FS to TO
   B. rTO to lTO
   C. lFS to rTO
   D. TO to FS

108. Which of the following does not happen as a runner increases speed?
   A. Stance time decreases.
   B. First, stride length increases at lower speeds, and then stride rate increases in order to increase speed further.
   C. The amount of knee flexion following footstrike decreases.
   D. Faster runners can achieve a higher stride rate.

109. Which statements about the kinematics of gait are correct?
   A. A stride can be defined as the interval from right footstrike to right footstrike.
   B. Initial increases in running speed are produced by concurrent increases in stride length and stride rate.
   C. Knee angle reaches 3.14 rad during a normal running gait (i.e., as seen on knee-thigh angle-angle diagrams).
   D. Stride rate is measured in Hz/s.

110. What is the typical pattern of arm motion during running?
   A. Concurrent shoulder and elbow flexion
   B. Shoulder extension and elbow flexion
   C. Concurrent shoulder and elbow extension
   D. Shoulder flexion and elbow extension

111. The shoulder joint is not capable of rotation about which axis?
   A. External-internal rotation
   B. Pronation-supination
   C. Flexion-extension
   D. Abduction-adduction

112. What are the significant differences between a throw and a kick?
   A. The duration of contact
   B. The rigidity of the limb
   C. The proximal-to-distal sequence of body segment activity
   D. The purpose of altering the momentum of the projectile

113. Which of the following are not characteristic of projectile motion in which the effects of air resistance are negligible?
   A. Horizontal distance depends on horizontal velocity at release and flight time.
   B. Flight time depends on the horizontal velocity at release and the height of release above the landing surface.
   C. The vertical velocity of the projectile remains constant during flight.
   D. The horizontal acceleration experienced by the projectile is zero.

114. What is the optimum angle of release when the landing height is lower than the release height?
   A. 0.785 rad
   B. Greater than 0.785 rad
   C. 1.57 rad
   D. Less than 0.785 rad

PROBLEMS

115. Determine the height in appropriate SI units of a person who is 5 ft 8 in.
116. According to the March 1994 issue of *Track and Field News*, ''Records Section,'' the world-record performances for selected race distances are as follows:

Distance	Women	Men
100 m	10.49 s	9.86 s
200 m	21.34 s	19.72 s
400 m	47.60 s	43.29 s
800 m	1 min 53.28 s	1 min 41.73 s
1,500 m	3 min 50.46 s	3 min 28.86 s
1 mile	4 min 15.61 s	3 min 46.39 s
3,000 m	8 min 06.11 s	7 min 28.96 s
5,000 m	14 min 37.33 s	12 min 58.39 s
10,000 m	28 min 31.78 s	26 min 58.38 s
Marathon	2 hr 21 min 08.00 s	2 hr 06 min 50.00 s

Calculate the average speed at which these performances were run.

117. Select the appropriate SI unit of measurement for each variable.

Acceleration	_______	A. kg · m/s
		B. rad/s
Area	_______	C. ft
		D. N/m
Mass	_______	E. μm^3
		F. kg
Height	_______	G. cm^2
		H. N
Weight	_______	I. lb
		J. m/s^2
Velocity	_______	K. ft/s
		L. m
Volume	_______	M. rad
		N. degree
Direction	_______	O. s

118. A constant force (e.g., gravity) applied to an unsupported object produces a constant acceleration, and the absence of a force means that the object is at rest or traveling at a constant velocity (i.e., zero acceleration). Draw a velocity-time graph for these two conditions.
119. A diver performs a dive from a handstand position off the 10-m tower. Assume that the effects of air resistance are negligible.
    A. Use Equation 1.4 to calculate the time it takes for the diver to reach the water.
    B. Determine the position (Equation 1.4) of the diver at 0.1-s intervals, and plot the results on a position-time graph.
    C. What is the velocity of the diver at the time contact is made with the water?
    D. What is the diver's acceleration at the 5-m mark?
120. The velocity-time graph shows four qualitatively different acceleration profiles associated with a movement. Briefly describe what happened to acceleration in each instance. In which instances was acceleration zero?

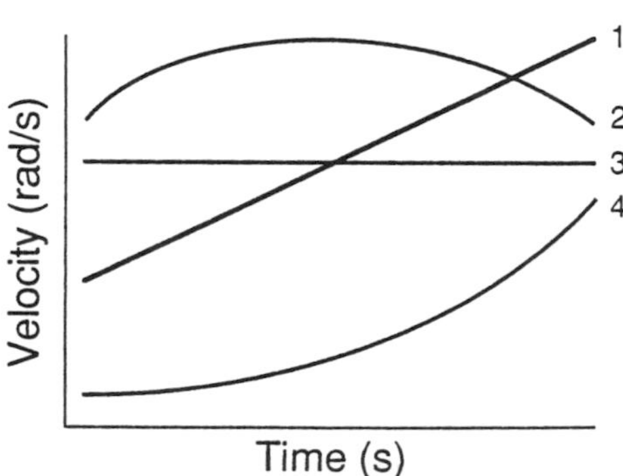

121. Suppose that an athlete ran a 100-m race in 9.98 s. What was his horizontal velocity when he crossed the finish line? Assume that the velocity of the athlete increased continuously from zero (initial) to some final value. What was the average acceleration of the athlete during the race?

122. What is the magnitude of the average acceleration (in m/s^2) required to slow a cyclist from 22 to 7 mph in 51 s?

123. Derive the acceleration-time graph from the velocity-time graph.

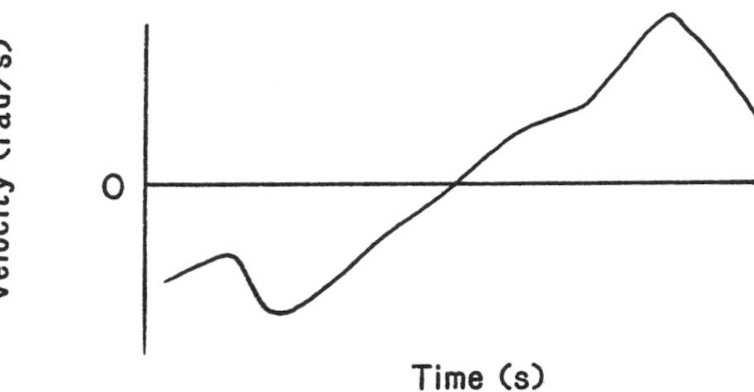

124. A ball is held 1.23 m above the ground and is dropped. It reaches the ground 0.5 s later.
   A. What was the average velocity (direction, magnitude, and unit of measurement) of the ball during the fall?
   B. What was the average acceleration experienced by the ball while it was falling?
   C. Suppose that the ball rebounded from the ground to a height of 1.0 m, where it was caught. Qualitatively graph the position-time relationship of the ball from its release until it was caught. Label the axes with the appropriate variables and their units of measurement.

125. The following position-time data represent the angular motion of the leg during a punt. The angle indicates the position of the thigh relative to the right horizontal. Calculate and graph the associated angular velocity and acceleration curves.

Angle (rad)	Time (s)
2.57	0.115
2.59	0.130
2.62	0.145
2.79	0.161
3.11	0.176
3.46	0.191
3.89	0.206
4.45	0.221 (contact)
5.06	0.236
5.49	0.252
5.83	0.267
5.90	0.282
5.93	0.297
5.93	0.312
6.02	0.327

126. The motions involved in the overarm throw are quite complex. Figure 1.22 shows the velocity-time profile for the ball during a throw.
   A. Qualitatively graph both the position- and acceleration-time curves for the ball based on Figure 1.22.
   B. What feature of the throwing motion could cause the phase of negative acceleration that occurs about 100 ms prior to release?

127. The following position-time data of a runner were obtained during a 50-m sprint.

Distance (m)	0	10	21	29	40
Time (s)	0	3.1	4.3	5.5	6.6

Determine the following values for that runner:
   A. Average speed over 40 m
   B. Acceleration at 21 m
   C. Acceleration at 5.5 s

128. Based on the force-motion relationship, which diagram correctly represents the motion of the object once it is released (M. McCloskey, 1983)? Explain the reason for your choice.

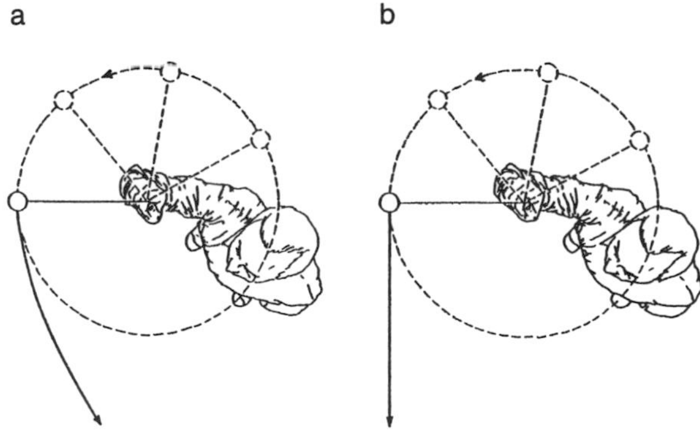

Note. From "Intuitive Physics" by M. McCloskey, 1983, *Scientific American*, **248**, p. 128. Copyright 1983 by Scientific American, Inc. Reprinted by permission.

129. A ball was dropped from a certain height with an initial velocity of zero and hit the ground 0.18 s later. Calculate the following variables related to the motion of the ball:
   A. Final velocity
   B. Average velocity
   C. Height from which the ball was dropped
   D. Average acceleration

130. An individual performed a throw in darts by holding the elbow joint stationary and rotating the forearm-hand (one rigid segment) about the elbow joint. Suppose that the distance from the elbow joint to the dart in the hand was 41 cm and that the forearm-hand had an angular velocity of 18 rad/s at release.
   A. Calculate the linear velocity of the dart at release.
   B. What was the magnitude of the acceleration component that accounts for the change in direction of linear velocity?
   C. If the angle of the linear velocity vector at release was 0.64 rad above horizontal, how far would the dart travel horizontally before it reached a landing height at the same level as the release height?

131. The following data were obtained from a film of an individual who was running at about 5.6 m/s. The measured angles (rad) are for the thigh and leg relative to the right horizontal.

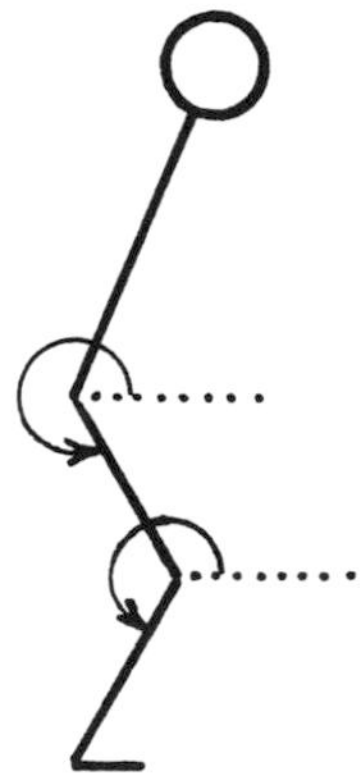

The time between each frame of film is 0.0067 s.

Frame	Thigh	Leg	Knee	Event
1	5.73	4.29	1.70	lTO
4	5.70	4.47	1.91	
7	5.65	4.75	2.24	
10	5.60	4.90	2.44	
13	5.51	4.95	2.58	
16	5.40	4.94	2.68	
19	5.30	4.81	2.65	rFS
22	5.24	4.59	2.49	
25	5.20	4.40	2.34	
28	5.12	4.24	2.26	
31	4.94	4.08	2.28	
34	4.71	3.96	2.39	
37	4.49	3.89	2.54	
40	4.25	3.90	2.79	
43	4.12	3.91	2.93	rTO
46	4.10	3.77	2.81	
49	4.12	3.53	2.55	
52	4.17	3.32	2.29	
55	4.27	3.08	1.95	
58	4.39	2.85	1.60	
61	4.49	2.72	1.37	lFS
64	4.65	2.67	1.16	
67	4.88	2.66	0.92	
70	5.09	2.81	0.86	
73	5.30	3.10	0.94	
76	5.53	3.39	1.00	
79	5.67	3.74	1.21	
82	5.70	4.10	1.54	
85	5.71	4.27	1.70	

Plot a knee-thigh angle-angle diagram.

132. A javelin is thrown at an angle of 0.78 rad to the horizontal with a release velocity directed along the line of projection of 22.5 m/s. What are the horizontal and vertical components of the release velocity?

133. A punted soccer ball reached a peak height of 23 m in an indoor stadium. If we assume that the effects of air resistance were negligible, how fast was the ball going when it hit the ground?

134. A young child hits a T-ball from a tee that is 0.75 m above the ground. The ball is hit so that it is displaced at an angle of 0.7 rad above the horizontal. The resultant velocity along the line of projection is 4 m/s. Assume that the effects of air resistance are negligible.
   A. How long does the ball stay in the air before it hits the ground?
   B. How far (horizontally) does the ball travel?
   C. Why is it reasonable to neglect air resistance?

135. At the start of a football game, the ball was kicked off with a resultant velocity of 27.0 m/s at an angle of 1.15 rad with respect to the horizontal. The ball had a hang time of 4.9 s before the receiver caught it. Assume that the effects of air resistance were negligible and that the ball reached a peak height at about one half of hang time.
   A. What was the peak height reached by the ball?
   B. What horizontal distance did the ball travel?

136. An athlete is about to jump with an initial takeoff velocity of 2.9 m/s.
   A. How high can she raise her center of gravity if she jumps straight up?
   B. What height will she reach if her takeoff velocity is at an angle of 0.70 rad to the ground?

137. In an intramural croquet championship, Ralphie Roundtree is playing Mindy Malaprop in the final. Mindy will win the championship if she can hit the final post on this shot. Suppose that Mindy contacts the ball, which has a mass of 1.76 kg, so that its initial velocity is 9.0 m/s. As the ball travels through the grass, it has an acceleration of –4.0 m/s^2. If Mindy hits the ball straight at the post, what is the greatest distance that she can be from the post when she hits the ball and still have the ball hit the post?

138. A basketball player drives in for a layup and leaves the ground at an angle of 0.91 rad to the horizontal with a velocity of 5.7 m/s. A defensive player goes straight up to block the shot with a velocity of 4.8 m/s.
   A. Which player jumps higher?
   B. How long is each player in the air? Assume that they do not contact each other.

139. A golf ball was driven horizontally from a point 3 m above a level fairway. If the ball landed (and stopped) on the fairway at a point 150 m horizontally from the starting point, what was the velocity of the ball immediately after it was hit?

140. An athlete who was performing a long jump had a vertical velocity of 3.4 m/s and a horizontal velocity of 9.5 m/s at takeoff. Assume that the effects of air resistance were negligible, that the height of the center of gravity was 1.09 m at takeoff, and that the height of the center of gravity at landing was 0.42 m.
   A. What was the velocity (magnitude and direction) at takeoff?
   B. How high did the athlete raise his center of gravity?
   C. What was the horizontal distance of the jump?

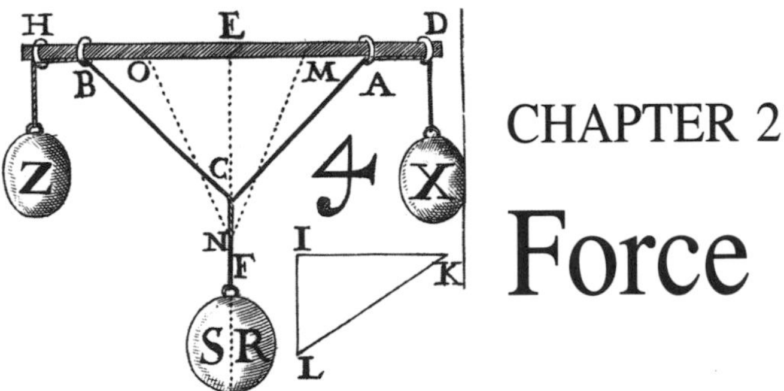

CHAPTER 2

Force

Force is a concept that is used to describe the *interaction* of an object with its surroundings, including other objects. It can be defined as an agent that produces or tends to produce a change in the state of rest or motion of an object. For example, a ball sitting stationary (zero velocity) on a pool table will remain in that position unless it is acted upon by a force. Similarly, a person gliding on ice skates will maintain a constant velocity unless a force changes the motion—that is, the speed (magnitude) or the direction. The study of motion that includes consideration of force as the cause of movement is called **kinetics**.

Because force is a fixed vector, the complete specification of its action must include magnitude, direction, and point of application. As a vector, force has both magnitude and direction and can be treated graphically or trigonometrically to determine the resultant effect of several forces (**composition**) or alternatively to resolve a resultant effect into several components (**resolution**). These characteristics are useful in the analysis of human movement because they can be used to determine (a) the net effect of several forces acting upon an object or body segment and (b) the functional effect (e.g., rotation vs. stabilization) of a single force acting upon a system.

Figure 2.1 illustrates how composition is used to determine the resultant effect of coactivation of both the clavicular and sternal portions of the pectoralis major muscle. The graphic technique involves manipulating forces represented as arrows (vectors), where arrow length indicates force magnitude and the orientation of the arrow represents force direction. A feature of vectors that is useful in this procedure is that they can slide along their line of action and still have the same effect. The technique involves (a) careful measurement of arrow lengths and angles and (b) a scale factor with which length is related to force. Figure 2.1a indicates the direction and magnitude of the force exerted by each of the clavicular ($\mathbf{F}_{m,c}$) and sternal ($\mathbf{F}_{m,s}$) portions of the muscle pectoralis major. The clavicular component exerts a force of 224 N, which is directed at an angle of 0.55 rad above the horizontal, and the sternal component has a magnitude of 251 N and acts 0.35 rad below the horizontal. The process of graphically determining the net effect involves adding the components head to tail by sliding either one of the vectors along its line of action (Figure 2.1b) and then joining the open ends (open tail to open head) to produce the resultant (Figure 2.1c). The magnitude of the resultant is obtained by measuring the length of the arrow (3 cm) and then converting this measurement to newtons (400 N) with the indicated scale. Direction is obtained by measuring the angle with a protractor. The resultant effect ($\mathbf{F}_m$), therefore, has a magnitude of 400 N and a direction of 0.1 rad above horizontal.

This analysis can also be applied to systems that comprise several force vectors. Figure 2.2 shows a leg with a fractured femur that is placed in a Russell traction, which consists of several cables and pulleys that suspend the limb. If we are interested in determining the resultant force exerted on the femur, we can perform a graphic analysis and add the vectors head to tail. To determine the resultant, we must draw each vector with the correct magnitude (length) and direction (angle). Three forces are applied to the leg, one through the strap at the knee and two through the pulley attached to the base of the cast. In a pulley-cable system, the tension in the cable

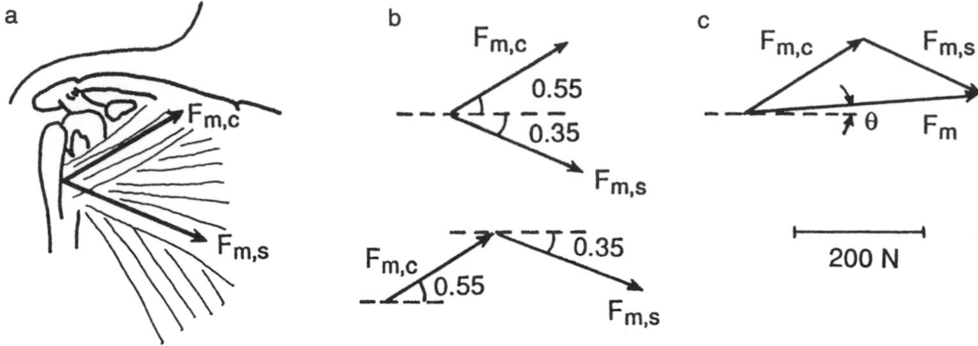

Figure 2.1 Graphical composition of the resultant force ($\mathbf{F}_m$) associated with the activation of both the clavicular ($F_{m,c}$) and sternal ($F_{m,s}$) components of the muscle pectoralis major. (Angles are measured in radians.)

is constant throughout the length of the cable, and the pulleys simply serve to alter the direction of the force vector. For this reason, the force vectors **P**, **Q**, and **S** in Figure 2.2 have the same magnitude but different directions. The resultant (**R**) has a magnitude indicated by its length and a direction of 0.35 rad above the horizontal.

The opposite procedure is to graphically resolve a resultant into components. This procedure (Figure 2.3) simply involves constructing a parallelogram (see Appendix B) such that the resultant represents the diagonal and thus the sides of the parallelogram indicate the magnitude and direction of the components. Again, the magnitude can be determined by use of a scale factor.

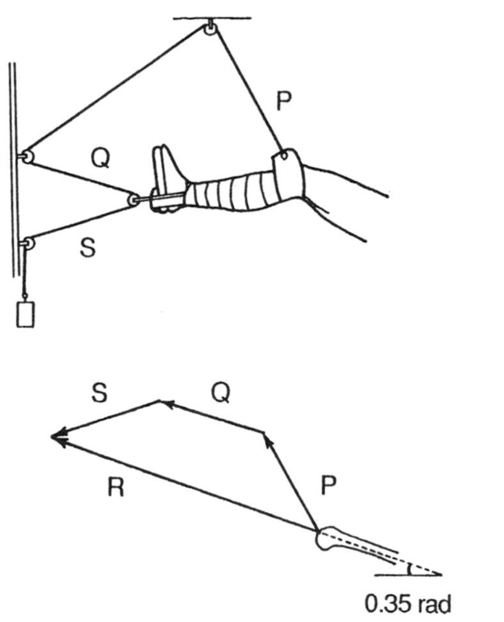

Figure 2.2 Leg with a fractured femur in a Russell traction.

Note. From *Biomechanics of Human Motion* (p. 75) by M. Williams and H.R. Lissner, 1962, Philadelphia: W.B. Saunders. Copyright 1962 by W.B. Saunders Company. Adapted by permission.

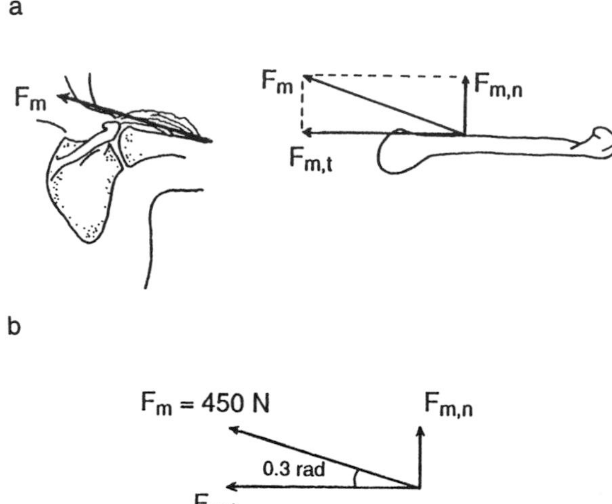

Figure 2.3 Graphic resolution of the resultant force ($\mathbf{F}_m$) produced by activation of middle deltoid muscle: (a) the normal ($F_{m,n}$) and tangential ($F_{m,t}$) components; (b) the trigonometric relations between $F_{m,n}$ and $F_{m,t}$.

In many instances, especially focusing on muscle force, the parallelogram becomes a rectangle so that we can identify components that are oriented at right angles (1.57 rad, or 90°) to each other. In the analysis of muscle force, these **orthogonal** (perpendicular) components are referred to as the normal and tangential components. The **normal component** represents the proportion of muscle force that acts perpendicular to the long bone of the segment and therefore produces the rotation of the body segment. The **tangential component** indicates the proportion directed along the bone toward the joint and thus contributes to joint stabilization or compression.

Composition and resolution can also be accomplished using trigonometry—the advantage being that less graphic precision is required. Suppose that the resultant force in Figure 2.3 has a magnitude of 450 N and its line of action is 0.3 rad (17.2°) with respect to the humerus. Let us calculate the magnitude of the components. First, we reconstruct the parallelogram, add the given data (Figure 2.3b), and use known trigonometric identities to solve for $F_{m,n}$ and $F_{m,t}$.

$$\sin 0.3 = \frac{F_{m,n}}{450}$$

$$F_{m,n} = 450 \times \sin 0.3$$

$$= 450 \times 0.2955$$

$$F_{m,n} = 133 \text{ N}$$

$$\cos 0.3 = \frac{F_{m,t}}{450}$$

$$F_{m,t} = 450 \times \cos 0.3$$

$$= 450 \times 0.9553$$

$$F_{m,t} = 430 \text{ N}$$

Thus the normal component has a magnitude of 133 N, and the tangential component has a magnitude of 430 N. Both of these values seem reasonable because (a) the angle of pull of most muscles is rather shallow and thus most of the resultant force is generated in the tangential direction, and (b) the resultant represents the diagonal of the parallelogram and therefore should be the largest of the three values.

To verify that the graphic and trigonometric procedures are comparable, you can graphically resolve the resultant in Figure 2.3 to determine the magnitude of the normal and tangential components. Redraw the resultant ($\mathbf{F}_m$) on graph paper with a scale of 1 cm = 25 N and with the arrow directed 0.3 rad above the horizontal. Next, draw the rectangle around the resultant so that the resultant appears as the diagonal. Measure the lengths of the two sides ($F_{m,n}$ and $F_{m,t}$) and use the scale factor to convert these distances to forces. How similar are the trigonometric ($F_{m,n}$ = 133 N and $F_{m,t}$ = 430 N) and the graphic results?

Newton's Laws of Motion

Isaac Newton (1642–1727) addressed the relationship between force and motion and provided three statements, which are known collectively as the *laws of motion*. These laws are referred to as the laws of inertia, acceleration, and action-reaction.

Law of Inertia

> Every body continues in its state of rest or uniform motion in a straight line except when it is compelled by external forces to change its state.

More simply, *a force is required to stop, start, or alter motion*. This law is easily demonstrated by astronauts as they perform maneuvers in a weightless (microgravity) environment (e.g., tossing objects to one another or performing gymnastics stunts). In a gravitational world, however, forces act continuously upon bodies, and a change in motion occurs when there is a *net imbalance* of forces. In this context, the term *body* can refer to the entire human body, just part of the human body (e.g., thigh, hand, torso), or even some object (e.g., shot put, baseball, Frisbee).

To appreciate fully the implications of this law, we need to understand what is meant by the term **inertia**. The concept of inertia relates to the difficulty with which an object's velocity is altered. **Mass**, expressed in grams (g), is a measure of the amount of matter constituting an object and is a quantitative measure of inertia. Consider two objects of different mass but with the same amount of motion (similar velocities). It is more difficult to alter the motion of the more massive object, and hence the more massive object is described as having a greater inertia. Because motion is described in terms of velocity, the inertia of an object is a property of matter that is revealed only when the object is being accelerated, that is, when there is a change in velocity.

According to the law of inertia, an object in motion will continue in uniform motion (constant velocity) unless acted upon by a force. This means that the tendency of an object in motion is to travel in a straight line. For example, consider the situation in Problem 128 in chapter 1. An individual swings a ball tied to a string over his head. When the string is released, will the ball follow the trajectory depicted in Panel a or b? To answer this question we simply need to invoke Newton's law of inertia. If no force is acting on an object, the object will, due to its inertia, travel in a straight line. Once the individual releases the string, no horizontal force is acting on the ball and the ball will travel in a straight line (b). But then, how can a pitched ball in baseball or softball be made to vary from a straight line? Quite simply, according to the law of inertia, other forces (e.g., air resistance, gravity) must be acting on the ball once it has been released. Remember, the trajectory of a projectile will be a straight line unless it is influenced by forces, such as gravity and air resistance.

Because uniform motion is represented as a constant velocity, both in magnitude and direction, a force must be present when an object travels along a curved path. This force prevents the object from following its natural tendency of traveling in a straight line. This can be demonstrated by considering the motion of the ball in Problem 128 at two instances in time. For these two positions, the motion is indicated by a velocity vector (Figure 2.4).

The length of the arrow (magnitude of the vector) is the same in the two positions, so the ball is traveling at a constant velocity (magnitude). But the direction of the vector differs. This change in direction represents a change in one feature of motion and hence can be

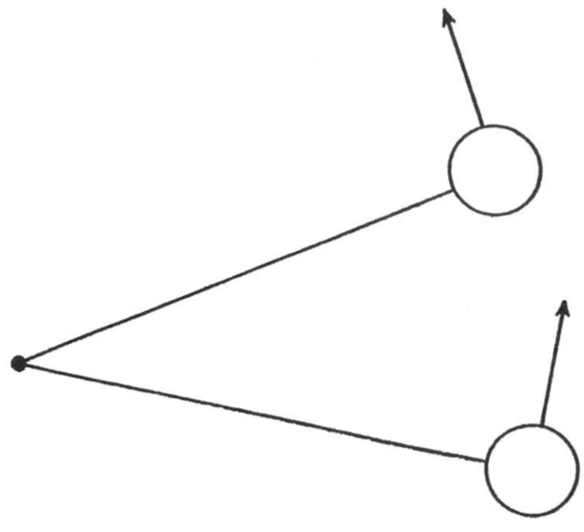

Figure 2.4 Ball on a string that is being displaced at a constant speed. The magnitude of the velocity vector remains constant, but its direction changes.

caused only by the presence of a force. The inwardly directed force that changes the direction but not the magnitude (speed) of velocity during angular motion is known as **centripetal force**. Centripetal force ($\mathbf{F}_c$) is defined as

$$\mathbf{F}_c = \frac{m\mathbf{v}^2}{r} \tag{2.1}$$

where m = mass, $\mathbf{v}$ = velocity, and r = the radius of the curved path. In Equation 2.1, what would happen to $\mathbf{F}_c$ if the speed of the ball ($\mathbf{v}$) remained constant, the length of the string (r) stayed the same, and the mass of the ball increased? Centripetal force ($\mathbf{F}_c$) would have to increase to satisfy these conditions. Conversely, $\mathbf{F}_c$ would decrease if the ball mass (m) and velocity ($\mathbf{v}$) remained constant while the length of the string (r) increased. Consider the problem recently resolved by track-and-field officials. Athletes performing the hammer throw event were throwing the hammer too far and were endangering the athletes running around the track. How could the officials manipulate the implement to decrease the distance of the throw?

Because the distance of a throw depends on the velocity at release, we rearrange Equation 2.1 so that velocity is on the left-hand side and then determine what can be changed:

$$\mathbf{v} = \sqrt{\frac{\mathbf{F}_c r}{m}}$$

This equation indicates that velocity (and hence the displacement experienced by the hammer) is directly related to $\mathbf{F}_c$ and r and inversely related to m. As $\mathbf{F}_c$ and r get smaller, so does $\mathbf{v}$, but as m gets smaller, $\mathbf{v}$ gets larger. The options available to the officials are to increase the mass of the hammer (m) or to decrease its length (r).

Law of Acceleration

> The rate of change of momentum of a body is proportional to the applied force and takes place in the direction in which the force acts.

The term **momentum** (**G**) describes the quantity of motion possessed by a body and is defined as the product of mass (m) and velocity ($\mathbf{v}$). A runner with a mass of 60 kg moving at a horizontal speed of 8 m/s possesses a momentum of 480 kg · m/s. Thus

$$\mathbf{G} = m\mathbf{v} \tag{2.2}$$

and the rate of change in momentum can be written as

$$\frac{\Delta \mathbf{G}}{\Delta t} = \frac{\Delta(m\mathbf{v})}{\Delta t}$$

The quantity Δm is of negligible concern in the analysis of human movement (i.e., mass is constant); therefore, according to the law of acceleration, the applied force (**F**) is proportional to the rate of change in momentum,

$$\mathbf{F} = \frac{\Delta \mathbf{G}}{\Delta t} = m\frac{\Delta \mathbf{v}}{t}$$

Further, as $\Delta\mathbf{v}/\Delta t$ represents the time rate of change in velocity, it is equivalent to acceleration (**a**), and

$$\mathbf{F} = m\mathbf{a} \tag{2.3}$$

Thus Equation 2.3 is the algebraic expression of Newton's law of acceleration and states that force is equal to mass times acceleration. Conceptually this is a cause-and-effect relationship. The left-hand side (**F**) can be regarded as the cause because it represents the interactions between a system and its environment. In contrast, the right-hand side reveals the effect because it indicates the kinematic effects ($m\mathbf{a}$) of the interactions on the system.

Law of Action-Reaction

> To every action there is an equal and opposite reaction.

The law of action-reaction implies that every effect one body exerts on another is counteracted by an effect that the second body exerts on the first. This interaction between bodies is described as a force, which has an effect on both bodies. In a strict sense it is misleading to imply that one body exerts an action, and the other body responds with a reaction. Rather, the two bodies interact simultaneously and this interaction is referred to as a force.

For example, consider a person performing a jump shot in basketball. During the act of jumping, the person exerts a force against the ground, and the ground responds simultaneously with a reaction force (the ground reaction force) on the jumper. The law of action-reaction indicates that the forces between the jumper and the ground (i.e., the effect of the jumper on the ground and the ground on the jumper) are equivalent in magnitude but opposite in direction. The consequence of this interaction, as specified by the law of acceleration (**F** = *m***a**), is that each body (i.e., the basketball player and the ground) experiences an acceleration that depends on its mass. If the average force was 1,500 N and the person had a mass of 75 kg, the person would experience an acceleration of 20 m/s^2. Due to the large mass of the planet Earth, however, the acceleration experienced by the ground would be imperceptible; that is, the product of mass and acceleration cannot exceed 1,500 N. Because the mass of the ground is large (compared to the basketball player), the acceleration of the ground would be small.

Free Body Diagram

In the analysis of human motion, many variables influence performance. The **free body diagram** reduces the complexity of a chosen analysis. It is a simplified diagram, usually a stick figure, of the system isolated from its surroundings that notes all the interactions between the system and its surroundings. The free body diagram provides a view of the body or a part of the body as an isolated entity in space (in this context a body can be anything that occupies space and has inertia). The free body diagram is a powerful analytical technique; it defines the extent of an analysis. Because force is the concept used to denote interactions between a system and its surroundings, *a free body diagram is a simplified figure upon which all the external forces that influence the system are indicated as arrows.* The arrows represent the forces as vectors. The free body diagram is a rigid unit, or a collection of interconnected parts that can be regarded as a single rigid system, which is isolated except for the force vectors acting on it. In general, the term *system* refers to an entity that can be characterized by at least one input and one output variable, whose internal behavior determines the relationship between the output and the input (Lamarra, 1990). For example, the heart is a system whose output is blood flow and whose inputs include mechanical, neural, and chemical stimuli. In a free body diagram, the input is the forces acting on the system, and the output is the motion of the system. Identifying an appropriate system, which may vary from one problem to another, is one of the greatest difficulties that beginning students encounter in the analysis of human movement.

Dempster (1961) traces the development of the free body diagram and indicates that the earliest precedents appear in the works of Archimedes, Leonardo da Vinci, and Giovanni Borelli. Examples of Borelli's work are shown in Figure 2.5. Borelli (1680) appears to have understood the concept of body levers and the need to balance muscle force and load in order to maintain equilibrium. Dempster's historical perspective indicates

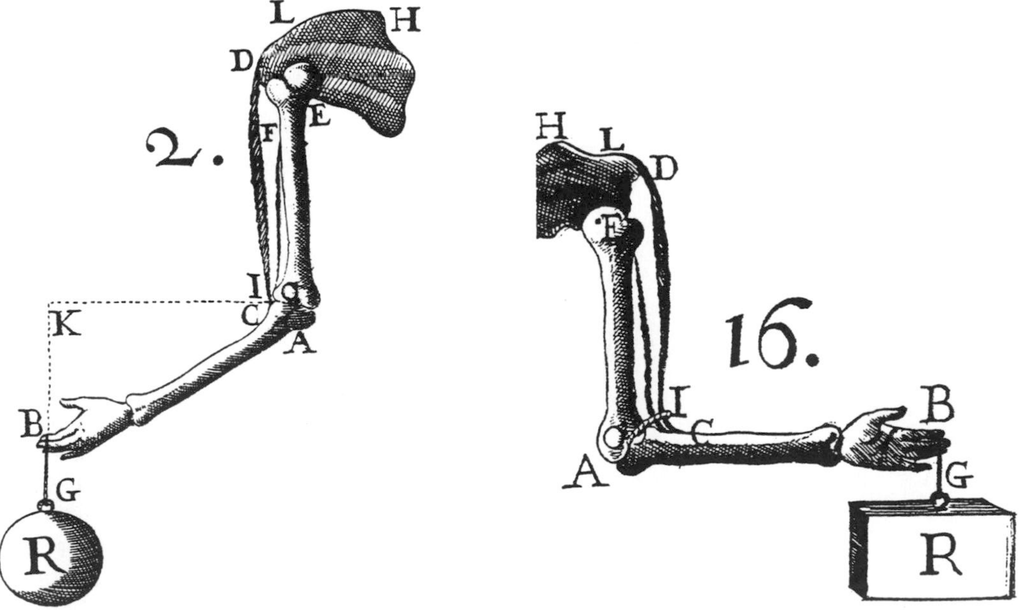

Figure 2.5 Representative equilibrium diagrams, among the earliest precedents of free body diagrams, from the work of Borelli.

that although the free body diagram as a technical method has developed over the past 100 years, its origins are based on more than 350 years of advances in classical physics.

In the remainder of this section, we examine several free body diagrams and consider the ways in which it is possible to define a system as a basis for the analysis of human movement. The actual forces that can be included in a free body diagram will be described in more detail later in the chapter. In general, however, when a free body diagram includes only two forces, the forces must be equal, opposite, and collinear. When the free body diagram includes three or more forces, the forces must be concurrent (i.e., their lines of action must converge); otherwise one of the forces would exert an unbalanced effect on the system (Meriam, 1987).

For the first example, suppose that we wanted to determine the magnitude and direction of the ground reaction force experienced by a runner (Figure 2.6a). According to the law of action-reaction, when the runner pushes on the ground, the ground pushes back; we want to know how hard the ground pushes back. The first step is to specify and draw the system (Figure 2.6b). The next step is to identify all the external forces acting on the system with arrows of correct length (magnitude) and direction and to label them appropriately (Figure 2.6c). The forces shown in Figure 2.6c represent air resistance ($\mathbf{F}_a$), weight ($\mathbf{F}_w$), and ground reaction force ($\mathbf{F}_g$). In general, the forces include the weight of the system and forces arising from contact with other bodies (e.g., ground reaction force, air resistance).

Free body diagrams show only the forces acting on the system and not those within the system. For example, the free body diagram in Figure 2.6c does not show any muscle forces across the knee joint even though the quadriceps femoris muscle is undoubtedly contracting; this is not shown because the muscle force is internal to this system. If, however, the aim of an analysis is to examine muscle activity across a joint during a movement, then a different system must be defined. If we consider a particular interaction, such as the muscle activity associated with a specific movement, then we must define the system so that the interaction occurs between the system and its surroundings. We need to specify the system so that we can see the agent causing the interaction. In the case of determining muscle activity, this involves figuratively cutting through the desired joint (e.g., knee joint). Around each joint there are a variety of tissues that include muscle, joint capsule, ligaments, and so forth. We usually mechanically distinguish the effects of muscle from those of the other tissues. Consequently, when we draw a free body diagram that involves cutting through a joint, we identify a net muscle force (the resultant muscle force) and a force that accounts mainly for the bone-on-bone contact (the joint reaction force) of adjacent body segments. Let us return to the example outlined in Figure 2.6. For examining the effects of the muscles across the knee joint at the instant shown in Figure 2.6b, the appropriate free body diagram is shown in Figure 2.7.

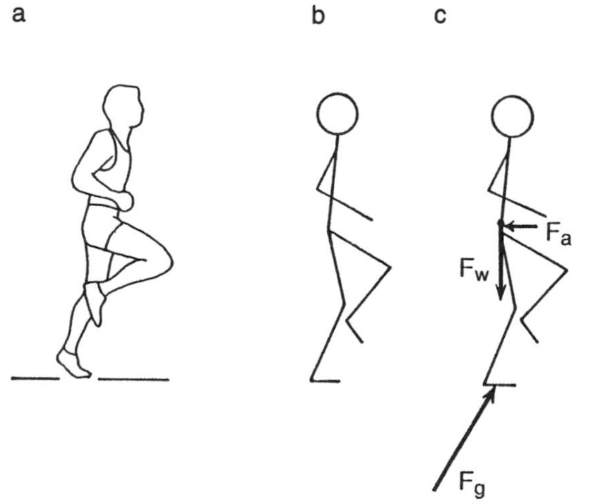

Figure 2.6 Derivation of a free body diagram from (a) real-life figure to (b) the identification of the system and (c) the inclusion of the forces imposed on the system by the surroundings.

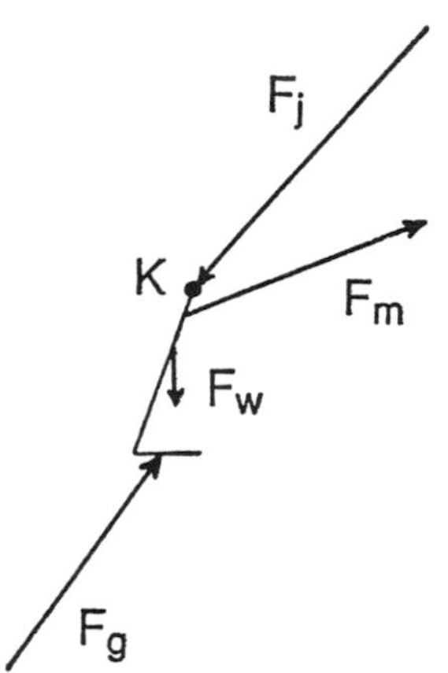

Figure 2.7 Free body diagram of a system (leg + foot) upon which four external forces are acting. $\mathbf{F}_g$ = ground reaction force; $\mathbf{F}_j$ = joint reaction force, $\mathbf{F}_m$ = resultant muscle force; $\mathbf{F}_w$ = weight; K = knee joint.

Suppose the situation arose in which a weight lifter, whose training program includes power cleans (the barbell is lifted from about knee height to the chest in one continuous motion), began to develop back pains after several months of training. An appropriate analysis might include determining how much force the muscles that extend the back and hip exert during the movement. Because this force is greatest at the beginning of the movement, the analysis would focus on the weight lifter at the beginning position (Figure 2.8a). The object of the analysis is to calculate a muscle force, so the system must be defined to identify the muscle effect on the free body diagram. This involves figuratively cutting through the hip joint so that the system becomes either the part of the weight lifter above the hip joint plus barbell or the part of the weight lifter below the hip joint. In the former instance, the stick figure would simply include a circle for the head and a line for the trunk and arms (Figure 2.8b).

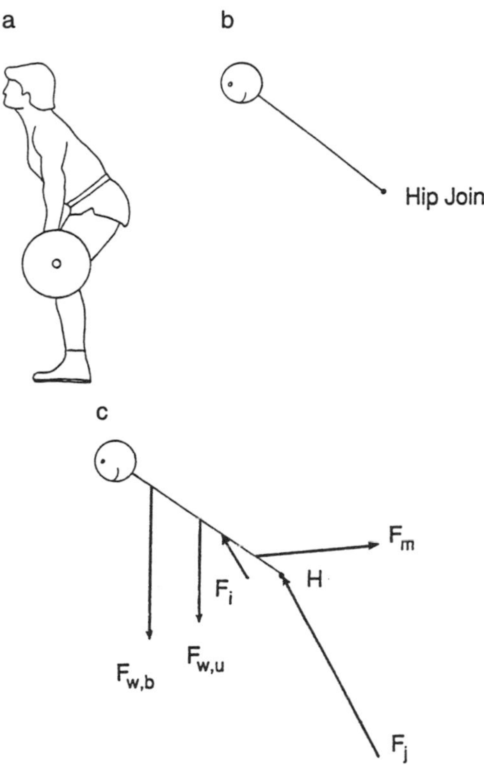

Figure 2.8 The derivation of a free body diagram to analyze part of a weight lifting movement: (a) whole-body diagram of the weight lifter; (b) system used to expose the forces across the hip joint; (c) free body diagram showing the interactions between the system and its surroundings.

The next step would be to illustrate all the interactions, drawn as force vectors (arrows), between the system (upper body of the weight lifter plus the barbell) and the surroundings. First, one of the most obvious forces is that due to gravity—that is, the weight of the barbell ($\mathbf{F}_{w,b}$) and the weight of the upper body of the lifter ($\mathbf{F}_{w,u}$). The direction of the weight vectors is always vertically downward. Second, because we are cutting through a joint, we must show both the resultant muscle force ($\mathbf{F}_m$) and the joint reaction force ($\mathbf{F}_j$) on the free body diagram. The joint reaction force ($\mathbf{F}_j$) represents the forces due to ligaments, joint capsules, and bone-on-bone contact. Although the direction of $\mathbf{F}_m$ is reasonably straightforward in that it opposes the rotation produced by the load (i.e., the weight vectors), the direction of $\mathbf{F}_j$ is usually unknown and is arbitrarily drawn in any direction. But if the system is in equilibrium, the forces directed to the right must be balanced by forces directed to the left, forces directed upward must be counteracted by forces that go downward, and forces that produce clockwise rotations must be balanced by forces that cause counterclockwise rotation. In other words, the forces must be concurrent. Another force appropriate for this free body diagram is that due to the fluids in the abdominal cavity (intraabdominal pressure, $\mathbf{F}_i$), which exerts a force that tends to cause extension about the hip joint (*H*) (Andersson, Örtengren, & Nachemson, 1977; Eie & Wehn, 1962; Rab, Chao, & Stauffer, 1977). This reflexively controlled force functions as a protective mechanism and has a significant role in lifting activities (Marras, Joynt, & King, 1985). These five forces ($\mathbf{F}_{w,b}$, $\mathbf{F}_{w,u}$, $\mathbf{F}_j$, $\mathbf{F}_m$, $\mathbf{F}_i$) represent the major interactions between this system and its surroundings. The appropriate free body diagram is shown in Figure 2.8c.

In the weight-lifter example (Figure 2.8), the muscle forces acting about the hip were simplified into a single resultant muscle force ($\mathbf{F}_m$). In reality, many muscles cross the hip joint and are active during lifting tasks. It is possible to draw more detailed free body diagrams that identify the individual muscles. Figure 2.9 provides

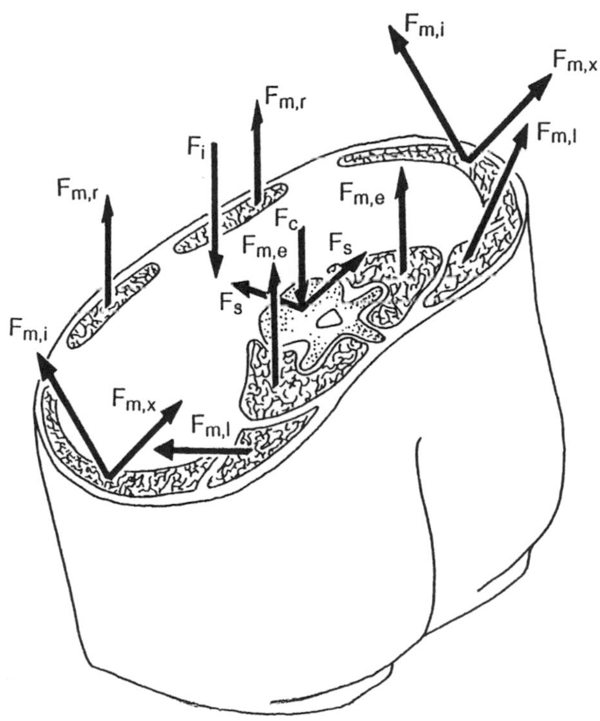

Figure 2.9 Schematic free body diagram (space diagram) of the lumbar trunk cross-sectional model.

Note. From "Biomechanics of the Human Spine and Trunk" by J.A. Ashton-Miller and A.B. Schultz. In *Exercise and Sport Sciences Reviews* (Vol. 16) (p. 188) by K.B. Pandolf (Ed.), 1988, New York: MacMillan. Copyright 1988 by Macmillan. Adapted by permission.

an example of this approach (Ashton-Miller & Schultz, 1988). The free body diagram represents a cross section through the trunk at the level of the L3 vertebra. The free body diagram includes 10 muscles ($\mathbf{F}_m$), 5 on each side; the intraabdominal pressure ($\mathbf{F}_i$); and the compressive ($\mathbf{F}_c$) and shear ($\mathbf{F}_s$) forces on the vertebra. The muscles represented in the diagram are rectus abdominus ($\mathbf{F}_{m,r}$), internal oblique ($\mathbf{F}_{m,i}$), external oblique ($\mathbf{F}_{m,x}$), latissimus dorsi ($\mathbf{F}_{m,l}$), and erector spinae ($\mathbf{F}_{m,e}$). The force exerted by each muscle is represented by a single equivalent force vector based on the cross-sectional area of the muscle, with the vector located at the centroid of the area (Németh & Ohlsén, 1986). This free body diagram can be used to determine the effects of the muscles on the net force and moment of force about the center of the L3 intervertebral disk (Ashton-Miller & Schultz, 1988).

Despite its elegance, the example shown in Figure 2.9 does not represent a true free body diagram. A free body diagram must include a correct quantitative graphical solution in which both the direction and magnitude of the involved forces are drawn to scale. A schematic of the free body diagram, as in Figure 2.9, that does not include the proper force magnitudes is called a **space diagram** (Dempster, 1961).

Human Movement Forces

Once we define a system with a free body diagram, we can consider the interactions (forces) that are likely to occur between a system and its surroundings. In general, the forces included on a free body diagram are those due to contact with the surroundings as well as those due to gravitational, electrical, and magnetic effects. In the analysis of human movement, eight forces are frequently included on a free body diagram: weight, ground reaction force, joint reaction force, muscle force, intraabdominal pressure, fluid resistance, elastic force, and inertial force.

Weight

Newton characterized **gravity** in a statement known as the **law of gravitation**:

> All bodies attract one another with a force proportional to the product of their masses and inversely proportional to the square of the distance between them.

That is,

$$\mathbf{F} \propto \frac{m_1 m_2}{\mathbf{r}^2} \tag{2.4}$$

where m_1 and m_2 are the masses of two bodies and $\mathbf{r}$ is the distance between them.

These forces of attraction between objects are generally regarded as negligible in the study of human movement with the exception of the attraction between Earth and various objects. The magnitude of this attraction, a force known as weight, depends on the mass of the objects involved and the distance between them. **Weight** is an expression of the amount of gravitational attraction between an object and Earth. As a force, it is measured in newtons (N). Of course, weight varies proportionately with mass—the greater the mass, the greater the attraction—but the two are separate quantities. Weight is a force (a derived quantity), whereas mass (a base measure in the SI system) is a measure of the amount of matter. Weight represents the interaction between an object and Earth, and the magnitude of the interaction depends on their respective masses and the distance between them; therefore, gravity decreases as altitude increases above sea level. This accounts for one of the advantages of performing at high altitude in events in which the contestant must overcome gravity (e.g., long jump, shot put, weight lifting). Reductions in gravity, especially microgravity conditions, have interesting implications for the control of movement (Davis and Cavanagh, 1993).

The *magnitude of the total-body weight vector* is determined quite readily by reading the value off a bathroom scale. The validity of this procedure can be demonstrated with a simple analysis based on Newton's law of acceleration (Equation 2.3),

$$\Sigma\mathbf{F} = m\mathbf{a}$$

which states that the sum (Σ = sigma) of the forces (**F**) produces an acceleration (**a**) of the system that depends on the mass (m) of the system. The quantities of force and acceleration are boldfaced to indicate that they are vectors. Because weight acts only in the vertical direction (z), the analysis can be confined to those components that act vertically: Weight ($\mathbf{F}_w$) is directed downward and indicated as negative, and the vertical component of the ground reaction force ($F_{g,z}$), which is provided by the scale, is directed upward and indicated as positive. The decision to label $\mathbf{F}_w$ negative and $F_{g,z}$ as positive is quite arbitrary; however, it is essential to distinguish between these differences in direction. To meet these needs, each free body diagram should be accompanied by a reference axis that shows the positive directions of each component (Figure 2.10): positive vertical (z), horizontal (x), and rotation directions.

The appropriate free body diagram is shown in Figure 2.10, and the analysis is as follows:

$$\Sigma\mathbf{F}_z = ma_z$$

$$-\mathbf{F}_w + F_{g,z} = ma_z$$

$$\mathbf{F}_w = F_{g,z} - ma_z$$

Because the person is stationary, $a_z = 0$. Thus

$$\mathbf{F}_w = F_{g,z} - 0$$

$$\mathbf{F}_w = F_{g,z}$$

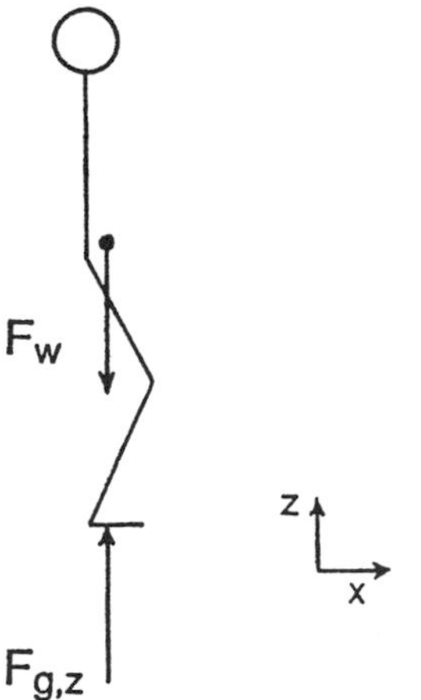

Figure 2.10 Free body diagram of the whole body.

The *direction of the weight vector* is always vertically downward toward the center of the Earth; the vector originates from a point referred to as the center of gravity. The **center of gravity** (CG) represents a balance point, a location about which all the particles of the object are evenly distributed. It is an abstract point that moves when the body segments are moved relative to one another. The CG is not confined to the physical limits of the object; for example, the CG for a doughnut is undoubtedly located within the inner hole. In fact, many successful high jumpers can project the CG outside the body or at least toward the outer limits (see Figure 2.11b). As Figure 2.11 shows, the height that a jumper can clear comprises three components (Hay, 1978): (a) the height (H_1) of the CG above the ground at takeoff, (b) the height (H_2) that the jumper can raise the CG above H_1, and (c) the difference between the maximum height reached by the CG and the height of the crossbar (H_3). With regard to H_3, although a person may jump 2.28 m (7 ft 6 in.), this does not imply that the CG was raised to that height but merely that the body passed over the bar—the two conditions are not synonymous. Figure 2.11b shows a theoretical high jump technique that maximizes the H_3 height. A somewhat similar rationale applies to pole vaulting (Fletcher & Lewis, 1960; Hubbard, 1980). Once the jumper has left the ground, the path that the CG will follow has been predetermined (recall the details of projectile motion in chapter 1). Consequently, the ability to project the CG out of the body (i.e., attain a positive H_3) serves as a means to acquire extra clearance height.

Segmental Analysis. Like many forces encountered in the analysis of human movement, weight is a distributed force with the net effect shown at a central location. For example, in the free body diagram shown in Figure 2.9, the force exerted by each muscle is represented by a single vector located at the center of the cross-sectional area of the muscle. Also, the contact between a player's head and a soccer ball would be shown to occur at a point on a free body diagram, but, in fact, the contact would be distributed over a large part of the forehead. Similarly, the center of gravity represents the center of the mass of the object and is the point about which the mass of the object is evenly distributed. Accordingly, when the mass of the object is redistributed, which usually happens in human movement, the location of the CG also moves. In Figure 2.11b the location of the CG shifts when the position of the high jumper changes.

In order to determine the location of the CG in such movements as high jumping, biomechanists have developed a procedure known as a *segmental analysis.* This procedure involves estimating the mass and the location of the CG for each body segment and then using this information to determine CG location for the whole body. This procedure will be discussed in chapter 3. At this point we introduce the database on segmental masses and CGs, which represents the magnitude and location of the weight vector for various segments of the human body.

Several groups of investigators have dissected cadavers to derive simple mathematical expressions with which to estimate various **anthropometric**

a

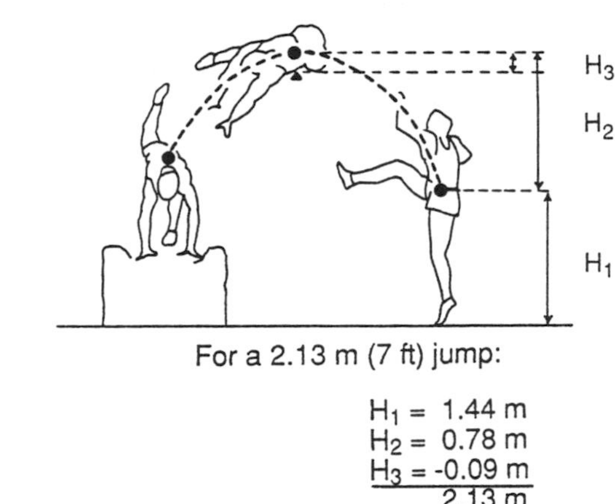

b

Figure 2.11 Contributions to the height recorded in the high jump: (a) schematic high jump with some values for a 2.13-m jump; (b) theoretical technique in which the jumper projects the center of gravity out of the body.
Note. From *The Biomechanics of Sports Techniques* (2nd ed.) (pp. 426, 433) by James G. Hay, 1978, Englewood Cliffs, NJ: Prentice Hall, Inc. Copyright 1978 by Prentice Hall, Inc. Adapted by permission.

(measurements of the human body) segmental dimensions (e.g., Chandler, Clauser, McConville, Reynolds, & Young, 1975; Dempster, 1955; Hanavan, 1964, 1966; D.I. Miller & Morrison, 1975). As described by D.I. Miller and Nelson (1973), this database is not extensive and mainly consists of measurements made on older Caucasian males. Some of the data derived from one of the most comprehensive cadaver studies ($n = 6$), that performed by Chandler et al., are reported in Tables 2.1 and 2.2. These data consist of regression equations to estimate segmental weights and CG locations (Table 2.1) and actual values for segmental moments of inertia about the three axes (Table 2.2).

The mathematical expressions that specify the relationships between total-body weight and segment weight are known as regression equations. In general, regression equations allow us to estimate the value of one parameter based on values for another parameter. Table 2.1 presents formulas for estimating segment weight from total-body weight and for estimating the location of the CG for each segment based on the measured segment length. For example, if an individual weighed 750 N, then

$$\text{trunk weight} = 0.532 \times 750 - 6.93$$
$$= 392 \text{ N}$$

which accounts for 52% of total-body weight. Similarly,

$$\text{hand weight} = 0.005 \times 750 + 0.75$$
$$= 4.5 \text{ N}$$

which represents about 0.06% of total-body weight. (Exactly what Chandler et al. [1975] meant by the designations of trunk, hand, and so forth is indicated in Appendix D.) Table 2.1 can also be used to estimate CG location. If the length of an individual's thigh (hip-to-knee distance) is 36 cm, then, according to Chandler et al., the CG for that thigh would be located at 39.8% of that distance from the hip joint:

$$\text{CG location} = 36 \times 0.398$$
$$= 14.3 \text{ cm from the hip joint}$$

Table 2.1 Regression Equations Estimating Body Segment Weights and Locations of Center of Gravity

Segment	Weight (N)	CG location (%)
Head	0.032 BW + 18.70	66.3
Trunk	0.532 BW − 6.93	52.2
Upper arm	0.022 BW + 4.76	50.7
Forearm	0.013 BW + 2.41	41.7
Hand	0.005 BW + 0.75	51.5
Thigh	0.127 BW − 14.82	39.8
Shank	0.044 BW − 1.75	41.3
Foot	0.009 BW + 2.48	40.0

Note. See Appendix D for the body segment organization used by Chandler et al. (1975). Body segment weights are expressed as a percentage of total-body weight (BW), and the segmental center of gravity (CG) locations are expressed as a percentage of segment length as measured from the proximal end of the segment.

Table 2.2 Whole-Body Moments of Inertia (kg·m²) About the Somersault, Cartwheel, and Twist Axes

Position	Somersault	Cartwheel	Twist
Layout	12.55	15.09	3.83
Open pike (arms out to side)	8.38	8.98	4.79
Closed pike (fingers touching toes)	8.65	6.60	3.58
Tuck	4.07	4.42	2.97

Because human motion is based on the rotation of body segments about one another, the distribution of mass in a segment is as important to know as its mass. The distribution of mass can be quantified by a parameter known as the moment of inertia. The **moment of inertia** is the angular equivalent of inertia (mass) and represents a measure of the resistance that an object offers to a change in its motion about an axis. The moment of inertia (I) is defined as

$$I = \sum_{i=1}^{n} m_i r_i^2 \tag{2.5}$$

where n indicates the number of elements (particles) in the system, m_i represents the mass of each element of the system, and r_i is the distance of each element from the axis of rotation. The moment of inertia is a measure of the mass distribution about an axis. In characterizing the distribution of mass, we do so with reference to three orthogonal (perpendicular) axes that are referred to as the principal axes of rotation. In the analysis of human movement, these axes are usually defined as the somersault (side-to-side), cartwheel (front-to-back), and twist (longitudinal) axes. Table 2.2 indicates estimates of the moment of inertia about the three principal axes for the human body in different positions (D.I. Miller & Nelson, 1973). Note that the moment of inertia declines as the mass of the body is brought closer to the axis of rotation; for example, for each body position the moment of inertia is least about the twist axis.

In addition to estimates of mass and CG location, the cadaver studies have yielded segmental moment of inertia values. These can be obtained relatively easily with the oscillation timing technique (Lephart, 1984). In contrast to the regression equations of Table 2.1, the data for the segmental moments of inertia were reported by Chandler et al. (1975) as mean values (Table 2.3). However, Zatsiorsky and Seluyanov (1983, Tables 5–7) have used a radiation scanning technique (described later in this section) to determine multiple-regression equations with which to estimate the segmental moments of inertia about the principal axes. As with the whole-body measures, the moment of inertia for each segment (except the head) is similar about the somersault and cartwheel axes and least about the twist axis.

Table 2.3 Segmental Moments of Inertia (kg · m^2) About the Somersault, Cartwheel, and Twist Axes

Segment	Somersault	Cartwheel	Twist
Head	0.0164	0.0171	0.0201
Trunk	1.0876	1.6194	0.3785
Upper arm	0.0133	0.0133	0.0022
Forearm	0.0065	0.0067	0.0009
Hand	0.0008	0.0006	0.0002
Thigh	0.1157	0.1137	0.0224
Leg	0.0392	0.0391	0.0029
Foot	0.0030	0.0034	0.0007

To supplement the segmental data derived from cadavers, some investigators have used *mathematical modeling* procedures. In this approach, the human body is represented as a set of geometric components, such as spheres, cylinders, and cones (Hall & DePauw, 1982; Hanavan, 1964, 1966; Hatze, 1980, 1981a, 1981b; D.I. Miller, 1979). One of the first to use this approach was Hanavan (1964, 1966). Figure 2.12 shows the Hanavan model, which divides the human body into 15 simple geometric solids of uniform density. The advantage of this model is that it requires only a few simple anthropometric measurements (e.g., segment lengths and circumferences) to personalize the model and to predict the center of gravity and moment of inertia for each body segment. However, three assumptions typically used in modeling body segments limit the accuracy of the estimates: Segments are assumed to be rigid, the boundaries between the segments are assumed to be distinct, and segments are assumed to have a uniform density. In reality, there can be substantial displacement of the soft tissue during movement, the boundaries between segments are fuzzy, and the density varies within and between segments.

Based on the same approach, Hatze (1980) has developed a more detailed model of the human body (Figure 2.13). Hatze's hominoid consists of 17 body segments and requires 242 anthropometric measurements for individualization. The model subdivides the segments into small mass elements of different geometrical structures, allowing the shape and density fluctuations of a segment to be modeled in detail. Furthermore, no assumptions are made regarding bilateral symmetry, and the model differentiates between male and female subjects, adjusting the densities of certain segmental parts according to the value of a special subcutaneous fat indicator. The model is able to account for changes in body morphology, such as those due to obesity and pregnancy, and can accommodate children. Indeed, the model has been used to estimate the mass, center of mass, and moment of inertia for the limb segments of infants (Schneider & Zernicke, 1992). Although acquisition of the input for the model (242 anthropometric measurements) is time-consuming, the output of the model provides accurate estimates of volume, mass, CG location, and moments of inertia for the identified body segments. Typical estimates for two male subjects are shown in Table 2.4. One noteworthy feature of the model is the approach used to represent the trunk and shoulder segments (Figure 2.13).

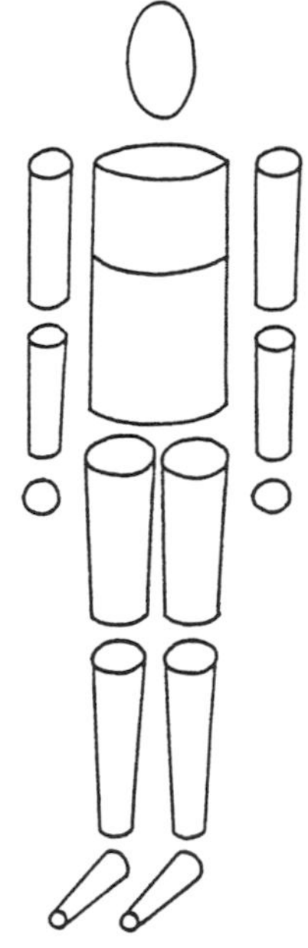

Figure 2.12 The Hanavan model of the human body.

Another technique available to determine body segment parameters is based on the recently developed imaging procedures, such as computed tomography and magnetic resonance imaging (Engstrom, Loeb, Reid, Forrest, & Avruch, 1991; P.E. Martin, Mungiole, Marzke, & Longhill, 1989). An example of this approach is the use of *radioisotopes*. With this approach, various properties of body segments are determined based on the measurement of the intensity of a gamma-radiation beam before and after it passes through a segment. The principle involves scanning a subject's body and

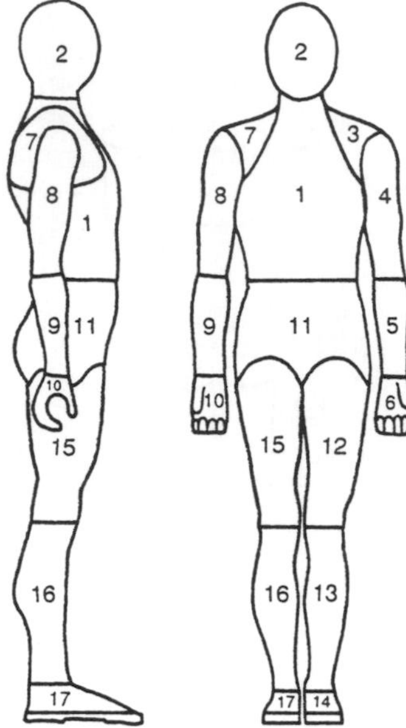

Figure 2.13 Diagrammatic representation of a 17-segment hominoid. The 17 segments are 1, abdomino-thoracic; 2, head; 3, left shoulder; 4, left upper arm; 5, left forearm; 6, left hand; 7, right shoulder; 8, right upper arm; 9, right forearm; 10, right hand; 11, abdomino-pelvic; 12, left thigh; 13, left shank; 14, left foot; 15, right thigh; 16, right shank; and 17, right foot.

Note. From "A Mathematical Model for the Computational Determination of Parameter Values of Anthropomorphic Segments" by H. Hatze, 1980, *Journal of Biomechanics*, **13**, p. 835. Copyright 1980 by Pergamon Press, Ltd. Adapted by permission.

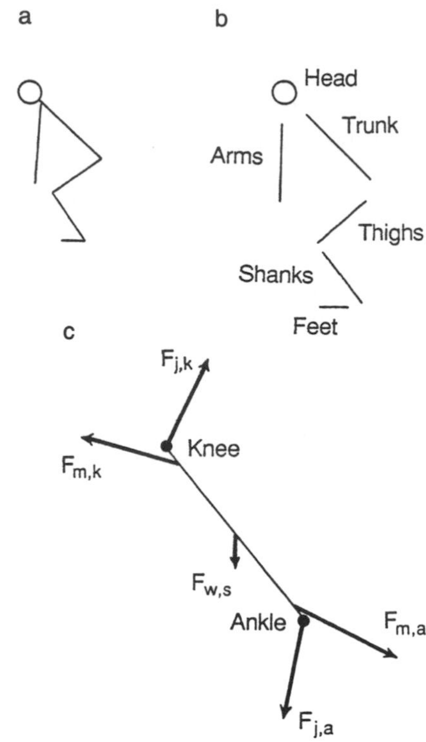

Figure 2.14 Free body diagram of a weight lifter: (a) whole-body stick figure; (b) segmental components; (c) free body diagram of the leg. $\mathbf{F}_{j,a}$ and $\mathbf{F}_{j,k}$ = ankle and knee joint reaction forces; $\mathbf{F}_{m,a}$ and $\mathbf{F}_{m,k}$ = ankle and knee resultant muscle forces; $\mathbf{F}_{w,s}$ = weight of the shank.

obtaining the surface density and coordinates of the body segments affected by the radiation. Zatsiorsky and Seluyanov (1983) performed this procedure on 100 male subjects (age = 23.8 ± 6.2 years; height = 1.74 ± 0.06 cm; weight = 730 ± 91 N) and derived regression equations (Table 2.5) with which to estimate body segment parameters for other subjects.

EXAMPLE

Let us briefly consider a segmental analysis of the weight lifter depicted previously in the Free Body Diagram section. We want to determine the contribution of the muscles crossing the knee to the movement. First, we need to decide how many segments to use to represent the lifter (Figure 2.14a). The movement shown at the top of Figure 1.7 (p. 11) indicates that the arms do not bend at the elbow, so we can consider the arms as one segment. And because the movement is confined to the sagittal plane, we assume that the left and right sides of the body are doing more or less the same thing. These simplifications allow us to reduce our free body diagram from the maximum of 15 to 17 segments (based on Hanavan's and Hatze's models) down to 6 segments. Furthermore, the whole-body stick figure will become six separate systems, one for each segment (Figure 2.14b); this is the essence of a segmental analysis.

Next we need to draw the appropriate free body diagram (Figure 2.14c). This includes identifying the system and determining the location (CG) and magnitude of the weight vector ($\mathbf{F}_{w,s}$). Because we want to determine the muscle forces across the knee joint, we need to use the shank as the system. Both the CG location and the magnitude of the weight vector can be estimated from the data in Table 2.1. According to Table 2.1, the CG for the shank is located at 41.3% of the shank length from the knee joint. If the shank of the weight lifter measured 36 cm, the CG would be 14.9 cm (36 × 0.413 = 14.9) from the knee. Similarly, the magnitude of the weight vector is determined as a function of total-body weight. For a 800-N weight lifter, the weight of one shank would be 33.5 N (0.044 × 800 − 1.75 = 33.5). Finally, the other forces acting on the system, which we are about to describe, must be added to the free body diagram.

Ground Reaction Force

The **ground reaction force** is a concept that is used to describe the reaction force provided by the supporting

Table 2.4 Segmental Parameter Values for Two Male Subjects Computed With the Hatze Model

	Subject F.B. (23 years)			Subject C.P. (26 years)		
Segment	Volume	CG location	I_{xx}	Volume	CG location	I_{xx}
Abdomino-thoracic	19.111	0.439	0.3117	19.803	0.444	0.3302
Head-neck	4.475	0.517	0.0303	4.537	0.516	0.0337
Left shoulder	1.438	0.727	0.0047	2.042	0.706	0.0080
Right shoulder	1.890	0.711	0.0071	2.121	0.699	0.0084
Left arm	2.110	0.432	0.0196	2.123	0.437	0.0203
Right arm	2.021	0.437	0.0168	2.340	0.428	0.0229
Left forearm	1.023	0.417	0.0067	1.223	0.413	0.0086
Right forearm	1.190	0.404	0.0079	1.313	0.412	0.0093
Left hand	0.453	0.515	0.0011	0.416	0.533	0.0010
Right hand	0.446	0.531	0.0011	0.417	0.524	0.0010
Abdomino-pelvic	8.543	0.368	0.0399	9.614	0.395	0.0541
Left thigh	8.258	0.479	0.1475	8.744	0.473	0.1653
Right thigh	8.278	0.480	0.1415	8.729	0.466	0.1702
Left shank	3.628	0.412	0.0615	3.856	0.420	0.0798
Right shank	3.686	0.417	0.0663	3.798	0.417	0.0747
Left foot	0.887	—	0.0041	1.032	—	0.0051
Right foot	0.923	—	0.0042	1.055	—	0.0051

Note. Volume = computed segment volume (L); CG location = distance from proximal end of segment (proportion of segment length, where 0.500 indicates 50% of segment length); I_{xx} = moment of inertia about the somersault (xx) axis through the CG. Segment densities were abdomen = 1,000 + 30 i_m, head = 1,120, neck = 1,040, and arms and legs = 1,080 + 20 i_m kg/m^3, where i_m = 1 for males and 0 for females.

Table 2.5 Regression Equations Estimating Body Segment Masses and Locations of Centers of Gravity

	Mass (kg) coefficients			CG location (% segment length) coefficients		
Segment	B_0	B_1	B_2	B_0	B_1	B_2
Foot	−0.8290	0.0077	0.0073	3.7670	0.0650	0.0330
Shank	−1.5920	0.0362	0.0121	−6.0500	−0.0390	0.1420
Thigh	−2.6490	0.1463	0.0137	−2.4200	0.0380	0.1350
Hand	−0.1165	0.0036	0.0017	4.1100	0.0260	0.0330
Forearm	0.3185	0.0144	−0.0011	0.1920	−0.0280	0.0930
Upper arm	0.2500	0.0301	−0.0027	1.6700	0.0300	0.0540
Head	1.2960	0.0170	0.0143	8.3570	−0.0025	0.0230
Upper torso	8.2144	0.1862	0.0584	3.3200	0.0076	0.0470
Middle torso	7.1810	0.2234	−0.0663	1.3980	0.0058	0.0450
Lower torso	−7.4980	0.0976	0.0490	1.1820	0.0018	0.0434

Note. The multiple regression equations are in the form $Y = B_0 + B_1X_1 + B_2X_2$, where Y = predicted segment mass or CG location, X_1 = total body mass (kg), X_2 = height (cm), and B_0, B_1, B_2 = coefficients given in the table. Body-segment mass and CG locations are expressed as functions of body mass and height. Adapted from Zatsiorsky and Seluyanov, 1983.

horizontal surface. It is derived from Newton's law of action-reaction *to represent the reaction of the ground to the accelerations of all the body segments*. The ground reaction force can be measured with an instrument that is known as a force platform, which essentially operates like a sophisticated scale for measuring weight. Researchers began using this technique in the 1930s (Elftman, 1938, 1939; Fenn, 1930; Manter, 1938), although the idea had been proposed earlier (Amar, 1920; Marey, 1874).

One important difference between a force platform and a weight scale is that the force platform can measure the ground reaction force in three dimensions, and it can do so quickly with minimal distortion of the signal. The resultant ground reaction force can be resolved into three components whose directions are functionally defined as vertical (upward-downward), forward-backward, and side-to-side. Because the ground reaction force measured during the stance phase represents the reaction of the ground to the actions of the runner that are transmitted through the support leg, these components represent the acceleration of the system in these respective directions. The extent to which a segment influences the ground reaction force depends on its mass and the acceleration of its CG. D.I. Miller (1990) estimates that the trunk and head account for about 50% of the runner's acceleration, whereas each leg contributes about 17% and the arms about 5%.

An instructive example of the association between the ground reaction force and the associated movement kinematics is shown in Figure 2.15. The movement is a vertical jump. The height of a vertical jump depends on the magnitude of the vertical velocity at takeoff, which, in turn, is determined by the vertical component of the ground reaction force. As Figure 2.15 shows, the subject begins from an upright position and then lowers the center of gravity by approximately 0.2 m before changing direction (velocity goes from negative to positive values) and moving upward to the takeoff position. Takeoff occurs when the vertical component of the ground reaction force falls to zero (time = 0.53 s). The subject was in the air for about 0.41 s, during which time the center of gravity was raised 0.49 m from the starting position. Recall that Figure 2.15 indicates the net effect of the interaction between the subject and the ground; accordingly, the kinematics describe the motion of the total-body center of gravity.

As Figure 2.15 shows, the peak downward velocity (acceleration = 0) occurred about midway through the downward movement (time = 0.19 s), and the peak upward velocity (positive values; acceleration = 0) occurred just prior to takeoff. During the flight phase of the jump, the ground reaction force was zero and the vertical component of acceleration had a value of –9.81 m/s^2 (i.e., the effect of gravity at the jumper's center of gravity). As noted previously in the context of Newton's law of acceleration, the ground reaction force and the acceleration graphs paralleled each other but were offset. Initially, the acceleration of the system (total-body center of gravity) was zero, and the ground reaction force was equal to body weight. As the body segments began to accelerate, the system acceleration changed, and the ground reaction force changed in parallel but about the body weight line. The ground reaction force record consisted of four phases: (a) an initial phase, where the ground reaction force was less than body weight (negative acceleration); (b) a phase where the ground reaction force was greater than body weight (positive acceleration); (c) the flight phase, where the ground reaction force was zero (acceleration = –9.81 m/s^2); and (d) the impact phase, where the jumper returned to the ground.

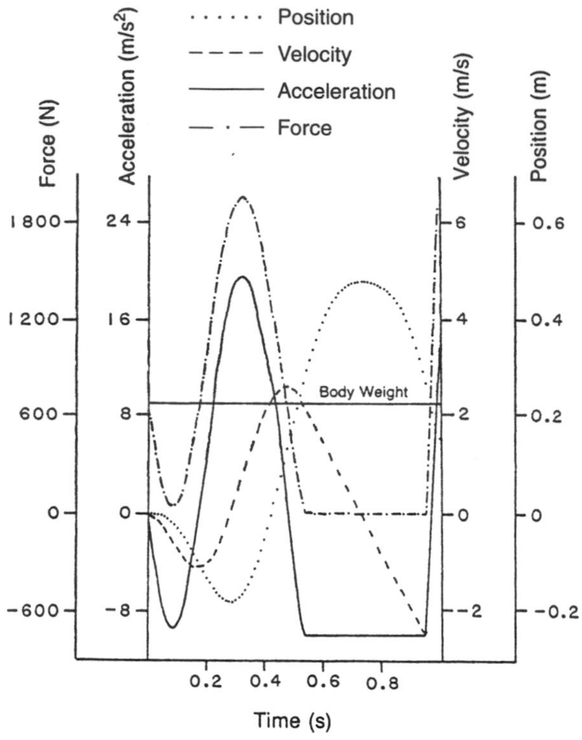

Figure 2.15 The vertical components of the kinematics and kinetics associated with a vertical jump.
Note. From "A Biomechanical Analysis of the Contribution of the Trunk to Standing Vertical Jump Takeoffs" by D.I. Miller. In *Physical Education, Sports and the Sciences* (p. 357) by J. Broekhoff (Ed.), 1976, Eugene, OR: Microform Publications. Copyright 1976 by Microform Publications. Adapted by permission.

Figure 2.16 illustrates the vertical component ($F_{g,z}$) of the ground reaction force that is associated with walking and with running (for a more detailed treatment of the differences in the ground reaction force in walking and running, consult Alexander, 1984b). These data indicate the manner in which $F_{g,z}$ changed from the instant of foot contact with the ground (time zero on the abscissa) until the instant that the same foot left the ground (the time at which $F_{g,z}$ returned to zero). In the analysis of locomotion, this epoch is known as the stance or support phase. $F_{g,z}$ is nonzero only when the foot is in contact with the ground. Note that $F_{g,z}$ changes continuously throughout the period of support. Recall from the discussion of weight that the magnitude of the weight vector is equivalent to $F_{g,z}$ when the system (body) is not accelerating in the vertical direction. Accordingly, when $F_{g,z}$ differs from body weight, the system is experiencing a vertical acceleration; when $F_{g,z}$ is greater than body weight, the vertical acceleration of the CG is upward, and when $F_{g,z}$ is less than body weight, the vertical acceleration of the CG is downward.

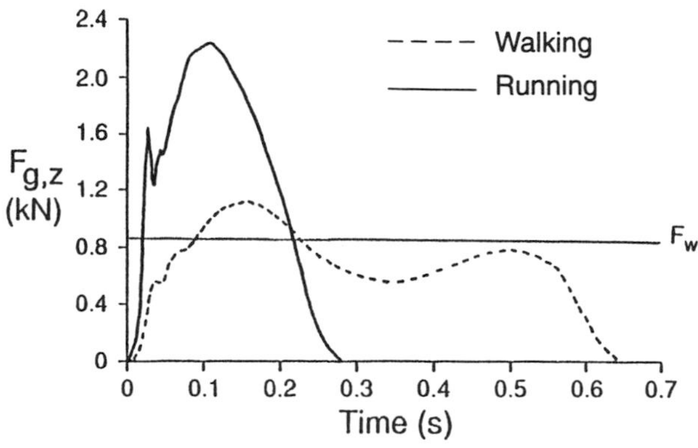

Figure 2.16 Vertical component of the ground reaction force ($F_{g,z}$) during the period of support in walking and running.

Another difference in $F_{g,z}$ between walking and running is that the walking record has two peaks, whereas that for running has a single peak. Indeed, Alexander (1984b) has used this difference as an index to quantify changes in $F_{g,z}$ with speed. Given this fluctuation in $F_{g,z}$ during walking, when does the CG of the individual move downward during the stance phase of walking? As mentioned in the discussion on kinematics in chapter 1, it is not possible to determine the direction of displacement directly from a force-time graph. Because force is proportional to acceleration (**F** = *m***a**), we cannot determine displacement from acceleration.

Because the ground reaction force represents the reaction of the ground to the action (acceleration of the CG) of the runner, the movement of the runner while the foot is on the ground is reflected in the ground reaction force (Figure 2.17). A runner in the nonsupport phase of a stride experiences a downward acceleration (due to gravity) of 9.81 m/s^2, which means that when the foot contacts the ground there is an upward-directed vertical component ($F_{g,z}$) to counteract the downward motion of the runner. Furthermore, when the foot contacts the ground, it is initially in front of the runner's CG, which causes the ground to respond with a backward-directed (braking) horizontal component ($F_{g,y}$). As the runner's CG passes over the support foot, the horizontal component changes direction so that it is acting forward (propulsion). The side-to-side component ($F_{g,x}$) is more difficult to explain, but it has a lesser magnitude and is more variable than $F_{g,z}$ and $F_{g,y}$. However, $F_{g,x}$ has been shown to be correlated ($r = 0.71$) with the position of the foot relative to the midline of progression during contact with the ground (K.R. Williams, 1985).

In chapter 1, we discussed some kinematic changes that occur as a runner increases speed. Because these changes are caused by changes in the actions of the body segments, they are reflected in the ground reaction force. Munro, Miller, and Fuglevand (1987) characterized some of these changes as the running speed of 20

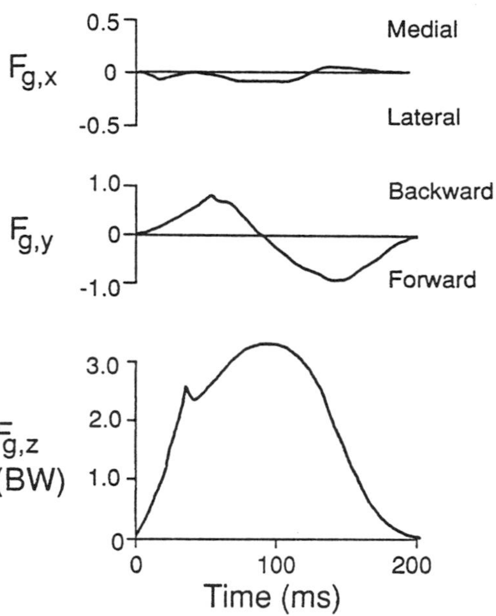

Figure 2.17 Generalized force-time curves for the three components of the ground reaction force during the support phase of a running stride. The forces are expressed relative to body weight (BW).

male subjects increased from 3.0 to 5.0 m/s. They found that, on average, the stance time decreased from 270 to 199 ms, the peak $F_{g,z}$ increased from 2.51 to 2.83 $\mathbf{F}_w$ (where $\mathbf{F}_w$ equals body weight), and the average force during the stance phase increased from 1.4 to 1.7 $\mathbf{F}_w$. These results indicate that the increase in running speed from 3.0 to 5.0 m/s involved a greater $F_{g,z}$ applied over a shorter duration (Figure 2.18). The average force for both the braking and propulsion directions of $F_{g,y}$ also increased with running speed. Similar results were reported by Nilsson and Thorstensson (1989), but these

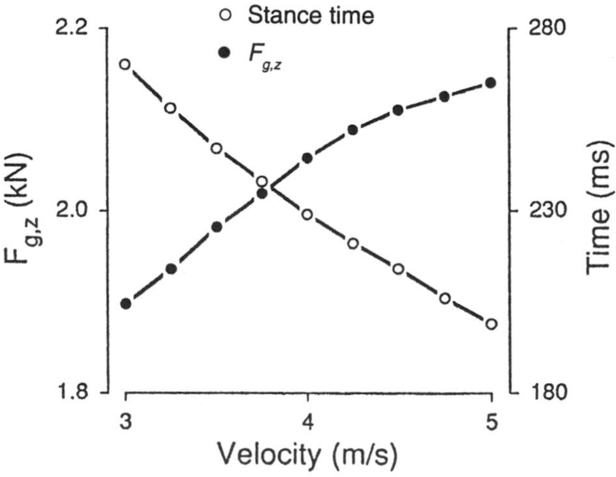

Figure 2.18 Change in the maximum $F_{g,z}$ and stance time with running speed.

covered the range from 1.0 to 6.0 m/s and included the transition from walking to running.

Center of Pressure. Although the ground reaction force reflects the action of all body segments, it is possible to gain some insight into the action of the support limb during the stance phase with ground reaction force measurements. These measurements have indicated that the kinematics of the foot are relatively complex during the stance phase (Cavanagh, 1990; Cavanagh, Andrew, Kram, Rodgers, Sanderson, & Henning, 1985; Cavanagh, Pollock, & Landa, 1977; Chao, 1986; Nigg, 1986; Rodgers, 1993; Vaughan, 1984). The part of the foot that initially contacts the ground varies with speed, but it is generally the lateral border, usually between the midfoot and the heel, with the foot in a supinated position (adduction and inversion of the foot, where *inversion* refers to rotation of the sole toward the midline of the body). After this initial footstrike, the foot is pronated (abduction and eversion of the foot, where *eversion* refers to the rotation of the sole away from the midline of the body), and the ankle is dorsiflexed (ankle flexion). As the runner's CG passes over the foot, both the knee and the ankle joints begin to extend. By the time the foot leaves the ground, the ground reaction force vector has moved from the lateral border of the foot at footstrike to a point at or near the base of support of the big toe.

This motion of the foot-ankle-shank can be recorded in at least two ways related to the ground reaction force; by measuring the center of pressure and the pressure distribution under the foot. The **center of pressure** represents the point of application of the resultant ground reaction force. It is simply the central point of the pressure exerted on the foot. Cavanagh and LaFortune (1980) characterized runners as either midfoot or rearfoot strikers; this distinction referred to the initial location of the center of pressure on the foot (Figure 2.19). For midfoot strikers the initial location of the center of pressure was on the lateral border in the middle of the foot. (This does not mean that neither the forefoot nor the rearfoot contacted the ground, merely that the center of the pressure was in the middle of the foot.) Similarly, for rearfoot strikers the initial location of the center of pressure was on the rear of the foot.

The actual distribution of the pressure (i.e., area over which the ground reaction force was exerted) has been measured with the use of photoelastic material or miniature load cells (Cavanagh & Ae, 1980; Hennig, Cavanagh, Albert, & Macmillan, 1982). Like weight, the ground reaction force is a distributed force, and the region over which it is distributed shifts during the stance phase of walking or running and with different types of footwear. An example is shown in Figure 2.20. Sanderson and Hennig (1992) compared the pressure distributions in a cycling shoe and a running shoe by carefully positioning 12 piezoceramic transducers in each shoe. Figure 2.20 shows the pressure distributions

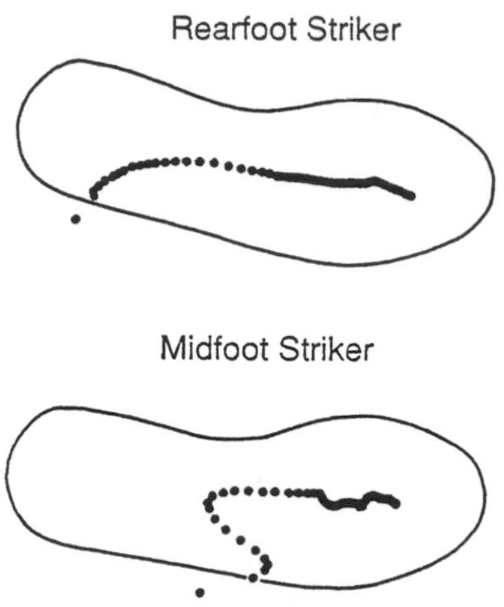

Figure 2.19 The displacement of the center of pressure (point of application of the ground reaction force vector) during the stance phase of a running stride (4.5 m/s).
Note. Adapted from "Ground Reaction Forces in Distance Running" by P.R. Cavanagh and M.A. LaFortune, 1980, *Journal of Biomechanics*, **13**, p. 401, Copyright 1980, with kind permission from Pergamon Press Ltd., Headington Hill Hall, Oxford 0X3 OBW, UK.

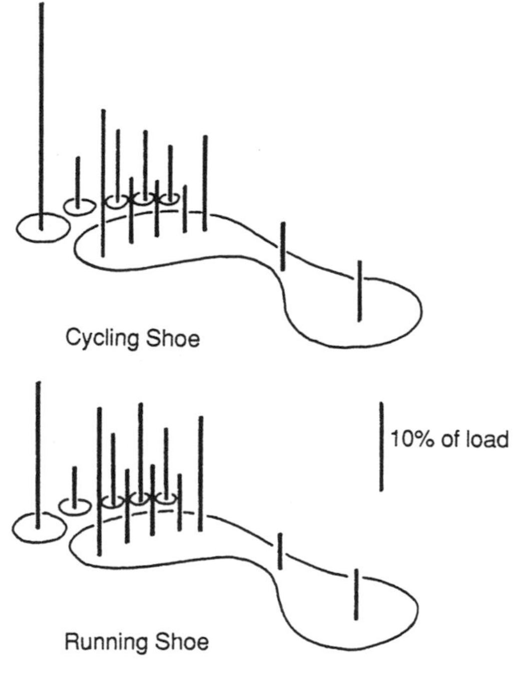

Figure 2.20 A comparison of pressure distributions in a cycling shoe and a running shoe during cycling.
Note. Data provided by David J. Sanderson, Ph.D.

when the crank angle was 1.57 rad past the top vertical position; the pressures are expressed as a percentage of the load. The greatest proportion of the load was applied to the first metatarsal and the hallux. In addition, higher pressures were applied to the lateral forefoot with the running shoe.

Shoe Design. Both the kinematics of locomotion and the ground reaction force can be altered by modifying the shoes worn by a runner. For example, both knee flexion velocity (Frederick, 1986) and the rate of pronation (Luethi & Stacoff, 1987) increase following footstrike when the soles of running shoes are hard. Well-designed running shoes serve a dual purpose—to minimize the shock associated with the foot contacting the ground (i.e., cushioning) and to provide stability for the ankle joint (especially the subtalar joint) during the stance phase. Unfortunately, an increase in cushioning is associated with a decrease in stability, and vice versa.

Normal foot motion during stance includes a supination-pronation sequence that distributes the impact force throughout the foot and up the support limb. *The task of a running shoe is to attenuate the impact force and to control the normal motion of the foot.* The ability of a shoe to provide appropriate stability is usually assessed in terms of rearfoot motion. Such an assessment involves determining the angle between the shank and the calcaneus throughout the stance phase. Figure 2.21 shows the changes in rearfoot angle for three runners. The runner with a rearfoot angle of –0.14 rad has a more vertical heel marker (pronation), and the runner with an angle of 0.07 rad has a more vertical shank marker (supination). Although the range of motion of the rearfoot angle covers about 0.31 rad during normal stance (Figure 2.21), running in running shoes appears to cause greater pronation than does barefoot running (Frederick, 1986). Luethi and Stacoff (1987) have demonstrated that a poorly designed shoe can have a negative effect on stability by increasing the amount of pronation during the stance phase by about 0.09 rad (5°). This effect is undesirable because overpronation may be one of the most important causes of running injuries (Clement, Taunton, Smart, & McNicol, 1981).

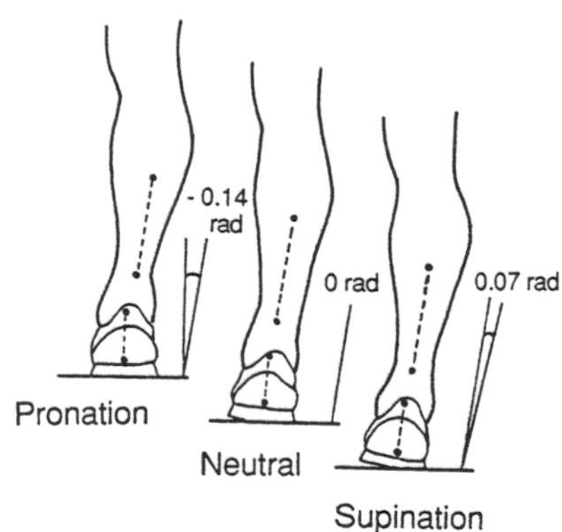

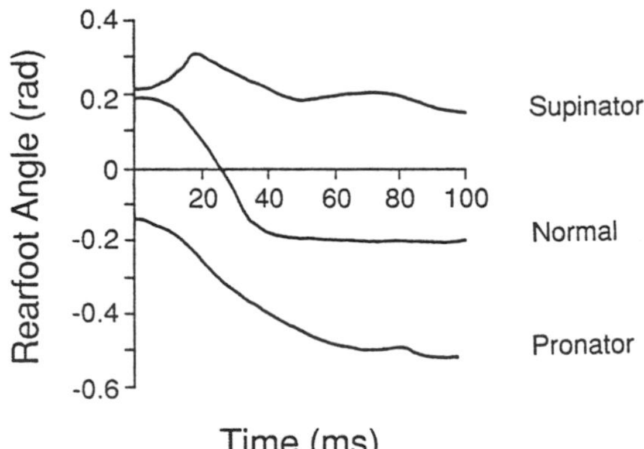

Figure 2.21 Changes in rearfoot angle during the support phase of running for three runners.
Note. From "Rearfoot Motion in Distance Running" by C.J. Edington, E.C. Frederick, and P.R. Cavanagh. In *Biomechanics of Distance Running* (p. 149), by P.R. Cavanagh (Ed.), 1990, Champaign, IL: Human Kinetics. Copyright 1990 by Human Kinetics Publishers, Inc. Adapted by permission.

The features of shoe design that influence stability are sole hardness, heel height and lift, and the angle of flare of the sole (Frederick, 1986). Soft soles allow more pronation and rearfoot motion. Heel height (distance from the ground to the heel) and heel lift (height of the heel above the forefoot) have been found to reduce pronation by some investigators (Bates, Osternig, Mason, & James, 1978; Stacoff & Kaelin, 1983) but not by others (Clarke, Frederick, & Hamill, 1983). Increases in heel flare up to 0.26 rad (15°) result in a significant reduction in pronation and rearfoot motion (Frederick, 1986).

Friction. The resultant of the two horizontal components of the ground reaction force ($F_{g,y}$ and $F_{g,x}$) represents the friction, or **shear force** ($\mathbf{F}_s$), between the shoe and the ground. Friction is important in locomotion because it provides the basis for the horizontal progression of the CG. Recall that the three components of the ground reaction force represent the acceleration of the total body CG, which is the sum of the body segment CG accelerations, in the three directions. Accordingly, $\mathbf{F}_s$ corresponds to the acceleration of the total body CG in the horizontal plane (i.e., the forward-backward and side-to-side directions).

In general, the maximum friction force is determined by the magnitude of the force that is normal (perpendicular) to the surface, which in the case of the ground reaction force is $F_{g,z}$, and a coefficient (μ) that characterizes the contact (rough or smooth, dry or lubricated, static or dynamic) between the two objects:

$$\mathbf{F}_{s,\max} = \mu F_{g,z} \qquad (2.6)$$

For a given shoe-ground contact, the friction coefficient varies depending on whether the shoe is stationary relative to the ground (static, μ_s) or is moving (dynamic, μ_d). Because μ_s is greater than μ_d, $\mathbf{F}_s$ reaches greater values when the shoe does not move or slide on the ground. This explains why it is more difficult to make a rapid turn once the shoe begins to slip on the ground. The magnitude of μ between shoes and surfaces is determined experimentally by calculating the ratio of $\mathbf{F}_s$ to

$F_{g,z}$ at the point in time just before the shoe moves relative to the ground. With this approach, μ_s has been reported to range from 0.3 to 2.0, with 0.6 for a cinder track and 1.5 for grass (Nigg, 1986; Stucke, Baudzus, & Baumann, 1984), and μ_d has ranged from 0.003 to 0.007 when racing on ice skates (de Koning, de Groot, & van Ingen Schenau, 1992). When $\mathbf{F}_s$ is less than the maximum, however, friction is simply equal to the resultant force in the horizontal plane.

As described by D.I. Miller and Nelson (1973), there are three regions to the friction domain: (a) just prior to the moment of the shoe slipping on the ground, when $\mathbf{F}_s$ is equal to $\mu_s F_{g,z}$ (this is the maximum value of friction); (b) once the shoe slips on the ground, when $\mathbf{F}_s$ is equal to $\mu_d F_{g,z}$ (this value is less than the maximum friction force); and (c) as the friction force increases from the moment that the shoe contacts the ground up until the maximum ($\mathbf{F}_{s,\max}$), when friction is less than the maximum and can be determined as the resultant force in the horizontal plane ($\mathbf{F}_s = \sqrt{F_{g,x}{}^2 + F_{g,y}{}^2}$). These characteristics indicate that, for a given shoe-ground condition (i.e., μ), friction increases with $F_{g,z}$, and because $F_{g,z}$ is predominantly influenced by body weight ($\mathbf{F}_w$), the amount of friction is often greater for heavier individuals. Furthermore, as long as the shoe does not slip on the ground, the friction force is smaller than the maximum possible ($\mathbf{F}_{s,\max}$).

Joint Reaction Force

When a system for a free body diagram is defined so that it passes through a joint, the concept of a joint reaction force ($\mathbf{F}_j$) is invoked *to account for the* net *forces generated by bone-on-bone contact between adjacent body segments.* The joint reaction force represents the net effects that are transmitted from one segment to another and is due to the muscle, ligament, and bony contact forces that are exerted across a joint. If a pair of agonist-antagonist muscles is coactive, then the joint reaction force represents the difference in the activity of the two muscles—that is, the net effect. Although the joint reaction force does not represent the absolute bone-on-bone force, it can have substantial magnitudes; values of 3 kN have been estimated at the humeroradial joint for some activities. Knowledge of these effects is essential in the design of prostheses if, for example, the prosthesis and the patient-prosthesis interface are to withstand the forces that will be encountered (Goodfellow & O'Connor, 1978; Rydell, 1965).

Perhaps the most intuitive factor underlying the joint reaction force is that due to the tissues surrounding the joint, especially the ligaments. The contribution of the ligaments to $\mathbf{F}_j$, however, appears controversial; some investigators have found it to be relatively small except at the extremes of the range of motion and under some loading conditions (Amis, Dowson, & Wright, 1980), whereas others report ligament (cruciate) forces three times that of body weight during walking (Collins & O'Connor, 1991). In contrast, $\mathbf{F}_j$ is definitely known to be influenced by the muscle force (recall that muscle force [Figure 2.3] can be resolved into normal and tangential components and that the tangential component is transmitted to the joint as a compressive force). Similarly, because the human body comprises a set of rigid segments that are connected together, a force that acts on one segment can be transmitted to all other body segments. For this reason, the ground reaction force exerted on the foot is distributed throughout the entire body and influences $\mathbf{F}_j$. Any contact force acting on the system can affect $\mathbf{F}_j$. Furthermore, an effect due to the motion of other body segments, called the inertia force, can be transmitted between segments.

It is extremely difficult to measure $\mathbf{F}_j$ experimentally. The magnitude of $\mathbf{F}_j$ is usually estimated by determining all the other forces on a free body diagram and assuming that the remaining effect is due to $\mathbf{F}_j$. This can be done, for example, if the system is in equilibrium, which means that all the forces acting on the system must be balanced. We will use this approach in chapter 3. It also is possible to use various mathematical procedures, such as minimizing muscle stress, to estimate the magnitude of $\mathbf{F}_j$. An, Kwak, Chao, and Morrey (1984) have used such an approach on the elbow (humeroulnar) joint when a load is applied at the wrist so that it is perpendicular to the forearm. Values for $\mathbf{F}_j$ were 6 to 16 times greater than the load when the joint was moved over a range of motion from complete extension to a right angle. When the loads encountered in normal daily activities were considered, this meant that values for $\mathbf{F}_j$ equal to 0.3 to 0.5 times body weight were commonly encountered at the elbow joint. Similarly, the magnitude of $\mathbf{F}_j$ can be calculated by using the muscle architecture and limb geometry to estimate muscle force. Amis et al. (1980) used this procedure to estimate the variation in $\mathbf{F}_j$ at the humeroradial, humeroulnar, and humerocoronoid joints during isometric flexion (Figure 2.22a) and extension tasks (Figure 2.22b) at the elbow joint. The calculated joint reaction forces ($\mathbf{F}_j$) shown in Figure 2.22 are those associated with the muscle activity necessary to resist the indicated forces at hand.

Values for $\mathbf{F}_j$ have been reported for such activities as standing, moving from sitting to standing, walking, running, weight lifting, and landing from a drop (summarized in Harrison, Lees, McCullagh, & Rowe, 1986). When an individual ran at 4.5 m/s, Harrison et al. found that the maximum values for $\mathbf{F}_j$ occurred at midstance (Figure 2.23) and reached a peak compression force of 33 times body weight at the knee joint but only a peak compressive force of 9 times body weight and a peak shear force of 4 times body weight at the ankle joint. Even the common task of going from an erect posture to a squat position and then rising again is associated with large joint reaction forces. For this task, the maximum tibiofemoral joint reaction force that was normal

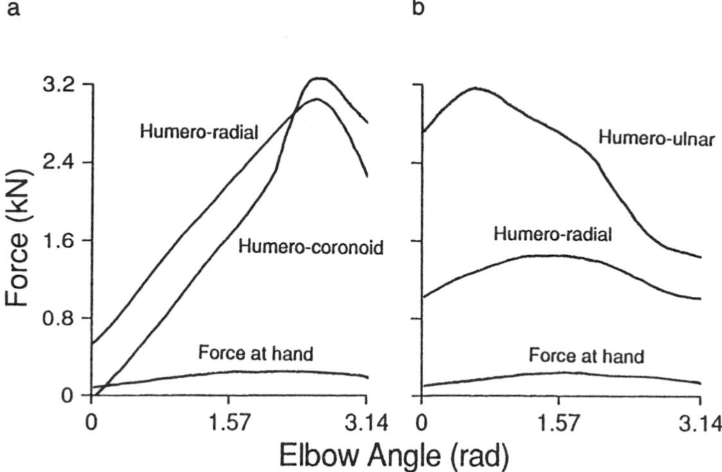

Figure 2.22 Joint reaction forces at the elbow joint during isometric flexion (a) and extension (b) efforts. An angle of 0 rad indicates complete extension.
Note. Adapted from "Elbow Joint Force Predictions For Some Strenuous Isometric Actions" by A.A. Amis, D. Dowson, and V. Wright, 1980, *Journal of Biomechanics*, **13**, pp. 772, 773, 1980, with kind permission from Pergamon Press Ltd., Headington Hill Hall, Oxford 0X3 OBW, UK.

to the surface (compressive) ranged from 4.7 to 5.6 times body weight, whereas the component that was tangential to the surface (shear) ranged from 3.0 to 3.9 times body weight (Dahlkvist, Mayo, & Seedhom, 1982). Power lifters experience maximum compressive forces of 17 times body weight and maximum shear forces of 2.3 times body weight at the L4-L5 joint during the performance of the dead lift (Cholewicki, McGill, & Norman, 1991). Results such as these emphasize that the joint reaction force varies over the range of motion and can have a substantial magnitude, especially in comparison with the loads encountered in daily activities.

Muscle Force

Previously, we described human movement as the consequence of the mechanical interactions between muscle and its surroundings. In this scheme, the role of muscle is to exert a force that is transmitted through the tendon to the bone and to cause body segment rotation. However, the generalized concept of force (i.e., a push or pull that tends to alter an object's state of motion) cannot be extrapolated to muscle, because a muscle cannot push—it can only pull. In mechanical terms, muscle can exert a **tensile** (pulling) **force** but not a **compressive**

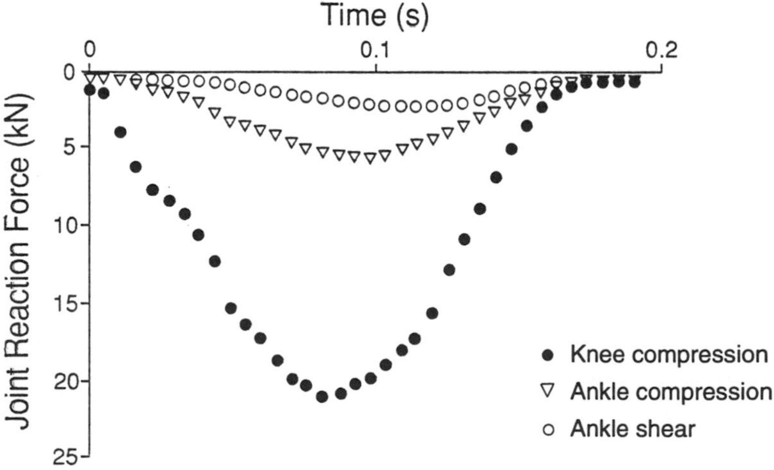

Figure 2.23 Joint reaction forces at the knee and ankle during the stance phase of running (4.5 m/s).
Note. From "A Bioengineering Analysis of Human Muscle Force and Joint Forces in the Lower Limbs During Running" by R.N. Harrison, A. Lees, P.J.J. McCullagh, and W.B. Rowe, 1986, *Journal of Sports Sciences*, **4**, p. 213. Copyright 1986 by E. & F.N. Spon Ltd. Adapted by permission.

(pushing) **force**. However, the tensile force exerted by muscle can be transmitted along a segment and exert a compressive bone-on-bone force.

Because of a muscle's unidirectional capability of force exertion, movement about a joint is controlled by an opposing set of muscles. For example, in the absence of other forces, the movement of elbow extension-flexion is controlled by one muscle group causing extension (elbow extensors) and another group controlling flexion (elbow flexors). However, when other forces, such as gravity, are acting on the system, then the elbow flexors can control extension, and the extensors can control flexion. For this reason, *the function of a muscle depends critically on the context in which it is activated.* Nonetheless, control of human movement requires a minimum of one pair of opposing muscles to control each anatomical motion (e.g., flexion-extension, abduction-adduction).

Because force is a vector, muscle force can be represented as an arrow and described in terms of the vector characteristics of magnitude and direction. In this context, there are four common assumptions related to the analysis of human movement. First, most skeletal muscles exert forces across a joint and can cause body segments to rotate, which allows us to consider many of the human body functions as inanimate machines. Crowninshield and Brand (1981) attribute this assumption to Leonardo da Vinci (1452–1519). Second, the human body can be regarded as a series of rigid body segments; the deformation of soft tissue and the movement of body fluids within the body segment have a negligible effect on the movement (Braune & Fischer, 1889). Third, the direction of the muscle force vector is a straight line between the proximal and distal attachments, with the application of force represented as a point (Brand et al., 1982). In reality, the attachment of muscles does not occur at a point but is spread over a finite (i.e., measurable) area. As discussed previously, every force is distributed over a finite area. However, when the size of the area is negligible compared to the other dimensions of the system, the force may be considered to be acting at a point. If the attachment site is substantial (e.g., trapezius, pectoralis major), a muscle may be represented by several lines of action (van der Helm & Veenbaas, 1991). The fourth assumption, an extension of the idea that movement in a gravitational world is determined by the net imbalance of forces (law of inertia), is that mechanical analyses focus on the net effect of muscles crossing a joint. For this reason, most of our calculations enable us to determine the **resultant muscle force** rather than the force exerted by individual muscles.

Magnitude of Muscle Force. Both the magnitude and direction of the muscle force vector are difficult to measure. To measure the magnitude of muscle force directly, we must be able to measure the force transmitted by the tendon. In an isolated-muscle experiment, this measurement involves connecting the tendon to a force transducer (Ralston, Inman, Strait, & Shaffrath, 1947). In human experiments, where the tendon is not detached from the bone, muscle force can be measured by placing a buckle transducer on the tendon. Gregor, Komi, and Järvinen (1987) used the buckle transducer technique to measure the force in the Achilles tendon during cycling; they measured peak tendon forces of about 700 N in the right Achilles tendon when subjects were pedaling at a rate of 90 rpm (revolutions per minute) while producing 265 W of power.

Most human subjects, however, are not willing to volunteer for such an invasive procedure, and it is necessary to use more indirect techniques to assess muscle force. Most information on the magnitude and direction of a muscle force has been derived from indirect estimates. One common approach has been to determine the **cross-sectional area** of muscle from sections that are made perpendicular to the orientation of the muscle fibers and to use this information to estimate muscle force (Fick, 1904). Cross-sectional area is a measurement of the end-on view of the area at the level that the section (cut) has been made. These measurements can be obtained from cadavers or from imaging procedures (ultrasound, computed tomography [CT], magnetic resonance imagery) (Narici, Roi, & Landoni, 1988). In addition to the data available on the muscles that cross the elbow joint (Table 2.6), there are data for most other muscle systems, such as the hip, leg, and wrist (Brand, Pedersen, & Friederich, 1986; Clark & Haynor, 1987; Häkkinen & Keskinen, 1989; Lieber, Fazeli, & Botte, 1990).

The capacity of muscle to produce force is related to the cross-sectional area of the muscle by a constant of about 30 N/cm^2. This constant is referred to as **specifictension** and varies from 16 to 40 N/cm^2 in careful measurements (Edgerton, Apor, & Roy, 1990; Kanda & Hashizume, 1992; McDonagh & Davies, 1984; Nygaard, Houston, Suzuki, Jørgensen, & Saltin, 1983; Roy & Edgerton, 1992). In other words, muscle can generate 30 N of force for each square centimeter of cross-sectional area. *Specific tension, therefore, is a measure of the capability of muscle to exert force that is independent of the amount of muscle.* The relation for estimating muscle force ($\mathbf{F}_m$) is given as

$$\mathbf{F}_m = \text{specific tension} \times \text{cross-sectional area} \quad (2.7)$$

According to Table 2.6, biceps brachii with a cross-sectional area of 5.8 cm^2 can develop a maximum force of 174 N. The maximum force that could be exerted for the elbow flexor muscles (biceps brachii, brachialis, brachioradialis) acting together would be 456 N for the cross-sectional data listed in Table 2.6. However, Edgerton et al. (1990) report that the largest cadaver in their sample of four had a cross-sectional area of 34.7 cm^2 and therefore could have exerted a maximum force with the elbow flexor muscles of 1,297 N.

Table 2.6 Summary Data on Cross-Sectional Areas (CSA) and Moment Arms for the Elbow Flexor and Extensor Muscles

Muscle	CSA (cm^2)	Predicted force (N)	Moment arm (cm)	Torque (N·m)	Torque (% maximum)
Biceps brachii	5.8	174	3.8	6.6	32
Brachialis	7.4	222	2.9	6.4	31
Brachioradialis	2.0	60	6.1	3.7	18
Pronator teres	3.6	108	1.6	1.2	6
Extensor carpi radialis longus	3.1	93	3.0	2.8	14
Triceps brachii	23.8	536	—	—	—

Note. Predicted force was estimated by multiplying the CSA values by a specific tension of 30 N/cm^2. Torque was determined as the product of predicted force and moment arm. The % maximum data indicate the contributions of the respective elbow flexor muscles to the total elbow flexor torque.
Data are from Edgerton, Apor, and Roy (1990).

What accounts for the spread in the specific-tension constant that relates force capability to area (i.e., 16 to 40 N/cm^2)? The factors that might influence specific tension include muscle fiber type, gender, and muscle architecture. Some evidence suggests that the force capability of muscle depends on the percentage of fast-twitch fibers (e.g., Schantz, Fox, Norgren, & Tydén, 1981), but the specific tension of fast-twitch fibers (25.4 N/cm^2) has not been found to differ from that of slow-twitch fibers (23.8 N/cm^2) (Lucas, Ruff, & Binder, 1987). Reports of nonconstant specific tension for the different muscle fiber types are based largely on indirect estimates from studies on motor units (Bodine, Roy, Eldred, & Edgerton, 1987; McDonagh, Binder, Reinking, & Stuart, 1980). One important difference between the muscle fiber and motor unit measurements is that the force exerted by an isolated muscle fiber can be measured directly, whereas the force exerted by a motor unit is measured at the tendon and is affected by all the connective tissue structures between the crossbridges and the tendon. For this reason, it seems that specific tension is constant for the different muscle fiber types, but the transmission of force from the crossbridges to the tendon varies among motor unit types.

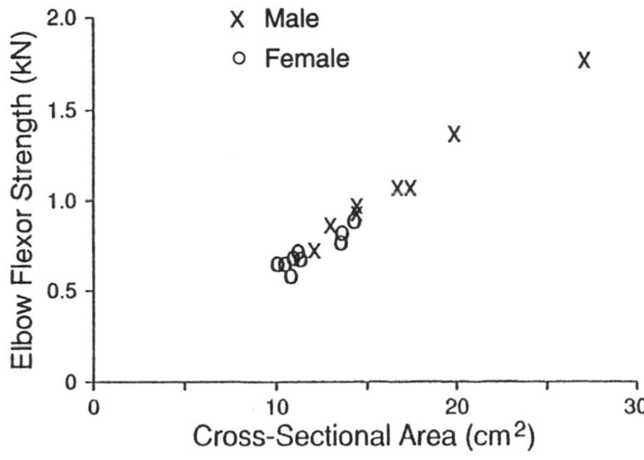

Figure 2.24 Relationship between elbow flexor strength and the cross-sectional area of the flexors for 129 male and 126 female subjects.
Note. Data from Ikai & Fukunaga, 1968.

Antigravity muscles—those involved with the maintenance of an upright posture (e.g., knee extensors)—are about twice as strong as their opposing muscles, but this is due to differences in size rather than specific tension. Similarly, males are generally stronger than females (when strength is defined as the capacity to produce force in an isometric contraction), but this is due to differences in muscle mass (Figure 2.24). The cause of these differences is hormonal; testosterone (male) is better than estrogen (female) at stimulating the protein synthesis that results in muscle growth. For the population shown in Figure 2.24, Ikai and Fukunaga (1968) determined a mean constant of 61.1 N/cm^2 for the elbow flexors (this specific-tension value was artificially high because the measurement of cross-sectional area included only biceps brachii). The specific tension of 61.1 N/cm^2 shown in the slope of Figure 2.24 means that, on average, an individual is capable of generating 61.1 N of force for each square centimeter of cross-sectional area. When the cross-sectional areas of all the elbow flexor muscles are included, the specific tension is 30 to 37 N/cm^2 (Edgerton et al., 1990).

These observations suggest that the magnitude of muscle force is not affected by muscle fiber type or gender but varies with cross-sectional area, whereas specific tension remains constant. One factor that does appear to affect this relationship, however, is muscle architecture. If we consider a muscle with a simple structure in which the fibers are aligned parallel to one another and extend from one end of the muscle to the other (e.g., biceps brachii), then the absolute force at each point along the length of the muscle is essentially identical. Consequently, the force exerted at one end of the muscle is the same as that in the middle of the muscle and at the other end of the muscle. Because muscles have a fusiform rather than a cylindrical shape, cross-sectional area varies along the length of the muscle. If muscle force is constant along its length but cross-sectional area varies, then specific tension also must vary in order to keep the product (Equation 2.7) constant; that is, specific tension must vary in a reciprocal manner with cross-sectional area. Although much remains to be learned about how force is transmitted from the crossbridges to the tendon, this example suggests that muscle architecture does influence muscle force and probably contributes to the variability in the values reported for specific tension.

In addition to cross-sectional area, two other techniques can be used to estimate the magnitude of muscle force; these are EMG and intramuscular pressure. The EMG enables us to measure the electrical activity in muscles, which is a direct response to the activation signals from the nervous system. Under isometric conditions, the magnitude of the EMG is highly correlated with muscle force (Bigland & Lippold, 1954; Lawrence & De Luca, 1983). Although this association is less direct for nonisometric conditions (Calvert & Chapman, 1977; Milner-Brown & Stein, 1975), there are algorithms with which it is possible to estimate the magnitude of the muscle force from the EMG (Hof, Pronk, & van Best, 1987; Hof & van den Berg, 1977; Marras & Sommerich, 1991a, 1991b). In contrast, the measurement of *intramuscular pressure* with a catheter seems to provide a less variable index of muscle force than does the EMG (Parker, Körner, & Kadefors, 1984; Sadamoto, Bonde-Petersen, & Suzuki, 1983). The disadvantage of the intramuscular pressure measurement is that it is an invasive procedure that involves inserting a pressure transducer into the muscle. Like the EMG, intramuscular pressure increases linearly with muscle torque, but the slope of the relationship differs among muscles (Figure 2.25).

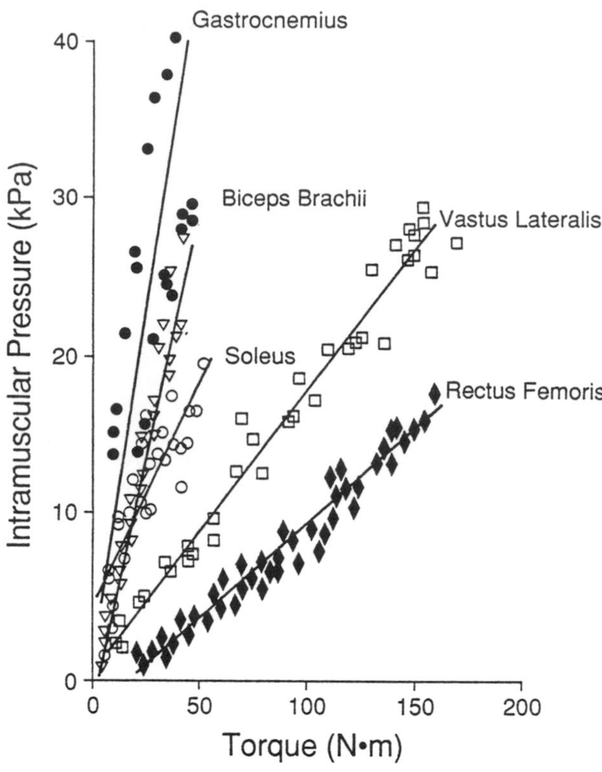

Figure 2.25 Association between intramuscular pressure and muscle torque during isometric contractions for five muscles in one subject.

Note. From "Skeletal Muscle Tension, Flow, Pressure, and EMG During Isometric Contractions in Humans" by T. Sadamoto, F. Bonde-Petersen, and Y. Suzuki, 1983, *European Journal of Applied Physiology*, **51**, p. 403. Copyright 1983 by Springer-Verlag, Inc. Adapted by permission.

Direction of Muscle Force. When we include a muscle force on a free body diagram, we must first determine which muscle group (e.g., flexors or extensors) is exerting the force and then draw the force vector so that it is directed back across the joint. For example, consider a free body diagram that shows the Achilles tendon force measured by Gregor et al. (1987) during cycling. Figure 2.26 indicates that the system consists of the foot (from the ankle to the toes) and that it interacts with its surroundings with the following forces: $\mathbf{F}_g$ (ground reaction force—actually the force exerted by the pedal), $\mathbf{F}_w$ (weight), $\mathbf{F}_j$ (joint reaction force), and $\mathbf{F}_m$ (muscle force). Notice that $\mathbf{F}_m$ is directed back across the ankle joint, in much the same position as the Achilles tendon.

Although the angle of pull of a muscle is generally shallow, the angle does change over the range of motion and can become substantial. An et al. (1984) measured the angle of pull of the muscles that cross the elbow joint. With the forearm in a horizontal position, these angles are 1.4 rad for biceps brachii, 1.2 rad for brachialis, 0.4 rad for brachioradialis, 0.2 rad for pronator teres, and 0.05 rad for triceps brachii. Furthermore, the lines of action of most muscle force vectors are actually curvilinear rather than linear. Figure 2.27 shows the lines of action for six muscles that cross the elbow joint. The lines of action were estimated from the measurement of the centroid location in serial cross sections of each muscle (An, Hui, Morrey, Linscheid, & Chao, 1981).

Intraabdominal Pressure

Because the intraabdominal cavity contains mainly liquid and viscous material, it can be considered an incom-

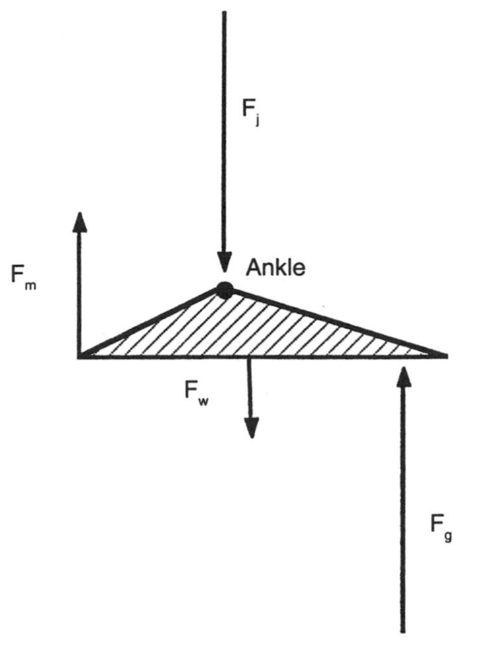

Figure 2.26 Free body diagram of the foot showing the measurement of the Achilles tendon force during cycling.

pressible element that can transmit forces from the muscles encasing the cavity to the supporting structures of the trunk. The muscles surrounding the abdominal cavity include the abdominal muscles anteriorly (rectus abdominus, external and internal obliques, and transversus abdominus), the diaphragm above, and the muscles of the pelvic floor below. Voluntary pressurization of the abdominal cavity is referred to as the **Valsalva maneuver**. Intraabdominal pressure ($\mathbf{F}_i$), measured in pascals (Pa; 1 Pa = 1 N/m^2), can be increased by closing the epiglottis and activating various trunk and abdominal muscles (Cresswell, Grundström, & Thorstensson, 1992). When the trunk muscles are activated, this causes the **intrathoracic pressure** to increase in addition to the intraabdominal pressure. During lifting and jumping activities intrathoracic and intraabdominal pressures tend to change in parallel; the intraabdominal pressure is usually greater (Harman, Frykman, Clagett, & Kraemer, 1988).

Intraabdominal pressure has been proposed as a mechanism to reduce the load on back muscles during lifting tasks (Bartelink, 1957; Cresswell, 1993; McGill & Norman, 1993; J.M. Morris, Lucas, & Bressler, 1961). Figure 2.28a shows this effect during the lifting of a 91-kg mass. The system consists of the upper body above the lumbosacral joint. This system interacts with its surroundings with three weight vectors—weight of the head-neck-arms ($\mathbf{F}_{w,h}$), weight of the trunk ($\mathbf{F}_{w,t}$), and a 91-kg load ($\mathbf{F}_{w,l}$)—that all act vertically downward, a resultant muscle force due to the hip and back extensor muscles, a joint reaction force, and an extensor force due to the intraabdominal pressure. When intraabdominal pressure is not considered, the back and hip muscles must exert a force ($\mathbf{F}_m$) of 8,223 N, and the joint reaction force ($\mathbf{F}_j$) must be 9,216 N just to support the load (Figure 2.28a). However, when an intraabdominal pressure of 19.7 kPa (force of 810 N) is included in the calculation, $\mathbf{F}_m$ is reduced to 6,403 N and $\mathbf{F}_j$ becomes 6,599 N (Figure 2.28b).

Intraabdominal pressure, which can be measured with a catheter pressure transducer inserted into the abdominal cavity, varies for most activities, especially those including jumping, lifting, and the Valsalva maneuver (Harman et al., 1988). Figure 2.29 provides an example

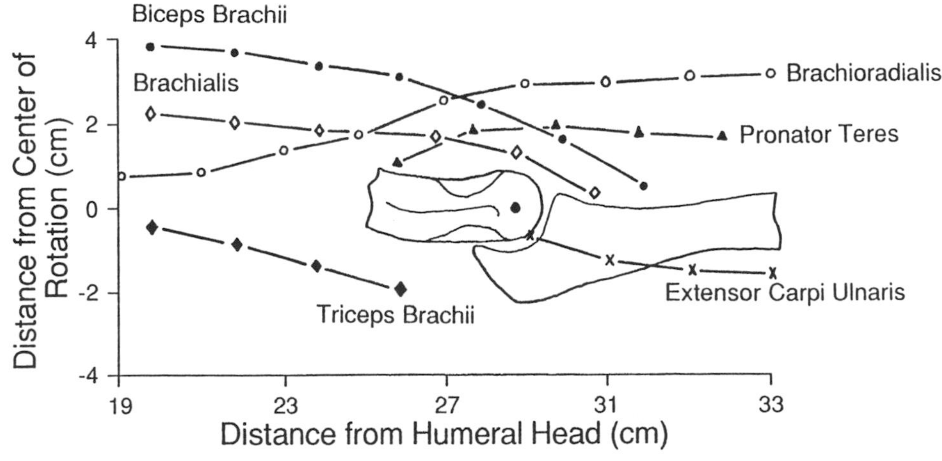

Figure 2.27 Graphical representation of the centroid line of action for six muscles that cross the elbow joint. The elbow joint (humeroulnar) is shown extended and in a neutral position.

Note. Adapted from "Muscles Across the Elbow Joint: A Biomechanical Analysis" by K.N. An, F.C. Hui, B.F. Morrey, R.L. Linscheid, and E.Y. Chao, 1981, *Journal of Biomechanics*, **14**, p. 663, Copyright 1981, with kind permission from Pergamon Press, Headington Hill Hall, Oxford 0X3 OBW, UK.

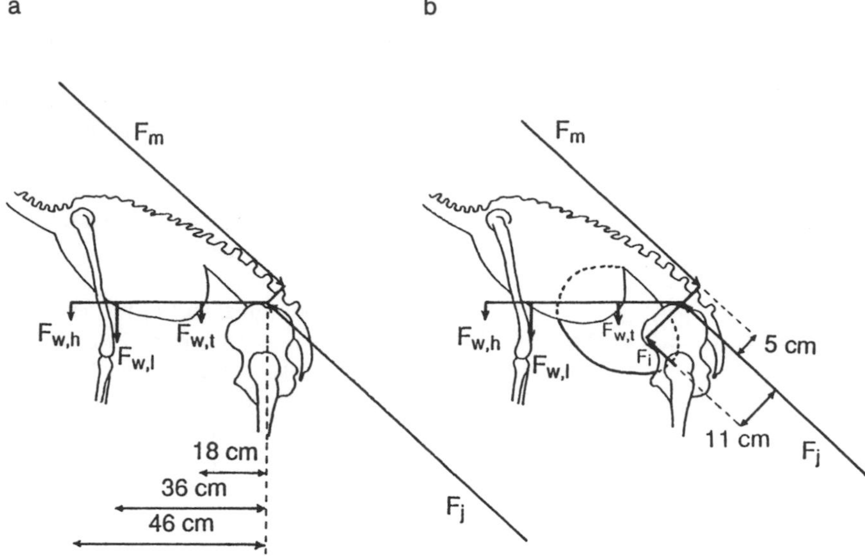

Figure 2.28 Effect of intraabdominal pressure on the muscle and joint reaction forces involved in lifting a 91-kg load.
Note. From "Role of the Trunk in Stability of the Spine" by J.M. Morris, D.B. Lucas, and B. Bressler, 1961, *Journal of Bone and Joint Surgery*, **43A**, pp. 344, 345. Copyright 1961 by The Journal of Bone & Joint Surgery, Inc. Adapted by permission.

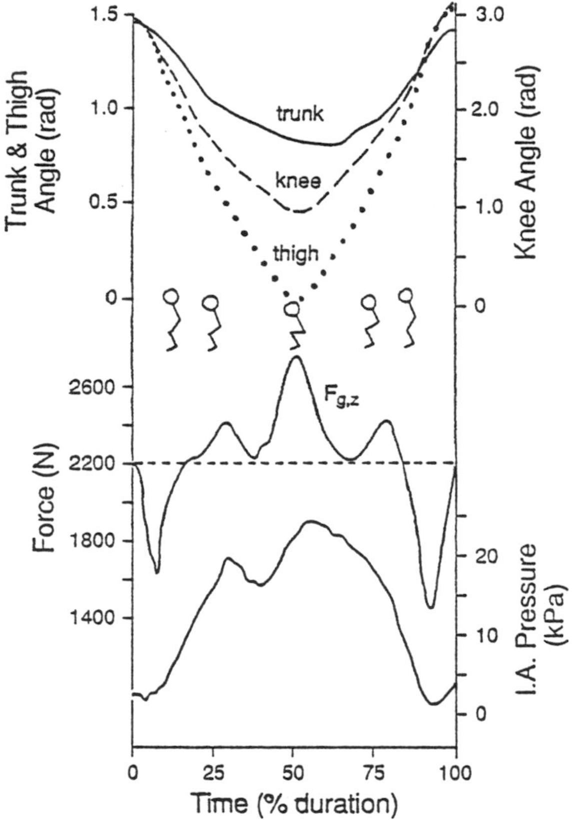

Figure 2.29 Changes in joint angles, the vertical component of the ground reaction force, and intraabdominal pressure during a squat lift by a weight lifter.
Note. From "Biomechanics of the Squat Exercise Using a Modified Center of Mass Bar" by J.E. Lander, B.T. Bates, and P. DeVita, 1986, *Medicine and Science in Sports and Exercise*, **18**, p. 473. Copyright 1986 by Williams and Wilkins. Adapted by permission.

of the change in intraabdominal pressure during the squat lift, an event in the sport of power lifting. In this lift, the lifter begins from an erect position and lowers the load until the thighs are parallel to the ground (the trunk, knee, and thigh angles reach minimum values), after which the lifter returns to an erect posture. The stick figures in Figure 2.29 indicate the extent of the movement and that the barbell is supported on the shoulders. Because the load (weight of the lifter and the barbell) in this example is 2,200 N, the vertical component of the ground reaction force ($F_{g,z}$) fluctuates about this value. Values for $F_{g,z}$ that are less than 2,200 N indicate that the system is accelerating downward, and values greater than 2,200 N depict an upward acceleration of the system. The increase in intraabdominal pressure tends to coincide with values greater than 2,200 N when the load on the trunk muscles is greatest. Other investigators have shown that the use of a weight-lifting belt can increase the intraabdominal pressure during lifting activities (Harman, Rosenstein, Frykman, & Nigro, 1989; Lander, Hundley, & Simonton, 1992; McGill & Norman, 1993).

Despite the correlation between various movements and changes in intraabdominal pressure, there is some controversy over the functional role of this mechanical effect (Marras & Mirka, 1992). It has been proposed, for example, that one effect of intraabdominal pressure is to reduce the compressive forces that act on the intervertebral disks. However, Nachemson, Andersson, and Schultz (1986) have shown that although the Valsalva maneuver increases the intraabdominal pressure, it can also increase the pressure on the nucleus of the L3 disk for some moderate tasks. Nonetheless, for the most strenuous task in which the subjects leaned forward 0.53 rad while holding an 8-kg load in outstretched arms, a

Valsalva maneuver increased the intraabdominal pressure from 4.35 kPa to 8.25 kPa and reduced the **intradiscal pressure** from 1,625 kPa to 1,488 kPa.

As a pressure in a confined volume, the intraabdominal pressure exerts a force over the surface area of the abdominal cavity. The force that the intraabdominal pressure exerts on the trunk is usually estimated as the product of intraabdominal pressure and the surface area of the diaphragm, which J.M. Morris et al. (1961) estimated to be about 0.0465 m^2 for an adult. If this estimate is combined with the peak intraabdominal pressure of 25 kPa shown in Figure 2.29, then the force acting on the diaphragm due to the intraabdominal pressure would be approximately 1,163 N during the squat lift. Clearly, this is not an insignificant force in terms of human movement.

Fluid Resistance

Both human motion (e.g., ski jumping, cycling, swimming, skydiving) and projectile motion (e.g., the flight of a discus or golf ball) can be profoundly influenced by the fluid (gaseous or liquid) medium in which they occur. As an object passes through a fluid, energy transfers from the object to the fluid. This phenomenon is referred to as **fluid resistance**. In essence, this transfer depends on the extent to which the object disturbs the fluid. Consequently, the transfer (fluid resistance) increases as the speed of the object increases.

The interaction between the fluid and the object is typically schematized as shown in Figure 2.30, where the fluid is shown flowing around a stationary object. The schematized lines of fluid flow are **streamlines**; they conceptually represent consecutive layers of particles in the fluid. One streamline, or layer, is adjacent to the object, another streamline lies on top of this, and so forth. The streamline closest to the object is the **boundary layer**. When a dye is placed in the fluid, it is apparent that the streamlines can move at different velocities. Typically, the streamline touching the object has the lowest velocity because it is slowed down by the effect of friction between the layer of particles and the object; this effect is known as *friction drag*.

When the object has an asymmetrical shape, such as an airfoil, or when it is spinning, the streamline traveling the greater distance around the object travels at a greater velocity, and the pressure on that side of the object is less than on the other side. This effect is known as **Bernoulli's principle**, which states that fluid pressure is inversely related to fluid velocity. When the pressure on the top side of the airfoil is less because the velocity of the streamlines is greater, a force pushes the airfoil upward. This effect is known as **lift** ($\mathbf{F}_l$) and represents the fluid resistance force that acts perpendicularly to the direction of fluid flow (Figure 2.31).

When the streamlines move around an object, they may remain uniform, which is referred to as **laminar flow** (Figure 2.30a), or they may become nonuniform, which is called **turbulent flow** (Figure 2.30b). The

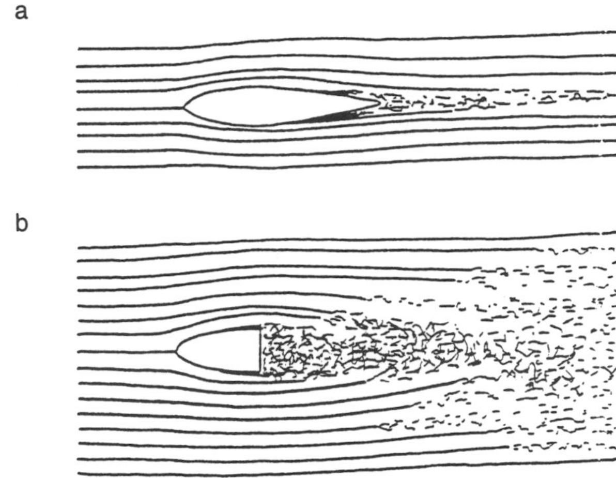

Figure 2.30 Tracings of smoke streamlines indicating the air traveling around a streamlined object (a) and around the same object when it is cut in half (b). The airflow is much more turbulent around the object cut in half; the smoke streamlines break up.

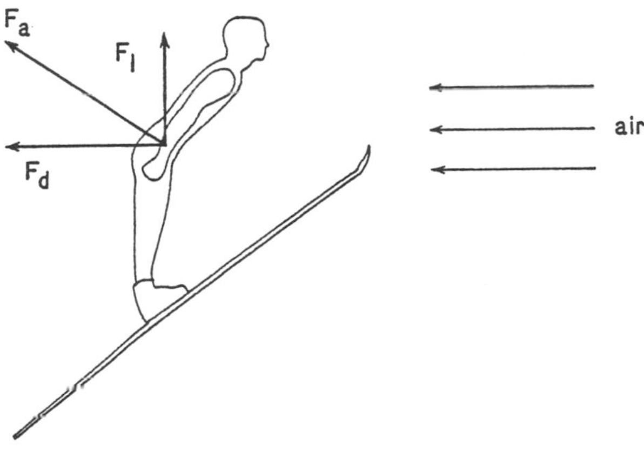

Figure 2.31 Air resistance encountered by a ski jumper. The air is shown moving relative to the jumper. The resultant air resistance vector ($\mathbf{F}_a$) is resolved into two components, drag ($\mathbf{F}_d$) and lift ($\mathbf{F}_l$).

major factor influencing fluid flow, and whether it is laminar or turbulent, is the shape and texture of the object. When an object (e.g., a cycling helmet,) is streamlined, its shape maintains a laminar flow of the streamlines for longer than normal. If the flow becomes turbulent, this generates a fluid resistance effect called **drag** ($\mathbf{F}_d$). The direction of the drag vector is parallel to the direction of fluid flow (Figure 2.31). The drag due to turbulent flow is sometimes referred to as **pressure drag**, or form drag, to distinguish it from the drag due to friction of the boundary layer (**surface drag**). In general, drag opposes the forward motion of the object, whereas lift produces vertical motion (up or down). But this is not always the case; in swimming, lift generated by the hands and legs contributes to forward motion. In fact, the key to a successful swimming stroke is to move the hands through the water so that both the drag and lift components contribute to the forward motion of the swimmer (Schleihauf, 1979). As Figure 2.32 shows, the orientation of the swimmer's hand (shown as an airfoil) changes continuously so that the resultant fluid resistance force ($\mathbf{F}_f$) is directed forward throughout the stroke. Both drag ($\mathbf{F}_d$) and lift ($\mathbf{F}_l$) contribute to $\mathbf{F}_f$ in this direction. Drag acts parallel to the direction of fluid flow, lift acts perpendicularly.

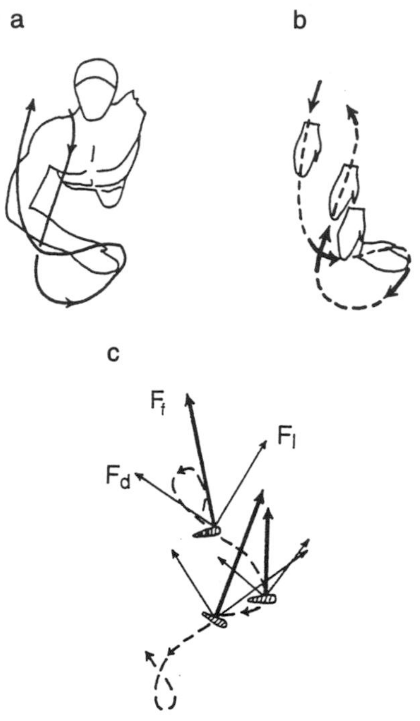

Figure 2.32 Path and orientation of the right hand during a freestyle stroke: (a) front view of the path followed by the hand; (b) flow of the water across the hand at four points during the stroke; (c) view from above of the path followed by the hand; hand enters the water at the top and leaves at the bottom.
Note. From "A Hydrodynamic Analysis of Swimming Propulsion" by R.E. Schleihauf. In *Swimming III* (pp. 85, 91) by J. Terauds (Ed.), 1979, Baltimore: University Park Press. Copyright 1979 by University Park Press.

It is frequently useful to distinguish factors that can influence the drag component of fluid resistance. Two of these factors have already been mentioned; the nature of the fluid flow around the object (pressure drag) and the quantity of friction between the boundary layer of the fluid and the object (surface drag). Because pressure drag increases as the fluid flow shifts from laminar to turbulent, designers often streamline an object to minimize the tendency for the fluid flow to become turbulent. In recent years (since the 1984 Los Angeles Olympics), cyclists' helmets have become more pointed in the rear because such a design has been shown to reduce airflow turbulence. Furthermore, the surface texture of an object, such as the dimples in a golf ball, can influence the width of the turbulent wake and hence decrease the pressure drag experienced by the object. The golf ball dimples decrease the width of the wake and pressure drag by delaying the point along the object at which the streamlines become turbulent. The net effect is that for a given golf swing, a dimpled golf ball can be driven 239 m, whereas a smooth golf ball would only travel 46 m (Townend, 1984).

The texture of the object affects not only the width of the turbulent wake but also the velocity of the fluid layers close to the object. The rougher the surface of the object, the greater the friction and hence the surface drag. Concern for surface drag is evident in many sports, including swimming, cycling, and rowing. For this reason, many swimmers shave body hair and wear caps, and cyclists experiment with exotic materials and designs for clothing. Surface drag, however, has a much smaller effect than pressure drag.

In addition to pressure and surface drag, a third factor that affects the drag component of fluid resistance is **wave resistance**. This effect sets the upper limit on the speed of surface ships and may be important at the higher velocities associated with competitive swimming. The resistive effects of waves are probably due to the uneven density of the fluid (water vs. air) that the swimmer encounters. The presence of waves probably decreases the average proportion of the swimmer's body that is out of the water and therefore increases fluid density. The current practice in pool design is to minimize wave turbulence with carefully designed gutters and lane markers.

Magnitude of Fluid Resistance. The magnitude of the fluid resistance vector can be determined from

$$\mathbf{F}_f = kA\mathbf{v}^2 \tag{2.8}$$

where k is a constant, A represents the *projected area* of the object, and $\mathbf{v}$ refers to the velocity of the fluid relative to the object. Projected area refers to a silhouette view of the frontal area of the object as it moves through the fluid. The constant k is an abbreviation for the term 0.5 ρC_D for drag and 0.5 ρC_L for lift; ρ accounts for fluid density, C_D distinguishes the effects of laminar and turbulent flow, and C_L is proportional to the angle between fluid flow and the orientation of the object.

Consider the action of the air resistance vector on the runner (Hill, 1928) in Figure 2.6. The constant k is about 0.55 kg/m^3 for a runner moving through air. The projected area of the runner, that is, the front view of the body on which the air acts, can be estimated as 0.15 multiplied by the square of standing height for the 1.55-m runner in Figure 2.6 (A = 0.36 m^2). This estimate, however, seems a little low compared to those obtained with a skier in semisquat (0.27 m^2) and upright (0.65 m^2) positions (Spring, Savolainen, Erkkilä, Hämäläinen, &

Pihkala, 1988). Nonetheless, if the runner had a speed of 6.5 m/s and experienced a tailwind of 0.5 m/s, the relative velocity (**v**) would be 6.0 m/s. Thus

$$\mathbf{F}_f = 0.55 \times (0.15 \times 1.55^2) \times 6.0^2$$
$$= 7.14 \text{ N}$$

Calculations by Shanebrook and Jaszczak (1976) have indicated that, at middle-distance running speeds (approximately 6 m/s), up to 8% of the energy expended by the runner is used to overcome air resistance, whereas at sprinting speeds this value can be up to 16% of the total energy expenditure. Drafting (shielding) can abolish 80% of this cost: When running at 6.0 m/s, drafting behind another runner is equivalent to reducing speed by about 0.1 m/s. Ward-Smith (1985) has calculated the effects of head winds and tailwinds on performance times in a 100-m sprint:

Wind speed (m/s)	Head wind (s)	Tailwind (s)
1	+0.09	−0.10
3	+0.26	−0.34
5	+0.38	−0.62

With a 3-m/s head wind, the 100-m time would be increased by 0.26 s, whereas a 3-m/s tailwind would decrease the time by 0.34 s. The effects of the head winds and tailwinds are not symmetrical for 3 and 5 m/s (i.e., there are not similar gains and losses), because the issue is not the force associated with the wind but rather the work done against it. Similarly, 80% of the power generated by a cyclist traveling on level ground at 8.05 m/s (18 mph) is used to overcome air resistance (Gross, Kyle, & Malewicki, 1983). Cyclists can reduce power output by about 30% and still sustain the same speed when drafting behind another cyclist.

The drag experienced by swimmers has been estimated by measuring the force needed to tow passive subjects through the water. With this technique, however, the limbs do not actually perform swimming movements. The drag acting on a swimmer also has been estimated by measuring the propulsive force exerted by a subject swimming at a constant velocity. Under these conditions, the acceleration of the swimmer is zero, and hence the forces acting on the swimmer are balanced. In the horizontal direction, therefore, the propulsive force is equal to the resistive force (drag). With this method, the average drag experienced by a swimmer doing the front crawl is 53 N at a constant speed of 1.48 m/s (van der Vaart et al., 1987).

The magnitude of the fluid resistance vector depends on the extent to which the fluid is disturbed by the object. Equation 2.8 indicates that this disturbance is proportional to the size of the projected area and the speed of the object relative to the fluid. Because the projected area is a function of the orientation of the object to the fluid, variation of the orientation can be used to alter the lift and drag experienced by the object as it passes through the medium. For example, consider a person water-skiing behind a boat. The ability of the skier to remain on top of the water is due to the force exerted by the fluid on the skis. The lift component of this force ($\mathbf{F}_f$) must be at least as great as the downward forces, the latter being due mainly to the weight of the skier. As the boat goes faster (i.e., as the speed of the skis relative to the water increases), the magnitude of the projected area can decrease to maintain the necessary lift. In fact, this trade-off between projected area and relative velocity occurs throughout the skiing run. At the start, the skier in the water points the skis almost vertically. In this position the projected area of the skis to the water is maximized. As the boat begins to move and tow the skier, the skis disturb the water and create a lift force as the water flows around the skis. At the beginning of the run, the projected area must be great because it combines with a low relative speed to generate the necessary $\mathbf{F}_f$ and to lift the skier to the surface. As the speed of the boat increases, the skis can be lowered because a smaller projected area is then needed. Once the skier releases the tow rope, the lift force eventually becomes insufficient, and the skier sinks into the water.

Terminal Velocity. Because the relative velocity of the object is a squared term in Equation 2.8, it is the single most important factor influencing fluid resistance. A sky diver represents an interesting example of this phenomenon. After the diver jumps from the airplane, the speed of the sky diver will increase up to some terminal value. How do we determine the value of this terminal velocity? Once again, the analysis should begin with an appropriate free body diagram (Figure 2.33).

When the sky diver jumps out of the airplane, the system (sky diver, parachute, and associated equipment)

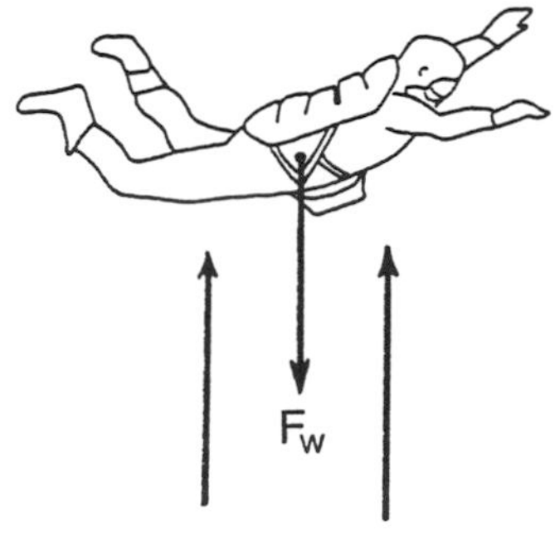

Figure 2.33 The forces experienced by a sky diver during free fall.

is accelerated, due to gravity ($\mathbf{F}_w$), toward the ground. But the downward acceleration of the system is only briefly equal to the value for gravity (9.81 m/s^2), because as system speed increases, the opposing effect of air resistance increases. We know from Equation 2.8 that the magnitude of air resistance increases as the square of relative velocity; thus, as $\mathbf{v}$ increases, so does the force due to air resistance. The speed of the system will continue to increase (i.e., acceleration will be nonzero) until the force due to air resistance is equal to the weight of the system (i.e., the acceleration due to gravity). After these two forces become equal, speed will remain constant, and the system will have reached a terminal velocity. Because velocity remains constant (i.e., zero acceleration) and the system is in equilibrium at terminal velocity, the forces must be balanced. Thus

$$\mathbf{F}_w = \mathbf{F}_f$$

And because we have defined $\mathbf{F}_f$ (Equation 2.8),

$$\mathbf{F}_w = kA\mathbf{v}^2$$

We can rearrange this relationship to determine the terminal velocity:

$$\mathbf{v} = \sqrt{\frac{\mathbf{F}_w}{kA}}$$

If k = 0.55 kg/m^3, A = 0.36 m^2, and $\mathbf{F}_w$ = 750 N, then

$$\mathbf{v} = \sqrt{\frac{750}{0.55 \times 0.36}}$$

terminal velocity = 61.6 m/s (138 mph)

Despite being in free-fall conditions, experienced sky divers can maneuver reasonably well, performing somersaults, cartwheels, and other movements. How can they do this? As we know from the free body diagram (Figure 2.33), sky divers experience two forces as they fall: weight and air resistance. Both of these forces are distributed forces; that is, they are distributed over the entire system but are drawn as acting at one or two points. The weight vector is always drawn as acting at the CG. Similarly, the force due to air resistance has a central balance point: the center of air resistance. If the line of action of the air resistance vector does not pass through the CG, then the air resistance exerts a rotary force (torque) about the CG and causes the sky diver to experience angular motion. By moving the limbs, the sky diver can alter the projected area and shift the center of air resistance so that the vector does not pass through the CG.

Projectile Aerodynamics. In chapter 1 we considered the case where air resistance had a negligible effect on the trajectory of projectiles. But for many human movements, air resistance does exert a significant effect. One of the most notable contributions is the effect of air resistance on the trajectory of baseballs, golf balls, volleyballs, cricket balls, and other such projectiles. Air resistance provides a force, in addition to gravity, that can alter the trajectory of a projectile. Air resistance causes baseballs to curve, golf balls to slice, and tennis balls to drop. These effects on the trajectory of a projectile are due to the *lift component of air resistance.* According to Bernoulli's principle, there is an inverse relationship between the pressure exerted by a fluid and its velocity. Fluid pressure decreases as velocity increases. Consequently, when fluid flows around an object, there can be a pressure gradient across the object if there is a difference in the velocity of the fluid flow on the two sides (top and bottom) of the object.

The difference in velocity of the streamlines as they flow around the object can be caused by the shape of the object (e.g., airfoil) or the spin that the object possesses. If the shape is not symmetrical, the streamlines will have different distances to travel, and the velocity of a streamline will be greater on the side where it has to travel farther. For an airfoil, the distance and velocity of airflow are greater on the top side, and therefore the pressure is less on the top; this causes a pressure gradient in the bottom-to-top direction (i.e., lift). Similarly, a spinning projectile will create a pressure gradient across itself by influencing the velocity of the streamlines. For example, if a baseball has a counterclockwise spin as it passes through the air, the streamlines flowing in the counterclockwise direction will have a greater velocity (due to reduced surface drag) and hence there will be a lower pressure on the left side of the ball. This will create a pressure gradient across the baseball from right to left and cause the baseball to curve to the left (Figure 2.34).

Elastic Force

A pulling force can increase the length of a material or tissue. This stretch is possible because of the molecular composition and organization of the material. The extent of the stretch depends on the nature of the material and the magnitude of the pulling force. For an ideal spring, the relation between the applied force and the amount of stretch (deformation of the dimensions of the spring) is given by the simple expression

$$\mathbf{F} = kx \tag{2.9}$$

where $\mathbf{F}$ = pulling force, k = spring stiffness, and x = amount of stretch or deformation. As k increases, the spring stiffness increases, and it is necessary to exert a greater pulling force to accomplish the same Δx. The constant k represents the slope of the relationship between $\mathbf{F}$ and x. The stiffer the spring, the steeper the slope; the more compliant the spring, the lower the slope (Figure 2.35a).

If the force applied to the spring is increased indefinitely, the spring will experience permanent changes in

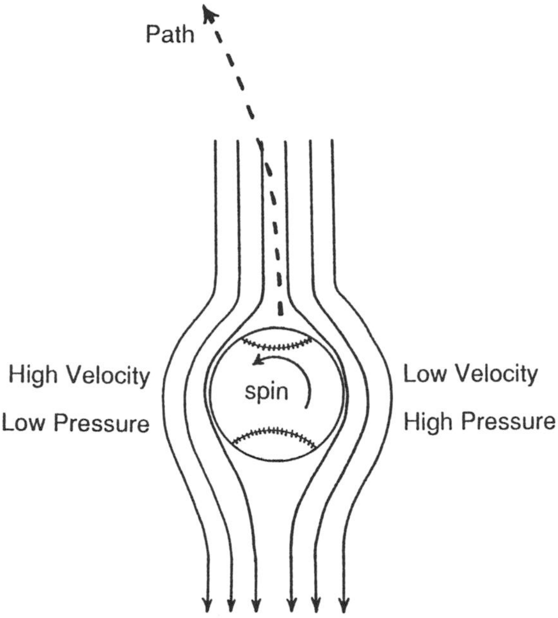

Figure 2.34 The effect of the lift component of air resistance on the trajectory of a baseball. The baseball has a spin in the counterclockwise direction and will curve to the left.

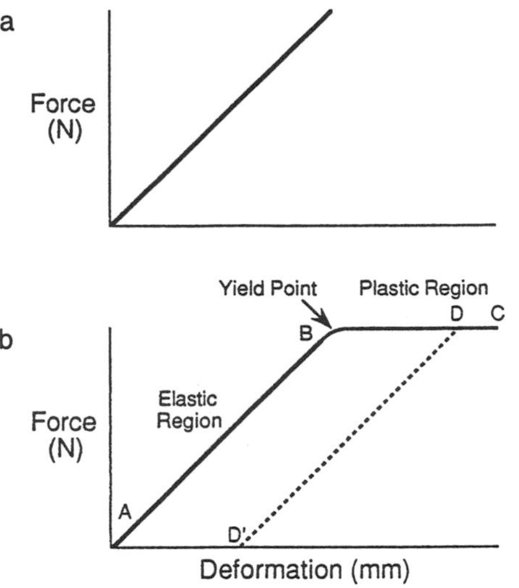

Figure 2.35 Force-deformation relationship for a spring stretched within its elastic range (a) and to failure (b). *Note*. Graph B is from "Biomechanics of Bone" by M. Nordin and V.H. Frankel. In *Basic Biomechanics of the Musculoskeletal System* (p. 6) by M. Nordin and V.H. Frankel (Eds.), 1989, Philadelphia: Lea & Febiger. Copyright 1989 by Lea & Febiger. Adapted by permission.

its dimensions and will eventually break. This behavior is shown by the force-deformation graph in Figure 2.35b. Biological tissue (e.g., bone, muscle, ligament, skin) exhibits similar behavior. If tissue is stretched within its elastic limits (points *A* to *B* in Figure 2.35b) and then released, the tissue will return to its original length due to its elastic properties. The elasticity of tissue, which differs with such factors as tissue type, age, and health status, is a property of its molecular structure (Figure 2.36). The deformation that a tissue experiences when it is stretched within its elastic limits is transient. But if the deformation extends beyond the yield point of the tissue (point *B* in Figure 2.35b) and into the plastic region, there will be long-term changes in the structure of the tissue. For example, if the tissue is stretched to point *D* (Figure 2.35b) and then released, the tissue will assume a resting length (point *D'*) that will be longer than the initial length (point *A*) due to a change in its structure. The tissue will fail if the deformation continues to point *C*.

To compare the force-deformation curve across conditions, the variables force and deformation must be normalized and expressed as a stress-strain relationship. **Stress** (Pa) represents the force applied per unit area of the tissue, where area is measured in the plane that is perpendicular to the force vector. **Strain** (%) indicates the change in length of the tissue relative to its initial length. The upper limit for physiological strain in tendons and ligaments is 2% to 5%, with tendons breaking at about 8% (Alexander & Ker, 1990). Because the peak force in the Achilles tendon of a 70-kg human running

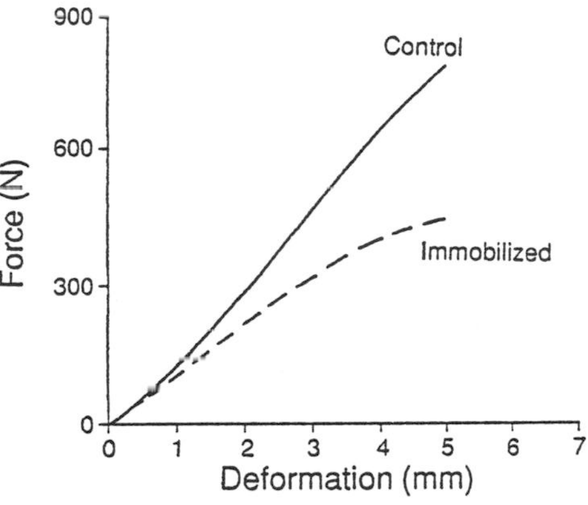

Figure 2.36 Force-deformation curve for the anterior cruciate ligament. The graphs compare the stiffness (slope) of the relationship in healthy knees and in knees that had been immobilized in a plaster cast for 8 weeks. *Note*. From "Functional Properties of Knee Ligaments and Alterations Induced by Immobilization" by F.R. Noyes, 1977, *Clinical Orthopaedics and Related Research*, **123**, p. 219. Copyright 1977 by J.B. Lippincott Company. Adapted by permission.

at moderate speed is about 4,700 N and the cross-sectional area of the tendon is about 89 mm, the peak stress is about 53 MPa (Alexander & Ker, 1990). The elastic region of a stress-strain relationship can be characterized by the **modulus of elasticity** (*E*), which is defined as the ratio of stress (σ) to strain (ε).

$$E = \frac{\sigma}{\varepsilon} \tag{2.10}$$

At least two features of the force-deformation relationship are important in kinesiology. First, the extent to which the force-deformation characteristic of a tissue resembles an ideal spring indicates the potential that the tissue has to store elastic energy (**strain energy**) during a stretch and to use this energy subsequently to do work on its surroundings. At the peak of a stretch, biological tissue possesses a certain amount of strain energy due to the forces exerted by the molecular structure of the tissue. It is this effect that we ascribe to **elastic force**; which is the potential that a tissue has to return to its original length (Equation 3.23). For example, when a trampolinist bounces on a trampoline, the bed is stretched to an extent that depends on the magnitude of the pushing force provided by the trampolinist's body. Toward the bottom of the bounce, the downward pushing force is overcome by the upward-directed elastic force exerted by the bed, and the trampolinist is accelerated upward. The ability of biological tissue, particularly muscle and tendon, to store strain energy and to subsequently use it by exerting an elastic force is thought to have significant effects in movement. For example, prosthetic feet are designed to mimic intact limbs and are able to store and return elastic energy during the stance phase of gait (Czerniecki, Gitter, & Munro, 1991).

Second, a stretch into the plastic region causes at least a partial reorganization of the structure of the material that is associated with some weakening of the material. When a tissue is stretched into the plastic region and then released, the final length of the tissue will be longer than its initial length. By this mechanism, flexibility exercises induce plastic changes in biological tissues that increase the range of motion about a joint. As we will discuss later (chapter 8), considerable research has been conducted to determine the most effective procedures for causing length increases in the tissues that limit joint mobility. These procedures need to provide a compromise between the change in length of the tissue and the weakening that accompanies the structural reorganization.

Inertial Force

Newton told us that a moving object continues to move in a straight line and at a constant speed unless it is acted on by a force (the law of inertia). The object's resistance to any change in its motion is called its *inertia. An object in motion can, due to its inertia, exert a force on another object.* To demonstrate this effect, place one of your forearms in a vertical position (hand pointing upward) with the upper arm horizontal. Relax the muscles of your forearm that cross the wrist joint. Slowly begin to oscillate your forearm in a small arc in a forward-backward direction. As the movement becomes more vigorous, you should notice that your hand begins to flail, particularly if your hand is relaxed. You suspect that the motion of your hand has been caused by the motion of the forearm, and you are correct. The forearm (due to its motion) has exerted an inertial force on the hand and caused the motion of the hand. This simple example emphasizes the mechanical coupling that exists between our body segments.

The hammer throw provides another example of the inertial force. We learned in the section on projectile motion (chapter 1) that the distance an athlete can throw the hammer depends on the hammer's speed and the angle and height at which it is released. For the world-record performance of 86.3 m, the hammer had a release velocity of 29.1 m/s (65 mph) if the angle of release was optimum (0.78 rad, or 45°). To achieve this release speed, an athlete needs to perform five revolutions while traveling across a 2.13-m diameter ring. The athlete applies two forces to the hammer during this procedure (Figure 2.37): (a) a pulling force that provides an angular acceleration to increase the speed of rotation and (b) a centripetal force to maintain the angular nature of the motion. Brancazio (1984) estimated that to achieve a release velocity of 29.1 m/s, the athlete must provide an average pulling force of about 45 N throughout the event and a centripetal force of about 2.8 kN at release.

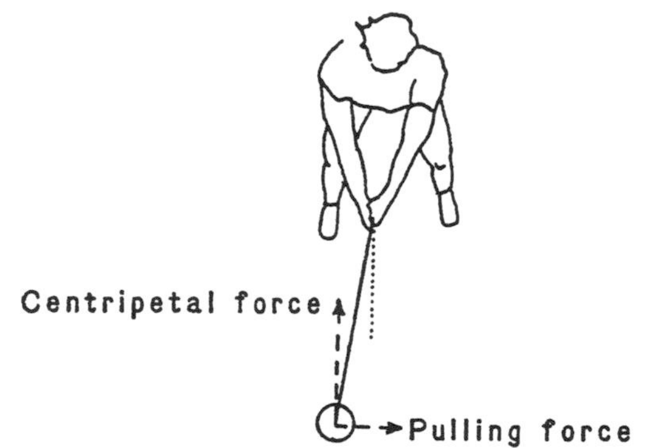

Figure 2.37 Forces exerted by the thrower on the hammer.

Note. From *Sport Science: Physical Laws and Optimum Performance* (p. 296) by P.J. Brancazio, 1984, New York: Simon and Schuster. Copyright 1984 by Peter Brancazio. Adapted by permission.

Because forces act in pairs (action-reaction), the hammer must exert a force on the thrower (centrifugal force) of 2.8 kN at release. This inertial effect of the hammer can be quite easily visualized by imagining that the athlete forgot to let go of the hammer but instead held on past the release point. As soon as the centripetal force exerted by the athlete on the hammer declined below 2.8 kN, the hammer, due to its motion, would exert an inertial force on the athlete. The net effect of this force would cause the athlete to be dragged out of the throwing circle.

Although the example of the hammer throw is an extreme account of inertial effects, such effects are significant in everyday events like running and kicking. In particular, the motion of one body segment (e.g., thigh) can quite readily affect the motion of another body segment (e.g., shank). Figure 2.38 shows the changes that occur in knee angle and resultant muscle torque (rotary force exerted by muscle) about the hip and knee joints during the swing phase of a running stride (Phillips & Roberts, 1980). The net muscle torque (rotary force) about the hip joint is essentially biphasic: It has a flexor direction for the first half of the swing phase and then acts in an extensor direction for the second half of the swing phase. The flexor hip torque accelerates the thigh in a forward direction while the extensor torque accelerates the thigh in a backward direction. Similarly, the resultant muscle torque about the knee joint (dashed line in Figure 2.38) appears biphasic with an intermediate period of zero torque. However, whereas the hip torque has a flexor-extensor sequence, the resultant muscle torque about the knee has an extensor-flexor sequence. In Figure 2.38, positive torque indicates flexor torque for the hip but extensor torque for the knee. Thus a net hip flexor torque is associated with a net knee extensor torque, and in the latter part of the swing phase there is an association between the net hip extensor and knee flexor torques.

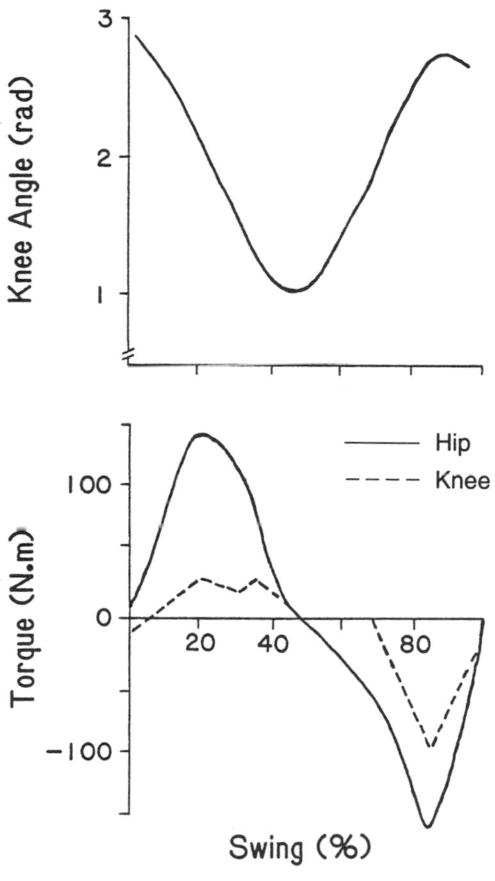

Figure 2.38 Knee angle and resultant muscle torques about the hip and knee joints during the swing phase of a running stride. The subject was a Masters runner who was running at 5.1 m/s.

Note. From "Muscular and Non-Muscular Moments of Force in the Swing Limb of Masters Runners" by S.J. Phillips and E.M. Roberts. In *Proceedings of the Biomechanics Symposium* (pp. 265-266) by J.M. Cooper and B. Haven (Eds.), 1980, Bloomington, IN: Indiana State Board of Health. Adapted by permission.

The role of the resultant muscle torque about the knee joint is made clearer when we consider the way in which knee angle changed during the swing phase. We can do this by comparing the two graphs in Figure 2.38. Knee angle first decreased and then increased during the swing phase (knee angle graph in Figure 2.38). When the knee joint flexed (first half of the swing phase), this was controlled by a resultant extensor torque about the knee joint; that is, *flexion was controlled by the extensor muscles*. This condition represents a lengthening of the active muscles (extensors, quadriceps femoris) and is referred to as an eccentric contraction. Similarly, during the phase of knee joint extension, there was a resultant flexor (hamstrings) torque about the knee joint. In other words, the displacement at the knee joint was controlled by eccentric contractions; knee flexion was controlled not by the knee flexors but by the knee extensors, and vice versa. The cause of the shank motion was the inertial force exerted by the thigh on the shank. By using the muscles that cross the hip joints, subjects can cause the knee joint to flex and to extend. The muscles that cross the knee joint are used to control the effect of the thigh inertial force on the shank; the eccentric contractions serve to brake the forward and backward rotation of the shank. This example illustrates that the inertial force exerted by one body segment on another is an important consideration in the analysis of human movement (Bizzi & Abend, 1983; Phillips, Roberts, & Huang, 1983; Putnam, 1991). We will return to this concept in chapter 3 when we discuss intersegmental dynamics.

Torque

All human motion involves the rotation of body segments about their joint axes. These actions are produced by the interaction of forces associated with external loads and muscle activity. In particular, *human movement is the consequence of an imbalance between the components of the forces that produce rotation*. The capability of a force to produce rotation is referred to as **torque** or **moment of force**. Torque represents the rotational effect of a force with respect to an axis: the tendency of a force to produce rotation.

Suppose an individual is exercising on a Cybex. A number of exercises can be performed on the Cybex, such as extension and flexion about the knee joint. Figure 2.39a illustrates the position of the subject for a knee extension task. The knee joint is positioned in line with an axis about a lever that rotates. The subject pushes the lever and causes it to rotate. One feature of the Cybex is that it can measure the pushing effort of the subject, but there is often confusion as to what the Cybex actually measures. To measure force, it would be necessary for the transducer to be located along the line of action of the force vector. As Figure 2.39b shows, the transducer is actually located at a distance from the line of action of the force exerted by the subject on the lever; thus *the Cybex measures torque and not force.*

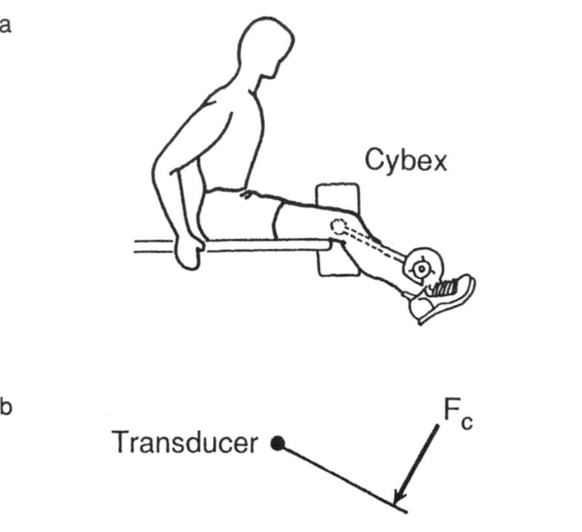

Figure 2.39 The measurement capability of a Cybex: (a) subject performing a knee extension exercise; (b) the line of action of the pushing force (contact force—F_c) that the subject exerts on the lever.

Torque is a vector that is equal to the magnitude of the force times the perpendicular distance between the line of action of the force and the axis of rotation. This distance, known as the **moment arm**, may be expressed as a scalar or a vector quantity. As a scalar quantity we simply indicate the magnitude of the distance between the line of action and the axis of rotation. As a vector quantity, however, we must also be concerned with its direction, and this is specified as being directed from the axis of rotation to the line of action of the force. Algebraically,

$$\mathbf{T} = \mathbf{F} \times \text{moment arm} \tag{2.11}$$

where **T** = torque and **F** = force. Because the moment arm is a perpendicular distance, it is the shortest distance from the line of action of the force to the axis of rotation. Both **F** and the moment arm indicated in Equation 2.11 can be vector quantities; therefore, the direction of the torque vector is perpendicular to the plane in which the force and the moment arm exist. As explained in Appendix B, when two vector quantities are multiplied and the product is a vector, the direction of the product is always perpendicular to the plane in which the other two vectors are located; this procedure is called the **cross product** or **vector product**. Recall from basic geometry that a plane is defined by two lines (such as vectors). Torque, as a vector product, will always be directed out of or into the page.

For graphic convenience torque is often represented as a curved arrow in the same plane as the moment arm and force vectors (e.g., Figure 2.41). The direction of the curved arrow can be determined by applying the **right-hand-thumb rule**, as shown in Figure 2.40. This involves drawing the vectors to be multiplied so that their tails are connected. To determine a torque vector, one must translate the moment-arm and force vectors of the free body diagram (Figure 2.40a) so that their tails contact without changing the directions of the two vectors (Figure 2.40b). Then the right hand, with thumb extended, is used to perform the vector product (Figure 2.40c). In vector terminology (Appendix B), torque is equal to the moment arm (**r**) crossed into force (**T** = **r** × **F**). Take the right hand and extend it in the direction of the moment-arm vector with the palm facing the force vector; the extended thumb will indicate the direction of the torque vector. When this is done for the moment-arm and force vectors shown in Figure 2.40, the product is a torque coming out of the page. To show this graphically, we used a curved arrow and define counterclockwise rotations as positive (see the reference axis in Figure 2.40). If we take the right hand again, curl the fingers in a counterclockwise direction and extend the thumb, this is the positive direction of the torque vector. Consequently, a counterclockwise direction corre-

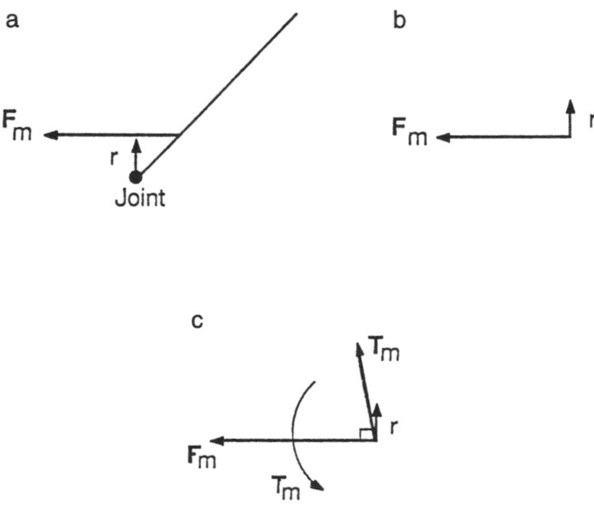

Figure 2.40 Use of the right-hand rule to determine the direction of the torque vector.

sponds to a vector coming out of the page, and we call this direction positive.

Torque, or moment of force, is always determined with respect to a specific axis (as indicated by the moment arm) and therefore must be expressed with reference to the same axis. For example, to discuss the rotary effect of activating a muscle group (e.g., elbow flexors), we need to indicate the point about which the rotation will occur (e.g., elbow joint). Because torque is calculated as the product of a force and a distance, the units of measurement are newtons times meters (N·m).

EXAMPLE

A person is recovering from knee surgery by doing seated knee extension exercises with a weighted boot. The exercise involves sitting at the end of a trainer's bench and raising the leg (shank + foot) from a vertical (knee angle = 1.57 rad) to a horizontal position and then lowering the leg again. What torques about the knee joint (K) are involved in this exercise? Figure 2.41 depicts the appropriate system, from the knee joint down to the toes and the four forces—joint reaction force ($\mathbf{F}_j$), resultant muscle force ($\mathbf{F}_m$), limb weight ($\mathbf{F}_{w,l}$), and boot weight ($\mathbf{F}_{w,b}$)—that represent the interaction of this system with its surroundings (Figure 2.41a). The first step in determining the torques is to draw the moment arms (a, b, c), which we will treat as scalar quantities in this example. This involves extending the line of action for each force and then drawing a perpendicular line from the line of action to the axis of rotation (Figure 2.41b). As the line of action for the joint reaction force passes through the axis of rotation, its moment arm, and therefore the moment of the joint reaction force about the knee joint, is equal to zero. Thus, for the system indicated in Figure 2.41, there are three forces that produce a torque about the knee joint. Figure 2.41c shows the resultant muscle torque ($\mathbf{T}_m$), the torque due to the weight of the boot ($\mathbf{T}_{w,b}$), and the torque due to the weight of the leg ($\mathbf{T}_{w,l}$). The torque due to the total load ($\mathbf{T}_l$) is determined as the sum of $\mathbf{T}_{w,b}$ and $\mathbf{T}_{w,l}$. The direction of the curved torque arrow is the same as the rotation that the torque causes.

Calculating the moment of force (torque) is a straightforward procedure that often involves the use of trigonometric functions. Suppose the person illustrated in Figure 2.41a is using an 80-N weight boot and has a body weight of 700 N; a magnitude and direction of the resultant muscle force vector of 1,000 N and 0.25 rad, respectively; a leg length of 0.36 m; and a distance of 0.05 m from the knee joint to the point of application of the resultant muscle force vector. What are the magnitudes of the three torques? First, let us focus on the moment of force produced by the resultant muscle force and draw the appropriate free body diagram:

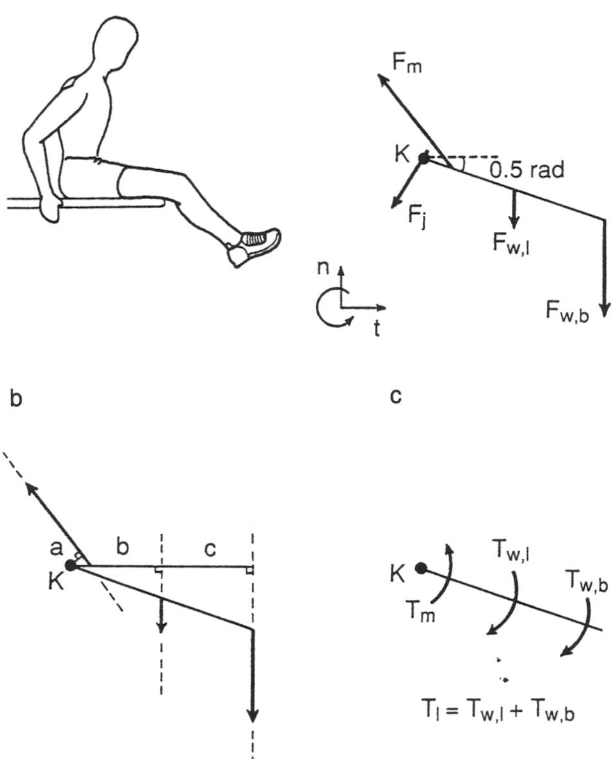

Figure 2.41 Free body diagram of the leg of a person performing leg extension exercises.

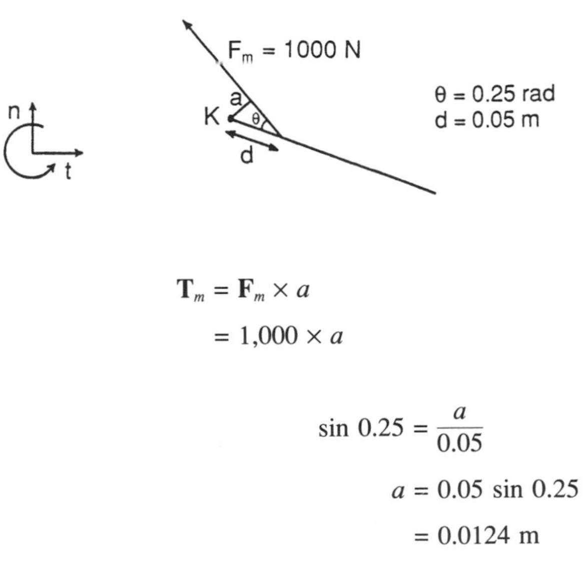

$$\mathbf{T}_m = \mathbf{F}_m \times a$$
$$= 1{,}000 \times a$$

$$\sin 0.25 = \frac{a}{0.05}$$
$$a = 0.05 \sin 0.25$$
$$= 0.0124 \text{ m}$$

$$\mathbf{T}_m = 1{,}000 \times 0.0124$$
$$= 12.4 \text{ N·m}$$

The moment of force due to the weight of the limb can be determined in a similar manner. The weight of the shank and foot can be estimated from the combined shank and foot regression equations in Table 2.1, with the center of gravity (i.e., point of application for the weight vector) for the leg (shank + foot) located at a distance of 43.4% of shank length as measured from the knee joint. Hence,

$\theta = 0.5$ rad

$$d = 0.36 \times 0.434$$
$$= 0.156 \text{ m}$$
$$\mathbf{F}_{w,l} = (0.044 \times 700 - 1.75) + (0.009 \times 700 + 2.48)$$
$$= 37.8 \text{ N}$$
$$\mathbf{T}_{w,l} = \mathbf{F}_{w,l} \times b$$
$$= 37.8 \times b$$
$$\cos 0.5 = \frac{b}{d}$$
$$b = 0.156 \cos 0.5$$
$$= 0.14 \text{ m}$$
$$\mathbf{T}_{w,l} = 37.8 \times 0.14$$
$$= 5.2 \text{ N·m}$$

Finally, what is the torque about the knee joint due to the weight boot? The boot is located at the end of the leg, 0.36 m from the knee joint, and has a magnitude of 80 N.

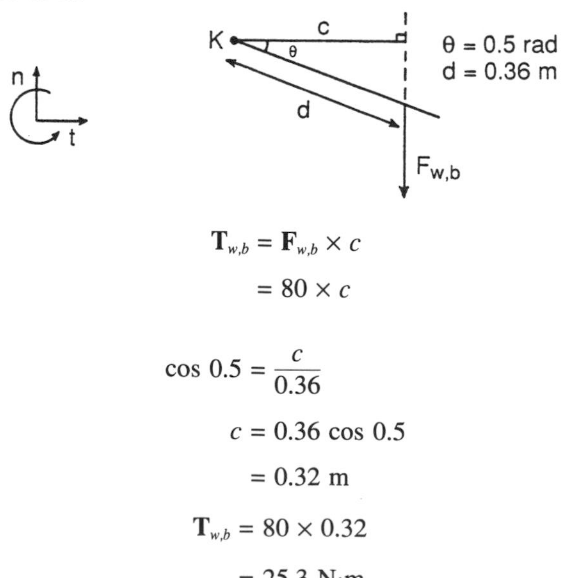

$$\mathbf{T}_{w,b} = \mathbf{F}_{w,b} \times c$$
$$= 80 \times c$$
$$\cos 0.5 = \frac{c}{0.36}$$
$$c = 0.36 \cos 0.5$$
$$= 0.32 \text{ m}$$
$$\mathbf{T}_{w,b} = 80 \times 0.32$$
$$= 25.3 \text{ N·m}$$

Note that the direction of the weight vector is always vertically downward. And because a moment arm is perpendicular to the line of action of a vector, *the moment arm of a weight vector is always horizontal.*

The net moment of force about the knee joint in this example can be determined by summing the magnitudes and taking into account the direction of each torque (i.e., counterclockwise indicated as positive). As discussed previously, the direction of each torque vector can be determined with the right-hand-thumb rule. The effect of each force on the system is considered separately by taking the right hand, curling the fingers into a loose fist, and extending the thumb. The right hand is positioned so that the curled fingers indicate the direction in which the force will cause the system to rotate. The extended thumb then indicates the direction of the torque vector that produces the rotation. Recall that *a torque can be represented by a vector that is directed perpendicular to the plane of this page.* If such a vector extends toward you, it will cause the system to rotate in a counterclockwise direction, which we usually call positive. In the present example, the forces $\mathbf{F}_{w,l}$ and $\mathbf{F}_{w,b}$ produce clockwise rotation and thus are identified as negative torques.

To determine the net effect of these forces on the system, we sum (Σ) the moments of force about the knee joint (M_K):

$$\Sigma M_K = (\mathbf{F}_m \times a) - (\mathbf{F}_{w,l} \times b) - (\mathbf{F}_{w,b} \times c)$$
$$= 12.4 - 5.2 - 25.3$$
$$\Sigma M_K = -18.1 \text{ N·m}$$

The net moment of force, therefore, has a magnitude of 18.1 N·m and acts in a clockwise direction. However, the fact that the torque acts in a clockwise direction says nothing about the direction of the limb displacement. For example, recall the elbow flexion-extension problem we considered in chapter 1 in the Acceleration and Muscle Activity section. As indicated in Figure 1.9, displacement (Δ position) and force (acceleration) are often offset, so the direction of one does not tell us directly about the direction of the other (see also Problem 2.140). We need information on position, velocity, and acceleration (torque) to get a complete description of the motion.

Because the torque is equal to the product of force and moment arm, the rotary effect of a force can be altered by either factor, singly or in combination. Many anatomical structures (e.g., calcaneus, patella, olecranon process, vertical spines of vertebrae) have the mechanical effect of altering the moment arms of muscles relative to joints. For example, consider the attachment of the patellar tendon to the tibia (Figure 2.42). If two individuals could generate the same amount of muscle force in the quadriceps femoris muscle group, the person with the greater perpendicular distance (due to a difference in the patella) from the line of action of the muscle force vector to the axis of rotation would be able to

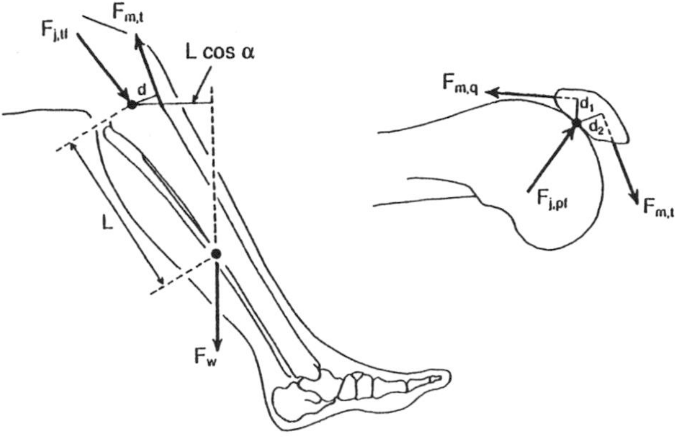

Figure 2.42 The knee extensor mechanism. The torque exerted by the quadriceps femoris muscle group is influenced by two moment arms—one from the center of the tibiofemoral joint to the patellar tendon (*d*) and the other from the patellofemoral joint to the patellar tendon (d_1) and the quadriceps muscles (d_2). $\mathbf{F}_{j,pf}$ = patellofemoral joint reaction force; $\mathbf{F}_{j,tf}$ = tibiofemoral joint reaction force; $\mathbf{F}_{m,t}$ = patellar tendon force; $\mathbf{F}_{m,q}$ = quadriceps muscle force; $\mathbf{F}_w$ = weight of the leg.
Note. From "Biomechanics of the Knee-Extension Exercise" by E.S. Grood, W.J. Suntay, F.R. Noyes, and D.L. Butler, 1984, *Journal of Bone and Joint Surgery*, **66-A**, p. 727. Copyright 1984 by The Journal of Bone and Joint Surgery. Adapted by permission.

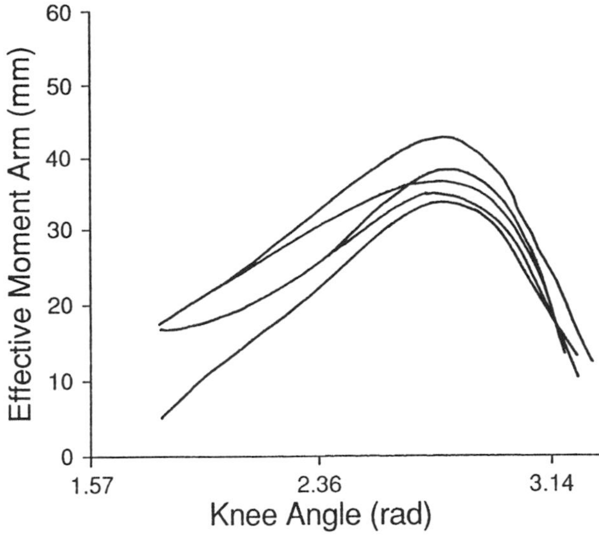

Figure 2.43 The effective moment arm for the knee extensor mechanism over the range of motion from a right angle to complete extension.
Note. From "Biomechanics of the Knee-Extension Exercise" by E.S. Grood, W.J. Suntay, F.R. Noyes, and D.L. Butler, 1984, *Journal of Bone and Joint Surgery*, **66-A**, p. 729. Copyright 1984 by The Journal of Bone and Joint Surgery. Adapted by permission.

generate the greater knee extensor torque. Because strength is a measure of the capacity to exert torque, it includes not simply a measure of muscle force but also the influence of the moment arm.

In general, muscles have a shallow angle of pull or are located close to the joint about which they exert a torque. Consequently, anatomical moment arms are typically short. But moment arms change throughout the range of motion (Figure 2.43). Data from An et al. (1981), for example, indicate that the moment arms for the major elbow flexor muscles double as the elbow goes from a fully extended position to 1.75 rad (100°) of flexion, whereas the moment arm for triceps brachii (elbow extensor) decreases by about one third over the same range of motion (Table 2.7). Similar observations have been made for muscles that cross the hip joint (Németh & Ohlsén, 1986) and the ankle and knee joints (Rugg, Gregor, Mandelbaum, & Chiu, 1990; Spoor, van Leeuwen, Meskers, Titulaer, & Huson, 1990).

Table 2.7 Moment Arms (cm) Associated With the Elbow Flexor and Extensor Muscles for Fully Extended and 1.75 rad of Flexion and for the Hand in Two Positions (Neutral and Supinated)

	Elbow extended		Elbow flexed	
Muscle	Neutral	Supinated	Neutral	Supinated
Flexor				
Biceps brachii	1.47	1.96	3.43	3.20
Brachialis	0.59	0.87	2.05	1.98
Brachioradialis	2.47	2.57	4.16	5.19
Extensor				
Triceps brachii	2.81	2.56	2.04	1.87

Note. Data are from An et al. (1981).

EXAMPLE

Let us examine the significance of the variation in moment arm over a range of motion. Consider an individual who performs push-ups to exhaustion. The prime mover for push-ups is the elbow extensor muscle, triceps brachii. Suppose this muscle was maximally active as the individual approached exhaustion and, based on the cross-sectional data presented previously (Table 2.6), the muscle was exerting a force of 550 N. According to the data in Table 2.7, the moment arm for the triceps brachii is about 2.81 cm with elbow extended (Figure 2.44a) and about 2.04 cm with elbow flexed to 1.75 rad (Figure 2.44b). This variation in moment-arm length can be thought of as a change in length of *d* in Figure 2.44c. Because of this variation, the maximum torque would be approximately 14.1 N·m and 10.2 N·m for the extended and flexed positions, respectively. That is, in the flexed position the torque due to the triceps brachii force, the prime mover for the exercise, is less than in the extended position.

Consequently, failure to perform any more push-ups is more likely to occur in the flexed position, where

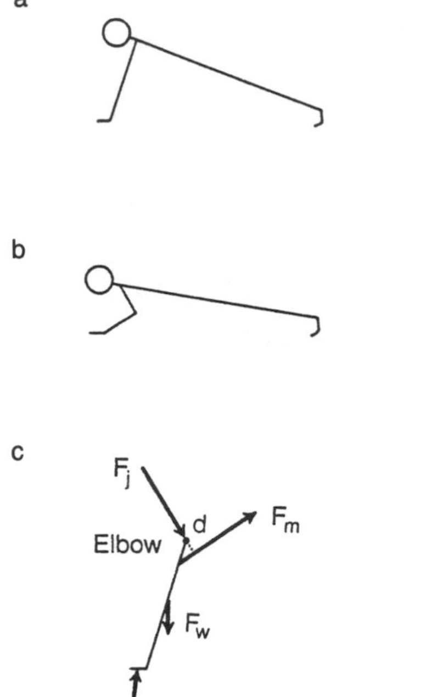

Figure 2.44 A person performing a push-up exercise: (a) straight-arm position; (b) bent-arm position; (c) free body diagram that isolates the resultant muscle force ($\mathbf{F}_m$) about the elbow joint. The net torque ($\mathbf{T}_m$) about the elbow joint is equal to the product of $\mathbf{F}_m$ and the moment arm (d).

there is a minimal amount of torque. In this case, failure is the inability to raise body weight, a constant load, up and down. A similar rationale applies to the point of failure during pull-ups to exhaustion. The moment arms for the elbow flexor muscles are minimal with complete elbow extension, and if the muscle force is reasonably constant throughout the range of motion, that is the point at which failure occurs.

Summary

This chapter describes the ways in which an individual can interact with the surrounding environment to initiate or alter a movement. As defined by Newton, these interactions are identified as forces. Eight forces are commonly encountered in the study of human movement: weight, ground reaction force, joint reaction force, muscle force, intraabdominal pressure, fluid resistance, elastic force, and inertial force. This chapter describes the characteristics of each force and its role in human movement. In addition, the chapter introduces the free body diagram, an engineering technique that is used to define the conditions for a biomechanical analysis. Examples are provided to indicate the conditions under which each of the eight forces should be included on a free body diagram. Furthermore, because human movement consists of body segments rotating about one another, the concept of a rotary force (torque) and the relevant calculations are introduced.

TRUE-FALSE QUESTIONS

1. Force is a vector. **TRUE FALSE**
2. It is not possible to determine the net effect of several forces. **TRUE FALSE**
3. A resultant force can be represented graphically as the diagonal of a trapezoid. **TRUE FALSE**
4. Orthogonal components are perpendicular to one another. **TRUE FALSE**
5. The tangential component of a muscle force acts along the body segment and contributes to joint compression. **TRUE FALSE**
6. The normal component is always less than the resultant. **TRUE FALSE**
7. The law of inertia states that a force is required to stop, start, or alter motion. **TRUE FALSE**
8. The unit of measurement of inertia is g. **TRUE FALSE**
9. Mass is a measure of an object's inertia. **TRUE FALSE**
10. A force must act on an object in order for it to follow a curved path. **TRUE FALSE**
11. The magnitude of the centripetal force influences the curvature (radius of the curved path) of a trajectory. **TRUE FALSE**
12. Momentum describes the quality of motion. **TRUE FALSE**
13. Momentum is a vector quantity. **TRUE FALSE**
14. The law of action-reaction states that $\mathbf{F} = m\mathbf{a}$. **TRUE FALSE**
15. In mechanics, a body is anything that occupies space and has inertia. **TRUE FALSE**
16. The inputs on a free body diagram are the forces acting on the system. **TRUE FALSE**
17. All forces acting on a free body diagram must be collinear. **TRUE FALSE**
18. Internal forces cannot be shown on a free body diagram. **TRUE FALSE**
19. The forces across a joint on a free body diagram include a resultant muscle force and a joint reaction force. **TRUE FALSE**

20. A space diagram is a schematic free body diagram. TRUE FALSE
21. Weight is a force. TRUE FALSE
22. The weight vector is always directed vertically downward. TRUE FALSE
23. The center of gravity is not a stationary point. TRUE FALSE
24. Regression equations provide precise values for a parameter. TRUE FALSE
25. The thigh segment has the greatest mass of any segment in the human body. TRUE FALSE
26. The moment of inertia is a measure of the mass distribution about an axis. TRUE FALSE
27. The unit of measurement for moment of inertia is $kg \cdot m^2$. TRUE FALSE
28. In the layout position, the moment of inertia is greatest about the somersault axis. TRUE FALSE
29. The ground reaction force represents the net velocities of all body segments. TRUE FALSE
30. $F_{g,z}$ has a nonzero value for about one half of the stance phase of a running stride. TRUE FALSE
31. The $F_{g,y}$ component of the ground reaction force is not affected by body weight. TRUE FALSE
32. An individual experiences an upward acceleration only when $F_{g,z}$ is greater than body weight. TRUE FALSE
33. The average value of $F_{g,z}$ during the stance phase decreases with running speed. TRUE FALSE
34. The foot usually contacts the ground in a pronated position during running. TRUE FALSE
35. The center of pressure represents the point of application of the weight vector. TRUE FALSE
36. The center of pressure is measured in Pa. TRUE FALSE
37. Running shoes that provide greater cushioning also provide greater stability. TRUE FALSE
38. The stability of the foot during the stance phase is assessed by the measurement of the supination-pronation range of motion. TRUE FALSE
39. The friction between a shoe and the ground is the resultant of $F_{g,x}$ and $F_{g,y}$. TRUE FALSE
40. The maximum friction force is proportional to the magnitude of the force that is normal to the contact plane. TRUE FALSE
41. The dynamic coefficient of friction is greater than the static coefficient. TRUE FALSE
42. The joint reaction force is mainly due to forces exerted by ligaments. TRUE FALSE
43. The joint reaction force is never greater than body weight. TRUE FALSE
44. It is difficult to measure the joint reaction force directly. TRUE FALSE
45. A shear force is tangential to the surface. TRUE FALSE
46. Muscle cannot exert a compressive (pushing) force. TRUE FALSE
47. The resultant muscle force represents the net effect of the muscle activity about a joint. TRUE FALSE
48. Muscle force can be measured directly by determining the force transmitted in a tendon. TRUE FALSE
49. The force that a muscle can exert is proportional to its volume. TRUE FALSE
50. The unit of measurement for specific tension is N. TRUE FALSE
51. Larger muscles have a greater specific tension. TRUE FALSE
52. Specific tension does not depend on muscle fiber type. TRUE FALSE
53. The specific tension of muscles in males is greater than that in females. TRUE FALSE
54. Muscle architecture can influence specific tension. TRUE FALSE
55. Muscle force can be estimated from measurements of intramuscular pressure. TRUE FALSE
56. The unit of measurement for pressure is J. TRUE FALSE
57. Muscle force usually has a shallow angle of pull. TRUE FALSE
58. Intraabdominal pressure exerts a force that tends to extend the trunk. TRUE FALSE
59. Intraabdominal pressure cannot be measured in humans. TRUE FALSE
60. Intraabdominal pressure can exert a force that is greater than body weight. TRUE FALSE
61. Layers of fluid particles flowing around an object are referred to as streamlines. TRUE FALSE
62. According to Bernoulli's principle, fluid pressure increases with fluid velocity. TRUE FALSE
63. The lift component of fluid resistance acts perpendicular to the direction of fluid flow. TRUE FALSE
64. The drag component of fluid resistance always retards forward motion. TRUE FALSE
65. Pressure drag depends on the texture of the object that causes the velocity of the boundary layer to decrease. TRUE FALSE
66. The unit of measurement for surface drag is N. TRUE FALSE
67. Waves decrease the velocity of a swimmer. TRUE FALSE

68. Pressure drag is the largest contributor to the drag component of fluid resistance. **TRUE FALSE**
69. The most significant factor in fluid resistance is the relative velocity between the object and the fluid. **TRUE FALSE**
70. The average drag experienced by a swimmer performing the front crawl at about 1.5 m/s is around 500 N. **TRUE FALSE**
71. At terminal velocity, a sky diver experiences a vertical acceleration of $-9.81\ m/s^2$. **TRUE FALSE**
72. The lift component of air resistance can cause a baseball to curve. **TRUE FALSE**
73. The constant *k* indicates the stiffness of a spring. **TRUE FALSE**
74. Tissue returns to its initial length when it is stretched within its plastic limits. **TRUE FALSE**
75. The unit of measurement for strain is %. **TRUE FALSE**
76. Stress indicates the change in length of a tissue. **TRUE FALSE**
77. Elastic force represents the potential that a tissue has to return to its initial length. **TRUE FALSE**
78. The unit of measurement for elastic force is N/cm^2. **TRUE FALSE**
79. Centrifugal force causes the path of an object to be linear. **TRUE FALSE**
80. The knee extensor muscles can control knee flexion. **TRUE FALSE**
81. Inertial force is the effect that one body segment can exert on another due to its mass. **TRUE FALSE**
82. The unit of measurement for moment of force is N·m. **TRUE FALSE**
83. A force will exert a torque about a point only if the moment arm has a nonzero value. **TRUE FALSE**
84. Torque is a scalar quantity. **TRUE FALSE**
85. The cross product is used to add vector quantities. **TRUE FALSE**
86. The unit of measurement for moment arm is N. **TRUE FALSE**
87. A curved arrow shows a vector perpendicular to the page. **TRUE FALSE**
88. Moment arm is a scalar quantity. **TRUE FALSE**
89. The moment arm for a weight vector is always vertical. **TRUE FALSE**
90. If a muscle exerts a torque of −25 N·m, the body segment will rotate in the negative direction. **TRUE FALSE**
91. The moment arm for the elbow flexors is greatest when the elbow is in an extended position. **TRUE FALSE**

MULTIPLE-CHOICE QUESTIONS

92. What technique is used to determine the net effect of several forces acting on a system?
    A. Resultant
    B. Composition
    C. Orthogonal
    D. Resolution
93. What effect does the normal component of a muscle force exert on a body segment?
    A. Compression
    B. Rotation
    C. Coactivation
    D. Stabilization
94. Which of Newton's laws of motion states that $\mathbf{F} = m\mathbf{a}$?
    A. Gravitation
    B. Inertia
    C. Action-reaction
    D. Acceleration
95. Which parameter is a quantitative measure of inertia?
    A. Mass
    B. Acceleration
    C. Force
    D. Weight
96. Which variable does not influence centripetal force?
    A. Mass
    B. Radius
    C. Acceleration
    D. Velocity
97. Momentum is calculated as the product of which two variables?
    A. Mass and acceleration
    B. Weight and velocity
    C. Weight and time
    D. Mass and velocity

98. Which statements about free body diagrams are correct?
   A. The lines of action of the forces are concurrent.
   B. The forces are not drawn to scale.
   C. Forces that are internal to the system are not shown.
   D. The forces are always collinear.

99. Why is weight described as a distributed force?
   A. When the system is stationary, $F_{g,z} = \mathbf{F}_w$.
   B. It acts at the center of gravity.
   C. The center of gravity is not a stationary point and is not even confined to the physical limits of the system.
   D. Each particle in the system has a weight.

100. Which body segment has its center of gravity closest to the midpoint of segment length?
   A. Head
   B. Shank
   C. Forearm
   D. Upper arm

101. Which statements about moment of inertia are correct?
   A. It is represented by the symbol I.
   B. It indicates the distribution of the mass of a system about its center of gravity.
   C. It is smallest about the twist axis.
   D. The unit of measurement is $kg \cdot m^2$.

102. Which techniques have been used to obtain anthropometric assessments of the body segments?
   A. Radiation scanning
   B. Cadaver studies
   C. Magnetic resonance imaging
   D. Mathematical modeling

103. What assumptions did Hanavan use in his model of the human body?
   A. Segments have a uniform density.
   B. The left and right sides are symmetrical.
   C. Segments are rigid.
   D. The boundaries between segments are distinct.

104. Which model divides the human body into the greatest number of segments?
   A. Hatze
   B. Chandler, Clauser, McConville, Reynolds, and Young
   C. Zatsiorsky and Seluyanov
   D. Hanavan

105. What is happening when $F_{g,z} = \mathbf{F}_w$?
   A. The displacement of the center of gravity is changing direction.
   B. The center of gravity is passing over the foot.
   C. The center of gravity has a zero acceleration in the vertical direction.
   D. All the body segments are stationary.

106. Which features of the ground reaction force change with an increase in running speed?
   A. Peak $F_{g,z}$
   B. Duration of $F_{g,y}$
   C. Peak $\mathbf{F}_w$
   D. Average $F_{g,z}$

107. What is the unit of measurement for center of pressure?
   A. cm
   B. Pa
   C. N
   D. N/cm^2

108. What are the objectives of well-designed running shoes?
   A. To maximize the center of pressure
   B. To provide stability for the ankle joint
   C. To minimize the shock associated with foot contact
   D. To store elastic energy

109. Which variable accounts for differences in the friction between a shoe and the ground among different people?
   A. μ_s
   B. $\mathbf{F}_{s,\max}$
   C. $F_{g,z}$
   D. $F_{g,x}$ and $F_{g,y}$

110. Why can joint reaction forces have large magnitudes?
   A. Ligaments exert high intersegmental forces.
   B. Contact forces are transmitted from one segment to another.
   C. Muscles have a shallow angle of pull and hence a large tangential component of muscle force.
   D. The shear component is usually larger than the compression component.

111. Which techniques are used to estimate the magnitude of a muscle force?
   A. EMG
   B. Cross-sectional area
   C. Intramuscular pressure
   D. Buckle transducer

112. How much muscle force would be transmitted through the Achilles tendon of an adult male cycling at a moderate intensity?
   A. 7 N
   B. 70 N
   C. 700 N
   D. 7,000 N

113. What is the average value of specific tension?
   A. 30 N/cm^2
   B. 30 kg/cm^2
   C. 30 N/m^2
   D. 30 Pa

114. Which variables account for the range of specific tension values reported in the literature?
   A. Gender
   B. Measurement technique (e.g., muscle fiber vs. motor unit)
   C. Muscle fiber type
   D. Muscle architecture
115. What is the unit of measurement for intraabdominal pressure?
   A. N
   B. Pa
   C. J
   D. N·m
116. Why does the use of a weight-lifting belt enable a lifter to increase intraabdominal pressure?
   A. Activates the Valsalva maneuver
   B. Causes a parallel increase in intrathoracic pressure
   C. Supports the activity of the abdominal muscles
   D. Increases the intradiscal pressure
117. What is Bernoulli's principle?
   A. Fluid flow around an object can be represented as layers of particles, with the layers referred to as streamlines.
   B. The streamline closest to the object is referred to as the boundary layer and has the lowest velocity.
   C. The change in fluid flow from laminar to turbulent causes an increase in fluid resistance.
   D. The pressure in a fluid is inversely related to the velocity of the particles.
118. What term is used to refer to the fluid resistance due to turbulent airflow?
   A. Surface drag
   B. Lift
   C. Pressure drag
   D. Wave resistance
119. Which variable exerts the greatest effect on the fluid resistance experienced by an object?
   A. Relative velocity
   B. Fluid density
   C. Orientation of the object
   D. Projected area
120. Why can a pitched baseball be made to curve toward or away from the batter?
   A. The pitcher releases it with a curved trajectory.
   B. The seams cause an increase in surface drag.
   C. Spin causes a pressure gradient across the ball.
   D. Gravity causes it to curve to one side or the other depending on how the ball is released.
121. How is a force-deformation relationship normalized?
   A. Force-length curve
   B. Stress-strain relationship
   C. Spring stiffness
   D. Energy-work relationship
122. What is the unit of measurement for stress?
   A. %
   B. N
   C. mm
   D. Pa
123. What term is used to describe the mechanical effect that the motion of one body segment will have on its neighbors?
   A. Elastic force
   B. Inertial force
   C. Resultant muscle torque
   D. Intersegmental pressure
124. Identify the incorrect statements about muscle torque.
   A. Maximal muscle torque remains constant over the range of motion at a joint.
   B. It is a vector located in the same plane as the moment arm and the muscle force vectors.
   C. The torque is equal to the muscle force when the moment arm is zero.
   D. Muscle torque is usually larger than body weight.

PROBLEMS

125. The medial (M) and lateral (L) heads of gastrocnemius both exert a force of 350 N that is directed 0.25 rad to the left and right of the midline, respectively. Determine the magnitude and direction of the resultant.

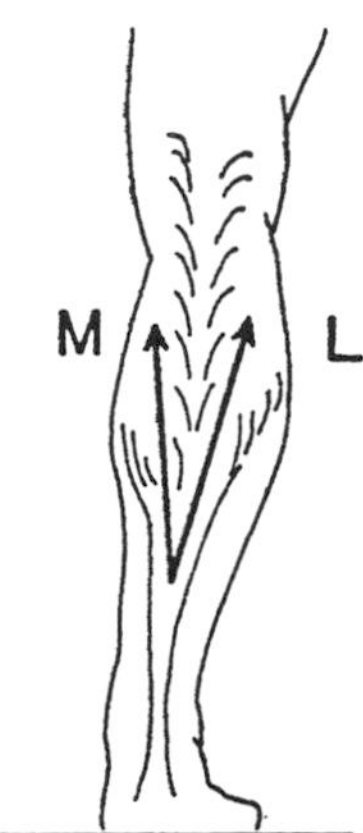

126. The free body diagram shows the forces acting on the head to maintain the flexed neck position.
   A. Use the regression equations in Tables 2.1 and 2.5 to estimate the weight of the head. Assume that the person weighs 619 N and is 1.62-m tall.
   B. When a system is in equilibrium, the forces acting on the system are concurrent, as is shown by the dashed lines. If $\mathbf{F}_m$ is directed at an angle of 0.58 rad to the horizontal, calculate the value of θ for these three forces to be concurrent.
   C. Assume that $\mathbf{F}_m$ has a magnitude of 54 N. Calculate the magnitude of the horizontal component of $\mathbf{F}_m$ and the horizontal and vertical components of $\mathbf{F}_j$, which acts at the atlantooccipital joint.

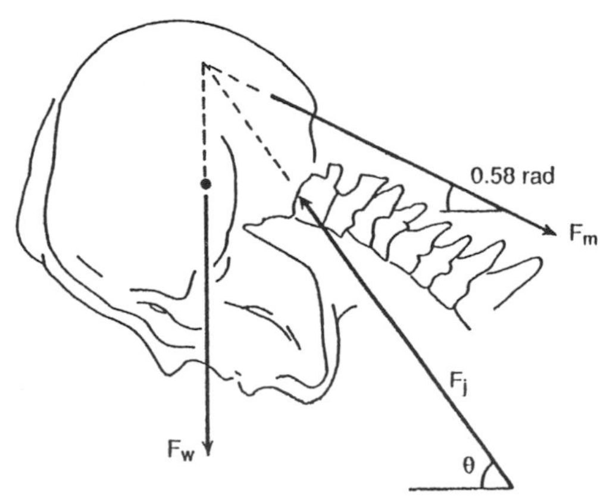

Note. From *Biomechanics of Human Motion* (p. 98) by M. Williams and H.R. Lissner, 1962, Philadelphia: W.B. Saunders. Copyright 1962 by W.B. Saunders Company. Adapted with permission.

127. The three elbow flexor muscles are biceps brachii, brachialis, and brachioradialis. Suppose that the elbow joint is flexed to 1.57 rad and that these muscles are generating 290, 440, and 95 N of force, respectively, and pulling at angles of 0.7, 0.2, and 0.3 rad, respectively. For simplicity, assume the three muscles share a common point of attachment on the radius. What are the direction and the magnitude of the resultant muscle force?

128. During knee extension the patella moves in the groove between the medial and lateral condyles of the femur. An imbalance between the four muscles that pull on the patella can result in the patella not tracking correctly (Grabiner, Koh, & Draganich, 1994). This can occur, for example, if the knee has been immobilized in a cast for a period of time, and the vastus medialis and lateralis muscles have not atrophied at the same rate. Given the following magnitudes and direction for the four muscle-force vectors, determine the resultant vector:

   Rectus femoris—130 N, 0.25 rad to the left of vertical
   Vastus lateralis—180 N, 0.50 rad to the left of vertical
   Vastus intermedius—160 N, 0.15 rad to the left of vertical
   Vastus medialis—210 N, 0.62 rad to the right of vertical

129. A person performing sit-ups is generating a 480-N force with the hip flexor muscles in the position indicated. Complete the free body diagram for this situation. What are the magnitudes of the normal and tangential components of the muscle force?

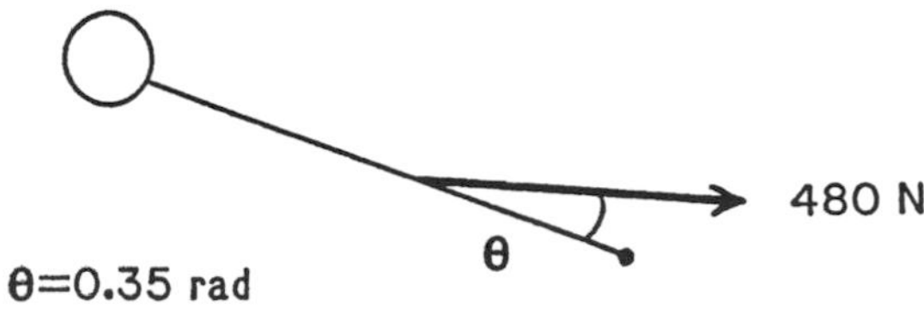

130. A manufacturer has designed and built a machine to strengthen the elbow flexor muscles. Draw the appropriate free body diagram to analyze the effect of the machine on the force developed by these muscles.

131. The kicking coach of a football team would like to know what force the knee extensors of his punter produce during a punt. How would you define your system to address this question? Draw your free body diagram.

132. Draw a free body diagram that will enable calculation of the forces at the knee joint for the leg (thigh + shank + foot) in traction shown in Figure 2.2.
133. An individual weighs 523 N. Based on the regression equations in Table 2.1 (Chandler et al., 1975), how much does this person's trunk weigh and what percentage is this of total-body weight?
134. A person who is 1.72-m tall has thigh and shank lengths of 36 cm each and upper arm and forearm lengths of 25 cm each. For which of these limb segments is the CG farthest from, and for which is the CG closest to, the proximal joint?
135. A patient with a total-body mass of 78 kg performs a knee extension exercise while wearing a deLorme boot (4.5 kg). The patient has a standing height of 1.83 m. Suppose that the (x,y) coordinates (cm) for various landmarks are as follows:

Landmark	Coordinates
Knee joint	0,0
Ankle joint	−1, −44
Shank CG	0, −19
Foot CG	8, −49
deLorme boot CG	6, −52

A. Use the regression equations of Zatsiorsky and Seluyanov to estimate the mass of the shank and the foot.
B. Use Equation 2.5 to estimate the moment of inertia of the system (shank + foot + deLorme boot) about the somersault axis of the knee.

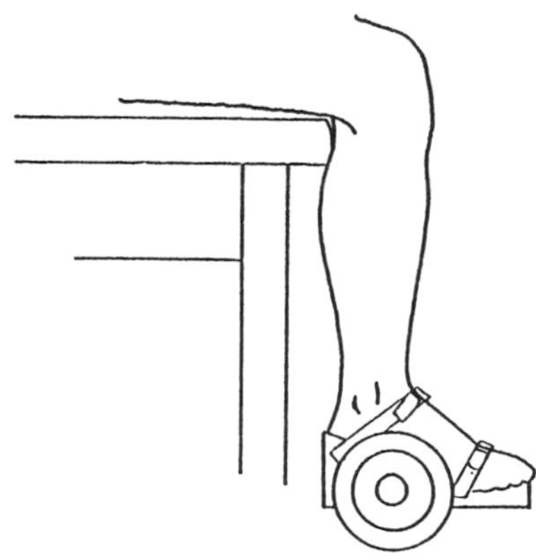

Note. From *Biomechanics of Human Motion* (p. 26) by M. Williams and H.R. Lissner, 1962, Philadelphia: W.B. Saunders. Copyright 1962 by W.B. Saunders Company. Reprinted with permission.

136. In the middle of the takeoff phase for a forward somersault, the center of gravity of a gymnast ($\mathbf{F}_w$ = 537 N) had an upward vertical acceleration of 65.5 m/s^2. What was the magnitude of $F_{g,z}$ at this point in the movement?

137. In the early part of the stance phase during running, the ground reaction force components have the following magnitudes (Figure 2.16): $F_{g,z}$ = 2.6 times body weight, $F_{g,y}$ = 1.0 times body weight, and $F_{g,x}$ = 0.02 times body weight. For a person who weighs 819 N, what is the magnitude of the resultant ground reaction force?
138. Based on the ground reaction force graphs shown in Figures 2.15 and 2.16, list the variables that a biomechanist would measure to describe the changes in the ground reaction force with speed.
139. At one point during a game of racquetball, the interaction between a shoe and the ground had the following characteristics: μ_s = 0.85, $F_{g,z}$ = 821 N, $F_{g,y}$ = 594 N, and $F_{g,x}$ = 332 N. Will the foot slide on the ground, or will it remain stationary? (To answer this question, calculate the maximum static friction force [$\mathbf{F}_{s,max}$] and the actual friction force [$\mathbf{F}_s$]. The shoe will not slide if the actual friction force is considerably less than the maximum static friction.)
140. The graph shows the kinematics (vertical position, velocity, and acceleration) of the center of gravity of the diver and the reaction force of the diving board during the takeoff phase for a three-and-a-half-backward somersault in the tuck position.

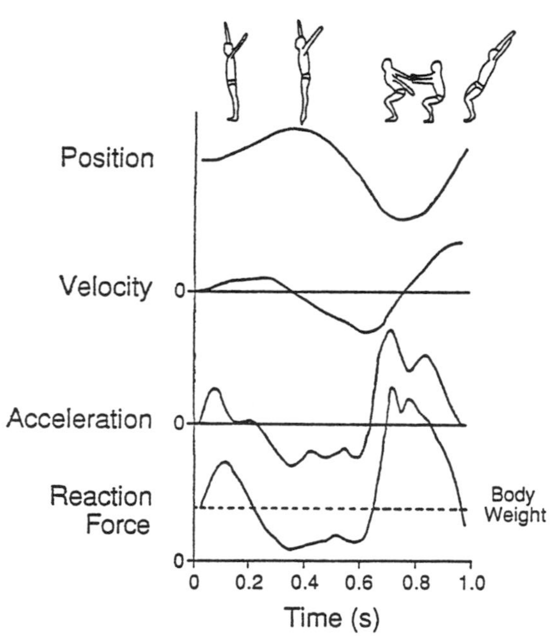

Note. From "Kinetic and Kinematic Characteristics of 10-m Platform Performances of Elite Divers: I. Back Takeoffs" by D.I. Miller, E. Hennig, M.A. Pizzimenti, I.C. Jones, and R.C. Nelson, 1989, *International Journal of Sport Biomechanics*, **5**, p. 66. Copyright 1989 by Human Kinetics Publishers, Inc. Adapted by permission.

A. When did the displacement of the diver change direction? How many times did this occur during the takeoff phase?
B. When did the center of gravity accelerate in the upward direction, and when did it accelerate in the downward direction?
C. What is the relationship between the acceleration and reaction force curves?
D. What is the relationship between the position and acceleration curves?

141. Eie (1966) measured the intraabdominal pressure of a weight lifter who held a 130-kg barbell at about knee height; the position was similar to that shown in Figure 1.7. In this position, intraabdominal pressure was measured as 225 mm Hg, and the cross-sectional area of the abdomen at the level of L5-S1 was measured as 608.5 cm^2. What extensor force did the intraabdominal pressure exert on the trunk?

142. The amount of fluid resistance experienced by an object depends on the extent to which the fluid is disturbed by the object. From the fluid resistance equation (Equation 2.8), which two variables have a substantial effect on this disturbance?

143. A sky diver experiences four phases during a jump: (a) an initial phase of acceleration from the moment of leaving the plane until reaching terminal velocity; (b) a phase of terminal velocity; (c) a phase of decreased vertical velocity immediately after the parachute opens; and (d) a final phase of free fall with the parachute opened and the sky diver preparing to land. Draw a velocity-time graph that represents these four phases. Draw an air resistance-time graph that shows how the magnitude of the air resistance changes to produce the velocity-time graph.

144. The figure shows the position adopted by a downhill speed skier in an attempt to minimize the effects of fluid resistance. Define the abbreviations *D*, *F*, *G*, *L*, and *N*, and explain the variables affecting each variable.

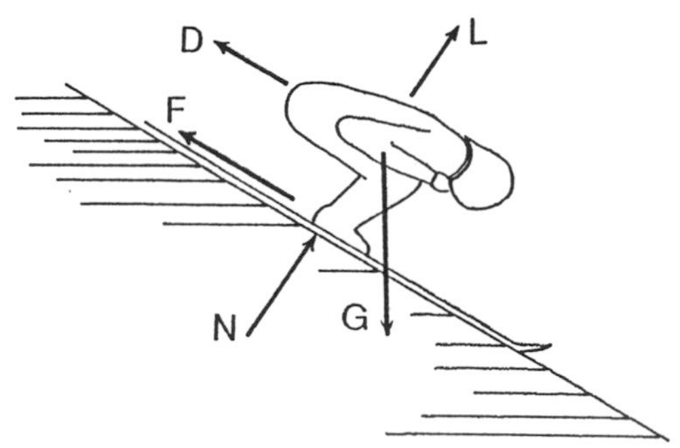

Note. From "Theoretical Drag Analysis of a Skier in the Downhill Speed Race" by S. Savolainen, 1989, *International Journal of Sport Biomechanics*, **5**, p. 27. Copyright 1989 by Human Kinetics Publishers, Inc. Reprinted with permission.

145. The behavior of a spring stretched within its elastic limits can be represented by the equation, $\mathbf{F} = kx$. What are the units of measurement for each of the variables in this equation?

146. The specific tension of muscle is about 30 N/cm^2.
A. The cross-sectional areas of the prime movers for elbow flexion and extension have been measured as follows:

Muscles	Cross-Sectional Area
Biceps brachii	4.6 cm^2
Brachialis	7.0 cm^2
Brachioradialis	1.5 cm^2
Triceps brachii	18.8 cm^2

Determine the maximum force that the elbow flexors (as a group of muscles) can exert.
B. For an individual whose triceps brachii has a cross-sectional area of 9.4 cm^2, what is the specific tension?
C. As the elbow joint is flexed from an extended position to an angle of 1.75 rad, the moment arm for the elbow flexors doubles in size. If the specific tension is 30 N/cm^2 in the extended position, what is it in the flexed position?

147. For each of the following four positions, draw in the moment arms for the two weight vectors relative to the knee joint.

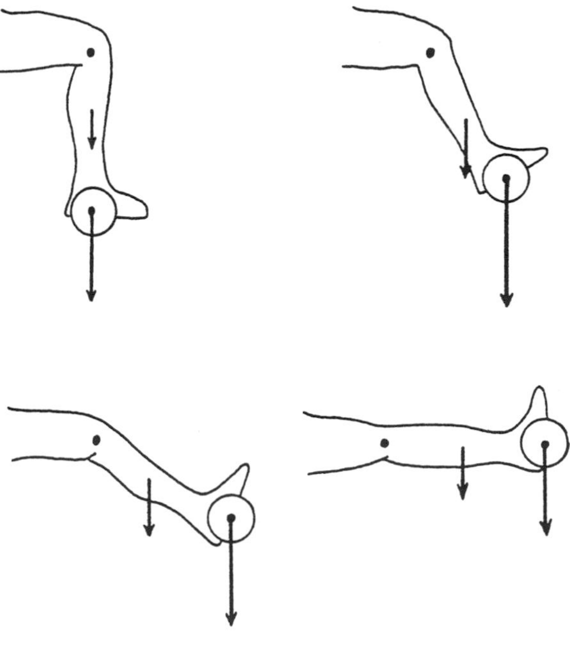

148. In each of the following cases, determine the moment produced by the force about point *B*. The rigid beam is pivoted at point *B* and is thus free to rotate.

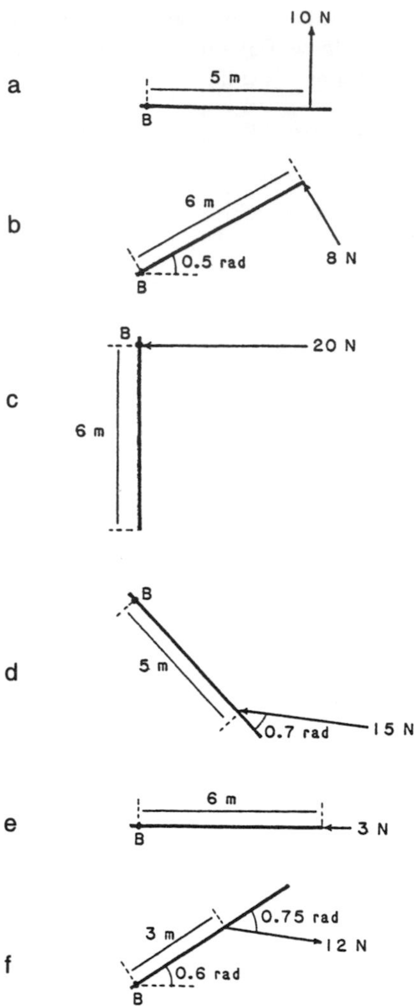

149. An individual in a wheelchair applied a force (137 N) to the hand rim on the wheel. In Condition a, the direction of the force was tangent to the rim; that is, perpendicular to the radius (26.7 cm). In Condition b, the force was applied at an angle of 1 rad to the tangent to the rim. What was the torque about the wheel axis in the two conditions?

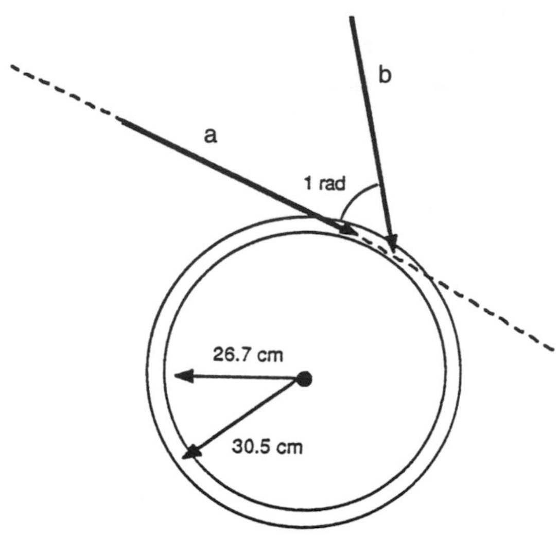

150. Draw a free body diagram for each of the whole-body positions shown. For each force included on the free body diagrams, draw in the moment arm from the force to the feet.

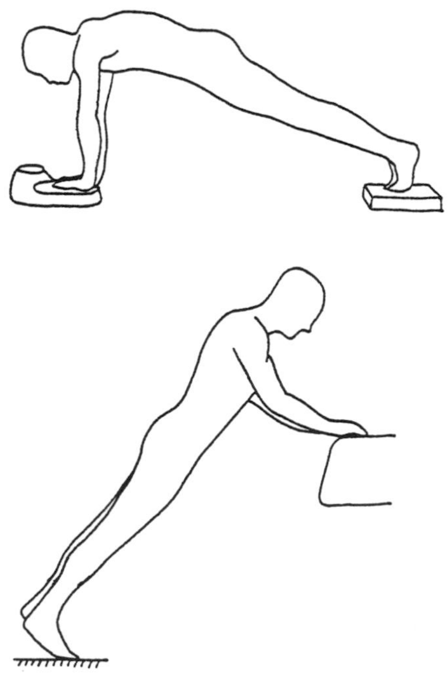

Note. From "Free-Body Diagrams as an Approach to the Mechanics of Human Posture and Motion" by W.T. Dempster. In *Biomechanical Studies of the Musculo-Skeletal System* (pp. 107, 112) by F.G. Evans (Ed.), 1961, Springfield, IL: Charles C Thomas. Copyright 1961 by Charles C Thomas. Adapted by permission.

151. Imagine doing pull-ups to failure. Draw a free body diagram (e.g., Figure 2.44) to help explain the point at which you would expect the failure to occur during the exercise. Given the following data on cross-sectional area and moment arm for the elbow flexors, estimate the torque exerted by the elbow flexors at the extremes of the movement (i.e., extended vs. flexed).

		Moment arm (cm)	
	CSA (cm^2)	Extended	Flexed
Biceps brachii	4.6	1.47	3.43
Brachialis	7.0	0.58	2.05
Brachioradialis	1.5	2.47	4.16

152. Williams and Lissner (1962) describe the following exercise, in which a patient performs a rehabilitation exercise for the quadriceps femoris muscles: The foot is connected via a pulley to a beam that has weights attached to it. Draw a free body diagram of the system (shank + foot) so that it is possible to calculate the forces at the knee joint. Describe what happens to the resistance encountered by the thigh muscles through the range of motion for this knee extension exercise.

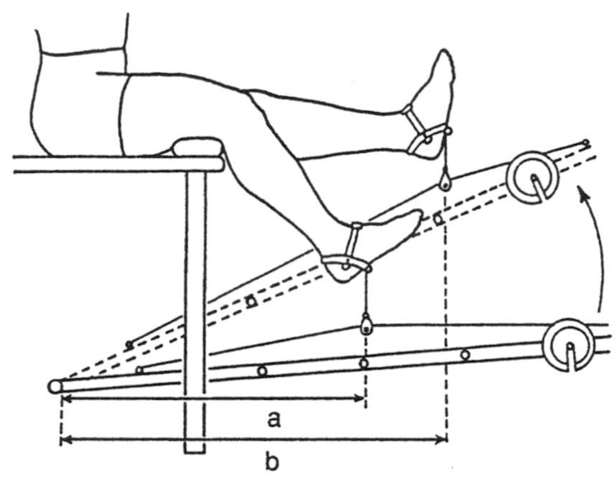

Note. From *Biomechanics of Human Motion* (p. 43) by M. Williams and H.R. Lissner, 1962, Philadelphia: W.B. Saunders. Copyright 1962 by W.B. Saunders Company. Reprinted with permission.

CHAPTER 3
Movement Analysis

The motion of an object—which is described in terms of position, velocity, and acceleration—is characterized as the consequence of the interaction between the object and its surroundings. This interaction is commonly represented as a force. The nature of the relationship between force and motion is outlined in Newton's laws of motion, a set of three laws that form the Newtonian approach to the analysis of motion. The particular algebraic relationship between force and motion that is used in an analysis depends on the objective of the analysis. In general, the alternatives are to focus on (a) the instantaneous value of force (force-mass-acceleration), (b) the effect of force applied over an interval of time (impulse-momentum), or (c) the application of a force that causes an object to move through some distance (work-energy).

Force-Mass-Acceleration Approach

When the force-mass-acceleration approach ($\mathbf{F} = m\mathbf{a}$) is used, a system may exist in one of two states; that is, acceleration (**a**) may be either zero (static) or nonzero (dynamic).

Static Analysis

When the focus of an analysis is on the value of force at one instant in time, the appropriate procedure is to use the force-mass-acceleration approach, so named because it is based on Newton's law of acceleration:

$$\Sigma\mathbf{F} = m\mathbf{a}$$

Recall that the left-hand side of this relationship ($\Sigma\mathbf{F}$) represents the interactions between a system and its surroundings, whereas the right-hand side indicates the kinematic effects of the interactions on the system. A force-mass-acceleration analysis can be further categorized by the magnitude of the acceleration (**a**), that is, whether it is zero or nonzero. **Statics** is a special case of Newton's law of acceleration, in which acceleration is zero because the object is either stationary or moving at a constant velocity. In a static analysis the system is in equilibrium, and the sum of the forces in any given direction is zero ($\Sigma\mathbf{F} = 0$). That is, the interactions are balanced, and therefore the system will not experience an acceleration. Similarly, the sum of the moments of force is zero ($\Sigma\mathbf{M}_o = 0$). At most, there are three independent scalar equations available to solve coplanar (i.e., confined to the same plane) statics problems:

$$\Sigma F_x = 0 \tag{3.1}$$

$$\Sigma F_y = 0 \tag{3.2}$$

$$\Sigma M_o = 0 \tag{3.3}$$

Equations 3.1 and 3.2 refer to the sum (Σ) of the forces in two linear directions (x and y), the directions being perpendicular to one another, whereas Equation 3.3 represents the sum of the moments of force about point o, which may or may not be the center of gravity of the system.

In what follows, we shall examine four examples of static analysis: determination of the magnitude and direction of an unknown force, calculation of resultant muscle torque, resolution of the magnitude and direction of a resultant muscle force and a joint reaction force, and location of the total-body CG using a segmental analysis.

EXAMPLE 1

Consider the system in Figure 3.1a, which is in equilibrium and on which are acting three known forces and one unknown. What is the magnitude of the unknown force? The first step is to identify the two directions, at right angles to one another, that will be used and the sense of each that will be denoted as positive. This includes specification of the positive angular direction. These have been indicated to the left of the system: The selected directions are horizontal (x) and vertical (y), and the positive senses are to the right and upward, respectively. Thus any horizontal force directed to the right will be regarded as positive, and any to the left will be indicated as negative. The declaration of the horizontal-vertical directions is also an indication that only forces in these directions can be handled in the analysis and thus any force (e.g., **R**) not in either direction must be *resolved* into such directions (e.g., Figure 2.3).

The next step in the analysis, therefore, is to resolve **R** graphically into horizontal (R_x) and vertical (R_y) components (Figure 3.1b). Once all forces are resolved into the x and y directions, the forces in each direction can be summed to determine the magnitude of the unknowns.

$$\Sigma F_x = 0$$

$$R_x - 3 = 0$$

$$R_x = 3 \text{ N}$$

$$\Sigma F_y = 0$$

$$5 + 10 - R_y = 0$$

$$R_y = 15 \text{ N}$$

Each calculation for determining the magnitudes of R_x and R_y began with the specification of one of two independent equations (Equation 3.1 or 3.2). It is of little consequence whether an inappropriate equation is chosen, for if the calculation proceeds logically, it will soon become evident that such was the case. For example, suppose $\Sigma M_o = 0$ is selected to determine R_x or R_y. It is not possible to sum the moments of force in this problem, because no distances are given with which to determine the torques. Thus one of the other two equations must be chosen.

With the calculated values of R_x and R_y, the resultant (**R**) can be determined by the Pythagorean relationship:

$$\mathbf{R} = \sqrt{R_x^2 + R_y^2}$$

$$= \sqrt{3^2 + 15^2}$$

$$\mathbf{R} = 15.3 \text{ N}$$

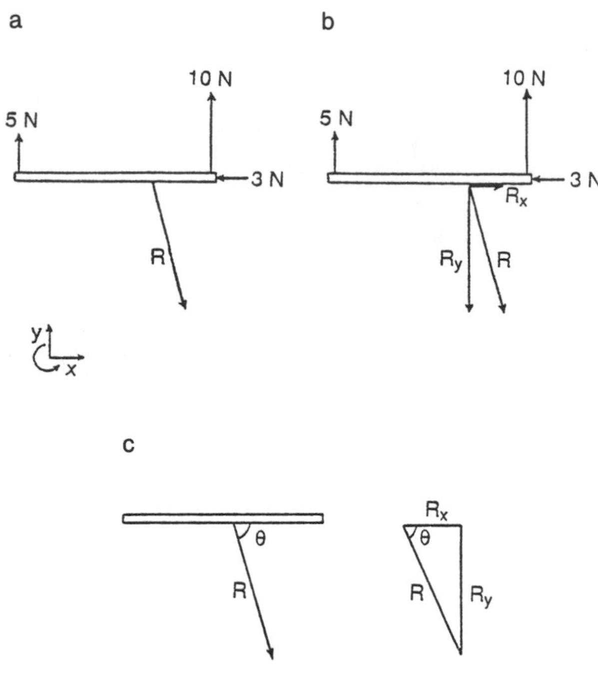

Figure 3.1 Distribution of forces acting on an object.

This result, however, specifies only the magnitude of the resultant and not its direction. We can actually indicate the direction of **R** relative to several references; for example, we could determine the angle relative to a horizontal (R_x) or a vertical (R_y) reference. Let us calculate and determine the angle of **R** relative to the system (Figure 3.1c). By the parallelogram rule, **R** is the diagonal of the rectangle, which has R_x and R_y as its sides. Thus R_x, R_y, and **R** represent the sides of a triangle, so we can determine θ as follows:

$$\cos\theta = \frac{R_x}{\mathbf{R}}, \qquad \sin\theta = \frac{R_y}{\mathbf{R}}, \qquad \tan\theta = \frac{R_y}{R_x}$$

Accordingly,

$$\cos\theta = \frac{R_x}{\mathbf{R}}$$

$$\theta = \cos^{-1}\frac{R_x}{\mathbf{R}}$$

$$\theta = \cos^{-1}\frac{3}{15.3}$$

$$\theta = \cos^{-1} 0.1961$$

$$\theta = 1.37 \text{ rad}$$

EXAMPLE 2

Another example of a static force-mass-acceleration analysis was indicated in Figure 2.41 (see Figure 3.2).

The object of the analysis was to determine the magnitude and direction of the resultant muscle torque about the knee joint (K), given a specific value for the resultant muscle force ($\mathbf{F}_m$). Figure 3.2 shows three moment arms ($a = -0.0124$, $b = 0.110$, and $c = 0.320$ m), the joint reaction force ($\mathbf{F}_j$), the leg weight ($\mathbf{F}_{w,l} = 40.6$ N), and the weight of the boot ($\mathbf{F}_{w,b} = 80$ N). Suppose the individual must hold the limb at an angle of 0.5 rad below the horizontal: What resultant muscle torque must the person generate to accomplish this task? First, reconstruct the free body diagram. Next, select the appropriate equation (Equations 3.1 to 3.3), and proceed to sum the forces or moments of force. If you choose Equation 3.3 and sum the moments of force about the somersault axis through the knee joint, you can ignore the joint reaction force; point o in Equation 3.3 is chosen to be the knee joint. Also, because the purpose of the analysis is to determine the resultant muscle torque, it is appropriate to use Equation 3.3. Consequently,

$$\Sigma \mathbf{M}_K = 0$$

Three forces will produce a rotation of the system about point K (the axis of rotation), and these need to be indicated. The calculation, therefore, proceeds as follows:

$$\Sigma \mathbf{M}_K = 0$$

$$(\mathbf{F}_m \times a) - (\mathbf{F}_{w,l} \times b) - (\mathbf{F}_{w,b} \times c) = 0$$

$$(\mathbf{F}_m \times a) = (\mathbf{F}_{w,l} \times b) + (\mathbf{F}_{w,b} \times c)$$

$$= (40.6 \times 0.11) + (80.0 \times 0.32)$$

$$= 4.63 + 25.27$$

$$(\mathbf{F}_m \times a) = 29.9 \text{ N·m}$$

The product of resultant muscle force ($\mathbf{F}_m$) and its moment arm equals the resultant muscle torque; therefore, the individual must exert a torque of 29.9 N·m to maintain the limb in the indicated position. What resultant muscle torque was the person exerting when the example (Figure 2.41) was solved previously? What is the magnitude of $\mathbf{F}_m$? Which muscle group would have to be dominant to obtain this direction?

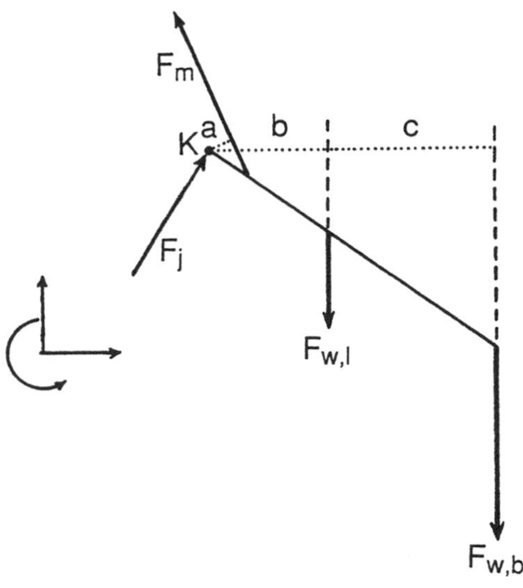

Figure 3.2 Free body diagram of the leg and foot of an individual performing a knee extension exercise (from Figure 2.41).

EXAMPLE 3

A student sits at the end of a trainer's bench and uses a rope-pulley apparatus to strengthen her quadriceps femoris muscle group with an isometric exercise. Her leg weighs 30 N, and a force of 100 N is applied to the rope pulley. The quadriceps muscle group attaches via the patellar tendon to the tibia at a distance of 7 cm from the tibiohumeral (knee) joint, the center of gravity for the limb is 20 cm from the knee joint, and the distance between the knee joint and the rope-pulley apparatus is 45 cm. The rope-pulley apparatus involves an ankle strap with a rope connecting the strap to a weight stack at the back of the bench. The weight stack can be set so that it does not move, and hence the individual performs an isometric contraction. The free body diagram of the system, from the knee (K) joint down to the toes, is shown in Figure 3.3. Figure 3.3 shows the weight of the leg ($\mathbf{F}_w = 30$ N), the load due to the rope-pulley apparatus ($\mathbf{F}_l = 100$ N), three distances ($a = 7$ cm, $b = 20$ cm, and $c = 45$ cm), and three angles ($\theta = 0.25$ rad, $\beta = 0.87$ rad, and $\gamma = 0.70$ rad). Also shown are the normal (n) and tangential (t) components of the joint reaction force ($\mathbf{F}_j$). Perform the necessary calculations to find

1. the magnitude of the net quadriceps femoris force ($\mathbf{F}_{m,q}$) necessary to maintain the illustrated position,
2. the magnitude of the normal ($F_{j,n}$) and tangential ($F_{j,t}$) components of the joint reaction force, and
3. the magnitude and direction of the resultant joint reaction force ($\mathbf{F}_j$).

The initial step is to identify the two independent directions in which the forces are to be summed and the positive sense for these linear and angular directions. Because this problem concerns the application of forces and torques to a body segment, it is most appropriate to choose the normal (n) and tangential (t) directions. These are indicated to the right of the system.

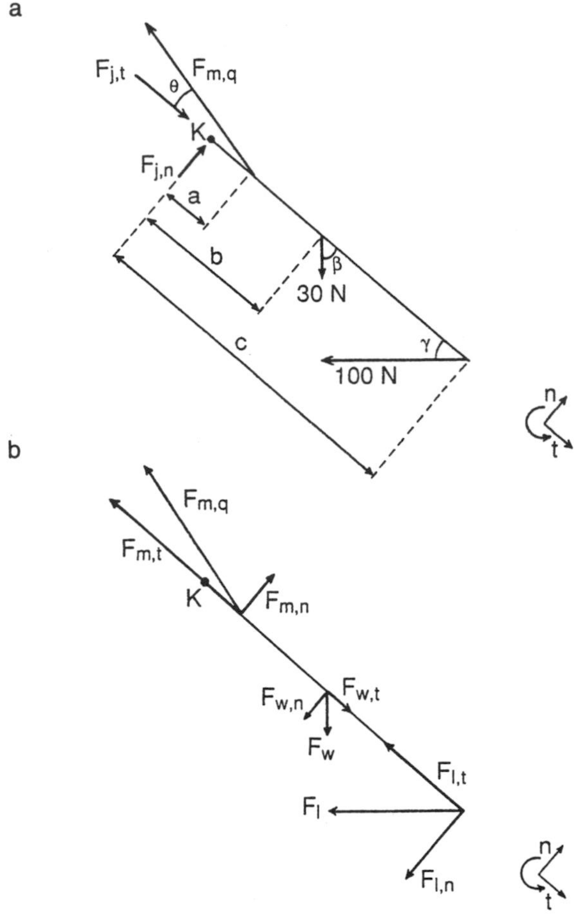

Figure 3.3 Free body diagram of the leg performing an isometric exercise to strengthen the quadriceps femoris muscle group.

The same three equations (Equation 3.1 through 3.3) can be used to determine the magnitude of $\mathbf{F}_{m,q}$, because the system is in equilibrium. Because three unknown forces are acting on the system ($F_{j,n}$, $F_{j,t}$, $\mathbf{F}_{m,q}$), it seems sensible to sum the moments about point *K* so that we can momentarily ignore $F_{j,n}$ and $F_{j,t}$ and solve for $\mathbf{F}_{m,q}$. This involves summing the moments of force about point *K* for the weight of the leg ($\mathbf{F}_w$) and the load exerted by the rope-pulley apparatus ($\mathbf{F}_l$) and setting these equal to the torque associated with the quadriceps femoris muscle activity ($\mathbf{F}_{m,q}$). To perform this calculation, we need to determine the moment arm for each force by carefully analyzing Figure 3.3; this involves combining the distances *a*, *b*, and *c* with the angles for each force, where

$$a \sin \theta = 7 \sin 0.25$$
$$= 1.73 \text{ cm}$$

$$b \sin \beta = 20 \sin 0.87$$
$$= 15.29 \text{ cm}$$

$$c \sin \gamma = 45 \sin 0.70$$
$$= 28.99 \text{ cm}$$

The calculation proceeds as follows:

$$\Sigma \mathbf{M}_K = 0$$

$$(\mathbf{F}_{m,q} \times a \sin \theta) - (\mathbf{F}_w \times b \sin \beta) - (\mathbf{F}_l \times c \sin \gamma) = 0$$

$$(\mathbf{F}_{m,q} \times 1.73) - (30 \times 15.29) - (100 \times 28.99) = 0$$

$$(\mathbf{F}_{m,q} \times 1.73) = (30 \times 15.29) + (100 \times 28.99)$$

$$(\mathbf{F}_{m,q} \times 1.73) = 458.7 + 2{,}899.0$$

$$\mathbf{F}_{m,q} = \frac{3{,}357.7}{1.73}$$

$$\mathbf{F}_{m,q} = 1{,}941 \text{ N}$$

To determine the magnitude of $F_{j,n}$ and $F_{j,t}$, we need to resolve each force into its normal and tangential components (Figure 3.3b) and then sum the forces in each direction:

$$F_{m,n} = \mathbf{F}_{m,q} \sin 0.25$$
$$= 1{,}941 \sin 0.25$$
$$= 480 \text{ N}$$

$$F_{m,t} = \mathbf{F}_{m,q} \cos 0.25$$
$$= 1{,}941 \cos 0.25$$
$$= 1{,}881 \text{ N}$$

$$F_{w,n} = \mathbf{F}_{w,l} \sin 0.87$$
$$= 30 \sin 0.87$$
$$= 23 \text{ N}$$

$$F_{w,t} = \mathbf{F}_{w,l} \cos 0.87$$
$$= 30 \cos 0.87$$
$$= 19 \text{ N}$$

$$F_{l,n} = \mathbf{F}_l \sin 0.70$$
$$= 100 \sin 0.70$$
$$= 64 \text{ N}$$

$$F_{l,t} = \mathbf{F}_l \cos 0.70$$
$$= 100 \cos 0.70$$
$$= 77 \text{ N}$$

These values represent the magnitudes of the normal and tangential components as they are shown in Figure 3.3b. The magnitude of the *tangential* component of the joint reaction force is determined as follows:

$$\Sigma F_t = 0$$

$$F_{j,t} - F_{m,t} + F_{w,t} - F_{l,t} = 0$$

$$F_{j,t} = F_{m,t} - F_{w,t} + F_{l,t}$$

$$= 1{,}881 - 19 + 77$$

$$F_{j,t} = 1{,}939 \text{ N}$$

The magnitude of the *normal* component can be found in a similar manner:

$$\Sigma F_n = 0$$

$$F_{j,n} + F_{m,n} - F_{w,n} - F_{l,n} = 0$$

$$F_{j,n} = - F_{m,n} + F_{w,n} + F_{l,n}$$

$$= - 480 + 23 + 64$$

$$F_{j,n} = - 393 \text{ N}$$

The magnitude of $F_{j,n}$ is determined as negative 393 N. This does not mean that it is a negative force but rather that its direction is incorrect. If the force is calculated as negative, this means that the direction of force is actually opposite to that indicated on the free body diagram. When the free body diagram was drawn, we were told that the system was in equilibrium. This means that the forces in each direction must add to zero. For the tangential direction, we assumed that $F_{j,t}$ was acting in the positive direction; this turned out to be correct. Also, we assumed that $F_{j,n}$ would have to act in the positive direction in order for the system to be in equilibrium. But this was incorrect, as indicated by the negative value that we calculated for $F_{j,n}$. In reality, $F_{j,n}$ acts in the negative direction in order for the system to be in equilibrium. Thus $F_{j,t}$ and $F_{j,n}$ and the associated resultant ($\mathbf{F}_j$) act on point *K* in the manner illustrated in Figure 3.4. Now that we know the magnitude of $F_{j,n}$ and $F_{j,t}$, we can determine the magnitude of the resultant joint reaction force ($\mathbf{F}_j$) with the Pythagorean relationship:

$$\mathbf{F}_j = \sqrt{F_{j,n}^{\ 2} + F_{j,t}^{\ 2}}$$

$$= \sqrt{393^2 + 1{,}939^2}$$

$$\mathbf{F}_j = 1{,}978 \text{ N}$$

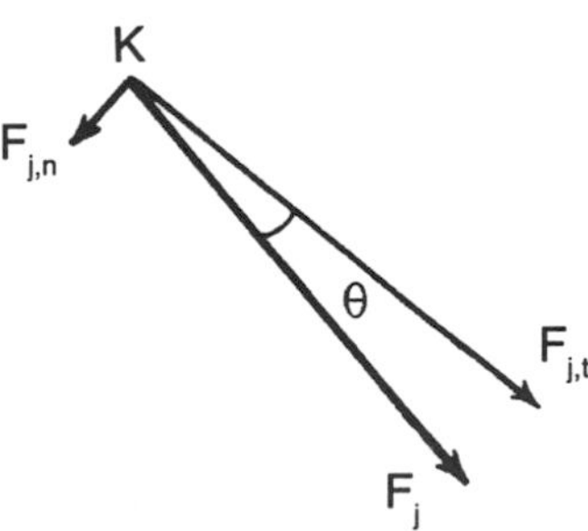

Figure 3.4 Determination of the resultant knee joint reaction force for the free body diagram outlined in Figure 3.3.

Thus, when a 100-N load is applied at the ankle, the knee joint experiences a joint reaction force that is almost 20 times larger than the load.

Finally, we can calculate the direction of the resultant joint reaction force with respect to the axis of the leg as

$$\tan \theta = \frac{F_{j,n}}{F_{j,t}}$$

$$\theta = \tan^{-1} \frac{F_{j,n}}{F_{j,t}}$$

$$\theta = \tan^{-1} \frac{393}{1{,}939}$$

$$\theta = 0.2 \text{ rad}$$

The resultant joint reaction force, therefore, has a magnitude of 1,978 N and is directed 0.2 rad below the leg segment.

EXAMPLE 4

Use the principles of statics to calculate the location of the total-body center of gravity (CG). Previously the CG of an object was defined as a balance point, which means that the mass of the system is in equilibrium about this point. Therefore, *the location of the total-body CG for multisegmented objects, such as the human body, can be determined using the static version of Newton's law of acceleration* ($\Sigma \mathbf{F} = 0$ and $\Sigma \mathbf{M}_o = 0$). This is accomplished by finding the point about which the body segments are balanced. In other words, the CG can be located by requiring the system to be in equilibrium and then finding the point (the CG) about which this is true. Because the body is actually in equilibrium (i.e., balanced) with respect to the CG, the sum of the moments of force due to segment weights about the CG is equal to zero. The only forces acting on the system in this example are the segmental weight vectors.

To demonstrate this technique, let us determine the location of the total-body CG of a gymnast about to perform a backward handspring (Figure 3.5). The necessary steps include the following:

1. *Identify the appropriate body segments.* The human body can be divided into 14 to 17 segments for a segmental analysis (these boundaries were discussed in chapter 2 in the section on segmental

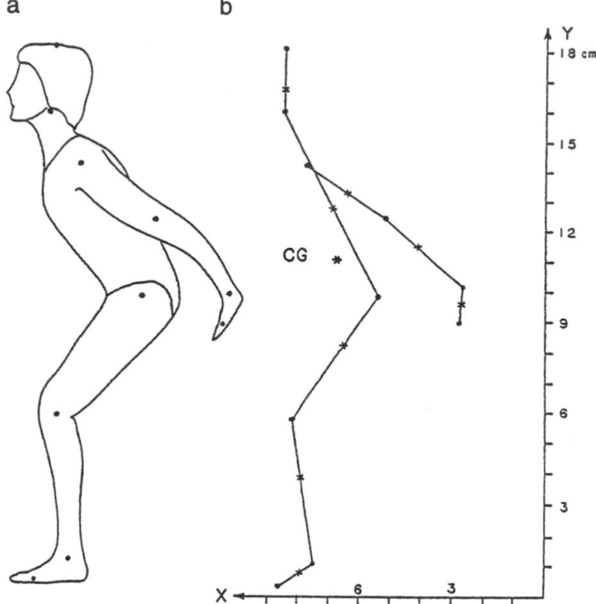

Figure 3.5 Location of total-body center of gravity (CG) as a function of the position of the body segments: (a) limits of the respective body segments; (b) location of the segmental CGs as a percentage of segment length.

analysis). The number of segments to be used in an analysis is determined by the number of joints that experience angular displacement during the movement. If there is no rotation about the elbow joint, for example, then the arm (upper arm + forearm) can be represented as a single segment. For this example, we will use 14 segments: head, trunk, upper arms, forearms, hands, thighs, legs (shanks), and feet. These segments are indicated in Figure 3.5a by the marks over the joint centers that represent the proximal and distal anatomical landmarks for each segment (Chandler et al., 1975) (Appendix D).

2. *Join the joint-center markers to construct a stick figure* (Figure 3.5b).
3. *From Table 2.1, determine the location of the CG for each segment as a percentage of segment length.* These lengths are measured from the proximal end of each segment (Figure 3.5b).

Segment	CG location (%)	Proximal end
Head	66.3	Top of the head
Trunk	52.2	Top of the neck
Upper arm	50.7	Shoulder
Forearm	41.7	Elbow
Hand	51.5	Wrist
Thigh	39.8	Hip
Leg (shank)	41.3	Knee
Foot	40.0	Ankle

4. *Estimate segmental weights as a function of body weight* ($\mathbf{F}_w$) *with the Chandler et al. (1975) regression equations* (Table 2.1). Suppose the gymnast weighs 450 N.

Segment	Equation	Weight
Head	$0.032 \times \mathbf{F}_w + 18.70 =$	33.10
Trunk	$0.532 \times \mathbf{F}_w - 6.93 =$	232.47
Upper arm	$0.022 \times \mathbf{F}_w + 4.76 =$	14.66
Forearm	$0.013 \times \mathbf{F}_w + 2.41 =$	8.26
Hand	$0.005 \times \mathbf{F}_w + 0.75 =$	3.00
Thigh	$0.127 \times \mathbf{F}_w - 14.82 =$	42.33
Leg (shank)	$0.044 \times \mathbf{F}_w - 1.75 =$	18.05
Foot	$0.009 \times \mathbf{F}_w + 2.48 =$	6.53

5. *Measure the location of segmental CGs relative to an xy-axis* (Figure 3.5b). The location of this axis is quite arbitrary and does not affect the location (with respect to the body) of the total-body CG; you can convince yourself of this by doing the calculation twice with the *xy*-axis in a different location each time.

Segment	*x*-coordinate (cm)	*y*-coordinate (cm)
Head	8.6	16.8
Trunk	6.9	12.8
Upper arm	6.5	13.3
Forearm	4.2	11.5
Hand	2.8	9.6
Thigh	6.6	8.2
Leg (shank)	8.0	3.9
Foot	8.0	0.8

6. *With the segmental weight* ($\mathbf{F}_{w,s}$) *and location* (x,y) *data, sum the segmental moments of force about the y-axis* ($\Sigma \mathbf{M}_y = \mathbf{F}_{w,s} \times x$) and the *x*-axis ($\Sigma \mathbf{M}_x = \mathbf{F}_{w,s} \times y$). Double the limb segmental weights to account for both limbs. Please refer to the following table.

Segment	x (cm)	y (cm)	$\mathbf{F}_{w,s}$ (N)	$\Sigma\mathbf{M}_y$ (N·cm)	$\Sigma\mathbf{M}_x$ (N·cm)
Head	8.6	16.8	33.10	284.7	556.1
Trunk	6.9	12.8	232.47	1,604.4	2,975.6
Upper arm	6.5	13.3	29.32	190.6	390.0
Forearm	4.2	11.5	16.52	69.4	190.0
Hand	2.8	9.6	6.00	16.8	57.6
Thigh	6.6	8.2	84.66	558.8	694.2
Leg (shank)	8.0	3.9	36.10	288.8	140.8
Foot	8.0	0.8	13.06	104.5	10.5
				3,018.0	5,014.7

7. *Find the location of the point that would produce the same moment of force for total-body weight about the x- and y-axes as that due to the sum of the segmental effects.* The net moment of force due to segmental weights about the x- and y-axes (3,018.0 N·cm and 5,014.7 N·cm, respectively) is the same as the net moment of force due to total-body weight. Thus

$$\mathbf{F}_w \times \text{moment arm} = \sum_{i=1}^{8} (\mathbf{F}_{w,s} \times \text{moment arm})_i$$

$$\text{moment arm} = \frac{\Sigma(\mathbf{F}_{w,s} \times \text{moment arm})}{\mathbf{F}_w}$$

To find the x- and y-coordinates for the total-body CG,

$$x = \frac{3{,}018 \text{ N·cm}}{450 \text{ N}} = 6.71 \text{ cm}$$

$$y = \frac{5{,}014.7 \text{ N·cm}}{450 \text{ N}} = 11.14 \text{ cm}$$

With these coordinates, the location of the total-body CG is indicated in Figure 3.5b relative to the x- and y-axes established at the beginning of the example. These coordinates (6.71, 11.14) represent the point about which the mass of the system is evenly distributed.

Dynamic Analysis

A static analysis is the most elementary approach to the kinetic analysis of human movement. In contrast, when the system is subjected to unbalanced forces, it will accelerate. Use of the force-mass-acceleration approach in situations in which acceleration is nonzero is referred to as dynamic analysis. The general form of the three independent planar equations (Equations 3.1 to 3.3) used in the static approach also applies to the dynamic problem, except the right-hand side of the equations is now equal to the product of mass and acceleration. Hence, the scalar components and the angular term may be written as

$$\Sigma F_x = ma_x \quad (3.4)$$

$$\Sigma F_y = ma_y \quad (3.5)$$

$$\Sigma \mathbf{M}_o = I_o\alpha + m\mathbf{a}d \quad (3.6)$$

where $\mathbf{a} = \sqrt{a_x^2 + a_y^2}$ and $\Sigma\mathbf{F} = \sqrt{(\Sigma F_x)^2 + (\Sigma F_y)^2}$ in the x-y plane. As in the static case, these equations are independent and represent two linear (x and y) and one angular direction. In expanded form, Equation 3.4 states that the sum (Σ) of the forces (F) in the x direction is equal to the product of the mass (m) of the system and the acceleration of the system's center of mass in the x direction (a_x). Equation 3.5 similarly addresses forces and accelerations in the y direction. Equation 3.6 states that the sum of the moments of force *about point o* is equal to two effects, one related to the angular kinematics of the system and the other related to the linear kinematics of the system. The term for angular kinematics includes the product of the relevant moment of inertia (I_o) and angular acceleration of the system about the axis of rotation (α). The linear kinematics term includes the product of the mass of the system, the linear acceleration of the system CG (**a**), and the distance (d) between point o and the system center of gravity.

The point o can be any point about which the moments are summed. If point o is the CG and it is a fixed point, then the acceleration of the system CG is zero, which results in the terms a_x, a_y, and $m\mathbf{a}d$ becoming zero and Equations 3.4 to 3.6 reducing to $\Sigma\mathbf{M}_o = I_o\alpha$. If the system comprises a single rigid body, such as one body segment that is rotating about a fixed axis, then the resultant effect of the forces acting on the system with regard to a normal-tangential reference frame can be calculated as follows (Meriam & Kraige, 1987):

$$\Sigma F_n = mr\omega^2 \quad (3.7)$$

$$\Sigma F_t = mr\alpha \quad (3.8)$$

$$\Sigma M_o = I_o\alpha \quad (3.9)$$

where $mr\omega^2$ is referred to as the centripetal force and $mr\alpha$ is identified as the centrifugal force.

Inertia was previously described as the resistance that an object offers to a linear change in its velocity. Likewise, the *moment of inertia* (*I*) is a quantity that indicates the resistance of an object to an angular change in its state of motion. It is a measure of the distribution of the mass of the object with respect to the axis about which it is rotating. As discussed previously, the unit of measurement for moment of inertia is kg·m^2. Can you tell how the moment of inertia is calculated from its unit of measurement? The greater the distribution of mass of the object about an axis of rotation, the greater the moment of inertia. Thus the moment of inertia is not a fixed quantity but, like the CG, may be altered by shifting the mass within the system. For example, a springboard diver performing somersaults rotates about the somersault axis. In the pike position (straight legs) the body segments are spread further from this axis than in the tuck position, and accordingly the moment of inertia of the diver about the somersault axis is greater in the pike position (Table 2.2). Similarly, the axis that passes from front to back through the center of gravity is known as the cartwheel axis, and the axis that passes from head to toe through the center of gravity is referred to as the twist axis. In general, the moment of inertia of the human body is least about the twist axis and greatest about the cartwheel axis (Table 2.2). The moment of inertia about the somersault axis, however, can be made to vary over a substantial range by going from a tuck to a layout position.

Because the right-hand sides of Equations 3.4 through 3.6 are nonzero, the free body diagram of the system can be equated to a **mass acceleration diagram**. (This has also been referred to as the resultant force diagram [Meriam & Kraige, 1987]). That is, by Newton's law of acceleration (**F** = *m***a**), force (free body diagram) equals mass times acceleration (mass acceleration diagram). In this context the free body diagram (FBD) represents the left-hand side of the equation and the mass acceleration diagram (MAD), the right-hand side. In other words, the free body diagram defines the system and how it interacts (forces shown with arrows) with its surroundings. The mass acceleration diagram shows the effects of these interactions on the system—that is, how the interactions alter the motion of the system.

EXAMPLE 1

A weight lifter raises a barbell to his chest. Suppose we want to determine the torque developed by the back and the hip extensor muscles when the barbell is about knee height. The typical strategy is to define a system that includes the torque about the hip joint as an external effect and then to account for all the other forces that act on the system and to determine the kinematics of the system. By accounting for all the other forces acting on the system, we assume that what remains unexplained (the residual effect) is due to the net torque exerted by the muscles across the hip joint. Torque calculated in this manner is referred to as the *residual moment of force*.

For the weight-lifter example, an appropriate system (as identified previously) would include the upper body from the lumbosacral (*LS*) joint to the head (Figure 3.6). We can identify five forces that act on this system: the resultant muscle torque ($\mathbf{T}_m$) about the lumbosacral joint, the joint reaction force ($\mathbf{F}_j$), the weight of the barbell ($\mathbf{F}_{w,b}$ = 1,003 N), the weight of the system ($\mathbf{F}_{w,u}$ = 525 N), and a force due to the intraabdominal pressure ($\mathbf{F}_i$ = 1,250 N). And from a film or video analysis of the event, we can determine the kinematics of the movement (Figure 1.7). The object of the analysis is to determine the magnitude of $\mathbf{T}_m$. As with the static situations, the obvious approach is to sum the moments about the LS joint, thereby eliminating $\mathbf{F}_j$ from consideration. The analysis proceeds as follows:

$$\Sigma \mathbf{M}_{LS} = I_{LS}\alpha + m\mathbf{a}d$$

$$(\mathbf{F}_{w,b} \times a) + (\mathbf{F}_{w,u} \times \mathrm{b}) - (\mathbf{F}_i \times c) - \mathbf{T}_m = I_{LS}\alpha + m\mathbf{a}d$$

$$\mathbf{T}_m = (\mathbf{F}_{w,b} \times a) + (\mathbf{F}_{w,u} \times b) - (\mathbf{F}_i \times c) - I_{LS}\alpha - ma_xd_y - ma_yd_x$$

$$\mathbf{T}_m = (1{,}003 \times 0.38) + (525 \times 0.24) - (1{,}250 \times 0.09) - (7.43 \times 8.7) - (155.8 \times 0.2 \times 0.40) - (155.8 \times [-0.1] \times 0.24)$$

$$\mathbf{T}_m = 381 + 126 - 11 - 64 - 12 + 4$$

$$\mathbf{T}_m = 424 \text{ N·m}$$

The main difference between static and dynamic analyses, therefore, is the inclusion of values for the moment

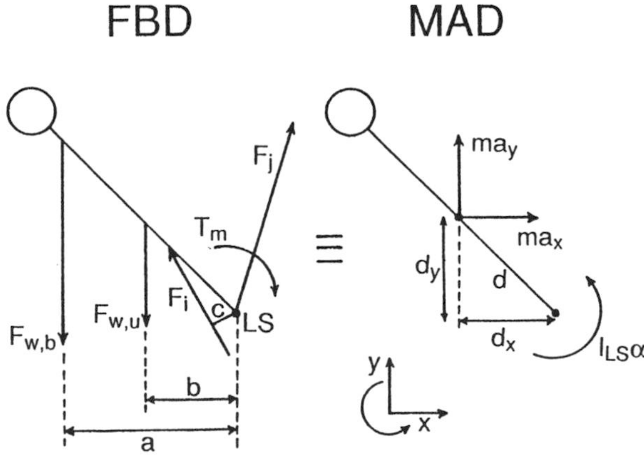

Figure 3.6 A dynamic analysis (free body diagram [FBD] = mass acceleration diagram [MAD]) of a weight lifter performing the clean lift. $\mathbf{F}_i$ = 1,250 N; $\mathbf{F}_{w,b}$ = 1,003 N; $\mathbf{F}_{w,u}$ = 525 N: a = 38 cm; b = 24 cm; c = 9 cm; d = 47 cm; I_{LS} = 7.43 kg·m^2; α = 8.7 rad/s^2.

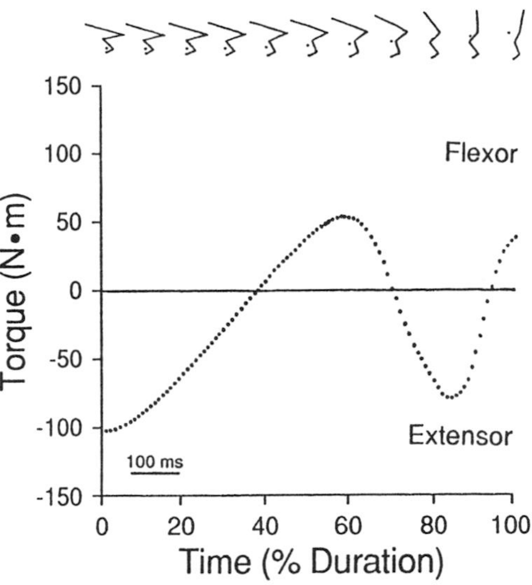

Figure 3.7 Resultant muscle torque about the knee joint during the clean lift in weight lifting.
Note. From "Muscular Control of a Learned Movement: The Speed Control System Hypothesis" by R.M. Enoka, 1983, *Experimental Brain Research*, **51**, p. 140. Copyright 1983 by Springer-Verlag, Inc. Reprinted by permission.

of inertia and angular acceleration in the dynamic analysis. Although these appear to have been obtained with minimal effort in this weight-lifting example, both measurements, particularly the angular acceleration, usually require substantial data processing that can be time-consuming. The value for angular acceleration used in this example (8.7 rad/s^2) was derived from position-time data obtained from a film of a weight lifter performing the movement; this derivation of acceleration was based on the same procedure as outlined in Table 1.1. The value of the moment of inertia needed for the dynamic analysis (7.43 kg·m^2) was estimated with regression equations that were based on measurements made on cadavers (Table 2.3).

The weight-lifting event discussed in this example takes an experienced athlete about 0.4 s to complete. If we film the event at 100 frames per second, we will have 40 frames of film that contain relevant information on the event. To completely describe the time course of $\mathbf{T}_m$, we need to perform the force-mass-acceleration calculation shown above for each frame of data. The result will be a set of instantaneous torques that can be plotted as a torque-time curve for the movement. An example for the knee joint is shown in Figure 3.7; the data represent the mean torque about the knee joint for 15 experienced weight lifters as each lifted a barbell (1,141 N, which was 1.5 times body weight) to chest height. The graph indicates that the resultant muscle torque about the knee joint reached maximal values of about 100 and 50 N·m in the extensor and flexor directions, respectively, and that direction fluctuated between extensor and flexor during the movement. Positive torque represents a net flexor torque. This torque-time graph corresponds to the angle-angle diagram shown in Figure 1.7.

EXAMPLE 2

An issue that often arises in the study of human movement is whether a dynamic analysis is necessary or whether the movement is slow enough that it can be assumed to be **quasistatic**—that is, whether we can analyze it using static techniques. Rogers and Pai (1990) examined this issue in a simple movement that involved a human subject moving from double- to single-leg support, as occurs when an individual is about to start walking. In the initiation of a step, body weight has to be shifted from two-legged support to single-leg support while the other leg flexes at the knee. This is accomplished by an increase in the activity of the lateral hip muscles of the flexing leg, an increase in the vertical component of the ground reaction force under the flexing leg, and a shift of the center of pressure toward the single-support leg (Rogers & Pai, 1990). Because the ground reaction force, as we discussed in chapter 2, represents the acceleration of the individual's center of gravity, Rogers and Pai sought to determine how well the change in the vertical component of the ground reaction force under the flexing leg could be determined using a static analysis.

If we assume that step initiation can be treated as a quasistatic event, then the system is in equilibrium and the forces acting on the system are balanced (Figure 3.8). We can sum the moments of force about the point of application of $F_{z,ss}$ (point o) and estimate the magnitude of $F_{z,f}$.

$$\Sigma M_o = 0$$

$$(F_{z,f} \times d_2) - (\mathbf{F}_w \times d_1) = 0$$

$$F_{z,f} = \frac{\mathbf{F}_w \times d_1}{d_2}$$

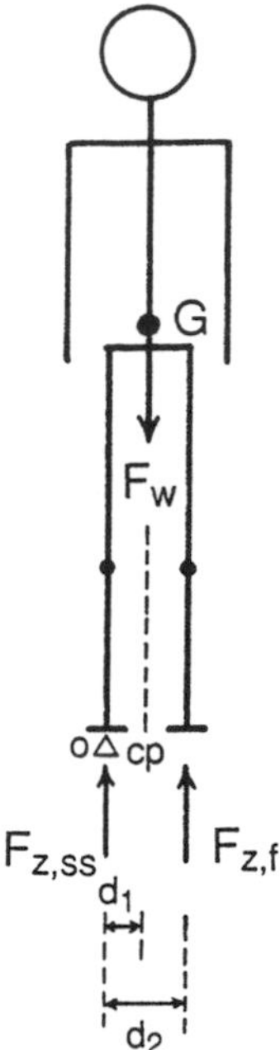

Figure 3.8 Quasistatic model used to predict the vertical component of the ground reaction force under the flexing leg during a transfer from two- to one-legged support. cp = center of pressure; d_1 = distance between cp and $F_{z,ss}$; d_2 = distance between o and $F_{z,f}$; $F_{z,f}$ = vertical component of the ground reaction force beneath the flexing leg; $F_{z,ss}$ = vertical component of the ground reaction force beneath the single support leg; $\mathbf{F}_w$ = body weight; G = center of gravity, and o = point of application of $F_{z,ss}$.
Note. From "Dynamic Transitions in Stance Support Accompanying Leg Flexion Movements in Man" by M.W. Rogers and Y.-C. Pai, 1990, *Experimental Brain Research*, **81**, p. 399. Copyright 1990 by Springer-Verlag, Inc. Reprinted by permission.

Because $\mathbf{F}_w$ is constant, $F_{z,f}$ can be predicted as the ratio of changes in d_1 and d_2. To obtain the data necessary to test this prediction, Rogers and Pai (1990) had subjects stand on two force platforms, one foot on each, which would measure the vertical component of the ground reaction force and its point of application (center of pressure) on each foot. In addition, they used a motion-analysis system and a segmental analysis to determine the location of the total-body CG; d_1 represents the distance between the vertical projection of $\mathbf{F}_w$ and the point of application of $F_{z,ss}$. The results are shown in Figure 3.9. Focus on the graphs that show the change in $F_{z,f}$ with time; the solid line represents the actually measured $F_{z,f}$, the dashed line indicates the estimate of $F_{z,f}$ with the quasistatic analysis, and the dotted area represents the difference between the two. In Figure 3.9, the graphs on the left are for a rapid step initiation (leg flexion), whereas those on the right are for a slow step initiation. The results indicate that a quasistatic analysis is appropriate for this task when it is performed slowly but not when it is performed at normal or fast speeds. Similarly, slow lifts can be analyzed with a quasistatic analysis (Toussaint, van Baar, van Langen, de Looze, & van Dieën, 1992).

EXAMPLE 3

A dynamic analysis can, in general, proceed in either of two directions. Given information on the forces and torques, we can determine the associated kinematics, or given the kinematics, we can determine the underlying forces and torques. These two approaches are referred to as **forward dynamics** and **inverse dynamics**, respectively. The *inverse dynamics approach* involves obtaining the derivative of position-time data to yield velocity- and acceleration-time data—for example, measuring position and calculating joint forces. The principal disadvantage of this method is that errors embedded in the position-time data are greatly magnified by the time the data have been processed to yield acceleration. In contrast, the *forward dynamics approach* involves the integration (in the calculus sense) of forces and torques (or accelerations) to produce the related kinematic information—for example, measuring acceleration and calculating velocity and position. The main difficulty associated with this technique is the need to accurately specify initial conditions. An alternative technique involves the use of features from both the inverse and forward approaches. In this method, measurements of position, linear acceleration, and angular velocity are combined to provide reliable estimates of link kinematics and joint loads (Ladin & Wu, 1991; Wu & Ladin, 1993).

Calculation of the net torque exerted about the hip by the weight lifter in Example 1 used the inverse dynamics approach; in that example we proceeded from the kinematics of the movement, along with some force information, to determine the unknown force. It has been suggested that the nervous system uses inverse dynamics when it plans movements (Hollerbach & Flash, 1982). In this scheme, the nervous system determines the *desired* kinematics for performing a movement and then uses inverse dynamics to calculate the muscle torques needed to produce these kinematics. The complexity of these calculations, however, makes it unlikely that the nervous

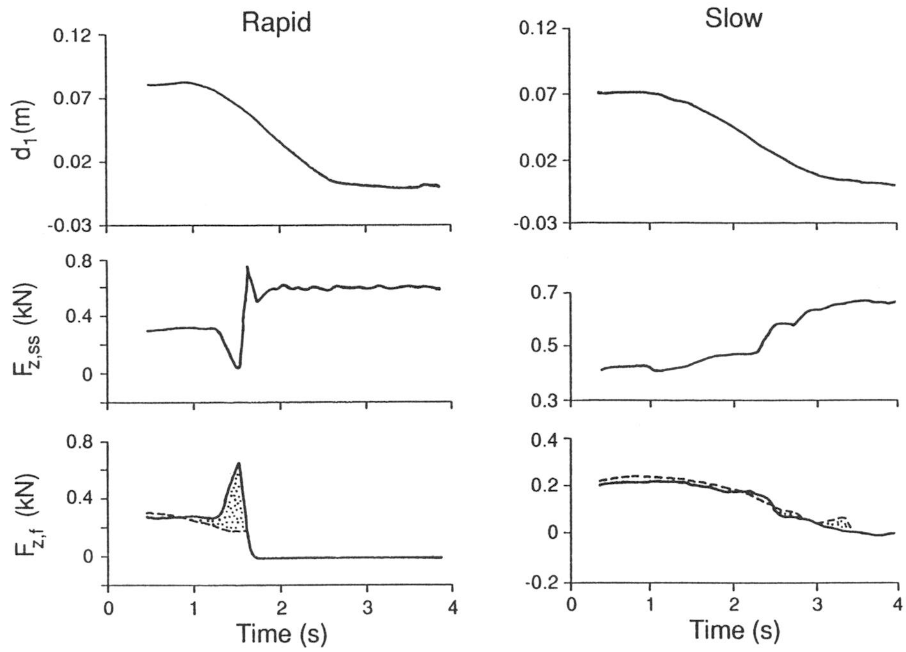

Figure 3.9 Changes in d_1, $F_{z,ss}$ and $F_{z,f}$ with time during step initiation when the task is performed rapidly (left) and slowly (right).

Note. From "Dynamic Transitions in Stance Support Accompanying Leg Flexion Movements in Man" by M.W. Rogers and Y.-C. Pai, 1990, *Experimental Brain Research*, **81**, p. 399. Copyright 1990 by Springer-Verlag, Inc. Reprinted by permission.

system organizes movements in this manner (Hasan, 1990).

Bobbert, Schamhardt, and Nigg (1991) used inverse dynamics with reasonable accuracy to estimate the magnitude of the vertical component of the ground reaction force during the support phase of running. They used four video cameras to obtain the kinematic information needed to determine the vertical acceleration of the center of gravity of each body segment and combined this with data on body segment mass (Clauser, McConville, & Young, 1969) to calculate the vertical inertial force (mass × acceleration) for each segment. The inertial forces for each segment were summed to calculate $F_{g,z}$, which was then compared to the measured $F_{g,z}$. With this technique, the magnitude of the initial peak in the $F_{g,z}$-time graph was estimated with less than 10% error and its time of occurrence within 5 ms. To achieve such accuracy in the estimation of forces, however, we need extremely accurate kinematic data (Bobbert et al., 1991; Ladin & Wu, 1991).

Intersegmental Dynamics. Because human movement typically involves the motion of more than one body segment, the kinematics experienced by an anatomical landmark is determined by the cumulative motion of all the involved body segments. For example, consider an individual about to punt a football for distance. This task clearly involves the coordinated motion of the thigh, shank, and foot segments; the motion of the contact point on the foot (m) depends on the relative motion of the foot, shank, and thigh segments (Figure 3.10). The displacement of point m (s_m) is given by

$$s_m = s_{m/A} + s_{A/K} + s_{K/H} + s_H$$

where the expression $s_{m/A}$ refers to the displacement of point m relative to the ankle joint and the subscripts A, K, and H represent the ankle, knee, and hip, respectively. According to this equation, the displacement of point m depends on four displacement terms: its displacement relative to the ankle joint, the displacement of the ankle joint relative to the knee joint, the displacement of the knee joint relative to the hip joint, and the absolute displacement of the hip joint.

Similarly, if each segment also has an angular velocity (ω) and acceleration (α), then the linear velocity (**v**) and acceleration (**a**) of point m depend on the relative kinematics of each segment in combination with the absolute kinematics of the hip joint. The angular velocity and acceleration of each segment are shown in Figure 3.10a, where the subscripts 1, 2, and 3 represent the foot, shank, and thigh, respectively. A parallel equation to the one for s_m can be developed for $\mathbf{v}_m$:

$$\mathbf{v}_m = \mathbf{v}_{m/A} + \mathbf{v}_{A/K} + \mathbf{v}_{K/H} + \mathbf{v}_H$$

Because linear and angular velocity are related by the equation $\mathbf{v} = r\omega$,

$$\mathbf{v}_m = r_1\omega_1 + r_2\omega_2 + r_3\omega_3 + \mathbf{v}_H$$

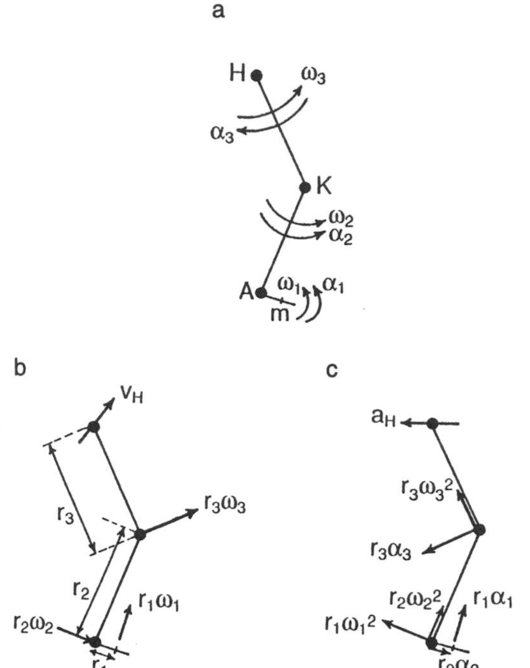

Figure 3.10 Kinematics of the lower extremity during a kick. (a) The foot, shank, and thigh each experience an angular velocity (ω) and acceleration (α) as the extremity moves to contact the ball at point *m*. The foot rotates about the ankle (*A*), the shank about the knee (*K*), and the thigh about the hip (*H*) joint. (b) The kinematic factors that influence the linear velocity of point *m*. (c) The kinematic variables that affect the linear acceleration of point *m*.

where r_1 corresponds to the distance from the ankle to point *m*, r_2 indicates the length of the shank, and r_3 represents the length of the thigh. The direction of each term is tangent to the path traced by the relevant anatomical landmark (Figure 3.10b): $r_1\omega_1$ is tangent to the path of point *m*, $r_2\omega_2$ is tangent to the path of the ankle (the end point of the shank), and $r_3\omega_3$ is tangent to the path of the knee (the end point of the thigh).

The acceleration of point *m* ($\mathbf{a}_m$) also depends on the acceleration of the involved segments (Figure 3.10c) and, based on the relationship between linear and angular acceleration ($\mathbf{a} = r\omega^2 + r\alpha$), can be written as

$$\mathbf{a}_m = r_1\omega_1^2 + r_1\alpha_1 + r_2\omega_2^2 + r_2\alpha_2 + r_3\omega_3^2 + r_3\alpha_3 + \mathbf{a}_H$$

where $r\omega^2$ represents the change in direction of **v** for a segment and $r\alpha$ indicates the change in magnitude of **v**. An important feature of this relationship is that *the acceleration of each segment in this system is influenced by the acceleration of* all *the other segments*. For example, rearrangement of the above expression for $\mathbf{a}_m$ indicates that the acceleration of the shank during the football punt is dependent on the angular acceleration of the foot and thigh and the linear acceleration of the hip:

$$r_2\omega_2^2 + r_2\alpha_2 = \mathbf{a}_m - r_1\omega_1^2 - r_1\alpha_1 - r_3\omega_3^2 - r_3\alpha_3 - \mathbf{a}_H$$

Miller and Munro (1984) use this approach to determine the body segment contributions to the height achieved during the flight phase of a springboard dive.

Kinematic coupling between segments occurs because of the dynamic interactions between segments. Based on what we know about the relationship between force and acceleration (**F** = *m***a**), these interactions between the accelerations of the body segments indicate that there must be *interactive forces between body segments during human movement*. The study of intersegmental dynamics examines these **motion-dependent interactions** between segments (Hoy & Zernicke, 1986; Phillips et al., 1983; Putnam, 1991). This is a difficult area of study, however, because of the complexity of the equations that are needed to explicitly identify the motion-dependent effects. Putnam (1991) has provided a thorough description of these procedures by deriving the equations for a two-segment lower extremity (thigh and shank) during a kick. The derivation begins with the standard Newtonian analysis in which the equation for the shank (the distal segment), isolated from the thigh, can be written as

$$\mathbf{F}_{j,k} + \mathbf{T}_k + \mathbf{F}_{w,s} = m_s\mathbf{a}_s$$

where $\mathbf{F}_{j,k}$ represents the joint reaction force at the knee, $\mathbf{T}_k$ indicates the resultant torque about the knee, $\mathbf{F}_{w,s}$ refers to the weight of the shank, m_s denotes the mass of the shank, and $\mathbf{a}_s$ accounts for the acceleration of the center of gravity of the shank. The next steps in the derivation are to replace $\mathbf{a}_s$ with an expression consisting of $r\omega^2$ and $r\alpha$ terms, as was done above for the football-punt example, and then to sum the moments of force about the knee joint. The result is an expression in which the torque about the knee joint is equal to a moment due to $\mathbf{F}_{w,s}$ and torques due to several motion-dependent effects that include $r\alpha$ for the thigh, $r\omega^2$ for the thigh, $r\alpha$ for the shank, and the linear acceleration of the hip.

An important goal of this analysis is to determine the magnitude of the motion-dependent effects. Figure 3.11a provides such a comparison by showing the time course of the resultant muscle torque about the knee joint and the resultant motion-dependent torque exerted by the thigh on the shank during the kick. This graph shows that the effect of thigh motion on the shank is as large as that due to the muscles that cross the knee joint. The motion-dependent torque exerted by the thigh on the shank is negative for about the first 60 ms of the kick; the negative direction means that the thigh accelerated the shank in a backward-rotation direction. For most of the kick, the thigh motion accelerates the shank in a forward direction.

By a similar analysis it is possible to identify the effect of shank motion on the thigh. As described by Putnam (1991), the derivation would again begin with a free body diagram for the thigh and proceed to substituting a set of $r\omega^2$ and $r\alpha$ terms for $\mathbf{a}_s$ and summing the torques about the hip joint. The effect of shank motion on the thigh is not simply a mirror image of the effect of thigh motion on the shank. The shank is the distal segment in this two-segment system, whereas the thigh is the proximal segment with the shank on one end and the hip joint at the other. The shank experienced four motion-dependent effects; the thigh is subjected to six. As with the thigh-on-shank interaction, we are interested in the magnitude of the shank-on-thigh motion-dependent effect. Figure 3.11b shows the net torque acting on the thigh due to the motion of the shank. Once again, the magnitude of the motion-dependent torque is large compared to the resultant muscle torque about the hip joint. Furthermore, the muscle and motion-dependent torques act in opposite directions for most of the kick. When all the torques acting on the thigh are taken into account, the resultant torque has a smaller magnitude and causes an acceleration of the thigh in the forward-rotation direction for the first 100 ms of the kick and then a backward-rotation acceleration for the remainder of the kick.

One conclusion that can be made from the study of intersegmental dynamics is that the motion of a body segment can exert significant torques on the other segments in the system. These effects are substantial for whole-limb, rapid movements such as kicking and throwing but are also significant even in finger movements (Darling & Cole, 1990). Consequently, the control of movement by the nervous system must accommodate these motion-dependent effects (Smith & Zernicke, 1987). For example, the magnitude of the resultant muscle torque about the hip joint (Figure 3.11b) must be sufficient to overcome the shank-on-thigh motion-dependent torque and to accelerate the thigh in the forward direction initially and then in the backward direction.

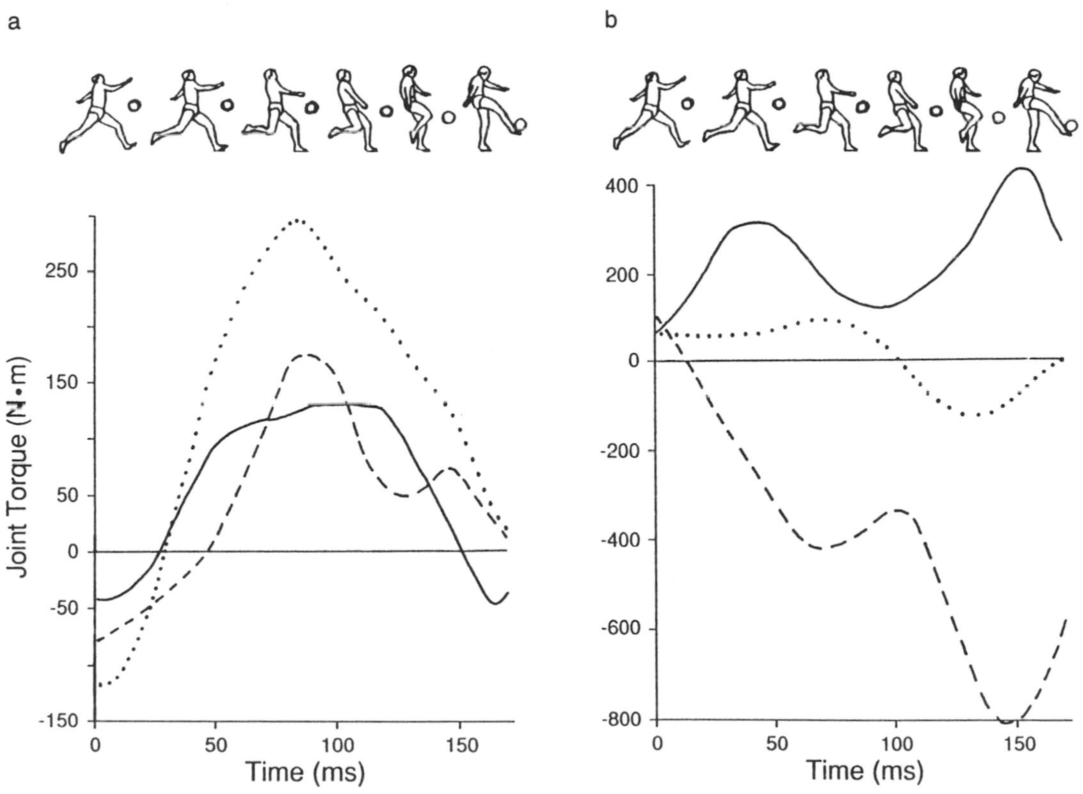

Figure 3.11 Motion-dependent effects during a kick: (a) the resultant muscle torque about the knee joint (solid line), the motion-dependent torque exerted by the thigh on the shank (dashed line), and the net effect (dotted line) of all torques acting on the shank; (b) the resultant muscle torque about the hip joint (solid line), the motion-dependent torque exerted by the shank on the thigh (dashed line), and the net effect (dotted line) of all the torques acting on the thigh.

Note. From "Interaction Between Segments During a Kicking Motion" by C.A. Putnam. In *Biomechanics VIII-B* (pp. 691, 692) by H. Matsui and K. Kobayashi (Eds.), 1983, Champaign, IL: Human Kinetics. Copyright 1983 by Human Kinetics Publishers, Inc. Adapted by permission.

The means by which the nervous system knows about these motion-dependent effects, however, appear complex (Koshland & Smith, 1989a, 1989b).

Impulse-Momentum Approach

Many human movements can be considered collisions in which the human body, either directly or indirectly, collides with an object in its surroundings. A convenient perspective from which to analyze collisions is the study of momentum. The linear *momentum* (**G**) of a moving object is defined as the product of its mass and velocity. The unit of measurement for momentum is kg·m/s (mass × velocity). Momentum is a vector quantity whose direction is the same as the velocity. According to Newton's law of acceleration, the momentum of a system is altered when a force acts on the system. And because forces are not applied instantaneously but rather over intervals of time, we generally consider the effect of a force as an **impulse** (force × time). Figure 3.12, which illustrates the vertical component of the ground reaction force ($F_{g,z}$) during running, provides an example of an application of a force over a period of time. An impulse is determined graphically as the area under a force-time curve (Figure 3.12) or numerically as the product of the average force (N) and time (s). If the average force ($F_{g,z}$) in Figure 3.12 is 1.3 kN and the time of application is 0.29 s, the impulse will be 377 N·s. It should be apparent that the magnitude of the impulse could be altered by varying, singly or in combination, either average force or time of application.

If Newton's law of acceleration is interpreted to focus on epochs rather than instants of time, the law indicates that *the application of an impulse will result in a change in the momentum of the system*. This is the basis of the impulse-momentum approach to the analysis of motion. Momentum was previously used to describe the quantity of motion possessed by a system. Linear momentum (**G**) is defined as mass times velocity (Equation 2.2), and angular momentum (**H**) is calculated as the product of moment of inertia and angular velocity. The impulse-momentum relationship is derived from Newton's law of acceleration:

$$\Sigma \mathbf{F} = m\mathbf{a}$$

$$\Sigma \mathbf{F} = m\frac{(\mathbf{v}_f - \mathbf{v}_i)}{t}$$

$$\Sigma \mathbf{F}t = m(\mathbf{v}_f - \mathbf{v}_i)$$

$$\Sigma \mathbf{F}t = \Delta m\mathbf{v} \quad \text{or} \quad \bar{\mathbf{F}}\cdot t = \Delta m\mathbf{v} \qquad (3.10)$$

where the term $\Sigma \mathbf{F}t$ represents the area under a force-time curve (Figure 3.12) and is equivalent to the product of the average force and its time of application ($\bar{\mathbf{F}}\cdot t$). Equation 3.10 suggests that if the magnitude of the impulse is known, its effect on the momentum of the system can be calculated. Conversely, if the change in momentum can be measured, it is possible to determine the applied impulse.

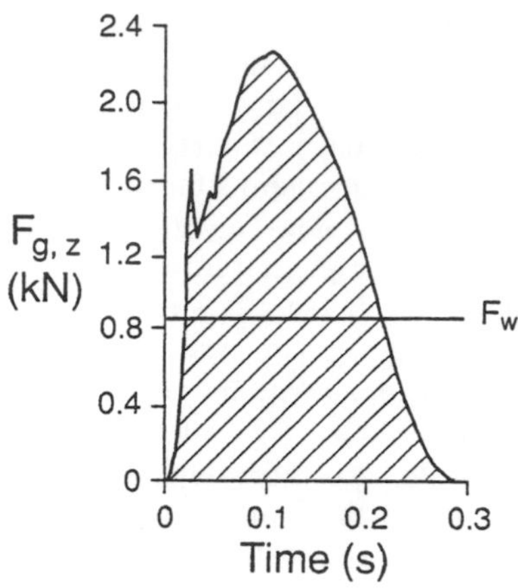

Figure 3.12 Graphic representation of an impulse as an area (shaded) under a force-time curve.

EXAMPLE 1

By filming a person spiking a volleyball and measuring the mass (m) of the ball, we can determine the impulse applied to the ball. From the film it would be necessary to measure both the velocity of the ball before ($\mathbf{v}_b$) and after ($\mathbf{v}_a$) contact and the total time the hand is in contact with the ball (t_c). For one performance of this skill, we obtained

$$\mathbf{v}_b = 3.6 \text{ m/s}$$

$$\mathbf{v}_a = 25.2 \text{ m/s}$$

$$m = 0.27 \text{ kg}$$

$$t_c = 18 \text{ ms}$$

Equation 3.10 is applied as follows:

$$\bar{F}\cdot t = \Delta m\mathbf{v}$$

$$= m\Delta\mathbf{v}$$

$$= m(\mathbf{v}_b - \mathbf{v}_a)$$

$$= 0.27(25.2 - 3.6)$$

$$\bar{F}\cdot t = 5.83 \text{ N·s}$$

Because we know the contact time (t_c), we can determine the average force ($\bar{F}$) exerted by the spiker during the contact:

$$\bar{F}\cdot t = 5.83 \text{ N·s}$$

$$= \frac{5.83}{t_c}$$

$$= \frac{5.83}{0.018}$$

$$\bar{F} = 324 \text{ N}$$

Thus, although the impulse appeared to be small (5.83 N·s), the brief duration of the force application resulted in forces that were quite substantial ($\bar{F}$ = 324 N). Incidentally, the time of contact with the volleyball in this example is quite similar to those (10 to 16 ms) recorded during the kicking of a ball (Asami & Nolte, 1983).

EXAMPLE 2

In most contact events, such as the spiking of a volleyball, the momentum of an object is altered by applying relatively high forces for brief periods of time. There are instances, however, in which the change in momentum is accomplished by applying smaller forces for longer periods. A good example (Brancazio, 1984) of this is the distinction between the consequences of a person (mass = 71 kg) jumping off a 15-m building onto the pavement and the consequences of another person diving off a 15-m cliff into the ocean. In both instances the individual will have a speed of about 17.3 m/s just prior to contact (linear momentum = $m \times \mathbf{v}$ = 1,228 kg·m/s). Eventually, however, the speed (and thus momentum) of each person will reach zero. The jumper will experience large forces (probably fatal) for a brief interval upon contact with the pavement, whereas the diver will encounter smaller forces, due to contact with the water, over a longer period of time, but the change in momentum for each individual will be the same.

EXAMPLE 3

Whenever an impulse is applied to a system, the momentum of the system will change in proportion to the net impulse. Furthermore, the effect on momentum will be confined to the direction of the impulse. Consider the forward-backward component ($F_{g,y}$) of the ground reaction force during the support phase of running (Figure 2.17). The graph illustrates that the runner experiences two horizontal impulses during support. Initially $F_{g,y}$ is directed backward, creating a retarding or braking impulse; then $F_{g,y}$ changes direction, eliciting a propulsive impulse. Because these impulses act in opposite directions, the change in momentum that the runner (or the system) experiences in the y-direction depends on the difference between the braking and propulsion impulses. When the individual is running at a constant speed (no change in momentum), the two impulses are equal. For a runner to increase speed, however, the propulsive impulse must exceed the braking impulse; to decrease speed, the converse is true. These relationships are shown in Figure 3.13. The change in momentum due to the ground reaction force is equal to the difference between the initial (before stance) and final (after stance) values for momentum.

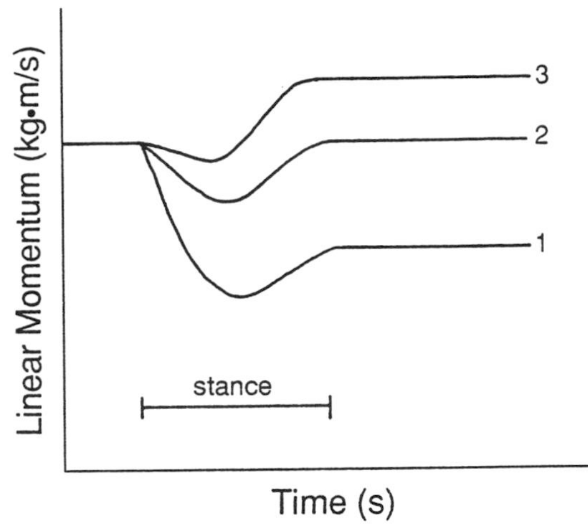

Figure 3.13 Change in the forward-backward linear momentum of a runner in response to (1) a braking impulse that is larger than the propulsive impulse; (2) equivalent braking and propulsive impulses; and (3) a propulsive impulse that is larger than the braking impulse.

Because impulses in locomotion and other such activities are influenced by the weight of the individual, it is common practice to normalize the impulses and to express them as a proportion of the body weight impulse. For example, if a runner weighs 630 N and experiences a braking impulse of 15.8 N·s that lasts for 0.1 s, then the braking impulse was 0.25 times the body weight impulse (15.8/[630 × 0.1] = 0.25). Munro et al. (1987) have reported braking and propulsive impulses of 0.15 to 0.25 times body weight impulse with runners traveling at 3.0 to 5.0 m/s. These impulses increased with running speed.

Collisions

Whereas the linear momentum of a diver or a gymnast performing an airborne stunt is not conserved, linear momentum is conserved during collisions. Many sports activities are based on collisions such as those between players (e.g., rugby, boxing), between a participant and an inanimate object (e.g., handball, soccer), and between inanimate objects (e.g., badminton, golf). *A collision does not create or destroy momentum; the sum of the momentum of the colliding objects remains constant.* Consider, for example, a head-on tackle between a running back (mass = 100 kg; velocity = 7.2 m/s to the right) and a defensive back (mass = 85 kg; velocity = −5.9 m/s to the left):

$$\begin{aligned}\text{momentum after tackle} &= \text{momentum before tackle}\\ &= (100 \times 7.2) + (85 \times [-5.9])\\ &= 720 - 502\\ &= 218 \text{ kg·m·s}^{-1}\end{aligned}$$

Because momentum is a vector with the same direction as that of velocity, the positive sign for the momentum (218 kg·m·s^{-1}) indicates that after the head-on tackle the running back continues to move to the right, albeit with a lesser momentum than before the tackle. What is the speed of the two players who are now joined in a tackle? The change in velocity experienced by each player will be inversely proportional to his mass—hence the advantage of size (mass) in contact sports.

When two objects collide, they exert equal and opposite forces on each other. This occurs because a force represents a physical interaction in the form of a push or pull. When two objects are involved in the interaction (collision), they both experience the same force. And because the contact associated with the collision is the same for the two objects, the impulse applied to each object and the change in momentum experienced by each object are the same. So in a collision there is a conservation of linear momentum. This means that the sum of the linear momentums of the two objects (*A* and *B*) before the collision is the same as the sum of the linear momentum after the collision, as described above in the football example:

$$(m_A\mathbf{v}_A)_{\text{before}} + (m_B\mathbf{v}_B)_{\text{before}} = (m_A\mathbf{v}_A)_{\text{after}} + (m_B\mathbf{v}_B)_{\text{after}} \tag{3.11}$$

The left-hand side of Equation 3.11 represents the momentum of the system before the collision, and the right-hand side indicates the momentum after the collision. Because mass does not usually change in human movements, this equation can be rearranged to show the change in momentum of object *A* is equal to the change in momentum of object *B*:

$$(m_Av_A)_{\text{before}} - (m_Av_A)_{\text{after}} = (m_Bv_B)_{\text{after}} - (m_Bv_B)_{\text{before}}$$

$$m_A(v_{\Delta,\text{before}} - v_{\Delta,\text{after}}) = m_B(v_{B,\text{after}} - v_{B,\text{before}})$$

$$m_A\Delta v_A = m_B\Delta v_B$$

$$\frac{\Delta v_A}{\Delta v_B} = \frac{m_B}{m_A} \tag{3.12}$$

Equation 3.12 indicates that the change in velocity of object *A* with respect to object *B* is inversely proportional to the mass of the two objects. For example, consider hitting a tennis ball with a racket. After contacting the racket (the collision), the velocity of the ball is usually much greater than the velocity of the racket. The difference in the velocity of the ball and the racket is determined by the ratio of their masses.

One important feature of collisions is whether or not they are elastic; that is, do the objects bounce off one another or do they remain joined after the collision? If the collision is elastic, then each object preserves some kinetic energy (velocity after the collision is nonzero). The *elasticity* of a collision is indicated by the **coefficient of restitution** (e). A perfectly elastic collision has a coefficient of restitution equal to one; the velocity after the collision is identical to that before the collision. However, for most collisions in human movement the coefficient of restitution is less than one:

$$\text{coefficient of restitution} = \frac{\text{speed after collision}}{\text{speed before collision}}$$

The coefficient of restitution of a ball is usually measured by dropping the ball from a known height to the ground and measuring how high it rebounds. If a ball is dropped from a height of 1.0 m and rebounds to 0.5 m, then the coefficient of restitution is 0.5. Selected coefficients of restitution for various balls at an impact speed of 24.6 m/s (55 miles/hr) include softball, 0.40; tennis ball, 0.55; golf ball, 0.58; basketball, 0.64; soccer ball, 0.65; and Superball, 0.85. The coefficient of restitution tends to decrease as the speed of the collision increases.

The coefficient of restitution quantifies the extent to which a perfect collision is modified by material properties of the objects involved in the collision. After a bat contacts a ball, the velocities of the ball and the bat depend not only on the mass of each but also on the coefficient of restitution. The coefficient of restitution represents the constant of proportionality between the speed before the collision and the speed after the collision. For example, consider the contact of a baseball and a bat, which is described by the equations that define the coefficient of restitution and the conservation of momentum:

$$\text{speed after collision} = -e(\text{speed before collision})$$

$$v_{B,a} - v_{b,a} = -e(v_{B,b} - v_{b,b}) \tag{3.13}$$

where *B* represents the bat and *b* indicates the ball. If we assume that the velocity vectors of the ball and the bat are collinear (lie in the same line, as shown in Figure 3.14), rearrangement of Equation 3.13 allows us to specify the velocity of the bat ($v_{B,a}$) and the ball ($v_{b,a}$) after contact:

$$v_{b,a} = v_{B,a} + e(v_{B,b} - v_{b,b}) \tag{3.14}$$

$$v_{B,a} = v_{b,a} - e(v_{B,b} - v_{b,b}) \tag{3.15}$$

In order to determine the velocity of the bat after the collision ($v_{B,a}$), we next substitute Equation 3.14 into the expression for the conservation of linear momentum (Equation 3.11); the uppercase symbols are for the bat, and the lowercase symbols are for the ball.

$$m_Bv_{B,b} + m_bv_{b,b} = m_Bv_{B,a} + m_bv_{b,a}$$

$$m_Bv_{B,b} + m_bv_{b,b} = m_Bv_{B,a} + m_b[v_{B,a} + e(v_{B,b} - v_{b,b})]$$

$$v_{B,a}(m_b + m_B) = m_bv_{b,b}\,(1 + e) + v_{B,b}\,(m_B - m_be)$$

$$v_{B,a} = \frac{m_bv_{b,b}\,(1 + e) + v_{B,b}\,(m_B - m_be)}{m_b + m_B} \tag{3.16}$$

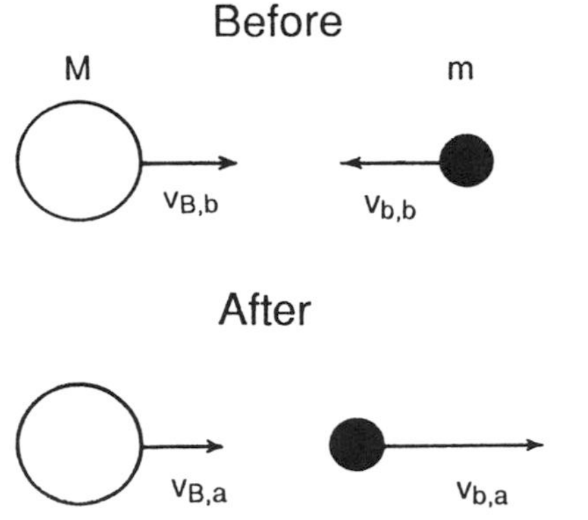

Figure 3.14 Mass and velocity of the bat (B) and ball (b) before ($v_{B,b}$, $v_{b,b}$) and after ($v_{B,a}$, $v_{b,a}$) a collision.

Similarly, we can substitute the expression for $v_{B,a}$ with Equation 3.14 and solve for $v_{b,a}$:

$$m_B v_{B,b} + m_b v_{b,b} = m_B v_{B,a} + m_b v_{b,a}$$

$$m_B v_{B,b} + m_b v_{b,b} = m_B[v_{b,a} - e(v_{B,b} - v_{b,b})] + m_b v_{b,a}$$

$$v_{b,a}(m_b + m_B) = m_B v_{B,b}\,(1 + e) + v_{b,b}\,(m_b + m_B e)$$

$$v_{b,a} = \frac{m_B v_{B,b}\,(1 + e) + v_{b,b}\,(m_b + m_B e)}{m_B + m_b} \quad (3.17)$$

Equations 3.16 and 3.17 represent the general case for the velocity of a bat ($v_{B,a}$) and ball ($v_{b,a}$) after a collision. Typically, however, ball velocities are rarely collinear before ($v_{b,b}$) and after ($v_{b,a}$) impact, and the directions of $v_{b,b}$ and $v_{b,a}$ are usually opposite (i.e., toward and away from the batter). The noncollinearity of the ball velocities is accommodated by inserting a cosine term next to $v_{b,b}$ in Equations 3.16 and 3.17 (see Townend, 1984, p. 142, for an example). Similarly, the difference in the direction of $v_{b,b}$ with respect to the other velocities must be included in the equations, as shown in the following example.

Given the following conditions,

mass of bat (m_B) = 0.93 kg (30 oz)

mass of ball (m_b) = 0.16 kg (5 oz)

velocity of ball
before impact ($v_{b,b}$) = −38 m/s (85 miles/hour)
velocity of bat
before impact ($v_{B,b}$) = 31 m/s (70 miles/hour)
coefficient of restitution
(specified in rules) = 0.55

we can determine the velocity of the baseball and bat after a hit as follows:

$$v_{B,a} = \frac{m_b v_{b,b}\,(1 + e) + v_{B,b}\,(m_B - m_b e)}{m_b + m_B}$$

$$= \frac{-6.08\ (1 + 0.55) + 31\ (0.93 - 0.088)}{0.16 + 0.93}$$

$$= 15.3 \text{ m/s}$$

$$v_{b,a} = \frac{m_B v_{B,b}(1 + e) + v_{b,b}\,(m_b + m_B e)}{m_B + m_b}$$

$$= \frac{28.8\ (1 + 0.55) + 38\ (0.16 + 0.51)}{0.16 + 0.93}$$

$$= 64.3 \text{ m/s}$$

Although these calculations of ball and bat velocities after impact rely on the application of standard techniques from physics, they may actually underestimate the effect of the collision. The critical variables affecting velocity after a collision are the coefficient of restitution, the velocity before contact, and mass. Whereas the above example used the actual mass of the ball and the bat, the mass of the bat may be larger due to Plagenhoef's concept of the *effective striking mass*. There does not seem to be any information available on whether the mass of the baseball bat is supplemented due to its connection to the rigid links of the athlete.

Another aspect of ball-bat collisions, which is somewhat related to the effective striking mass, is the notion of a sweet spot. Contact in a ball-bat collision is said to occur at the sweet spot when no reaction force is felt at the hands. Three theories seek to account for the sweet spot: the center of percussion, the natural frequency node, and the coefficient of restitution (Brancazio, 1984). Of these, the explanation provided by the **center of percussion** seems to have gained the most acceptance. When a ball contacts a bat, the location along the bat where no reaction force ($\mathbf{F}_r$) is felt at the hands is referred to as the center of percussion (Figure 3.15c). When a force ($\mathbf{F}_c$) is applied at the CG of the bat, the bat experiences only linear motion (Figure 3.15a). However, when the force is not applied at the CG, the bat experiences both linear and angular motion (Figure 3.15b). In this latter case, there is a point where the linear translation and the angular rotation cancel, and the bat does not move. The contact point that produces a stationary pivot point is the center of percussion (Figure 3.15c). It is located distal on the bat to the CG. The location of the ball contact on the bat relative to the centers of gravity and percussion determines the nature of the reaction force at the hands (Figure 3.15c).

Angular Momentum

When we analyze angular motion, we do so in terms of the torques that are acting on the system. Torque represents the rotary effect of a force and is calculated as the product of force and moment arm. A synonym for torque is moment of force. Similarly, angular momentum is derived as the moment of linear momentum. Angular momentum (**H**) describes the quantity of angular motion.

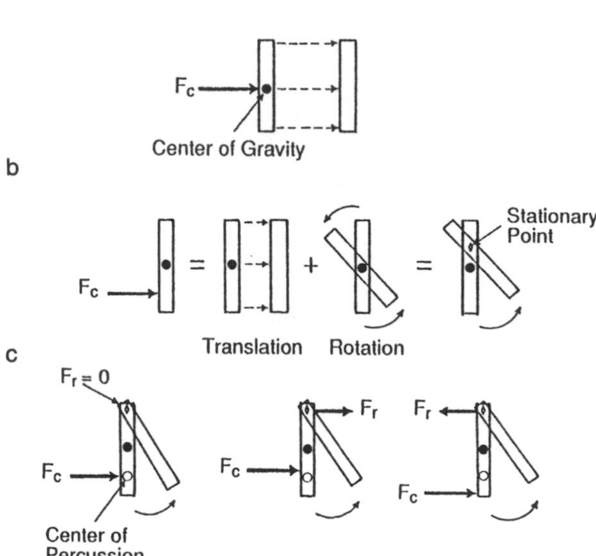

Figure 3.15 The effect of a force that is applied to a bat on the motion of the bat: (a) a contact force ($\mathbf{F}_c$) applied at the center of gravity; (b) $\mathbf{F}_c$ applied at a place other than the center of gravity; (c) $\mathbf{F}_c$ applied at the center of percussion.

Note. From *Sport Science: Physical Laws and Optimum Performance* (p. 236) by P.J. Brancazio, 1984, New York: Simon and Schuster. Copyright 1984 by Peter Brancazio. Adapted with permission.

It is calculated as the product of moment of inertia and angular velocity, and its unit of measurement is kg·m²/s.

Recall that statics presents a special case of the force-mass-acceleration approach in situations in which one of the quantities of the equation is zero; the same is true for equations in the impulse-momentum context. Specifically, if there is no impulse applied to a system, the left-hand side of Equation 3.10 is zero, a state in which momentum is said to be conserved. That is, *if there is no change in momentum, the momentum of the system must remain constant.* This can occur in linear and angular motion and represents the principle of the **conservation of momentum**. However, conservation in both the linear and angular directions does not always occur at the same time. This is particularly evident in free fall, found in such activities as gymnastics (e.g., Dainis, 1981; Nissinen, Preiss, & Brüggemann, 1985) and diving (e.g., Bartee & Dowell, 1982; Frohlich, 1980; Stroup & Bushnell, 1970; Wilson, 1977) and is classically demonstrated by the cat (and other animals) in the air-righting reaction that results in the cat always landing on its feet when it is dropped from a low height (Kane & Scher, 1969; Laouris, Kalli-Laouri, & Schwartze, 1990; Magnus, 1922; Marey, 1895).

EXAMPLE 1

Once a springboard diver has left the board, the diver may experience two appreciable forces, weight and air resistance (Figure 3.16). (Generally, however, air resistance is really only a factor in tower dives.) Let us consider the effect of these forces on the linear (**G**) and angular (**H**) momentum of the diver. From the impulse-momentum relationship, we have

$$\text{impulse} = \Delta \text{ linear momentum}$$

$$\bar{\mathbf{F}} \cdot t = \Delta \mathbf{G}$$

$$\bar{\mathbf{F}} \cdot t = \Delta m\mathbf{v}$$

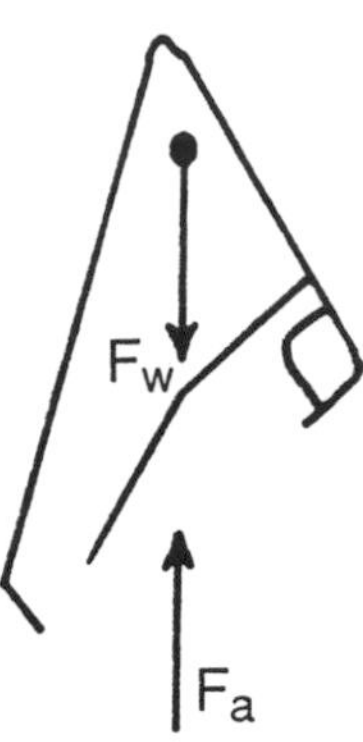

Figure 3.16 Forces experienced by a diver during free fall. $\mathbf{F}_a$ = air resistance; $\mathbf{F}_w$ = weight.

Because the average force includes the effect of weight ($\mathbf{F}_w$) and air resistance ($\mathbf{F}_a$), $\bar{\mathbf{F}}$ is not zero and there is a change in linear momentum; that is, $\Delta m\mathbf{v} \neq 0$. Therefore, linear momentum does not remain constant. In contrast, consider the effect on angular momentum:

$$\text{angular impulse} = \Delta \text{ angular momentum}$$

$$(\bar{\mathbf{F}} \times d) \cdot t = \Delta \mathbf{H}$$

$$(\bar{\mathbf{F}} \times d) \cdot t = \Delta I\omega$$

The angular impulse is equal to the product of the average torque ($\bar{\mathbf{F}} \times d$) and its time of application, where d is the moment arm from the line of action of each force to the axis of rotation (the CG). Because both $\mathbf{F}_w$ and $\mathbf{F}_a$ act through the CG, the moment arm for each is equal to zero, and there is no torque about the CG during the dive. The absence of a torque means that the right-hand side of the equation ($\Delta I\omega$) is equal to zero, and therefore momentum does not change but remains constant. This represents an example of the conservation of momentum.

Because angular momentum remains constant and is equal to the product of moment of inertia and angular velocity, any change in one parameter (i.e., moment of inertia or angular velocity) will be accompanied by a complementary change in the other parameter. For example, suppose a diver performs a multisomersault in the pike position. If, during the dive, it became apparent that the diver would not make the appropriate number of revolutions, then one alternative (because **H** remains constant) would be to assume a tuck position, which would be accompanied by an increase in the speed of rotation. The moment of inertia (I_g) of the diver about the somersault axis passing through the CG is about 7.5 kg·m^2 in the pike position, as opposed to 4.5 kg·m^2 in the tuck position. If, in the pike position, the diver had an angular velocity (ω) of 6 rad/s, then on changing to a tuck the speed would increase to 10 rad/s, and the product of the two parameters ($\mathbf{H} = I_g\omega$) would remain constant ($\mathbf{H}$ = 45 kg·m^2·s^{-1}). Specifically,

Pike

$$\mathbf{H} = 7.5 \text{ kg·m}^2 \times 6 \text{ rad/s}$$

$$\mathbf{H} = 45 \text{ kg·m}^2\text{·s}^{-1}$$

Tuck

$$\mathbf{H} = 4.5 \text{ kg·m}^2 \times 10 \text{ rad/s}$$

$$\mathbf{H} = 45 \text{ kg·m}^2\text{·s}^{-1}$$

and thus angular momentum (**H**) remains constant. The diver could also slow the speed of rotation by increasing the moment of inertia, in this case by assuming a greater layout position. This exchange between angular velocity and moment of inertia in order to conserve angular momentum during a dive is illustrated in Figure 3.17.

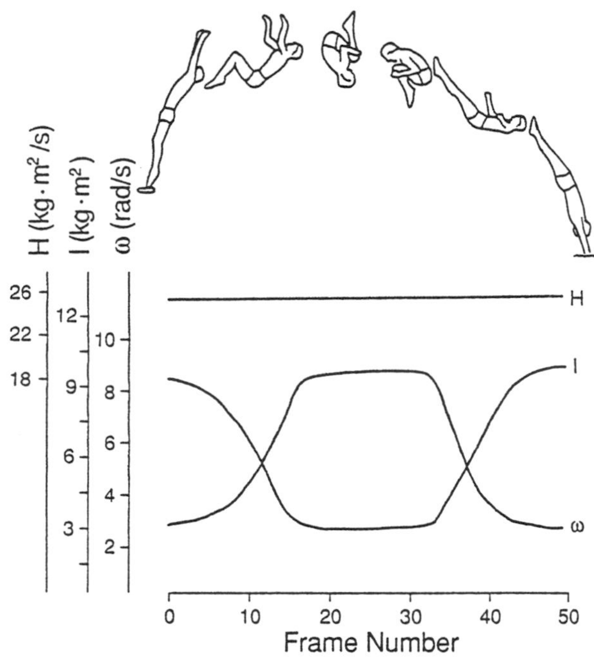

Figure 3.17 The conservation of angular momentum is accompanied by reciprocal changes in moment of inertia and angular velocity during a backward one-and-a-half somersault dive.

Note. From *The Biomechanics of Sports Techniques* (p. 149) by J.G. Hay, 1978, Englewood Cliffs, NJ: Prentice Hall. Copyright 1978 by Prentice Hall, Inc. Adapted with permission.

EXAMPLE 2

The impulse-momentum relationship can also be used to explain how a diver is able to *initiate a twist* even though no force is available to assist in the maneuver. Figure 3.18 shows a diagram of a diver performing a somersault in a layout position; the angular momentum of the diver ($\mathbf{H}_g$) is acting at the center of gravity (Figure 3.18a). According to the right-hand rule, in which direction is the diver performing the somersault? To initiate the twist, the diver rotates the arms about a cartwheel axis passing through the chest; the right arm goes above the head and the left arm moves across the trunk (Figure 3.18b). Because the arms rotate about a cartwheel axis, this does not alter $\mathbf{H}_g$ about the somersault axis. However, rotation of the arms generates an angular momentum in one direction about a cartwheel axis ($\mathbf{H}_{g,\text{arms}}$) that is counteracted (to keep $\mathbf{H}_g$ about the cartwheel axis zero) by an equivalent angular momentum of the trunk in the opposite direction about the cartwheel axis ($\mathbf{H}_{g,\text{trunk}}$) ($\mathbf{H}_{g,\text{arms}} + \mathbf{H}_{g,\text{trunk}} = 0$). Following this action-reaction maneuver, the arms are displaced as shown in Figure 3.18b, and the trunk is inclined to the vertical by an angle θ. Because the diver is in free fall, the angular momentum remains constant at the value that the diver possessed when leaving the board. Now, however, because the orientation of the diver has changed, $\mathbf{H}_g$ has components about both the somersault and twist axes of the diver; the direction of $\mathbf{H}_g$ remains constant, but the axes move with the diver (Figure 3.18c). The component of $\mathbf{H}_g$ about the twist axis ($H_{g,t}$) is equal to $\mathbf{H}_g$ sin θ, and the component about the somersault axis ($H_{g,s}$) can be determined as $\mathbf{H}_g$ cos θ.

For example, consider the tilt angle (θ = angle that the trunk is inclined to the vertical) necessary to execute a forward layout dive with a full twist from a 3-m diving board. Assume that no twist is initiated from the board. In this example, the diver has an angular momentum of 30 kg·m^2/s about the somersault axis at the moment of takeoff, $I_{g,s}$ represents the moment of inertia of the diver about the somersault axis passing through the CG, $I_{g,t}$ indicates the moment of inertia of the diver about the twist axis passing through the CG, and t is the

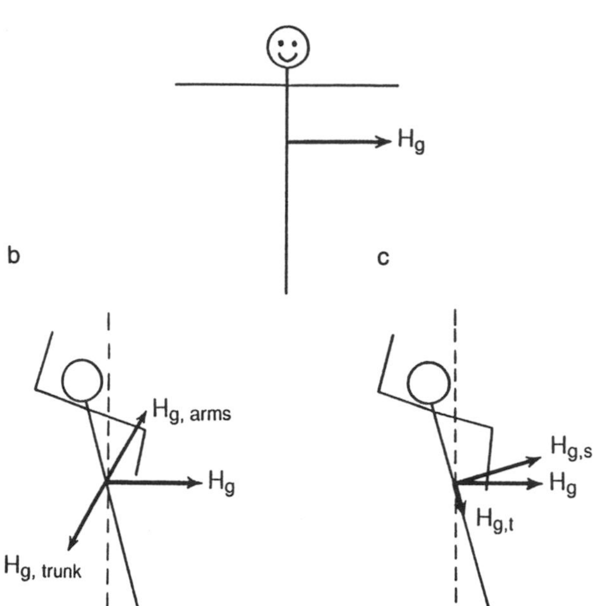

Figure 3.18 Initiation of a twist during a dive.

time taken to perform the dive. The values for these variables are

$$\mathbf{H}_g = 30 \text{ kg}\cdot\text{m}^2/\text{s}$$

$$I_{g,s} = 14 \text{ kg}\cdot\text{m}^2$$

$$I_{g,t} = 1 \text{ kg}\cdot\text{m}^2$$

$$t = 1.5 \text{ s}$$

$$H_{g,t} = I_{g,t}\omega_t$$

$$30 \sin\theta = 1 \times \frac{6.28}{1.5}$$

$$\theta = \sin^{-1}\frac{4.19}{30}$$

$$\theta = 0.14 \text{ rad}$$

Given these initial conditions, the diver will need to tilt by 0.14 rad to complete a full twist in the dive. How much rotation about the somersault axis would accompany a tilt of 0.14 rad? The calculations are as follows:

$$H_{g,s} = I_{g,s}\omega_s$$

$$30 \cos 0.14 = 14\ \omega_s$$

$$29.7 = 14\ \omega_s$$

$$\omega_s = 2.12 \text{ rad/s}$$

$$\frac{\Delta\theta}{t} = 2.12 \text{ rad/s}$$

$$\Delta\theta = 3.18 \text{ rad}$$

Thus, a tilt angle (θ) of 0.14 rad will result in a full-twisting, one-half (3.18 rad) somersault dive.

EXAMPLE 3

The type of angular momentum analysis described in the previous two examples has limited application in the study of human movement, because such analysis focuses on the changes about the center of gravity when the body is in a relatively stationary layout position. For most movements, however, the body segments participate in more substantial movements relative to one another. These conditions require a **linked-system analysis**, in which we calculate the angular momentum of each segment about its center of gravity (local angular momentum) and then determine the angular momentum of the center of gravity for each segment about the system (whole body) center of gravity (remote angular momentum). This relationship has the following form:

$$\begin{aligned}\mathbf{H}^{S/CS} &= \mathbf{H}^{B1/C1} + \mathbf{H}^{B2/C2} \\ &+ \mathbf{H}^{B3/C3} + \ldots \quad \text{(local terms)} \\ &+ \mathbf{H}^{C1/CS} + \mathbf{H}^{C2/CS} \\ &+ \mathbf{H}^{C3/CS} + \ldots \quad \text{(remote terms)}\end{aligned} \tag{3.18}$$

where

$\mathbf{H}$ = angular momentum
$B1$ = body segment 1
$C1$ = center of gravity for segment 1
CS = center of gravity for the system
S = system
/ = with respect to

Local terms have the form $I_g\omega$, whereas remote terms consist of the vector (cross) product $\mathbf{r} \times m\mathbf{v}$, which represents the moment of linear momentum relative to the system center of gravity. Figure 3.19 shows an example of the time course of angular momentum of the whole body as determined by a linked-system analysis (Hay, Wilson, Dapena, & Woodworth, 1977). In this example, a gymnast performed a vault that included periods where he contacted either the ground or the vaulting horse and times when he was not in contact with either. As we would expect from our discussion of impulse and momentum, the angular momentum of the gymnast changed when he contacted his surroundings and remained constant during the flight phases. The change in angular momentum, of course, must have been due to an angular impulse experienced by the gymnast.

Miller and Munro (1985) analyzed body segment contributions to performance of a springboard dive. They calculated local and remote angular momentum values for each body segment during the takeoff phase

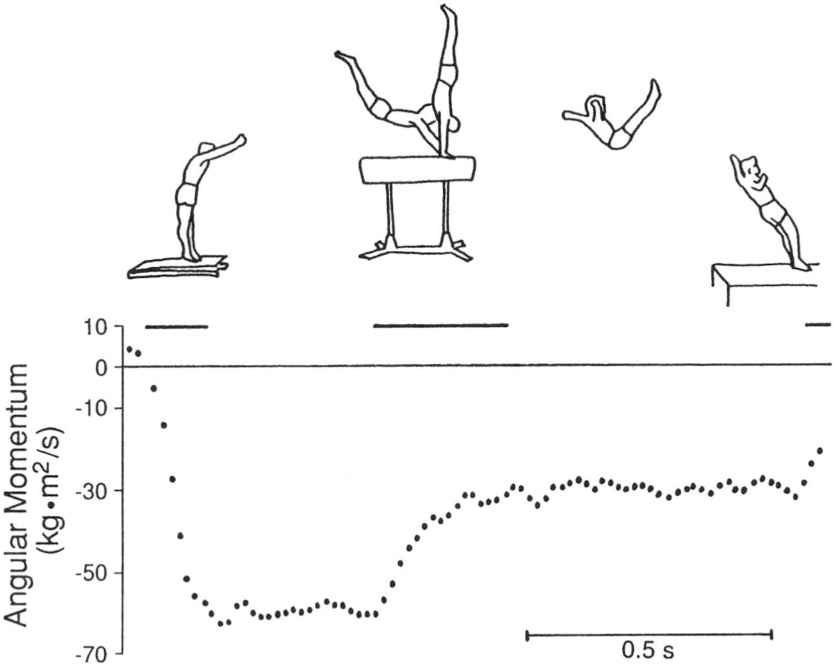

Figure 3.19 Angular momentum of a gymnast during a vault. The lines at the top of the graph indicate when the gymnast was in contact with the ground or the vaulting horse.
Note. From "A Computational Technique to Determine the Angular Momentum of a Human Body" by J.G. Hay, B.D. Wilson, J. Dapena, and G.G. Woodworth, 1977, *Journal of Biomechanics*, **10**, p. 271. Copyright 1977 by Pergamon Press. Reprinted by permission.

of various dives performed by Greg Louganis. The dives ranged from forward to reverse somersault and from a straight (layout) dive to multiple somersaults (up to three and a half). Because angular momentum is conserved during the flight phase of a dive, the quantity of angular momentum generated during the takeoff phase determines the performance options available to the diver. From 79% to 93% of the angular momentum needed for each dive (–70 to 57 kg·m^2/s) was obtained from remote angular momentum; that is, it was due to the motion of the body segments about the center of gravity of the diver. The contribution from local angular momentum was more substantial in forward dives than in reverse dives. And the contribution of the body segments differed across the dives, with distinctions between forward- and reverse-somersault dives and between straight and multiple-somersault dives.

EXAMPLE 4

The impulse-momentum relationship can also be used to explain the controversial air-righting reaction, the so-called cat twist—the ability of cats (and other animals) to land on their feet when dropped from an upside-down position. This is an interesting issue because it raises the question of how a cat can "acquire" angular momentum and perform the twist if there is no initial angular momentum and no angular impulse to cause a change in angular momentum. Biomechanists cannot agree on how the cat does this. One explanation (Hay, 1978) suggests that the twist can be performed by sequentially altering the moment of inertia of different body parts during the fall (Figure 3.20). In this scheme, the cat first rotates the upper body by bringing the forelimbs close to the trunk to reduce the moment of inertia about the longitudinal axis. In reaction to the upper body rotation, the lower body rotates in the opposite direction (conservation of angular momentum). But because the lower limbs are extended, the moment of inertia is large and the associated displacement of the lower body is much less than that of the upper body in the opposite direction. The cat then performs the converse, rotating the lower body with a low moment of inertia (larger displacement) and eliciting a reaction of the upper body, where the moment of inertia is large and hence the displacement is small. In the final phase, both sets of limbs are extended to maximize the moment of inertia and to minimize the angular velocity. This explanation is based on the conservation of angular momentum and relies on the differential alteration in the upper and lower body moment of inertia.

The other explanation is based on a muscle contraction-induced conservation of angular momentum for the upper and lower body (Hopper, 1973). As shown in Figure 3.21a, the cat is modeled as a two-segment system that comprises an upper (G_2) and lower (G_1) body that are linked by a set of muscles (*PQ*). In this scheme, the twist is initiated by a contraction of the linking muscles (*PQ*), which causes the two segments (G_1 and

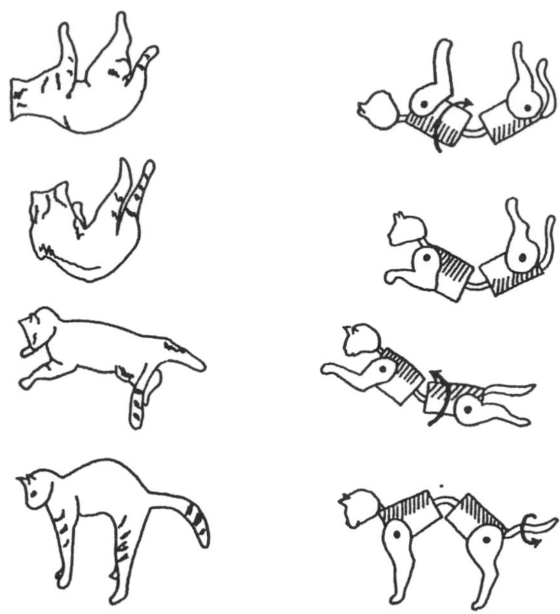

Figure 3.20 The air-righting reaction in which a falling cat twists by 3.14 rad and lands on its feet.
Note. From *The Biomechanics of Sports Techniques* (p. 158) by J.G. Hay, 1978, Englewood Cliffs, NJ: Prentice Hall. Copyright 1978 by Prentice Hall, Inc. Reprinted by permission.

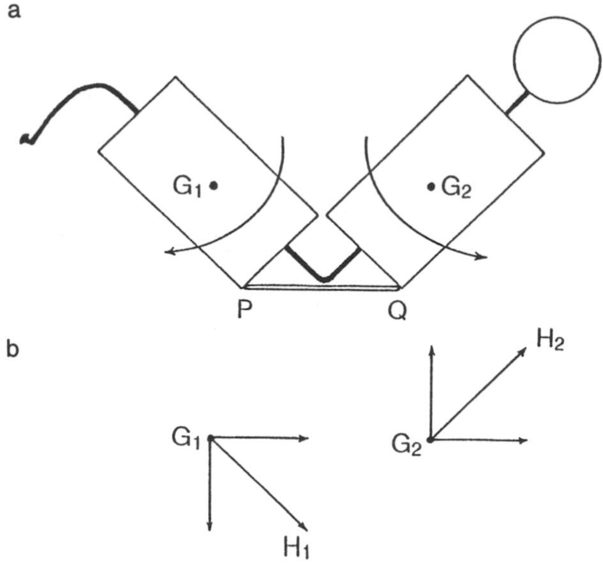

Figure 3.21 Model of a cat performing the air-righting reaction.
Note. From *The Mechanics of Human Movement* (p. 167) by B.J. Hopper, 1973, New York: Elsevier. Copyright 1973 by B.J. Hopper. Adapted by permission.

G_2) to rotate (as indicated by the arrows) about their respective longitudinal axes. These rotations can be represented by angular momentum vectors, one for each segment ($\mathbf{H}_1$ and $\mathbf{H}_2$), as shown in Figure 3.21b. By the right-hand rule, $\mathbf{H}_1$ is directed diagonally downward and $\mathbf{H}_2$ is directed diagonally upward. Because $\mathbf{H}_1$ and $\mathbf{H}_2$ are directed diagonally, they have components in both the horizontal and vertical directions. The angular momentum components in the vertical direction cancel, but those in the horizontal direction sum to nonzero and indicate a positive angular momentum to the right, which reveals a rotation that will enable the cat to twist and to land on its feet.

EXAMPLE 5

Consider the ability to walk and run, a fundamental motor skill. Although the objective is simply translating (forward-backward direction) the center of gravity, this is accomplished by the coordinated angular displacement of the body segments. The net result of the segmental motion is that the body also experiences translation in the upward-downward and side-to-side directions and rotations about the somersault, cartwheel, and twist axes. Perhaps the most obvious of these motions are the forward-backward translation of the center of gravity, the rotation of the body segments about their somersault axes, and the rotation of the upper and lower body about the twist axis. The motion of the runner about the twist axis can be modeled as a two-segment system comprising the upper and lower body (Figure 3.22a). When the foot pushes against the ground during running, the ground responds with a ground reaction force that, over time, applies an angular impulse to the body about the twist axis. To prevent whole-body angular motion (nonzero angular momentum) about the twist axis, the runner responds to the angular impulse provided by the ground reaction force to the lower body by contracting the trunk muscles and generating an opposing angular momentum for the upper body; that is, the arms and trunk rotate in the opposite direction to the legs. This interaction is shown in Figure 3.22b: For the first 40% of the stride, the left foot is in contact with the ground, producing a positive angular momentum ($\mathbf{H}_z$) for the lower body while concurrently the upper body rotates with a negative $\mathbf{H}_z$. In the second part of the stride, the converse occurs while the right foot is in contact with the ground. The primary function of arm motion during running, therefore, is to counteract the angular momentum of the legs about the twist axis (Hinrichs, 1987; Hinrichs, Cavanagh, & Williams, 1987).

Work-Energy Approach

The third type of analysis, the work-energy approach, addresses *the ability to move objects*. If an object or system is moved some distance as the result of a force,

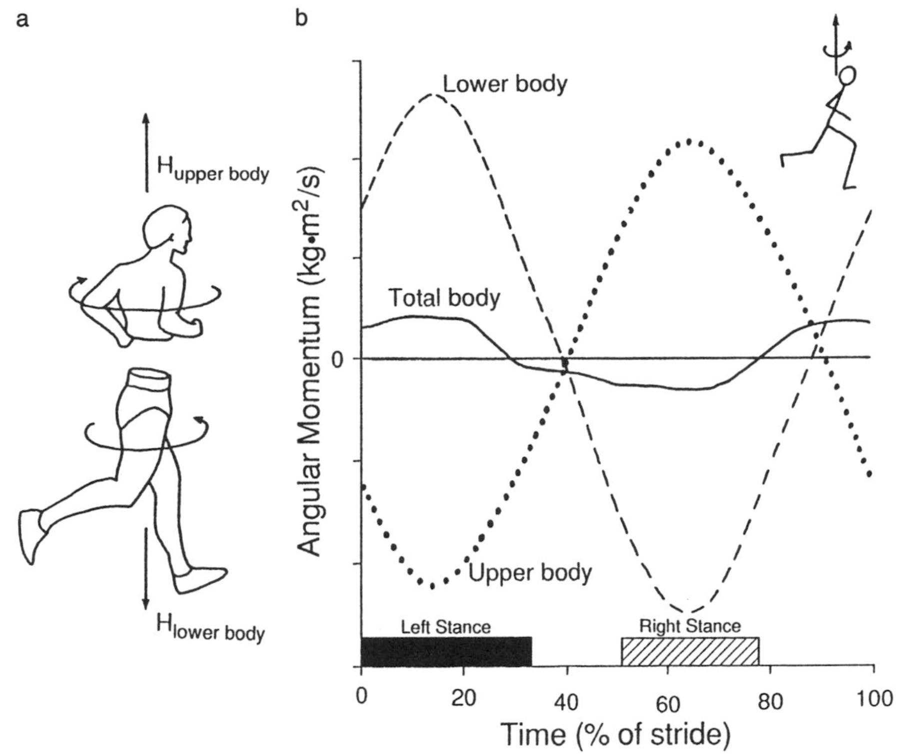

Figure 3.22 Angular momentum about the twist axis during a single running stride: (a) two-segment model of a runner; (b) angular momentum for the upper body, lower body, and total body.

Note. From "Upper Extremity Function in Running. II: Angular Momentum Considerations" by R.N. Hinrichs, 1987, *International Journal of Sport Biomechanics*, **3**, pp. 259 and 261. Copyright 1987 by Human Kinetics Publishers, Inc. Adapted by permission.

work is performed on it. Work (U) is a scalar quantity; it is calculated as the product of the displacement experienced by the system and the component of the force acting in the direction of the displacement. Work can be represented graphically as the area under a force-displacement (distance or length) curve. For example, a person rehabilitating from knee surgery uses a machine that can measure the force applied to a lever (e.g., Cybex). Figure 3.23 depicts the force exerted by the leg against the lever of the machine over a 1.57-rad range of motion. The work ($\bar{F} \cdot d$) performed by the leg on the machine can be calculated by measuring the area under the force-displacement curve. If the average force applied to the lever is 50 N·m, then work can be determined as follows:

$$\text{work} = \text{force} \times \text{displacement}$$

$$\bar{F} \cdot d = 50 \text{ N·m} \times 1.57 \text{ rad}$$

$$= 78.5 \text{ J}$$

where the quantity of work is measured in joules (J; 1 J = 1 N·m).

The work-energy principle states that the work (output) done by a system is equal to the change in energy (input) of the system. In other words, *the performance of work requires the expenditure of energy*. In a work-energy analysis, the difference in system energy between two positions represents the work done on or by the system. If energy is transferred from the system to its surroundings, the work is done by the system. If the energy transfer proceeds in the opposite direction, work is done by the surroundings on the system. For example, when a person lifts a barbell by using the elbow flexor muscles, the muscles shorten (concentric contraction) and perform work on the barbell, a performance that physiologists refer to as **positive work**. When the person lowers the barbell, however, the activated elbow flexor muscles are lengthened (eccentric contraction), and the barbell does work on the muscles; this energy transfer is known as **negative work**. The metabolic energy required to perform negative work is much less than that required to perform an equivalent amount of positive work.

According to the work-energy principle, energy can be described as the capacity to do work and can be regarded as a measure of the fuel available to the system for the performance of work. In fact, many processes or changes that occur in nature can be considered in terms of the transformation of energy from one form to another. In the analysis of human movement, the

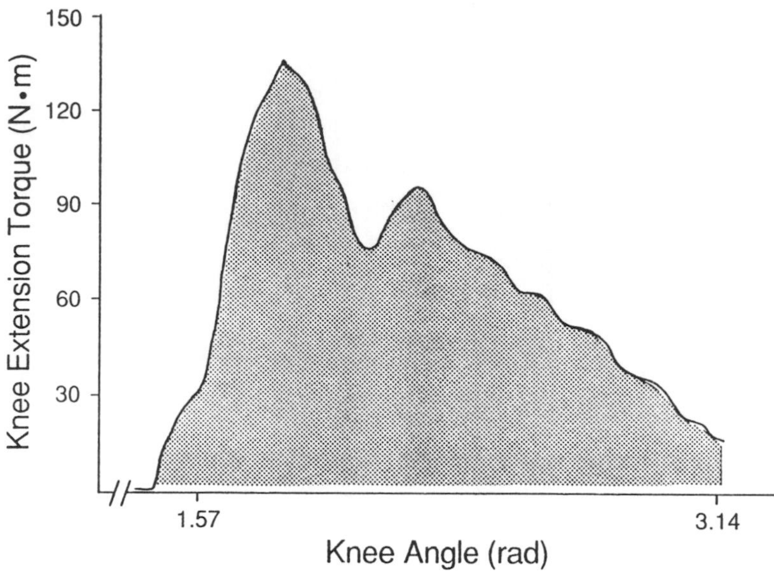

Figure 3.23 Idealized force-displacement curve for a single leg extension exercise, as measured with a Cybex II isokinetic dynamometer.

significant forms of energy include potential energy due to gravity ($E_{p,g}$), translational (linear) kinetic energy ($E_{k,t}$), rotational kinetic energy ($E_{k,r}$), potential energy due to strain ($E_{p,s}$), energy that is degraded to heat (E_h), and liberated metabolic energy (E_m). The relation between work and energy can be stated as

$$\text{work} = \Delta \text{ energy}$$

$$U = \Delta E \tag{3.19}$$

$$U = \Delta E_{p,g} + \Delta E_{k,t} + \Delta E_{k,r} + \Delta E_{p,s} + \Delta E_h + \Delta E_m$$

However, in the analysis of whole-body motion there are, as yet, no adequate descriptions of $\Delta E_{p,s}$ or ΔE_h (van Ingen Schenau & Cavanagh, 1990).

In the study of the work-energy relationship, it is often convenient to characterize a process by expressing the amount of energy used to perform a given quantity of work. This characterization has as its basis the **first law of thermodynamics**, which describes the association between the performance of work and the change in energy. This energy balance can be represented as

$$\Delta E_m = \Delta E_h + U \tag{3.20}$$

In biological systems, the exchange of energy for work done is not a completely efficient process, and as a result not all of the energy transformation is associated with the performance of work. As indicated in Equation 3.20, some of the metabolic energy (ΔE_m) is degraded to heat (ΔE_h). Muscle is often described as having an efficiency of about 25% (de Haan, van Ingen Schenau, Ettema, Huijing, & Lodder, 1989; van Ingen Schenau & Cavanagh, 1990), which means that 25% of the liberated metabolic energy is available to perform work, and the remaining 75% is converted into heat or used in recovery processes. The ratio of the work done to the change in energy describes the **efficiency** of the process. When work is accomplished with a minimum expenditure of energy, this represents the most economical performance of the task. The terms *efficiency* and *economy* are not synonyms. Figure 3.24 shows the distinction: Efficiency corresponds to the constant-energy condition, and economy is associated with the constant-work condition. The most efficient performance (point 2) occurs when the greatest amount of work is done for a given change in energy. The most economical performance (point 3) occurs when the smallest change in energy is used to perform the work.

When work is done on an object to counteract the effect of another force, the object is described as possessing **potential energy**. Once the force maintaining the object in the new position is removed, the object will tend to return to its original position and, in the process, will perform work. Thus the energy that an object has due to its position is referred to as potential energy (i.e., the object has the potential to do work). Analysis of human movement commonly involves two types of potential energy: gravitational energy and elastic, or strain, energy. Gravitational energy ($E_{p,g}$) is due to the location of the object or system in a gravitational field above some baseline (e.g., the height of an object above the ground). Elastic, or strain, energy ($E_{p,s}$) is due to the stretch of an object (e.g., tennis racket) or tissue (e.g., muscle) beyond its resting length. The magnitude of the gravitational potential energy can be determined by

$$E_{p,g} = mgh \tag{3.21}$$

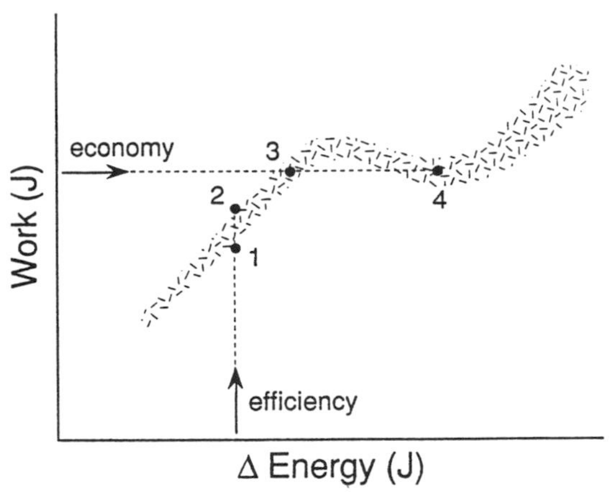

Figure 3.24 An arbitrary work-energy relationship for a task.

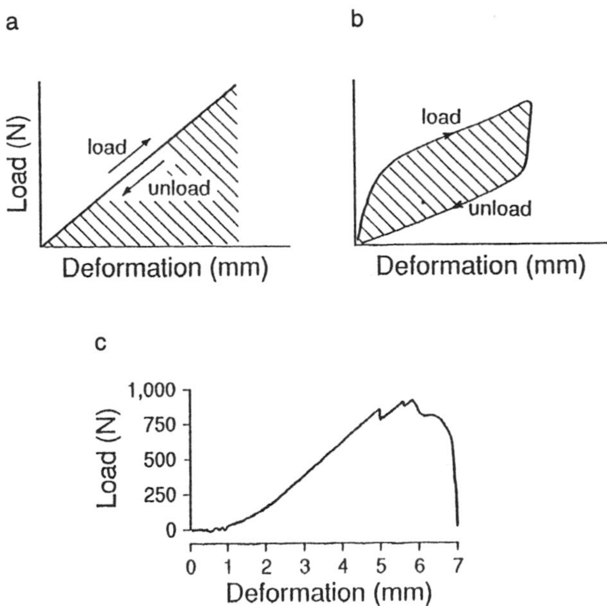

Figure 3.25 The load-deformation relationship of a linear spring (a), a viscoelastic element (b), and an anterior cruciate ligament (c).

Note. Panel c is from "Functional Properties of Knee Ligaments and Alterations Induced by Immobilization" by F.R. Noyes, 1977, *Clinical Orthopaedics*, **123**, p. 216. Copyright 1977 by J.B. Lippincott Company. Adapted by permission.

where m represents mass, g indicates the acceleration due to gravity (9.8 m/s^2), and h is the height of the system above the baseline (e.g., ground). Because g is constant, $E_{p,g}$ can be changed only by varying the mass of the system or its height above baseline.

Strain, or elastic, energy represents the energy a system possesses because of its tendency to return to its prestretch position. **Strain energy** represents a restorative capability of elastic material. In some analyses of human performance, consideration of the strain-energy capability of an object (e.g., diving board, archery bow, tennis racket) or tissue (e.g., muscle, connective tissue) is essential. Biological tissue can sustain small stretches without damage and can perform work in the process of returning to the initial length. Whether or not the tissue recovers its prestrain form depends on the magnitude of the stretch or strain (*the increase in length relative to an initial length*). If the magnitude is small and the tissue can return to its initial length, the strain occurred within the elastic capabilities of the tissue. But if the amount of the strain exceeds the elastic capability of the material, and the tissue cannot recover its prestrain length, the strain will induce plastic changes in the structure of the tissue. Figure 3.25c shows the effect of increasing load on the deformation of an anterior cruciate ligament; when the stretch exceeds about 5 mm, there are structural changes in the ligament. When people use stretching techniques to increase muscle length, they are attempting to cause plastic changes in the connective tissue (Sapega, Quedenfeld, Moyer, & Butler, 1981).

The amount of strain experienced by a system in response to an applied force depends on its material properties. For a perfectly elastic system, such as an ideal spring, the change in length (x) is proportional to the applied force (**F**):

$$\mathbf{F} = kx \tag{3.22}$$

Equation 3.22 describes a linear relationship (Figure 3.25a) in which the change in length (deformation) increases with the magnitude of the applied force (load). The slope of the load-deformation relationship, which is referred to as the spring stiffness, is defined by k; the slope of the relationship increases for greater values of k, which indicates a stiffer spring. The amount of energy stored in a stretched spring ($\Delta E_{p,s}$) is given by integrating Equation 3.22, which yields

$$\Delta E_{p,s} = \frac{1}{2}kx^2 \tag{3.23}$$

For a spring that is stretched within its elastic limits, all the energy that is stored in the spring when it is stretched can be used to perform work when the spring returns to its initial length; that is, no energy is lost in the load-unload cycle (shaded area in Figure 3.25a). This is illustrated in Figure 3.25a by the superimposed lines (arrows) for the loading and unloading conditions.

In contrast to a spring and its linear elastic characteristics, muscle and connective tissue are best characterized as viscoelastic elements; that is, they exhibit both elastic and viscous properties. The load-deformation relationship of a viscoelastic element is not linear, it contains a **hysteresis loop** (Figure 3.25b), which indicates that the force exerted by the element is different during the loading and unloading phases. The significance of a hysteresis loop is that some energy is lost in the process of loading and unloading, so not all of the energy stored in the system during loading (stretching) is available to do work during the unloading (shortening) phase. This energy loss is shown in Figure 3.25b as the shaded area. Nonetheless, as viscoelastic elements, muscle and connective tissue can store energy when they are stretched and use this energy to do work when they subsequently return to their initial length (Cavagna & Citterio, 1974; Shorten, 1987).

The capacity of an object to perform work because of its motion is referred to as **kinetic energy**. An object acquires kinetic energy when a force does work against its inertia and changes its motion. Kinetic energy (E_k) consists of both linear and angular terms. Its magnitude can be determined by

$$E_k = E_{k,t} + E_{k,r} \tag{3.24}$$

$$E_k = \frac{1}{2}mv^2 + \frac{1}{2}I\omega^2$$

where m and I represent the mass and moment of inertia of the system, respectively, and v and ω indicate its linear and angular velocity. The quantity of E_k possessed by a system is most influenced by its velocity, because this term is squared. In the analysis of human movement, the angular term is often most significant for the limbs, and the linear term dominates for the trunk and head.

The term E_m (see Equation 3.19) is intended to account for the proportion of the food that we eat that provides *metabolic energy* for biological functions. In human performance, this energy serves two general functions. It provides energy (ATP) for muscle contractions, and it provides energy for sustaining the physiological activity of the cardiorespiratory system, the nervous system, and the muscular activity associated with the maintenance of posture. Exercise physiologists often determine the metabolic energy used in human performance by measuring the oxygen consumed during the activity (Daniels, 1985; di Prampero, 1986; Joyner, 1991), especially if the activity involves large muscle groups and is sustained for at least a few minutes. Based on the performance of a 70-kg, 175-cm male subject, Table 3.1 compares the energy cost for aerobic events of comparable duration (di Prampero, 1986). The average speed achieved during an event is inversely proportional to its energy cost per unit of distance (km).

Table 3.1 World-Record Times (1986) and Average Speeds for Aerobic Events of Comparable Duration

Event	Distance (km)	Time (s)	Speed (m/s)	Energy cost (kJ/km)
Swimming (freestyle)	1.5	895	1.68	1,400
Race walking	3.0	788	3.81	356
Running	5.0	780	6.41	300
Speed skating (ice)	10.0	874	11.44	175
Cycling	10.0	713	14.02	160

Note. From "The Energy Cost of Human Locomotion on Land and in Water" by P.E. di Prampero, 1986, *International Journal of Sports Medicine*, **7**, 55-72.

EXAMPLE 1

If we assume that the universe is a closed system, then we can apply the concept that energy is not created or destroyed but rather is conserved. Therefore, when energy is used to perform work, it is not lost but changed into another form. For example, consider a 60-kg (mass) acrobat about to leap off a 10-m tower. The acrobat has a potential energy of 5.88 kJ. The instant the acrobat leaves the tower, there will be no supporting surface to provide a ground reaction force, and because of gravity the acrobat will fall. Mechanically, this process can be explained as a conversion of energy from one form to another. The fall of the acrobat can be described as a change of energy from potential ($E_{p,g}$) to kinetic ($E_{k,t}$). Because the acceleration due to gravity is constant, the speed of falling increases with the distance covered in the fall. This correlates with an increasing conversion from potential to kinetic energy.

Consider the system energy after the acrobat has fallen 4 m. At this position, the potential energy of the acrobat is 3.53 kJ, a decrease of 2.35 kJ. The kinetic energy ($E_{k,t}$) of the system can be determined by the equation

$$E_{k,t} = \frac{1}{2}mv^2$$

To calculate the kinetic energy of the acrobat after falling 4 m, we need to determine the acrobat's velocity (v) at that point. Velocity can be calculated by using Equation 1.4 to determine the time (0.9045 s) it took the acrobat to fall 4 m and then using the definition of acceleration:

$$a = \frac{\Delta v}{\Delta t}$$

$$a = \frac{v_f - v_i}{t}$$

$$9.81 = \frac{v_f - 0}{0.9045}$$

$$v_f = 8.86 \text{ m/s}$$

where v_i and v_f refer to the initial (at the beginning of the fall) and final (at 4 m) velocities, respectively. The final velocity value (8.86 m/s) can then be used in Equation 3.24 to determine the acrobat's kinetic energy at 4 m.

Alternatively, because $E_{k,t} = \frac{1}{2}mv^2$ and we know how much $E_{p,g}$ has been reduced (2.35 kJ) due to the law of conservation of mechanical energy $[(E_{p,g} + E_{k,t})_f = (E_{p,g} + E_{k,t})_i]$, we can determine v by rearranging this relation as follows:

$$E_{k,t} = \frac{1}{2}mv^2$$

$$v = \sqrt{\frac{2 \times E_{k,t}}{m}}$$

$$v = \sqrt{\frac{2 \times 2{,}350}{60}}$$

$$v = \sqrt{78.33 \text{ m}^2/\text{s}^2}$$

$$v = 8.85 \text{ m/s}$$

(Note that $\frac{\text{J}}{\text{kg}} = \frac{\text{N·m}}{\text{kg}} = \frac{\text{kg·m·m}}{\text{kg·s}^2}$)

Using either method, we obtain a final velocity of about 8.85 m/s. Thus the kinetic energy changed from zero as the acrobat began to fall to a value of 2.35 kJ after 4 m, an increase that matched the decrease in potential energy.

EXAMPLE 2

Although the first law of thermodynamics provides a simple formal statement of the energy balance associated with the performance of work, it is difficult to apply this law to such complex tasks as human locomotion (Tucker, 1975; Ward-Smith, 1983). Figure 3.26 represents one such effort, a flow diagram of the energy exchanges that occur during sprinting (Ward-Smith, 1984). In this scheme, ATP (chemical energy) is provided by metabolic processes to the muscles for the performance of work. As mentioned, some of the ATP is converted to heat ($H1$). Although the net result is the performance of work by the runner on the environment, this is accomplished by the angular motion of the body segments. Accordingly, the work (positive and negative) done by the muscles provides kinetic (both translational and rotational) and potential energy to the limbs. In the process, however, some of the mechanical energy is degraded to heat ($H2$) due to frictional effects within joints and muscles. The mechanical energy (kinetic and potential) acquired by the limbs is transferred into potential and kinetic energy of the runner's center of gravity and into work against air resistance ($W3$). Functionally, the mechanical energy of the center of gravity can be partitioned into two effects: (a) $W1$, the addition of kinetic energy to its horizontal component, and (b) $W2$, an addition to the vertical component, which amounts to doing work against gravity. The kinetic and potential energy of the limbs and of the center of gravity moving against gravity ($W2$) results in the storage of strain

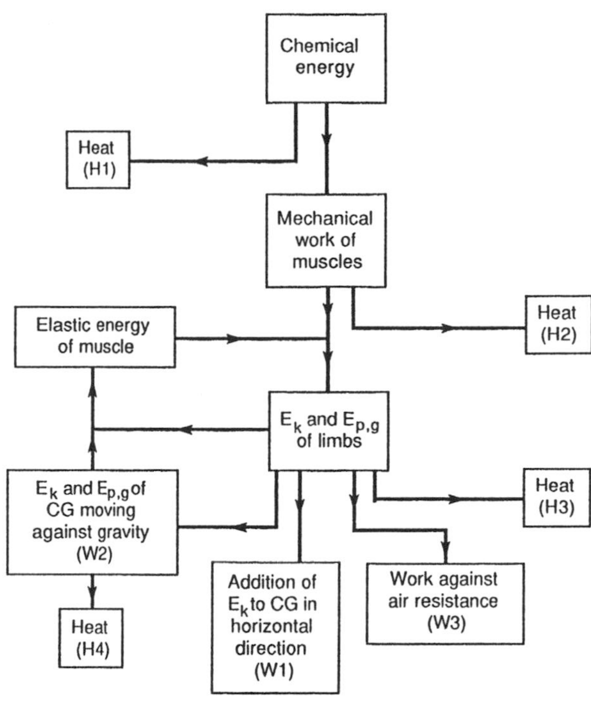

Figure 3.26 Flow diagram of the energy exchanges that occur during running.
Note. Adapted from "Air Resistance and Its Influence on the Biomechanics and Energetics of Sprinting at Sea Level and at Altitude" by A.J. Ward-Smith, 1984, *Journal of Biomechanics*, **17**, p. 340, Copyright 1984, with kind permission from Pergamon Press Ltd., Headington Hill Hall, Oxford 0X3 OBW, UK.

(elastic) energy in the muscles (including tendons) that can be recycled. This phenomenon is referred to as the storage and utilization of elastic energy, or the stretch-shorten cycle. Over a number of strides, the net vertical displacement of the runner's center of gravity is zero, and thus no work is done against gravity. Consequently, the ATP supplied to the muscles is ultimately converted into heat ($H1 + H2 + H3 + H4$) or used to do work ($W1 + W3$). The process of running then, at least in terms of the work-energy relationship, can be described as a repetitive sequence of Equation 3.20.

EXAMPLE 3

The speed that a swimmer can achieve depends on how much energy is available to perform the task and on how much work the swimmer can perform for a given amount of energy. Most of the work done by a swimmer is used to overcome drag. The energy cost for world-record performances in the front crawl ranges from 1,900 kJ/km for the 100-m event to 1,170 kJ/km for the 1,500-m event. The energy cost at a given swimming speed is least for the front crawl, slightly greater for the back stroke, and much greater for the breast stroke (di Prampero, 1986). The quantity of mechanical work done in relation to the energy expended is referred to as efficiency. Swimming efficiency, which ranges from 1% to 10%, is influenced by such variables as skill, stroke, buoyancy, and velocity. Efficiency can be estimated from the following equation:

$$\text{efficiency} = \frac{\text{work output}}{\text{energy input}} \times 100$$
$$= \frac{F \times d}{\text{VO}_2} \times 100$$

Because it is usually more convenient to measure the rate at which work is done (i.e., power), this relationship is also commonly expressed as

$$\text{efficiency} = \frac{\text{work output / time}}{\text{energy input / time}} \times 100$$
$$= \frac{(F \times d) / t}{\text{VO}_2 / t} \times 100$$
$$= \frac{F \times v}{\dot{\text{V}}\text{O}_2} \times 100$$

where $\dot{\text{V}}\text{O}_2$ is the rate of oxygen consumption, v is the velocity of the swimmer, and F represents the horizontal component of the propulsive force exerted by the water on the swimmer. When swimming speed is constant, F is equivalent to drag.

EXAMPLE 4

When a person drops onto the ground from a low height, the person must perform work on the ground in order to reduce the $E_{k,t}$ to zero. This can be accomplished by using various landing strategies (McNitt-Gray, 1991), such as a soft or a stiff landing; the stiff landing involves a more erect posture and less flexion at the hip and knee joints. DeVita and Skelly (1992) examined this task (vertical drop of 59 cm) and found that the hip and knee muscles absorbed (negative work) more energy in the soft landing (hip, –0.60 vs. –0.39 J/kg; knee, –0.89 vs. –0.61 J/kg), whereas the ankle muscles absorbed more in the stiff landing (–0.88 vs. –1.00 J/kg). The muscles of the lower extremity absorbed 19% more of the body's kinetic energy during the soft landing, thereby reducing the impact stress on other body tissues. Most of the energy was absorbed by the plantarflexor muscles (44% of the total muscle work), followed by the knee extensors (34%) and the hip extensors (22%).

Power

In many short-duration events (e.g., sprinting, Olympic weight lifting, arm wrestling, vertical jump), the rate at which muscles can produce work, referred to as *power production*, is often the variable limiting performance. Power production is measured as the amount of work done per unit of time. The duration that an activity can be sustained is inversely related to the power requirements of the activity (Figure 3.27). **Power** can be determined as work ($F \cdot d$) divided by the amount of time (Δt), or as the product of force (F) and velocity ($\mathbf{v}$). Because distance is synonymous with displacement ($\Delta\mathbf{r}$) in this context, the two expressions can be demonstrated to be equivalent as follows:

$$\frac{F \cdot d}{\Delta t} = F \cdot \mathbf{v}$$

$$\frac{F \cdot \Delta\mathbf{r}}{\Delta t} = F \cdot \mathbf{v}$$

$$F \cdot \frac{\Delta\mathbf{r}}{\Delta t} = F \cdot \frac{\Delta\mathbf{r}}{\Delta t}$$

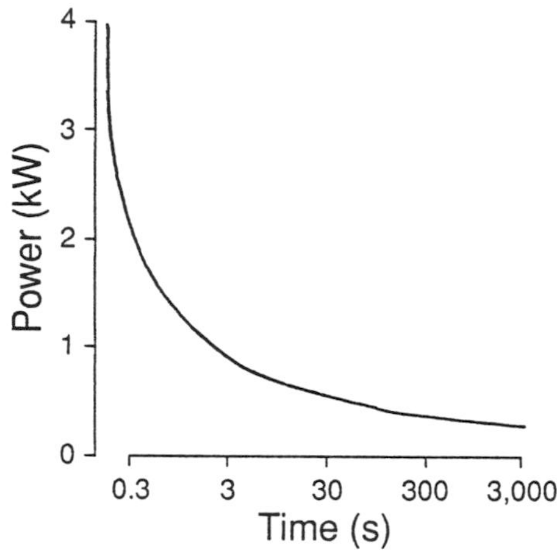

Figure 3.27 Maximum power production decreases with the duration of an activity.
Note. From "Measurement of Human Power Output in High Intensity Exercise" by H.K.A. Lakomy. In *Current Research in Sports Biomechanics* (p. 47) by B. VanGheluwe and J. Atha (Eds.), 1987, Basel: Karger. Copyright 1987 by S. Karger AG. Reprinted by permission.

As a measure of the rate of work performance, power is a scalar quantity that is measured in watts (W); 1 kW (1.36 horsepower) is the metabolic power corresponding to an O_2 consumption of about 48 ml/s. Because work represents a change in the energy of the system, power can also be considered *the rate of energy utilization.* There are at least three different ways of calculating total power production (DeLooze, Bussman, Kingma, and Toussaint, 1992).

EXAMPLE 1

Recall the elbow extension-flexion example (Figure 1.9) considered previously. Because power can be determined as the product of force and velocity and because acceleration and force are proportional, the power-time profile associated with the movement can be obtained by multiplying the velocity and acceleration curves. When velocity and acceleration have the same sign (positive or negative), *power is positive and represents an energy flow from the muscles to the arm*. Conversely, *when power is negative* (i.e., velocity and acceleration have opposite signs), *energy flows from the arm to the muscles*. These conditions are known as **power production** and **power absorption**, respectively, indicating energy flow from and to the muscles. With regard to the elbow extension-flexion movement, positive velocity (greater than zero) indicates elbow extension, positive acceleration represents an extension-directed force, and positive power (production) represents energy flowing from the appropriate muscles to the system (forearm-hand). The power-time curve is determined by multiplying the velocity and acceleration (force) graphs. Whenever the velocity or acceleration graph crosses zero (i.e., changes sign) so does the power curve. The resulting power-time graph for the elbow extension-flexion movement is depicted as a four-epoch event (Figure 3.28). The first two epochs, power production and absorption, respectively, occur during elbow extension (see the velocity-time graph) and represent periods of positive and negative work. During positive work, the muscles do work on the system; during negative work, the system (due to its inertia) does work on the muscles. A similar sequence (power production, then absorption) occurs during the flexion phase of the movement.

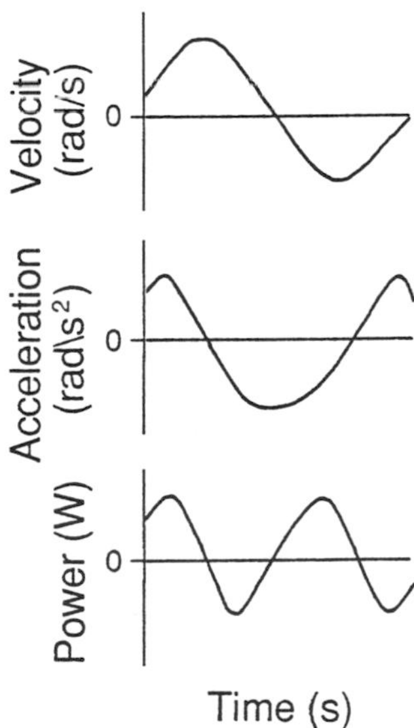

Figure 3.28 Power-time profile associated with the muscles across the elbow joint during the elbow extension-flexion movement shown in Figure 1.9. Positive velocity represents an extension movement; positive acceleration indicates an acceleration in the direction of extension; and positive power (production) refers to the power produced by the system.

This example emphasizes the correlation between the concepts of positive and negative work, power production and absorption, and concentric and eccentric muscle activity. As shown in the lower panel of Figure 3.28, the elbow extension-flexion movement is associated with a four-phase power-time profile. The epochs of power production correspond to periods of positive work and, therefore, concentric muscle activity. Recall that in chapter 1 we determined that this movement involved a sequence of activity: concentric extensor, eccentric flexor, concentric flexor, and eccentric extensor activity. According to this scheme, therefore, power absorption is related to negative work and eccentric muscle activity.

Such an analysis, however, is only a first approximation, because in reality not all of the power observed during the production phase (especially the second production phase) is produced by the use of E_m. Some power is probably due to the elastic capabilities of muscle and tendon ($E_{p,s}$). We shall consider this possibility in more detail in Part II.

EXAMPLE 2

The power-time data shown in Figure 3.28 are typically obtained by calculating the product of angular velocity and resultant muscle torque about a joint at selected intervals during a movement. This procedure, however, involves substantial data analysis because both velocity and torque have to be derived from measurements made during the movement. Angular velocity can be determined by obtaining a video or film record of the performance and then using the numerical analysis techniques outlined in Table 1.1. The number of velocity measurements depends on the frame rate set on the camera; for most human movements, rates of 50 to 200 frames per second are adequate. The resultant muscle torque can be estimated by performing a dynamic analysis and proceeding from some external measurement of force, such as the ground reaction force, backward through the system to the joint of interest (Figure 3.6). The result of the dynamic analysis is a set of instantaneous values for the resultant muscle torque throughout the movement (Figure 3.7). The number of torque values depends on the rate at which the computer sampled the ground reaction force signal; this usually varies from 500 to 1,000 Hz. The final step in calculating power is to match each angular velocity measurement with a value for resultant muscle torque (the closest one in time) and then determine the product. This calculation is done for each angular velocity value throughout the movement.

Huijing (1992) performed such an analysis on the final phase of the vertical jump (Figure 3.29). In this analysis, Huijing focused on the ankle joint and the power produced by the muscles that cross this joint. The results shown in Figure 3.29 are for the final 300 ms of the jump, when the subject went from the position of minimum knee angle to takeoff. At the minimum knee angle (–300 ms), the resultant muscle torque was

a

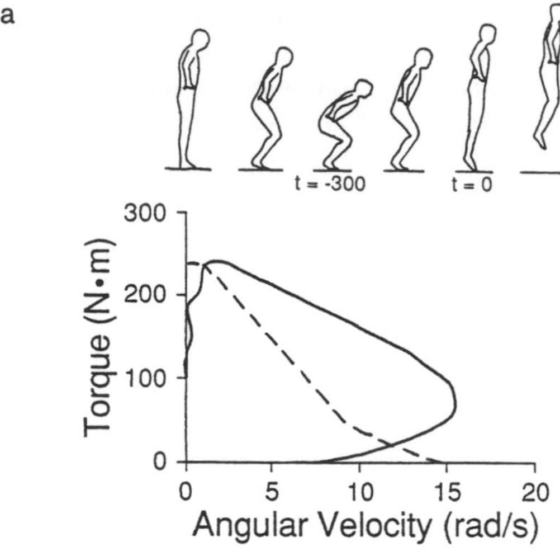

b

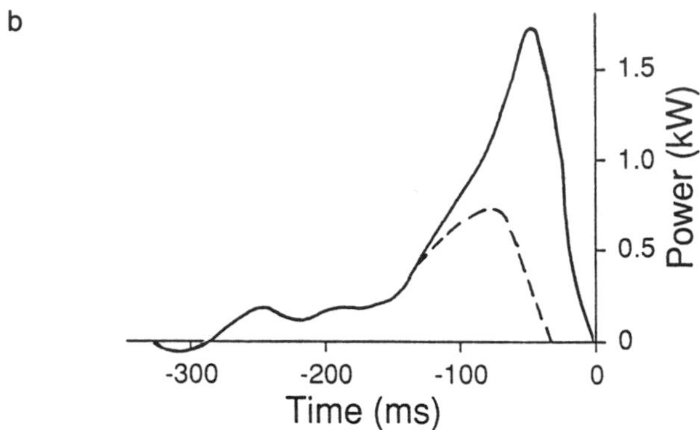

Figure 3.29 Power production at the ankle joint during a vertical jump: (a) resultant muscle torque about the ankle joint as a function of the angular velocity at the ankle joint (the theoretical torque-velocity relationship resembles the Hill relationship, which is shown by the dashed line); (b) power produced by the muscles that cross the ankle joint. *Note.* From "Elastic Potential of Muscle" by P.A. Huijing. In *Strength and Power in Sport* (p. 166) by P.V. Komi (Ed.), 1992, Oxford: Blackwell Scientific Publications, Ltd. Copyright 1992 by the International Olympic Committee. Adapted by permission.

about 100 N·m and the angular velocity was 0 rad/s; in this position the ankle joints were producing no power. From this initial position, the jump comprised three phases: For the next 150 ms (from −300 to about −150 ms), the resultant muscle torque increased to a maximum (240 N·m), and the angular velocity remained low (<2 rad/s), which meant the power production (positive) remained low—about 200 W. Then, although resultant muscle torque declined to about 75 N·m, both angular velocity (15 rad/s) and power production (1,700 W) increased to a maximum. Finally, angular velocity, resultant muscle torque, and power all declined to zero at takeoff. This example demonstrates that the maximum power a muscle can produce does not occur when the torque is maximum; it depends on the combination of torque and velocity and on velocity's having a significant effect on the product. This issue is examined in more detail in chapter 6.

Summary

This chapter uses the definitions provided in the first two chapters to describe the biomechanical techniques for studying movement. These techniques consist of three formulations of Newton's law of acceleration. The particular formulation chosen for an analysis depends on the details of the movement to be analyzed and the information desired from the analysis. The options are to focus on the *instantaneous* value of force (force-mass-acceleration approach), to examine the effect of force applied over an *interval of time* (impulse-momentum approach), or to consider the application of a force that causes an object to be displaced some *distance* (work-energy approach). These three approaches allow us to calculate force-time relationships, determine the effect of a force on the quantity of motion possessed by a system, and estimate the energy requirements of a task.

TRUE-FALSE QUESTIONS

1. The force-mass-acceleration approach allows us to determine instantaneous values of force. **TRUE FALSE**
2. The SI unit of measurement for force is kg. **TRUE FALSE**
3. A static analysis can be performed only on a system that is stationary. **TRUE FALSE**
4. In the expression ΣM_o, the abbreviation M refers to torque. **TRUE FALSE**
5. There are no forces acting on a system when it is in equilibrium. **TRUE FALSE**
6. The term *moment of force* is a synonym for *torque*. **TRUE FALSE**
7. The magnitude of the resultant force is always greater than any of the components. **TRUE FALSE**
8. A curved arrow means the vector is perpendicular to the page. **TRUE FALSE**
9. The symbol Σ stands for *change in*. **TRUE FALSE**
10. The direction of a force vector can be defined as an angle. **TRUE FALSE**

11. Provided that the moment arm is known, it is possible to calculate $\mathbf{F}_m$ by summing the moments about a joint. **TRUE FALSE**
12. The moment arm for a muscle with respect to a joint axis is usually less than 10 cm. **TRUE FALSE**
13. When calculating the location of the CG for a system, the *y*-coordinate is obtained from the expression ΣM_y. **TRUE FALSE**
14. $F_{m,n}$ is usually less than $F_{m,t}$. **TRUE FALSE**
15. The tangential component of the joint reaction force acts in parallel to the long axis of the body segment. **TRUE FALSE**
16. The joint reaction force is never greater than the weight of the system. **TRUE FALSE**
17. The location of the center of gravity can be determined by using a static analysis. **TRUE FALSE**
18. Intraabdominal pressure exerts a force that causes trunk flexion. **TRUE FALSE**
19. The moment of force can be measured in N·cm. **TRUE FALSE**
20. A system subjected to unbalanced forces will accelerate. **TRUE FALSE**
21. The symbol *I* represents inertia. **TRUE FALSE**
22. The term *quasistatic* refers to slow movements. **TRUE FALSE**
23. The unit of measurement for center of pressure is newtons. **TRUE FALSE**
24. A quasistatic analysis is appropriate for a slow, but not a rapid, performance of step initiation. **TRUE FALSE**
25. Inverse dynamics are used to calculate the forces from the kinematics of a movement. **TRUE FALSE**
26. The unit of measurement for ω is m/s. **TRUE FALSE**
27. The term $r\omega^2$ accounts for the change in magnitude of **v**. **TRUE FALSE**
28. Motion-dependent interactions refer to the effects of external forces (e.g., ground reaction force, air resistance) on the motion of the system. **TRUE FALSE**
29. The unit of measurement for the torque due to a motion-dependent effect is N·m. **TRUE FALSE**
30. Intersegmental dynamics are important in rapid finger movements. **TRUE FALSE**
31. Momentum is a vector quantity. **TRUE FALSE**
32. An impulse is apparent as the area under a force-time graph. **TRUE FALSE**
33. The unit of measurement for impulse is N·m. **TRUE FALSE**
34. An impulse causes a change in the momentum of a system. **TRUE FALSE**
35. The impulse-momentum relationship is derived from Newton's law of acceleration. **TRUE FALSE**
36. The product of mass and velocity is referred to as an impulse. **TRUE FALSE**
37. The abbreviation for linear momentum is **H**. **TRUE FALSE**
38. 1 J = 1 N·m. **TRUE FALSE**
39. A runner will increase speed when the braking and propulsive impulses are equal. **TRUE FALSE**
40. Angular impulse = $\Delta I_g\omega$. **TRUE FALSE**
41. The momentum of an object involved in a collision does not change before and after the collision. **TRUE FALSE**
42. Contact forces associated with kicking a ball or spiking a volleyball last about 0.2 s. **TRUE FALSE**
43. The coefficient of restitution for balls is always less than one. **TRUE FALSE**
44. The coefficient of restitution is dimensionless. **TRUE FALSE**
45. The mass of a baseball bat will influence the velocity of the ball and bat after a collision. **TRUE FALSE**
46. The unit of measurement for center of percussion is N. **TRUE FALSE**
47. Angular momentum is the moment of linear momentum. **TRUE FALSE**
48. The principle of conservation of momentum refers to conditions when a system has zero momentum. **TRUE FALSE**
49. The product of angular velocity and moment of inertia remains constant during the flight phase of a dive. **TRUE FALSE**
50. The moment of inertia of a diver is least about the somersault axis. **TRUE FALSE**
51. In a linked-system analysis, the angular momentum of a segment about its center of gravity is referred to as the remote angular momentum. **TRUE FALSE**
52. The main function of arm motion during running is to increase linear momentum in the horizontal direction. **TRUE FALSE**
53. Momentum defines the quantity of motion possessed by a system. **TRUE FALSE**

54. Work is a vector quantity. **TRUE FALSE**
55. The unit of measurement for work is N·m. **TRUE FALSE**
56. The performance of work requires the expenditure of energy. **TRUE FALSE**
57. When the surroundings do work on a system, this is referred to as negative work. **TRUE FALSE**
58. The unit of measurement for energy is J. **TRUE FALSE**
59. The first law of thermodynamics describes the balance between work and energy. **TRUE FALSE**
60. Potential energy has both linear and angular components. **TRUE FALSE**
61. An object has potential energy when work is done on it to alter its momentum. **TRUE FALSE**
62. Muscle can store strain energy. **TRUE FALSE**
63. Connective tissue loses some energy during loading and unloading so that it does less work on shortening than when it was stretched. **TRUE FALSE**
64. An object can possess both linear and angular kinetic energy. **TRUE FALSE**
65. The term $\dot{V}O_2$ represents the rate of energy input to the body. **TRUE FALSE**
66. The unit of measurement J/kg represents a measure of work normalized to body mass. **TRUE FALSE**
67. Power = work × velocity. **TRUE FALSE**
68. Angular power is easier to measure than linear power. **TRUE FALSE**
69. Power can be calculated as the rate of energy utilization. **TRUE FALSE**
70. Muscle performs an eccentric contraction when it produces power. **TRUE FALSE**
71. The energy used to produce power comes only from E_m. **TRUE FALSE**
72. The maximum power a muscle can produce occurs at the maximum torque. **TRUE FALSE**
73. The maximum power a muscle can produce is less than 1 kW. **TRUE FALSE**

MULTIPLE-CHOICE QUESTIONS

74. Rank the following four forces, from easiest to most difficult, in terms of how easy they are to measure:
    A. Joint reaction force
    B. Segment weight
    C. Ground reaction force
    D. Muscle force
75. What is the unit of measurement for moment of inertia?
    A. kg
    B. kg·m/s
    C. kg·m^2
    D. N·m/s
76. In which direction does the cartwheel axis pass?
    A. Back to front
    B. Head to toe
    C. Side to side
    D. Medial to lateral
77. Which statement or statements about the force-mass-acceleration approach to the study of human movement are incorrect?
    A. The mass acceleration diagram shows all the external forces.
    B. According to the law of acceleration ($\mathbf{F} = m\mathbf{a}$), 1 N = 1 kg·m/s^2.
    C. The right-hand side of the law of acceleration is nonzero in a dynamic analysis.
    D. A quasistatic analysis assumes that system acceleration is zero.
78. Which term best describes the technique used to determine the kinematics experienced by a system given a set of forces and system dimensions?
    A. Inverse dynamics
    B. Quasistatic
    C. Forward dynamics
    D. Residual analysis
79. Which motion-dependent effect does not exert a torque about the knee joint?
    A. $r\alpha$ for the thigh
    B. $r\omega^2$ for the thigh
    C. $r\alpha$ for the shank
    D. $r\omega^2$ for the shank
80. The change in velocity of object A with respect to object B due to a collision depends on which of the following?
    A. Relative mass of the two objects
    B. Initial velocity of both objects
    C. Angle between the two objects
    D. Initial momentum of the two objects

81. Identify the correct statements about the center of percussion.
   A. The center of percussion is located distal to the center of gravity on a baseball bat.
   B. No reaction force is felt at the hands when a ball contacts a bat at its center of percussion.
   C. A larger center of percussion will result in the ball being hit further with the bat.
   D. When a ball-bat contact occurs at the center of percussion, the bat will rotate about a stationary pivot point.

82. What is the unit of measurement of angular momentum?
   A. kg·rad/s
   B. $kg \cdot m/s^2$
   C. $kg \cdot m^2/s$
   D. kg·rad·m/s
   E. $kg \cdot rad/s^2$

83. What happens after a diver has left the board and before the diver contacts the water?
   A. Angular momentum is conserved.
   B. The moment of inertia is constant.
   C. No impulse acts on the diver.
   D. Linear momentum is conserved.

84. What is the unit of measurement for angular impulse?
   A. N·m·s
   B. $N \cdot m/s^2$
   C. $N \cdot m^2/s$
   D. N·m·rad/s

85. The impulse-momentum relationship describes which of the following?
   A. The effect of energy on the quantity of motion
   B. The need to apply a force to stop, start, or alter motion
   C. The change in the quantity of motion due to the application of a force over an interval of time
   D. The effect of the momentum of an object on its surroundings
   E. The change in the momentum of an object when a force displaces the object

86. Work can be represented as the area under which type of graph?
   A. Force-time
   B. Force-velocity
   C. Force-displacement
   D. Force-energy

87. In the analysis of whole-body motion, least is known about which of the following?
   A. $E_{p,g}$
   B. $E_{k,r}$
   C. $E_{k,t}$
   D. $E_{p,s}$

88. Which phrase best defines the term *efficiency*?
   A. The maximum amount of energy a system can produce
   B. The performance of work with the minimum expenditure of energy
   C. The maximum work done per unit of energy used
   D. The maximum amount of work a system can perform

89. Which of the following energy forms is not important in human movement?
   A. Translational kinetic energy
   B. Strain energy
   C. Heat
   D. Shear energy
   E. Chemical energy

90. What is the unit of measurement for strain?
   A. %
   B. mm
   C. rad
   D. J

91. Identify the incorrect statements regarding work and power.
   A. The performance of work requires the expenditure of energy.
   B. The unit of measurement for power is the watt.
   C. A muscle performs positive work when it does an eccentric contraction.
   D. A musclc absorbs power when it does positive work.
   E. Power can be calculated as the rate of change in energy.

PROBLEMS

92. The following systems are in equilibrium, acceleration is equal to zero, and thus

$$\Sigma F_x = 0$$

$$\Sigma F_y = 0$$

$$\Sigma \mathbf{M}_o = 0$$

Determine the magnitude of the unknown forces.

a

3 N
6 N
E
4 N
F

b

40 N
H
25 N
100 N
J

c

5 N
20 N
M
0.3 rad
L

d

5 N
3 m
10 m
A
B
S
T

93. Select the appropriate SI unit of measurement for each variable.

1. Potential energy	________	A.	kg·m/s
		B.	rad/s
2. Force	________	C.	N·s
		D.	N·m
3. Impulse	________	E.	W
		F.	kg
4. Angle	________	G.	kg·m²/s
		H.	N
5. Weight	________	I.	N/s
		J.	m/s²
6. Momentum	________	K.	J·s
		L.	J/s
7. Work	________	M.	rad
		N.	degree
8. Power	________	O.	J

94. Find the resultant elbow flexor force ($\mathbf{F}_m$) that is required to keep the forearm and hand in a horizontal position. The forearm and hand together weigh 30 N, and the center of gravity for the system is located 12 cm from the axis of the elbow joint (*E*). The resultant muscle force is applied 5 cm along the segment from *E* and pulls at an angle of 0.4 rad. If a 100-N weight is added to the system and held in the hand (25 cm from the elbow joint axis), what resultant muscle force ($\mathbf{F}_m$) is required to maintain equilibrium?

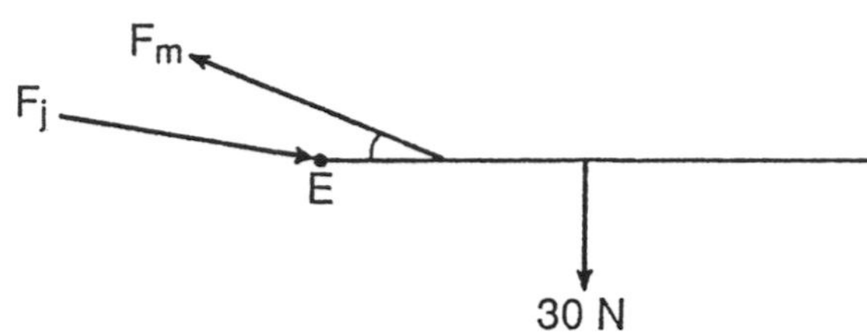

95. What resultant muscle force ($\mathbf{F}_m$), pulling at an angle of 0.6 rad to the forearm-hand and acting 5 cm from the elbow (*E*) joint axis, would be required to maintain the forearm-hand in a position 0.2 rad above the horizontal? The forearm-hand weighs 45 N, with the point of application of the weight vector located 10 cm from *E*. Before beginning the calculation, draw an accurate free body diagram of the system. How much of the resultant muscle force ($\mathbf{F}_m$) tends to cause rotation, and how much causes stabilization?

96. In this figure, an individual is shown performing a leg extension exercise on a Cybex machine.
    A. Draw and label a free body diagram that would enable you to calculate the net muscle torque about the knee joint.

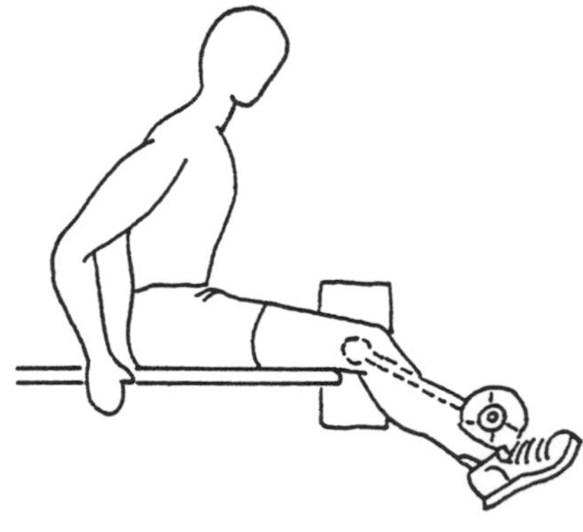

    B. Calculate the resultant muscle torque about the knee joint that the individual must exert to hold the shank in a position 0.24 rad below the horizontal. The following system dimensions apply.

body weight = 785 N

leg length (knee to ankle) = 39 cm

contact force (acts perpendicular to shank) = 972 N

contact force location = 31 cm from the knee

limb (shank + foot) CG = 41.3% of shank length from the knee joint

limb (shank + foot) weight = 0.045 $\mathbf{F}_w$ − 1.75

97. Borelli, one of the first biomechanists, observed in 1680 that a man can support a mass (R) of 20 libra (probably a weight of 67 N) from the tip of his thumb. What resultant muscle force ($\mathbf{F}_m$) must the muscles (flexor and extensor pollicis longus) across the distal phalangeal joint (P) exert to support the weight? The appropriate free body diagram is shown in Panel b, where $\mathbf{F}_w$ represents the phalanx weight (0.5 N) and $\mathbf{F}_j$, the joint reaction force.

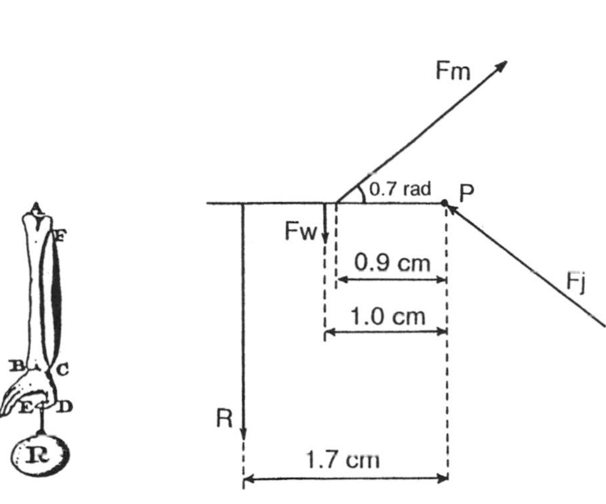

Note. From "Mechanics of Skeleton and Tendons" by R.M. Alexander. In *Handbook of Physiology: Sec. I, The Nervous System: Vol. II. Motor Control. Part I* (p. 28) by V.B. Brooks (Ed.), 1981, Bethesda, MD: American Physiological Society. Copyright 1981 by the American Physiological Society. Adapted by permission.

98. Consider the illustrated free body diagram of the forearm-hand, oriented vertically, with the following dimensions: The cross-sectional area of the elbow flexors ($\mathbf{F}_{m,e}$) is 14 cm, the weight ($\mathbf{F}_w$) of the system is 40 N, the joint reaction force ($\mathbf{F}_j$) has a magnitude of 850 N, and the $\mathbf{F}_{m,e}$ vector acts at an angle of 0.3 rad to the forearm-hand.
   A. If muscle has a specific tension of 27 N/cm^2, determine the maximum torque that $\mathbf{F}_{m,e}$ can exert about the elbow joint (E) in the position indicated in the free body diagram.
   B. Determine the magnitude of the normal and tangential components of the force exerted by $\mathbf{F}_{m,e}$.
   C. If the forearm-hand has a length of 43 cm, and the center of gravity (g) is located at 32% of that length from the proximal end of the segment, how far is it from E to g?
   D. Calculate the magnitude of the net muscle torque required to maintain the system in a vertical position.

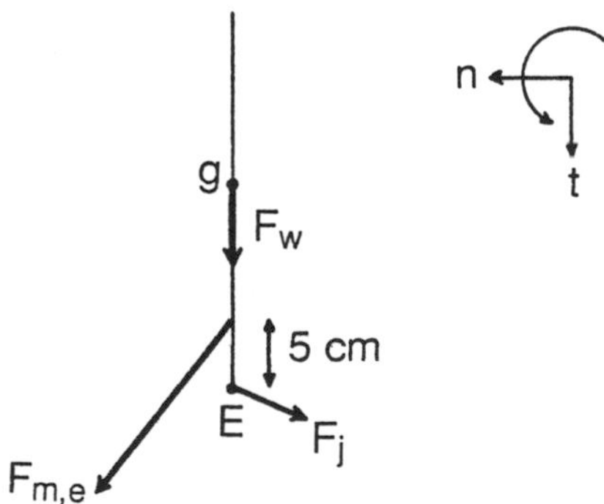

99. A subject performed the curl exercise, which involved an elbow flexion-extension movement with a weight (dumbbell) held in the hand. The hand was in a supinated position, and at one point in the movement the system dimensions were as follows: The subject weighed 721 N; the dumbbell weighed 25 N; the forearm + hand had a length of 40 cm and a weight of 51 N; the location of the center of gravity for the system was 14 cm from the elbow joint; the elbow flexor muscles exerted a force of 1,208 N; the distance to the attachment of the elbow flexor muscles was 3 cm; and the angle of pull of the elbow flexor muscles was 0.29 rad.
   A. Calculate the net torque (magnitude and direction) when the forearm-hand was 0.53 rad below the horizontal.
   B. Was the elbow being flexed or extended?

100. The following are *x*- and *y*-coordinates for the lower limb of an individual who weighs 562 N:

Landmark	*x*	*y*
Hip	0.47	0.74
Knee	0.71	0.47
Ankle	0.63	0.09
Toes	0.73	0.00

Plot these points on graph paper and join them to form a stick figure of the lower limb. Use the regression equations in Table 2.1 to find the location of the center of gravity for the lower limb. Repeat the procedure using the regression equations in Table 2.5, and compare the locations to those obtained with the Table 2.1 data.

101. One common technique to determine the location of the whole-body CG is to place the individual in a supine position on a rigid board and measure the ground reaction force ($F_{g,z}$) at a known distance (x) from the feet. Use Equation 3.3 to derive an expression for determining CG location relative to the feet. Suppose your subject has a weight ($\mathbf{F}_w$) of 652 N and a height of 172 cm. The value for $F_{g,z}$ has a magnitude of 311 N and a point of application (x) of 2.0 m from the feet. Calculate the location of the CG from the feet as a percentage of height.

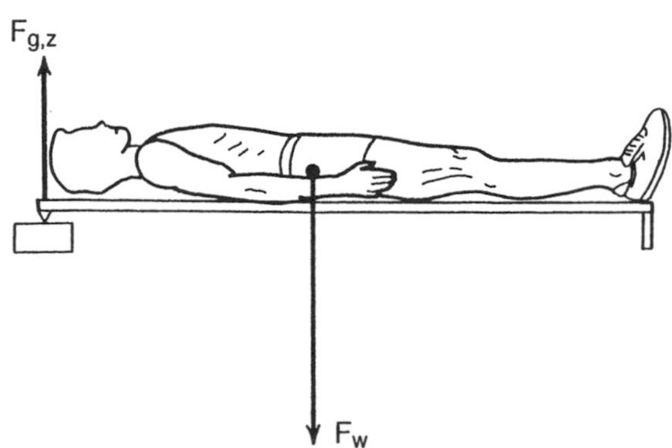

102. Use the regression equations in Table 2.1 and the procedures described under Example 4 of the Static Analysis section to locate the CG for this individual ($\mathbf{F}_w$ = 583 N).

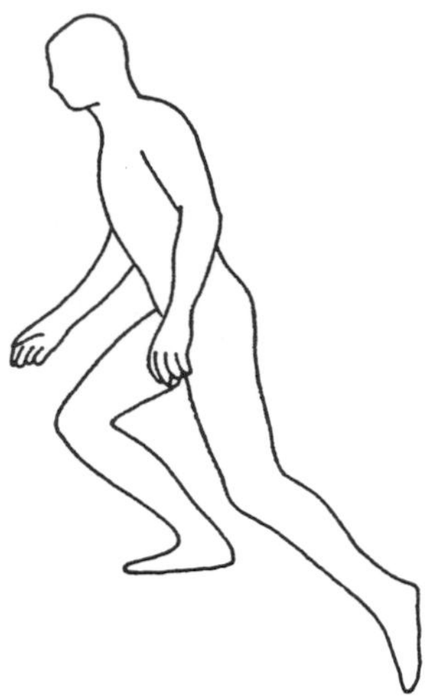

103. A photograph was taken of an athlete performing rehabilitation from knee surgery. The exercise was performed with the knee joint stationary so that the center of gravity of the system did not experience a linear acceleration (**a**). The shank-foot ($\mathbf{F}_{w,s}$) weighed 120 N, and an ankle weight ($\mathbf{F}_{w,a}$) of 250 N was attached.
   A. What was the magnitude of the load torque in this position?
   B. Suppose the leg is being lowered.
      i. What muscle group is controlling the movement?
      ii. What is the maximum torque ($\mathbf{T}_m$) that the muscle group may exert and still allow the leg to be lowered?
   C. Suppose the moment of inertia of the system about the knee joint (K) has a magnitude of 0.11 kg·m. What magnitude of torque ($\mathbf{T}_m$) must the athlete exert to obtain a system acceleration of 9.4 rad/s? (Hint: Perform a dynamic analysis, including the drawing of free body and mass acceleration diagrams.)

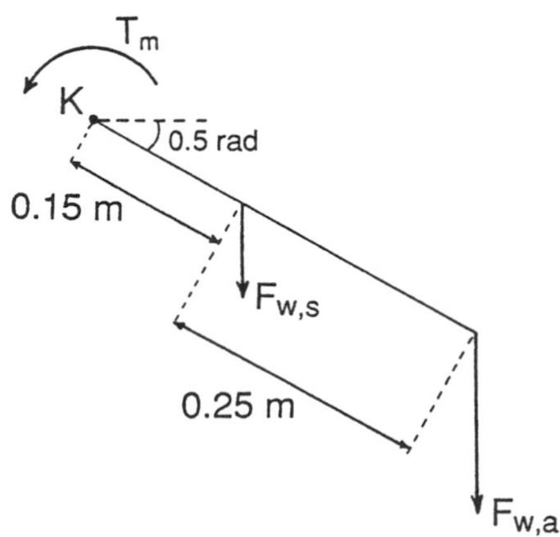

104. The figure shows the resultant muscle torque about the hip (solid line), knee (dashed line), and ankle (dot-dash line) joints during the stance phase of an individual running at 4.5 m/s.
   A. Which direction (positive or negative torque) indicates an extensor torque?
   B. Did the quadriceps femoris exert a greater torque than the plantarflexor muscles?
   C. Why might the peak torques occur in a hip-knee-ankle sequence?
   D. How could angular impulse be calculated with these data?

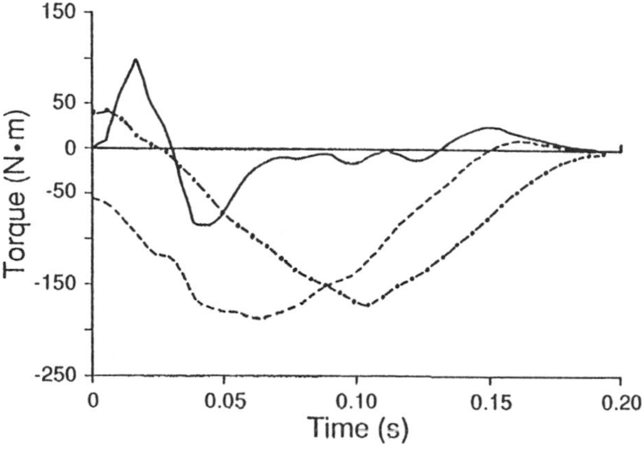

Note. From "A Bioengineering Analysis of Human Muscle and Joint Forces in the Lower Limbs During Running" by R.N. Harrison, A. Lees, P.J.J. McCullagh, and W.B. Rowe, 1986, *Journal of Sports Sciences*, **4**, p. 212. Copyright 1986 by E. & F.N. Spon Ltd. Adapted by permission.

105. Luhtanen (1988) measured angular velocities at the moment that the ball is contacted during an overhead volleyball serve. Luhtanen reported angular velocities of 7, 18, and 38 rad/s for the upper arm, forearm, and hand, respectively. Assume that the shoulder had a linear velocity of 0.2 m/s and that segment lengths were 26 cm for the upper arm, 25 cm for the forearm, and 7 cm from the wrist to the center of the contact area on the hand. Calculate the linear velocity of the hand as it contacts the ball.

106. A golf ball has a mass of 46 g. When a golf ball is hit with a driver, its speed will change from 0 to about 60 m/s.
   A. What is the change in momentum of the ball?
   B. If the time of contact of the club with the ball is 0.5 ms, what is the average force on the ball?
   C. If the club head has a mass of 200 g and a speed of 28 m/s just prior to contacting the ball, what is the speed of the club head after hitting the ball?

107. In walking across a tightrope sideways, a circus performer begins to fall backward. According to the conservation of momentum principle, in which direction (forward or backward) will the performer's arms be circled so that balance is maintained? Explain the reason for your choice. (Hint: Draw a stick figure of the performer with an angular momentum vector for the arms and another for the remainder of the body.)

108. A soccer ball (weight = 4.17 N) was traveling at 7.62 m/s until it was contacted by the head of a soccer player and sent traveling at 12.8 m/s in the opposite direction. If the ball were in contact with the player's head for 22.7 ms, what was the average force applied to the ball?

109. According to Stevenson (1985), a diver (mass = 58 kg) entering the water from a 10-m tower experiences a change in vertical velocity from 16.8 m/s to 5.2 m/s in 133 ms. What was the average force exerted by the water on the diver to get this change in momentum?

110. The total-body angular momentum of a diver changes from zero to some significant value (positive or negative depending on the type of dive) during the takeoff phase. Miller and Munro (1985) reported that Greg Louganis had an angular momentum at takeoff of -70 kg·m^2/s in the forward three-and-a-half somersault dive performed in the pike position, an angular momentum of 54 kg·m^2/s in the reverse two-and-a-half somersault dive in the pike position, and an angular momentum of -17 kg·m^2/s in a straight forward dive.
   A. What does a negative angular momentum indicate? Why does the magnitude of the angular momentum differ for the three dives?
   B. Use the moment of inertia values in Table 2.2 for an individual in a pike position to calculate the average angular velocity for each of the three dives.
   C. The durations of the takeoff phase were 0.42, 0.44, and 0.39 s, respectively. Calculate the angular impulse and the average torque about the center of gravity for each of the dives.

111. A diver wants to perform a double-twisting, one-and-a-half somersault forward dive. The diver is able to rotate the trunk by 0.15 rad from the vertical by rotating the arms about a cartwheel axis through the shoulder girdle. The diver is in the air for 1.62 s and has a moment of inertia about the somersault axis through the center of gravity of 14.3 kg·m^2. How much $\mathbf{H}_g$ does the diver require at takeoff to perform this dive?

112. A baseball weighs 1.43 N and has a speed of 24.4 m/s just before the catcher catches it. What is the kinetic energy of the ball?

113. A skier (weight = 653 N) leaves from the top of a ski jump that is 90 m above the takeoff area of the jump.
   A. Neglecting friction and air resistance, determine the jumper's speed at takeoff.
   B. Given $k = 0.55$ kg/m^3 and $A = 0.047$ m^2, what is the air resistance encountered by the skier immediately after takeoff?

114. If a diving board behaves like a linear spring, or at least close to it, the amount of strain energy that it can store is given by Equation 3.23. A 575-N diver depressed the board by 0.76 m during a dive off a 3-m board, and the $E_{p,s}$ of the board was 727 J at maximum depression.
   A. What is the value of k for the board?
   B. What is the kinetic energy of the diver at last contact with the board?
   C. What is the vertical velocity of the diver at last contact with the board?
   D. How high will the center of gravity of the diver reach?
   E. What is the vertical velocity of the diver at the time that the hands first contact the water?

115. A box of exercise equipment weighs 89 N and is located on a flat surface. Suppose that you push the box ($\mathbf{F}_p$) with a force of 396 N.
   A. Calculate the work you would need to perform to push the box 2 m along the horizontal surface in the direction of the applied force.
   B. Suppose that $\mu_s = 1.4$. Is it possible to move the box with $\mathbf{F}_p = 396$ N?

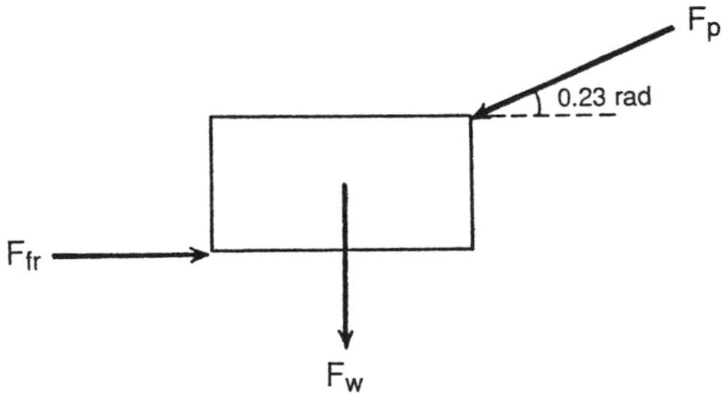

116. A swimmer performs the front crawl at a constant speed of 1.4 m/s and encounters a drag force of 93 N. The swimmer consumes oxygen at a rate of 2.3 L/min.
   A. If 1 L of oxygen produces 20.9 J of energy, what is the rate (J/s) of production of metabolic energy at this rate of oxygen consumption?
   B. What is the efficiency of the swimmer?
   C. What are the units of measurement for efficiency?

117. Vaughan (1980) obtained the following position-, velocity-, and acceleration-time histories for the center of gravity of a person bouncing on a trampoline. Use these data to graph qualitatively the power-time relationship during this event; use the same approach as in Figure 3.28. Once you have drawn the graph, explain what it means. The person is in contact with the trampoline bed between touchdown and takeoff.

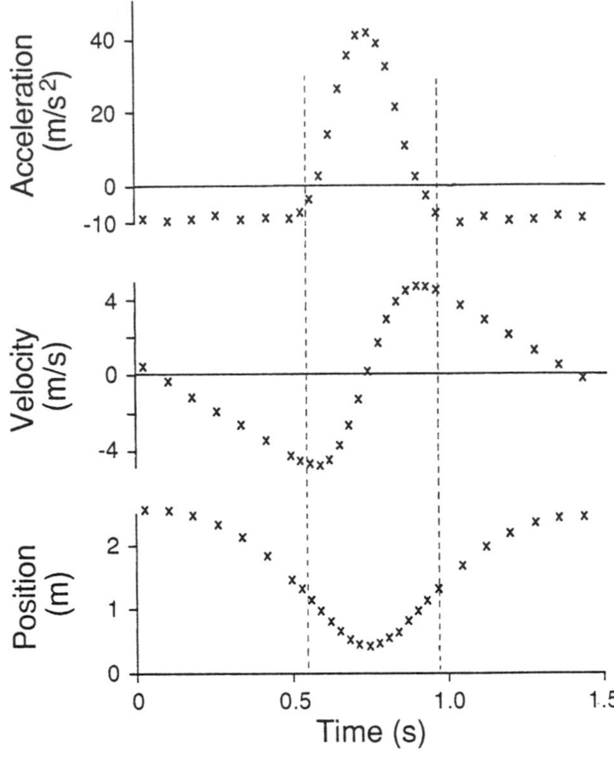

Note. From "A Kinetic Analysis of Basic Trampoline Stunts" by C.L. Vaughan, 1980, *Journal of Human Movement Studies*, **6**, p. 241. Copyright 1980 by Henry Kimpton (Publishers) Ltd. Adapted by permission.

118. The power produced by muscle is a significant determinant of performance.
   A. What is the unit of measurement for power?
   B. Define power production. How do we vary the power that a muscle can produce?
   C. It is possible to derive (simply!) an algebraic expression for power as a function of energy. If work = Δ energy, what is the relationship between power and energy?

119. Draw a power-time graph for an elbow flexion-extension movement performed in a vertical plane. Assume that the initial position has the elbow fully extended.

Summary of Part I

At the beginning of Part I, a number of specific objectives were listed to help us achieve the goal of defining the mechanical bases of movement. Now that you've completed Part I, you should

- understand the definitions of and relationships (numeric and graphic) between the kinematic variables position, velocity, and acceleration;
- know how to read a graph carefully and how to interpret the relationship between the two or more variables shown on the graph;
- appreciate the relationships between linear and angular motion;
- be able to provide a kinematic description of gait, throwing, and kicking;
- realize that many of the details of projectile motion can be determined from the definitions of position, velocity, and acceleration;
- consider force as a concept used to describe an interaction between two objects and understand that the magnitude of the interaction can be determined using Newton's laws, particularly the law of acceleration;
- be able to use a free body diagram to define conditions of an analysis and the free body and mass acceleration diagrams to provide a graphic version of Newton's law of acceleration;
- be able to identify the ways in which the human body interacts with its surroundings to influence movement;
- conceive of torque as the rotary effect of a force for which torque is defined as the product of force and moment arm;
- realize that force acting over time (impulse) causes a change in the momentum (quantity of motion) of a system;
- understand that the performance of work (force × distance) requires the expenditure of energy and that the work can be done by the system (positive) or on the system (negative); and
- consider power as a measure of the rate of doing work or the rate of using energy.

PART II

The Single Joint System

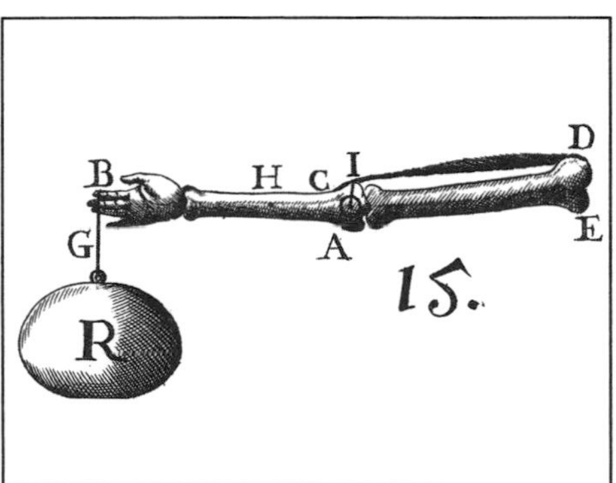

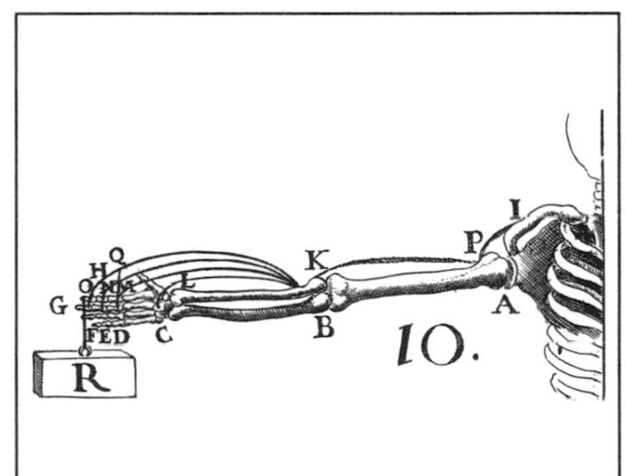

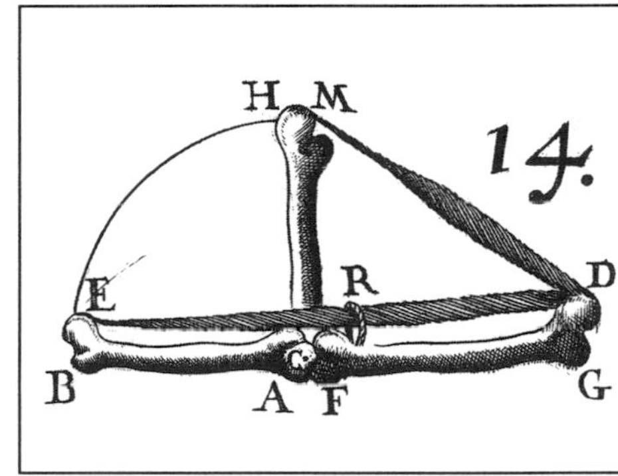

Human movement is a complex phenomenon. To examine the neuromechanical bases of movement, we need to make several simplifying assumptions. For example, in Part I, ''The Force-Motion Relationship,'' we assumed that the force exerted by muscle acts on rigid body segments, and we ignored the complexity and variety of structures and attachment sites. These simplifications allowed us to address some of the basic concepts associated with the study of movement.

Part II continues this focus by describing the human body as a system of rigid links that are rotated by muscles about pinned frictionless joints. The activation of muscle is controlled (neurons) and monitored (sensory receptors) by the nervous system. Five elements (i.e., rigid link, synovial joint, muscle, neuron, and sensory receptor) make up the basic apparatus for the production for human movement and thus form a *biological model* that is called the **single joint system** (Figure II.1). Note that Figure II.1 is just a model. Although the model includes only one neuron, there are generally hundreds that innervate each muscle and provide both motor commands (motor neurons) and sensory information (sensory receptors). Similarly, about each joint there are generally groups of muscles, each of which controls movement in a limited number of directions.

In Part II, we shall first review simplified descriptions of these five elements (chapter 4, ''Single Joint System Components''), then consider several interactions between the elements (chapter 5, ''Single Joint System Operation''), and finally examine the properties of the single joint system that enable us to control precisely muscle force (chapter 6, ''Single Joint System Activation'').

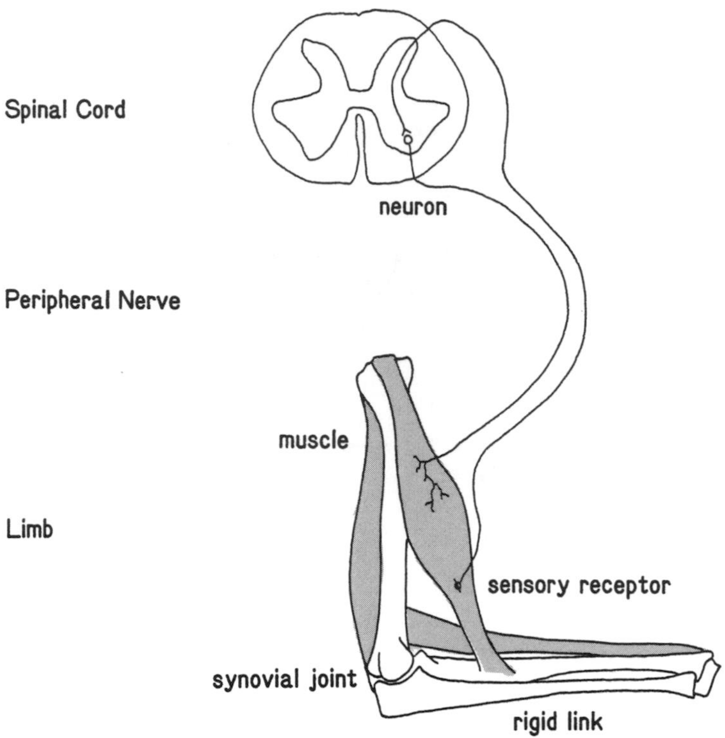

Figure II.1 The five components of the single joint system.

Objectives

As stated previously, the goal of this text is to decribe movement as the interaction of a biological model (a simplified version of humans) with the physical world. In Part I, we examined the biomechanics of movement, focusing on the relationship between force and motion. The aim of Part II is to define a biological model and to describe the interactions among its components. Specific objectives include the following:

- To describe the single joint system and to explain why it is an adequate model for the study of the basic features of movement
- To list the details of the structural characteristics of the five components of the single joint system
- To define the basic functional unit of the system
- To explain the means by which information is transmitted rapidly throughout the system
- To indicate the theoretical bases of techniques used to monitor the activation of the single joint system
- To outline the link between the neural signal and muscle contraction
- To examine the neuroanatomical basis and the role of afferent information in the operation of the system
- To describe the options used by the nervous system to control muscle force
- To define the mechanical characteristics of the force-generating units
- To evaluate the effects of different architectural arrangements of the force-generating units on the force exerted by muscle

CHAPTER 4

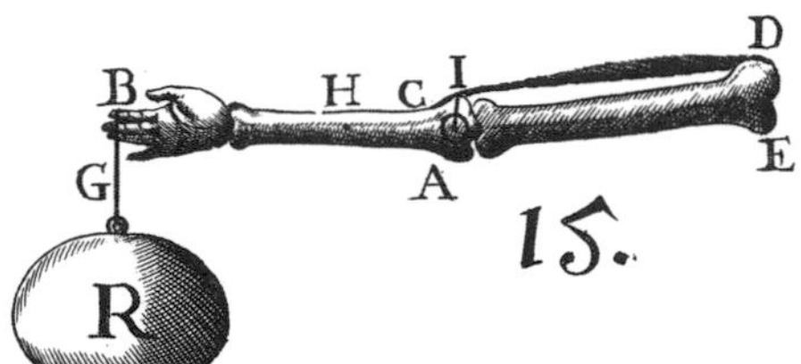

Single Joint System Components

Although it is anticipated that you have encountered most of the material in this chapter in introductory human anatomy and physiology courses, the material is repeated here to emphasize a functional focus. Indeed, perhaps the most important task in chapter 4 is to emphasize that *the morphological and mechanical features of the single joint system elements are determined largely by the functions they serve*. To this end, the chapter focuses on the morphological and mechanical characteristics of the single joint system components with some mention of their adaptive capabilities. Because the components are all living tissue, they adapt to the demands that they experience; for example, muscle size increases with strength training, bone strength decreases during spaceflight, and sensory receptors become less sensitive with age. The interactions among the elements of this system in the production of movement are considered in subsequent chapters.

Rigid Link

Many different tissues provide a structural framework for the single joint system. These tissues, known as connective tissue, comprise living (cells) and nonliving (intercellular material) substances that are bathed in tissue fluid. The cells (e.g., fibroblasts, macrophages, fat cells, mast cells) perform functions necessary for the maintenance of the tissue. The intercellular material, which forms the matrix in which the cells live, includes the proteins collagen, elastin, and reticulum and determines the physical characteristics of the tissue. The principal connecting elements that constitute the rigid link of the single joint system include bone, tendon, and ligament.

Bone

Bone performs several functions essential to movement production. It (a) provides mechanical support as the central structure of each body segment, (b) produces red blood cells, and (c) serves as an active ion reservoir for calcium and phosphorus. Bone is a living tissue consisting of a protein matrix (mainly collagen) upon which calcium salts (especially phosphate) are deposited. These minerals give bone its solid consistency. Water accounts for 20% of the wet weight of bone; the protein matrix, which is mainly osteocollagenous fibers, represents 35%; and the bone salts account for 45%. The osteocollagenous fibers determine the strength and resilience of bone. The basic structural unit of bone is the **osteon** (also called the haversian system), which consists of a series of concentric layers of mineralized matrix that surround a central canal containing blood vessels and nerves. A typical osteon has a diameter of about 200 μm. Although bone is often classified as cancellous or cortical, the biomechanical properties of the two types are similar and differ only in the degree of porosity and density.

In the study of bone biomechanics, bone is examined as a material, as a structure, and as a system (Roesler, 1987). The material properties of bone are generally characterized by the *load-deformation relationship* (chapter 2). In this scheme, a load is applied to the tissue and the ensuing deformation (change in length) is measured. Different loads can be applied to bone (Figure 4.1) to identify such features as strength, stiffness, and ability to store energy. Rather than measure the absolute load and deformation, however, researchers must normalize these measurements so that the biomechanical properties can be determined independently from the geometry (e.g., size) of the tissue. Load is normalized as **stress** (force per unit area); the unit of measurement is MPa or MN/m^2. Deformation is normalized as **strain** (change in length as a function of initial length); strain is a dimensionless quantity.

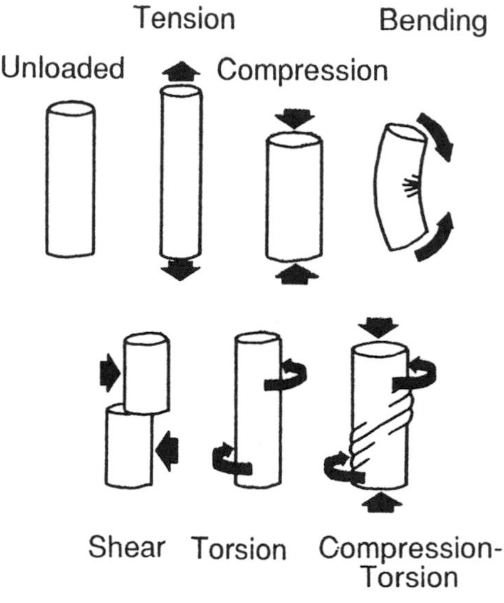

Figure 4.1 The types of loads that can be applied to a tissue, such as bone.
Note. From "Biomechanics of Bone" by M. Nordin and V.H. Frankel. In *Basic Biomechanics of the Musculoskeletal System* (p. 10) by M. Nordin and V.H. Frankel (Eds.), 1989, Philadelphia: Lea & Febiger. Copyright 1989 by Lea & Febiger. Adapted with permission.

Idealized stress-strain curves for human cortical bone tested with tension and compression loads are shown in Figure 4.2. The material properties of bone are characterized by the measurement of the yield stress (σ_{yield}), the stress at which the relationship changes from the elastic to the plastic region; the ultimate stress or strength of bone (σ_{ult}); the ultimate strain (ε_{ult}); the slope of the elastic region (elastic modulus); and the slope of the plastic region (plastic modulus). Figure 4.2 shows how these properties are represented on the stress-strain curve by peak values, inflection points, and slopes. In addition, the energy that bone can absorb is measured

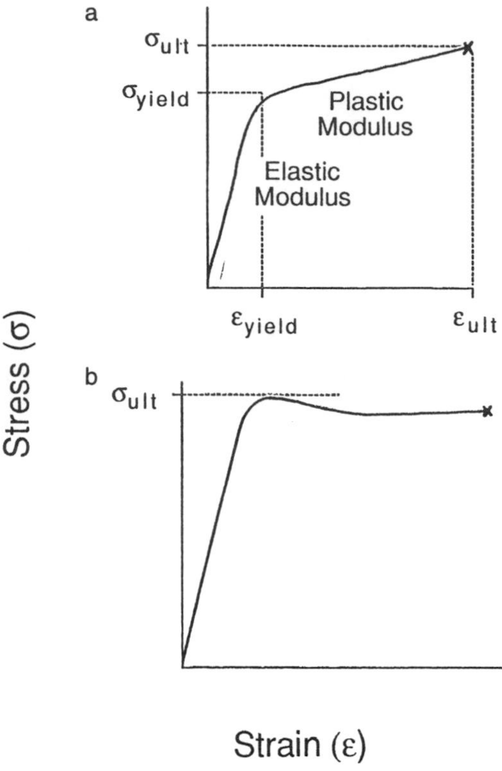

Figure 4.2 Idealized stress-strain relationship for human cortical bone subjected to tension (a) and compression (b) loads.
Note. From "Mechanical Properties and Composition of Cortical Bone" by D.R. Carter and D.M. Spengler, 1978, *Clinical Orthopaedics and Related Research*, **135**, p. 199. Philadelphia: Lippincott. Copyright 1989 by J.B. Lippincott and Company. Adapted with permission.

as the area under the stress-strain curve. Table 4.1 lists typical values for these parameters for some human bones as a function of age (Burstein, Reilly, & Martens, 1976).

The second focus in the study of bone biomechanics is to treat *bone as a structure*. In this focus, the observed form and structure of a bone are explained by the function that it serves and the environmental conditions that it experiences. For example, the human tibia experiences a greater strain in the compressive direction (847 μstrain) than in the tensile direction (578 μstrain) when a person is jogging at 2.2 m/s (Lanyon, Hampson, Goodship, & Shah, 1975). In animal studies, peak strain has been found to increase with the speed of locomotion. Furthermore, the femur more commonly experiences longitudinal than perpendicular forces. The long bones of animals of different sizes typically experience compressive forces of 40 to 80 MPa during locomotion (Biewener, 1991). For these reasons, the femur offers the greatest resistance to compression loads, is weakest in response to shear loads, and has an intermediate strength for tension loads. Measurements indicate that the human femur has a strength (σ_{ult}) of 132 MPa when

Table 4.1 Stress-Strain Characteristics for Human Bone Subjected to a Tension Load

Bone	Age (years)	σ_{yield} (MPa)	σ_{ult} (MPa)	Elastic modulus (GPa)	Plastic modulus (GPa)	ε_{ult}	Energy (MPa)
Femur	20–29	120	140	17.0	0.75	0.034	3.85
	30–39	120	136	17.6	0.64	0.032	3.55
	40–49	121	139	17.7	1.00	0.030	3.19
	50–59	111	131	16.6	0.89	0.028	2.84
	60–69	112	129	17.1	0.98	0.025	2.65
	70–79	111	129	17.1	0.98	0.025	2.65
	80–89	104	120	15.6	1.08	0.024	2.23
Tibia	20–29	126	161	18.9	1.17	0.040	4.36
	30–39	129	154	27.0	0.91	0.039	5.77
	40–49	140	170	28.8	1.39	0.029	4.09
	50–59	133	164	23.1	1.21	0.031	4.19
	60–69	124	147	19.9	1.20	0.027	3.05
	70–79	120	145	19.9	1.18	0.027	3.27
	80–89	131	156	29.2	1.43	0.023	2.96

Note. Modified from Burstein, Reilly, and Martens (1976).

a tensile force is applied along the length of the bone (longitudinal) but a strength of only 58 MPa for tensile forces acting perpendicular (normal) to the bone. Similarly, the femur has a compressive strength of 187 and 132 MPa for longitudinal and perpendicular forces, respectively (Cowin, 1983). The absolute strength of bone varies depending on the stress and strain it experiences. In engineering terms, bone has a **safety factor** of between two and five; that is, bones are two to five times stronger than the forces they commonly encounter in the activities of daily living (Alexander, 1984a; Biewener, 1991).

From the perspective of bone as a structure, these observations emphasize that function has a major effect on the cellular organization and hence the mechanical characteristics of bone. As the stress-strain patterns vary, so do the mechanical properties and cellular organization of bone. **Wolff's law** characterizes this relationship as follows: ''Every change in the . . . function of bone . . . is followed by certain definite changes in . . . internal architecture and external confirmation [sic] in accordance with mathematical laws'' (Carter, 1984, p. 1). Some studies related to this law have focused on bone as a structure and have attempted to identify appropriate parameters by which the structure and geometry of bone can be described. Other studies, however, have focused on the processes by which bone adapts to its environment; these represent the study of *bone as a system*. The processes experienced by bone include growth, reinforcement, and resorption, which are collectively termed **remodeling** (Lanyon & Rubin, 1984). The time required for one complete cycle of remodeling (replacement of all structures) seems to be about 10 to 20 years for limb bones of the adult human (Alexander, 1984a).

Remodeling represents a balance between bone absorption by osteoclasts and bone formation by osteoblasts. The balance between these processes changes continually and is influenced by such factors as physical activity (stress-strain loads), age, and disease. Several reports in the literature have documented the remodeling of bone with changes (increases and decreases) in activity. For example, Shumskii, Merten, and Dzenis (1978) found a greater deposition and density of bone in the tibia of athletes than in that of control subjects. Similarly, Dalén and Olsson (1974) found that bone mineral content for cross-country runners (25 years' experience) was 20% greater in appendicular sites (distal radius, ulna, calcaneus) but less than 10% greater in axial sites (lumbar vertebrae and head of femur) compared with that of control subjects. Even weight lifters (6 years' experience) have been reported to show an increase in bone mineral density at weight-bearing sites (lumbar spine, trochanter, femoral neck) but not at non-weight-bearing sites (midradius) (Colletti, Edwards, Gordon, Shary, & Bell, 1989). Furthermore, bone mineral content can increase even after 6 weeks of exercise (Beverly, Rider, Evans, & Smith, 1989).

In contrast, it appears that a major problem experienced by astronauts during extended flights is the loss of bone tissue (Zernicke, Vailis, & Salem, 1990). The microgravity conditions of spaceflight cause bone **demineralization**, the excessive loss of salts from the skeleton (Anderson & Cohn, 1985; Morey, 1979). This raises two concerns: Bone is weaker and therefore more susceptible to fracture during strenuous activities (e.g., extravehicular activity), and bone seems to have difficulty recovering from an episode of demineralization when the astronauts return to a normal gravity environment. For these reasons,

NASA has been interested in exercise programs that might limit the loss of bone tissue during **weightlessness**. Studies indicate that bone remodeling is best induced by loads that are applied intermittently rather than in a sustained, continuous manner (Lanyon & Rubin, 1984). Furthermore, the magnitudes of the joint forces have a greater effect on bone mass than do the number of loading cycles (Whalen, Carter, & Steele, 1988). Whereas few of us will experience a loss of bone tissue due to weightlessness, our bones may decline in mass and strength as a function of age (Snow-Harter & Marcus, 1991). This effect, known as **osteoporosis**, involves an increase in the porosity of bone that results in a decrease in its density and strength and an increase in its vulnerability to fracture.

Attempts to determine the mechanisms that control remodeling have included the study of the electrical properties of bone (Bassett, 1968; Singh & Katz, 1989). Apparently, when a stress is sufficient to cause bone collagen fibers to slip relative to one another, this action will generate electrical potentials in bone. This generation of electrical potentials due to pressure has been referred to as the **piezoelectric** effect of bone. Scientists have attempted to establish a relationship between the mechanical stress experienced by bone and the associated electrical effects as the mechanism underlying Wolff's law. It has been shown, for example, that weak electrical currents can induce the formation of a callus (Marino, 1984). One practical consequence of this phenomenon has been the use of electric and magnetic stimulation to enhance fracture healing, especially for those fractures where nonunion of the bones seems doubtful (Brighton, 1981). Some investigators have even used the electrical stimulation of bone to attempt to impede the onset of osteoporosis.

Tendon and Ligament

As connecting elements, tendon links muscle to bone, and ligament connects bone to bone. The principal difference between tendon and ligament is the organization of the collagen fibril, which is specialized to accommodate these respective functions. Because the function of tendon is to transmit muscle force to bone or cartilage, the structure of tendon is such that it is least susceptible to deformation from **tensile** forces (i.e., the pulling force exerted by muscle). In contrast, only small longitudinal-pushing (**compressive**) and side-to-side (**shear**) forces are required to deform tendon. Although ligament also mainly encounters tensile forces, its primary function is joint stabilization, and hence it is designed to provide multidirectional stability and to accommodate tensile, compressive, and shear forces (Fu, Harner, Johnson, Miller, & Woo, 1993).

Tendons and ligaments are dense connective tissue that contain collagen, elastin, proteoglycans, water, and cells (fibroblasts). The proteins collagen and elastin are synthesized and secreted by the fibroblasts. Approximately 70% to 80% of the dry weight of tendon and ligament consists of Type I collagen, which is a fibrous protein that has considerable mechanical stability (Burgeson & Nimni, 1992). The collagen **fibril** (Figure 4.3) is the basic load-bearing unit of both tendon and ligament. The structure of the fibril from the alpha chains of the triple helix to the packing of the collagen molecules in the microfibril is the same for tendon and ligament. The major distinction between the two connecting elements concerns *the way in which the fibrils are arranged*. In tendon, the fibrils are arranged longitudinally in parallel to maximize the resistance to tensile forces. In ligament, the fibrils are generally aligned in parallel with some oblique or spiral arrangements to accommodate forces in different directions.

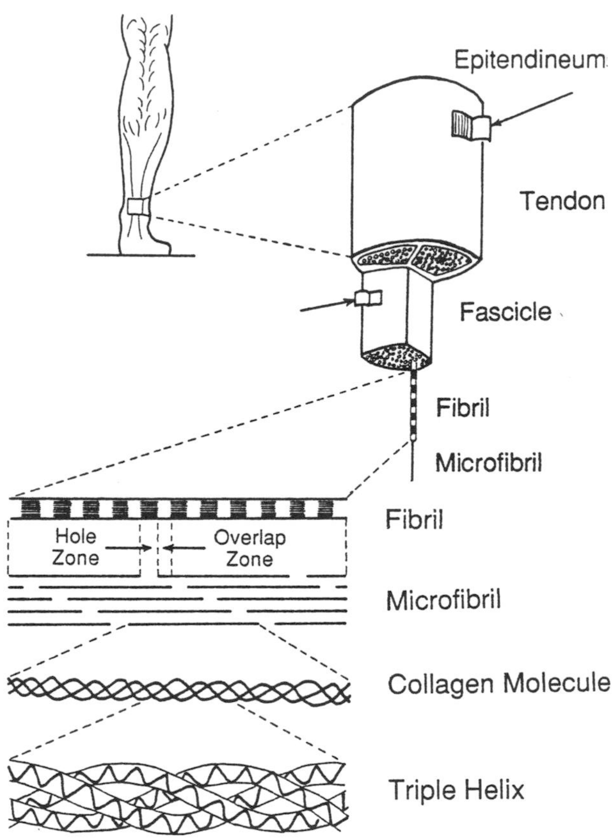

Figure 4.3 The hierarchical organization of a tendon to the level of the triple helix of collagen.

The **triple helix** structure of collagen indicates that the molecule consists of three intertwined polypeptide chains. Each polypeptide chain includes a sequence of about 1,000 amino acids (mainly proline, hydroxyproline, and glycine) and is known as an *alpha chain*. The three-stranded collagen molecules are then arranged end to end (in series), and five such rows are stacked in parallel (side by side) to form the **microfibril**. The collagen fibril, which is the basic load-bearing unit of tendon and ligament, consists of bundles of microfibrils held together by biochemical bonds (cross-linkages) between the collagen molecules; these cross-links occur both within and between the rows of collagen molecules in the microfibril (Figure 4.3). Because these **cross-links** bind the microfibrils together, *the number and state of the cross-links are thought to have a substantial effect on the strength of the connective tissue* (Bailey, Robins, & Balian, 1974). Thus the functional focus of tendon and ligament is the fibril, a collection of units (microfibrils) bound together by cross-links. The strength of the fibril depends on these cross-links. The number and state of cross-links are thought to be determined by such factors as age, gender, and activity level.

In addition to collagen and elastin, the extracellular matrix of tendon and ligament includes proteoglycans and water. The water typically binds to the proteoglycans to form a gel whose viscosity decreases with activity. This property is known as **thixotropy**. The resistance that a tissue offers to lengthening at a given velocity depends on its viscosity; the higher the viscosity, the greater the resistance to stretch. As viscosity decreases, therefore, the tissue is able to accommodate higher velocity stretches before it is damaged. The viscosity of thixotropic tissue changes due to prior activity, such as a warm-up or a sustained period of inactivity.

The biomechanical properties of tendon and ligament are frequently characterized as a load-deformation relationship in response to a tensile load (Figure 4.4). In these experiments, a specimen (e.g., ligament, tendon, ligament-bone) is obtained from a cadaver and is mounted on a device that stretches the tissue at a prescribed rate (strain rate) to failure and measures the displacement (elongation) and force. Incidentally, clinical observations on the failure of connective tissue suggests that tissue tear is more common than avulsion from the bone. Figure 4.4 shows the variation in stretch and peak force for different specimens. For example, a medial patellar tendon-bone specimen was stretched 10 mm and exerted a peak tensile force (σ_{yield}) of about 3 kN before it began to fail, whereas an anterior cruciate ligament-bone specimen was stretched 15 mm and exerted a peak tensile force (σ_{yield}) of around 1.5 kN before failure. The data shown in Figure 4.4 are taken from a study that compared the mechanical properties of various collagenous tissues for use in reconstruction of articular cartilage at the knee joint (Noyes, Butler, Grood, Zernicke, & Hefzy, 1984). The gracilis tendon represents the tissue between the muscle and the tibial insertion. The fascia lata specimen was 7- to 10-cm wide and was taken from the middle of the thigh just proximal to the lateral femoral condyle. The data indicate that the patellar tendon-bone specimen was stronger than the anterior cruciate ligament-bone specimen, but both the patellar tendon and anterior cruciate specimens were stronger than those of the gracilis tendon and the fascia lata (Figure 4.4).

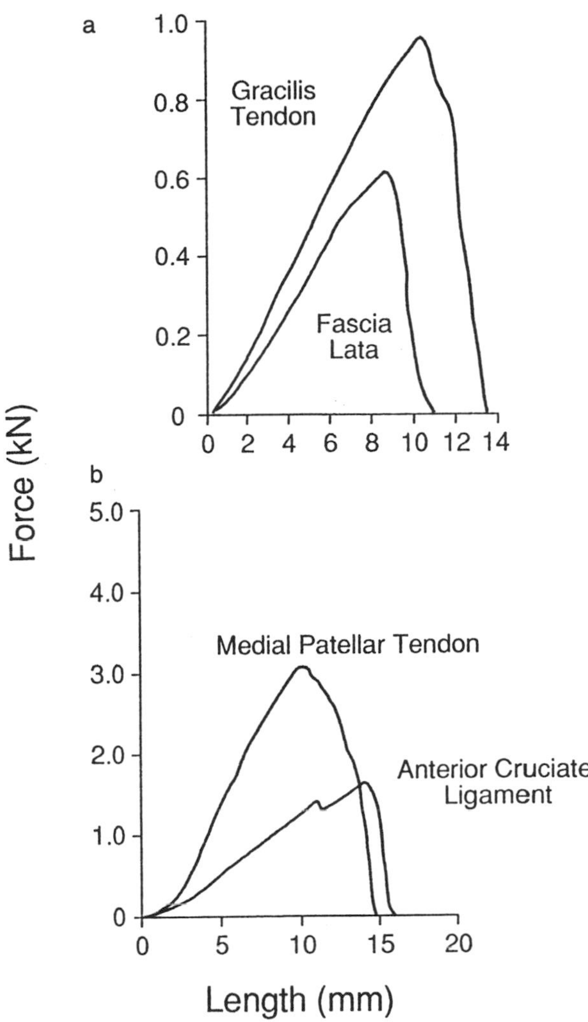

Figure 4.4 Load deformation relationships for connective tissue specimens stretched to failure.
Note. From "Biochemical Analysis of Human Ligament Grafts Used in Knee-Ligament Repairs and Reconstructions" by F.R. Noyes, D.L. Butler, E.S. Grood, R.F. Zernicke, and M.S. Hefzy, 1984, *Journal of Bone and Joint Surgery*, **66-A**, pp. 346, 347. Copyright 1984 by The Journal of Bone and Joint Surgery, Inc. Adapted by permission.

When the load and deformation are normalized so that load is expressed per unit of cross-sectional area and deformation is described as a percentage of the initial length, the biomechanical properties of tissues can be compared as stress-strain relationships. Figure 4.5 represents an idealized stress-strain relationship for collagenous tissue, such as tendon and ligament. The stress-strain relationship comprises three regions: toe, linear, and failure. The *toe region* corresponds to the initial part of the relationship, in which the collagen fibers are stretched and straightened from a resting zig-zag pattern. The *linear region* represents the elastic capability of the tissue; the slope of the relationship in this region is referred to as the elastic modulus and is steeper for stiffer tissues. Beyond the linear region, the slope decreases as some of the fibers are disrupted in the *failure region*. When connective tissue experiences a strain of this magnitude, the tissue will undergo plastic changes and there will be a change in its resting length. From the stress-strain relationship, the tissue can be characterized by the measurements of ultimate stress (σ_{ult}), the ultimate strain (ε_{ult}), the elastic modulus, and the energy absorbed (area under the stress-strain curve). These properties tend to decline with such conditions as reduced use (e.g., immobilization, bed rest), age, and steroid use but increase with chronic exercise (Butler, Grood, Noyes, & Zernicke, 1978; Noyes, 1977). In addition, tendon properties can vary with the function of muscle (Shadwick, 1990).

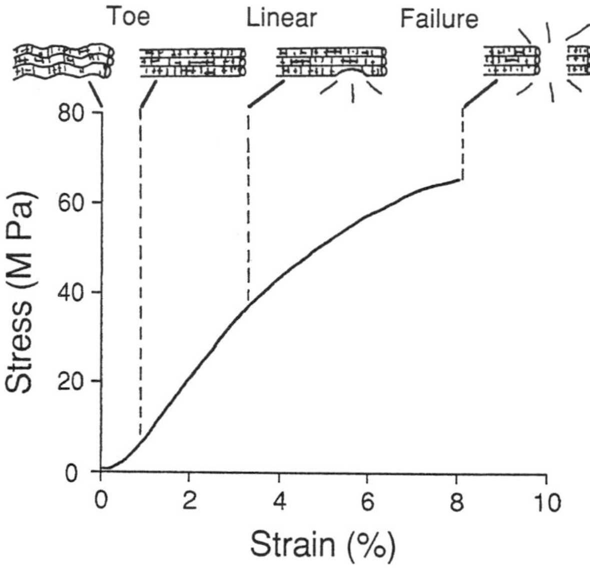

Figure 4.5 Idealized stress-strain relationship for collagenous tissue. The tissue can experience only a small change in length before it sustains some damage.
Note. From "Biomechanics of Ligaments and Tendons" by D.L. Butler, E.S. Grood, F.R. Noyes, and R.F. Zernicke. In *Exercise and Sport Sciences Reviews* (Vol. 6) (p. 145) by R.S Hutton (Ed.), 1978, Philadelphia: The Franklin Institute Press. Copyright 1978 by The Franklin Institute Press. Adapted by permission.

Synovial Joint

There are about 206 bones in the human body, and these bones form approximately 200 joints. Joints are generally classified as belonging to one of three groups: the fibrous joint, which is relatively immovable (e.g., sutures of skull, interosseus membrane between radius and ulna or between tibia and fibula); the cartilaginous joint, which is slightly movable (e.g., sternocostal, intervertebral disks, pubic symphysis); or the synovial joint, which is freely movable (e.g., hip, elbow, atlantoaxial). Because the synovial joint most closely approximates the frictionless pinned joint of the rigid-link assumption, it is taken as the joint component of the single joint system.

The synovial joint serves two functions: It provides mobility of the skeleton by permitting one body segment to rotate about another, and it transmits forces from one segment to another. These interactions, which involve the contact of adjacent bones, are controlled by a number of structural features, including articular cartilage, joint capsule, synovial membrane, and the geometry of the bones. The surfaces of the bones that form the joint are lined with **articular cartilage**, a dense white connective tissue. Articular cartilage has no blood vessels, lymph channels, or nerves. It contains chondrocytes, cells that maintain the organic component of the extracellular matrix, and a dense network of fine collagen fibrils in a concentrated solution of proteoglycans. Water is the most abundant component of articular cartilage; most of the water is located near the surface. The collagen fibrils, proteoglycans, and water determine the biomechanical behavior of articular cartilage.

Articular cartilage is protected by two forms of lubrication (Mow, Proctor, & Kelly, 1989): (a) boundary lubrication, absorption of the glycoprotein lubricin by the surface of the articular cartilage, and (b) fluid film lubrication, a thin film of lubricant that causes separation of the articulating surfaces. Boundary lubrication seems to be important when the contact surfaces sustain high loads for long periods of time. In contrast, fluid film lubrication seems more important when the loads are low and variable and the contact surfaces move at high speeds relative to one another. Synovial joints appear to be capable of self-lubrication, whereby fluid is exuded in front of and beneath the moving contact of the articular cartilage but is reabsorbed as the load passes (Mow et al., 1989). The flux of fluid across the articular cartilage probably serves to provide nutrients for the chondrocytes.

The function of articular cartilage is to allow relative motion of opposing joint surfaces with minimal friction and wear and to modify the shape of the bone to ensure better contact with its neighbor (C.G. Armstrong & Mow, 1980; Askew & Mow, 1978). This is accomplished by articular cartilage behaving as a water-filled sponge. Articular cartilage is a **viscoelastic** material, which means that when it is subjected to a constant load or a constant deformation, its response (mechanical behavior) changes over time. This response can include changes in the thickness of articular cartilage due to the stress-related flux of water. Articular cartilage is thicker in more active individuals and increases in thickness when an individual goes from a resting to an active state. Articular cartilage can become specialized, such as at the temporomandibular and knee joints, and can develop intraarticular disks or menisci that enlarge the contact areas of the articulating surfaces. The loads experienced by articular cartilage are supported by the collagen-proteoglycan matrix and by the resistance offered by fluid flow through the matrix.

The articulating surfaces of the synovial joint are enclosed in a **joint capsule**, which attaches to the bones of the joint and thus separates the joint cavity from surrounding tissues. The internal aspect of the capsule and those areas of the articulating bones that are not covered with articular cartilage are lined with **synovial membrane**, a vascular membrane that secretes synovial fluid into the joint cavity. The synovial fluid provides nourishment for and lubricates the articular cartilage. The joint capsule is a loose structure that surrounds the entire joint and, in some places, fuses with capsular ligaments, which are thought to keep the articulating surfaces in close proximity. Like articular cartilage, the joint capsule and associated ligaments adapt to alterations in the pattern of activity. For example, one of the greatest dangers associated with the immobilization of joints for rehabilitative purposes is the adaptations that occur in the joint capsule and ligaments. Apparently, connective tissue has a tendency to adapt to the shortest functional length; when a joint is immobilized, the capsule and ligaments shrink, and new tissue is synthesized to accommodate the shorter length. These changes reduce the mobility at the joint. Such immobilization may contribute to the development of osteoarthritis (Videman, 1987).

The geometry of synovial joints (i.e., architecture of the articulating surfaces, design of ligaments) largely determines the quality of motion between two adjacent body segments. A synovial joint may permit rotation about one to three axes. These axes could pass through the joint from side to side, from front to back, or from end to end. Each axis is referred to as a **degree of freedom**. The only motion possible at the humeroradial joint (elbow) is in the flexion-extension plane (Engin & Chen, 1987), which is rotation about an axis that passes side to side through the joint. In contrast, the hip and shoulder (glenohumeral) joints have three degrees of freedom (Engin & Chen, 1986, 1988; Högfors, Sigholm, & Herberts, 1987), which means that the joint structure permits rotation about each of the axes. Motion at these joints can occur in the flexion-extension plane, in the abduction-adduction plane, and in rotation about a longitudinal axis. A summary of the quality and quantity of motion possible at some major joints in the human body is provided in Table 4.2.

Muscle

Muscles are molecular machines that convert chemical energy, initially derived from food, into force. The properties of muscle include (a) irritability, the ability to respond to a stimulus; (b) conductivity, the ability to propagate a wave of excitation; (c) contractility, the ability to modify its length; and (d) adaptability, a limited growth and regenerative capacity. In histology, there are three types of vertebrate muscle: cardiac, smooth, and skeletal. Only skeletal muscle is considered in the single joint system analysis of human movement. Cardiac muscle, which makes up most of the heart wall, is composed of a network of individual cells that show obvious striations. Smooth muscle, which is found in the viscera and in the walls of blood vessels, consists of individual cells that are separated by a thin cleft and that lack apparent striations. Skeletal muscle comprises fused cells in which the striations are well defined (Figure 4.6 shows the organization of skeletal muscle). With the exception of some facial muscles, skeletal muscles act across joints to produce rotation of body segments. The following comments refer specifically to skeletal muscle and its role as the generator of human movement. The role of muscle in the single joint system is to exert a force that interacts with forces imposed by the surroundings to produce rotation of the rigid links.

Muscle contains a great many identifiable elements. As you develop an understanding of this system, focus on the function of muscle and the process by which that function is carried out. In particular, focus on the elements related to *two critical processes of muscle function: (a) the relationship between the sarcolemma and the sarcoplasmic reticulum and (b) the components of the sarcomere*. As we shall note in chapter 5, these processes are associated with the connection between the nervous system and muscle (sarcolemma-sarcoplasmic reticulum) and the force that muscle can exert (sarcomere).

Gross Structure

Muscle fibers are linked together by a three-level network of collagenous connective tissue. **Endomysium** surrounds individual muscle fibers, **perimysium** collects bundles of fibers into fascicles, and **epimysium** ensheathes the entire muscle (Figure 4.6, a and b). This connective tissue matrix, which exists throughout the entire muscle and not just at its ends, connects muscle fibers to tendon and hence to the skeleton (Tidball, 1991). Due to this relationship, muscle fibers and connective tissue (including tendon) operate as a single functional unit. Sometimes this relationship is emphasized by referring to the **musculotendinous unit**; however, in this text the term *muscle* denotes both the contractile tissue and the associated connective tissue (including tendon).

Muscle fibers vary from 1 to 400 mm in length and from 10 to 60 μm in diameter. The cell membrane encircling each set of myofilaments that comprises a muscle fiber is known as the **sarcolemma**. As a plasma membrane, the sarcolemma provides active and passive selective membrane transport, an essential property of excitable membranes. Because of this property, the sarcolemma allows some material to pass through

Table 4.2 Motion at the Major Joints of the Human Body

Joint	Articulating surfaces	Degrees of freedom	Range of motion
Spine	Atlantooccipital	Flexion-extension	0.3 rad
		Lateral flexion	0.2 rad
	Atlantoaxial	Flexion-extension	0.2 rad
		Rotation (longitudinal)	0.8 rad
	C3–C7	Flexion-extension	0.7 rad flexion, 0.4 rad extension
		Lateral flexion	1.7 rad
		Rotation (longitudinal)	1.5 rad
	Thoracic	Flexion-extension	0.06–0.2 rad
		Lateral flexion	0.06–0.16 rad
		Rotation (longitudinal)	0.16 rad
	Lumbar	Flexion-extension	0.3 rad
		Lateral flexion	0.1 rad
		Rotation (longitudinal)	0.03 rad
Shoulder	Glenohumeral, acromioclavicular, sternoclavicular, scapulothoracic	Flexion (sagittal plane)	3.14 rad
		Extension (sagittal plane)	0.9 rad
		Flexion (transverse plane)	2.4 rad
		Extension (transverse plane)	0.9 rad
		Abduction (frontal plane)	3.14 rad
		Internal-external rotation	3.14 rad (arm abducted and elbow at 1.57 rad)
Elbow	Humeroulnar, humeroradial	Flexion-extension	2.4 rad
	proximal radioulnar, humeroradial	Pronation-supination	1.2 rad pronation, 1.5 rad supination
Wrist	Radiocarpal, intercarpal	Flexion-extension	1.5 rad flexion, 1.4 rad extension
		Radial-ulnar deviation	0.3 rad radial, 0.6 rad ulnar
Hip	Femoral head-acetabulum	Flexion-extension	2.4 rad flexion, 0.3 rad extension
		Abduction-adduction	0.5 rad abduction, 0.4 rad adduction
		Internal-external rotation	1.2 rad internal, 1.5 rad external
Knee	Tibiofemoral	Flexion-extension	2.4 rad
		Internal-external rotation	0.7 rad internal, 0.5 rad external (knee at 1.57 rad)
		Abduction-adduction	<0.1 rad (knee at 0.5 rad)
Ankle	Tibiotalar, fibulotalar, distal tibiofibular	Flexion-extension	0.2–0.4 rad dorsiflexion, 0.4–0.6 rad plantarflexion
Foot	Subtalar	Inversion-eversion	0.3 rad inversion, 0.1 rad eversion
	Intertarsal and tarsometatarsal	Dorsiflexion, plantarflexion	<0.1 rad dorsiflexion, 0.3 rad plantarflexion
	Metatarsophalangeal	Flexion-extension	0.5 rad flexion, 1.5 rad extension

Note. The sagittal plane divides the body into left-right, the frontal plane distinguishes front-back, and the transverse plane separates the body into upper-lower. The range of motion data represent average values for young adults.

(passive transport) and actually helps other material to pass through (active transport). The sarcolemma is about 7.5 nm thick. The fluid enclosed within the fiber by the sarcolemma is referred to as **sarcoplasm**. Within the sarcoplasm are fuel sources (e.g., lipid droplets, glycogen granules), organelles (e.g., nuclei, mitochondria, lysosomes), enzymes (e.g., myosin ATPase, phosphorylase), and the contractile apparatus (bundles of myofilaments arranged into myofibrils).

In addition, the sarcoplasm contains an extensive, hollow, membranous system that is functionally linked to the surface sarcolemma and assists the muscle in conducting commands from the nervous system. This membranous system includes the sarcoplasmic reticulum, lateral sacs (terminal cisternae), and transverse (T) tubules (Figure 4.7). The **sarcoplasmic reticulum** runs longitudinally along the fiber, parallel to and surrounding the myofibrils. At specific locations along the

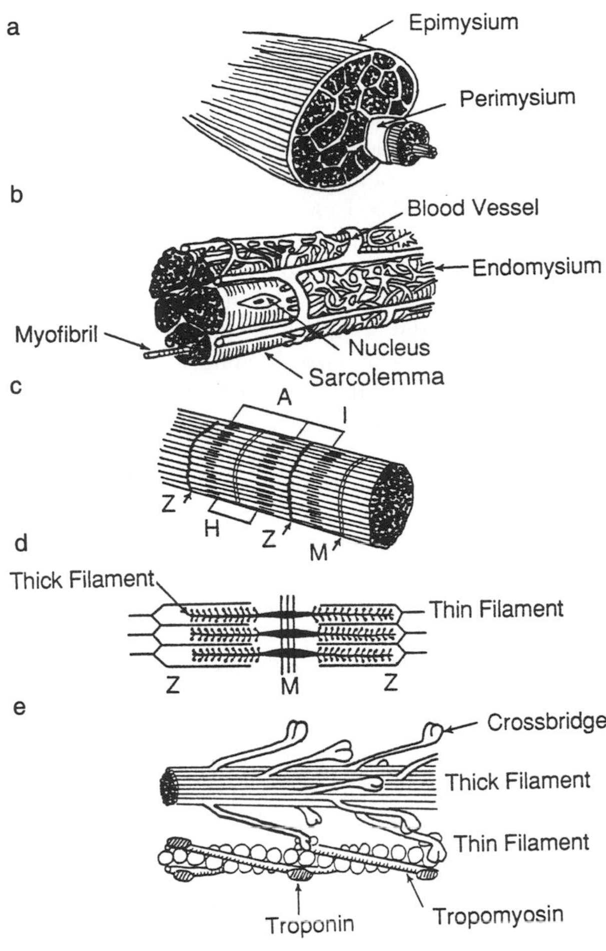

Figure 4.6 Organization of skeletal muscle from the gross to the molecular level. Note the levels of connective tissue within muscle, the bands and zones that comprise the sarcomere, and the molecular components of the thick and thin filaments: (a) whole muscle; (b) a group of muscle fibers; (c) one myofibril; (d) a sarcomere; (e) one thick and one thin filament.

Note. From "Biomechanics of Skeletal Muscle" by M.I. Pitman and L. Peterson. In *Basic Biomechanics of the Musculoskeletal System* (p. 90) by M. Nordin and V.H. Frankel (Eds.), 1989, Philadelphia: Lea & Febiger. Copyright 1989 by Lea & Febiger. Reprinted by permission.

myofibril, the sarcoplasmic reticulum bulges into **lateral sacs**. Perpendicular to the sarcoplasmic reticulum and associated with the lateral sacs are **transverse tubules**, branching invaginations of the sarcolemma. The term **triad** refers to one transverse tubule plus two terminal cisternae of the sarcoplasmic reticulum, one on each side of the T tubule. As we shall consider in more detail later, this triad aids in the rapid communication between the sarcolemma and the contractile apparatus.

Sarcomere

Skeletal muscle fibers can be regarded as a series of repeating units, each of which comprises the same characteristic banded structure. Histologically, this unit is defined as the **sarcomere** and represents the zone of a myofibril from one Z band to another (Figure 4.6d and Figure 4.7). A **myofibril** is a series of sarcomeres added end to end. The sarcomere is the basic contractile unit of muscle and comprises an interdigitating set of thick and thin contractile proteins (Figure 4.6d). Because a sarcomere has a length of about 2.5 µm in resting muscle, a 10-mm myofibril represents 4,000 sarcomeres added end to end. Each myofibril is composed of bundles of myofilaments (thick and thin contractile proteins) and has a diameter of about 1 µm. In many muscles, however, myofibrils are difficult to identify, and hence it is more appropriate to emphasize the myofilaments (Hoyle, 1983).

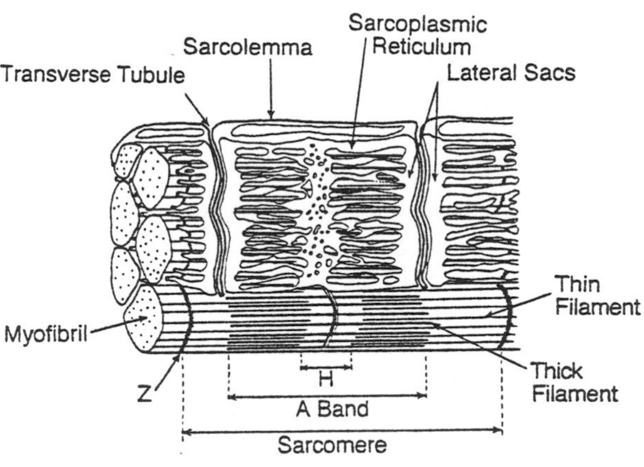

Figure 4.7 Alignment of the transverse tubules and sarcoplasmic reticulum with respect to the myofibrils. The figure shows six myofibrils that belong to a single muscle fiber.
Note. From *A Textbook of Histology* (9th ed.) (p. 281) by W. Bloom and D.W. Fawcett, 1968, Philadelphia: W.B. Saunders. Copyright 1968 by W.B. Saunders. Adapted by permission.

The obvious striations of skeletal muscle are due to the differential refraction of light as it passes through the contractile proteins. The thick-filament zone (Figure 4.6c), which includes some interdigitating thin filaments, is doubly refractive (i.e., the formation of two refracted rays of light from a single incoming ray) and comprises the dark band, called the **A band** (anisotropic). Within the A band is a zone that contains only thick filaments. Because this zone is clear of thin filaments, it is known as the **H band** (Hellerscheibe, or clear disk). The area between the A bands contains predominantly thin filaments and, because it is singly refractive, is called the **I band** (isotropic).

Each set of filaments (thick and thin) is attached to a central transverse band; the thick filaments attach to the **M band** (Mittelscheibe, or middle disk—the band located in the middle of the A band), and the thin filaments connect to the **Z band** (Zwischenscheibe, or between disk). A cross section through the A band shows that each thick filament (myosin with six protruding crossbridges) is surrounded by six thin filaments (small open circles), whereas a single thin filament can interact with only three thick filaments.

Myofilaments

The **myofilament** contains thick and thin filaments, both of which are composed of several proteins (Figure 4.8). The structure of the thin filament is dominated by actin but also includes the regulatory proteins tropomyosin and troponin. Each thin filament is composed of two helical strands of fibrous actin (**F-actin**) (Figure 4.8a). Each F-actin strand is a polymer (i.e., chemical union of two or more molecules) of some 200 globular actin (**G-actin**) molecules (Figure 4.8b). A G-actin molecule is a protein containing about 374 amino acids.

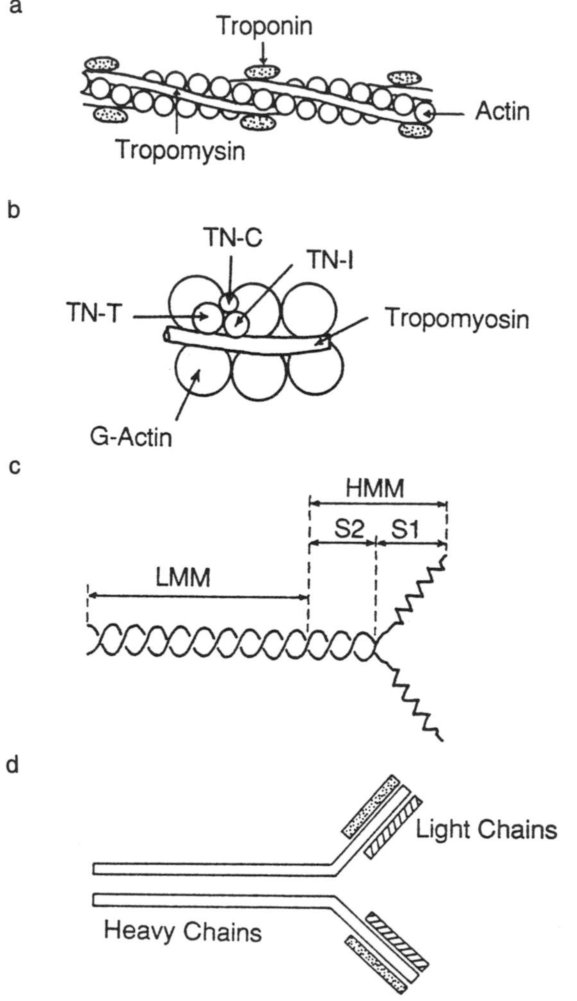

Figure 4.8 Organization of the myofilaments: (a) the thin filament; (b) troponin (TN) elements; (c) fragments of the myosin molecule; (d) myosin heavy and light chains.
Note. Part D is from "Myosin Isoenzymes as Molecular Markers for Muscle Physiology" by R.G. Whalen, 1985, *Journal of Experimental Biology*, **115**, p. 46. Copyright 1985 by The Company of Biologists Limited. Adapted by permission.

Located in the groove of the F-actin helix are two coiled strands of **tropomyosin** (Figure 4.8, a and b). The structure of tropomyosin is referred to as a two-chain coiled-coil; each of these chains contains approximately 284 amino acids. The **troponin** (TN) complex has a globular structure that includes three subunits (Figure 4.8b); the TN-T unit binds troponin to tropomyosin, TN-I inhibits four to seven G-actin molecules from binding to myosin when tropomyosin is present, and TN-C can reversibly bind Ca^{2+} ions as a function of calcium concentration (see Appendix F for TN-T, TN-I, and TN-C). TN-C has four binding sites, two for Ca^{2+} and two for Ca^{2+} or Mg^{2+}. Thus the thin filament has as its backbone two strands of actin molecules (F-actin) upon which are superimposed (wrapped around or attached to) two-strand (tropomyosin) and globular (troponin) proteins. These proteins operate in much the same way during a muscle contraction; tropomyosin and troponin influence the activity of actin.

A set of thin filaments that projects longitudinally into one sarcomere connects in the Z-band region to another set, which projects in the opposite direction into an adjacent sarcomere. At this connection, each thin filament appears to be linked to its four closest neighbors. A region of considerable flexibility, the Z band changes its shape under different conditions. Z-band width can vary from one muscle fiber to another (e.g., different muscle fiber types), and it probably also varies as a consequence of training (Sjöström, Ängquist, et al., 1982; Sjöström, Kidman, Larsén, & Ängquist, 1982).

Of the thick filament proteins, most is known about the myosin molecule (Cooke, 1990). It is a long, two-chain, helical tail that terminates in two large globular heads (Figure 4.8c). With the aid of an enzyme (protease), the myosin molecules can be decomposed into **light meromyosin** (LMM) and **heavy meromyosin** (HMM) fragments, the latter of which is further subdivided into **subfragments 1** and **2** (S1 and S2). LMM has a relatively low molecular weight (135 kilodaltons [kDa]), whereas HMM is heavier (335 kDa). The globular heads of the myosin molecule, one of which contains an ATP- and the other an actin-binding side, are known as S1 (115 kDa). The remaining portion of HMM is called S2 (60 kDa). The role of these four components (HMM, LMM, S1, and S2) in a muscle contraction has been examined extensively by muscle biologists.

Because LMM binds strongly to itself under physiological conditions, approximately 400 myosin molecules aggregate to form the dominant element of the thick filament (Pepe & Drucker, 1979). The union is not random but structured. The molecules are aligned in pairs, and the S1 element of each molecule is oriented to its partner at 3.14 rad (180°). The next pair is displaced by a translation of about 0.0143 μm and a rotation of 2.1 rad (120°). The result is an ordered alignment of myosin molecules in which the HMM projections (crossbridges) encircle the thick filament (Figure 4.6e). Each sarcomere actually contains two such sets of myosin molecules; however, because the S1 elements of the two sets point in opposite directions, the LMM fragments unite in the M band (Figure 4.6c) to form a single filament.

The myosin molecule contains two hinge regions (i.e., zones of relatively greater flexibility). These occur at the LMM-HMM and S1-S2 junctions. In the resulting alignment, the HMM fragment can extend from the thick filament to within close proximity of the thin filament (Figure 4.6e). Due to the ability of S1 to interact with actin, the HMM extension has been called the **crossbridge**. Each thick filament is surrounded by and can interact with six thin filaments because the crossbridges encircle the thick filament. There are about 1,600 thick filaments/μm^2 in the human quadriceps femoris muscle (Claassen, Gerber, Hoppeler, Lüthi, & Vock, 1989).

The proteins that comprise the contractile apparatus can be distinguished as the products of different genes. Eight multigene families contribute the major components of the sarcomere: myosin heavy chain, alkali light chain, DTNB light chain, actin, tropomyosin, troponin C, troponin I, and troponin T (Gunning & Hardeman, 1991; Tsika, Herrick, & Baldwin, 1987). The first three of these components form the myosin molecule. Because proteins comprise sequences (chains) of amino acids, a protein with a high molecular weight (200 kDa) is referred to as a **heavy chain**. Conversely, a protein with a low molecular weight (<30 kDa) is considered a **light chain**. And a given protein that is synthesized with a slightly different amino acid composition is known as an **isoform**; that is, there can be different isoforms (amino acid compositions) of the same protein that may or may not be the product of different genes (Babij & Booth, 1988). Isoforms are sometimes referred to as *isoenzymes*. There can be different isoforms of the myosin heavy chain and different isoforms of the light chain (Table 4.3). Although the different heavy chain isoforms appear to have physiological significance, less is known about the functional consequences of differences in other contractile-protein isoforms. However, isoforms of the contractile proteins do change during development and as a consequence of altered physical activity.

The myosin molecule consists of two coiled heavy chains with light chains attached to the myosin heads (Figure 4.8d). The isoforms differ in cardiac, smooth, and skeletal muscle. The heavy chains in skeletal muscle have a molecular weight of 200 kDa. There is a strong relationship between the maximum velocity at which muscle fibers can shorten and the ATPase activity of the heavy chain isoform contained in the fiber. There appear to be one slow and five fast heavy chain isoforms (Table 4.3). Four light chains are attached to the globular heads, and these are distinguishable by molecular weight (16, 18 to 20, and 25 to 27 kDa), by whether they can be phosphorylated, and by the experimental agent (alkali or DTNB) that separates them from the heavy chain.

Table 4.3 Isoforms of the Contractile Proteins in Adult Skeletal Muscles

	Skeletal muscle	
Gene family	Slow	Fast
Myosin heavy chain	S	F_{2A}, F_{2B}, F_{2X}, F_{EO}, F_{SF}
Alkali light chain	1_{Sa}, 1_{Sb}	1_F, 3_F
Dithionitrobenzoic Acid light chain	2_S, $2_{S'}$	2_F
Actin	α_{sk}	α_{sk}
Tropomyosin	β, α_S	β, α_F
Troponin C	S	F
Troponin I	S	F
Troponin T	S	F

Note. S = slow, F = fast; F_{2A}, F_{2B} = correspond to the two fast-twitch fibers defined by histochemistry; F_{2X} = defined by antibody staining and protein analysis; F_{EO} = found in adult extraocular muscle; F_{SF} = super-fast contractile proteins of jaw muscle.

From "Multiple Mechanisms Regulate Muscle Fiber Diversity" by P. Gunning & E. Hardeman, 1991, *FASEB Journal*, **5**, p. 3065. Copyright 1991 by Federation of American Societies for Experimental Biology. Reprinted by permission.

As with the heavy chains, there appear to be different isoforms of the light chains for fast- and slow-twitch muscle fibers (Table 4.3). The light chains probably modulate the interaction between actin and myosin, although the specific function is unknown.

Cytoskeleton

Since the initial proposal of the sliding filament theory of muscle contraction, it has become obvious that there must exist a set of structures that determines the organization of sarcomeres both within a single myofibril and between adjacent myofibrils. The set of structures that provides the physical framework for the interaction of the contractile proteins has been termed the **cytoskeleton** (Cooke, 1985). The cytoskeleton has been described as consisting of two lattices; the **exosarcomeric** cytoskeleton maintains the lateral (side-by-side) alignment of the myofibrils, and the **endosarcomeric** cytoskeleton maintains the orientation of the thick and thin filaments within the sarcomere (Waterman-Storer, 1991).

The exosarcomeric cytoskeleton consists of **intermediate filaments** that are arranged longitudinally along and transversely across sarcomeres (Figure 4.9). The intermediate filaments—which consist of such proteins as desmin, vimentin, and synemin—are localized at the Z bands and connect each myofibril to its neighbor and to the sarcolemma. The connection of the intermediate fibers to the sarcolemma and subsequently to the surrounding tissue probably involves specialized structures (Thornell & Price, 1991), especially at the junction between muscle and tendon (Tidball, 1991). For example, the transmembrane protein integrin may serve to connect myofibrils to the extracellular matrix of connective tissue (collagen). Some cytoskeletal filaments also span between M bands in adjacent myofibrils. The endosarcomeric cytoskeleton acts as a third filament system along with actin and myosin. This system is extensible and consists of the proteins titin and nebulin. Titin is thought to be responsible for resting muscle elasticity, whereas nebulin maintains the lattice array of actin (Waterman-Storer, 1991).

Neuron

We have thus far described three elements of the single joint system: the rigid link that forms the structural basis of the system, a joint about which rigid links rotate, and an organ known as muscle that is capable of exerting a force on the rigid links. Next we consider the nervous system and its cellular components, which represent the means by which we activate muscle.

Despite years of study, not much is known about the function of the nervous system. Understandably, we know more about relatively simple functions such as reflexes than about more complex processes such as learning or the control of movement. Furthermore, we know more about the properties of single nerve cells than about the behavior of groups of cells. Our model (single joint system) includes only a few rudimentary elements of the nervous system so that we can focus on the basic features related to the control of movement.

There are only two cell types in the nervous system: **neurons** and **neuroglia**. The neuron is characterized by a distinctive cell shape, an outer membrane (axolemma) capable of generating and conducting an electrical signal, and a unique structure (synapse) for the transfer of information. Less is known about the neuroglia, which are nine times more numerous than neurons. Neuroglia are primarily thought to provide structural and metabolic repair following injury and protective support for the neurons (Somjen, 1987; Varon & Somjen, 1979). For example, three prominent functions known to be performed or assisted by glial cells are myelination, phagocytosis, and metabolism. Myelination is accomplished by oligodendrocytes in the central nervous system and by Schwann cells in the peripheral nervous system. In this process, the surface membrane of the glial cell (oligodendrocyte or Schwann cell) wraps around the axon, a branch of the neuron that is involved in sending out commands (Figure 4.10). One consequence of this myelination is that the commands sent by the neuron travel at a much greater speed. In phagocytosis, glial cells (microglia) are known to proliferate around damaged neurons (injured or degenerating) and to transform into large macrophages that remove the debris. The contributions of neuroglia to metabolism involve modulation of the ions, transmitters, and metabolites that are necessary for the normal function of neurons.

Although neurons are a morphologically diverse group of cells, their common function is performed in three distinct phases: (a) the reception of information (input), (b) an evaluation of the input to determine whether an output signal should be transmitted, and (c) transmission of the output signal. A typical neuron has four morphological regions—dendrites, soma, axon,

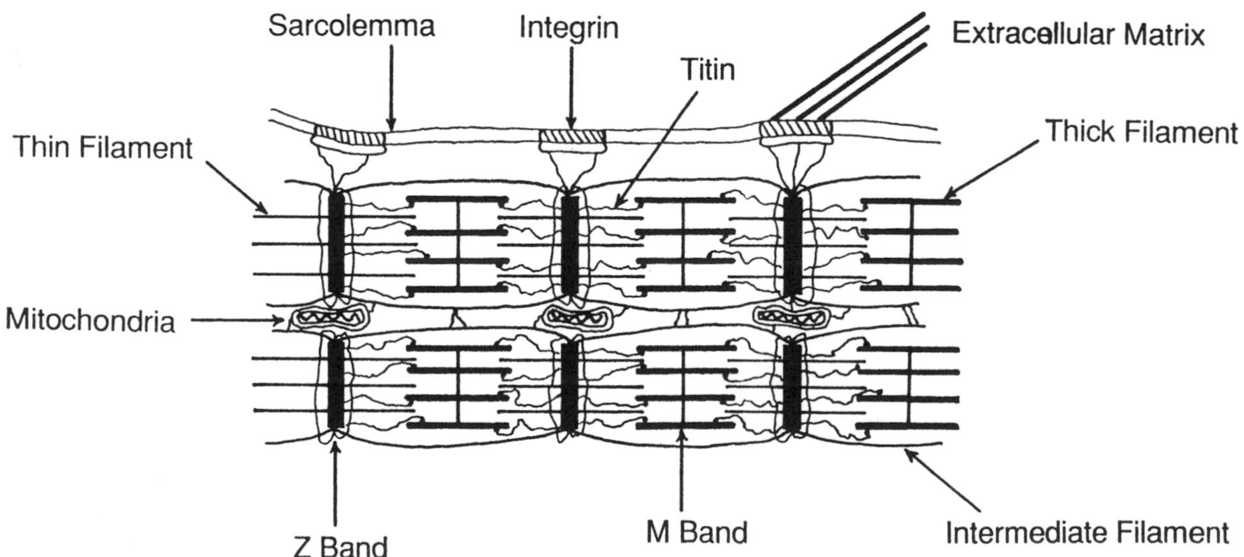

Figure 4.9 Organization of the cytoskeleton in skeletal muscle. Cytoskeletal filaments connect adjacent sarcomeres (thick and thin myofilaments) to one another in series and in parallel and to the extracellular matrix of connective tissue.
Note. Adapted from Thornell and Price (1991) and Waterman-Storer (1991).

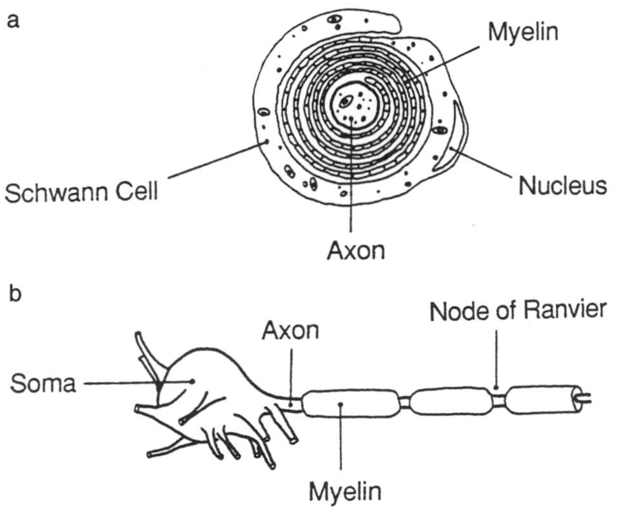

Figure 4.10 Myelination of an axon by a Schwann cell: (a) cross-sectional view of a Schwann cell ensheathing an axon; (b) an axon ensheathed by many Schwann cells, which provide an interrupted covering of myelin. The gaps in the myelin are known as nodes of Ranvier.

and presynaptic terminal (Figure 4.11)—which interact to accomplish these tasks (Palay & Chan-Palay, 1987). The **soma** (cell body) contains the apparatus (e.g., nucleus, ribosomes, rough endoplasmic reticulum, Golgi apparatus) needed for the synthesis of macromolecules. The **axon** is a tubular process that arises from the soma at the axon hillock. One of its functions is to serve as a cable for transmitting the output signal (Rall, 1987), which is an electrical event known as the **action potential**. The axon hillock is the most excitable portion of the axon and represents the site of initiation of the action potential. The axon usually gives off branches that are referred to as **collaterals**. Large axons are surrounded by a fatty insulating sheath called **myelin**, which is provided by Schwann cells in peripheral nerves. The myelin increases the rate at which the action potential is transmitted along the axon (conduction velocity). Near its end, the axon divides into many fine branches that form functional contacts with the receptive surface of cells. These contacts are referred to as **synapses**. The ending of the axon involved in the synapse is identified as the **presynaptic terminal** and includes the means for transferring the output signal from the neuron to the effector cell. In neuron-to-neuron interactions, the most common receptive site is the **dendrites**, the other processes extending from the soma. Although synapses can occur between any neuronal parts (e.g., axodendritic, axosomatic, axoaxonic, dendrodendritic, dendrosomatic), about 80% of the input sites (synapses) are located on the dendrites.

Three functional classes of neurons are important in the single joint system: afferent, interneuron, and efferent (Figure 4.12). This scheme represents the flow of

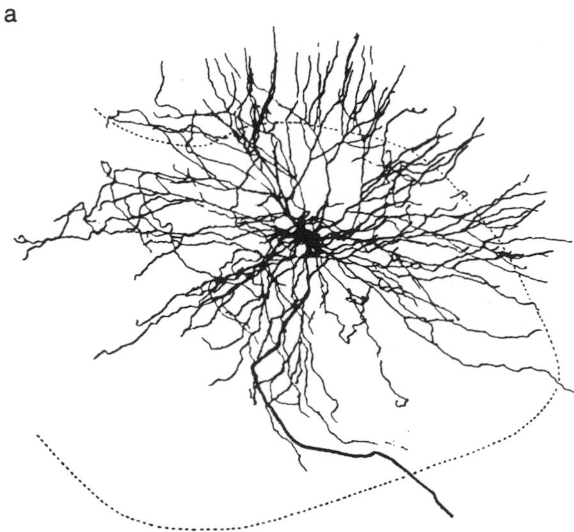

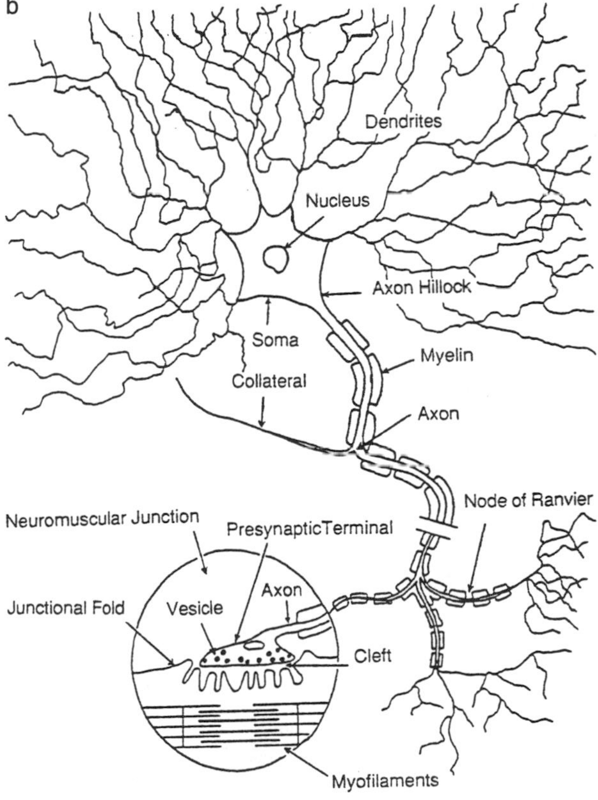

Figure 4.11 Morphological features of neurons. (a) A reconstruction of a motor neuron from the spinal cord of a cat shows a large dendritic tree surrounding a small soma and an axon exiting at the bottom of the figure. (b) An idealized (not to scale) neuron showing the four morphological regions: dendrites, soma, axon, and presynaptic terminal. *Note.* Part a is from "Electrophysiological and Morphological Measurements in Cat Gastrocnemius and Soleus α-Motoneurones" by B. Ulfhake and J.-O. Kellerth, 1984, *Brain Research*, **307**, p. 170. Copyright 1984 by Elsevier Science Publishers B.V. Reprinted by permission.

information from the surroundings (outside the single joint system) into the system (central nervous system) and ends with a response. **Afferent** neurons convey sensory information (action potentials) from the surroundings into the central nervous system. We will examine the role of the different sensory modalities in the control of movement when we discuss sensory receptors, the fifth element of the single joint system. Afferent signals enter the central nervous system and act at a local level; they are also distributed throughout the central nervous system. Simple reflexes are an example of local effects. An afferent signal enters the spinal cord and results in a response (output) signal from the same level of the spinal cord. **Interneurons** account for 99% of all neurons and represent the central nervous system component that modulates the interaction between input (afferent) and output (efferent) signals. Interneurons can elicit excitatory and inhibitory responses in other neurons. This modulation (excitation and inhibition) can occur directly, with the interneuron forming part of the circuit between afferent and efferent, or it can occur indirectly, whereby the interneuron can alter the excitability of the connection between the afferent and efferent (Figure 4.12). In addition, the efferent neuron can receive input from other structures within the central nervous system, and the interneuron can also modulate this interaction.

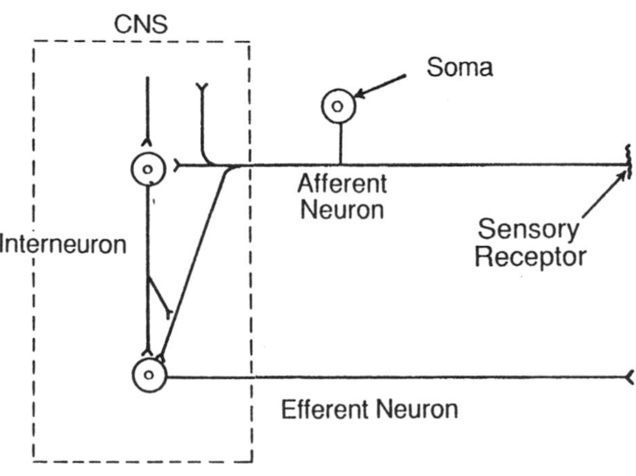

Figure 4.12 The three functional classes of neuron include the afferent neuron, interneuron, and efferent neuron.

Efferent neurons transmit the output signal (action potentials) from the central nervous system to the effector organ. In the single joint system, the effector organ is muscle. Efferent neurons that innervate muscle are referred to as **motor neurons**. The somas of these neurons are located in the brain and in the gray matter of the spinal cord, and their axons exit the cord and are bundled together into peripheral nerves that course to the target muscles. Forty-three pairs of nerves (12 cranial and 31 spinal) in the human body leave the central nervous system (CNS) and form the peripheral nervous system (Figure 4.13). The spinal cord is often described as a segmented structure in which the segments correspond to the vertebrae. Between each pair of vertebrae, a set of axons exits and another set enters on each side (left and right) of the cord. The axons belonging to efferent neurons exit the spinal cord in the **ventral** (front) **roots**, whereas the axons of the afferent neurons enter through the **dorsal** (back) **roots** (Figure 4.14). Motor neurons have large-diameter, myelinated axons that traverse from the spinal cord directly to skeletal muscle. The somas of the motor neurons are located in the ventral horn of the spinal cord.

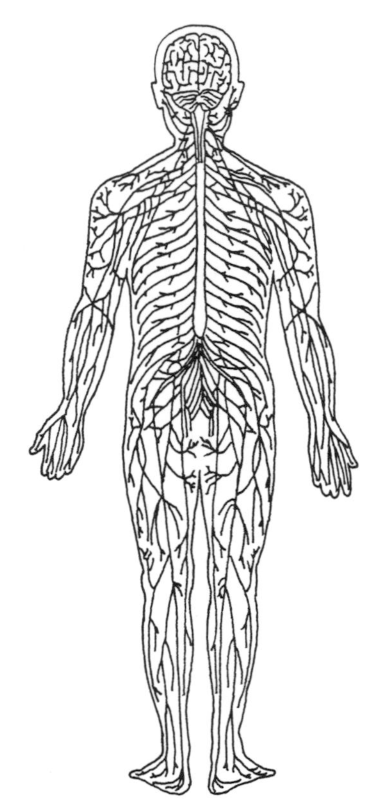

Figure 4.13 The human nervous system. The central nervous system consists of the brain and spinal cord. The peripheral nervous system consists of the nerves that exit from the brain and spinal cord and innervate the entire body.
Note. From *Human Physiology: The Mechanisms of Body Function* (p. 168) by A.J. Vander, J.H. Sherman, and D.S. Luciano, 1990, New York: McGraw-Hill, Copyright 1990 by McGraw-Hill. Reprinted by permission.

In contrast to motor neurons that innervate skeletal muscle, some efferent neurons connect to cardiac muscle, smooth muscle, and glands. These neurons form the **autonomic nervous system**. These neurons control such physiological processes as arterial blood pressure, gastrointestinal motility and secretion, perspiration, and body temperature. Most of these functions are not under voluntary control but instead are regulated by autonomic

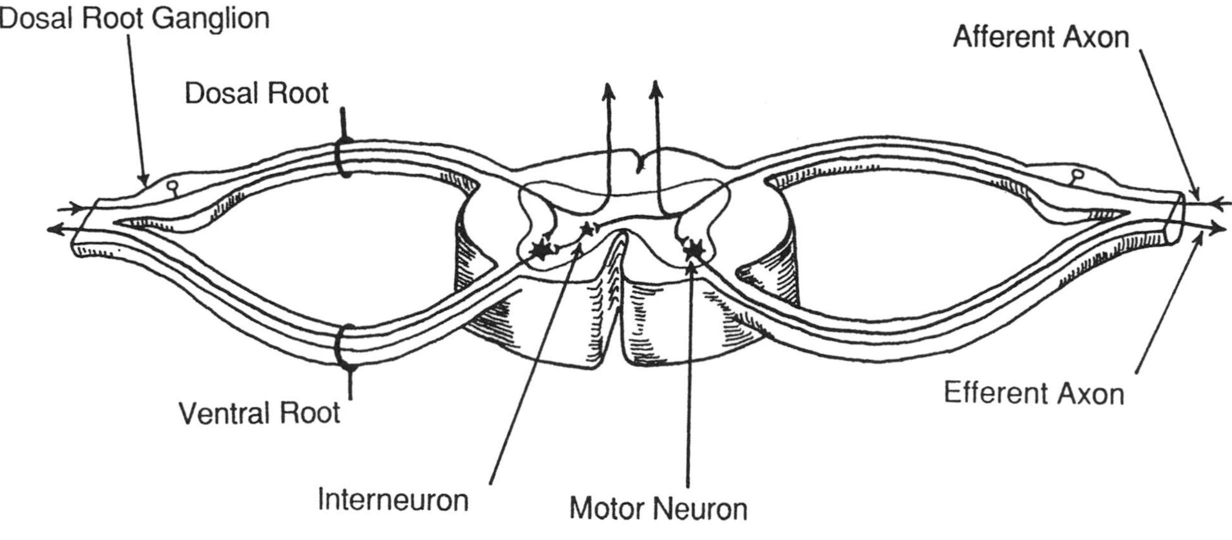

Figure 4.14 Segmental organization of the spinal cord. At the level of each vertebra in the spinal column, the spinal cord gives off a pair of dorsal and ventral roots to each side (right and left) of the body.

reflexes. Whereas motor neurons directly innervate skeletal muscle, there are two neurons in the autonomic nervous system between the CNS and the effector cell. One of these neurons has its soma in the CNS, and the other has its soma outside the CNS in cell clusters known as **autonomic ganglia**. Based on anatomical and physiological differences, the autonomic nervous system can be subdivided into sympathetic and parasympathetic divisions. In the **sympathetic** division, the cell bodies of the first neurons are located in the thoracic spinal cord, and the autonomic ganglia containing the cell bodies of the second neurons are usually located close to the spinal cord. In the **parasympathetic** division, the cell bodies of the first neurons are located in the brain and the sacral region of the spinal cord, and the autonomic ganglia contain the cell bodies of the second neurons, located within the effector organ.

As the axons of motor neurons exit the spinal cord in the ventral root, they form peripheral nerves that also contain axons of afferent neurons and the autonomic nervous system. When the nerve reaches the muscle, it subdivides first into primary nerve branches and then into smaller branches until single axons contact single muscle fibers (Figure 4.15). The connection (synapse) between an axon and a muscle fiber is known as a **neuromuscular junction**; sometimes it is also referred to as a *motor end plate*. At the neuromuscular junction, the presynaptic membrane (axon) is separated from the postsynaptic membrane (muscle) by a 1- to 2-µm cleft. An action potential generated by the motor neuron is transmitted across this cleft by an electrochemical process in which the electrical energy embodied in the nerve action potential is converted to chemical energy in the form of a neurotransmitter. The excitation associated with the nerve action potential generates the release of a chemical neurotransmitter by the presynaptic terminal; this process is considered in more detail in chapter 5. At the neuromuscular junction the neurotransmitter is **acetylcholine** (ACh) (Iversen, 1987). The neurotransmitter, in turn, causes a change in the permeability and the electrical status of the postsynaptic membrane such that the signal is converted to a muscle (sarcolemmal) action potential. Thus the energy contained in the action

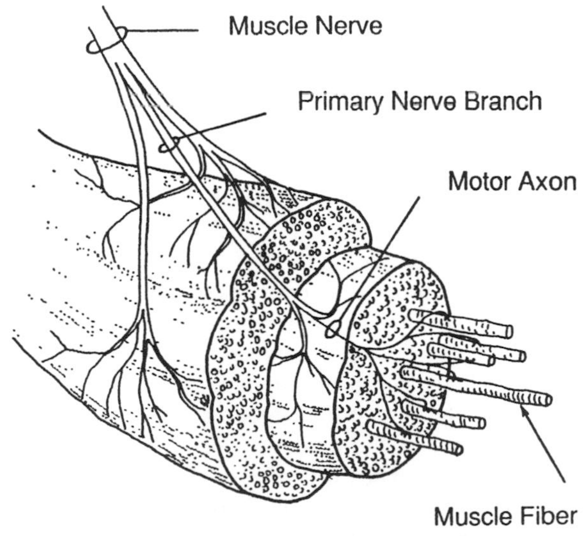

Figure 4.15 Subdivision of the muscle nerve down to the level of single axons that innervate single muscle fibers. Primary nerve branches can produce a dissociation in the activation of different parts of a muscle.
Note. From "Structure and Function in Vertebrate Skeletal Muscle" by S.E. Peters, 1989, *American Zoologist*, **29**, p. 222. Copyright 1989 by the American Society of Zoologists. Adapted by permission.

potential of the motor neuron is converted into chemical energy by the release of neurotransmitter and then back to electrical energy by the generation of the muscle action potential. By this process, the motor neuron represents the fourth element of the single joint system and provides the ability to control the force exerted by muscle.

Sensory Receptor

Now we consider the sensory receptor, the fifth and final element of the single joint system. The reason for including the first four elements in this biological model for the control of movement is intuitively obvious, but why include sensory receptors? The basic function of sensory receptors is to provide information to the system on its own state and that of its surrounding environment. This type of information flow, from the sensory receptors to the CNS, is sometimes referred to as **feedback**; it represents the transfer of information back to the CNS. *It appears as a general principle, both in engineering design and in biological systems, that the more maneuverable a system, the more feedback it requires to maintain its stability* (Hasan & Stuart, 1988). As a biological system, the human body is highly maneuverable in that we can perform all sorts of movements, which require considerable feedback to control. In fact, the number of afferent neurons that provide feedback information is much greater than the number of efferent neurons involved in activating muscle.

Sensory receptors convert energy from one form to another through a process known as **transduction**. Energy can exist in a variety of forms, such as light, pressure, temperature, and sound, but the common output of sensory receptors is electrochemical energy in the form of action potentials. The action potentials are transmitted centrally and used by the CNS to monitor the status of the musculoskeletal system. The human body contains many types of sensory receptors, which can be distinguished on the basis of their location (exteroceptors, proprioceptors, interoceptors), function (mechanoreceptors, thermoreceptors, photoreceptors, chemoreceptors, nociceptors), and morphology (free nerve endings, encapsulated endings).

The single joint system needs at least two types of information to control movement. It needs to know where it is and when it is disturbed by something that happens in its environment. This information is provided by **proprioceptors**, which detect stimuli generated by the system itself, and by **exteroceptors**, which detect external stimuli (Sanes, 1987). With this information, the single joint system is able to organize a rapid response to a disturbance, to determine its position, and to distinguish between self-generated and imposed movements. Proprioceptors include muscle spindles, tendon organs, and joint receptors. Exteroceptors include the eyes, ears, and skin receptors that respond to temperature, touch, and pain.

Muscle Spindle

Provided a muscle operates across a joint and is subject to unexpected loads, it will contain a variable number of muscle spindles (6 to 1,300) distributed throughout the muscle (Hasan & Stuart, 1984; Matthews, 1972). The spindles are fusiform-shaped and lie in parallel with the skeletal muscle fibers (Figure 4.16). Although the muscle spindle is a morphologically complex sensory receptor, it is essentially a collection of miniature skeletal muscle fibers (2 to 12) enclosed in a connective tissue capsule (Figure 4.17). These smaller muscle fibers are referred to as **intrafusal fibers**; those outside the muscle spindle are called **extrafusal fibers**. Because of

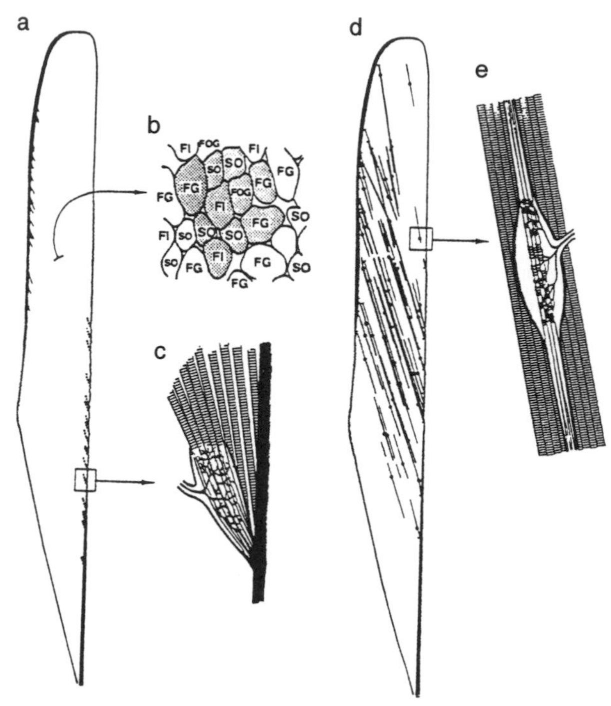

Figure 4.16 Distribution of the muscle spindle and tendon organ in the medial gastrocnemius muscle of the cat: (a) a longitudinal section of the muscle with tendon organs stained (the tendon organs are associated with the aponeurosis); (b) a cross section through the muscle showing the mixture of muscle fiber types; (c) enlarged view of a single tendon organ in series with skeletal muscle fibers; (d) a longitudinal section of the muscle showing the muscle spindle distribution throughout the belly of the muscle; (e) enlarged view of a muscle spindle showing its location in parallel with the skeletal muscle fibers.
Note. From "Functional Anatomy of the Association Between Motor Units and Muscle Receptors" by B.R. Botterman, M.D. Binder, and D.G. Stuart, 1978, *American Zoologist*, **18**, p. 136. Copyright 1978 by the American Society of Zoologists. Reprinted by permission.

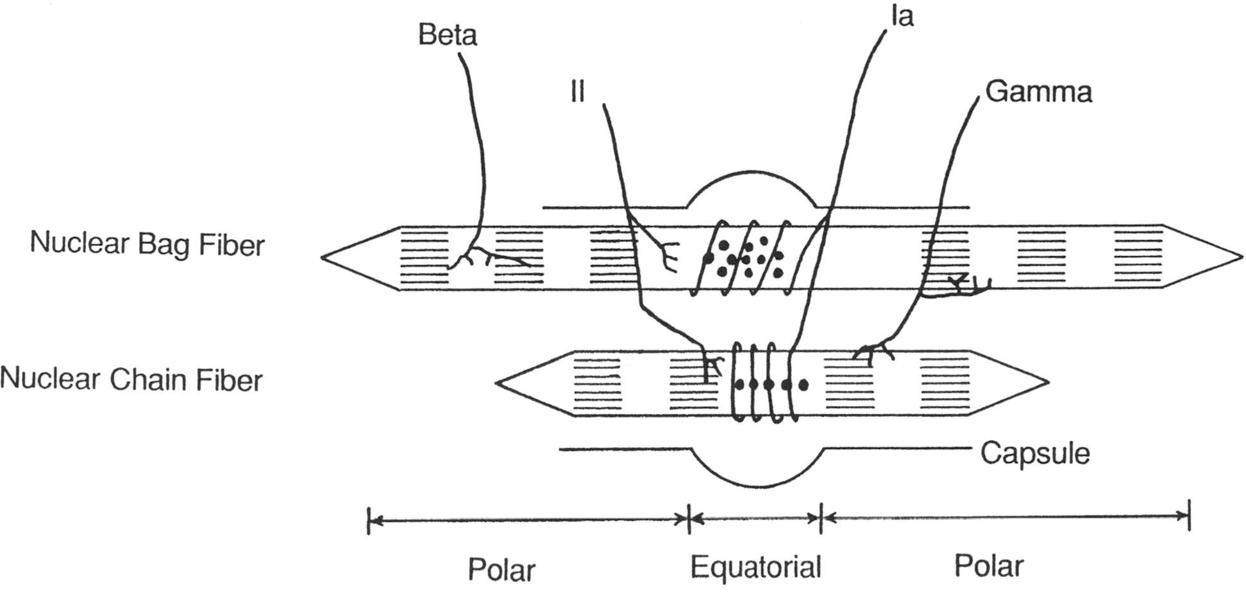

Figure 4.17 A schematized muscle spindle.

the greater myofilament content (cross-sectional area) of the extrafusal fiber, it can generate approximately 36 times more force than an intrafusal fiber. There are two types of intrafusal fiber, which differ as to the arrangement of their nuclei. The nuclei of the **nuclear chain fiber** are arranged end to end as the links in a chain, whereas those in the **nuclear bag fiber** cluster in a group. Both types of fiber, however, are devoid of myofilaments in the *equatorial* (central) region. The nuclear bag fiber is the longer of the two, extending at both ends beyond the capsule.

As a sensory receptor, the muscle spindle has an afferent supply over which the action potentials are transmitted to the CNS. In general, afferent axons are classified into four groups, primarily due to differences in axonal diameter. Group I axons have the greatest diameters, and Group IV, the smallest; the larger the axon diameter, the faster the action potentials can be conducted (Table 4.4). Each muscle spindle has a variable number (8 to 25) of Group I and Group II afferents. The larger of the two afferent axons, the **Group Ia afferent**, has an ending that spirals around the equatorial regions of both the nuclear chain and bag fibers. The **Group II afferent** has a nonspiral ending that connects principally to the chain fibers. Not all muscle spindles have Group II afferents, but they all have Group I afferents. The somas of Group I and Group II afferent neurons are located in the dorsal root ganglion, close to the spinal cord (Figures 4.12 and 4.14).

In addition to an afferent system, the intrafusal fibers of the muscle spindle receive efferent input. In general, skeletal muscle fibers are innervated by three groups of motor neurons (α, β, γ), which can be distinguished by size and by the fibers that they innervate (Table 4.4).

Table 4.4 Classification of Nerve Fibers

Fiber type	Function[a]	Fiber diameter (μm)	Conduction velocity (m/s)
Efferent			
Aα	Skeletal muscle	15	100
Aβ	Skeletal muscle + muscle spindle	8	50
Aγ	Muscle spindle	5	20
B	Sympathetic preganglionic	3	7
C	Sympathetic postganglionic (unmyelinated)	1	1
Afferent			
Ia	Muscle spindle	13–20	80–120
Ib	Tendon organ	13–20	80–120
II	Muscle spindle	6–12	35–75
III	Deep pressure sensors in muscle	1–5	5–30
IV	Pain (unmyelinated), temperature	0.2–1.5	0.5–2

[a]Examples of functions served by the different classes of nerve fibers.

The **alpha (α) motor neurons** are the largest and innervate the extrafusal muscle fibers; **gamma (γ) motor neurons** are the smallest and connect exclusively to intrafusal muscle fibers; and **beta (β) motor neurons** are intermediate in size and innervate both the extrafusal and intrafusal muscle fibers. The beta and gamma motor neurons contact the intrafusal fibers in their myofilament-rich *polar* regions. When an action potential is initiated at the axon hillock of a beta or gamma motor neuron and transmitted to a muscle spindle, the net effect is a contraction (shortening) of the intrafusal fibers in the polar regions and a stretch of the equatorial region and of the muscle spindle afferents (Groups I and II). The stretch results in activation of the afferents by the generation of action potentials that are transmitted to the CNS.

Communication within the afferent element of the single joint system begins with the development of a **generator potential** by the sensing elements of the sensory receptors. For mechanoreceptors (e.g., muscle spindle, tendon organ, joint receptor), the generator potential is created by mechanical deformation of the sensory terminal. The generator potential is conducted along the axon to the trigger zone, and if the signal is large enough, an action potential is initiated. The most excitable portion of the afferent axon, the trigger zone, is located close to the sensing elements, usually within the capsule of the muscle spindle and the tendon organ. The action potential is transmitted centrally, where it impinges on various motor neurons and interneurons.

Muscle spindles provide information on *changes in muscle length*. Because muscle spindles are arranged in parallel with skeletal muscle fibers, there are two ways that the sensory receptor can be activated and thereby provide an afferent signal related to muscle length. The first, described above, is by the gamma or beta motor neurons causing the intrafusal muscle fibers to contract and stretch the equatorial region, where some of the Group I and Group II afferent endings are located. The stretch reversibly deforms the afferent endings and results in the generation of an action potential (the axon hillock is located near the sensory receptor) that is transmitted back to the CNS. Because the gamma and beta motor neurons can activate the muscle spindle (fusiform shape), they are sometimes referred to as **fusimotor** neurons. The second way to activate the spindle is by passive stretch of the entire muscle.

Figure 4.18 shows how muscle spindle activity was found to work in one study (Al-Falahe, Nagaoka, & Vallbo, 1990). The experiment measured the discharge of a single muscle spindle in the extensor digitorum muscle in the forearm of a human subject while a finger was moved sinusoidally either by the experimenter (Figure 4.18a) or by the subject (Figure 4.18b). The muscle spindle activity was recorded by placing a probe (an electrode) in the muscle nerve; this technique is known as microneurography. The bottom trace in the figure shows the action potential discharge by the muscle spindle during the two conditions. The second-to-bottom trace shows the discharge rate of the muscle spindle, which reached peak values of about 25 Hz. The third trace from the bottom shows the metacarpophalangeal joint angle, with an upward deflection indicating flexion and stretch of the extensor digitorum muscle. The top trace shows the electrical activity (EMG, electromyogram) in the extensor digitorum muscle; the muscle was electrically silent when the movement was imposed by the investigator (a) but active when performed by the subject (b). These data show that for both types of movement, imposed and active, the muscle spindle discharge increased when the extensor digitorum muscle was stretched during the flexion (upward) phase of the movement.

The information on muscle stretch that is provided by the muscle spindle can serve at least two purposes: This information can help the CNS determine the posi-

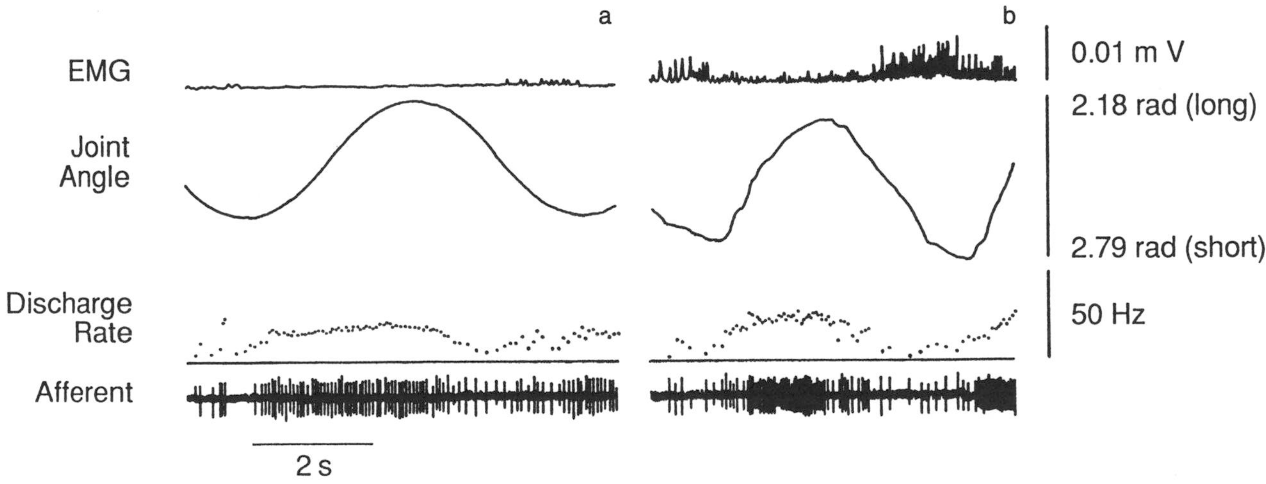

Figure 4.18 Discharge of a human muscle spindle during a passive (a) and active (b) stretch (upward joint angle) of a forearm muscle.
Note. From "Response Profiles of Human Muscle Afferents During Active Finger Movements" by N.A. Al-Falahe, M. Nagaoka, and A.B. Vallbo, 1990, *Brain*, **113**, p. 330. Copyright 1990 by Oxford University Press. Adapted by permission of Oxford University Press.

tion and orientation of the single joint system, and it can signal a disturbance imposed on the system by its surroundings. These processes are examined in chapter 5.

Tendon Organ

In contrast to the muscle spindle, the tendon organ is a relatively simple sensory receptor; it includes a single afferent and no efferent connections (Hasan & Stuart, 1984). Few tendon organs are located in the tendon proper. Most are arranged around a few extrafusal muscle fibers as they connect with an aponeurosis of attachment (Figure 4.16). An *aponeurosis* refers to the tendinous sheaths that usually extend along and deep into the belly of the muscle. Due to this location, the tendon organ is described as being in series with skeletal muscle fibers. The sensory terminal of the afferent neuron is contained within a capsule and branches to encircle several strands of collagen that comprise the aponeurosis (Figure 4.19a). It is estimated that about 10 skeletal muscle fibers are included in a typical tendon organ capsule and that each of these muscle fibers is innervated by a different alpha motor neuron. The afferent neuron associated with the tendon organ is referred to as the Group Ib afferent (Table 4.4).

When a muscle and its connective tissue attachments are stretched, either by pulling the muscle (passive stretch) or by activating the skeletal muscle fibers (active stretch), the strands of collagen pinch and excite the Group Ib afferent (Figure 4.19b). Because the tendon organ is activated in this way, it is described as a *monitor of muscle force*. The level of force necessary to excite a tendon organ depends upon the mode of activation. Passive stretch requires a muscle force of 2 N, whereas the activity of a single muscle fiber (30 to 90 μN) is sufficient in active force conditions (Binder, Kroin, Moore, & Stuart, 1977).

An example of the discharge of a tendon organ is shown in Figure 4.20 (Al-Falahe et al., 1990). The microneurography technique was used to record the discharge of a single tendon organ located in the extensor digitorum muscle of a human volunteer. The subject performed a finger movement against a zero load (Figure 4.20a) and against a light load (Figure 4.20b). In each panel, the top trace records the angle of the metacarpophalangeal joint, with flexion of the joint and stretch of the muscle shown as an upward deflection. The middle trace indicates the discharge rate of the tendon organ, which achieved peak rates of about 40 Hz. The bottom trace shows the EMG (electrical activity) of the muscle performing the finger movement. Figure 4.20 shows how the discharge of a tendon organ parallels the EMG. Because there is a close association between muscle EMG and force, these data indicate that tendon organ discharge monitors the force exerted by muscle.

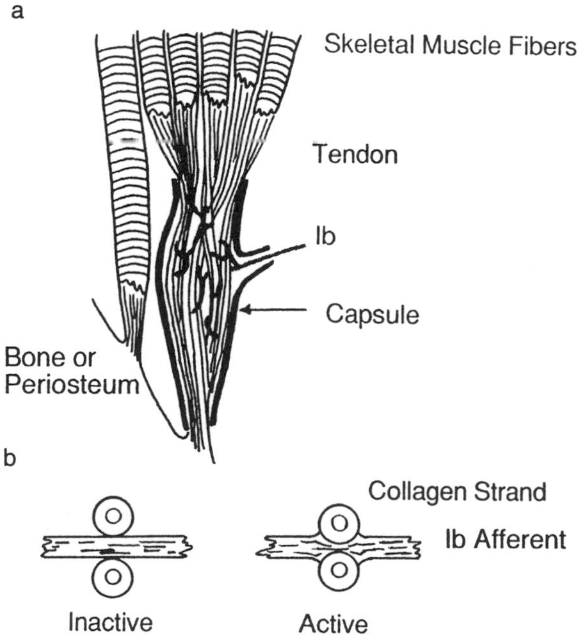

Figure 4.19 The tendon organ. (a) The encapsulated Group Ib afferent encircles the tendons of several skeletal muscle fibers. (b) Contraction of the skeletal muscle fibers causes the collagen strands (tendon) to squeeze and thus activate the Group Ib afferent.

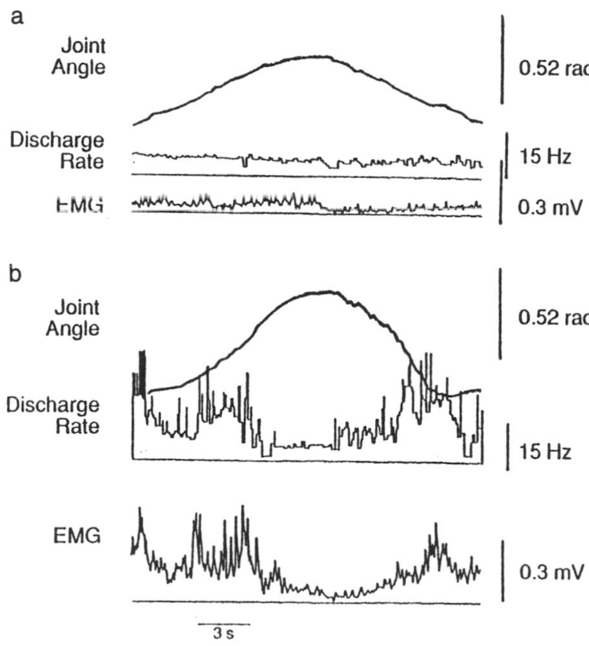

Figure 4.20 Discharge of a human tendon organ during finger movements against a zero load (a) and a light load (b).

Note. From "Response Profiles of Human Muscle Afferents During Active Finger Movements" by N.A. Al-Falahe, M. Nagaoka, and A.B. Vallbo, 1990, *Brain*, **113**, p. 339. Copyright 1990 by Oxford University Press. Adapted by permission of Oxford University Press.

Joint Receptors

As with the muscle spindle and the tendon organ, joint receptors can function as mechanoreceptors and provide information for the single joint system that is necessary for the control of movement (Burgess, Horch, & Tuckett, 1987). In contrast to the muscle spindle and the tendon organ, however, the joint receptor is not a single, well-defined entity. Rather, joint receptors vary in location (e.g., joint capsule, ligament, loose connective tissue), type (e.g., Ruffini ending, Golgi ending, Pacinian corpuscle, free nerve endings), and presumably function. These receptors are served by Group II, III, and IV afferent neurons.

The **Ruffini endings** typically consist of two to six thinly encapsulated, globular corpuscles with a single myelinated parent axon that has a diameter of 5 to 9 μm. These receptors may be categorized as static or dynamic mechanoreceptors and are capable of signaling joint position and displacement, angular velocity, and intraarticular pressure (Johansson, Sjölander, & Sojka, 1991). **Pacinian corpuscles** are thickly encapsulated with a parent axon of 8 to 12 μm diameter. These receptors have low thresholds to mechanical stress and apparently detect acceleration of the joint (Bell, Bolanowski, & Holmes, 1994). **Golgi endings** are thinly encapsulated, fusiform corpuscles that are similar to tendon organs. The axon of the afferent neuron connected to the Golgi ending has a diameter of 13 to 17 μm. These receptors have high thresholds, and monitor tension in ligaments, especially at the extremes of the range of motion. **Free nerve endings** are widely distributed and constitute the joint nociceptive system. They have small-diameter axons (0.5 to 5 μm) and are active when a joint is subjected to an abnormal mechanical stress or to chemical agents. These four types of joint receptors are able to provide the single joint system with information about the position, displacement, velocity, and acceleration of movement as well as noxious stimuli experienced by the joint (Johansson et al., 1991). For the knee joint, the receptor types appear to be located in distinct anatomical regions, and therefore are capable of providing unique afferent information (Zimny & Wink, 1991).

The significance of joint receptors for the normal function of the single joint system has been convincingly demonstrated by the effect of joint pathology on muscle activation. For example, when normal subjects are given a large experimental effusion in the knee joint (fluid injected into the joint space), the ability to activate the quadriceps femoris is greatly reduced, even in the absence of pain (Stokes & Young, 1984; Young, Stokes, & Iles, 1987). This inhibition of muscle activation depends on the volume of fluid added but can cause a 30% to 90% reduction in the maximal voluntary activation of quadriceps. Conversely, the removal of fluid from the knee joint, such as after a meniscectomy, can markedly improve the use of the muscle by a patient. In the absence of pain, chronic joint effusion can cause weakness and atrophy of muscle. Similarly, patients with old cruciate ligament tears typically exhibit a decrease in strength of both the quadriceps femoris and hamstrings (Grabiner, Koh, & Andrish, 1992; Johansson et al., 1991).

The axons innervating joint receptors distribute their information extensively throughout the CNS, from the spinal cord up to the brain. At the level of the single joint system, however, joint receptors seem to have a more potent effect on fusimotor neurons (gamma motor neurons) than on alpha motor neurons. Consequently, joint receptors are more likely to exert an effect on the single joint system indirectly by modulating the activity of the muscle spindle than by directly influencing the output of alpha motor neurons.

Cutaneous Mechanoreceptors

In contrast to the three sensory receptors discussed previously in this section, cutaneous mechanoreceptors provide information exclusively on external events that affect the single joint system. The cutaneous mechanoreceptors in the hand and foot certainly provide important information on how we interact with our surroundings. For example, the decline in somatosensory acuity, presumably involving cutaneous mechanoreceptors, contributes to the reduced control of posture among elderly subjects and among patients with various pathologies (Horak, Shupert, & Mirka, 1989). In addition to mechanoreceptors, the skin includes thermoreceptors and nociceptors that can also influence the function of the single joint system, even of the alpha motor neurons that are activated during the performance of a given task.

Four types of cutaneous mechanoreceptors have been identified: Merkel disks, Meissner corpuscles, Ruffini endings, and Pacinian corpuscles (Rothwell, 1987). The first two are found close to the surface of the skin, whereas the latter two are deeper. The **Merkel disk** is sensitive to local vertical pressure and does not respond to lateral stretch of the skin. This receptor responds with a rapid initial discharge of action potentials that is quickly reduced to a slow steady rate. **Meissner corpuscles** are innervated by 2 to 6 axons, and each axon may innervate more than one corpuscle. The Meissner corpuscle is sensitive to local, maintained pressure, but its response (discharge of action potentials) fades rapidly. **Ruffini endings** are innervated by a single axon and respond to stretch of the skin over a wide area. This sensitivity, however, depends on the direction of the stretch; the ending will be excited with stretch in one direction and inhibited by stretch at a right angle to the preferred direction. The response of Ruffini endings adapts slowly to a sustained stretch. **Pacinian corpuscles** are the largest receptors in the skin and are innervated by a single axon. The Pacinian corpuscle detects a rapidly changing pressure stimulus. These four cutaneous mechanoreceptors provide the single joint system with the ability to detect stimuli applied to the skin over small and large areas, and for brief and sustained durations.

Summary

Because of the complexity of the human body, we have developed a simplified model of the components of the motor system that are essential for movement. This simplified model is called the single joint system. This chapter reviews the properties of the five components of the single joint system: rigid link, synovial joint, muscle, neuron, and sensory receptor. The single joint system is an extension of the concepts encountered in Part I on the free body diagram, with the addition of components to control (muscle and neuron) and monitor (sensory receptor) the rotation of the rigid body segments. Because these five components are comprised of living tissue, they adapt to the physical stress encountered in the activities of daily living. The result is a dynamic system with continually changing properties.

TRUE-FALSE QUESTIONS

1. Connective tissue includes living cells. **TRUE FALSE**
2. Collagen is a protein. **TRUE FALSE**
3. Water accounts for 80% of the wet weight of bone. **TRUE FALSE**
4. The basic structural unit of bone is the fibroblast. **TRUE FALSE**
5. Stress is a measure of force normalized to the strength of the tissue. **TRUE FALSE**
6. Strain is a dimensionless quantity. **TRUE FALSE**
7. Strain is a measure of the change in length. **TRUE FALSE**
8. A stress-strain curve represents a normalized load-deformation relationship. **TRUE FALSE**
9. The stress-strain relationship is linear in the elastic region. **TRUE FALSE**
10. The slope of the stress-strain relationship beyond σ_{yield} represents the plastic modulus. **TRUE FALSE**
11. σ_{ult} indicates the energy that bone can absorb. **TRUE FALSE**
12. The tibia experiences a greater strain in tension than in compression when an individual jogs. **TRUE FALSE**
13. The femur offers the greatest resistance to compression loads. **TRUE FALSE**
14. Bone is designed to withstand forces two to five times greater than it commonly encounters each day. **TRUE FALSE**
15. Wolff's law describes the relationship between structure and function of bone. **TRUE FALSE**
16. The process of bone replacement is referred to as *remodeling*. **TRUE FALSE**
17. Bone resorption is performed by osteoclasts. **TRUE FALSE**
18. Demineralization describes the loss of salts from the skeleton. **TRUE FALSE**
19. Bone remodeling is best induced with continuous loads. **TRUE FALSE**
20. Bone density increases with age in a process known as osteoporosis. **TRUE FALSE**
21. Electrical stimulation has been used to heal fractures. **TRUE FALSE**
22. Bone can generate electrical potentials when it is stressed. **TRUE FALSE**
23. Ligament connects bone to bone. **TRUE FALSE**
24. Both tendon and ligament are designed to mainly resist tensile forces. **TRUE FALSE**
25. The unit of measurement for compressive force is MPa. **TRUE FALSE**
26. Tendon and ligament mainly consist of collagen. **TRUE FALSE**
27. The fibrils are all arranged longitudinally in parallel in ligament. **TRUE FALSE**
28. The microfibril represents groups of fibrils bound together by cross-links. **TRUE FALSE**
29. The cross-links are made out of elastin. **TRUE FALSE**
30. The strength of tendon and ligament increases with the number of cross-links. **TRUE FALSE**
31. The viscosity of ligament changes after a warm-up. **TRUE FALSE**
32. Tensile stress of tendon increases linearly with strain. **TRUE FALSE**
33. The stress-strain relationship does not differ among tendons from different muscles. **TRUE FALSE**
34. The synovial joint is freely movable. **TRUE FALSE**
35. Articular cartilage is innervated by Group III afferent neurons. **TRUE FALSE**

36. Boundary lubrication refers to the absorption of a lubricant by the surface of the articular cartilage. TRUE FALSE
37. Self-lubrication refers to the thin film of lubricant that separates the articulating surfaces. TRUE FALSE
38. Articular cartilage is a thixotropic tissue. TRUE FALSE
39. The thickness of articular cartilage can change with activity. TRUE FALSE
40. The joint capsule secretes synovial fluid. TRUE FALSE
41. A synovial joint has from one to three degrees of freedom. TRUE FALSE
42. The hip joint has one degree of freedom. TRUE FALSE
43. Cardiac muscle consists of individual cells that are separated by a thin cleft and that lack apparent striations. TRUE FALSE
44. Not all skeletal muscles cross joints. TRUE FALSE
45. The endomysium surrounds individual muscle cells. TRUE FALSE
46. The sarcolemma is an excitable membrane. TRUE FALSE
47. A skeletal muscle fiber is a single cell. TRUE FALSE
48. A single contractile protein is known as a myofibril. TRUE FALSE
49. The sarcoplasmic reticulum is an invagination of the sarcolemma. TRUE FALSE
50. A sarcomere represents the zone of a myofibril from one Z band to another. TRUE FALSE
51. Sarcomere length is about 2.5 μm. TRUE FALSE
52. The dark band is the A band. TRUE FALSE
53. The H band is located in the middle of the A band. TRUE FALSE
54. The thick filaments connect to the M band. TRUE FALSE
55. A single thin myofilament is surrounded by six thick filaments. TRUE FALSE
56. A thin filament contains two strands of G-actin. TRUE FALSE
57. Actin is a protein. TRUE FALSE
58. Troponin has three subunits. TRUE FALSE
59. The abbreviation TN-T represents the component of troponin that binds to tropomyosin. TRUE FALSE
60. Tropomyosin is a two-strand protein bound to actin. TRUE FALSE
61. The Z band is highly flexible and adaptable. TRUE FALSE
62. The light meromyosin fragment of myosin can be further subdivided into subfragments 1 and 2. TRUE FALSE
63. Subfragment 1 connects light and heavy meromyosin. TRUE FALSE
64. The light meromyosin extension is called the crossbridge. TRUE FALSE
65. Each thick filament contains one crossbridge. TRUE FALSE
66. Molecular weight is measured in daltons. TRUE FALSE
67. A heavy chain has a high molecular weight. TRUE FALSE
68. A protein that is synthesized with a slightly different amino acid composition is known as an isoform. TRUE FALSE
69. Isoforms of contractile proteins do not vary with muscle fiber type. TRUE FALSE
70. Each myosin molecule contains one heavy chain. TRUE FALSE
71. There are four light chains attached to the globular heads of a myosin molecule. TRUE FALSE
72. The perimysium forms the cytoskeleton in skeletal muscle. TRUE FALSE
73. The exosarcomeric cytoskeleton maintains the lateral alignment of the myofibrils. TRUE FALSE
74. The intermediate fibers are the major component of the endosarcomeric cytoskeleton. TRUE FALSE
75. Neuroglia are passive structures that only provide structural support for neurons. TRUE FALSE
76. The cell body of a neuron is known as a *soma*. TRUE FALSE
77. Most input to a neuron occurs on the axon. TRUE FALSE
78. An action potential is generated at the collateral. TRUE FALSE
79. The functional contact of a neuron with a target cell is referred to as the *axon hillock*. TRUE FALSE
80. Dendrites form the presynaptic terminal. TRUE FALSE
81. Afferent neurons convey sensory information to the CNS. TRUE FALSE
82. Afferent signals are transmitted directly to efferent neurons in the spinal cord. TRUE FALSE
83. Efferent neurons that innervate skeletal muscle are known as *motor neurons*. TRUE FALSE
84. The axons of efferent neurons enter the spinal cord in the dorsal roots. TRUE FALSE
85. All reflexes are controlled through the autonomic nervous system. TRUE FALSE

86. Each muscle fiber is innervated by a single efferent axon. **TRUE FALSE**

87. A synapse between an afferent and an efferent neuron is known as a neuromuscular junction. **TRUE FALSE**

88. Acetylcholine is a neurotransmitter. **TRUE FALSE**

89. Muscle fibers can generate an action potential. **TRUE FALSE**

90. Sensory receptors provide feedback to the CNS by way of afferent neurons. **TRUE FALSE**

91. A sensory receptor cannot be both a mechanoreceptor and a proprioceptor. **TRUE FALSE**

92. Intrafusal fibers are smaller than extrafusal fibers. **TRUE FALSE**

93. The nuclear bag fiber is devoid of myofilaments. **TRUE FALSE**

94. The Group Ia axon serves an afferent function. **TRUE FALSE**

95. The Group II afferent connects to the nuclear chain fiber. **TRUE FALSE**

96. The Aα fiber has the largest diameter of the efferent axons. **TRUE FALSE**

97. Gamma motor neurons can cause intrafusal fibers to contract. **TRUE FALSE**

98. The equatorial region of an intrafusal fiber is stretched when the polar regions shorten. **TRUE FALSE**

99. Stretch of an intrafusal fiber causes the development of a generator potential. **TRUE FALSE**

100. Muscle spindles provide information on changes in muscle length. **TRUE FALSE**

101. Alpha motor neurons are known as fusimotor neurons. **TRUE FALSE**

102. The tendon organ is innervated by β motor neurons. **TRUE FALSE**

103. The tendon organ is located in series with skeletal muscle fibers. **TRUE FALSE**

104. Discharge rates of tendon organs are measured in Hz. **TRUE FALSE**

105. Tendon organs do not respond to passive changes in muscle force. **TRUE FALSE**

106. Microneurography is used to measure action potentials in a nerve. **TRUE FALSE**

107. Ruffini endings detect acceleration of the joint. **TRUE FALSE**

108. Golgi endings signal tension in ligaments. **TRUE FALSE**

109. Muscle activity is inhibited when fluid accumulates in a joint. **TRUE FALSE**

110. Joint receptors have a potent effect on fusimotor neurons. **TRUE FALSE**

111. Cutaneous mechanoreceptors are important in the control of posture. **TRUE FALSE**

112. Meissner corpuscles sense local vertical pressure. **TRUE FALSE**

113. Pacinian corpuscles detect rapidly changing pressure. **TRUE FALSE**

114. The Merkel disk responds with an initial rapid discharge of action potentials. **TRUE FALSE**

MULTIPLE-CHOICE QUESTIONS

115. Which of the following is not a function of bone?
   A. Provides mechanical support
   B. Secretes synovial fluid into the joint cavity
   C. Produces red blood cells
   D. Serves as an active ion (Ca^{2+}, Mg^{2+}) reservoir

116. Which material accounts for the greatest proportion of the wet weight of bone?
   A. Water
   B. Bone salts
   C. Osteoclasts
   D. Osteocollagenous fibers

117. What is the basic structural unit of bone?
   A. Osteoblast
   B. Osteon
   C. Protein matrix
   D. Osteoclast

118. What is the unit of measurement for stress?
   A. MPa
   B. Pa/cm^2
   C. N
   D. %

119. Which parameter indicates the maximum strength of bone?
   A. σ_{ult}
   B. Elastic modulus
   C. ε_{ult}
   D. σ_{yield}

120. For which loads does the human femur offer the greatest resistance?
   A. Compression
   B. Torsion
   C. Tension
   D. Shear

121. What is the normalized version of the load-deformation relationship?
   A. Wolff's law
   B. Force-length relationship
   C. Stress-strain relationship
   D. Osteoclast-osteoblast activity ratio
122. Why are the bones of astronauts weaker following a spaceflight?
   A. A decrease in the piezoelectric potentials
   B. A reduction in the safety factor
   C. The development of osteoporosis
   D. Increased demineralization
123. Which term describes side-to-side forces?
   A. Shear
   B. Tensile
   C. Orthogonal
   D. Compressive
124. What do cross-links bind together?
   A. Microfibrils
   B. Triple helix
   C. Alpha chains
   D. Amino acids
125. What property of tendon and ligament influences its velocity-dependent resistance to stretch?
   A. Elasticity
   B. Piezoelectric
   C. Thixotropy
   D. Viscosity
126. In which region of the stress-strain relationship of tendon and ligament does disruption of the collagen fibers begin to occur?
   A. Toe
   B. Failure
   C. Linear
   D. Plastic
127. What measure indicates the slope of the stress-strain relationship in the linear region?
   A. Plastic modulus
   B. Energy absorbed
   C. σ_{yield}
   D. Elastic modulus
128. Which joint is included in the single joint system?
   A. Cartilaginous
   B. Loose
   C. Synovial
   D. Fibrous
129. What is the lubrication process whereby a lubricant separates the articulating surfaces?
   A. Self-lubrication
   B. Boundary lubrication
   C. Fluid film lubrication
   D. Viscous lubrication
130. Which statement best describes why the thickness of articular cartilage changes due to the stress-related flux of water?
   A. Articular cartilage behaves as a water-filled sponge.
   B. There are no blood vessels or nerves in articular cartilage.
   C. Articular cartilage is capable of self-lubrication.
   D. The glycoprotein lubricin can be absorbed by the surface of the articular cartilage.
131. Which of the following is not a property of muscle?
   A. Conductivity
   B. Transduction
   C. Irritability
   D. Contractility
132. What level of the hierarchy of muscle is known as a muscle cell?
   A. Fascicle
   B. Sarcomere
   C. Myofibril
   D. Fiber
133. What is the length of a human muscle fiber?
   A. 50 nm
   B. 50 μm
   C. 50 mm
   D. 50 cm
134. Which part of a muscle cell is an extension of the sarcolemma?
   A. Sarcoplasmic reticulum
   B. Terminal cisternae
   C. Transverse tubule
   D. Endomysium
135. Which band contains only the thick myofilaments?
   A. Z
   B. H
   C. A
   D. I
136. Which band refers to the zone where two sets of thick filaments of opposite polarity (pointing in the opposite direction) are joined together?
   A. M
   B. H
   C. I
   D. Z
137. What are the regulatory proteins in muscle?
   A. Actin and myosin
   B. Troponin and tropomyosin
   C. Light and heavy meromyosin
   D. Subfragments 1 and 2
138. What does the abbreviation TN represent?
   A. Titin
   B. Troponin
   C. Transverse network
   D. Tropomyosin

139. Which part of the myofilaments is the crossbridge?
   A. Heavy meromyosin extension
   B. Light chains
   C. Subfragments 1 and 2
   D. Myosin molecule
140. Which factor is not used to distinguish between the light chains attached to the myosin globular heads?
   A. Whether or not they can be phosphorylated
   B. Molecular weight
   C. Experimental agent used to separate them from the heavy chain
   D. Amino acid sequence
141. What are the intermediate fibers?
   A. Part of the endosarcomeric cytoskeleton
   B. A myofilament (contractile protein)
   C. Fibers in a muscle spindle
   D. The principal component of the exosarcomeric cytoskeleton
142. Which function is not performed by neuroglia?
   A. Phagocytosis
   B. Myelination
   C. Ion transport
   D. Generation of action potentials
143. What are the four morphological regions of a neuron?
   A. Axon, presynaptic terminal, dendrite, soma
   B. Soma, synapse, axon, dendrite
   C. Dendrite, collateral, presynaptic terminal, soma
   D. Collateral, soma, dendrite, axon
144. Which part of the neuron generates the output signal?
   A. Axon hillock
   B. Dendrite
   C. Soma
   D. Collateral
145. Which component is not one of the three functional classes of neurons?
   A. Interneuron
   B. Efferent
   C. Sensory receptor
   D. Afferent
146. Which neurons are called motor neurons?
   A. Efferent
   B. Group II
   C. Type C fiber
   D. Interneuron
147. Which physiological process does the autonomic nervous system not control?
   A. Body temperature
   B. Perspiration
   C. Contraction of skeletal muscle
   D. Blood pressure
148. Which element provides the efferent innervation of the muscle spindle?
   A. Intrafusal fiber
   B. Gamma motor neuron
   C. Nuclear chain fiber
   D. Group Ia neuron
149. Where is the sensory receptor of the Group Ia axon located?
   A. Polar region of the intrafusal fibers
   B. Dorsal root ganglion
   C. Equatorial region of the intrafusal fibers
   D. Along the extrafusal muscle fibers
150. Which fibers are innervated by alpha motor neurons?
   A. Skeletal muscle fibers
   B. Nuclear chain fibers
   C. Collagenous fibers of the tendon organ
   D. Nuclear bag fibers
151. What do tendon organs signal?
   A. Local pressure applied to the skin
   B. Muscle force
   C. Joint displacement
   D. Changes in muscle length
152. What afferent axon innervates the tendon organ?
   A. Group II
   B. Group Ib
   C. Beta motor neuron
   D. Fusimotor neuron
153. Which sensory receptor does not function as a joint receptor?
   A. Merkel disk
   B. Golgi ending
   C. Free nerve ending
   D. Pacinian corpuscle
154. What parameters are not sensed by joint receptors?
   A. Muscle force
   B. Joint position
   C. Noxious stimuli
   D. Angular velocity of the joint

PROBLEMS

155. Consider the biological model we call the single joint system.
   A. Why is it necessary to use such a model?
   B. What are the five elements of the model?
   C. Why does the model not include such elements as the lungs and heart?
156. When a bone specimen (0.9 cm × 0.3 cm × 1.8 cm) was subjected to a tensile load of 1,890 N, it experienced a stretch of 0.8 mm; the specimen had a length of 1.8 cm. What stress and strain were applied to the specimen?

157. Use the data in Table 4.1 to plot the stress-strain relationship for the femur and tibia of a 30- to 39-year-old subject. Label the graph with the correct variables and units of measurement.

158. The human femur has a strength of 58 MPa for forces acting perpendicular to its long axis.
   A. What mechanical variable do the units MPa represent?
   B. Approximately what perpendicular load (in MPa) does this suggest that the human femur experiences in activities of daily living?
   C. What types of loads does the femur experience in everyday life?

159. You have been hired to conduct an exercise class for a group of women (45 years and older). What activities would you prescribe to minimize the decline in bone strength due to osteoporosis?

160. Identify the basic structural unit of bone, the basic load-bearing unit of tendon and ligament, the basic contractile unit of muscle, and the element of the nervous system that activates skeletal muscle.

161. The strength of tendon and ligament depends on the state of the cross-links. What are cross-links, and why does the strength of tendon and ligament depend on them?

162. Why does prior activity influence the resistance that tendon and ligament offers to a stretch?

163. Compare the stress-strain relationships in Table 4.4 for bone with that in Figures 4.4 and 4.5 for tendon and ligament.
   A. Which tissue (bone or tendon-ligament) has the greater elastic modulus? What does the difference in the modulus indicate?
   B. Which tissue has the greater strength? Which variable indicates the strength of the tissue?
   C. Describe the qualitative differences in the stress-strain relationships for the two tissues.

164. Why does articular cartilage increase in thickness when an individual goes from resting to an active state?

165. Articular cartilage is part of the synovial joint element of the single joint system.
   A. How does lubrication help articular cartilage perform its function?
   B. What is the function of articular cartilage?
   C. What happens to articulating surfaces when articular cartilage is damaged?

166. Immobilization represents a condition of reduced use of the single joint system.
   A. Why is it necessary to perform passive range of motion activities about an injured or immobilized joint?
   B. Immobilization causes bone to become weaker and its stiffness to decrease. Draw the stress-strain relationship for a bone that has been immobilized for 6 weeks, and compare it to one that has experienced normal activity.

167. For a joint with three degrees of freedom, describe the types of motion that can occur and indicate the axis of rotation for each motion and the plane in which it occurs.

168. What is the difference between the sarcolemma and the endomysium?

169. Consider the organization of sarcomeres in a muscle fiber.
   A. How many sarcomeres (average width = 2.5 μm) are there in a 7.2-mm myofibril?
   B. If a muscle fiber had a cross-sectional area of 1,809 μm^2, a myofibril had a diameter of 1 μm, and the myofibrils occupied 82% of the muscle fiber cross-sectional area, how many myofibrils would there be in the muscle fiber? If each myofibril had a length of 7.2 mm, how many sarcomeres would there be in the fiber?
   C. Are most sarcomeres in a muscle fiber arranged in series or in parallel?

170. How many strands of F-actin and tropomyosin are there in a single thin filament?

171. Table 4.3 indicates some of the isoforms that have been identified in skeletal muscle.
   A. What is an isoform?
   B. Why is this concept used to describe muscle structure?
   C. What is known about the functional significance of contractile-protein isoforms?

172. What does the cytoskeleton in skeletal muscle accomplish? Why is it necessary?

173. Neurons form a network of cells that make rapid communication possible.
   A. What are the three phases of communication that a neuron performs?
   B. What morphological regions of a neuron are involved in each of these phases?
   C. Identify three differences between afferent and efferent neurons.

174. How does a sensory receptor, such as a muscle spindle, detect a mechanical stimulus, convert the stimulus to an electrical event, and transmit the information to the nervous system?

175. Why do intrafusal fibers not contribute a significant force to whole-muscle force?

176. According to axon counts, the number of afferent axons is much greater than the number of efferent axons. What does this suggest about the control of the single joint system?

177. Gamma motor neurons provide an efferent innervation of muscle spindles.
   A. How do the fusimotor neurons change the excitability of the muscle spindle?
   B. Draw a graph of the discharge rate of a muscle spindle over time when a muscle is given a certain stretch. Label the axes with the correct variables and units of measurement. Draw the response of the muscle spindle for two conditions: (1) normal steady-state conditions, and (2) after the muscle spindle has been made hyperexcitable by fusimotor neurons.
   C. How can fusimotor neurons be activated?

178. Although a force of 2 N delivered by passive stretch is required to activate a tendon organ, an activated muscle fiber that exerts a force of 50 μN can activate a tendon organ.
   A. How many times smaller is the 50 μN compared with 2 N?
   B. What mechanical reason can you think of to explain the difference?
   C. What advantage is there for the single joint system to have these two different thresholds?

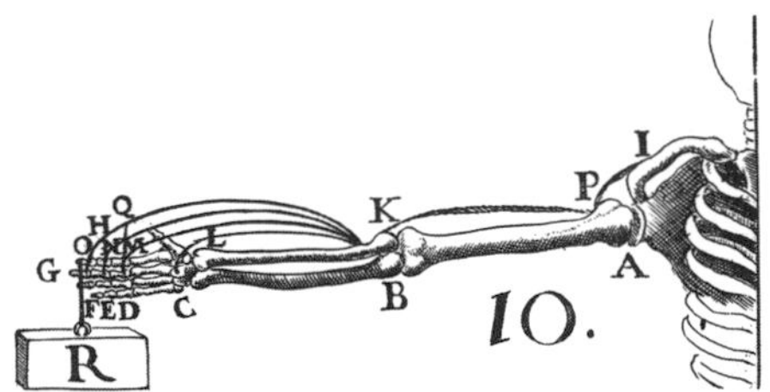

CHAPTER 5

Single Joint System Operation

In chapter 4 we discussed the morphological characteristics of the five elements (rigid link, synovial joint, muscle, neuron, sensory receptor) of the single joint system. In chapter 5 we consider how these elements interact to produce movement. Human movement, particularly elite performance, can be impressive. When we consider that each movement is the consequence of a multitude of minute interactions between nerve and muscle cells, movement becomes all the more spectacular. Chapter 5 addresses four key features of the operation of the single joint system: (a) the motor unit, the basic functional unit of the single joint system; (b) excitable membranes, the means by which information is transmitted rapidly throughout the system; (c) excitation-contraction coupling, the link between the neural signal and muscle contraction; and (d) sensory receptor feedback, the role of afferent information in the operation of the system.

Motor Unit

In Part I, "The Force-Motion Relationship," we discussed the concept that activated muscle exerts a force on a rigid link and that, depending on the relationship between the load on the link and the force exerted by the muscle, the link either rotates or remains stationary. In this scheme, muscle is characterized as able to exert a force that can vary in magnitude, a variation that is controlled by the nervous system. In a functional sense, therefore, the analysis of movement as the consequence of muscle activation must include consideration of the associated neural factors. The concept of a motor unit includes such a perspective. Specifically, a **motor unit** is defined as *the cell body and dendrites of a motor neuron, the multiple branches of its axon, and the muscle fibers that it innervates*. The anatomical and physiological features of a motor unit can vary markedly; these differences enable a muscle, which usually comprises a few hundred (100 to 1,000) motor units (McComas, 1991; Stein & Yang, 1990), to meet a variety of needs and to adapt to various acute and chronic demands.

Neural Component

The neural component of the motor unit essentially consists of the motor neuron and its dendrites (Figure 4.11). The features of the neural component that vary among motor units include morphology, excitability, and distribution of input. The morphological feature of the neural component that has received the greatest attention is motor neuron size. This feature is thought to be important because of its role in the activation of motor neurons. The size of a motor neuron can be indicated by the diameter of the soma, the surface area of the cell body, the number of dendrites arising from the soma, and the diameter of the axon. Based on measurements of motor neurons in the cat, these properties seem to be correlated so that the largest motor neuron has the greatest value for each of these properties (Stuart & Enoka, 1983). Furthermore, motor neurons supplying fast-twitch muscle fibers appear to be larger than those innervating slow-twitch muscle fibers.

As initially reported by Henneman (1957), there is strong correlation between a motor neuron's size and its excitability. Typically, the excitability of a motor neuron is assessed by placing a microelectrode inside the cell and measuring selected biophysical properties of the membranes. As summarized by Stuart and Enoka (1983) for cat motor neurons, these properties include input resistance (MΩ), rheobase (nA), afterhyperpolarization (ms), and axonal conduction velocity (m/s). **Input resistance** is a measurement of the electrical resistance that a cell offers to current that is injected by the intracellular microelectrode. Small motor neurons tend to have a high input resistance, which results in a greater response (i.e., they are more excitable) to a given input. **Rheobase** is a direct measure of excitability that indicates the amount of current that must be injected into the motor neuron to generate an action potential. The rheobase for small motor neurons is much less than that for larger motor neurons. **Afterhyperpolarization** refers to the duration of the trailing part of the action potential when the membrane is less excitable than in normal resting (steady-state) conditions. The duration of the afterhyperpolarization is thought to influence the maximum rate at which a motor neuron can generate action potentials. The afterhyperpolarization is much briefer in motor neurons that supply fast-twitch muscle fibers. The velocity at which action potentials are propagated is influenced by the size (diameter) of the axon. Because the axon diameter varies with motor neuron size, the larger motor neurons, which innervate fast-twitch muscle fibers, have high axonal **conduction velocities**. These associations indicate that small motor neurons are more excitable but generate and propagate action potentials at a slower rate than larger motor neurons.

To understand the function of the motor unit, we need to combine the information on the excitability of motor neurons with the details of the inputs that they receive. Motor neurons have extensive dendritic trees, which receive about 80% of the input directed to the cell (Figure 4.11). It appears, however, that the input from different sources can have a variable effect on the generation of an action potential by the motor neuron. This difference can be due to the *number and location of synapses associated with each input system*. One way to assess the effect of different inputs is to use a microelectrode to measure the **effective synaptic current** (nA) generated in the motor neuron in response to a given input (Heckman & Binder, 1991; Powers, Robinson, Konodi, & Binder, 1992). Presumably this measurement indicates the net effect of activating an input system and represents the signal that will be transmitted to the axon hillock where the action potential is generated. It appears that not all inputs are distributed uniformly to populations of motor neurons and that there are three patterns of distribution (Heckman & Binder, 1990): (1) *least input* (smallest effective synaptic current) to the largest motor neurons—input from the Group Ia afferent of the muscle spindle; (2) *uniform input* to all motor neurons—inhibitory input from a muscle spindle located in an antagonist muscle and from an interneuron (Renshaw cell); and (3) *greatest input* to the largest motor neurons—input from a brain stem nucleus (red nucleus) and from a nerve (sural) containing information from cutaneous receptors. These observations suggest that the activation of a motor neuron depends not only on its intrinsic excitability but also on the type (distribution) of input that it is receiving. Furthermore, the distribution of input influences the rate at which the motor neurons in a population are recruited (Kernell & Hultborn, 1990).

Muscle Component

Although a muscle fiber is innervated by a single motor neuron, each motor neuron innervates more than one muscle fiber (Figure 5.1). The number of muscle fibers innervated by a single motor neuron is referred to as the **innervation ratio** and varies from about 1:1,900 (e.g., gastrocnemius, tibialis anterior) to 1:15 (e.g., extraocular muscles). That is, one motor neuron may innervate from 15 to 1,900 muscle fibers. For example, the muscle first dorsal interosseus (a hand muscle) contains about 41,000 muscle fibers and 120 motor units; it has, on average, an innervation ratio of about 1:342. In contrast, the medial gastrocnemius has about 1,120,000 muscle fibers and 580 motor units—an average innervation ratio of 1:1,931 (Feinstein, Lindegård, Nyman, & Wohlfart, 1955). Each time a motor neuron is activated in the central nervous system, it elicits one or more action potentials in all of its muscle fibers (except under some fatiguing conditions); hence, the lower the innervation ratio, the finer the control of muscle force in terms of motor unit activation. The innervation ratio is also indicative of the number of times that an axon must branch to contact all of its muscle fibers.

Figure 5.1 shows the location of the fibers belonging to a single Type FR motor unit in the medial gastrocnemius muscle of a cat. The motor unit appeared to include 500 muscle fibers. Such measurements have shown that fibers of a single motor unit occupy a specific region of the muscle. This territory can extend to up to 15% of the volume of the muscle with a density of 2 to 5 muscle fibers per 100 belonging to the same motor unit (Burke, 1981). This means that a given region of a muscle contains muscle fibers from 20 to 50 different motor units. However, the density of muscle fibers belonging to the same motor unit increases with age. This occurs because innervation ratio changes as some motor

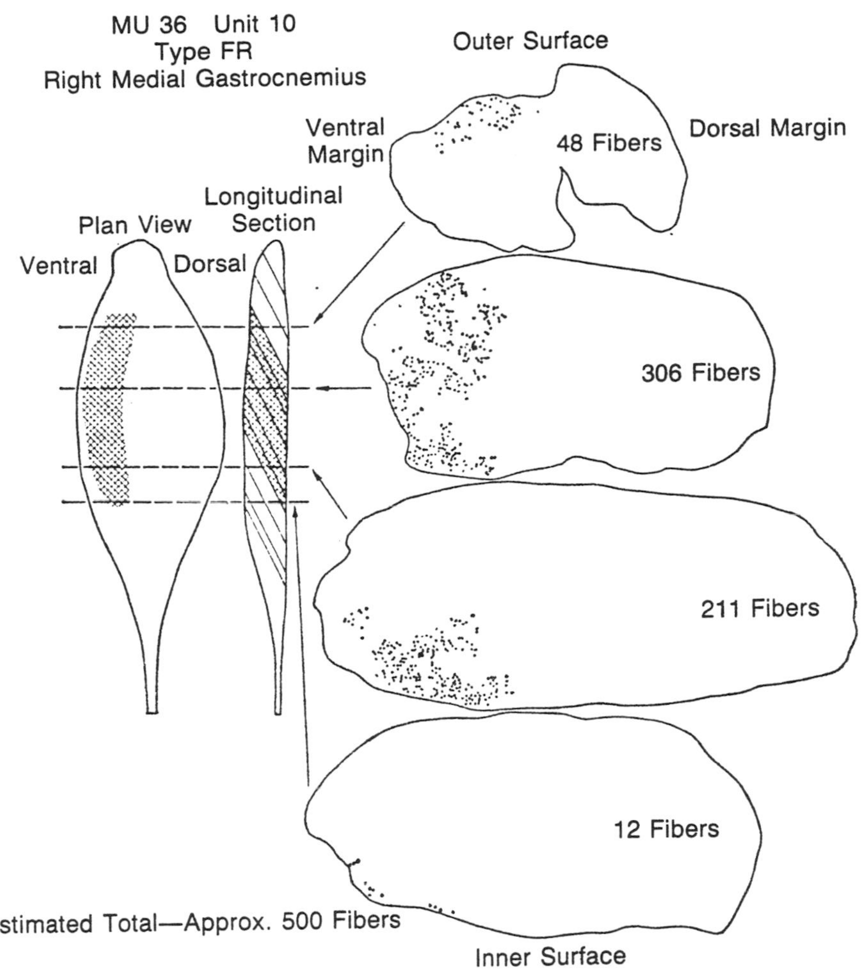

Figure 5.1 The distribution of muscle fibers is shown in longitudinal section on the left and in the cross-sectional views on the right. The muscle fibers belonging to the unit were confined to a subvolume of the muscle.
Note. From "Anatomy and Innervation Ratios in Motor Units of Cat Gastrocnemius" by R.E. Burke and P. Tsairis, 1973, *Journal of Physiology*, **234**, p. 755. Copyright 1973 by Cambridge University Press. Adapted with permission.

neurons degenerate and die, which causes the axons of some surviving motor units to develop more branches and reinnervate muscle fibers that have lost their axonal connection (Campbell, McComas, & Petito, 1973; Kanda & Hashizume, 1989; Stålberg & Fawcett, 1982). Thus, as humans age, the number of motor units declines but the size (number of muscle fibers) of the surviving motor units increases. Similar effects have been observed in such disease states as amyotrophic lateral sclerosis, spinal muscular atrophy, poliomyelitis, and diabetic neuropathies (McComas, 1991).

Not only is the territory of a single motor unit limited to a specific part of a muscle, but it appears that different parts of a muscle can contain distinct populations of motor units. This observation has given rise to the concept of a **neuromuscular compartment** (Peters, 1989; Windhorst, Hamm, & Stuart, 1989). A *compartment* is defined as the volume of muscle supplied by a primary branch of the muscle nerve. A compartment contains a unique population of motor units; the muscle fibers belonging to one motor unit are confined to a single neuromuscular compartment. The proportion of muscle fiber types can differ between the compartments of a single muscle. Neuromuscular compartments have been found in some, but not all, muscles. Because compartments can be activated independently, a single muscle, which is an anatomical entity, can consist of several distinct regions that each serve a different physiological function. Also, the existence of compartments suggests that it can be misleading to infer the function of a muscle based on the location of its attachments. The analysis of muscle function must consider both its architecture and its innervation pattern.

Motor units can be compared to one another based on a number of physiological properties, including the discharge characteristics, the speed of contraction, the magnitude of force, and the resistance to fatigue. Most comparisons are based on the properties of the muscle component of the motor unit. Two methodologies are commonly used for evaluating these parameters, one direct and the other indirect. *Direct* evaluation refers to the physiological measurement of the motor unit properties. Indirect assessment is based on histochemical and biochemical measurements. A direct assessment can be made by evaluating either the discharge characteristics of motor units (Gydikov & Kosarov, 1974) or the mechanical response of motor units to different inputs (Burke, Levine, Tsairis, & Zajac, 1973). We will consider the discharge characteristics in chapter 6. For now, we focus on the mechanical behavior of motor units (Figure 5.2). The quantal output of a motor unit is a **twitch**. This is the force-time response to a single input (Figure 5.2a). Although the output of a motor unit rarely consists of individual twitch responses, its output consists of a series of twitch responses that summate so that the force is greater than that associated with a single twitch. The force-time profile that consists of summated twitch responses is known as a tetanus (Figure 5.2c). A twitch represents a unit response of muscle to a single input and can be characterized by three measurements: the time from force onset to peak force (**contraction time**), the magnitude of the peak force, and the time it takes for the force to decline to one half of its peak value (**half relaxation time**) (Figure 5.2a). Contraction time is used as a measure of the speed of the contractile machinery. If contraction time is long, the motor unit is described as *slow twitch*; the term *fast twitch* refers to a twitch response that involves a brief contraction time (Figure 5.2b).

Scientists are unable to identify distinct populations of fast- and slow-twitch motor units in most human muscles. Instead, twitch contraction times of motor units have been reported to extend along a continuum from 20 to 72 ms (mean = 35 ms) for masseter, 35 to 85 ms (mean = 59 ms) for thenar muscles, and 40 to 110 ms (mean = 76 ms) for medial gastrocnemius (Garnett, O'Donovan, Stephens, & Taylor, 1978; Nordstrom & Miles, 1990; Thomas, Johansson, & Bigland-Ritchie, 1991). Differences in contraction speed among motor units are thought to be due largely to variations in the enzyme myosin ATPase (bound to one of the S1 heads), the rate at which Ca^{2+} is released from and taken up by the sarcoplasmic reticulum (Kugelberg & Thornell, 1983), and the architecture of the muscle.

When a series of stimuli are given, each stimulus elicits a single twitch response so that a series of twitch responses is produced. As the stimuli occur closer together, the twitches begin to sum, and the force profile becomes known as a tetanus. As the time between the stimuli decreases further, the tetanus changes from a sawtooth (**unfused tetanus**) (Figure 5.2c) to a smooth plateau (**fused tetanus**) (Figure 5.2d). When the stimuli are applied at a submaximal rate such that the interval between stimuli is 1.25 times the contraction time, the unfused tetanus may reach an initial peak and subsequently **sag**, or decline, before returning again to the peak value (Figure 5.2c shows a slight sag after the

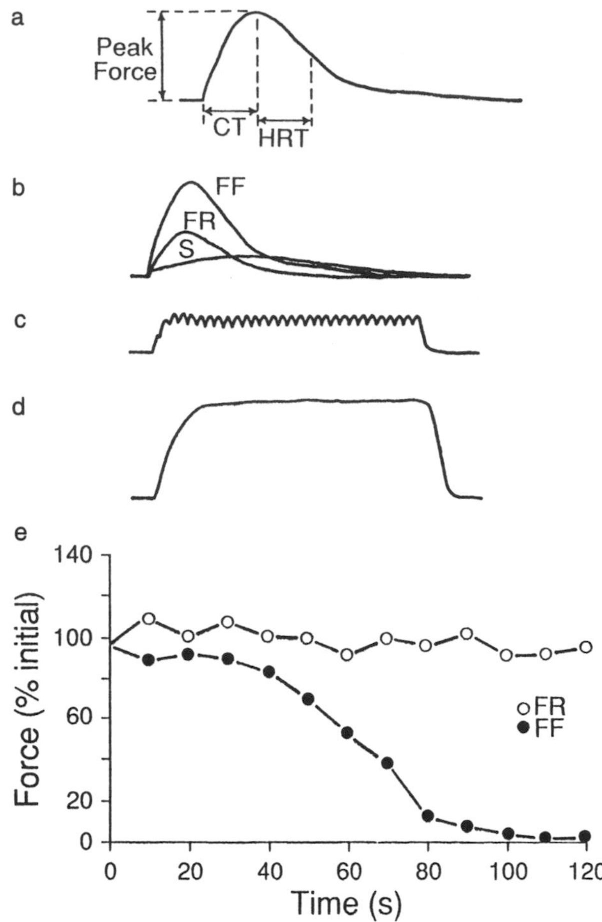

Figure 5.2 Twitch and tetanic responses of motor units in a cat hindlimb muscle: (a) Type FF motor unit twitch (contraction time [CT] = 24 ms; half relaxation time [HRT] = 21 ms; peak force = 0.03 N); (b) twitch responses for Types FF, FR, and S; (c) an unfused tetanus (stimulated at 28 Hz) for the FF unit (peak force = 0.42 N); (d) a fused tetanus (stimulated at 200 Hz) for the motor unit in panel c (peak force = 1.30 N); (e) fatigue test for the Type FF and Type FR motor units in panel b.

fourth stimulus). Fast-twitch motor units exhibit this sag profile in the unfused tetanus, whereas slow-twitch units do not; however, the sag has not been demonstrated in human motor units (Thomas, Johansson, & Bigland-Ritchie, 1991).

The capability of the motor unit to exert force is not measured from the twitch; it is assessed from the peak force of a single fused tetanus. That is, the maximum force exerted by a motor unit (and a muscle) occurs during a fused tetanus. The difference between the peak twitch force and the maximum force in a fused tetanus (referred to as P_o) is known as the twitch-tetanus ratio and generally varies from 1:1.5 to 1:10. Thus, in a fused tetanus the force may be 1.5 to 10 times greater than the twitch force. Peak tetanic force does not vary as a function of twitch contraction time; rather, P_o can vary over a substantial range for the same twitch contraction time, and conversely twitch contraction time can double with little change in P_o (Figure 5.3).

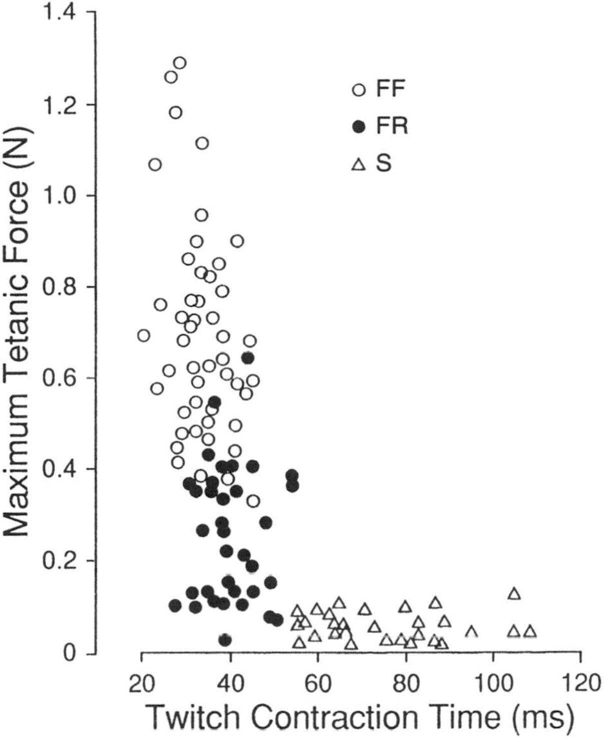

Figure 5.3 Maximum tetanic force and twitch contraction time for motor units of the medial gastrocnemius muscle of a cat.
Note. From "The Correlation of Physiological Properties and Histochemical Characteristics in Single Muscle Units" by R.E. Burke and P. Tsairis, 1974, *Annals of the New York Academy of Sciences*, **228**, p. 152. Copyright by The New York Academy of Sciences. Adapted with permission.

The force produced in a single tetanus declines over time if the motor unit is required to produce a series of tetani. The ability of a motor unit to prevent such a decline is considered a measure of its resistance to fatigue. The time between the onset of activation and the beginning of the decline in force differs markedly among motor units and can be assessed by various fatigue tests. A standard fatigue test (Burke et al., 1973) designed for cat hindlimb motor units involves eliciting tetani for 2 to 6 min at a rate of one tetanus each second—each tetanus lasts for 330 ms and includes 13 stimuli. The ratio of the peak force exerted after 2 min of this stimulus protocol compared with that exerted in the initial tetanus is referred to as the fatigue index (Figure 5.2e). This fatigue test stresses the connection between the electrical signal from the nervous system and the contraction of the muscle (Enoka & Stuart, 1992; Jami, Murthy, Petit, & Zytnicki, 1983). This test distinguishes fatigue-resistant types of motor units (Types S and FR) that have a fatigue index greater than or equal to 0.75 from fatigable motor units (Type FF) that usually have an index of less than 0.25. An index of 0.25 indicates that after 2 min the force exerted by the unit will be only 25% of that measured at the beginning of the test. These data suggest that activation of the Type S and Type FR motor units is more appropriate for sustained contractions because they are more resistant to fatigue.

Motor Unit Types

On the basis of two of these parameters, sag in the unfused tetanus and fatigue resistance, mammalian motor units can be classified into three groups (Burke, 1981): slow-contracting, fatigue-resistant (**Type S**); fast-contracting, fatigue-resistant (**Type FR**); and fast-contracting, fast-to-fatigue (**Type FF**). The Type S motor units produce the least force, and the Type FF produce the greatest (Table 5.1). This difference in force is due to variations in the number of muscle fibers belonging to a motor unit (i.e., the innervation ratio) and the size of individual muscle fibers (i.e., the quantity of contractile proteins per muscle fiber). Although there are differences between muscles and species, Type FF

Table 5.1 Motor Unit and Muscle Fiber Characteristics for Three Cat Hindlimb Muscles

	MG	FDL	TA
Number of muscle fibers	170,000	26,000	—
Number of motor units	270	130	—
Innervation ratio			
S	611	180	93
FR	553	132	197
FF	674	328	255
Mean tetanic force (mN)			
S	76	11	40
FR	287	53	101
FF	714	300	208
Muscle fiber area (μm^2)			
I	1,980	1,023	2,484
IIa	2,370	1,403	2,430
IIb	4,503	2,628	3,293
Muscle fiber types (%)			
I	25	10	11
IIa	20	37	50
IIb	55	53	39

Note. MG = medial gastrocnemius; FDL = flexor digitorum longus; TA = tibialis anterior.
Modified from Bodine, Roy, Eldred, & Edgerton (1987), Burke (1981), and Dum & Kennedy (1980).

motor units generally have the greatest innervation ratios and the largest muscle fibers (Table 5.1).

Motor units can also be classified *indirectly* based on histochemical and biochemical measurements. Both techniques involve determining the enzyme content of the muscle fibers, but one technique is qualitative (histochemistry) and the other is quantitative (biochemistry). Because enzymes are the catalysts for chemical reactions, measuring the amount of enzyme provides an index of the speed or quantity of the reaction. Thus, *the histochemical and biochemical techniques attempt to measure mechanisms responsible for the various physiological properties* (e.g., contraction speed, magnitude of force, fatigue resistance). Once a correlation can be determined between a chemical reaction and a physiological response, the quantity of enzyme can be interpreted as a correlate of the physiological response. Typically, three types of enzymes are measured: One type indicates contractile speed (myosin ATPase is an index of the maximum velocity of shortening) (Bárány, 1967), whereas the other two represent the metabolic basis (aerobic vs. anaerobic) by which the muscle fiber produces its energy for contraction. Commonly assayed enzymes for aerobic metabolism are succinic dehydrogenase (SDH) and nicotinamide nucleotide dehydrogenase tetrazolium reductase (NADH-TR); for anaerobic capabilities the enzymes are phosphorylase and alpha-glycerophosphate dehydrogenase (alpha GPD).

Based on these enzyme assays of muscle fibers (note that the histochemical and biochemical schemes refer only to the muscle component of the motor unit), it is possible to classify the fibers with a tripartite scheme. Two such schemes are commonly encountered in the literature. One scheme, which classifies fibers solely on the basis of myosin ATPase, uses the nomenclature *Types I*, *IIa*, and *IIb*. The distinction between **Type I** and **Type II muscle fibers** is based on the amount of ATPase activity remaining in the muscle fibers after preincubation in a bath with a pH of 9.4. Type I represents the slow-twitch, and Type II represents the fast-twitch muscle fibers. Type II muscle fibers can be further separated into two groups (IIa and IIb) after preincubation in baths with pHs of 4.3 (IIa) and 4.6 (IIb) (Brooke & Kaiser, 1974). The distinction between the muscle fiber types is shown in Figure 5.4 with a myosin ATPase stain of a thin cross section of a cat hindlimb muscle. The other scheme—which uses enzymes for contraction speed (myosin ATPase), aerobic capacity (SDH or NADH-TR), and anaerobic capacity (phosphorylase and alpha-GPD)—employs the terms slow-twitch, oxidative (**Type SO**), fast-twitch, oxidative-glycolytic (**Type FOG**), and fast-twitch, glycolytic (**Type FG**).

In general, Type I fibers are the same as Type SO fibers, Type IIa fibers correspond to Type FOG, and Type IIb fibers are the same as Type FG fibers. There are, however, some species-dependent differences (Nemeth & Pette, 1981). Furthermore, the SO muscle fibers belong to a **Type S motor unit**, the FOG fibers to a **Type FR motor unit**, and the FG fibers to a **Type FF motor unit**. Because all the muscle fibers in a motor unit are the same type, these fibers possess the same properties and are therefore **homogeneous** (Edström & Kugelberg, 1968; Nemeth et al., 1986). For example, a Type S motor unit has only Type SO muscle fibers, and these fibers all have the same physiological and biochemical characteristics. The same relationship applies to the other two motor unit types and their muscle fibers.

Figure 5.4 Photomicrograph of muscle fiber types in the tibialis posterior muscle of the cat hindlimb. The thin cross section of muscle was stained for myosin ATPase to show the distribution of Type I (dark), Type IIa (white), and Type IIb (gray) fibers. In the middle of the photomicrograph is a muscle spindle, with its capsule and small intrafusal muscle fibers.
Reprinted by permission of Robert C. Callister, Ph.D.

Each human muscle contains a mixture of all three muscle fiber types (Edgerton et al., 1975). The composition is usually assessed by using a muscle biopsy to extract muscle tissue, subjecting the extracted tissue to a myosin ATPase stain, and then classifying the fibers using the I-IIa-IIb scheme. It is difficult to measure the physiological properties of human motor units, although a few such experiments have been reported (Stuart & Enoka, 1983). One technique, known as microneurography, involves carefully placing a tungsten microelectrode in the nerve of a human volunteer and attempting to stimulate a single axon belonging to a muscle (Westing, Johansson, Thomas, & Bigland-Ritchie, 1990). With such an approach, it has proved difficult to identify distinct populations of motor units in human muscle that are similar to those found in other mammalian muscle (Thomas, Johansson, & Bigland-Ritchie, 1991). However, there are differences between muscle fiber types in human muscle. These differences, which include fiber type proportions and cross-sectional area, vary with the training history and gender of the individual and with the muscle examined (Table 5.2). In general, the Type II muscle fibers have a larger cross-sectional area than Type I fibers; the exceptions include muscles of runners and the middle trapezius and vastus lateralis muscles of females (Table 5.2).

Figure 5.5 summarizes the physiological and histochemical differences among the Type FF, FR, and S motor units. In this diagram, each motor unit is shown as a soma, which gives rise to an axon that descends vertically and branches out to innervate four muscle fibers. The differences in size (e.g., for soma, axon diameter, muscle fiber cross-sectional area) indicate general morphologic differences between the unit types. Beneath the motor units are shown the respective force-time (twitch) and fatigability profiles. Accordingly, the Type FF motor unit exerts the greatest twitch force, has the fastest (shortest) contraction time, and is the most fatigable. Figure 5.5 also illustrates the variable effect of the Ia afferent input from the muscle spindle onto the motor neuron; Type FF units receive fewer Ia afferents and hence produce a lesser response (excitatory postsynaptic potential [EPSP]) to Ia input. The Ia EPSP represents a voltage-time record, and the larger the EPSP, the more easily the motor neuron is excited by input from the muscle spindle.

In addition to the physiological features, Figure 5.5 indicates the histochemical profile for the muscle fibers of the different motor units. Each fiber attached to a motor unit illustrates the histochemical staining pattern for a different enzyme. The four enzyme systems tested were myosin ATPase (M-ATPase), myofibrillar ATPase after acid preincubation (Ac-ATPase) (this is used to differentiate between Type IIa and IIb fibers), oxidative (Ox), and glycolytic (Glycol.). The staining patterns are dark (indicating a high density of enzyme), intermediate, and clear. Myosin ATPase is correlated with the maximum velocity of shortening; therefore, the muscle fibers that stain dark for myosin ATPase suggest that the motor unit has a fast contractile speed. The first of the four muscle fibers for each motor unit shows the myosin ATPase stain; Types FG and FOG stain dark for myosin ATPase. Types SO and FOG stain dark for oxidative enzymes, and Types FG and FOG show a high glycolytic capability.

Functional Implications

Why are there different types of muscle fibers and motor units? The variety of muscle fiber and motor unit types in a muscle probably increases the range of demands that the muscle can accommodate. The two mechanisms that determine the proportion of fiber types in a muscle are heredity and usage. Although there are some molecular differences between muscle fibers that affect certain physiological properties (Groves, 1989), other characteristics depend on the amount of activity that the fibers experience. The evidence presented in Table 5.2 supports the effect of usage. Table 5.2 lists the proportions and cross-sectional areas of muscle fiber types for different athletes. Furthermore, training studies have shown that these properties can be changed for individual subjects with several weeks of intense training (Saltin & Gollnick, 1983), and patients with chronic heart failure exhibit an increase in the proportion of Type IIb fibers

Table 5.2 Fiber-Type Proportions and Cross-Sectional Area (CSA) of Selected Human Muscles

Muscle	*n*	Type I		Type II	
		%	CSA (μm^2)	%	CSA (μm^2)
Deltoid[a]					
Students	12	50	5,710	50	6,490
Wrestlers	8	61	5,700	49	7,010
Kayakers	9	71	5,170	29	7,430
Runners	9	49	5,800	51	5,850
Lifters	7	54	5,060	46	8,910
Tibialis anterior					
Female[b]	5	60	2,216	40	3,338
Male[c]	10	66	3,463	34	4,959
Trapezius (middle)					
Female[d]	5	63	2,746	37	2,270
Male[e]	5	67	3,853	33	4,647
Vastus lateralis[f]					
Female	203	51	4,044	49	3,409
Male	215	46	4,591	54	4,814

Note. *n* = number of subjects.
[a]Tesch and Karlsson (1985); [b]Henriksson-Larsén (1985); [c]Henriksson-Larsén, Fridén, and Wretling (1985); [d]Lindman, Eriksson, and Thornell (1991); [e]Lindman, Eriksson, and Thornell (1990); [f]Simoneau and Bouchard (1989).

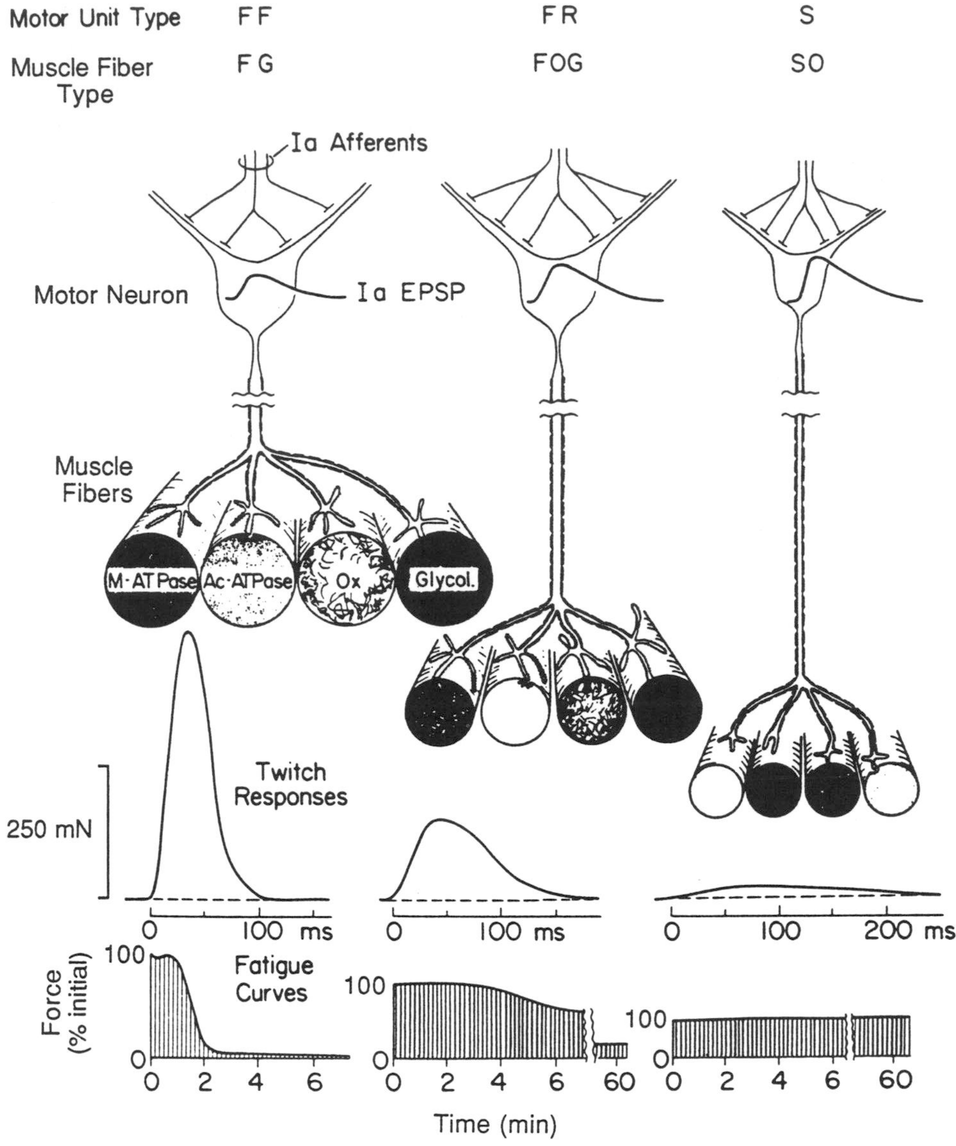

Figure 5.5 Physiological and histochemical features of motor units in the medial gastrocnemius muscle of the cat.
Note. From "Motor Unit Properties and Selective Involvement in Movement" by R.E. Burke and V.R. Edgerton. In *Exercise and Sport Sciences Reviews* (Vol. 3) (p. 36) by J.H. Wilmore and J.F. Koegh (Eds.), 1975, New York: Academic Press. Copyright 1975 by Academic Press. Adapted by permission.

due to deconditioning rather than pathological changes (Sullivan, Green, & Cogg, 1990). These proportions would not change if fiber type proportions did not depend on usage.

Given that the properties of muscle fibers and motor units depend on the patterns of usage, what variables related to activity cause the muscle to change? The simple answer seems to be that any physiological property—such as the magnitude of force, contractile speed, and fatigability—can adapt to make the muscle more able to meet the demands that it encounters. For example, Monster, Chan, and O'Connor (1978) examined the extent of usage for 15 muscles during a normal 8-hr working day and found that muscles with a higher proportion of Type I fibers tended to be used more frequently. Presumably, muscles with a higher proportion of Type I fibers have a greater oxidative capability and are less likely to fatigue due to metabolic factors. Another difference between muscle fiber types is the power that they can produce. Type II muscle fibers can produce more force, have a higher maximum velocity of shortening, and are able to produce more power. But although Type I fibers produce less power, they do so more efficiently (Brooks & Faulkner, 1991; Rome et al., 1988).

Whereas a case can be made for the need for different fiber types, performance during some tasks does not correlate well with the proportion of muscle fiber types.

When subjects sustained an isometric force to exhaustion at either 20%, 50%, or 80% of maximum with the knee extensors, the endurance time was not correlated with the proportion of fiber types in vastus lateralis (Maughan, Nimmo, & Harmon, 1985). Similarly, the maximum velocity of knee extension was not correlated with the proportion of fast-twitch muscle fibers; however, peak acceleration was weakly correlated with the proportion of fast-twitch muscle fibers (Houston, Norman, & Froese, 1988). In contrast, there appears to be a positive correlation between the proportion of fast-twitch fibers in biceps brachii and the maximum elbow flexor force (Nygaard et al., 1983) and between the proportion of Type I fibers and peak oxygen consumption (Mancini et al., 1989). These observations indicate that although human performance capabilities can be affected by the proportion of muscle fiber and motor unit types, other factors must also play a role in prescribing these capabilities.

Excitable Membranes

The motor unit represents the basic functional unit of movement control. How does the motor neuron command the muscle fibers into activity? In other words, what is the mechanism by which the two elements communicate? Interaction between the motor neuron and its muscle fibers occurs at a rapid, electrical level (*action potentials*) and at a slower, chemical level (*neurotrophism*). The latter is often neglected in the study of movement, yet it may represent a significant mechanism by which the properties of a motor unit are specified.

Neurotrophism

Neurotrophism refers to the sustaining influence that one biological element (e.g., the neuron) exerts directly on another (e.g., muscle fibers). Although an axon is typically represented as a simple cylinder, the internal structure of the axon is complex and supports fast **axonal transport** systems for moving materials at a rate of 250 to 400 mm per day. Transportation of the material can occur from the soma to the end plate (*orthograde transport*) and in the reverse direction (*retrograde transport*) (Figure 5.6). The transported material includes structural elements, proteins, RNA, and amino acids. For example, an important step in the electrochemical transfer of information from the neuron to the muscle is the release of the neurotransmitter by the vesicle into the synaptic cleft. After a number of cycles, the vesicle membrane requires repair and is transported back to the soma, where the membrane repair occurs; the vesicle is subsequently sent back to the neuromuscular junction via orthograde transport.

The transport mechanisms underlying neurotrophism consist of guiding structures (**microtubules**) along

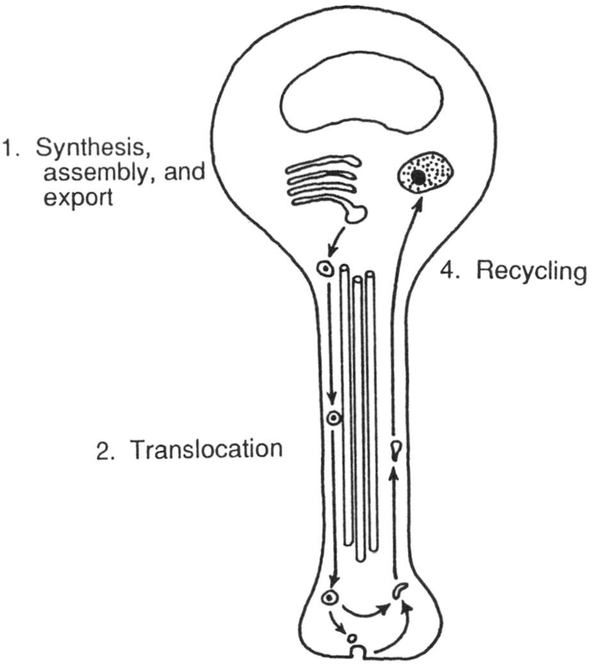

Figure 5.6 Diagram of axonal transport systems.
Note. From "Synthesis and Distribution of Neuronal Protein" by J.H. Schwartz. In *Principles of Neural Science* (2nd ed.) (p. 42) by E.R. Kandel and J.H. Schwartz (Eds.), 1985, New York: Elsevier Science Publishing Co., Inc. Copyright 1985 by Elsevier Science Publishing Co., Inc. Reprinted by permission of Appleton & Lange, Norwalk, CT.

which molecular "motors" can move organelles. The motors, which include the enzymes kinesin and cytoplasmic dynein, are attached to the microtubules and appear to function in a qualitatively similar manner to the crossbridge in muscle (Sheetz, Steuer, & Schroer, 1989). Each motor appears to translocate organelles in only one direction. Kinesin drives orthograde transport, and cytoplasmic dynein supports retrograde transport. On the basis of measurements made with a laser beam, it has been estimated that a single kinesin molecule can exert a force of about 2 pN and move organelles at about 0.5 to 2.0 µm/s.

There is substantial experimental evidence for both nerve-muscle and muscle-nerve trophism. For example, studies on the effects of denervation provide support for the existence of nerve-muscle trophism. When the nerve to a muscle is cut (**denervation**), extensive changes occur in the neuromuscular system: Muscle mass is lost (atrophy); muscle fibers degenerate as nuclei migrate to the center of the fiber and the mitochondria disintegrate; force output decreases and the time course of the twitch slows down; the resting membrane potential depolarizes within 2 hr of denervation; the resting membrane resistance increases, reflecting a decrease in membrane permeability; the sodium-channel structure is altered; the sarcolemma becomes capable of spontaneously generating action potentials; there is a synthesis of extrajunctional ACh receptors; the enzyme acetylcholinesterase is reduced; and intact motor axons are sprouted. Denervation causes the removal of both action potentials and axonal transport, but the onset of these changes has been found to depend on how far the nerve is cut from the muscle. This observation strongly suggests a significant role for axonal transport in regulating the normal properties of the nerve-muscle system. Furthermore, **reinnervation** (reconnection of the nerve to the muscle) results in a reversal of these changes.

Similarly, disturbance of the axon has been used to demonstrate the existence of muscle-nerve trophism (Lowrie & Vrbová, 1992). This can be shown by changes in an electrophysiological property of the motor neuron: the afterhyperpolarization phase of the motor neuron action potential (Czéh, Gallego, Kudo, & Kuno, 1978). When a drug (tetrodotoxin) that blocks axonal propagation but does not impair axonal transport is applied to an axon, the afterhyperpolarization phase changes if the axon is stimulated on one side of the block but not on the other. When the axon is artificially stimulated for 14 weeks on the motor neuron side of the block so that no action potential reaches the muscle, the afterhyperpolarization phase decreases in duration. In contrast, when the stimulation is applied on the muscle side of the block and the muscle is activated, the afterhyperpolarization phase does not change. This observation is interpreted as evidence that activation of the muscle, albeit with artificial stimulation, maintains the health of the motor neuron as displayed in the afterhyperpolarization phase of the action potential. Other electrophysiological properties of the motor neurons (e.g., action potential overshoot, resting membrane potential, axonal conduction velocity) also change after the axon is cut (Huizar, Kuno, Kudo, & Miyata, 1978).

As a physiological process, axonal transport can adapt to altered patterns of usage. For example, Jasmin, Lavoie, and Gardiner (1987) examined the transport of the enzyme acetylcholinesterase from the soma to the neuromuscular junction in the motor neurons of rats following 8 weeks of participation in either a swimming or running program. (**Acetylcholinesterase**, an enzyme found at the neuromuscular junction, terminates the activity of the neurotransmitter acetylcholine.) After 8 weeks of exercise, the axonal transport of acetylcholinesterase had increased in the runners but not in the swimmers. This finding suggests that axonal transport can adapt to chronic changes in activity, but the adaptations are specific to the type of activity; motor neuron activity increased in both the runners and swimmers, yet only the runners exhibited a change in the axonal transport of this enzyme.

The rate of axonal transport can also decrease, as appears to happen with aging (Frolkis, Tanin, Marcinko, Kulchitsky, & Yasechko, 1985). Older rats transport proteins inside the axon at a slower rate (200 mm/day) than do adult rats (380 mm/day). Furthermore, when axonal transport is halted, there is less change in the resting membrane potential and excitability of single muscle fibers in older rats. This means that axonal transport plays less of a role in specifying some properties of muscle fibers in older rats. The slowing of axonal transport with age is a primary factor involved in neuronal aging.

Similarly, axonal transport appears to be an important mechanism by which disease can invade the nervous system. Disease agents (viruses, bacteria) are taken up by vesicles during pinocytosis (the closure and release of the vesicle from the membrane after exocytosis) and become internalized inside the axon. Retrograde transport has been implicated in the movement of viruses (poliomyelitis and herpes) and the tetanus toxin (due to bacterial infection in the skin) from the periphery to the cell body (Ochs, 1987).

Axonal transport, therefore, represents a mechanism by which muscle can exert a neurotrophic effect on its motor neurons and by which motor neurons can influence the properties and health of the muscle fibers. As with all physiological processes, axonal transport, and hence neurotrophic effects, vary with activity and age.

Electrical Potentials

Of the two levels of nerve-muscle interaction, most is known about the rapid electrical form. This type of interaction is possible due to the excitable nature of the axolemma and sarcolemma, which are the membranes surrounding the axon and the muscle fiber, respectively. To describe the generation and function of the electrical signals, we will examine the reason that membranes are excitable (have a non-zero steady-state potential), the ionic basis of the action potential, and the generation of the action potential.

Steady-State Potential. Excitable membranes comprise a lipid bilayer on or in which proteins are located; they are semipermeable in that some lipid-soluble substances and smaller molecules can move through them. These membranes are referred to as **plasma membranes**. The membrane proteins—which account for about 50% to 70% of the membrane structure and serve structural, enzymatic, receptor, channel, and pump functions—are critical to the capability of the membrane to store, transmit, and release energy. Figure 5.7 shows some of the molecules associated with a plasma membrane. The membrane is surrounded by intracellular (e.g., sarcoplasm, axoplasm) and extracellular fluids (Figure 5.8), which contain variable concentrations of ions—notably sodium (Na^+), potassium (K^+), and chloride (Cl^-). The concentration ratio (outside the cell with respect to the inside) differs for each of these ions (e.g., for the muscle cell of a warm-blooded animal, $Na^+ = 12.08$, $K^+ = 0.03$, $Cl^- = 30.00$) such that the net effect is a charge distribution (voltage or **electrical potential**) across the membrane.

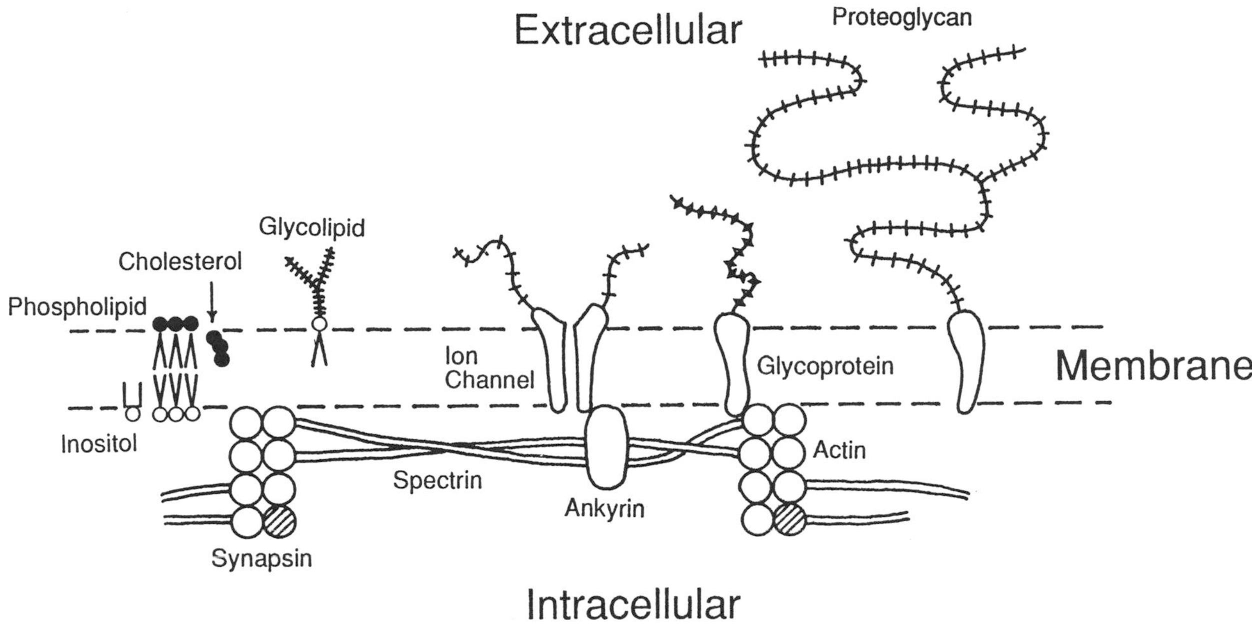

Figure 5.7 Some of the molecules associated with an excitable membrane.
Note. From *Neurobiology* (p. 45) by G.M. Shepherd, 1988, New York: Oxford. Copyright 1988 by Oxford University Press, Inc. Adapted by permission.

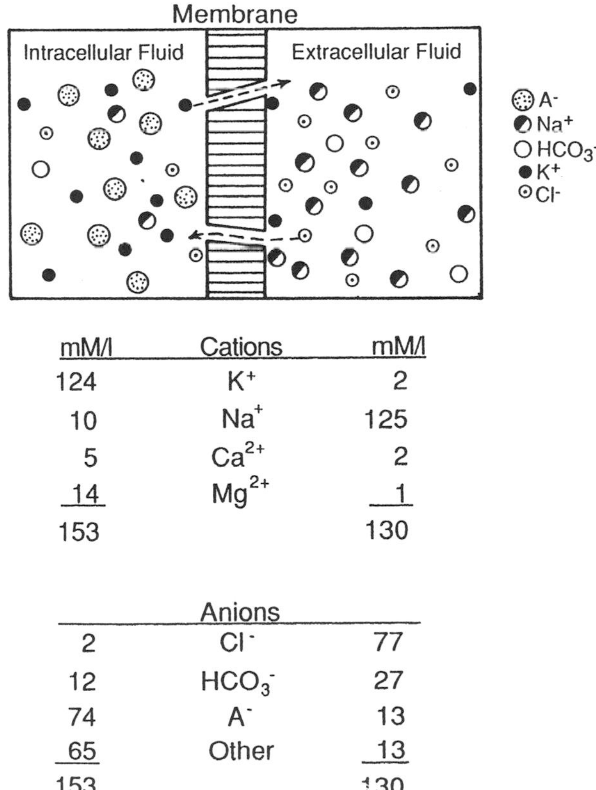

mM/l	Cations	mM/l
124	K^+	2
10	Na^+	125
5	Ca^{2+}	2
14	Mg^{2+}	1
153		130

	Anions	
2	Cl^-	77
12	HCO_3^-	27
74	A^-	13
65	Other	13
153		130

Figure 5.8 Distribution of ions about a mammalian excitable membrane. The membrane has channels through which K^+ and Cl^- can move easily, Na^+ with difficulty, and A^- (various negatively charged anions) not at all.

The particular distribution of an ion represents an equilibrium between two forces, the electrical and concentration gradients. The electrical force is due to the attraction of unlike charges and the repulsion of like charges. The concentration-gradient force causes ions to move from an area of high concentration to one that is lower.

When a membrane is at rest or in a steady-state condition (i.e., not conducting a signal), the electrical potential across the membrane remains relatively constant and is called the **resting membrane potential.** The resting membrane potential is usually about 60 to 90 mV, and the inside is negative with respect to the outside. The ions involved in establishing the electrical potential represent just a few of those included in the intracellular and extracellular fluids. The ionic distribution across the membrane represents a balance between the positive and negative ions and the concentration-gradient force experienced by each ion (Figure 5.9). In the resting state, Na^+ is prevented from remaining in the cell by the pumping activity of a membrane-bound protein, known as the **Na^+-K^+ pump**, that transports Na^+ from the intracellular to the extracellular fluid. Coupled to the localization of Na^+, K^+ is held internally by both the activity of the Na^+-K^+ pump and an electrical attraction to various organic anions (A), which are negatively charged ions such as amino acids and proteins that are too large to cross the membrane. Thus Na^+-K^+ pump activity keeps Na^+ in the extracellular fluid and K^+ in the intracellular fluid. The charges provided by Na^+ (attraction) and A^- (repulsion) cause Cl^- to be concentrated in the extracellular fluid. *The distribution of*

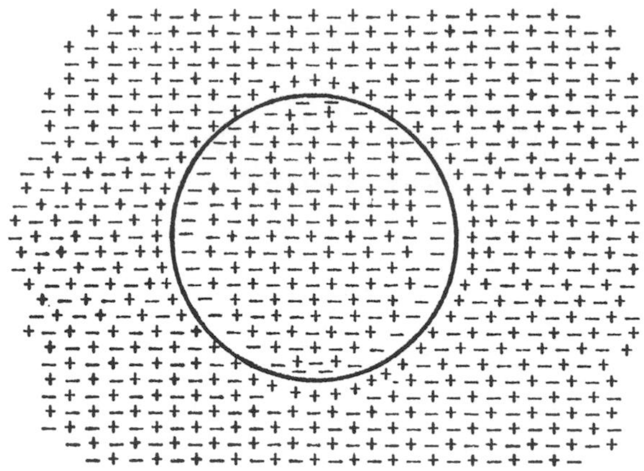

Figure 5.9 Conceptual view of the ionic charge distribution inside and outside a cell.
Note. From "Resting Membrane Potential and Action Potential" by J. Koester. In *Principles of Neural Science* (2nd ed.) (p. 50) by E.R. Kandel and J.H. Schwartz (Eds.), 1985, New York: Elsevier Science Publishing Co., Inc. Copyright 1985 by Elsevier Science Publishing Co., Inc. Reprinted by permission of Appleton & Lange, Norwalk, CT.

these ions (Na^+ and Cl^- externally, K^+ and A^- internally) largely determines the steady-state potential that exists across an excitable membrane.

The forces acting on these ions (Na^+, K^+, and Cl^-), and the flux of the ions across the membrane in a steady-state condition, are schematized in Figure 5.10. As mentioned previously, the forces driving the ions across the membrane (driving force) are due to chemical and electrical effects. The chemical force is associated with the concentration gradient. The magnitude of each force for the different ions is illustrated in Figure 5.10 by the length of its arrow. For example, Na^+ has a greater external (out) concentration, as indicated by the larger Na^+ symbol on the outside of the membrane, and hence it experiences a concentration-gradient force (chemical) that tends to drive it inward. The membrane is electrically negative inside (in) with respect to the outside (out). Positive ions (Na^+ and K^+) experience an inward-directed electrical force, and the negative ions (Cl^-) experience a force that drives them out. The net driving force (net DF) experienced by each ion can be determined by summing the chemical and electrical effects. For example, Na^+ is subject to inward chemical and electrical effects, which result in a large, inward net DF for Na^+. By similar reasoning, K^+ experiences a small, outward net DF, whereas the chemical and electrical components for Cl^- cancel, resulting in a zero net DF.

The net driving force for each ion combines with its permeability (P) to determine the net magnitude and direction of the ion's flux at a steady-state membrane potential. The ease with which an ion crosses a membrane (i.e., the permeability of the membrane to an ion) is indicated by the size of the symbol; excitable membranes are most permeable to K^+ (P_K) and least to Na^+ (P_{Na}). Due to these differences in permeability, the amount of ion that crosses a membrane (net flux) depends upon the combination of the net driving force experienced by that ion and its permeability. The result (Figure 5.10, net flux column) indicates that, *in steady-state conditions, approximately equal quantities of Na^+ and K^+ cross the membrane, whereas Cl^- remains internal.* However, once Na^+ and K^+ cross the membrane, they are quickly returned to their original sides of the membrane by the activity of the Na^+-K^+ pump such that the ionic distribution, and hence the electrical potential, remain relatively constant under steady-state conditions.

Ionic Basis of an Action Potential. Despite the apparent stability of the resting membrane potential, it is possible to disrupt the ionic distribution and to change the membrane potential. For example, the movement of negative ions across the membrane to the outside or vice versa (positive ions to the inside) produces a decrease in the membrane potential, a change referred to as **depolarization**—that is, the membrane becomes less polarized or closer to zero. In contrast, the movement of negative ions internally or the deposition of additional positive ions externally causes the membrane to become **hyperpolarized** (more polarized).

The shift in ions across the membrane is due to changes that occur within the membrane. These changes can be induced chemically (e.g., by a neurotransmitter) or electrically (e.g., by variations in membrane potential) and are described as gated changes in membrane permeability. For example, although Na^+ crosses the membrane in the resting state with considerable difficulty, the permeability of the membrane to Na^+ can be altered by about 500-fold, which makes this crossing much easier. This change is referred to as an increase in Na^+ **conductance**, or permeability. Once Na^+ can cross the membrane, it does so, moving down its concentration gradient (Figure 5.10) from the extracellular to the intracellular fluid. As the positive ions move inside the cell, the membrane becomes depolarized, which causes an additional increase in sodium conductance leading to a further depolarization, and so on. Such changes in Na^+ conductance are **voltage-gated**, because changes in the electrical potential (voltage) lead to changes in conductance, which in turn lead to further changes in the electrical potential, and so on. The function of voltage- and **transmitter-gated** channels is shown in Figure 5.11. In this model, the passage of Na^+ and K^+ through their respective voltage-gated channels is physically prevented by a gating molecule. The orientation of the gating molecule depends on the potential across the surrounding membrane. In the resting state (upper diagram in Figure 5.11a), the molecule is positioned so that it blocks the channel. As the potential across the membrane becomes depolarized, the molecule shifts its position and the channel is opened (lower diagram in Figure 5.11a). Similarly, the transmitter-gated channel is closed in steady-state conditions but

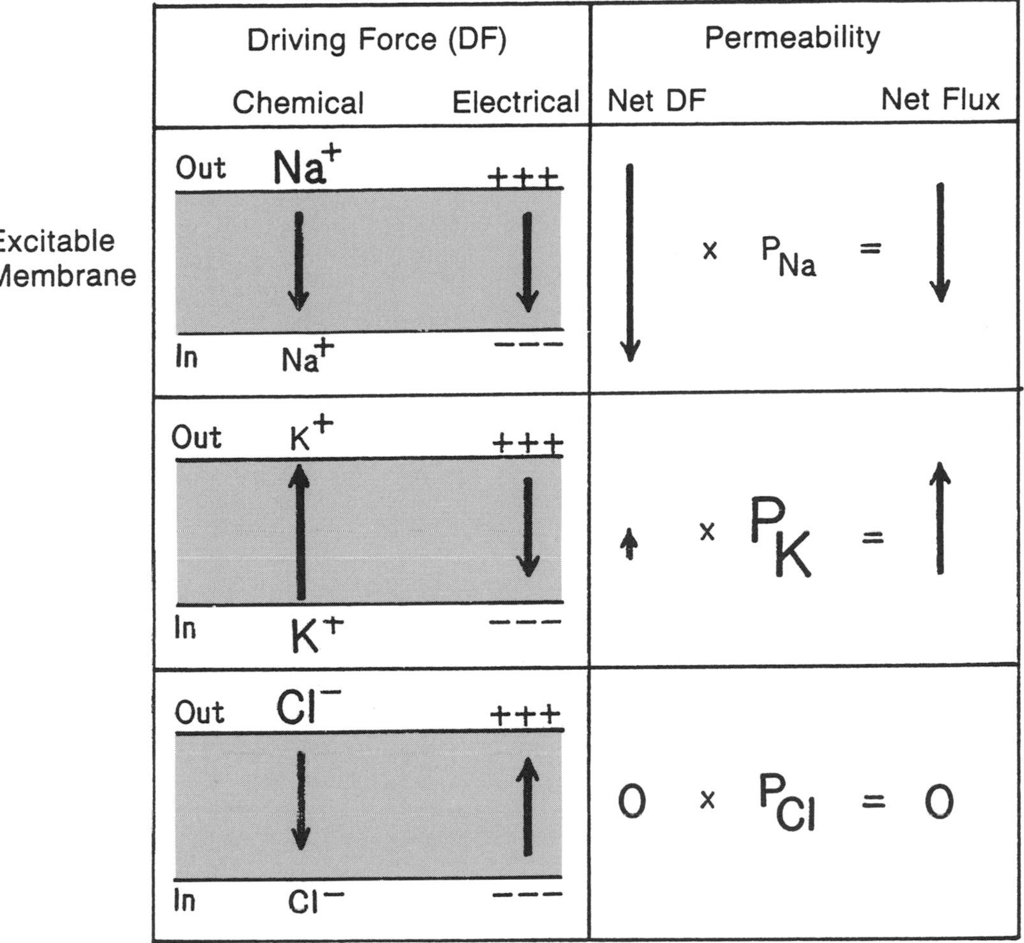

Figure 5.10 The net driving force and flux for Na^+, K^+, and Cl^- ions across a membrane with a steady-state membrane potential of –60 mV. Symbol size indicates the magnitude of the effect.
Note. From "Resting Membrane Potential and Action Potential" by J. Koester. In *Principles of Neural Science* (2nd ed.) (p. 55) by E.R. Kandel and J.H. Schwartz (Eds.), 1985, New York: Elsevier Science Publishing Co., Inc. Copyright 1985 by Elsevier Science Publishing Co., Inc. Reprinted by permission of Appleton & Lange, Norwalk, CT.

opens when a neurotransmitter attaches to the channel and causes it to change its structure. Both Na^+ and K^+ can pass through the same transmitter-gated channel. In general, the transmitter-gated channels initiate the changes in membrane potential that lead to the involvement of the voltage-gated channels.

Communication within the neuromuscular system is accomplished by transmitter- and voltage-gated changes in membrane potential. In general, these alterations are of two types: **synaptic potentials** and **action potentials**. Both types of potentials involve the same mechanism—an ionic shift across the membrane—but the synaptic potential essentially functions as a precursor to the action potential. Synaptic potentials decrease in amplitude as they are conducted along the membrane, whereas the action potential involves a self-regenerative process, known as **propagation**, that minimizes changes in amplitude. To distinguish between the processes of conduction (passive) and propagation (active), consider the ripples in a pond after a stone has been thrown into the pond. From the place that the stone contacts the water, ripples are set up in concentric circles, and these ripples move across the surface of the water away from the point of contact. Focus for a moment on a single ripple. As it travels across the surface it decreases in height, a phenomenon that is comparable to conduction of an electrical potential. If, however, the ripple were able to maintain the same height (a characteristic that would require the addition of energy to the ripple and therefore be an active process), then we would say that the ripple was being propagated.

Physically, an action potential is a patch of membrane where the potential has been reversed so that the inside is positive with respect to the outside. Like the wave in the pond, this patch of reversed polarity travels along the excitable membrane. It is possible to record an action potential in an experiment by placing a probe (an electrode) inside a cell (intracellular) and recording

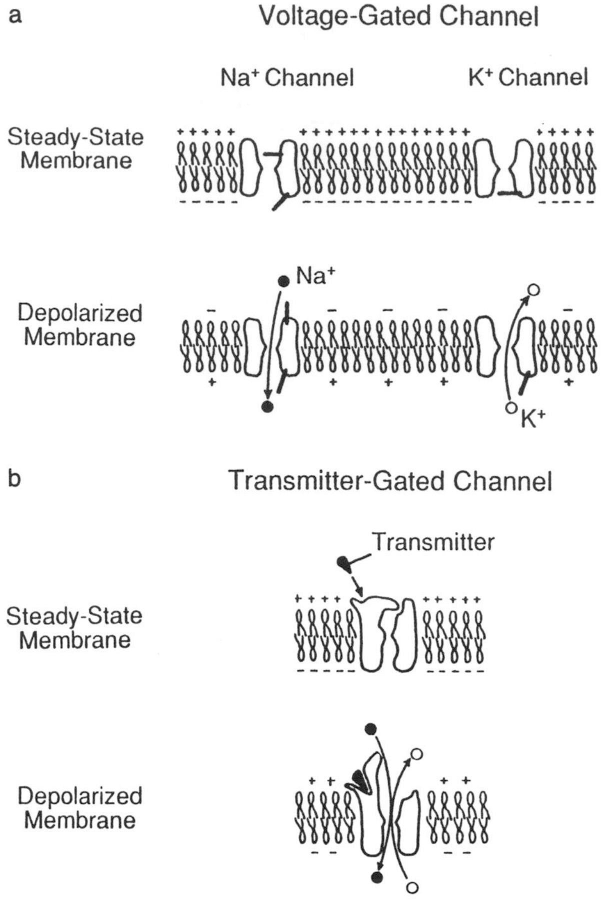

Figure 5.11 Function of gated channels in excitable membrane.
Note. From "Principles Underlying Electrical and Chemical Synaptic Transmission" by E.R. Kandel and S. Siegelbaum. In *Principles of Neural Science* (2nd ed.) (p. 99) by E.R. Kandel and J.H. Schwartz (Eds.), 1985, New York: Elsevier Science Publishing Co., Inc. Copyright 1985 by Elsevier Science Publishing Co., Inc. Reprinted by permission of Appleton & Lange, Norwalk, CT.

the potential inside the cell before, during, and after the action potential passes the electrode. With such an intracellular measurement, the action potential is recorded as a voltage-time event (Figure 5.12). An action potential is characterized by a rapid depolarization from the resting level to a complete reversal of potential (overshoot—the inside becomes positive with respect to the outside), followed by a **repolarization** (return to a polarized state) beyond the resting potential to a hyperpolarized state, and then a slow return to the resting membrane potential. These changes in membrane potential are the consequence of alterations in Na^+ and K^+ conductances (g_{Na} and g_K) that are temporally offset (Figure 5.12b). Because Na^+ and K^+ are both positive ions, their simultaneous movement across the membrane does not affect the membrane potential. However, because the net Na^+ influx occurs before the K^+ efflux and because the increased Na^+ conductance lasts for only a short time, the membrane experiences first a depolarization and then a repolarization. At any instant in time, therefore, the membrane potential (V_m) represents the imbalance in the Na^+ and K^+ conductances. The number of ions of one species (e.g., Na^+, K^+) that cross the membrane depends on the driving force and the permeability experienced by that ion (i.e., Na^+, K^+, Cl^-). At the conclusion of the action potential, for example, sodium conductance has returned to resting levels and potassium conductance has remained greater than normal, the net effect being a hyperpolarization of the membrane. Because it occurs after the action potential, the effect is referred to as the **afterhyperpolarization**. The Na^+-K^+ pumps gradually reverse the ionic distribution, returning Na^+ to the extracellular fluid and K^+ to the inside; this causes the membrane potential to return to steady-state levels.

Once a neuronal action potential has been initiated, it is propagated along the axon at speeds of up to 120 m/s. Because an action potential represents the reversal of the ionic distribution across the membrane (Area 2 in Figure 5.13b), an action potential and its propagation can be depicted as in Figure 5.13, where the electrical potential (Figure 5.13a) is shown relative to the ionic distribution (Figure 5.13b). Area 1 represents the membrane ahead of the action potential (inactive region), and Area 2 corresponds to the active region of the membrane; Area 2 shows the location of the action potential because it is the patch of membrane with the

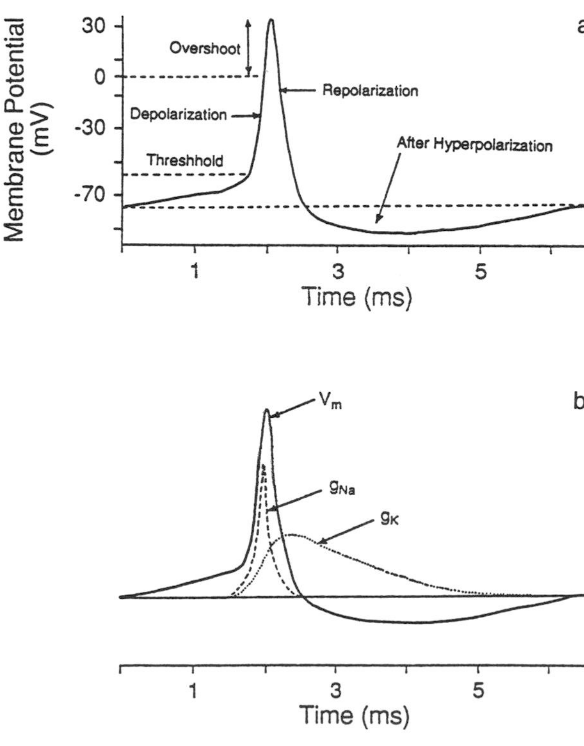

Figure 5.12 A schematic diagram of the action potential indicating (a) its phases and (b) its conductance changes.

reversed polarity. The depolarization spreads (positive charges on the inside indicate current flow) from Area 2 to Area 1 in the direction of action potential propagation. Although the depolarization also spreads from Area 2 to Area 3, the change in K^+ conductance in the membrane behind the action potential offsets any accumulation of positive charges on the inside.

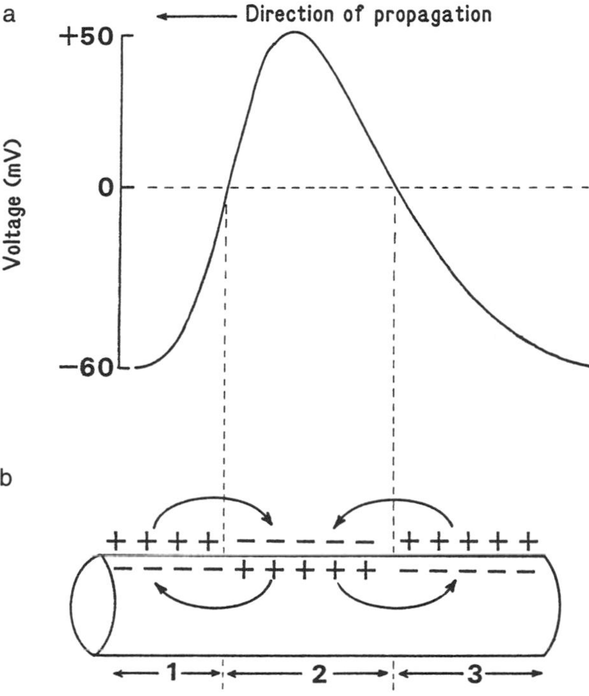

Figure 5.13 Propagation of an action potential and the associated ionic distribution.
Note. From "Functional Consequences of Passive Membrane Properties of the Neuron" by J. Koester. In *Principles of Neural Science* (2nd ed.) (p. 72) by E.R. Kandel and J.H. Schwartz (Eds.), 1985, New York: Elsevier Science Publishing Co., Inc. Copyright 1985 by Elsevier Science Publishing Co., Inc. Reprinted by permission of Appleton & Lange, Norwalk, CT.

The speeds at which action potentials are propagated vary with the diameter of the axon, mainly because much of the surface area of the large axons is insulated with myelin and thus prevented from involvement with the ionic shifts. Instead the action potential jumps from one node of Ranvier (noninsulated axolemma) to another; this is known as **saltatory conduction**. The presence of myelin, however, is somewhat fragile, and a number of neurological diseases involve a focal or patchy destruction of myelin sheaths that is usually accompanied by an inflammatory response. These demyelinating diseases can affect myelin in the central nervous system (e.g., multiple sclerosis, encephalomyelitis, myelopathy) and in the peripheral nervous system (e.g., Guillain-Barré syndrome, some types of neuropathies) and impair the propagation of action potentials.

Generation of an Action Potential. As we have discussed, the physical appearance of an action potential is a patch of membrane where the polarity has been temporarily reversed, that the change in membrane potential is due to gated changes in membrane conductance (*g*) for Na^+ and K^+, and that this reversal of polarity can move along an excitable membrane. But how do cells generate an action potential? To address this question, we will first consider the generation of an action potential by a neuron (an axonal action potential) and then examine the generation of an action potential by the sarcolemma of a muscle fiber (a muscle action potential).

An action potential is an *all-or-none* electrical command that is issued by a cell in response to the input that it receives. Both neurons and muscle fibers can generate action potentials. The inputs received by a neuron that underlie the generation of an action potential occur in the form of synaptic potentials, which is why the synaptic potential was described as the precursor to the action potential. Synaptic potentials can be either excitatory or inhibitory; they can either depolarize or hyperpolarize the receiving (**postsynaptic**) cell. Inhibitory synaptic potentials are due to an increased conductance to smaller ions, such as K^+ and Cl^-. The size of a synaptic potential depends on several factors (Burke, 1981), but the average amplitude of excitatory synaptic potentials elicited by single Group Ia afferents in motor neurons has been reported to be 71 μV for Type FF motor units, 118 μV for Type FR, and 179 μV for Type S (Stuart & Enoka, 1983).

A neuron can receive up to 10,000 synapses from various sources, such as peripheral sensory receptors, interneurons, and supraspinal centers. These synapses are distributed over the soma and the dendritic tree. When a neurotransmitter is released at a synapse, it generates a synaptic potential in the postsynaptic neuron. The synaptic potential is conducted over the neuron, decreasing in amplitude as it travels along the membrane. The further the synapse is from the axon hillock, the less effective the synaptic potential in shifting the membrane potential toward the **action potential threshold** of the neuron (trigger level for generation of an action potential). The neuron does not generate an action potential each time it receives a synaptic potential; the generation of an action potential is based on the sum of the synaptic potentials at the axon hillock. The neuron receives many inputs (synaptic potentials) at the same time; some are excitatory and others are inhibitory. This continuous input causes fluctuations of the membrane potential about some average steady-state level. When the membrane potential in the region of the axon hillock deviates by 10 to 15 mV above the steady-state level, the threshold has been reached, and the neuron generates an action potential that is propagated along the axon to an effector cell. Thus, a neuron generates an action potential in response to the net effect of numerous continuous inputs (synaptic potentials).

In general, synaptic potentials are generated and conducted by transmitter-gated changes in membrane conductance, whereas action potentials rely on voltage-gated changes in conductance. This distinction can be illustrated by considering the second example of action potential generation, that by the sarcolemma of the muscle fiber. This occurs at the neuromuscular junction, where the neurotransmitter is **acetylcholine** (ACh). When the axonal action potential reaches the synaptic bouton (nerve ending enlargement), it accelerates the release of ACh by promoting fusion of the vesicles with the terminal membrane (inside the axon). The action potential causes an increase in Ca^{2+} conductance in the axolemma of the synaptic enlargement, and the increase in intraaxonal Ca^{2+} facilitates vesicle fusion. This process of vesicular fusion and release of neurotransmitter is known as **exocytosis**. When ACh is released into the synaptic cleft, it diffuses across the cleft and attaches to a receptor on a transmitter-gated Na^+-K^+ channel (Figure 5.14a). Once the channel is opened, Na^+ flows into the muscle cell and K^+ flows out, which leads to the generation of a synaptic potential that is known as the **end-plate potential**. Because the end-plate potential causes a modest depolarization of the sarcolemma, this leads to the opening of voltage-gated Na^+ channels (Figure 5.14b), which further depolarizes the sarcolemma and shifts it toward threshold. Eventually, the sarcolemma, in the region of the neuromuscular junction, will be depolarized enough so that threshold is reached, and a sarcolemmal action potential will be generated and propagated along the excitable membrane.

A unique feature of the neuromuscular junction is the strength of the end-plate potential. In contrast to most synaptic potentials, the end-plate potential is essentially obligatory in that its occurrence usually results in the generation of a sarcolemmal action potential. The only time that the one-to-one association between the end-plate and action potentials appears to break down is during conditions of fatigue. There is probably no other synapse in the nervous system for which a given input (synaptic potential) will invariably result in the generation of an action potential. End-plate potentials have an average peak-to-peak amplitude of 1.4 mV in the extensor digitorum longus of mice, and this increases to an average of 1.8 mV after 12 weeks of running (Dorlöchter, Irintchev, Brinkers, & Wernig, 1991). The average amplitude of the sarcolemmal action potential in the rat diaphragm is around 84 mV (Metzger & Fitts, 1986).

The neurotransmitter ACh is released not only in response to an axonal action potential; small amounts are also released continuously. This spontaneous release of ACh elicits **miniature end-plate potentials**, which have amplitudes of about 0.7 mV in fast-twitch muscle fibers and 1.9 mV in slow-twitch muscle fibers (Lømo & Waerhaug, 1985). Interestingly, the frequency of the miniature end-plate potentials decreases with age in soleus but not in extensor digitorum longus (Alshuaib & Fahim, 1990). Because of the miniature end-plate potentials, the steady-state potential of the postsynaptic sarcolemma fluctuates about an average value, as does the postsynaptic membrane of the neuron.

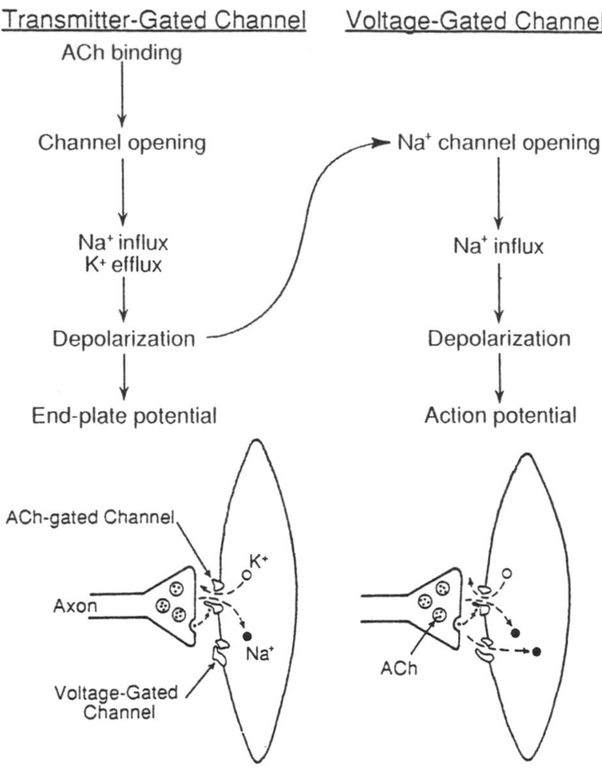

Figure 5.14 Activation of (a) transmitter-gated and (b) voltage-gated channels at the neuromuscular junction in response to the release of ACh.
Note. From "Principles Underlying Electrical and Chemical Synaptic Transmission" by E.R. Kandel and S. Siegelbaum. In *Principles of Neural Science* (2nd ed.) (p. 100) by E.R. Kandel and J.H. Schwartz (Eds.), 1985, New York: Elsevier Science Publishing Co., Inc. Copyright 1985 by Elsevier Science Publishing Co., Inc. Reprinted by permission of Appleton & Lange, Norwalk, CT.

Electromyography

Experimentalists have devised a technique with which to monitor the electrical activity of excitable membranes. The application of this technique to muscle is known as **electromyography** (EMG; *myo* = muscle). Because the EMG represents the measurement of sarcolemmal action potentials, it provides a window into the nervous system (Denny-Brown, 1949; Duchêne & Goubel, 1993; Hof, 1984; Loeb & Gans, 1986; Perry & Bekey, 1981; Person, 1963). In general, there will be no EMG unless it has been commanded by motor neurons. The most common approach to measuring EMG is to place an **electrode** (a type of probe that can measure voltage) near an excitable membrane and to record the action potentials as they pass the electrode. With this technique we can record an action potential as a voltage-time event. We used this procedure to measure the action potential for Figure 5.12; but in that instance the electrode was placed inside the cell. With EMG, we place the electrode outside the muscle cell and get an extracellular measurement of the change in voltage over time (Figure 5.15). The electrode, however, can be inside the muscle (intramuscular), although outside individual muscle cells, or outside the muscle. The most common approach is to place the electrode on the skin overlying the muscle; the electrode is outside the muscle.

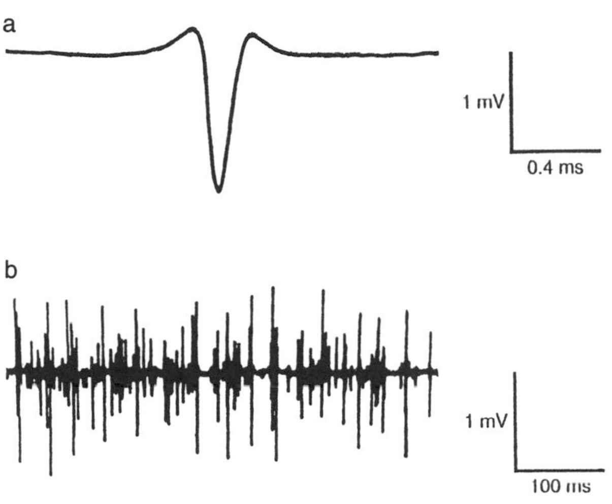

Figure 5.15 EMG records of (a) one and (b) many action potentials. Notice the difference in the time scale for the two records.

An extracellularly recorded action potential usually has a triphasic shape (Figure 5.15a). The reason for this shape is indicated in the diagram in Figure 5.16, which shows the association between the ionic events across the membrane and the recorded potential. An electrode is placed near an excitable membrane, outside the cell, to record the potential in the area (relative to some remote reference) as the action potential approaches, reaches, and passes by the electrode. In Figure 5.16, the electrode is on the right side of the diagram; this is the leading edge of the action potential. The central element of the action potential is the region where the polarity across the membrane has been reversed. This is caused by an influx of Na^+ ions and is apparent as the negative phase of the action potential; the negative potential corresponds to positive ions flowing into the cell. On each side of this negative phase, there is a single positive phase. The initial positive phase (the right side of the diagram) is due to passive local depolarization of the membrane (i.e., the membrane is less polarized). This happens because some Na^+ ions move forward from the region of reversed polarity in the intracellular fluid due to an electrical attraction (positive to negative), and some negatively charged ions in the extracellular fluid move backward. The local activity ahead of the region of reversed polarity is sensed by the electrode as the flow of positive ions out of the cell and is measured as a positive change in voltage. Similarly, the second positive phase of the potential (left side of the diagram) is caused by the efflux of K+ from the cell. The electrode, therefore, records the change in membrane potential with the movement of positive ions out of the cell registered as positive and the movement of positive ions into the cell recorded as negative.

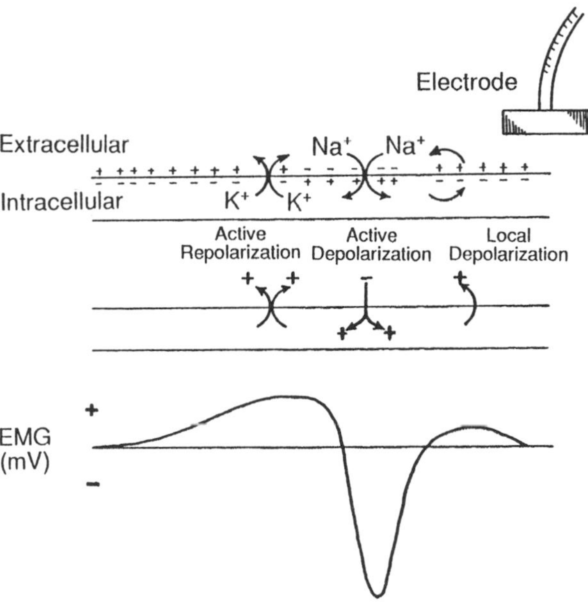

Figure 5.16 Association between the movement of ions across the sarcolemma (upper row), the corresponding current flow (middle row), and the shape of the extracellularly recorded action potential (bottom row).
Note. From *Electromyography for Experimentalists* (p. 46) by G.E. Loeb and C. Gans, 1986, Chicago: University of Chicago Press. Copyright 1986 by The University of Chicago. Adapted by permission.

The electrode serves as an antenna (Loeb & Gans, 1986) and can be arranged in either a monopolar or bipolar mode. With a monopolar mode, one electrode is placed over the signal source (e.g., muscle), and the signal that it senses (due to the influx of Na^+ and efflux of K^+) is compared with that at a distant location (ground). In contrast, a bipolar arrangement has two electrodes over the signal source and the output is the *difference* between the signals sensed by the two electrodes. Consequently, a monopolar arrangement is necessary to determine the absolute magnitude of the signal, whereas a bipolar configuration allows us to record selectively from a localized signal source.

The quantity of electrical activity that can be recorded in a muscle depends on the distance between the electrodes and the active motor units, the area of the recording surfaces of the electrodes, and the extent to which the electrodes are distributed over the muscle. In the analysis of human movement, an EMG is typically recorded with one pair of electrodes (each with a diameter of about 8 mm) that are separated by about 1.5 cm and placed on the skin over the muscle. Such electrodes will not record the entire EMG in a muscle but rather will record only muscle fiber action potentials within 1 to 2 cm of the electrodes (Fuglevand, Winter, Patla, & Stashuk, 1992). To obtain a more complete record of the EMG for an entire muscle researchers need to use either an array of electrodes distributed over the surface (Chanaud, Pratt, and Loeb, 1987; Emerson and Zahalak, 1981; Thusneyapan & Zahalak, 1989) or an electrode with a large recording area that is placed inside the muscle (Stålberg, 1980). The latter has been referred to as a macro EMG and is capable of recording most of the muscle fiber action potentials of an active motor unit. The macro EMG has been used to show that the amplitude of motor unit action potentials increases with age (Stålberg & Fawcett, 1982). For example, the maximum amplitudes of the action potentials for young subjects (younger than 60) were found to be 595 µV for biceps brachii, 1,068 µV for vastus lateralis, and 1,077 µV for tibialis anterior, whereas for older subjects (older than 60) the maximum amplitudes were 704 µV, 1,611 µV, and 962 µV, respectively. This change results from the loss of motor neurons with age and the peripheral reinnervation of muscle fibers by surviving motor neurons.

Typically, we are neither able to record nor interested in recording the action potential of a single muscle fiber or motor unit in the analysis of human movement. Most often, scientists measure the EMG of many motor units that are concurrently active. This type of EMG record is referred to as an **interference pattern** because it consists of many superimposed action potentials (see Figure 5.17; also, compare the time scale in Figure 5.17 to that for the single action potential in Figure 5.15). We are often interested in quantifying the interference EMG, and several procedures are available for this purpose (Davis and Vaughn, 1993; Loeb & Gans, 1986; Perry & Bekey, 1981). Two of the most common are rectification and integration. As shown in Figure 5.17, **rectification** consists of taking the absolute value of the EMG signal; an electronic module can be used to flip over the negative phases. Then the sharp peaks (high frequencies) that are present in the rectified EMG can be diminished by **integration**, an electronic process that consists of smoothing or filtering the EMG to reduce the high frequency content of the signal. After the interference EMG has been rectified and integrated, the EMG can be quantified by measuring the amplitude of the integrated EMG. The usefulness of this quantification procedure can be seen by comparing the changes in integrated EMG and force in Figure 5.17. To obtain the data in Figure 5.17, we had a subject slowly grade the force from zero to some target value, and we measured the interference EMG and force. After rectification and integration, the EMG paralleled the force record, which suggests that the EMG signal is a good index of muscle force under these conditions.

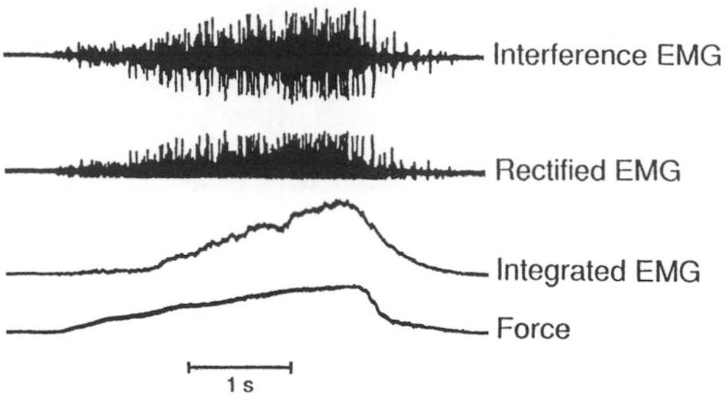

Figure 5.17 An interference EMG is rectified and integrated, and the resulting signal (integrated EMG) closely parallels the change in force.

When an integrated EMG is compared with the force exerted by muscle, there is a close association between the two (Bigland-Ritchie, 1981; Fuglevand, Winter, & Patla, 1993). But this association is limited to isometric conditions in which the muscle contracts without changing its overall length (Figure 5.18). Under these conditions, there can be a linear or a curvilinear relationship between the integrated EMG and force (Lawrence & De Luca, 1983). This means that the magnitude of the EMG provides a reasonable estimate of the force exerted by muscle. When muscle length changes (Figure 5.19), however, the location of the electrodes relative to the active muscle fibers changes, causing a change in the EMG that is unrelated to the input received by the muscle from the nervous system (Gerilovsky, Tsvetinov, & Trenkova, 1989; Inman, Ralston, Saunders, Feinstein, & Wright, 1952). For this reason, it is difficult to quantify EMG under nonisometric conditions (Bigland-Ritchie, 1981; Calvert & Chapman, 1977; Milner-Brown & Stein, 1975; cf., however, Hof & van den Berg 1981a, 1981b, 1981c, 1981d).

The interference EMG recorded during an activity is typically graphed as a function of time (Figure 5.17) and quantified by rectification and integration. This approach is referred to as a **time-domain analysis**. An alternative approach, the **frequency-domain analysis**, is to determine the frequency content of the EMG signal. Any signal (e.g., EMG, ground reaction force, velocity) can be characterized by its frequency content. To do so involves determining the combination of sine and cosine functions (Appendix B) that are needed to duplicate the

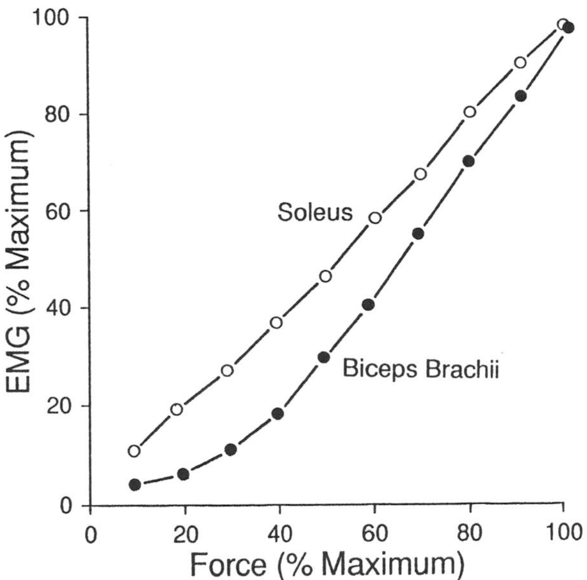

Figure 5.18 Integrated EMG-force relationship for soleus and biceps brachii muscles during an isometric contraction. Both EMG and force have been normalized to their respective maximum values.

Note. From "EMG/Force Relations and Fatigue of Human Voluntary Contractions" by B. Bigland-Ritchie. In *Exercise and Sport Sciences Reviews* (Vol. 9) (p. 78) by D.I. Miller (Ed.), 1981, Philadelphia: The Franklin Institute Press. Copyright 1982 by The Franklin Institute. Adapted with permission.

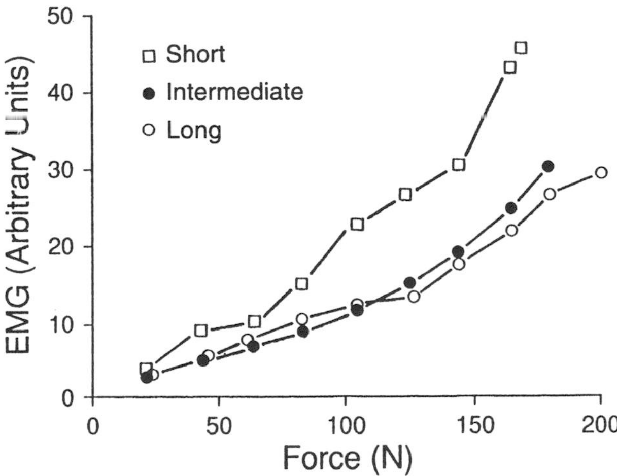

Figure 5.19 Measurement of EMG and force for triceps brachii at various lengths. The subject was a below-elbow amputee, and the muscle was attached to a force transducer.

Note. From "Relationships of the Surface Electromyogram to the Force, Velocity, and Contraction Rate of the Cineplastic Human Biceps" by C.W. Heckathorne and D.S. Childress, 1981, *American Journal of Physical Medicine*, **60**, p. 6. Copyright 1981 by The Williams & Wilkins Co. Reprinted by permission.

signal. Because the biological signals associated with human movement are not pure sine or cosine functions, it is necessary to combine several different sine-cosine functions to represent the signal. These functions will differ in frequency and amplitude. A more rapidly alternating sine-cosine function has a higher frequency. In general, the sharper the peaks and valleys in the biological signal, the higher the frequency content. Therefore, it is necessary to use higher frequency (more rapidly alternating) sine-cosine functions to represent the signal.

The most significant factor affecting the frequency content of the EMG is the shape of the constituent action potentials (Figure 5.20). When a motor unit sustains a discharge and fatigues, the shape of the action potential changes by decreasing in amplitude and increasing in duration (Figure 5.20a). These changes are evident in the frequency content of the signal (train of action potentials) at the beginning and end of the fatiguing contraction. The frequency-domain analysis essentially involves removing the time between successive action potentials so that the action potentials appear as periodic functions of time (Figure 5.20b). Each action potential sequence is then represented by a best-fitting combination of sine-cosine functions to characterize the frequency and amplitude of the signal, and the result is

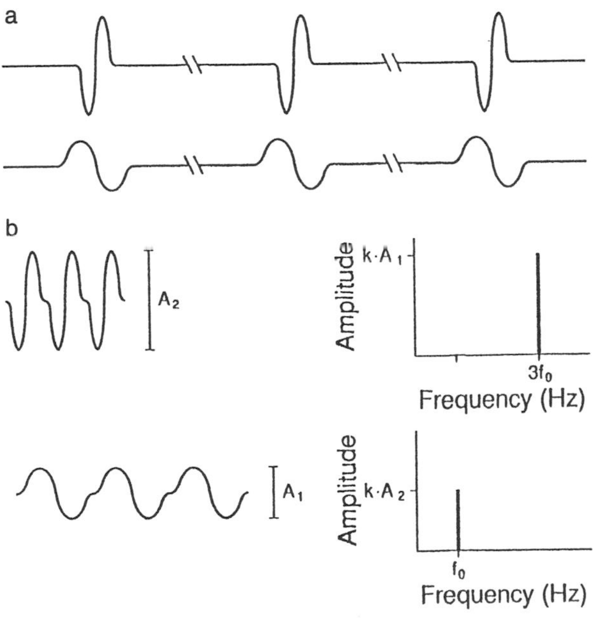

Figure 5.20 Frequency-domain analysis of the action potentials of a single motor unit at the beginning and end of a fatiguing contraction: (a) trains of action potentials at the beginning (upper trace) and end (lower trace) of a fatiguing contraction; (b) removal of the time between successive action potentials so that they appear as periodic functions that can be characterized in the frequency domain on a frequency-amplitude graph.

plotted on a frequency-amplitude graph. Because the train of action potentials for the single motor unit in Figure 5.20 can be represented by a single sine-cosine function (contains a single frequency), it can be characterized by a single line (amplitude and frequency) in the frequency domain. Notice that the action potential train recorded later in the fatiguing contraction has a smaller amplitude and frequency—three times smaller than the action potentials in the initial train. The lower frequency means that it can be represented by a less rapidly alternating sine-cosine function.

Because an interference EMG contains hundreds of action potentials with varying shapes and degrees of overlap, the frequency content is not confined to a single frequency but includes a range of frequencies. The frequencies span a given spectrum, and the amplitude of each frequency varies. Thus, to represent the interference EMG, we need to combine many sine-cosine functions of differing amplitude and frequency (rate of alternation). In general, the amplitude of each frequency is measured as power, and this type of graph is referred to as a **power spectrum**. An example of an interference EMG and the associated power spectrum is shown in Figure 5.21. A common index used to characterize the power spectrum is the **mean frequency**, which represents the central value of the frequency spectrum. Alternatively, the **median frequency** is the frequency that divides the spectrum into two halves based on the energy content of the signal. For most purposes, the mean and median frequencies provide comparable information, although the median frequency is the more sensitive measure during fatiguing contractions (Merletti, Knaflitz, & De Luca, 1990).

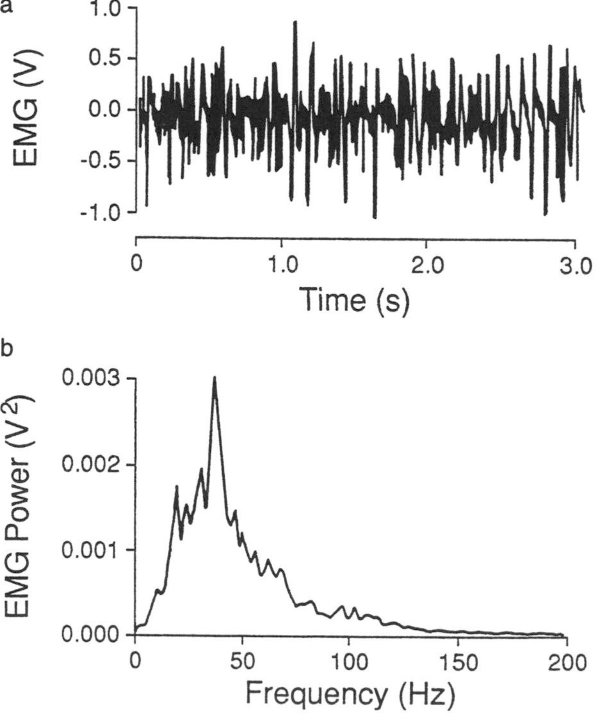

Figure 5.21 Power spectrum (b) for an interference EMG (a).

A power spectral analysis is commonly performed to document the changes in the EMG that occur with such activities as fatigue (Hägg, 1992). The two principal factors that cause an activity-dependent change in the frequency content of the interference EMG are *changes in action potential shape* (Figure 5.20) and a *reduction in motor unit discharge rates*. Action potential shape is most affected by changes in conduction velocity along the sarcolemma (De Luca, 1984; Lindström, Magnusson, & Petersén, 1970). As conduction velocity decreases, the duration of the action potential increases and its frequency content decreases, causing the mean and median frequencies to shift to lower values. Additionally, the discharge rate of motor units declines during sustained activity and at low rates (8 to 12 Hz), which can result in the grouping of action potentials into distinct bursts of activity (Taylor, 1962). This does not represent the synchronization of motor units, as some investigators have suggested. The effect of this grouping due to the reduction in discharge rate is to produce a prominent peak at this frequency (around 10 Hz) in the power spectrum (Krogh-Lund & Jøgensen, 1992). Because of these two effects (reductions in conduction velocity and discharge rate), the mean and median frequencies of the power spectrum decline to lesser values during fatiguing contractions. Furthermore, this leftward shift is one of the first events to occur as a muscle begins to fatigue.

Phonomyography. A muscle contraction is based on the electrical activation of muscle by the nervous system. As a muscle contracts, however, it makes sounds.

The first report of muscle sounds was made in 1663 (cited in Oster, 1984). We do not generally hear these sounds, because they occur at a low frequency, below the normal capacity of the human ear. The mean frequency is about 10 Hz at low forces and increases to about 22 Hz during a maximum voluntary contraction, compared with a mean frequency of about 60 Hz for an EMG measured during a moderate-force contraction (Maton, Petitjean, & Cnockaert, 1990; Orizio, Perini, Diemont, Figini, & Veicsteinas, 1990; Stokes & Cooper, 1992). However, these sounds can be heard with an appropriate microphone. The sounds are probably due to the lateral movements of muscle fibers as they contract. Because sound represents a series of pressure waves, it can be recorded and produces a signal that is similar to the interference EMG (Figure 5.17). Also, the sound signal can be analyzed with the same techniques used for the EMG (e.g., rectification and integration). The technique of recording muscle sounds is referred to as **phonomyography** (PMG); it has also been referred to as sound myography (SMG) and acoustic myography (AMG).

As with the EMG, the PMG is directly related to muscle force under isometric conditions; there appears to be a quadratic, not linear, relationship between PMG and muscle force (Maton et al., 1990; Stokes, Moffroid, Rush, & Haugh, 1988). Correlation analyses, however, show that EMG and PMG vary independently of one another as torque increases and then decreases and that PMG varies more than EMG. There are also significant differences between the EMG and PMG when submaximal contractions are sustained until exhaustion. For moderate-intensity contractions (60% to 80% MVC), the EMG at the end of the contraction is greater than at the beginning, whereas the final PMG values are less than at the beginning (Orizio, Perini, & Veicsteinas, 1989). These observations indicate that the EMG and PMG signals contain different information about a muscle contraction.

Excitation-Contraction Coupling

At the neuromuscular junction, the neurotransmitter acetylcholine (ACh) takes less than 100 μs to diffuse across the synaptic cleft and attach to receptors on the postsynaptic membrane. The attachment of ACh to the postsynaptic receptors results in the opening of the transmitter gated Na^+-K^+ channel and in the influx of Na^+ and the efflux of K^+ from the muscle fiber. The movement of Na^+ and K^+ across the sarcolemma results in the development of the end-plate potential (synaptic potential) that can trigger the generation of a sarcolemmal action potential. The processes involved in the conversion of an axonal action potential into a sarcolemmal action potential are referred to as **neuromuscular propagation**.

Once the axonal action potential has been transformed into a sarcolemmal action potential, several processes convert the command (action potential) that is issued by the motor neuron into a muscle fiber force. These processes are referred to as **excitation-contraction coupling**. The steps involved in excitation-contraction coupling are indicated in Figure 5.22: (1) propagation of the sarcolemmal action potential; (2) propagation of the action potential down the T tubule; (3) coupling of the action potential to the change in Ca^{2+} conductance of the sarcoplasmic reticulum; (4) release of Ca^{2+} from the sarcoplasmic reticulum; (5) reuptake of Ca^{2+} by the sarcoplasmic reticulum; (6) Ca^{2+} binding to troponin; and (7) interaction of the crossbridge (myosin) and actin. Most of these steps involve events that *permit* the interaction of actin and myosin (Ca^{2+} disinhibition), whereas only Step 7 actually *involves* the interaction of actin and myosin (crossbridge cycle).

Ca^{2+} Disinhibition

Under resting conditions, the thick and thin filaments are prevented from interacting by the regulatory action

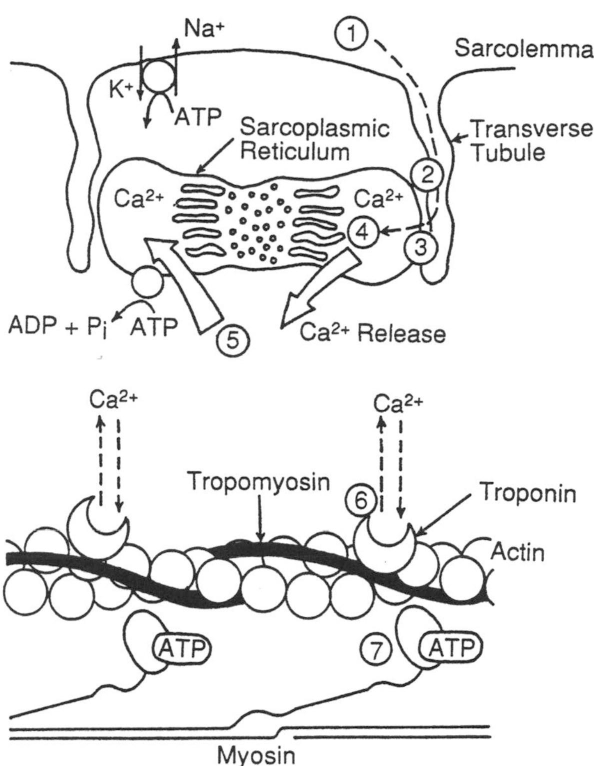

Figure 5.22 Events involved in excitation-contraction coupling.
Note. From "Mechanisms of Muscular Fatigue" by R.H. Fitts and J.M. Metzger. In *Principles of Exercise Biochemistry* (p. 214) by J.R. Poortmans (Ed.), 1988, Basel: Karger. Copyright 1988 by S. Karger AG. Adapted with permission.

of the proteins troponin and tropomyosin, and Ca^{2+} is largely stored in the sarcoplasmic reticulum. Steps 1 to 6 in Figure 5.22 identify the events involved in the removal of this inhibition. Essentially, these events enable the sarcolemmal action potential to trigger the release of Ca^{2+} from the sarcoplasmic reticulum and the subsequent inhibition of the regulatory action of troponin and tropomyosin. Because the net effect of this series of events is the removal, or inhibition, of inhibition, this result is referred to as **disinhibition**.

Ca^{2+} disinhibition begins with the muscle action potential as it is propagated along the sarcolemma at speeds of up to 6 m/s. Interestingly, the conduction (propagation) velocity appears to vary with training; it has been shown that body builders have higher conduction velocities for the biceps brachii muscle than do control subjects (5.5 vs. 2.8 m/s); (Kereshi, Manzano, & McComas, 1983). Figure 5.23 shows the role of Ca^{2+} in excitation-contraction coupling. In the resting state, the potential across the sarcolemma is negative with respect to outside, and most of the Ca^{2+} is stored in the terminal cisternae (Figure 5.23a). The action potential is propagated down the T tubule and triggers an increase in the Ca^{2+} conductance (g_{Ca}) of the terminal cisternae. In the absence of an action potential, g_{Ca} is normally low so that Ca^{2+} has difficulty crossing the membrane of the sarcoplasmic reticulum. The mechanism by which the action potential changes g_{Ca} may involve a chemical connection, a mechanical signal, or an effect due to changes in Ca^{2+} concentration; some evidence favors the mechanical connection between conformational (structural) changes in T-tubule molecules and the conductance of the sarcoplasmic reticulum (Ríos, Ma, & González, 1991; Ríos & Pizarró, 1988). Once g_{Ca} has been increased, Ca^{2+} is able to move from the terminal cisternae (lateral sacs) down its concentration gradient into the sarcoplasm (Figure 5.23b). The change in g_{Ca} is transient, and once the action potential has passed (Figure 5.23c), g_{Ca} returns to a resting level and Ca^{2+} is returned to the sarcoplasmic reticulum by Ca^{2+} pumps attached to the membrane of the sarcoplasmic reticulum. The rate at which Ca^{2+} is returned to the sarcoplasmic reticulum determines the rate of decline in force after the cessation of action potentials. For example, the rate of force decline for the twitch is quantified by the measurement of half relaxation time (Figure 5.2). After sustained activity that results in fatigue, there is a reduction in the rate of reuptake of Ca^{2+} (due to a decline in the activity of the Ca^{2+} pumps), which produces a decline in the rate of relaxation; this can be measured as an increase in the duration of the half relaxation time (Enoka & Stuart, 1992).

Once the Ca^{2+} concentration in the sarcoplasm is above a threshold level (10^{-7} M), Ca^{2+} binds to the TN-C element of the regulatory protein troponin (TN). The binding of Ca^{2+} to troponin is thought to cause a structural change in the thin filament such that the myosin-binding site on actin is uncovered and the two proteins (actin and myosin) are then able to interact (Figure 5.24). The uncovering of the binding site probably involves a transient rotation of the regulatory complex (troponin-tropomyosin). Once the action potential has passed, g_{Ca} of the sarcoplasmic reticulum returns to resting levels, and the Ca^{2+} pumps begin returning Ca^{2+} to the sarcoplasmic reticulum. This reuptake of Ca^{2+} lowers the concentration of Ca^{2+} in the sarcoplasm, which results in an inhibition of the activity of the enzyme (actomyosin ATPase) that regulates the interaction of actin and myosin.

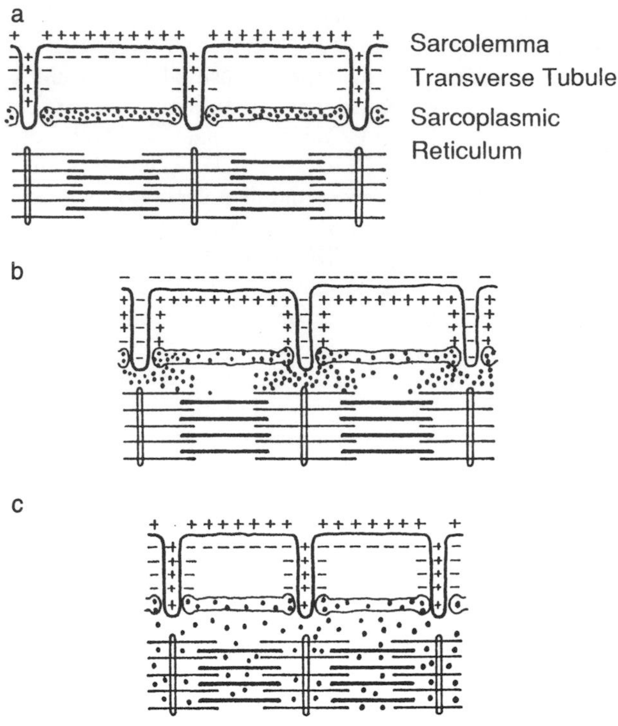

Figure 5.23 Role of Ca^{2+} in excitation-contraction coupling.
Note. From "Muscle" by J.C. Rüegg. In *Human Physiology* (p. 37) by R.F. Schmidt and G. Thews (Eds.), 1983, New York: Springer. Copyright 1983 by Springer-Verlag, Inc. Reprinted by permission.

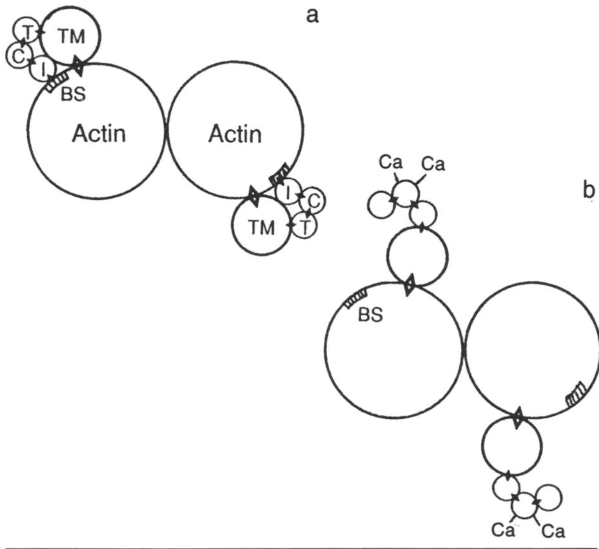

Figure 5.24 Scheme for the uncovering of the binding site (BS) on actin following the attachment of Ca^{2+} to the C component of troponin (TN): (a) arrangement of the thin filament in the resting state; (b) uncovering of the binding site.
Note. From "Some Aspects of the Role of the Sarcoplasmic Reticulum and the Tropomyosin-Troponin System in the Control of Muscle Contraction by Calcium Ions" by J. Gergely, 1974, *Circulation Research*, **34** (Suppl. III), p. 79. Copyright 1974 by the American Heart Association, Inc. Adapted by permission.

Crossbridge Cycle

The interaction of actin and myosin that occurs as a result of Ca^{2+} disinhibition is referred to as the crossbridge cycle and comprises three phases: attach, rotate, and detach (Figure 5.25). In the resting state, actin and myosin are prevented from interacting (Figure 5.25a). Once the binding sites on actin have been made available, the crossbridge (S1 globular heads) of myosin rapidly attaches to actin (Figure 5.25b). Each S1 subfragment includes two globular heads; one head has the binding site for actin-myosin interactions, and the other has the enzyme myosin ATPase for catalyzing the hy-

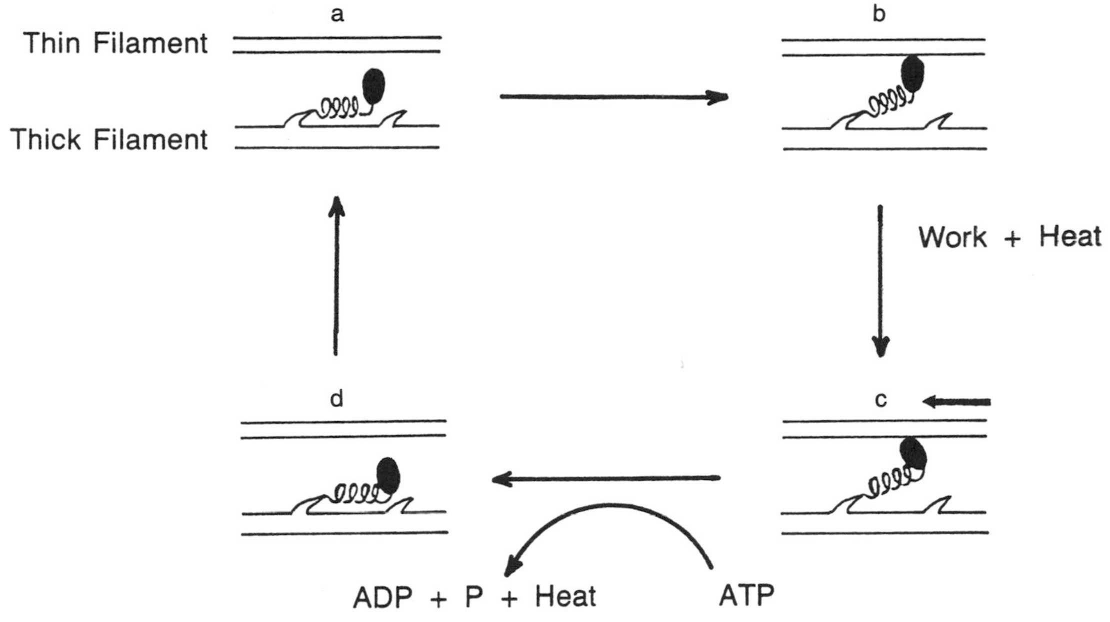

Figure 5.25 The crossbridge cycle.

drolysis of ATP (ATP → ADP + P_i + energy), which provides the energy for the crossbridge cycle (Cooke, 1990; Huxley, 1985; Lieber, 1992; Zahalak, 1990).

A muscle contraction represents the conversion of energy from a chemical to a mechanical form. Although the actual mechanism by which this transformation occurs is unknown, it is usually hypothesized that, following crossbridge formation, the splitting of a terminal phosphate from ATP provides energy for the crossbridge to rotate (Huxley, 1985). In this scheme, S1 rotates about S2 toward the LMM fragment (Figure 5.25c). Because the rotation occurs while actin and myosin are bound, the thick and thin filaments slide past one another and exert a force on the Z bands. The idea that filaments slide relative to one another represents the **sliding-filament theory** of muscle contraction. This movement occurs in each sarcomere along a myofibril, and the force is transmitted across the ends of each myofibril and through the connective tissue matrix around each myofibril and muscle fiber. The force exerted by muscle is due to the concurrent, but not synchronous, cycling of many crossbridges following Ca^{2+} disinhibition. We describe this activity as a muscle contraction and define **contraction** as *the muscle-activation state in which crossbridges are cycling in response to an action potential.*

Once in the rotated position (Figure 5.25c), the S1 binds another ATP molecule and myosin detaches from actin (Figure 5.25d). If no ATP is available (as in severe cases of starvation and following death), then detachment does not occur, and actin and myosin remain bound; this state is known as *rigor mortis*. Upon detachment, the crossbridges return to their prerotated position and are available when another binding site becomes uncovered. When the electrical events of the sarcolemma are finished, the permeability of the sarcoplasmic reticulum reverts to its normal low level, the Ca^{2+} pump (located in the membrane of the sarcoplasmic reticulum) returns Ca^{2+} to the sarcoplasmic reticulum, and the inhibitory effect of the troponin-tropomyosin complex is reestablished. Table 5.3 summarizes this sequence of events.

Although the crossbridge cycle is thought to involve only the attach-rotate-detach sequence, the rate of the different phases and the strain (change in length) experienced by the crossbridge varies with loading conditions. When the crossbridges in a sarcomere are active and cycling, the change in length of the sarcomere depends on the amount of force exerted by the crossbridges relative to the size of the load acting on the sarcomere. If sarcomere force and the magnitude of the load are the same, the sarcomere length will remain constant. Conversely, when the two are not equal, sarcomere length will change; it will shorten when sarcomere force is greater and lengthen when the load is greater. Average duration of the crossbridge cycle in the frog sartorius muscle has been estimated to be 0.34 s when sarcomere length does not change and 0.12 s when sarcomere length shortens (Hoyle, 1983). When sarcomere length increases, the stiffness and force exerted by each crossbridge increases, crossbridges can be forcibly detached, and the rate of crossbridge reattachment can be 200 times greater than in the normal crossbridge cycle (Lombardi & Piazzesi, 1990; Morgan, 1990).

Magnetic Resonance Imaging

Similar to the use of electromyography to measure muscle action potentials is the use of **magnetic resonance imaging** (MRI) to document the contractile activity of

Table 5.3 Sequence of Events in a Muscle Contraction

1. Initiation and propagation (active conduction) of the neural action potential (AP) from the axon hillock.
2. The neurotransmitter acetylcholine (ACh) is released from the neuromuscular junction (NMJ).
3. ACh increases the permeability of the postsynaptic membrane (sarcolemma) to Na^+ and K^+, which activates the transmitter-gated channels and causes the generation of an end-plate potential (EPP).
4. The EPP is conducted away from the NMJ and in the process activates the voltage-gated channels and depolarizes the sarcolemma. This results in the generation of a muscle (sarcolemmal) action potential.
5. The muscle action potential is propagated along the sarcolemma and into the T-tubule system.
6. The T-tubule action potential triggers an increase in the calcium conductance (g_{Ca}) of the sarcoplasmic reticulum and Ca^{2+} moves down its concentration gradient into the sarcoplasm.
7. When the Ca^{2+} concentration in the sarcoplasm is adequate, Ca^{2+} binds to troponin (TN) and disinhibits the regulatory proteins.
8. Subfragment 1 (S1) attaches to the actin-binding site, and the energy provided by the breakdown of ATP is transformed into crossbridge (XB) rotation.
9. The XB detaches once another ATP has bound to S1.
10. The attach-rotate-detach cycle of the XB continues while ATP is available and there is sufficient Ca^{2+} for disinhibition.
11. With the cessation of muscle APs, Ca^{2+} is pumped back into the sarcoplasmic reticulum and returned to the lateral sacs.
12. The removal of Ca^{2+} results in the resumption of inhibition by the regulatory proteins (TN and TM); XBs do not attach, and hence the myofilaments relax.

muscle. MRI, a technique that can localize atomic nuclei, can be used to study the structure and function of the single joint system. Conceptually, the technique involves three steps: (a) exposing an object to a magnetic field that aligns chemical elements with odd atomic weights; (b) perturbing the alignment; and (c) measuring the rate, called the relaxation rate, at which atomic species and atomic nuclei return to the initial alignment. The two principal relaxation rates are the longitudinal (T_1) and transverse (T_2) relaxation rates. Increases in both T_1 and T_2 relaxation times have been correlated with water content in response to muscle necrosis. As a result, MRI has been used to study muscle function. The technique has proven useful for the study of issues ranging from the task-related activation of muscle to muscle soreness and strains (Fleckenstein & Shellock, 1991).

In a typical experiment, a subject performs a prescribed task, and then the body part that was active is placed in the magnet for the measurement of the MRI signal. The MRI measurement takes several minutes. With this protocol, the MRI data can provide information on the cumulative effects of the exercise (crossbridge activity) but not a record of the moment-to-moment activity. The T_2 relaxation rate, for example, appears to document an exercise-induced change in the intracellular water content of muscle fibers following a muscle contraction (Fleckenstein, Canby, Parkey, & Peshock, 1988; Fisher, Meyer, Adams, Foley, & Potchen, 1990). Other factors that may influence the T_2 relaxation time include extracellular water, intracellular pH, level of oxygenation in the blood, fluid viscosity, temperature, and intramuscular fat (Yue, Alexander, Laidlaw, Gmitro, Unger, and Enoka, 1994). Nonetheless, the technique offers considerable potential for expanding our knowledge of muscle function.

Three observations underscore the usefulness of MRI for studying muscle function. First, there is a positive linear relationship between the intensity of the T_2 signal and the force of a muscle contraction. This relationship has been reported for shortening (concentric) and lengthening (eccentric) contractions and for several different muscle groups. For example, when a specific group of muscles moves a load and contracts concentrically, the oxygen consumption is greater and more motor units are recruited than when the muscles contract eccentrically to move the same load. The intensities of both the EMG and the T_2 signal are less during the eccentric contraction (Adams, Duvoisin, & Dudley, 1992; Fisher et al., 1990; Shellock, Fukunaga, Mink, & Edgerton, 1991). Second, the MRI measurement has enough spatial resolution to identify the subvolume of finger flexor muscles (flexor digitorum superficialis and flexor digitorum profundus) that control the activity of individual fingers (Fleckenstein, Watumull, Bertocci, Parkey, & Peshock, 1992). This capability enables researchers to determine the prevalence of functional compartmentalization within a single muscle and within a group of synergist muscles. It also makes it possible to determine whether a task involves coactivation of an agonist-antagonist set of muscles. Third, MRI signals may provide a technique to noninvasively determine the fiber-type composition of a muscle. A strength-training program for the knee extensor muscles resulted in a significant increase in the proportion of fast-twitch muscle fibers in the vastus lateralis muscle. This change in fiber-type proportion was highly correlated with increases in the T_1 and T_2 relaxation times (Kuno, Katsuta, Akisada, Anno, & Matsumoto, 1990). However, recent experiments question the validity of this approach (Parkkola, Alanen, Kalimo, Lillsunde, Komm, & Kormano, 1993).

Feedback From Sensory Receptors

In the operation of the single joint system, sensory receptors provide information that originates from a wide range of stimuli. Three types of afferent events are important in the control of movement: (a) exteroception; (b) proprioception; and (c) the consequences of action. This information is provided by two classes of sensory receptors: exteroceptors and proprioceptors. **Exteroceptors** detect external stimuli that impinge on the system and include, among others, the eyes, ears, and the skin receptors that respond to temperature, touch, and pain. This information is used to inform the system on the state of its external environment, including its location relative to its surroundings. **Proprioceptors** detect stimuli generated by the system itself (Sanes & Evarts, 1984), such as the mechanical variables associated with the activation of muscle, and are involved in the moment-to-moment control of movement. Proprioceptors include muscle spindles, tendon organs, and joint receptors. The information provided by exteroceptors and proprioceptors enables the system to organize a rapid response to a perturbation, to determine limb position, and to differentiate between self-generated and imposed movements. Afferent input from both exteroceptors and proprioceptors can affect motor output, but it appears that the proprioceptive effects are less intense than those generated by the exteroceptors (Sanes & Evarts, 1984).

Reflex Circuitry

The ability of sensory receptors to provide rapid responses to perturbations is based on the existence of short-latency connections between the input (afferent signal) and the output (motor response or efferent output). Such responses are termed reflexes and are defined as *a stereotyped motor response of an organism to a sensory stimulus*. The simplest neural circuit underlying a reflex involves a sensory receptor, its afferent innervation, and a group of motor units that receive input from the afferent. This circuit, however, can be embedded in the neural elements controlling a single muscle, be distributed among a group of synergists (muscles that exert a similar mechanical action), involve an interaction among an agonist-antagonist pair of muscles, or require the coordination of muscles in contralateral limbs.

Reflexes have evolved as mechanisms that can protect the system against unexpected disturbances. When the system is perturbed, such as by an unexpected stretch of a muscle, reflexes can generate a rapid response to counteract the perturbation. In this sense, reflexes are considered regulators because they compensate for disturbances and maintain a prescribed state (Houk, 1988). The neural circuits that enable input-output connections to compensate for disturbances perform a **negative feedback** function, in which the motor response tends to counteract the stimulus that initially activated the sensory receptor.

In the analysis of human movement, reflexes are measured to address clinical and experimental issues. Clinicians examine reflexes to determine the strength of the input-output connection and the ability of the patient to compensate for the disturbance. Measurements that differ from normal provide clues as to the deficits experienced by that individual and assist clinicians in formulating a diagnosis. Researchers in a laboratory setting use the reflex as a probe to determine the effect of various conditions and manipulations (e.g., treatment or training protocols) on an individual. In this section, we will consider some of the reflexes that are commonly used for both purposes.

Stretch Reflex. The two main functions of muscle are to generate power and to react to perturbations. Muscle needs to be springlike in order to react appropriately; the stretch reflex helps the muscle achieve this capability. When a muscle experiences a brief, unexpected increase in length (a stretch), the response is known as the stretch reflex. An example of a stretch reflex is shown in Figure 5.26. In this example, a human subject was grasping a handle that was unexpectedly displaced, resulting in a stretch of the extensor muscles that cross the wrist. The stretch reflex is indicated in the EMG elicited in the extensor muscles. As Figure 5.26 shows, the increase in EMG (response to stretch) begins soon after the onset of the handle displacement (stimulus). The stretch reflex consists of at least two components (Matthews, 1991). One component is the short-latency response (M1), which is thought to be mediated by a neural circuit limited to the spinal cord (Figure 5.27). The second component (M2) has a long latency and a more complex origin that may involve the motor cortex in the brain. A third component (M3) is occasionally observed. For comparison, Figure 5.26 shows that the various components of the stretch reflex all preceded the earliest voluntary EMG, which underscores the ability of reflexes to provide rapid responses to perturbations. Although the latencies for the stretch-reflex components vary, the M1 component generally has a latency of about 30 ms, the M2 has a latency of around 50 to 60 ms, and the earliest voluntary activity (EMG) begins at 170 ms.

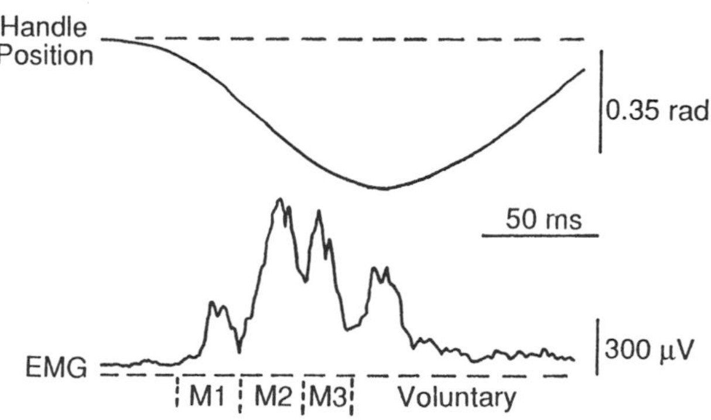

Figure 5.26 A stretch reflex that was elicited by an unexpected stretch (downward) of the extensor muscles that cross the wrist.

Note. From "The Human Stretch Reflex and the Motor Cortex" by P.B.C. Matthews, 1991, *Trends in Neurosciences*, **14**, p. 88. Copyright 1991 by Elsevier Science Publishers Ltd. Reprinted by permission.

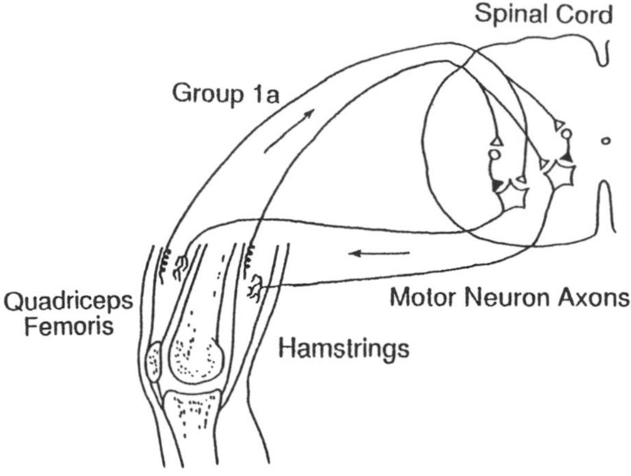

Figure 5.27 Neural circuits underlying the stretch and reciprocal inhibition reflexes.

The sensory receptor involved in the stretch reflex includes at least the muscle spindle, both its Group Ia and Group II afferents. The neural circuit involving the Group Ia afferent is shown in Figure 5.27. Action potentials are generated in the muscle spindle afferents in response to muscle stretch and are propagated centrally to the spinal cord, where they elicit synaptic potentials in the motor neurons innervating the muscle in which the spindle is located. In this instance, the motor neurons are described as **homonymous**, because they innervate the same muscle in which the sensory receptor (muscle spindle) is located. If the stretch is adequate, a sufficient number of synaptic potentials are generated in the motor neuron to elicit an action potential that results in a muscle contraction. *The stretch reflex seems most capable of accommodating small disturbances in muscle length.* The net effect of this input-output circuit is that the stretch (stimulus) elicits a contraction (response) that minimizes the stretch; this type of negative-feedback circuit has also been referred to as a **resistance reflex**.

The stretch reflex, however, does not function only as a stereotyped negative-feedback response but also as a flexible response that can adapt to changing conditions (Hammond, Merton, & Sutton, 1956). Because the stretch reflex can actually be elicited whenever there is an unexpected change in muscle length, it can also occur during movement. If, during movement, there is a mismatch between the expected and the actual muscle lengths, this difference may be sufficient to elicit a stretch reflex. Under these conditions, however, tendon organs may also be activated, and the Group Ib afferents will transmit their input (inhibitory) to the spinal cord (Matthews, 1990). In human movement, therefore, the unexpected stretch of a muscle can trigger input from many receptors, including the Group Ia, Ib, and II afferents. The net effect is an afferent barrage that can vary substantially from trial to trial, even to the extent of failing to elicit a short-latency response (such as occurs when a platform on which a subject is standing is suddenly displaced and a rapid response is required to restore balance) (Horak & Nashner, 1986; Nashner, 1977). Furthermore, because of the involvement of the cortex in the stretch reflex (M2 response), the nervous system is able to modulate the response and spread it to muscles that have not been stretched, even to antagonists, if the activity is mechanically appropriate (Matthews, 1991). The net effect of this flexibility is that both the short- and long-latency components of the stretch reflex can be modified to meet the demands of the task.

A response related to the stretch reflex is the **tendon tap reflex** (also known as the tendon jerk). This response is often elicited in clinical settings by striking the patellar tendon with an appropriate implement and assessing the vigor of the contraction. The tendon tap reflex represents a subset of the stretch reflex and involves only activation of the Group Ia afferent and the associated motor neuron output (Matthews, 1990). Because it is based on the monosynaptic connection of the Group Ia afferent, it is a rapid response and has a latency of 25 to 30 ms between striking the patellar tendon and recording the EMG in the quadriceps femoris muscles. The size of the tendon tap reflex is used as an index of the combined effect of muscle spindle responsiveness, motor neuron excitability, and the level of inhibition (presynaptic) acting on the Group Ia afferents in the spinal cord.

Reciprocal Inhibition Reflex. Whereas the unexpected stretch of a muscle can elicit several competitive (excitatory and inhibitory) inputs, the central actions of the input can also be diverse. An example of the divergence of the central actions is the effect that the Group Ia afferent input elicits in the motor neurons that innervate the antagonist muscle. As shown in Figure 5.27, the Group Ia afferent branches when it reaches the spinal cord. One of these branches synapses with an interneuron, the **Ia inhibitory interneuron**, which can generate inhibitory postsynaptic potentials in the motor neurons of the antagonist muscle (Day, Marsden, Obeso, & Rothwell, 1984; Katz, Penicaud, & Rossi, 1991). This connection, which serves to inhibit, or lower the excitability of, the motor neurons that innervate the antagonist muscle, is known as the **reciprocal inhibition reflex**. Therefore, activation of muscle spindles in the quadriceps femoris muscle (Figure 5.27) elicits an excitation of the homonymous (quadriceps femoris) motor neurons but inhibits those for the antagonist (hamstring) muscles. Because the net muscle activity about a joint is due to the difference in activity between an agonist-antagonist pair, the reciprocal inhibition reflex increases the likelihood that the stimulus sensed in one muscle will elicit a meaningful response in that muscle.

One way to test the strength of the reciprocal inhibition reflex in human subjects is to have a subject contract one muscle while an experimenter tests the excitability of an antagonist muscle. The excitability of a muscle can be tested by carefully shocking the nerve to the muscle and eliciting the Hoffmann reflex; a decrease in the reflex represents a decrease in the excitability of the motor neurons innervating the muscle. When a subject contracts the dorsiflexor muscles, there is an accompanying reduction in the Hoffmann reflex of the soleus muscle due to reciprocal inhibition (Shindo, Harayama, Kondo, Yanagisawa, & Tanaka, 1984). However, because the Ia inhibitory interneuron receives input from a variety of sources, including pathways from higher neural centers (e.g., pyramidal, rubrospinal, vestibulospinal) and proprioceptors, activation of the reciprocal inhibition reflex is not mandatory but depends on the state (level of excitability) of the interneuron.

Flexor-Withdrawal and Crossed-Extensor Reflexes. In addition to localized reflexes involving the muscle spindle and tendon organ, the repertoire of input-output connections includes those triggered by a variety of sensory receptors (e.g., cutaneous receptors, joint receptors, free nerve endings) with diverse responses. The flexor-withdrawal and crossed-extensor reflexes are examples of this diversity. When cutaneous receptors (Group III and IV afferents) sense a noxious stimulus, the response is to withdraw the site from the stimulus. If the stimulus is applied to the back of the leg, for example, the response will probably involve flexion of the leg away from the stimulus, hence the term **flexor-withdrawal reflex**. The response has a latency of about 100 ms. But because the sensory signals are transmitted over polysynaptic pathways, there is ample opportunity for the afferent input to be modified and for the response to vary from trial to trial.

The cutaneous afferent involved in the flexor-withdrawal reflex branches when it enters the spinal cord and distributes the input to various targets. These targets include the motor neurons that innervate the extensor muscles in the contralateral limb (Figure 5.28). The net effect is that the flexor-withdrawal reflex is often accompanied by a **crossed-extensor reflex**, which involves excitation of the motor neurons innervating extensor muscles and inhibition of the motor neurons of the flexor muscles in the contralateral limb. Such an association seems reasonable because if one leg is withdrawn (flexed) from a stimulus, it is probably desirable that the contralateral limb be extended to provide postural support.

Hoffmann Reflex. In contrast to the stretch, reciprocal inhibition, and crossed-extensor reflexes, which can occur under normal movement conditions, the **Hoffmann reflex**, or H reflex, is an artificially elicited response. The H reflex is a tool for testing the excitability of the neuromuscular system (Fisher, 1992; Hugon, 1973; Magladery & McDougal, 1950). It is elicited by applying a single electrical shock (stimulus) to a

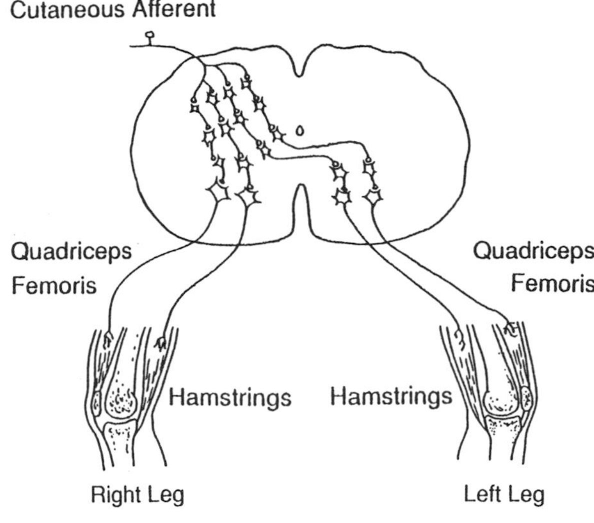

Figure 5.28 Neural circuit for flexor-withdrawal and crossed-extensor reflexes.

peripheral nerve. Because the stimulus is a single shock, the response is a twitch in the muscle innervated by the stimulated nerve; the response can be measured as the EMG or force associated with the twitch. The H reflex is most commonly examined in the soleus, but it can be elicited in such other muscles as quadriceps femoris, tibialis anterior, and muscles of the hand, foot, and forearm. For example, the procedure for eliciting the H reflex in the quadriceps femoris (Figure 5.29) involves placing the subject in a supine position and placing the stimulating electrode over the femoral nerve just below the inguinal ligament. Because the H reflex relies on the selective activation of the Group Ia afferents, the correct positioning of the electrodes with respect to the nerve is critical and can be difficult to achieve in some subjects; this is especially true for quadriceps femoris.

The neural circuit for the H reflex is shown in Figure 5.30. It involves the selective activation of the Group Ia afferents and the subsequent generation of action potentials in the motor neurons innervating the muscle. This is a population response that involves many Group Ia afferents as well as a number of motor units. The selective activation of the Group Ia afferents is achieved by beginning with a low-intensity stimulus and slowly increasing it until action potentials are generated in the largest diameter axons; in most human subjects these are the Group Ia afferents. With selective activation of the Group Ia afferents, action potentials are propagated centrally to the spinal cord (2 in Figure 5.30), where they elicit postsynaptic potentials in the motor neurons. The likelihood that these synaptic potentials will generate action potentials depends on the size of the synaptic potentials and the level of the membrane potential. The closer the membrane potential is to threshold for action potential generation, the more likely the stimulus (synaptic potentials) is to elicit an action potential. If the

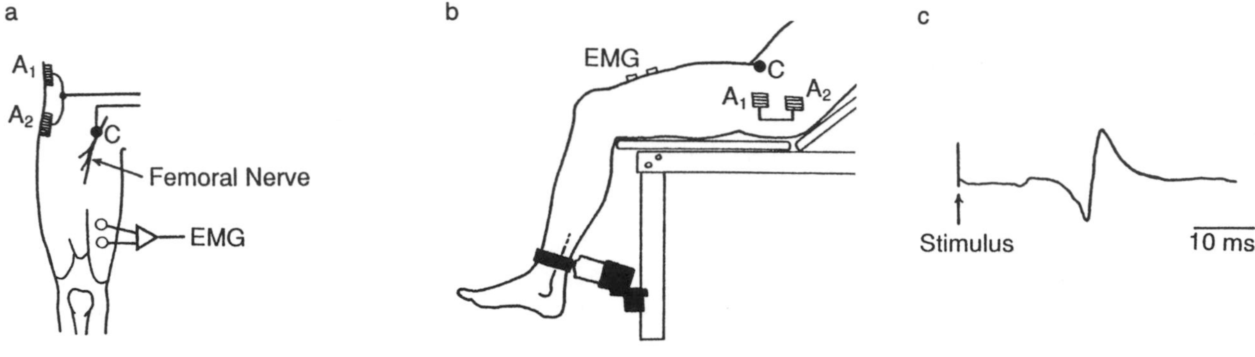

Figure 5.29 An H reflex elicited in quadriceps femoris: (a) location of the stimulating electrodes (*C*, cathode; A_1 and A_2, two anodes that are connected together); (b) position of the subject; (c) H reflex recorded as an EMG over vastus medialis.

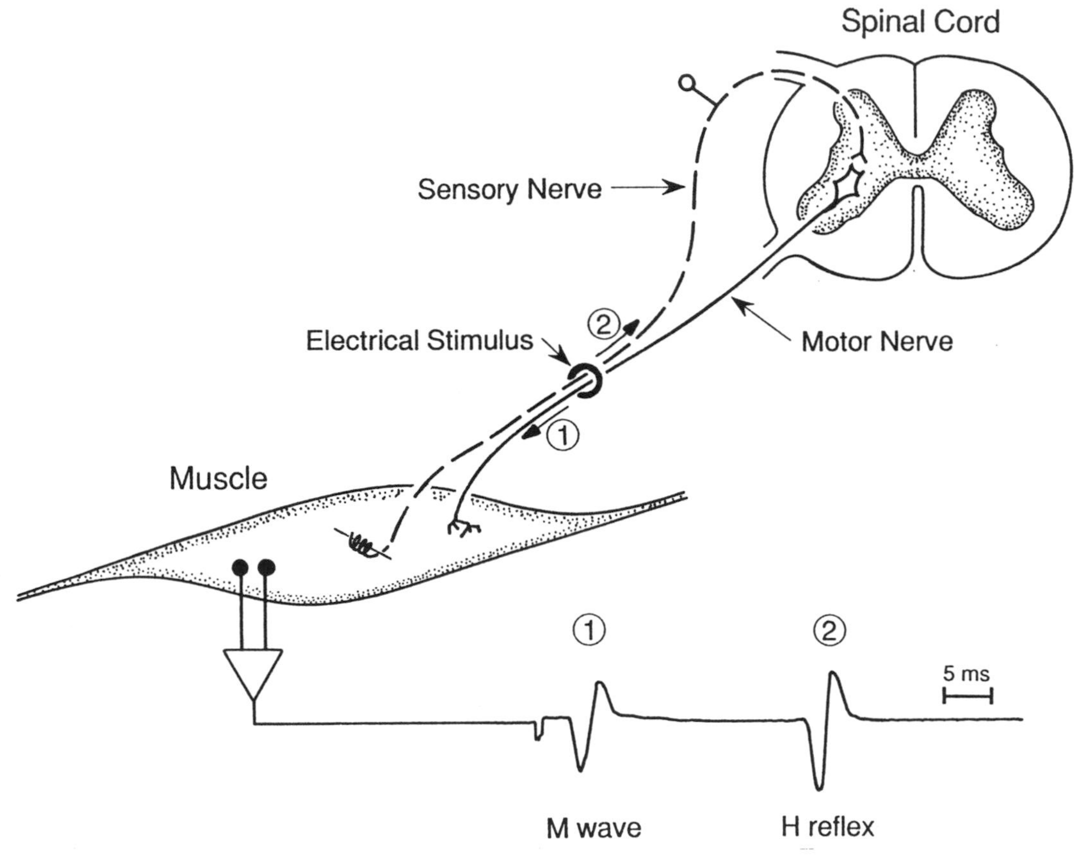

Figure 5.30 Circuit diagram for the H reflex. The electrical stimulus is applied over the peripheral nerve, and the response is measured in the muscle innervated by the nerve.

membrane potentials of more motor neurons are closer to threshold, then the stimulus will generate action potentials in more motor neurons, and the response (EMG and force) will be larger. For this reason, *the H reflex is used as a test of the level of excitability of the motor neuron pool.* The bigger the response, the greater the number of motor units that have contributed to the response, because of a higher level of excitability in the motor neuron pool.

Because the H reflex is a measure of the level of excitability in the motor neuron pool, even distant stimuli (e.g., loud and unexpected sounds) or the most remote muscle activity is sufficient to affect the response. For example, the H reflex elicited in the soleus muscle of a subject will increase in amplitude if the subject voluntarily activates other muscles, such as those involved in clenching the teeth. The reason for the increase in the amplitude of the H reflex is that the neural signal to the motor neurons that innervate the jaw-closing muscles to clench the teeth is distributed along the spinal cord, perhaps for postural needs, and generates excitatory postsynaptic potentials in many motor neurons.

Consequently, the membrane potentials of many soleus motor neurons are closer to threshold, so a constant electrical stimulus will cause more motor neurons to generate action potentials. The use of remote muscle activity to increase the excitability of a motor neuron pool is referred to as the **Jendrassik maneuver**. This effect has clinical applications. For example, a patient with weak leg muscles can be enabled to rise from a chair when a therapist provides resistance for the voluntary activation of arm and neck muscles by the patient. Activation of the arm and neck muscles elicits a Jendrassik effect in the leg muscles and can increase the force so that it is sufficient for the patient to stand.

Not only does the H reflex vary on a moment-to-moment basis due to such effects as the Jendrassik maneuver; it also varies as a result of chronic activity patterns (Nielsen, Crone, and Hultborn, 1993). For example, Sale, MacDougall, Upton, and McComas (1983) found that the H reflex elicited during a maximum voluntary contraction increased after a strength-training program. They interpreted this observation of reflex potentiation as an increase in motor neuron excitability that was due to the strength training. Similarly, Koceja and Raglin (1992) reported a decrease in the amplitude of the H reflex in a group of swimmers during an overtraining phase of their program but a return to normal during the tapering phase.

When the electrical stimulus used to elicit the H reflex is gradually increased, the stimulus begins to generate action potentials in axons of smaller diameter (Trimble & Enoka, 1991). After the Group Ia afferents, the class of axons with the next largest axon diameter is the alpha axons, which belong to the motor neurons. When action potentials are generated in the alpha axons, the response is a short-latency (5 ms) twitch, called the **M wave** (1 in Figure 5.30). The M wave is elicited experimentally to probe the integrity of the circuit between the site of the stimulus (muscle nerve) and the site of recording (usually the muscle EMG); that is, the M wave tests the integrity of neuromuscular propagation (Bigland-Ritchie, Kukulka, Lippold, & Woods, 1982; Enoka & Stuart, 1992). When the electrical stimulus is maximal, this results in the synchronous activation of all the muscle fibers and represents the summed response of all their action potentials. But when measured with a conventional electrode arrangement (bipolar, 8-mm diameter, 1.5-cm interelectrode distance), the M-wave record does not represent the action potentials in the entire muscle but only those from muscle fibers in close proximity to the electrodes (probably about 1 to 2 cm) (Fuglevand et al., 1992). This is the reason that M waves have a similar amplitude for small and large muscles (Hicks, Cupido, Martin, & Dent, 1992). Interestingly, M-wave amplitude declines with age, probably due to a reduction in the excitability of muscle fibers (Hicks et al., 1992).

Tonic Vibration Reflex. Like the H reflex, the tonic vibration reflex is an artificially elicited response that has both clinical and experimental uses (Lance, Burke, & Andrews, 1973). The neural circuit for this reflex is the same as that for the tendon tap reflex. It involves the muscle spindle and the homonymous motor units activated by excitation of the muscle spindles. However, it appears that vibration activates both mono- and polysynaptic pathways. The muscle spindle is extremely sensitive to small-amplitude vibration that has a frequency of 50 to 150 Hz. Application of vibration to a muscle with a standard clinical vibrator provides adequate stimulus to excite many muscle spindles and to activate enough motor units to cause a muscle contraction, which results in an increase in the force and EMG during a submaximal contraction (Figure 5.31). Furthermore, the vibration results in a decrease in the excitability of the motor neurons innervating the antagonist muscle through the reciprocal inhibition circuit.

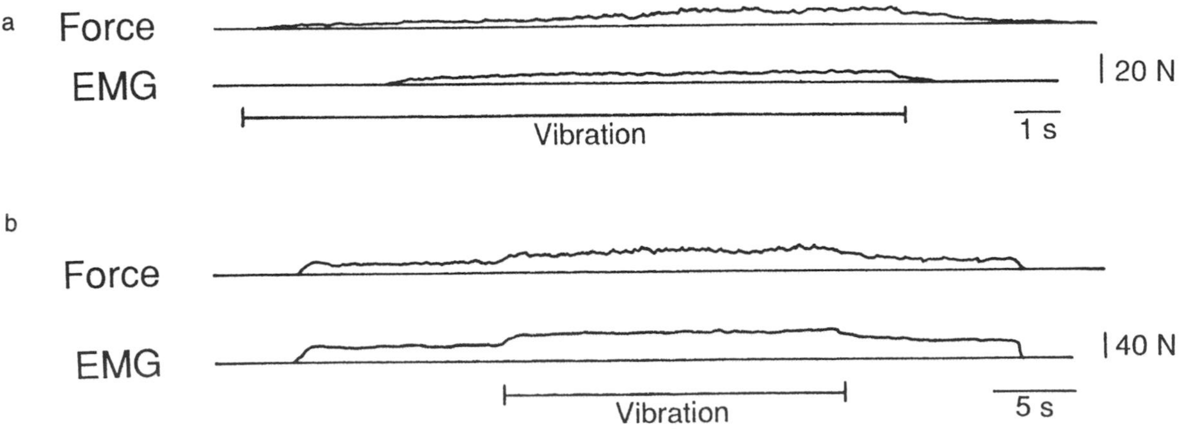

Figure 5.31 The tendon vibration reflex. (a) An active vibrator is placed over the tendon of biceps brachii and elicits an EMG in the muscle and a force that is exerted at the wrist. (b) The force and EMG associated with a submaximal contraction of biceps brachii can be enhanced by the use of a vibrator. Vibration was applied for the duration of the time indicated by the horizontal lines.

However, the effect of vibration is complex. Although vibration produces a reflex response (EMG and force), both the tendon tap and H reflexes are depressed during vibration.

The tonic vibration reflex has been used in the rehabilitation of stroke (cerebral vascular accident) patients, especially for restoring motor function in the upper extremity (Bishop, 1974, 1975a, 1975b). The typical approach is to apply vibration to the synergy (extensor or flexor muscles) that is not dominant following the accident. For example, generally the flexor synergy (scapular retraction and elevation, shoulder abduction and external rotation, elbow flexion and supination, and wrist and finger flexion) predominates, and one rehabilitation strategy is to elicit brief bursts of activity in the antagonist muscles (extensor synergy) with the use of vibration. The vibration enables the therapist to provide a powerful afferent input to the spinal cord and to manipulate the tonic activity between an agonist-antagonist set. With the appropriate selection of the stimulation site, it is possible to decrease the excitability of hyperactive motor neurons in one muscle and to increase the excitability of those that are hypoactive in the antagonist. The therapeutic effect of vibration is enhanced by placing the muscles to be vibrated in lengthened positions and by positioning the patient appropriately; a supine position facilitates the effect of vibration on the extensor synergy, whereas a prone position is better for the flexor synergy. Similarly, vibration of an antagonist muscle can be used to reduce the spasticity in an agonist muscle. The effect of vibration is to normalize movement in the clinic. However, after the patient leaves the clinic, these effects will disappear.

Reflexes and Movement

In the performance of even the simplest movement, motor neurons receive not only a barrage of input from many types of sensory receptors but also input from higher motor centers (the brain). All of this input must be organized carefully to achieve the desired activation of muscles (Fournier & Pierrot-Deseilligny, 1989; Windhorst, 1988). There is at least one important reflex-related principle that we need to consider concerning the management of this input. *The synthesis of this input does not occur at the level of the motor neuron but rather among the interneurons in the spinal cord.* This distinction is important because it has led to the observation that reflexes are not fixed and can be modified during movement.

The study of reflexes has, at least since the time of Sherrington (1857-1952), provided a dominant focus in the study of motor control. In this scheme, the motor neuron is represented as the central element to which afferent and supraspinal input is directed (Figure 5.32a). Although the motor neuron is still regarded as the **final common pathway**—the route through which the commands are issued to muscle—evidence suggests that the motor neuron is not the major integrating element for the control of movement. Millions of synaptic connections within the spinal cord form the basis of the interactions among the neural elements. These interactions are based on the summation of excitatory (open circle) and inhibitory (solid circle) postsynaptic potentials. It appears that the *interneuron* (I in Figure 5.32), rather than the motor neuron, serves as the focal point of this integration (Figure 5.32b) (Baldissera, Hultborn, & Illert, 1981; Pearson, 1993; Windhorst, 1988). Most of the neural traffic from afferent and supraspinal sources converges on a number of different interneurons rather than going directly to the motor neurons. As a consequence of this pre–motor neuron convergence, the input (e.g., Group

a

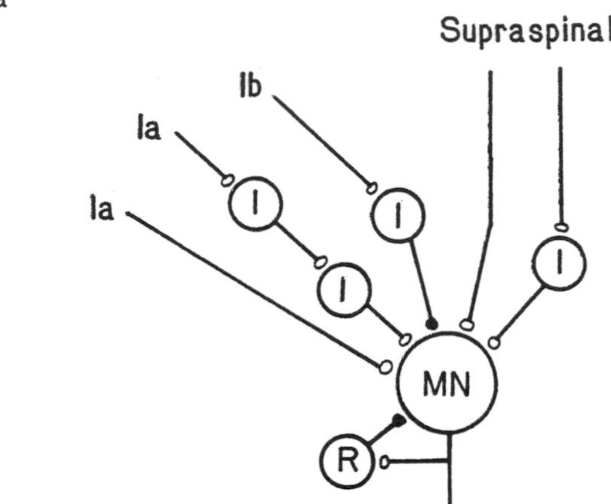

b

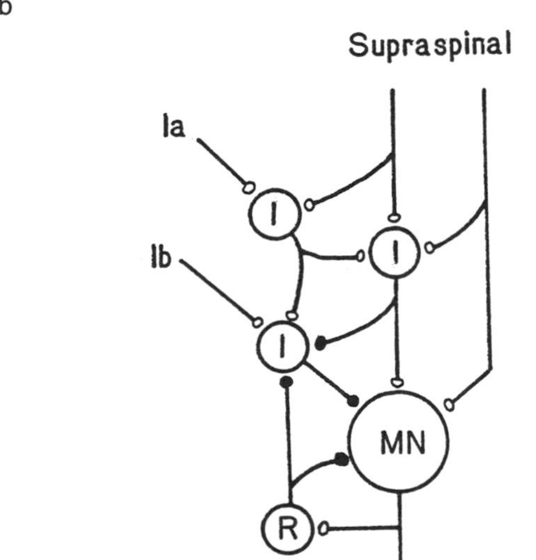

Figure 5.32 Schemes for the convergence of input in the spinal cord: (a) motor neuron focus; (b) interneuron focus. *Note.* From "Integration of Sensory Information and Motor Commands in the Spinal Cord" by R.E. Burke. In *Motor Control: From Movement Trajectories to Neural Mechanisms* (pp. 47, 53) by P.S.G. Stein (organizer), 1985, Class at the Annual Meeting of the Society for Neuroscience, Washington, DC. Adapted by permission.

Ia input caused by muscle stretch) can be altered before it is transmitted to the motor neurons. One effect of this arrangement is that input-output relationships (e.g., reflexes) become less stereotyped and can vary depending on the nature of the other input converging on the interneurons (Fournier & Pierrot-Deseilligny, 1989). The main advantage of this scheme is that it allows greater flexibility in input-output coupling.

The shift in focus to the interneuron as the key integrating element in the spinal cord largely came about because of technical developments and because of the systematic study of spinal cord connectivity. These studies have revealed a number of interesting features of nervous system operation (Stuart, 1987a, 1987b): Reflex circuitry involves alternate pathways, and therefore afferents have alternative excitatory and inhibitory pathways to motor neurons (Binder, Houk, Nichols, Rymer, & Stuart, 1982); there appear to be considerable differences among muscle systems, such as those between muscles that control limb, respiratory, and head-neck muscles (e.g., reciprocal inhibition occurs only in limb muscles); and the connections that occur between interneurons appear to be quite sophisticated in that they are capable of organizing most of the movements that we perform (Baldissera et al., 1981).

An example of the flexibility associated with interneuronal convergence is the variation in a reflex that can occur with movement (Nashner, 1982; Rossignol, Julien, Gauthier, & Lund, 1981; Stein & Capaday, 1988; Yang & Stein, 1990), such as the flexion reflex during walking (Crenna & Frigo, 1984). A noxious stimulus applied to the skin over the calf muscles will elicit a flexor-withdrawal reflex in the thigh muscles of the same leg (Figure 5.33). The response, which is intended to remove the limb from the painful stimulus, can be measured as an EMG in the thigh muscles (vastus lateralis and biceps femoris). When a subject is walking, however, these same muscles are also needed to support the body in selected phases of the stride cycle. Consequently, at these times it would be inappropriate to withdraw the limb in response to a noxious stimulus. The response to the painful stimulus must be modulated by the nervous system so that support is provided when necessary and the limb is withdrawn when it is mechanically possible. This modulation is shown in Figure 5.33 for the biceps femoris muscle. The response (magnitude of the EMG) consists of short- and long-latency components. The short-latency component (57 ms) was greatest during the transition from stance to swing, whereas the long-latency component (132 ms) was greatest at the end of the swing phase. This same type of modulation has also been reported for the H reflex during walking (Crenna & Frigo, 1987; Stein & Capaday, 1988).

Because of these types of observations, we must modify the definition of the reflex as a stereotyped response to emphasize that a reflex is a **state-dependent response**. This means that the response, despite a constant

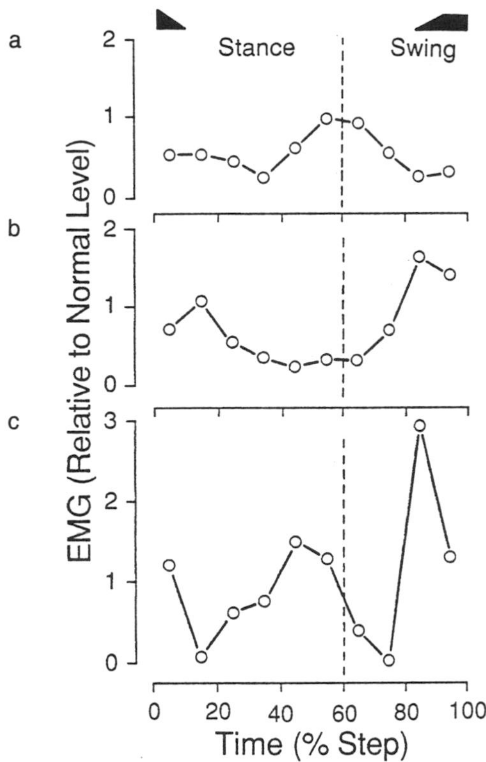

Figure 5.33 The EMG responses to a noxious stimulus are plotted relative to the normal level of EMG (1 = normal) during a walking step: (a) the short-latency response; (b) the long-latency response; (c) the combined response. *Note.* From "Evidence of Phase-Dependent Nociceptive Reflexes During Locomotion in Man" by P. Crenna and C. Frigo, 1984, *Experimental Neurology*, **85**, p. 341. Copyright 1984 by Academic Press. Reprinted by permission.

stimulus, depends on the subject's state of activity (e.g., stationary vs. active, swing vs. support). For example, the spinal stretch reflex is weak in most muscles during active moments, especially in comparison to the long latency reflex (Crago, Houk, and Hasan, 1976; Houk and Rymer, 1989; Marsden, Merton, and Morton, 1976; Melvill-Jones and Watt, 1971; Vallbo and Wessberg, 1993). In clinical and laboratory testing, therefore, it is necessary that the patient or subject assume a standardized position when the reflex is elicited.

Although there are many neural mechanisms that can modify a reflex, we know the most about two of these: recurrent inhibition and presynaptic inhibition. *Recurrent inhibition* is a local feedback circuit that can modify reflex responses through an interneuron known as the **Renshaw cell** (Windhorst, 1988). The Renshaw cell (R in Figure 5.32b) can be activated by supraspinal input, by Group III and Group IV muscle afferents, and by a collateral branch of the alpha motor neuron axon. In turn, the Renshaw cell generates an inhibitory postsynaptic potential in the same motor neurons and in other interneurons. The circuit involving the axon collateral, Renshaw cell, and motor neuron is referred to as **recur-**

rent inhibition (Figure 5.32). Recurrent inhibition from a given motor neuron pool is distributed to many other motor neuron pools but is weak or absent among pools innervating distal muscles (e.g., hand, foot) (Katz, Mazzocchio, Pénicaud, and Rossi, 1993). Activation of recurrent inhibition results in, among other effects, a decrease in the excitability of motor neurons. Because large motor neurons give off more collateral branches to Renshaw cells but receive fewer synapses from Renshaw cells (Stuart & Enoka, 1983), recurrent inhibition has a greater effect on the excitability of the smaller motor neurons. Furthermore, because recurrent inhibition in a motor neuron pool increases during weak contractions but decreases during strong contractions (Fournier & Pierrot-Deseilligny, 1989), recurrent inhibition is high when the smaller motor neurons are active and decreases as larger motor neurons are recruited. Recurrent inhibition may also increase during fatiguing contractions.

In addition to its effects on motor neurons, recurrent inhibition generates inhibitory postsynaptic potentials in gamma motor neurons and in the Ia inhibitory interneuron (Hultborn, Lindström, & Wigström, 1979). The connection to the gamma motor neuron means that recurrent inhibition can modulate the excitability of the muscle spindle and therefore influence the input-output relationship for the stretch reflex. The connection to the Ia inhibitory interneuron means that recurrent inhibition can inhibit (decrease the excitability of) the interneuron that mediates the reciprocal inhibition reflex; by this means recurrent inhibition is able to inhibit the inhibition, which is disinhibition. Recurrent inhibition, therefore, represents a significant element in the adaptability of reflexes.

Presynaptic inhibition also is able to modify reflexes given a constant input from a sensory receptor. Much has been learned about this mechanism by studying the adaptability of the stretch reflex (Matthews, 1990, 1991). The adaptability is achieved by imposing an inhibitory effect on the action potentials transmitted by the Group I and Group II afferents. This modification can occur by **presynaptic inhibition**, whereby an interneuron synapses on the afferent axon close to where it contacts the motor neuron and generates an inhibitory synaptic potential in the axon; thus, the interneuron modifies the afferent action potential before (presynaptic) it contacts the motor neuron. Presynaptic inhibition, which reduces the action potential and the subsequent amount of neurotransmitter released at the afferent–motor neuron synapse, modifies the input-output relationship so that for the same stimulus (muscle stretch), there can be variable amounts of response (muscle contraction) (Fournier & Pierrot-Deseilligny, 1989).

Acute and Chronic Adaptations. Not only can reflexes be modulated for moment-to-moment control; the input-output relationship also can be modified with activity. For example, 20 min of running on a treadmill can induce a reduction in the amplitude of the H reflex in the triceps surae muscles (Bulbulian & Bowles, 1992). And the reduction is greater during downhill running (−10% grade) than during level running; downhill running places a greater reliance on eccentric contractions of leg muscles and produces delayed-onset muscle soreness. One of the best examples of this plasticity resulting from long-term activity is the changes that can occur in the stretch reflex and the H reflex with appropriate training (Evatt, Wolf, & Segal, 1989; Wolpaw & Carp, 1990; Wolpaw, Herchenroder, & Carp, 1993). In an extensive set of experiments, Wolpaw and Carp have shown that it is possible to train (for 80 days) a nonhuman primate to increase or decrease the reflex response (stretch or H) to a constant stimulus. To a given stretch (unexpected change in muscle length), the reflex EMG can be trained to increase or to decrease without changes in the background EMG before the stretch is applied. These changes are probably those due to adaptations that occur in those elements of the neural circuit that are accessible by supraspinal input. For the H and stretch reflexes, this includes the Group Ia synapse on the motor neuron and the motor neuron itself. Because the Group Ia synapse can be inhibited by presynaptic inputs from several supraspinal sites, this seems to be the most likely candidate for the long-term adaptation of the stretch and H reflexes (Wolpaw & Carp, 1990).

Role of Proprioceptors. The sensory receptors discussed in chapter 4 not only contribute to reflexes but also serve a proprioceptive function. Proprioceptors include muscle spindles, tendon organs, and joint receptors. They respond to the mechanical variables associated with muscles and joints due to actions generated by the system itself. Hasan & Stuart (1988) have proposed that feedback from proprioceptors serves two roles in movement. One role concerns the interactions between the body and its surroundings, and the other modifies internally generated motor commands to accommodate musculoskeletal mechanics.

An important feature of the single joint system is its ability to generate movement in a form that is *appropriate for its surroundings*. One role of proprioceptors is to assist the system in meeting this need. It is not sufficient, for example, that the single joint system be capable of producing movement; it must do so within the constraints imposed by its surroundings. These constraints can include unexpected perturbations, limits on which limbs or muscles can be involved in a task, and the opportunity to facilitate a movement through segmental interactions (as described in the section on intersegmental dynamics in chapter 3). Accordingly, proprioceptors can provide information that enables the single joint system to accommodate the constraints imposed by the surroundings. This is accomplished by the use of assistance and resistance reflexes and by the selection of appropriate muscle synergies. **Resistance reflexes** are those responses that rely on proprioceptive feedback to resist unexpected disturbances due to some effect of the surroundings; the stretch reflex is a resistance reflex. In contrast, proprioceptors can provide information that is derived from the interaction between the single joint system and its surroundings to assist a movement. Because the performance of movement is preceded by postural adjustments, proprioceptors can provide feedback to stabilize the system relative to its surroundings and enhance the subsequent movement. Such adjustments are known as **assistance responses**. Furthermore, proprioceptors provide feedback that enables the system to select an appropriate group and sequence of muscle activity (muscle synergy) in order to accomplish a task. For the performance of any movement, the nervous system has many options (in terms of motor units, muscles, and segments) that may be used to accomplish the task. It is likely that proprioceptive information is used in selecting a particular muscle synergy for a given condition.

The second role served by proprioceptors is the *accommodation of musculoskeletal mechanics*. As will be discussed in chapter 6, the ability of the single joint system to produce movement depends not only on the excitation provided by the neural element but also on the mechanical properties of muscle and the geometry of the system. Proprioceptors can provide information that enables the nervous system to take into account these features of the system in order to generate an appropriate command for the performance of a movement. To achieve this goal, proprioceptors must provide information at the levels of a single muscle, a single joint, and several joints.

The force that a *single muscle* exerts is not constant for the same nervous system (motor) command. Muscle force depends on such mechanical properties as the length of the muscle and its rate of change. Unexpected changes in muscle length, therefore, disrupt the intended effect of a motor command. Proprioceptors can detect unexpected changes in muscle length and thereby minimize the effect of length on muscle force and enable the nervous system to control muscle force more precisely. A similar effect occurs at the level of a *single joint*. Because the moment arm for most muscles changes over its range of motion, with a peak value at a midrange position, muscle torque is frequently at its maximum at an intermediate position (Figure 2.42). Consequently, for a given motor command to generate a submaximal muscle force, the limb could move to either of two positions, one on either side of the maximum. To prevent this control problem, proprioceptive feedback may provide a signal that increases with joint angle so that the motor command increases continuously over the range of motion, and each limb position is defined uniquely. Finally, proprioceptors must provide feedback to assist with the control of *several joints*. Movement of a linked system, such as the human body, causes the motion of one body segment to exert an influence on all the other segments in the system (intersegmental dynamics). Proprioceptive feedback is critical for the system to interpret the nature of these interactions and to coordinate the activity of all the segments in the system.

Kinesthesia

Whereas reflexes represent the input-output connections by which the system can generate rapid responses to perturbations, sensory receptors also provide information that enables the system to determine the position of its limbs and to identify the agent (itself or something else) that caused it to move. This ability is called **kinesthesia**. Kinesthetic sensations are derived from the combination of input from different types of afferent fibers (e.g., skin, muscles and tendons, joint capsules, ligaments) and from centrally generated motor commands. In contrast to proprioceptive input, which is more general and includes vestibular sensations and inputs from muscles and joints that might not be perceived, kinesthetic sensations are perceived (McCloskey, 1987). Kinesthetic sensations include those concerning position and movement, effort and heaviness, and the perceived timing of movements.

Position and Movement. Although joint receptors were initially thought to provide the primary signal for the determination of joint position and movement, it is now thought that input from the muscle spindle is more important (Gandevia, McCloskey, & Burke, 1992; Matthews, 1987; McCloskey, 1987), except in the hand, where joint, cutaneous, and muscle feedback provide critical kinesthetic information. The importance of the muscle spindle can be demonstrated by vibrating a muscle of a blindfolded subject and asking the subject to indicate the nature of the sensation. The subject will report a rotation of the joint that far exceeds the rotation that actually occurs. Because the vibration selectively activates the muscle spindle afferents, the illusory rotation is attributed to the central effects of the input by the muscle spindle afferents. However, the use of muscle spindle input to determine joint position and movement is complicated by the existence of the fusimotor system and the possibility of activating the muscle spindle in the absence of a change in position. Therefore, in order to use the muscle spindle input to determine position and movement, the central nervous system must subtract the fusimotor activity from the muscle spindle signal. This is accomplished by supplying sensory centers with copies of the appropriate outgoing motor commands (**corollary discharges**). By this mechanism, the motor centers tell the sensory centers about the timing and magnitude of fusimotor activity and thereby enable the system to interpret muscle spindle feedback.

Effort and Heaviness. When a muscle is activated and exerts a force, tendon organs provide afferent information about the force exerted by muscle. It does not seem, however, that we typically use this information to provide an index of the intensity of muscle activity. For example, when a person supports a load (e.g., a briefcase) for a long period of time, it is necessary to increase the effort over time to support the unchanging load. Because the load does not change (a constant-force task), the force that is necessary to support the load does not change, and therefore the afferent signal from the force detectors (tendon organs) does not change. Why, then, does the load seem to become heavier? The sensation of an increase in the effort is derived not from peripheral afferents but from the centrally generated motor command associated with the task. A *corollary discharge* is sent from the motor centers to the sensory centers, and an estimate of the magnitude of the motor command is made by supraspinal sensory centers. This estimate is known as the **sense of effort**. The central basis for the sense of effort can be demonstrated by comparing the effort required for a constant-force task before and after weakening a muscle (e.g., by administering curare or by fatiguing the muscle). Although the force is the same in the two conditions, the effort is reported to be much greater in the weakened muscle due to an increase in the centrally generated motor command (McCloskey, Ebeling, & Goodwin, 1974). However, completely paralyzed humans (achieved by an intravenous dose of vecuronium, a neuromuscular blocker) are unable to perceive a focused effort associated with willed but unexecuted movement (Lansing & Banzett, 1993). This finding suggests that the sense of effort requires some form of sensory feedback related to the performance of the task and is not completely based on the centrally generated command.

Information from peripheral receptors, however, is not totally neglected and seems to be necessary to estimate the weight of an object. The **sense of heaviness** is probably based on an integration of proprioceptive feedback and the centrally generated command to move. As an individual prepares to lift an object, the central command gradually increases. At the onset of the lift (displacement of the object), proprioceptive feedback indicates a change in position, and the nervous system uses this signal to identify the magnitude of the central command necessary to lift the object and hence is able to estimate its weight (heaviness).

Timing of Movements. Human subjects are able to identify two moments in time related to the onset of movement: the dispatch of a command to move and the beginning of the actual movement. Because the perception that a command to move has been generated occurs before the onset of the EMG, it is probably derived from the centrally generated command to move. In contrast, the perception that a movement has begun occurs after the beginning of EMG activity and is probably based on proprioceptive signals.

Summary

This chapter describes four critical ways in which the five elements of the model of the motor system (the single joint system) interact to produce movement. These topics include the basic functional unit of the motor system (motor unit), the ability of the nerve and muscle cells to transmit signals rapidly (excitable membranes), the way in which the nervous system controls muscle activity (excitation-contraction coupling), and the role of sensory information in the operation of the single joint system (feedback from sensory receptors). At the most fundamental level, movement is the result of the nervous system's controlling motor unit activity by transmitting electrical signals that elicit a muscle contraction and using sensory feedback information to monitor the movement.

TRUE-FALSE QUESTIONS

1. The brain is one of the elements of the single joint system. TRUE FALSE
2. The motor neuron includes the motor unit and the muscle fibers that it innervates. TRUE FALSE
3. The word *innervate* refers to an anatomical connection and not to the process of activation. TRUE FALSE
4. A healthy human muscle can have as few as 10 motor units. TRUE FALSE
5. Small motor units tend to have a high input resistance. TRUE FALSE
6. About 80% of the input to motor neurons occurs on the dendrites. TRUE FALSE
7. The effective synaptic current is smallest in large motor neurons. TRUE FALSE
8. All input is distributed uniformly to populations of motor neurons. TRUE FALSE
9. The innervation ratio refers to the number of muscle fibers innervated by a single motor neuron. TRUE FALSE
10. The muscle fibers belonging to a single motor unit are generally confined to a single neuromuscular compartment. TRUE FALSE
11. A twitch is the response of a motor unit or muscle to a single action potential. TRUE FALSE

12. The maximum force that a motor unit can exert occurs in a fused tetanus. **TRUE FALSE**
13. The twitch-tetanus ratio is generally greater than 10. **TRUE FALSE**
14. Type FR motor units are fatigue-resistant. **TRUE FALSE**
15. Types I, IIa, and IIb are distinguished on the basis of the enzyme myosin ATPase. **TRUE FALSE**
16. NADH-TR is an enzyme that is important for aerobic metabolism. **TRUE FALSE**
17. The muscle fibers of a motor unit are all of the same type. **TRUE FALSE**
18. The cross-sectional areas of muscle fibers in a cat are much greater than those in a human. **TRUE FALSE**
19. The input from a tendon organ to a motor neuron is known as a Group Ia EPSP. **TRUE FALSE**
20. Fiber-type proportions cannot be changed with training. **TRUE FALSE**
21. Neurotrophism refers to the rapid electrical interaction between neurons. **TRUE FALSE**
22. The cutting of the nerve to a muscle is referred to as denervation. **TRUE FALSE**
23. The drug tetrodotoxin blocks axonal propagation but does not impair axonal transport. **TRUE FALSE**
24. The rate of axonal transport changes with age. **TRUE FALSE**
25. The sarcolemma is an electrically excitable membrane. **TRUE FALSE**
26. Under steady-state conditions, K^+ is mainly concentrated in the intracellular fluid. **TRUE FALSE**
27. The Na^+-K^+ pump is a protein located in the sarcolemma. **TRUE FALSE**
28. Ions do not move across the sarcolemma under steady-state conditions. **TRUE FALSE**
29. The sarcolemma is more permeable to Na^+ than to K^+. **TRUE FALSE**
30. The resting membrane potential is normally negative on the inside with respect to the outside. **TRUE FALSE**
31. Conductance is the inverse of permeability. **TRUE FALSE**
32. The symbol for conductance is g. **TRUE FALSE**
33. Voltage-gated changes in conductance are caused by changes in membrane potential. **TRUE FALSE**
34. Action potentials are propagated, whereas synaptic potentials are conducted. **TRUE FALSE**
35. Afterhyperpolarization is due to the efflux of K^+. **TRUE FALSE**
36. Action potentials can be propagated along the axon at 120 m/s. **TRUE FALSE**
37. An action potential is a graded response. **TRUE FALSE**
38. Action potential amplitude is measured in mV. **TRUE FALSE**
39. Synaptic potentials have a smaller amplitude than action potentials. **TRUE FALSE**
40. A neuron generates an action potential each time it receives a neurotransmitter that elicits a synaptic potential. **TRUE FALSE**
41. The neurotransmitter at the neuromuscular junction is potassium. **TRUE FALSE**
42. The end-plate potential is a synaptic potential. **TRUE FALSE**
43. An EMG is usually measured by placing an electrode inside a nerve. **TRUE FALSE**
44. EMG amplitude is measured in mV. **TRUE FALSE**
45. Integration involves obtaining the absolute value of an interference EMG by inverting the negative phases. **TRUE FALSE**
46. The integrated EMG is correlated to muscle force under isometric conditions. **TRUE FALSE**
47. EMG amplitude changes when muscle length changes. **TRUE FALSE**
48. Phonomyography is the measurement of the pressure in muscle. **TRUE FALSE**
49. The mean frequency of a PMG signal is less than that of an EMG. **TRUE FALSE**
50. The neurotransmitter at the neuromuscular junction is acetylcholine. **TRUE FALSE**
51. The excitable membrane in muscle is the endomysium. **TRUE FALSE**
52. The muscle fiber action potential is conducted rather than propagated. **TRUE FALSE**
53. The action potential results in the release of Ca^{2+} from the T tubule. **TRUE FALSE**
54. The term *disinhibition* describes the removal of inhibition. **TRUE FALSE**
55. Troponin and tropomyosin are described as disinhibitors because they prevent the interaction of actin and myosin in the resting state. **TRUE FALSE**
56. A muscle action potential causes an increase in the g_{Ca} of the terminal cisternae. **TRUE FALSE**
57. Ca^{2+} pumps attached to the sarcolemma return Ca^{2+} to the sarcoplasmic reticulum after an action potential has passed. **TRUE FALSE**
58. The rate at which muscle force returns to the resting value depends on the rate at which Ca^{2+} is removed from the sarcoplasm. **TRUE FALSE**

59. Ca^{2+} binds to troponin to produce disinhibition and activate the crossbridge cycle. **TRUE FALSE**
60. The crossbridge cycle involves the interaction of actin and myosin. **TRUE FALSE**
61. The term *contraction* refers to the shortening of muscle length. **TRUE FALSE**
62. ATP is necessary for the detachment phase of the crossbridge cycle. **TRUE FALSE**
63. The average duration of the crossbridge cycle is longer when muscle length changes than in isometric conditions. **TRUE FALSE**
64. Proprioceptors detect stimuli generated by the system itself. **TRUE FALSE**
65. Input from exteroceptors is more influential than that from proprioceptors. **TRUE FALSE**
66. *Regulation* refers to the process of maintaining a certain state or condition. **TRUE FALSE**
67. A negative feedback effect is one in which a reflex is negated or turned off. **TRUE FALSE**
68. The main function of the stretch reflex is to assist muscle in generating maximum power. **TRUE FALSE**
69. The M1 component of the stretch reflex is mediated by the spinal cord and has a long latency. **TRUE FALSE**
70. The term *homonymous* means that the elements (e.g., sensory receptor, motor neuron) innervate different muscles. **TRUE FALSE**
71. The stretch reflex is a negative feedback circuit. **TRUE FALSE**
72. The response involved in the stretch reflex is confined to the muscle that was stretched. **TRUE FALSE**
73. The tendon tap reflex involves the Group Ib afferents. **TRUE FALSE**
74. The reciprocal inhibition reflex represents the effect of Group Ia afferents on the motor neurons that innervate an antagonist muscle. **TRUE FALSE**
75. The Ia inhibitory interneuron prevents the Group Ia input from influencing muscles other than the stretched muscle. **TRUE FALSE**
76. The flexor-withdrawal reflex represents a response to a noxious stimulus. **TRUE FALSE**
77. The sensory signals involved in the flexor-withdrawal reflex are transmitted over polysynaptic pathways. **TRUE FALSE**
78. The crossed-extensor reflex is elicited by the same afferent signal that produces the flexor-withdrawal reflex. **TRUE FALSE**
79. The H reflex is an experimental tool and does not occur under natural conditions. **TRUE FALSE**
80. The H reflex relies on the selective activation of the Group Ia afferents. **TRUE FALSE**
81. The size of the H reflex depends on the level of excitability of the motor neuron pool. **TRUE FALSE**
82. The Jendrassik maneuver is used to increase the M wave. **TRUE FALSE**
83. A single stimulus to the nerve can elicit both an H reflex and an M wave. **TRUE FALSE**
84. The tonic vibration reflex involves the same neural circuit as the tendon tap reflex. **TRUE FALSE**
85. Muscle spindles are inhibited by small-amplitude vibration. **TRUE FALSE**
86. Vibration of a muscle during a submaximal contraction results in an increase in the EMG and force of the muscle. **TRUE FALSE**
87. Vibration of a muscle can elicit reciprocal inhibition in the antagonist. **TRUE FALSE**
88. Vibration causes an increase of the H reflex. **TRUE FALSE**
89. The term *final common pathway* describes the connection between a motor neuron and muscle. **TRUE FALSE**
90. The interneuron is the major integrating element in the spinal cord. **TRUE FALSE**
91. Because reflexes involve such few neural elements, they are stereotyped and cannot be changed. **TRUE FALSE**
92. The Renshaw cell is a specialized type of motor neuron. **TRUE FALSE**
93. The Renshaw cell participates in reciprocal inhibition. **TRUE FALSE**
94. Recurrent inhibition is weak among motor neurons innervating distal muscles. **TRUE FALSE**
95. Recurrent inhibition has a greater effect on the excitability of smaller motor neurons. **TRUE FALSE**
96. Recurrent inhibition can influence the stretch reflex by affecting either the gamma motor neuron or the Ia inhibitory interneurons. **TRUE FALSE**
97. Presynaptic inhibition prevents inhibitory input via Group Ib afferents from affecting the stretch reflex. **TRUE FALSE**
98. Reflexes can be modified by long-term training. **TRUE FALSE**
99. The stretch reflex is an example of an assistance reflex. **TRUE FALSE**
100. Kinesthetic sensation is derived from the input of many different types of afferent fibers. **TRUE FALSE**
101. Kinesthetic input is perceived. **TRUE FALSE**

102. Input from joint receptors provides the basis for the determination of joint position.
TRUE FALSE

103. A corollary discharge is a copy of an outgoing motor command. TRUE FALSE

104. The sense of effort is based on feedback from proprioceptors. TRUE FALSE

MULTIPLE-CHOICE QUESTIONS

105. The size of a motor neuron can be indicated by which of the following?
A. Axon diameter
B. Surface area of the cell body
C. Rheobase
D. Propagation

106. What is the unit of measurement for input resistance?
A. MΩ
B. m/s
C. A
D. ms

107. Small motor neurons are characterized by which of the following?
A. Brief afterhyperpolarization
B. Fast contraction time
C. High input resistance
D. High axonal conduction velocity

108. A twitch can be characterized by the measurement of which of the following?
A. The magnitude of the peak force
B. Half relaxation time
C. Contraction time
D. Degree of sag

109. Differences in contraction time among motor units are not due to which of the following?
A. Rate of Ca^{2+} release
B. Muscle architecture
C. Action potential amplitude
D. Variations in myosin ATPase

110. What is the fatigue index value of Type S motor units?
A. Less than 0.25
B. Equal to 0.50
C. Greater than 0.75
D. Equal to 1.0

111. Why do Type FF motor units produce the greatest force?
A. A greater innervation ratio
B. Larger cross-sectional area of individual muscle fibers
C. A greater quantity of myosin ATPase
D. A higher twitch-tetanus ratio

112. Myosin ATPase stains darkly in which type of muscle fibers?
A. FG
B. I
C. IIa
D. SO

113. Which neuronal component is not involved in axonal transport?
A. Kinesin
B. Microtubule
C. Action potential
D. Dynein

114. Which types of experiments have not been used to provide evidence for the existence of neurotrophism?
A. Reinnervation
B. Adaptability of fiber-type proportions with training
C. Denervation
D. Blockade of axonal propagation with tetrodotoxin

115. Membrane-bound proteins do not serve which function?
A. Channel
B. Structural
C. Storage
D. Receptor

116. What does Na^+ experience under steady-state conditions?
A. An inward concentration force
B. No net driving force
C. An outward electrical force
D. An inward flux of ions

117. When the potential across a membrane becomes closer to zero, which of the following best describes the membrane's condition?
A. Depolarized
B. Hyperpolarized
C. Repolarized
D. Afterhyperpolarized

118. What causes the rising phase of an action potential?
   A. K^+ efflux
   B. Voltage-gated change in g_K
   C. Na^+-K^+ pump
   D. Na^+ influx
119. Why can the EMG be used as a window into the nervous system?
   A. Action potentials are measured as they pass from the neuromuscular junction along the muscle.
   B. It is the only way to measure the negative and positive phases of the action potentials.
   C. There is a one-to-one relationship between axonal and muscle fiber action potentials.
   D. There is a high correlation between EMG and muscle force.
120. Which of the following does not describe an EMG measurement?
   A. Rectified
   B. Interference pattern
   C. Depolarized
   D. Integrated
121. The results of a frequency-domain analysis of the interference EMG are plotted on what type of graph?
   A. Amplitude-frequency
   B. Power-time
   C. Voltage-time
   D. Frequency-time
122. Which measurement is often used to characterize the results of a power spectral analysis?
   A. Median frequency
   B. Average amplitude
   C. Mean power
   D. Area under the amplitude-frequency graph
123. Where is Ca^{2+} released from during the process of excitation-contraction coupling?
   A. Sarcolemma
   B. Sarcoplasmic reticulum
   C. T tubule
   D. Terminal cisternae
124. What proteins regulate inhibition of the crossbridge cycle?
   A. Actin
   B. Troponin
   C. Kinesin
   D. Myosin
   E. Tropomyosin
125. What value best represents the conduction velocity of the muscle action potential?
   A. 0.5 m/s
   B. 5.0 m/s
   C. 50 m/s
   D. 100 m/s
126. Where are action potentials not propagated?
   A. Sarcolemma
   B. Sarcoplasmic reticulum
   C. T tubule
   D. Axolemma
127. What is the value for g_{Ca} under resting conditions?
   A. High
   B. Zero
   C. Low
   D. Negative
128. Which protein does Ca^{2+} bind to in the process of excitation-contraction coupling?
   A. Myosin
   B. Tropomyosin
   C. Actin
   D. Troponin
129. Which phrase best defines the term *contraction*?
   A. The shortening of muscle in response to an action potential
   B. The muscle-activation state in which crossbridges are cycling in response to an action potential
   C. The activation of muscle in response to the action potential–dependent release of Ca^{2+}
   D. The sliding of the myofilaments (actin and myosin) due to Ca^{2+}-mediated disinhibition
130. What happens to the crossbridge cycle when active muscle is lengthened?
   A. The duration of the cycle is shorter than for isometric.
   B. Fewer crossbridges are attached than during a shortening contraction.
   C. Reattachment of crossbridges occurs more rapidly than during a shortening contraction.
   D. The duration of the cycle is longer than for isometric.
131. What feature of contractile activity does magnetic resonance imaging measure?
   A. Crossbridge cycling
   B. Accumulation of intracellular water
   C. Change in muscle length
   D. Sarcolemmal action potentials
132. What three types of afferent events are important in the control of movement?
   A. Proprioception
   B. Consequences of action
   C. Exteroception
   D. Synergism
133. Reflexes can involve different types of responses, including which of the following?
   A. Coordination of muscles in contralateral limbs
   B. All-or-none activation of a single muscle
   C. Distribution among a group of synergists
   D. Interaction of an agonist-antagonist muscle pair

134. Which reflexes are capable of performing a negative feedback function?
    A. Stretch reflex
    B. Reciprocal inhibition reflex
    C. H reflex
    D. Tonic vibration reflex
135. What is the best definition of *homonymous*?
    A. A neural circuit that performs a negative feedback function
    B. A reflex that includes components mediated by both the spinal cord and the brain
    C. An anatomical association describing neural elements that belong to the same muscle
    D. A reflex response elicited by a single type of sensory receptor
136. Which sensory receptors are not involved in the stretch reflex?
    A. Joint receptors
    B. Group Ia afferents of the muscle spindle
    C. Group II afferents of the muscle spindle
    D. Group Ib afferents of the tendon organ
137. Which neuron mediates the reciprocal inhibition reflex?
    A. Renshaw cell
    B. Ia inhibitory interneuron
    C. Homonymous motor neuron
    D. Jendrassik cell
138. Which stimulus may elicit the flexor-withdrawal reflex?
    A. Unexpected stretch of muscle
    B. Noxious cutaneous
    C. An increase in muscle force in the contralateral limb
    D. Vibration of the muscle
139. The H reflex involves the selective activation of which axons?
    A. Alpha axons
    B. Group Ib axons
    C. Group III and Group IV axons
    D. Group Ia axons
140. The H reflex is used to test the excitability of which of the following?
    A. Neuromuscular junction
    B. Motor neuron pool
    C. Muscle spindles
    D. Muscle fibers
141. What is the approximate latency of the M wave?
    A. 100 ms
    B. 75 ms
    C. 30 ms
    D. 5 ms
142. What reflex involves the same neural circuit as the tonic vibration reflex?
    A. Stretch reflex
    B. Tendon tap reflex
    C. Reciprocal inhibition reflex
    D. H reflex
143. Which statement best defines the term *final common pathway*?
    A. The interneuron is the major integrating element in the spinal cord.
    B. Reflexes are one of the basic elements in the control of the neuromuscular system.
    C. The motor neuron serves as the route through which commands are issued to muscle.
    D. Afferents provide the input and efferents transmit the output.
144. Which term best describes the observation that reflex responses can be modified?
    A. State-dependent response
    B. Neuromuscular propagation
    C. Recurrent inhibition
    D. Presynaptic inhibition
145. Which neural elements receive input from Renshaw cells?
    A. Gamma motor neurons
    B. Motor neurons
    C. Ia inhibitory interneurons
    D. Group Ia afferents
146. What is one of the known effects of presynaptic inhibition?
    A. Inhibition of the Ia inhibitory interneuron
    B. Activation of the recurrent collateral onto the Renshaw cell
    C. Inhibition of a motor neuron before it has a chance to generate an action potential
    D. A decrease in an incoming signal in a Group Ia afferent
147. What is a resistance reflex?
    A. A postural adjustment or a movement that will enhance the effect of external forces
    B. A response to an unexpected disturbance
    C. A positive feedback function
    D. A reflex that is followed by voluntary activation of the muscle
148. What perceptions are based on kinesthetic sensations?
    A. Position or movement
    B. Heaviness
    C. Corollary discharge
    D. Timing of movements

PROBLEMS

149. Why is the motor unit referred to as the functional unit of the single joint system?

150. Draw a graph of effective synaptic current (nA) versus motor neuron size (small to large) that shows the three ways in which the input from a given source can be distributed within a motor neuron pool. Describe how these different distribution patterns influence the activation of motor neurons.

151. If the muscle fibers belonging to a single motor unit are distributed at a density of 5 muscle fibers per 100 and there are 870,536 muscle fibers in the cross section, what is the innervation ratio of the motor unit?

152. What causes contraction speed to vary among motor units?

153. Why do Type FF motor units exert the greatest force?

154. What is the difference between a Type I and a Type SO muscle fiber?

155. List four morphologic differences and six physiological differences between the three motor unit types.

156. Type I fibers produce less power than Type II fibers produce, but they do so more efficiently. What does this statement mean? What is efficiency? What type of experiments would be needed to confirm this statement?

157. A motor unit in a human biceps brachii muscle has a twitch contraction time of 42 ms, a peak twitch force of 18 mN, a fatigue index of 0.18, a twitch/tetanus ratio of 1:6, and an innervation ratio of 1:389.
   A. What type of motor unit is it?
   B. How many muscle fibers does the motor neuron innervate?
   C. What is the maximum force that the motor unit can exert?
   D. How much force can the motor unit exert at the end of the fatigue test?

158. What is the rate at which material is transported by axonal transport, and what is the rate at which action potentials are propagated? Why are the two rates so different?

159. It has been estimated that a single kinesin molecule can exert a force of about 2 pN. How big is this force? Express this force as a fraction of the body weight of a 632-N individual.

160. How do denervation experiments demonstrate the existence of nerve-muscle trophism?

161. A number of changes occur in the single joint system with age. Describe the changes that occur due to age and the functional implications of these changes in the following: motor neurons, neurotrophism, miniature end-plate potentials, muscle action potentials, and M waves.

162. The potential that exists across an excitable membrane is due to the unequal distribution of ions in the intra- and extracellular spaces.
   A. Which ions are involved in maintaining the resting membrane potential?
   B. What two forces are exerted on each of these ions?
   C. Why are the ions distributed unevenly across the membrane?
   D. Draw a graph of an action potential (label the axes with the correct variables and units of measurement) and explain how an action potential can be measured.
   E. If one type of positive ion enters the cell and another type of positive ion leaves the cell, the net change in charge is zero. An action potential is basically due to positive ions entering (influx) and leaving (efflux) the cell, so why does the membrane potential change during an action potential?

163. The distribution of ions across a membrane depends on the combined effects of the forces acting on each type of ion and the permeability of the membrane to the different ions. Why is the resting membrane potential attributed largely to the distribution of K^+ and the action potential described as being driven by Na^+?

164. In the generation of an action potential, why is the flow of Na^+ into the cell described as leading to depolarization, whereas K^+ efflux causes repolarization?

165. Communication within the single joint system is accomplished by transmitter- and voltage-gated changes in membrane potential.
   A. Describe the differences and similarities between synaptic potentials and action potentials.
   B. What causes action potentials to be propagated along an excitable membrane?

166. The EMG is often used in the analysis of human movement.
   A. What information does an EMG provide?
   B. What do the processes of rectification and integration accomplish?
   C. Why is it useful to rectify and integrate the EMG?

167. Some investigators have used the PMG as an alternative to the EMG.
   A. What is the PMG and what does it measure?
   B. Draw a graph to show the difference between PMG and EMG as a function of muscle force.
   C. Draw a graph to show how PMG and EMG change as a function of time during a sustained, submaximal contraction.
168. What is neuromuscular propagation?
169. What processes between the sarcolemma and crossbridges could be impaired and prevent a sarcolemmal action potential from causing the myofilaments to exert a force?
170. Why is a muscle contraction not defined as the shortening of a muscle due to the activity of crossbridges?
171. What are the three movement-related roles performed by exteroceptors and proprioceptors?
172. Sensory receptors contribute to the generation of reflexes.
   A. What is a reflex?
   B. How does a reflex perform a negative feedback function?
173. The stretch reflex is a flexible response that can adapt to changing conditions. Why then is it, or the tendon tap reflex, a useful tool in clinical settings?
174. Describe the functional roles of the following reflexes: stretch, reciprocal inhibition, flexor-withdrawal, and crossed-extensor.
175. The reciprocal inhibition reflex is not mandatory but depends on the level of excitability of the Ia inhibitory interneuron.
   A. Describe the role of the Renshaw cell in modulating the strength of the reciprocal inhibition reflex.
   B. Why is the role performed by the Renshaw cell important in the function of the single joint system?
176. The H reflex is not a naturally occurring reflex but a tool used by experimentalists.
   A. What is the rationale underlying the use of the H reflex to test the excitability of the single joint system? How do we know that the H reflex tests the excitability of the system?
   B. What is the difference between the H reflex and the M wave?
177. The interneuron is regarded as the integrating element of the spinal cord.
   A. What does this mean?
   B. How does the interneuron relate to the concept of the final common pathway?
   C. In what three ways does the role of the interneuron influence the function of the nervous system?
178. What is presynaptic inhibition, and how does it affect reflexes during movement?
179. Hasan and Stuart (1988) have identified six roles for proprioceptors in movement.
   A. What is the difference between a resistance reflex and an assistance response?
   B. Why is it necessary for proprioceptors to be involved in the selection of a synergy?
   C. Why is it necessary for proprioceptors to serve as mediators between commands generated by the nervous system and the mechanical properties of the single joint system?
180. What is the difference between kinesthesia and proprioception?
181. What is the sense of effort? On what information is the sensation based? What is the experimental evidence for the existence of a sense of effort?

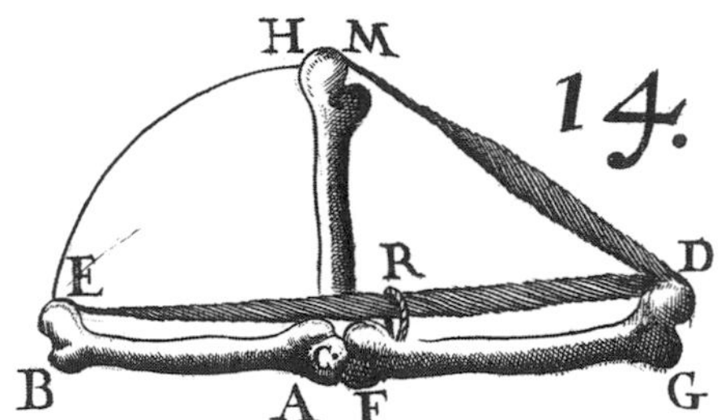

CHAPTER 6
Single Joint System Activation

The single joint system consists of five elements (rigid link, synovial joint, muscle, neuron, and sensory receptor) that represent a model of the motor system. Because movement is accomplished by activation of the motor system, we have discussed the characteristics of these elements (chapter 4) and four key interactions between these elements that enable us to move (chapter 5)—motor unit, excitable membranes, excitation-contraction coupling, and feedback from sensory receptors. Human movement, however, is characterized not by the haphazard activation of the motor system (single joint system) but rather by the careful and precise control of muscle force. In this chapter, we consider the properties of the single joint system that enable us to control muscle force so that we can perform a great variety of movements.

Neural Factors

The transformation of a muscle action potential into muscle force is referred to as excitation-contraction coupling. In this process, the electrical events that accompany the action potential provide the excitation for the muscle contraction. In a broader context, however, the muscle action potential is a product of nervous system input. Because the muscle action potential is usually related to an axonal action potential in a one-to-one manner, muscle action potentials represent the nervous system's *Go* command.

In chapter 5 we developed the concept that the motor unit represents the functional neuromuscular unit by which the single joint system controls muscle force. Because of this organization, muscle force is varied by altering the amount of motor unit activity (Kernell, 1992). This is accomplished by changing either the number of motor units that are active or the amount of activity of individual motor units once they have been activated. These two options are referred to as motor-unit **recruitment** (the process of motor unit activation) and modulation of **discharge rate** (action potential frequency).

Motor Unit Recruitment

In 1938, Denny-Brown and Pennybacker reported that the performance of a particular movement always appeared to be accomplished by the activation of motor units in a set sequence. This organization of motor unit activation has been called **orderly recruitment**. As the force exerted by a muscle increases, additional motor units are activated, or recruited, and once a motor unit is recruited, it remains active until the force declines. In this scheme, as shown in Figure 6.1, Motor Unit 1 is recruited first and remains active as long as the force does not decrease. The increase in force is accomplished, at least partly, by continuing to recruit motor units (four more in Figure 6.1). For the example shown in Figure 6.1, the force reaches a plateau value when additional motor units cease to be recruited and those that are active do not change the rate at which they discharge action potentials. As the force is reduced, motor units are sequentially inactivated, or **derecruited**, in the reverse order of recruitment; that is, the last motor unit recruited is the first derecruited. But Figure 6.1 is a special case because it represents the condition in which changes in force are due solely to a variation in the number of active motor units. Actual patterns of motor unit activity during the gradual change in force are shown in Figure 6.2; at low forces, these patterns include concurrent recruitment and modulation of discharge rate.

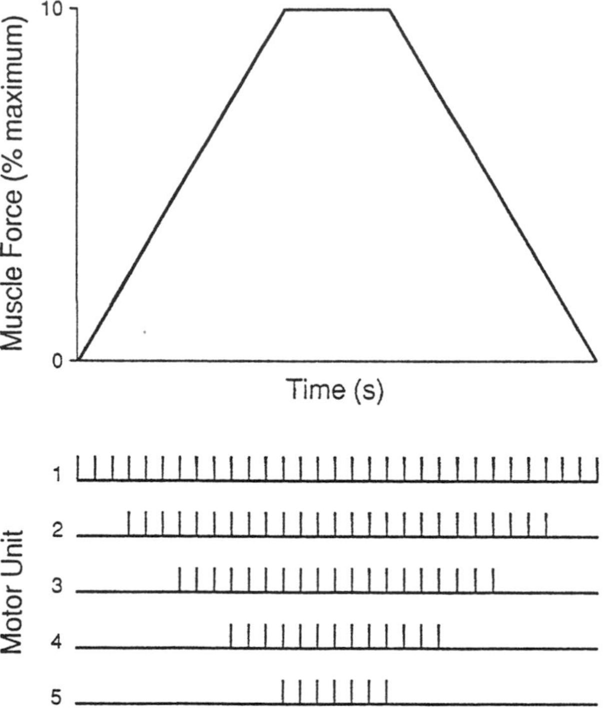

Figure 6.1 Idealized variation in muscle force due to the orderly recruitment and derecruitment of motor units. The motor unit records represent the action potentials generated by each motor unit as muscle force varies. For simplicity, the discharge rate of each motor unit is shown as regular, although the rate actually varies with muscle force.

The relative contribution of motor unit recruitment to muscle force varies between muscles. In some muscles (e.g., hand muscles such as adductor pollicis and first dorsal interosseus), all the motor units are probably recruited when the force reaches about 50% of maximum. Conversely, in other muscles (e.g., biceps brachii) motor unit recruitment continues up to 85% of the maximum force (Kukulka & Clamann, 1981). The increase in muscle force beyond that due to motor unit recruitment reflects the effect of varying discharge rate.

Mechanism. Orderly recruitment is a robust phenomenon. It has been demonstrated in a wide variety of muscle groups and animal species for many different tasks, but it appears to be the result of several physiological processes rather than of a single mechanism. One important factor underlying orderly recruitment appears to be motor neuron size, which is indicated by the surface area of the soma and dendrites. This effect, described as the **size principle**, suggests that *the orderly recruitment of motor units is due to variations in motor neuron size* such that the motor unit with the smallest motor neuron is recruited first and the motor unit with the largest motor neuron is recruited last (Binder & Mendell, 1990; Enoka & Stuart, 1984; Henneman, 1957, 1979). This organization is thought to exist among the motor neurons that innervate one particular muscle. Such a group of motor neurons is referred to as a **motor neuron pool**. For example, the human muscle tibialis anterior is innervated by approximately 445 alpha motor neurons; this represents the size of the motor neuron pool for the muscle. The size principle suggests that, within a motor neuron pool, the motor neurons are functionally organized according to size. When the muscle is required to exert a force, the smallest motor neuron in the pool is recruited first, and the largest is recruited last. For example, in tibialis anterior the largest alpha

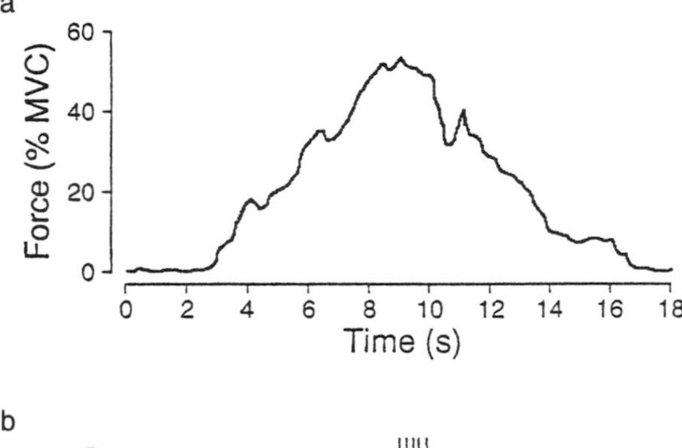

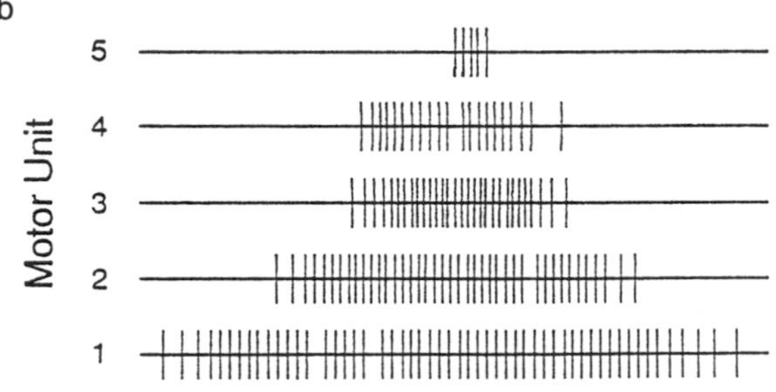

Figure 6.2 Recruitment and discharge pattern of five (of many) motor units during the performance of a graded contraction to 50% of the maximum force.
Note. From "Unusual Motor Unit Firing Behavior in Older Adults" by G. Kamen and C.J. DeLuca, 1989, *Brain Research*, **482**, p. 137. Copyright 1989 by Elsevier Science Publications. Reprinted by permission.

motor neuron (Number 445) will be recruited only when all the other motor neurons (Numbers 1 through 444) have been recruited. It appears that some morphological (e.g., number of dendrites, axon diameter, innervation ratio), biophysical (e.g., input resistance [R_N], rheobase [Rh], afterhyperpolarization [AHP]), and input (e.g., Renshaw cell pool [RCP], Group Ia afferent) characteristics vary with motor neuron size (Figure 6.3) such that the smallest motor neurons can be excited most easily.

The recruitment of motor units, however, does not depend solely on motor neuron size; it is also influenced by other motor neuron characteristics (intrinsic factors) and by the organization of synaptic input (extrinsic factors) on the dendrites and soma of the motor neurons in the pool (Burke, 1981). The *intrinsic factors* include motor neuron size, the sensitivity of the neurotransmitter receptors of the motor neuron, and the electrotonic characteristics of the motor neuron. *Electrotonic* refers to the electrical responses (voltage, current) of an excitable membrane due to changes in conductance (e.g., g_{Na}, g_K). Both the receptor sensitivity and the electrotonic characteristics of motor neurons can vary independently of motor neuron size. An increase in receptor sensitivity results in a more excitable cell to a given input, whereas changes in electrotonic characteristics result in a more effective conduction of synaptic inputs to the axon hillock. These effects make a motor neuron more responsive to inputs. The *extrinsic factors* include the number of synaptic terminals on a motor neuron from a given input system (e.g., Group Ia afferents, Renshaw cells, cutaneous afferents), the average amount of neurotransmitter liberated at each synapse, and the spatial distribution of the synapses over the soma and dendrites. Clearly, the greater the number of synaptic terminals or quantity of neurotransmitter released, the more excitable the motor neuron. And the closer the synapse is to the

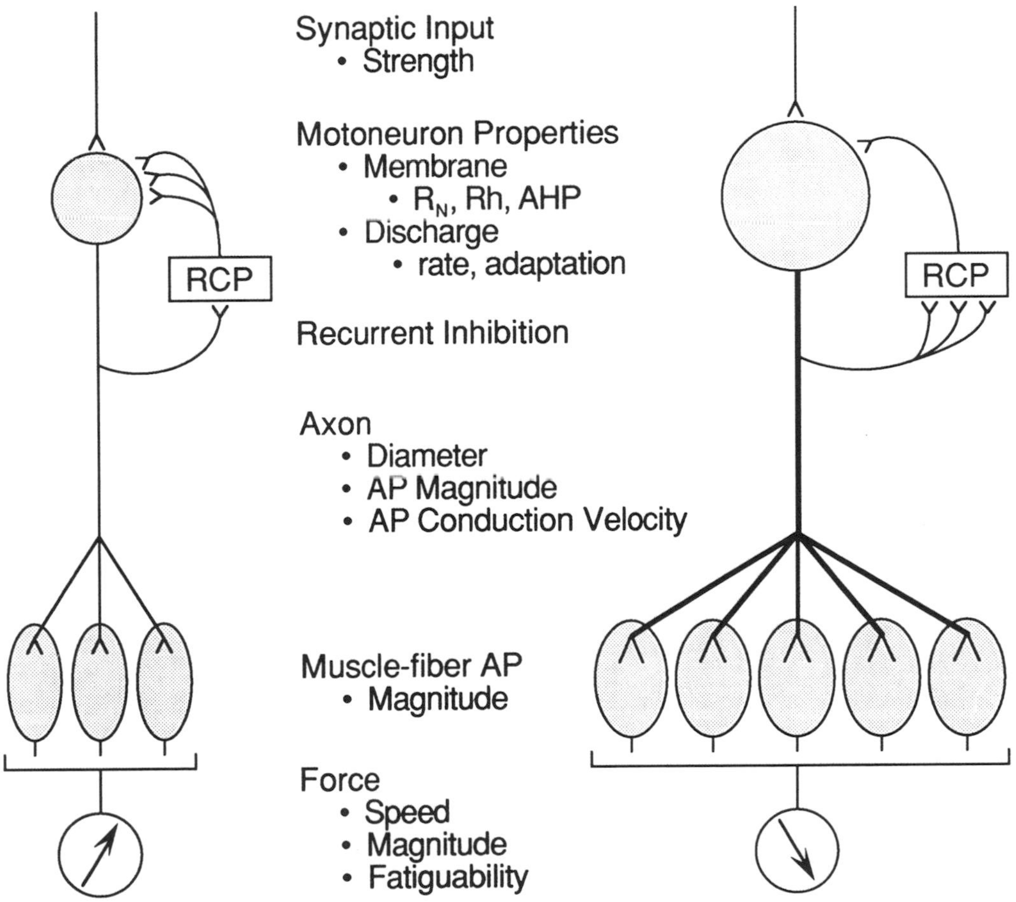

Figure 6.3 A model of two motor units that illustrates selected morphological and physiological measurements that correlate with motor neuron size. Morphological variations include the strength of synaptic input, axonal diameter, innervation ratio, and muscle fiber size. The functional consequences of such variations are examined physiologically by measuring the variables listed in the center of the figure. AP denotes action potential, and strength means amplitude of postsynaptic response. *Note*. From "Henneman's 'Size Principle': Current Issues" by R.M. Enoka and D.G. Stuart, 1984, *Trends in Neurosciences*, **7**, p. 227. Copyright 1984 by Elsevier Science Publications B.V. Reprinted by permission.

axon hillock, the more likely the postsynaptic potential is to elicit an action potential. Although we do not know exactly how each of these factors varies among motor neurons, it seems that there is a systematic variation, because the *effective synaptic current* in a motor neuron varies with input from different systems. The effective synaptic current is a measure of the net effect of the intrinsic and extrinsic factors on the excitability of motor neurons. In general, the effective synaptic current due to activation of an input system can be smallest in the large motor neurons, uniform in all motor neurons, or greatest in the large motor neurons (chapter 5).

Although the orderly recruitment of motor units has been shown to occur in many behaviors (Desmedt & Godaux, 1977), it is not a fixed pattern of activation, because it can vary in some conditions. One well-studied example of an alteration in recruitment order at the whole-muscle level is the **paw-shake response**. This behavior is elicited in animals, such as cats, when a piece of tape sticks to a paw and the animal's response is to vigorously shake the paw in an attempt to remove the tape. The behavior involves a rapid, alternating flexion and extension of the paw, which is controlled by the ankle flexor and extensor muscles. The extensor muscles include the slow-twitch muscle soleus and the fast-twitch muscle lateral gastrocnemius. In normal activities, such as standing, walking, running, and jumping, both the soleus and medial gastrocnemius are concurrently active, and the force exerted by soleus remains relatively constant across tasks while that exerted by medial gastrocnemius increases with the power demands of the task (Walmsley, Hodgson, & Burke, 1978). In the paw-shake response, however, the slow-twitch soleus is preferentially inactivated, and the behavior is produced by activation of the fast-twitch muscles (Smith, Betts, Edgerton, & Zernicke, 1980). This change in recruitment among a group of synergist muscles is necessary so that the animal can move the limb rapidly to dislodge the irritant.

At the motor unit level, some researchers have observed changes in recruitment order that result from manipulating sensory feedback (Garnett & Stephens, 1981; Kanda, Burke, & Walmsley, 1977). One way to demonstrate this effect is to determine the recruitment order of pairs of motor units in the absence and presence of a perceptible cutaneous sensation elicited by electrical stimulation of the skin. In the absence of the stimulus, one motor unit is recruited at a lower force than the other; that is, one has a lower **recruitment threshold**. In the presence of the cutaneous sensation, however, the unit with the higher recruitment threshold is activated first. Because the unit with the higher recruitment threshold probably exerts a greater force, this reversal of recruitment order represents the selective activation of stronger motor units. Such activation would be desirable if the cutaneous sensation required a rapid, forceful response.

Functional Implications. The advantage of orderly recruitment is that when a muscle is commanded to exert a force, the sequence of motor unit recruitment is predetermined and does not have to be specified by the brain. Therefore, the command generated by the brain need not include information on which motor units to activate; orderly recruitment relieves the brain of the need to be concerned with this level of detail in the performance of movement. However, because recruitment order is predetermined, largely by spinal mechanisms, it is not possible to selectively activate motor units in any order other than that specified by the intrinsic and extrinsic factors underlying orderly recruitment.

As a result of this predetermined order, the motor units recruited for a task depend on the proportion of the motor neuron pool that is needed. Figure 6.4 shows a hypothetical model of this relationship that was derived from the measurement of force in the triceps surae tendon of freely moving cats (Walmsley et al., 1978). In this scheme, the proportion of motor units recruited depends on the power demands of the task. For example, jogging at a slow speed represents an activity in which the muscle power requirements are minimal. The model indicates that jogging requires the recruitment of only some Type S and Type FR motor units. *Because recruitment order is fixed, the gradual increase in the power demands of a task involves the progressive recruitment of larger motor units.* However, motor unit size does not increase strictly according to motor unit type, so there is some overlap between Types S and FR and between Types FR and FF in both size and recruitment

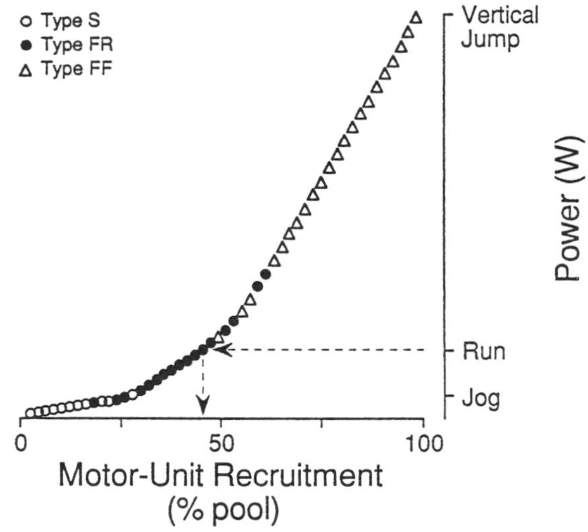

Figure 6.4 Hypothetical model of motor unit recruitment as dictated by the power demands of the task. Recruitment begins on the left (at 0% of the pool) and continues until enough units are recruited to produce the power needed to perform the task. For example, producing sufficient power to run would require the recruitment of about 48% of the motor units in the pool.

order (Stuart & Enoka, 1983). For this reason, it is not possible to selectively activate slow-twitch or fast-twitch motor units. When the power demands of the task are high, such as for a vertical jump, the model indicates that Type S motor units are recruited along with Types FR and FF, in a prescribed order.

Discharge Rate

The force exerted by muscle is due partly to variable combinations of the number of active motor units and the rate at which these motor units discharge action potentials. When a motor unit is recruited and the force exerted by the muscle continues to increase, the rate at which the motor unit discharges action potentials usually increases (Figure 6.2). Although each motor unit action potential will result in a motor unit twitch, when the action potentials occur close together, the twitches add together and exert a force that is greater than the twitch. The degree to which the twitches summate depends on the rate at which the action potentials are discharged (Figure 6.5). The relationship between action potential rate (frequency) and motor unit force is not linear; the increase in force due to increasing the action potential rate from 5 to 10 Hz is not the same as that due to increasing the rate from 20 to 25 Hz, although there is a difference of 5 Hz in each case. This effect can be shown by electrically stimulating a muscle or motor unit at different frequencies and measuring the elicited force. Such data can be plotted as a *force-frequency relationship* (Figure 6.6). The force-frequency relationship is sigmoidal (S-shaped), and the greatest increase in force (steepest slope) occurs at intermediate frequencies of stimulation (7 to 12 Hz). The actual relationship depicted in Figure 6.6 depends on muscle length; the curve shifts to the left for longer muscle lengths (Rack & Westbury, 1969). Although the relationship remains sigmoidal, the frequency for effecting the greatest change in force (i.e., the steepest part of the curve) is 3 to 7 Hz for long muscle lengths in contrast to 10 to 20 Hz for short muscle lengths. The force-frequency relationship is steepest for long muscle lengths.

The force-frequency relationship defines the variation in force as a function of activation rate, but this does not provide information on how the discharge rate actually varies during voluntary movements. Although it is technically difficult to measure the discharge of motor units over a wide range of forces, one group has done so and has classified units as either **tonic** or **phasic** based on features of discharge rate (Gydikov & Kosarov, 1973, 1974). One obvious difference between the two types is the relationship between muscle force and action potential rate, as shown in Figure 6.7. For tonic motor units, the relationship can be described as a ramp-and-plateau; discharge rate increases as muscle force increases at low levels, but at high forces the discharge

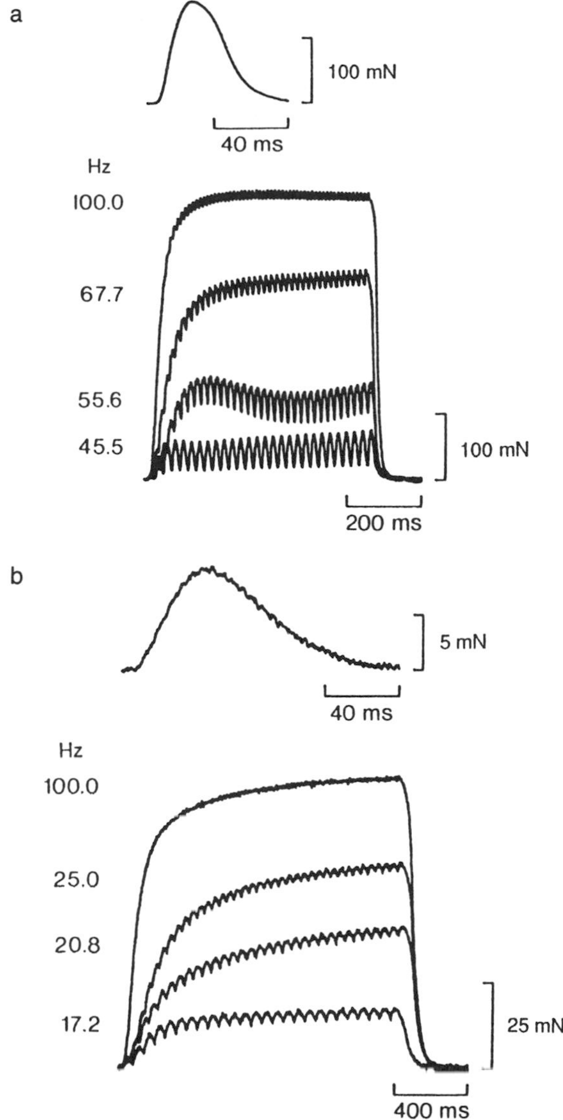

Figure 6.5 Effect of stimulus frequency on the summation of twitches for (a) a fast-twitch and (b) a slow-twitch motor unit. The slow-twitch unit had a longer contraction time and a smaller peak force; notice the different force scales. Four tetani were elicited in each motor unit. The tetani show the degree of summation of the twitch responses with various frequencies of stimulation (left column). Because the slow-twitch unit had a longer contraction time, there was greater summation at lower stimulus frequencies. *Note*. From "Gradation of Isometric Tension by Different Activation Rates in Motor Units of Cat Flexor Carpi Radialis Muscle" by B.R. Botterman, G.A. Iwamoto, and W.J. Gonyea, 1986, *Journal of Neurophysiology*, **56**, p. 497. Copyright 1986 by The American Physiological Society. Adapted by permission.

rate remains constant. In contrast, the discharge rate of phasic motor units increases over the entire range of muscle forces (i.e., a linear relationship). In addition, the tonic motor units generate smaller action potentials, are recruited at lower forces, and are less fatigable. The

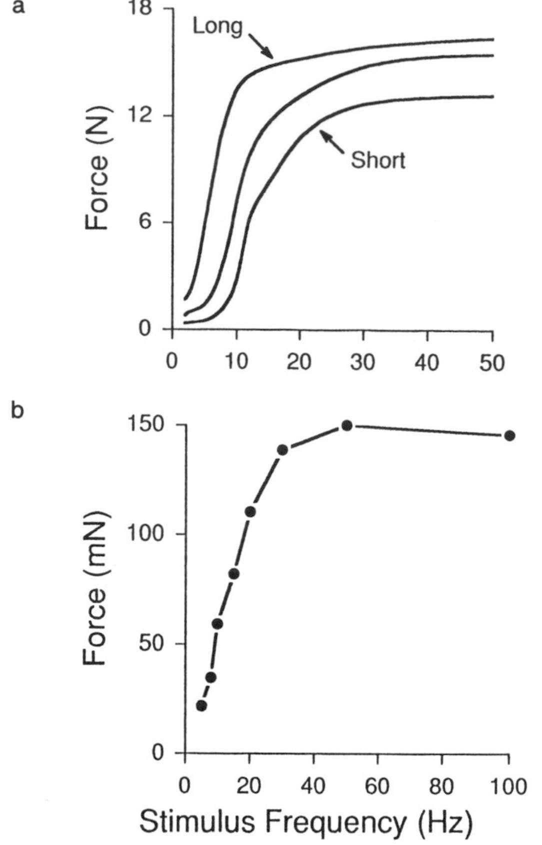

Figure 6.6 The force-frequency relationship. (a) Variation in whole-muscle force as a function of stimulus frequency for the cat soleus muscle at different lengths (short, intermediate, and long).

Note. From "The Effects of Length and Stimulus Rate on Tension in the Isometric Cat Soleus Muscle" by P.M.H. Rack and D.R. Westbury, 1969, *Journal of Physiology (London)*, **204**, p. 455. Copyright by the Cambridge University Press. Adapted by permission.

(b) Motor unit force in human thenar muscles as a function of stimulus frequency.

Note. From "Force-Frequency Relationships of Human Thenar Motor Units" by C.K. Thomas, B. Bigland-Ritchie, and R.S. Johansson, 1991, *Journal of Neurophysiology*, **65**, p. 1513. Copyright by the American Physiological Society. Adatped by permission.

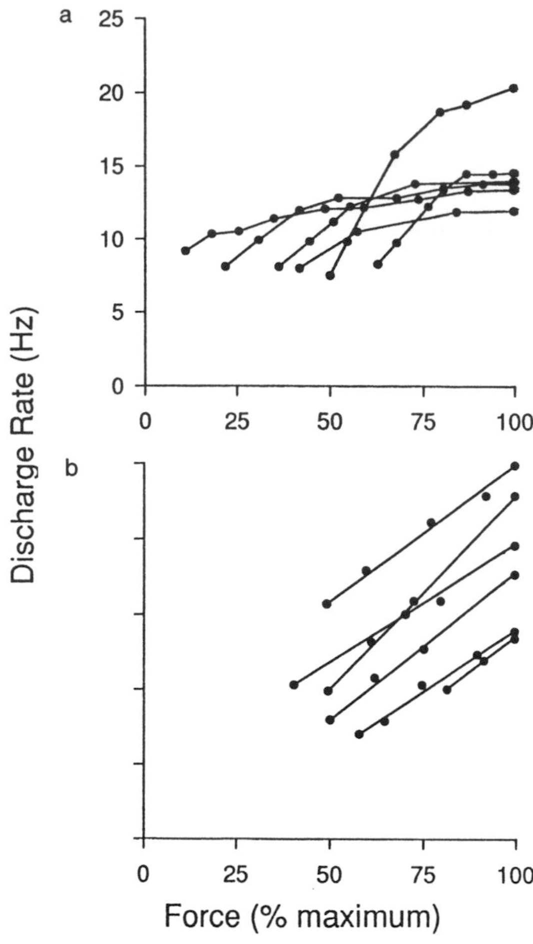

Figure 6.7 The relationship between muscle force (% maximum) and action potential rate of motor units for a muscle held at constant length: (a) ramp-and-plateau; (b) linear relationship.

Note. From "Some Features of Different Motor Units in Human Biceps Brachii" by A. Gydikov and D. Kosarov, 1974, *Pflügers Archiv*, **347**, p. 79. Copyright 1974 by Springer-Verlag, Inc. Adapted by permission.

phasic motor units seem to be important for dynamic conditions and contribute more to muscle force than do the tonic units. It is not known, however, how this tonic-phasic classification relates to the FF-FR-S scheme.

As is the case for motor unit recruitment, the extent of modulation in discharge rate appears to be muscle-specific. If motor unit recruitment for a particular muscle is complete at 50% of the maximum force (e.g., some hand muscles), then subsequent increases in force (i.e., 51% to 100%) must be due to variations in discharge rate. In contrast, for muscles in which recruitment continues up to 85% of the maximum force, only the final 15% of force (i.e., 86% to 100%) is due solely to variation in discharge rate. However, initial increases in force, at least up to 50% of the maximum force, are due to concurrent increases in recruitment and discharge rate (Monster & Chan, 1977; Person & Kudina, 1972). Figure 6.8 shows the activity of four motor units in rectus femoris while the force exerted by the knee extensor muscles was gradually increased up to 33% of maximum. The thick line shows the change in muscle force exerted by the knee extensor muscles, and the thin lines indicate the change in discharge rate of individual motor units. For each motor unit (thin lines), its discharge increased and then decreased with the change in force. Furthermore, while discharge rate was increasing other motor units were being recruited.

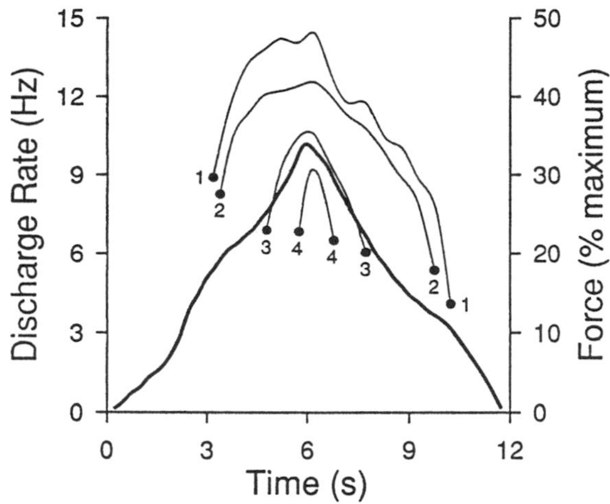

Figure 6.8 Motor unit activity (thin lines) during a gradual increase and then a decrease in force (thick line). The beginning of each thin line (left dot) indicates recruitment, and the end of the line (right dot) denotes derecruitment. The recruitment and derecruitment forces are determined by drawing a vertical line from the end of each thin line until it intersects the thick line. Motor Unit 1 was recruited at a force of 18% of maximum and derecruited at 10% of maximum (right ordinate); its discharge rate changed from 9 Hz at recruitment to 15 Hz at the peak force (left ordinate).
Note. From "Discharge Frequency and Discharge Pattern of Human Motor Units During Voluntary Contraction of Muscle" by R.S. Person and L.P. Kudina, 1972, *Electroencephalography and Clinical Neurophysiology*, **32**, p. 473. Copyright 1972 by Elsevier Publishing Co., Inc. Adapted by permission.

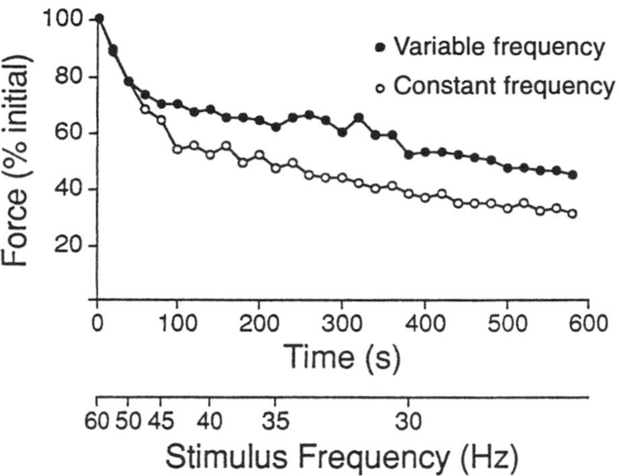

Figure 6.9 The submaximal force elicited in the knee extensor muscles by electrical stimulation.
Note. From "Preservation of Force Output Through Progressive Reduction of Stimulation Frequency in Human Quadriceps Femoris Muscle" by S.A. Binder-Macleod and T. Guerin, 1990, *Physical Therapy*, **70**, p. 622. Copyright 1990 by The American Physical Therapy Association, Inc. Adapted by permission.

Discharge Pattern. The force exerted by a muscle depends not only on motor unit recruitment and the frequency at which action potentials are discharged but also on the pattern of action potential activity (Windhorst, 1988). *Discharge pattern* refers to the relationship in time between an action potential and other action potentials generated by the same and by other motor units. There are at least three prominent effects of discharge pattern: muscle wisdom, double discharge, and motor unit synchrony. **Muscle wisdom** refers to the change in motor unit discharge that occurs during fatigue. When an individual sustains a fatiguing contraction, there is a decline in the rate at which motor unit action potentials are discharged (Bigland-Ritchie, Johansson, Lippold, Smith, & Woods, 1983; Dietz, 1978). This decline is not due to an impairment of the processes associated with action potential generation and propagation; it is an adaptation that matches the neural activity to the changing conditions in the muscle (Enoka & Stuart, 1992). Because it is an adaptive process that is controlled, at least partly, by the muscle, the fatigue-related decline in discharge rate has been referred to as muscle wisdom. The significance of this effect is shown in Figure 6.9, where the force elicited in the knee extensor muscles with a constant frequency of electrical stimulation is compared with a declining frequency of electrical stimulation. Over the 60-s interval, the force declined less when the frequency of stimulation was reduced than it did when frequency remained constant.

Another discharge pattern is that due to a **double discharge**, which refers to the discharge of two action potentials by a single motor unit within about 10 ms. Human motor units typically discharge over the range from 7 to 35 Hz, which means intervals of about 30 to 140 ms between consecutive action potentials. When a motor unit is electrically stimulated at about 12 Hz (82-ms interval) and then a double discharge (10-ms interval) is interposed in the train of stimuli, there is a substantial increase in the force exerted by the motor unit (Burke, Rudomin, & Zajac, 1970). In general, however, double discharges do not occur frequently in human motor units during voluntary movements. Although there have been few systematic studies on the topic, researchers have found that double discharges may vary between muscles and depend on the details of the task, such as whether the contractions are concentric or eccentric (Bawa & Calancie, 1983; Gydikov, Kossev, Kosarov, & Kostov, 1987).

In contrast to the double discharge, which refers to the action potentials discharged by a single motor unit, motor unit **synchrony** refers to the temporal relationship of the action potentials among motor units. If the action potentials generated by one motor unit are completely random (independent in time) with respect to those generated by another motor unit, they are asynchronous. However, when there is some relationship in time so that the discharges of action potentials by two motor units are not completely independent, they are synchronous. This effect is typically studied by recording the discharge of two motor units (Figure 6.10) and statistically comparing the temporal occurrence of action potentials. In most subjects, there is a slight degree of synchrony in motor unit discharge, which is due to shared input onto the motor neuron from some source (Bremner, Baker, & Stephens, 1991a, 1991b, 1991c; Nordstrom, Fuglevand, & Enoka, 1992). This synchrony exists between motor units located in the same muscle and between those in synergistic muscles. Furthermore, the amount of synchrony in motor unit discharge varies depending on the direction in which the muscle exerts a force. This variation suggests that the motor command differs under these conditions (Bremner et al., 1991c).

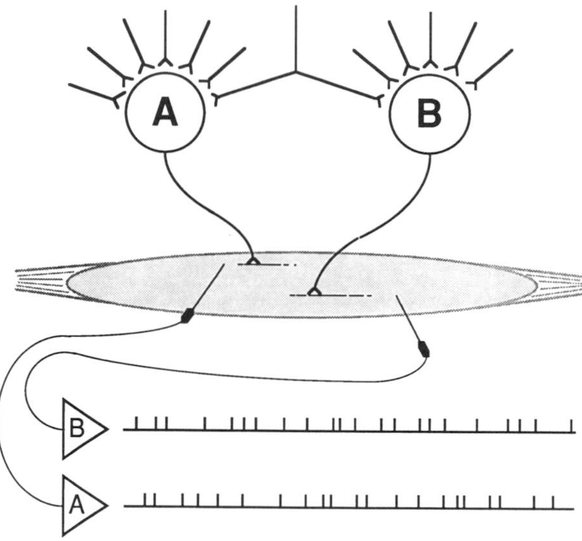

Figure 6.10 Measurement of motor unit action potentials. The motor neurons of two motor units (A and B) are shown receiving many input connections, including one input that is common to both motor neurons. The action potentials discharged by the two motor units (shown in the scales below the illustration) are recorded with wire electrodes inserted in the muscle. Although the action potentials discharged by the two motor units appear to be independent, there is a slight statistical correlation, and hence a degree of synchrony, that depends on the strength of the common input received by the two motor neurons.

An alternative method that has been used to test motor unit synchronization compares the discharge of a single motor unit with the EMG of the whole muscle. With this technique, which actually indicates the degree of correlation rather than synchronization of motor unit discharge, it has been reported that a strength-training program increases the degree of correlation in the discharge of the motor units belonging to the trained muscle (Milner-Brown, Stein, & Lee, 1975). Such an observation implies that an increase in motor unit synchronization will result in a greater muscle force. However, this technique contains methodological limitations that must be addressed before we can determine whether chronic activity alters motor unit synchronization (Yue, Fuglevand, Nordstrom, & Enoka, 1994).

Muscular Factors

Although the force that a muscle exerts clearly depends on the amount of excitation provided by the nervous system, it is perhaps less obvious that muscle force also depends on the properties of the muscle itself. These muscular factors include the mechanical properties of muscle and the structural effects due to differences in muscle architecture.

Muscle Mechanics

An action potential in a muscle fiber is an all-or-none event. But the force exerted by a muscle fiber following a muscle fiber action potential is not always the same. The force depends on internal factors, such as the discharge pattern, and on external factors, such as muscle fiber length and the speed of movement. **Muscle mechanics** is the study of the *external* mechanical variables (e.g., length, velocity, power, force) given the *internal* contractile state (e.g., discharge rate, availability of Ca^{2+}) of muscle.

Muscle Length. The sliding-filament hypothesis of muscle contraction states that the exertion of force by muscle is accompanied by the sliding of thick and thin filaments past one another. Although the actual mechanism causing this sliding is unknown (Pollack, 1983), three classes of theories attempt to provide an explanation: the filament-shortening, electrostatic, and independent force generator theories. The most popular explanation, the *crossbridge theory of muscle contraction*, belongs to the class of independent force generator theories. The crossbridge theory suggests that crossbridges extending from thick filaments are able to attach to the thin filaments and then undergo a structural-chemical transition (i.e., rotation of S1 as the muscle shortens), which exerts a tensile force. After the transition, the crossbridges detach and are free to repeat the cycle (Lieber, 1992).

According to this scheme, the development of force is dependent on these crossbridge attach-detach cycles; the greater the number of cycles occurring at the same time, the greater the force. And because force is exerted only during the attachment phase, the thick and thin filaments must be close enough to each other for the attachment to occur and thus for a force to be exerted. As the *length* of the muscle changes, and the thick and thin filaments slide past one another, the number of thin-filament binding sites available for the crossbridges changes. This leads to the observation that *tension varies with the amount of overlap between thick and thin filaments within a sarcomere*. The tension-striation spacing graph (Figure 6.11) shows the tension produced with different amounts of myofilament overlap. At a sarcomere length of 3.5 μm, there is minimal overlap of the thick and thin filaments; this does not happen under normal conditions. At a sarcomere length of about 2.0 μm, there is maximum overlap between the crossbridges (S1 extensions) and the binding sites on the thin filaments. As the filaments slide farther over one another and sarcomere length decreases to about 1.5 μm, the amount of **overlap of myofilaments** (i.e., the possibility of interaction between the crossbridges and the thin filaments) decreases. The net result of this change in the number of potential crossbridge attachments (due to variation in muscle length) is that the force that the muscle can exert will also vary with muscle length.

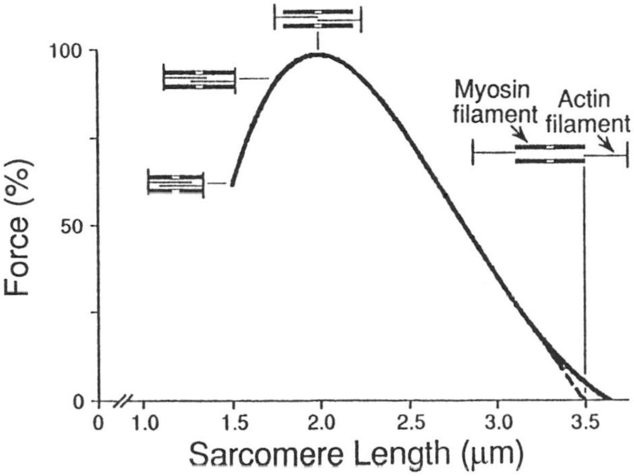

Figure 6.11 Relationship between contractile force, sarcomere length, and myofilament overlap. Force is expressed as a percentage of the maximum force exerted by the muscle fibers.

Note. From "Contractile Performance of Skeletal Muscle Fibres" by K.A.P. Edman. In *Strength and Power in Sport* (p. 103) by P.V. Komi (Ed.), 1992, Champaign, IL: Human Kinetics. Copyright 1992 by International Olympic Committee. Adapted by permission.

However, the force exerted by muscle is not dependent solely on the *active* process of crossbridge cycling. In addition, muscle includes a substantial amount of connective tissue (e.g., endomysium, perimysium, epimysium, tendon) and cytoskeletal components (e.g., intermediate filaments, titin, nebulin), which behave somewhat the way a stiff elastic band does. When stretched, these structures exert a *passive* force that combines with the active contribution due to crossbridge activity. Because of this interaction, the force exerted by *muscle* is due to both the contractile (myofilaments) and structural (connective tissue and cytoskeleton) elements; some researchers use the term *musculotendinous unit* to emphasize the contribution of the two components (active and passive) to the mechanical properties of muscle.

Figure 6.12 illustrates the contributions of the active and passive components to total muscle force as muscle length varies from the minimum contraction length to the maximum stretched length. This graph shows the **force-length relationship** of muscle. These data were obtained by the direct measurement of muscle force with a force transducer in patients who had a special type of below-elbow prosthesis. The experiment was performed by measuring two forces at each muscle length, one when the subject was resting and the other when the subject exerted a maximum voluntary force. When the subject remained relaxed, the passive force (due to the connective tissue and cytoskeletal elements) increased as the length of the muscle increased. When the subject performed a maximum voluntary contraction at each length, the force was due to both the passive (solid circle) and active (open circle) components and varied as shown in Figure 6.12. The dashed line represents the change in force due to the active component as a function of muscle length. At shorter muscle lengths, all of the force is due to the active component (crossbridge activity), whereas at the longer lengths most of the muscle force is due to the passive component. The profile for the active component indicates that the greatest overlap of the thick and thin filaments occurs at a muscle length that is about midway between the minimal and maximal lengths—this is typically the **resting length of muscle** (l_o).

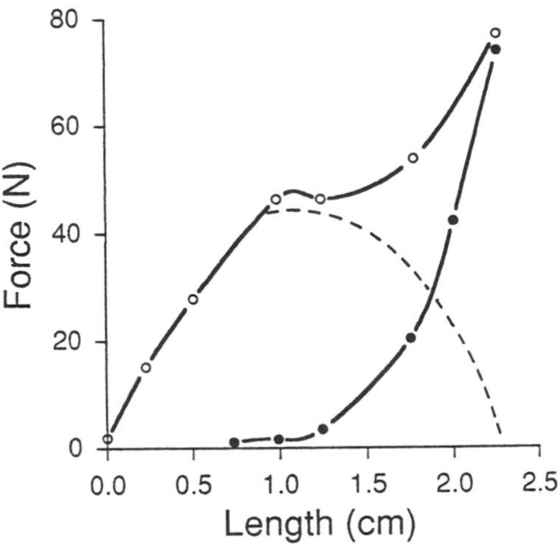

Figure 6.12 Contributions of the active and passive elements to the total muscle force as length varies. Force was measured isometrically.

Note. From "Mechanics of Human Isolated Voluntary Muscle" by H.J. Ralston, V.T. Inman, L.A. Strait, and M.D. Shaffrath, 1947, *American Journal of Physiology*, **151**, p. 615. Copyright 1947 by the American Physiological Society. Adapted by permission.

Muscle length also affects the submaximal force, even the twitch force (Rack & Westbury, 1969). In animal experiments, in which it is easier to measure directly the muscle force and to control precisely the excitation of muscle, it has also been shown that muscle force depends on muscle length. In the cat soleus muscle, for example, both the tetanic and twitch force reach a maximum at an ankle angle of about one rad (57.3°), which is a long muscle length (Figure 6.13a). Furthermore, the force due to the passive elements is not large in this muscle and is even less than peak twitch force. The magnitude of the force exerted by the passive elements varies substantially across muscles, in accordance with differences in muscle structure and the quantity of connective tissue (Woittiez, Huijing, & Rozendal, 1983). Muscle length affects not only the amplitude of the twitch but also its time course (Figure 6.13b); both contraction time and half relaxation time are longer with increases in muscle length. The effect of length on force is such that as length decreases, it is necessary to stimulate a muscle at higher frequencies in order to get the same force. For example, Rack and Westbury (1969) compared the stimulus frequency required for the soleus muscle to exert a comparable force at a long length and at one shorter by 10 mm. They found that it was necessary to increase the stimulus frequency from 10 Hz at the long length to 35 Hz at the shorter length to achieve the same force.

To determine the functional significance of the length-force relationship of muscle, we need to transform this characteristic into an *angle-torque relationship*. Recall that the contribution of muscle to movement depends on its ability to exert torque, which is the product of force and moment arm. However, the translation of the length-force relationship to an angle-torque relationship is confounded by at least three factors that may alter the length effect. First, the movement of most body segments (rigid links in our single joint system) is controlled by groups of muscles rather than by a single muscle. Often the fibers of each muscle in such a group are arranged differently (e.g., fusiform, pennate, bipennate), so at any joint position the fibers within the different muscles will probably be at different positions on their length-force curves. Second, because a number of muscles (e.g., rectus femoris, semitendinosus, gastrocnemius, biceps brachii) act across more than one joint, the length of these muscles, and therefore the force they exert, are not influenced solely by the position of one joint. For example, the knee extensor group includes a two-joint bipennate muscle (rectus femoris), two single-joint pennate muscles (vastus lateralis and vastus medialis), and a single-joint fusiform muscle (vastus intermedius). Third, variations in muscle force may not be reflected by similar changes in muscle torque, because torque is defined as the product of two variables (force and moment arm) and not simply the effect of length on force; it is also necessary to consider

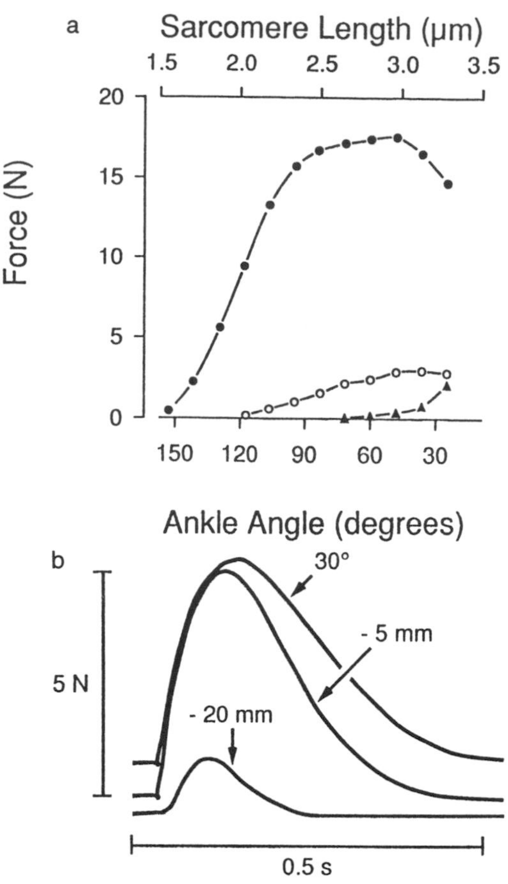

Figure 6.13 Effect of muscle length on the force exerted by the cat soleus muscle: (a) changes in tetanic (solid circles), twitch (open circles), and passive (triangles) forces as a function of the muscle length (muscle length increases to the right), (b) change in the twitch response as muscle length decreases from a long length (ankle angle of 30°) in 5-mm steps. 30° indicates the initial long length. –5 and –20 mm show shorter lengths.

Note. From "The Effects of Length and Stimulus Rate on Tension in the Isometric Cat Soleus Muscle" by P.M.H. Rack and D.R. Westbury, 1969, *Journal of Physiology (London)*, **204**, pp. 450, 451. Copyright 1969 by Cambridge University Press. Adapted by permission.

the effect of length (joint angle) on moment arm. Lieber (1992) has shown that the peak muscle force may not occur at the same joint angle as the peak moment arm, and hence the peak muscle torque will occur at a joint angle between the two. Generally, however, muscle force is the major determinant of muscle torque about a joint; exceptions include soleus, gluteus medius, and rectus femoris.

As a consequence of the effects of fiber architecture, the number of joints a muscle spans, and the influence of joint angle on moment arm, the angle-torque relationship has been found to vary among muscles (Kulig, Andrews, & Hay, 1984). These effects are characterized in Figure 6.14, which shows variations in muscle length represented as changes in joint angle. In general, the angle-torque relationship can have one of three forms: ascending, descending, or intermediate. In the ascending relationship (e.g., knee flexors, hip adductors), torque increases to a plateau as joint angle increases. In the descending relationship (e.g., hip abductors), torque is maximal at the minimum joint angle and decreases as joint angle increases. In the intermediate relationship (e.g., elbow extensors and flexors, knee extensors), torque is maximal at an intermediate joint angle; this is the most common form of the angle-torque relationship.

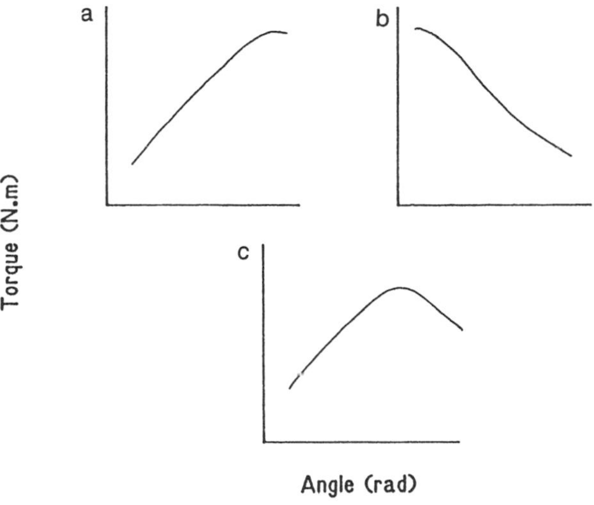

Figure 6.14 Three common forms of the angle-torque relationship: (a) ascending; (b) descending; (c) intermediate. *Note.* From "Human Strength Curves" by K. Kulig, J.G. Andrews, & J.G. Hay. In *Exercise and Sport Sciences Reviews* (Volume 12) (p. 422) by R.L. Terjung (Ed.,), 1984, Lexington, MA: Collamore. Copyright 1984 by D.C. Heath and Company. Adapted by permission.

Although the resultant muscle torque about a joint may have one of these three forms, the angle-torque relationship for individual muscles can be quite different. This has been demonstrated with a musculoskeletal model of the human leg (Hoy, Zajac, & Gordon, 1990). The model, which is based on musculoskeletal geometry and muscle parameters, defines the angle-torque relationship for 18 muscles or groups of muscles about the hip, knee, and ankle (Figure 6.15). The predictions of the model for resultant muscle torque can be compared with experimental measurements; in each graph compare the large dots (experimental) to the bold line. (The experimental torques were measured by different groups: hip extensor—Nemeth, Ekholm, Arborelius, Harms-Ringdahl, & Schuldt, 1983; hip flexor—Markhede & Grimby, 1980; knee extensor and flexor — Scudder, 1980; ankle plantarflexor—Sale, Quinlan, Marsh, McComas, & Belanger, 1982; and ankle dorsiflexor—Marsh, Sale, McComas, & Quinlan, 1981.) In general, the angle-torque relationship for individual muscles in each group was different. For example, the extensor torque about the hip was due to variable contributions by the hamstrings, adductor magnus, gluteals, and the combined effect of adductor longus, adductor brevis, and gracilis (Figure 6.15a). In contrast, the extensor torque about the knee was dominated by the contribution from the vasti, with a lesser contribution by rectus femoris, and the two angle-torque relationships were qualitatively similar (Figure 6.15c).

Change in Muscle Length. Once a muscle has been activated and begins to exert a torque, the tendency is for the filaments to slide past one another and for the muscle to shorten its length. But whether or not the length of the active muscle changes depends on the magnitude of the torque exerted by the muscle compared with that due to the load (Figure 6.16). This ratio (muscle/load) can have three distinct values: less than 1, equal to 1, or greater than 1. If the ratio has a value equal to 1, the muscle and load torques are equivalent and whole-muscle length will not change; this condition is referred to as an **isometric** contraction. The data shown in Figures 6.14 and 6.15 are for isometric contractions. Conversely, if the two torques are not equal (i.e., less than or greater than 1), then whole-muscle length will change. When the muscle torque is greater, the ratio will exceed 1 and the whole-muscle length will shorten; this is called a **concentric** contraction. When the torque due to the load is greater than the muscle torque, the ratio will be less than 1 and whole-muscle length will increase (while it is active); this is an **eccentric** contraction. Often these mechanical conditions (isometric, concentric, eccentric) are referred to as types of contractions. As we will discuss in chapter 9, there is a physiological as well as a mechanical rationale for making this distinction.

The angle-torque relationships shown in Figure 6.14 have been examined with isometric, concentric, and eccentric contractions. Obtaining these measurements requires a device that can apply a variable load to a human limb, and the load must be able to exceed the maximum voluntary muscle torque for eccentric contractions. Commercial products are available that meet these criteria, but they must be used with caution, especially on patient populations, because the device controls the movement during an eccentric contraction. In general, the angle-torque relationship remains qualitatively similar to that obtained with isometric contractions. The magnitude of the torque exerted in eccentric contractions, however, is greater than that exerted during concentric contractions, and that of the isometric torque falls between the eccentric and concentric values. Figure 6.17 shows data for concentric and eccentric angle-torque relationships for the leg and knee extensors; the leg extensor data concern the measured force exerted by the feet during the concurrent extension of

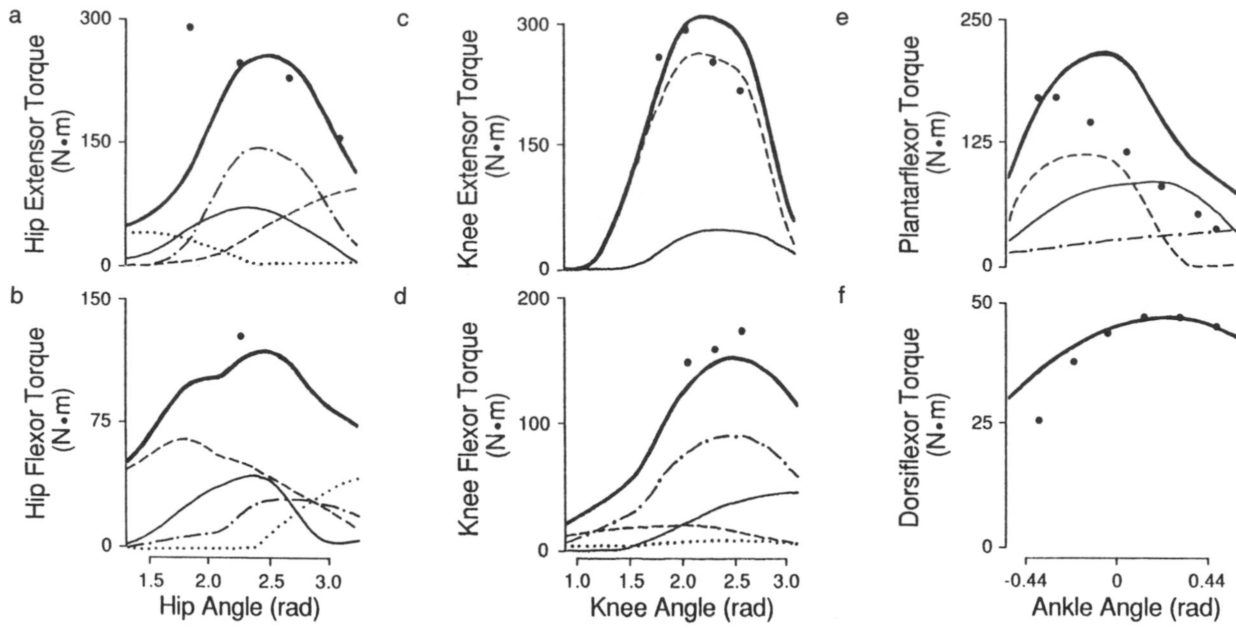

Figure 6.15 Maximal angle-torque relationships about the hip, knee, and ankle joints. The graphs indicate the resultant relationships (bold lines) as well as the contributions by individual muscles about each joint: (a) hip extensor—hamstrings (dash-dot line), adductor magnus (solid line), gluteals (dashed line), and adductor longus, adduction brevis, and gracilis (dotted line); (b) hip flexor—rectus femoris (solid line), iliopsoas, gluteus medius, and minimus (dashed line), gracilis, tensor fascia latae, pectineus, and sartorius (dash-dot line), and adductor longus and adductor brevis (dotted line); (c) knee extensor—vasti (dashed line) and rectus femoris (solid line); (d) knee flexor—short head of biceps femoris (dotted line), hamstrings (dash-dot line), gastrocnemius (solid line), and gracilis, sartorius, and tensor fascia latae (dashed line); (e) ankle plantarflexor—soleus (dashed line), gastrocnemius (solid line), and other plantarflexors (dash-dot line) (the negative ankle angle indicates dorsiflexion); (f) ankle dorsiflexor.

Note. Adapted from "A Musculoskeletal Model of the Human Lower Extremity: The Effect of Muscle, Tendon, and Moment Arm on the Moment-Angle Relationship of Musculotendon Actuators at the Hip, Knee, and Ankle" by M.G. Hoy, F.E. Zajac, and M.E. Gordon, 1990, *Journal of Biomechanics*, **23**, p. 165, Copyright 1990, with kind permission from Pergamon Press Ltd., Headington Hill Hall, Oxford OX3 DBW, UK.

the hip, knee, and ankle joints while subjects were in a seated position. For the leg extensors (Figure 6.17a), the eccentric torque was maximum at a knee angle of about 2.5 rad, whereas the concentric torque continued to increase up to the maximum knee angle. Over most of the range of motion, Westing, Seger, and Tharstenson (1990) found that the eccentric torque was greater than the concentric torque. For the knee extensors (Figure 6.17b), the eccentric and concentric torques were maximum at an intermediate knee angle (2.0 rad), and the eccentric torque was greater than the concentric torque at all knee angles. In addition, the knee extensor data compare the angle-torque relationships obtained with voluntary activation and electrical stimulation; the three conditions for both concentric and eccentric contractions were voluntary activation alone, electrical stimulation alone, and combined voluntary activation and electrical stimulation. For the concentric contractions, the three conditions produced similar torques at extended knee angles (2.0 to 3.0 rad), but the voluntary torque (curve d) was greater than the electrically stimulated (curve f) and combined (curve e) torques. For the eccentric contractions, the torque produced by the combined activation (curve a) was greater than that due to electrical stimulation (curve b) or maximal voluntary activation (curve c). Because the torque produced by voluntary activation was the least of the three conditions for the eccentric contractions, Westing et al. (1990) have proposed that the subjects did not maximally activate the knee extensors during the voluntary eccentric contractions.

This discussion of concentric and eccentric contractions has emphasized that the direction of the change (i.e., shorter or longer) in whole-muscle length depends on the ratio of the muscle and load torques; that is, it is defined by the mechanical conditions of the task. However, the change in whole-muscle length, which includes the muscle fibers and tendon, *does not necessarily coincide with the change in length experienced by the muscle fibers* (Fellows & Rack, 1987; Griffiths, 1991). When the medial gastrocnemius muscle of a cat is electrically stimulated to produce an isometric contraction, muscle fibers can shorten by as much as 28% of their resting length (Figure 6.18a). For this to happen, the tendon must lengthen by an equivalent amount so that whole-muscle length remains constant

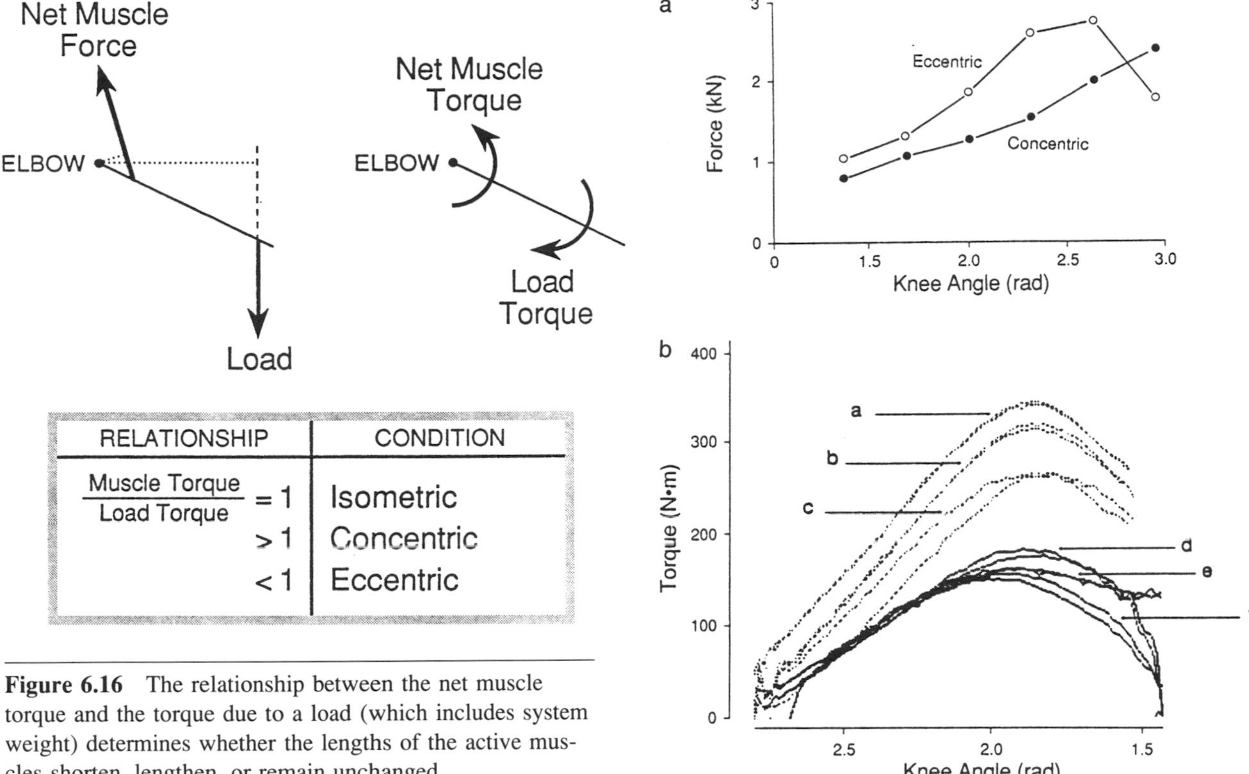

Figure 6.16 The relationship between the net muscle torque and the torque due to a load (which includes system weight) determines whether the lengths of the active muscles shorten, lengthen, or remain unchanged.

Figure 6.17 Angle-torque relationships for concentric and eccentric contractions. A knee angle of 3.14 rad represents complete extension. (a) The force exerted by the feet during leg extension as a function of knee angle.
Note. From "Function of the Quadriceps Femoris Muscle Under Maximal Concentric and Eccentric Contractions" by V. Eloranta and P.V. Komi, 1980, *Electromyography and Clinical Neurophysiology*, **20**, p. 162. Copyright 1980 by Nauwelaerts Publishing House. Adapted by permission.
(b) Knee extensor torque during voluntary activation (c, d), electrical stimulation (b, f), and combined activation (a, e) during eccentric (a, b, and c) and concentric (d, e, and f) contractions.
Note. From "Effects of Electrical Stimulation on Eccentric and Concentric Torque-Velocity Relationships During Knee Extension in Man" by A.H. Westing, J.Y. Seger, and A. Thorstensson, 1990, *Acta Physiological Scandinavica*, **140**, p. 19. Copyright 1990 by Scandanivian Physiological Society. Adapted by permission.

(i.e., isometric). This tendon **compliance** (mm/N), which is the inverse of stiffness, is most evident in muscles with long tendons. This effect occurs not only during electrical stimulation of muscle but also during voluntary movements such as walking. Figure 6.18b shows whole-muscle length, muscle fiber length, and EMG for the medial gastrocnemius muscle of a cat during the step cycle. Prior to the foot contacting the ground (at vertical line), both whole-muscle and muscle fiber length increased, then shortened; this is a standard concentric contraction. However, after foot contact there was an increase in whole-muscle length (an eccentric contraction) but a shortening of muscle fiber length. The increase in whole-muscle length must have been accomplished by a stretch of the tendon. This effect decreases with faster stretches and can reverse so that fast stretches can increase muscle fiber length. It seems, therefore, that *during isometric and slow eccentric contractions the change in muscle fiber length does not parallel the change in whole-muscle length in muscles with long tendons.*

One distinction that is often made between isometric and **anisometric** (concentric and eccentric) contractions is the inability of isometric contractions to perform work. Because work is defined as the product of force and distance, the failure to move an object, as in an isometric contraction, means that no work is done on the object. Recall from chapter 3 the relationship between work and energy: The amount of work done is determined by the change in energy of a system. For example, when a heavy briefcase is held in a stationary position, its energy (kinetic and potential) does not change, and therefore no work is done on the briefcase. However, for the briefcase to be held, the muscles that flex the fingers must be activated, albeit isometrically, and these muscles will consume energy (ATP). The fact that the muscles are doing work becomes quite apparent because after an interval of time the muscles tire and become unable to support the briefcase. If no work were being done, it would be possible to hold the briefcase indefinitely. Consequently, although the briefcase experiences no work, the finger flexor muscles must do work

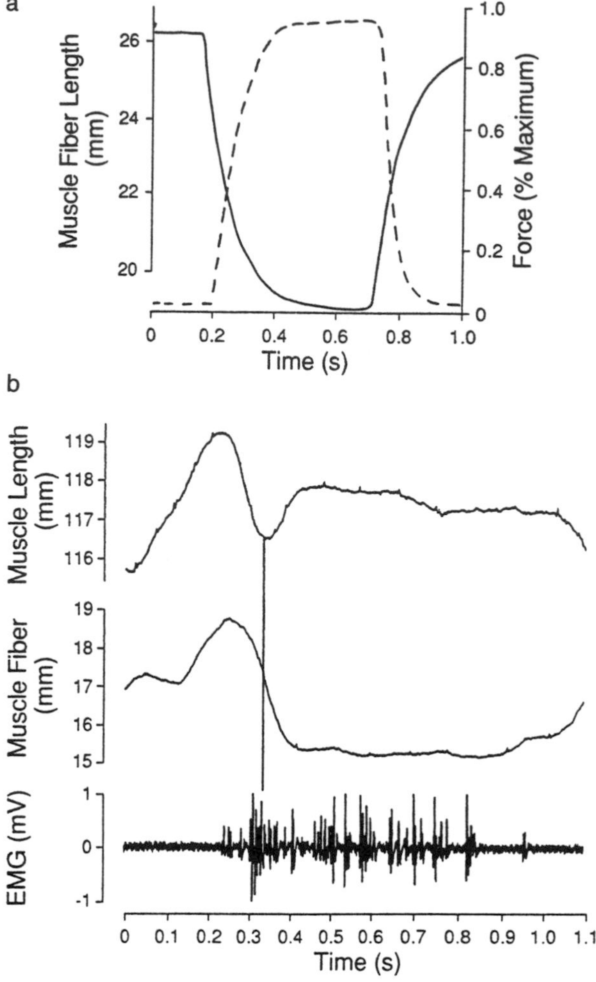

Figure 6.18 Change in muscle fiber length during isometric and eccentric contractions. (a) An isometric contraction was elicited in a medial gastrocnemius muscle. Whole-muscle length remained constant while muscle fiber length decreased and force increased. Muscle fiber length was measured with an ultrasound technique between two piezoelectric crystals. (b) An eccentric contraction in the medial gastrocnemius muscle following foot contact was associated with shortening of muscle fiber length. Footstrike occurred at the vertical line.
Note. From "Shortening of Muscle Fibres During Stretch of the Active Cat Medial Gastrocnemius Muscle: The Role of Tendon Compliance" by R.I. Griffiths, 1991, *Journal of Physiology (London)*, **436**, pp. 225, 229. Copyright 1991 by Cambridge University Press. Adapted by permission.

to lift the briefcase. To distinguish between the two, we can say that although the briefcase experiences no *mechanical work* the muscles do *metabolic work* to hold the briefcase in place.

Rate of Change in Muscle Length (Velocity). When the muscle torque does not equal the load torque, there is a change in whole-muscle length. The torque that the muscle can exert under these conditions depends on the magnitude and direction of the rate of change in length—the *velocity* of the muscle contraction. When the muscle shortens, it performs a concentric contraction, a mode of activity that is associated with a lesser muscle torque than for isometric (velocity = 0) contractions; thus, during maximum concentric contractions, the torque exerted by muscle is less than that for a maximum isometric contraction. In addition, *as the velocity of shortening increases, the amount of torque that the muscle can exert decreases* (Figure 6.19). Although the rate of crossbridge cycling (attach-rotate-detach) increases as the shortening speed increases, the average force exerted by each crossbridge decreases, and there may be fewer crossbridge attachments as the muscle shortens more quickly. Because the energy used by the muscle increases (due to increased crossbridge detachment) with shortening speed while the torque exerted decreases, the muscle becomes less *efficient* (work output/energy input) with increases in the discrepancy between the muscle and load torques.

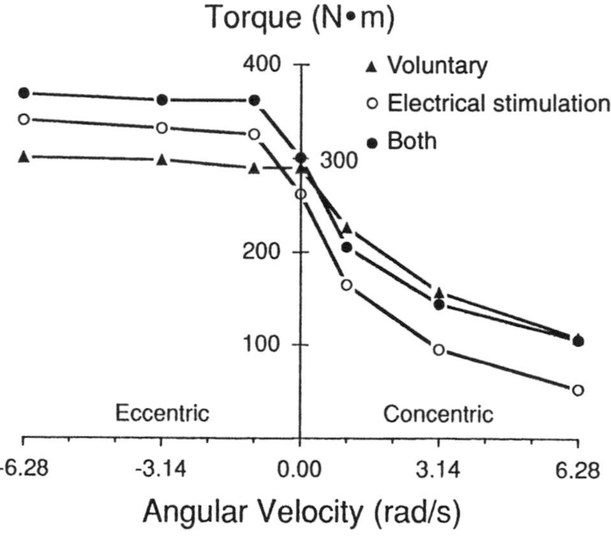

Figure 6.19 Velocity-torque relationship for the knee extensor muscles. Torque was measured at a knee angle of about 1.0 rad at various constant-velocity contractions.
Note. From "Effects of Electrical Stimulation on Eccentric and Concentric Torque-Velocity Relationships During Knee Extension in Man" by A.H. Westing, J.Y. Seger, and A. Thorstensson, 1990, *Acta Physiologica Scandinavica*, **140**, p. 19. Copyright 1990 by Scandanivian Physiological Society. Adapted by permission.

When the muscle torque is less than the load torque, whole-muscle length increases. The maximum torque that a muscle can exert when performing an eccentric contraction is greater than the isometric and concentric torques. As indicated in Figure 6.19, the maximum torque varies in the following order: Eccentric is greater than isometric and isometric is greater than concentric. Maximum torque during an eccentric contraction is about 1.5 to 1.8 times greater than the maximum isometric force. Why can a muscle exert the greatest torque during an eccentric contraction? Probably many factors contribute to dominance of the eccentric torque. One possibility is that the force required to break a crossbridge is greater than that required to hold an isometric contraction. The typical account given of crossbridge cycling is that the globular heads (S1) attach to actin binding sites and then undergo a conformational change, which is depicted as a rotation of S1. During an eccentric contraction, however, the mechanism responsible for crossbridge detachment is probably different, *presumably increasing the force exerted by the crossbridge.* Another possibility involves an enhancement of the contractile machinery activity by either *increasing the quantity of Ca^{2+} released* or *stretching the less completely activated sarcomeres within each myofibril* (Hoyle, 1983).

It must be emphasized, however, that the relationships depicted in Figure 6.19 represent a simplified version of the velocity effect, particularly with respect to eccentric contractions. This can best be demonstrated by explaining how the measurements were made for Figure 6.19. Each data point on the graph represents a measurement of torque for a given value of velocity; that is, torque is the dependent variable. The measurement of torque for each data point was made from a single contraction (concentric or eccentric) over a 1.4-rad range of motion. For example, a single measurement of torque was made at a knee angle of 2.62 rad from a contraction that went from an initial angle of 1.74 rad to a final angle of 3.14 rad at a speed of 3.14 rad/s. The next measurement involved the same conditions but with a different speed, and so on. Because the force that a muscle can exert varies with its length, the torque measurement was made at the same knee angle (2.62 rad) in each contraction. In effect, the velocity-torque relationships shown in Figure 6.19 were obtained under *constant-velocity conditions.*

This distinction is emphasized because it is known that muscle force varies in a complicated way when an active muscle goes from an isometric to an eccentric contraction while velocity increases (Rack & Westbury, 1974). An example of this interaction is shown in Figure 6.20. In this example, the length of the extensor digitorum longus muscle was controlled while the nerve to the muscle was stimulated and the muscle force was measured for concentric, isometric, and eccentric contractions (McCully & Faulkner, 1985). For each contraction, the muscle was stimulated at 150 Hz for 300 ms. After 100 ms of stimulation, muscle length was decreased, held constant, or increased. For the isometric contraction, length did not change while force rose to a plateau during the stimulation. When length did change, however, the change in force did not closely parallel the change in length, as would occur with a linear spring. During the constant-velocity (constant slope of the length curve) shortening contraction, force declined, at first rapidly (slope of the force curve) and then more slowly, than for the isometric value. In contrast, during the constant-velocity lengthening contraction, the force increased, first rapidly and then more slowly. For a constant-velocity contraction, the slope of the force record indicates the stiffness of the muscle. During the lengthening (eccentric) contraction, stiffness is initially high and then becomes less so; this high initial stiffness is referred to as the **short-range stiffness**.

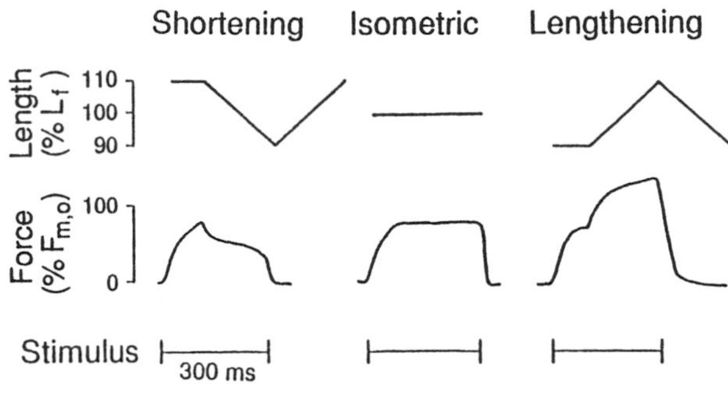

Figure 6.20 Change in muscle force during controlled changes in muscle length. Length is indicated as a percentage of muscle fiber length ($\%L_f$) at the resting length of muscle. Force is expressed as a percentage of the maximum isometric force ($\%F_{m,o}$). Stimulation was applied during the time indicated by the horizontal line at the bottom of each column.

Note. From "Injury to Skeletal Muscle Fibers of Mice Following Lengthening Contractions" by K.K. McCully and J.A. Faulkner, 1985, *Journal of Applied Physiology*, **59**, p. 120. Copyright 1985 by The American Physiological Society. Adapted by permission.

EXAMPLE 1

Despite these restrictions, the relationships among the maximum torque for the three contractions (concentric, isometric, and eccentric) can often be exploited in the prescription of exercise. For example, suppose chin-ups are part of an exercise program for a class. In many groups (e.g., physical education classes, adult fitness groups), several individuals are unable to perform a conventional chin-up, which involves elbow flexion to raise the chin to the bar followed by elbow extension to lower the body to a straight-arm hanging position. To avoid embarrassing participants, the instructor could have those unable to perform the complete chin-up do just the latter part—that is, begin with the chin touching the bar and slowly lower the body. The complete chin-up comprises two parts: concentric elbow flexor activity to raise the body and eccentric elbow flexor activity to lower the body. Because muscles can exert a greater torque eccentrically, most people can perform the lowering part of the exercise. However, the greatest torque that a muscle can generate eccentrically occurs at lower constant velocities, so the eccentric contractions need to be done slowly—this means a controlled lowering of the body and not just a collapse. Fortunately, the stress that a muscle experiences when it is exercised seems to induce similar adaptations whether the muscle is active eccentrically or concentrically (Komi & Buskirk, 1972). Thus any strength an individual gains by performing the eccentric part of the chin-up is also accessible in concentric contractions. Therefore, the individual who is unable to do a complete chin-up and begins a period of conditioning with the lowering chin-up will eventually strengthen the appropriate muscles and be able to do the complete chin-up.

EXAMPLE 2

Consider two individuals as they perform chin-ups; one person can do 20 chin-ups and the other can only do 2. Suppose they both did as many chin-ups as they could and as fast as they could. Intuition tells us that the person who can do 20 chin-ups would do the first chin-up much faster than the other person. Why is this? An explanation can be given on the basis of the velocity-torque relationship. To do this, let us make some simplifying assumptions: Both individuals weigh the same, the prime movers are the elbow flexors, and just the concentric (raising) phase is considered. Because of differences in strength, as indicated by the isometric torque (initial torque value) in Figure 6.21, the 20-chin-up person can generate the torque necessary to raise the body (horizontal line in Figure 6.21) at a higher velocity. In this example, the 20-chin-up person can perform the raising part of the chin-up at any speed up to about 4 rad/s, whereas the 2-chin-up person can only manage a maximum speed of about 2 rad/s. Thus, the first chin-up, before the effects of fatigue impair performance, can be done faster by the stronger individual.

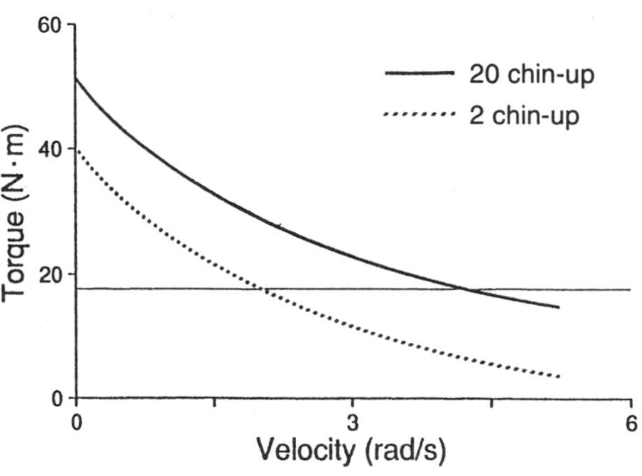

Figure 6.21 Velocity-torque relationships for the elbow flexor muscles of a 20-chin-up and a 2-chin-up person.

EXAMPLE 3

Although the measurement of muscle force in human subjects is a difficult procedure, Komi and colleagues have undertaken a series of extraordinary experiments in which they invasively attach a force transducer to the Achilles tendon of a volunteer and monitor the Achilles tendon force during various activities (Gregor, Komi, Browning, & Järvinen, 1991; Komi, 1990, 1992). In addition to measuring force, the researchers film the subjects while they walk, run, hop, jump, and bicycle so that muscle length and hence velocity can be estimated. The results of one such experiment for an individual running at 5.8 m/s are shown in Figure 6.22. The data represent the force-velocity time course during the stance phase; negative velocity indicates lengthening of the muscle (including tendon). When the foot contacted the ground, the gastrocnemius muscle initially lengthened, as indicated by the negative velocity, and force increased to a peak value of about 9,500 N. As the muscle shortened, force declined back to zero. This sequence of activity, which involves lengthening followed by shortening of active muscle, is referred to as an eccentric-concentric sequence or a stretch-shorten cycle.

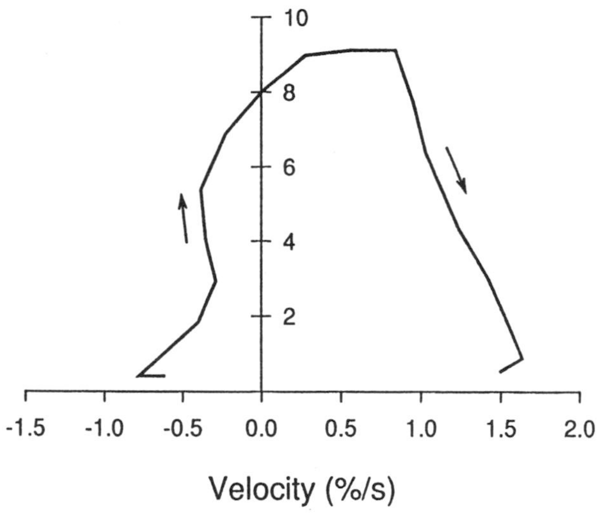

Figure 6.22 Achilles tendon force and gastrocnemius velocity during the stance phase of running at 5.8 m/s. Velocity is expressed as a percentage of the resting length per second (%/s). Footstrike occurred at the left endpoint, and takeoff occurred at the right endpoint of the graph.

Note. From "Stretch-Shortening Cycle" by P.V. Komi. In *Strength and Power in Sport* (p. 172) by P.V. Komi (Ed.), 1992, Oxford: Blackwell Scientific Publications, Ltd. Copyright 1992 by International Olympic Committee. Adapted by permission.

Muscle Power. In this chapter, we have described the mechanical characteristics of muscle in terms of the length-force and force-velocity relationships. The length-force relationship represents the static capability of muscle and indicates the force that muscle can exert at different muscle lengths. According to the crossbridge theory, muscle force varies in proportion to the amount of overlap between the thick and thin myofilaments. In contrast, the **force-velocity relationship** represents the dynamic capability of muscle that includes the effect of the rate of change in the amount of overlap on muscle force. Another way to characterize the dynamic capability of muscle is in terms of its ability to produce power; this is possible because *power is equal to the product of force and velocity or the rate of performing work* (Josephson, 1993). When a muscle shortens and the force it exerts causes an object to move, the muscle does *positive work*; that is, work is done by the muscle on the object. Conversely, a muscle does *negative work*, or has work done on it, when it lengthens while exerting a force on the object that is moved. A muscle produces power when it does positive work (concentric contraction) and absorbs power when it does negative work (eccentric contraction).

Consider an individual performing a squat exercise with a barbell (Figure 6.23). This exercise relies primarily on the knee extensor muscles. In going from the standing position to a squat, the knee extensors lengthen as they assist in lowering the body. In this instance the knee extensors are active eccentrically and are described as doing negative work; that is, work is done on the muscles by the weight of the system (barbell + lifter). When work is done on a muscle (negative work), the muscle *absorbs* power. When the individual rises from the squat position, the knee extensors perform positive work and hence *produce* power. In general, power production is limited by the rate at which energy is supplied for the muscle contraction (i.e., ATP production) the rate at which the myofilaments can convert chemical energy into mechanical work (Fitts, McDonald, & Schluter, 1991; Weis-Fogh & Alexander, 1977). Power production is often thought to limit human performance, especially if the event is of short duration (Enoka, 1988a; Ericson, 1988; Wilkie, 1960).

Power is determined as the product of the force and the velocity of the muscle contraction. Let us examine the relationship graphically using data from an isolated-muscle preparation. Figure 6.24 illustrates how the force that a muscle can exert declines as the speed of the shortening contraction increases. The force-velocity curve (solid line) contacts the axes at the two ends of the graph. At one end, velocity has a value of 0, and so the force value represents isometric force ($F_{m,o}$). At the other end, velocity has its highest value (known as v_{max}), and the force is equal to 0. At both of these locations, the product of force and velocity, and hence power, is equal to 0. Because both velocity and force are nonzero between these endpoints, power varies with a bell-shaped profile depending on the magnitude of the product.

The variation in power as a function of force and velocity can be examined rigorously with the Hill equation for the force-velocity relationship of muscle. Hill (1938) found that the decline in force as a function of

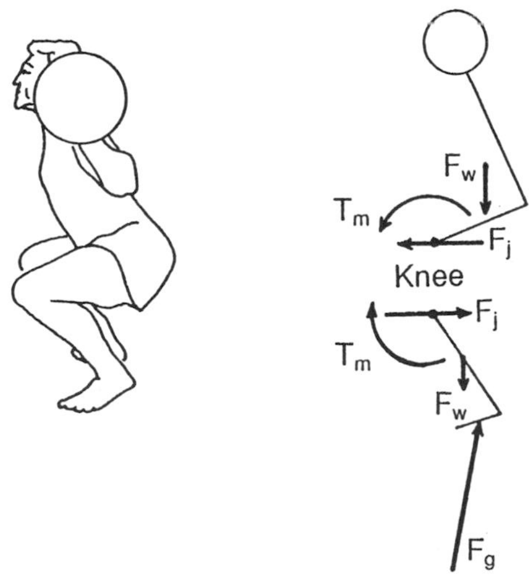

Figure 6.23 The squat exercise. The free body diagram isolates the net activity about the knee joint. The knee extensors ($\mathbf{T}_m$) are largely responsible for the execution of the movement.

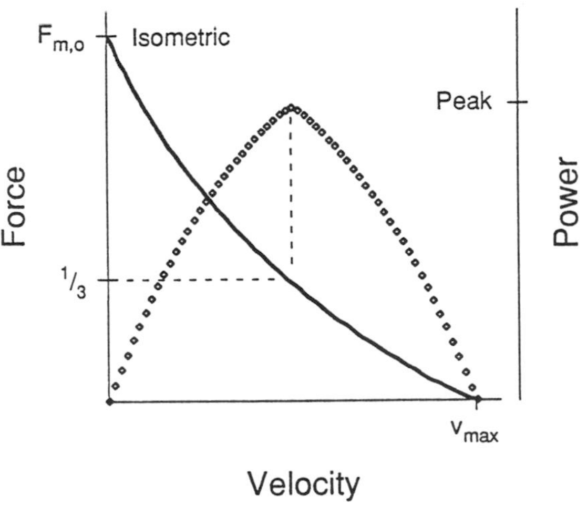

Figure 6.24 The force-velocity relationship (solid line) based on data from an isolated-muscle preparation. The power curve (right-hand vertical axis) is derived from the product of force and velocity.

shortening velocity could be represented by an equation for a rectangular hyperbola that had the following form:

$$v(F_m + a) = b\,(F_{m,o} - F_m) \quad (6.1)$$

$$F_m v + av = bF_{m,o} - bF_m$$

where v = velocity of shortening
v_o = maximum velocity of shortening at $F_{m,o}$
$F_{m,o}$ = maximal isometric force
F_m = force
a = coefficient of shortening heat ($0.15F_{m,o} - 0.25F_{m,o}$)
b = constant ($a \dfrac{v_o}{F_{m,o}}$)

In essence, Equation 6.1 expresses the dynamic capability of muscle as a work-energy relationship. The left-hand side of this equation ($F_m v + av$) corresponds to the rate of change in energy, and the right-hand side ($bF_{m,o} - bF_m$) indicates the rate of doing work. Because muscle has some viscosity (fluid friction), the rate of energy utilization is not linearly related to the rate of work production; that is, the power curve in Figure 6.24 is not a linear relationship. The term $F_m v$ represents the rate at which the contractile proteins do work (i.e., produce power) on the load while the term av represents a damping element (due to viscosity) that makes rapid movements wasteful. The interaction of these two terms ($F_m v$ and av) results in the increase and then decline in power production as the velocity of contraction increases—that is, the power curve shown in Figure 6.24. Hill's equation (Equation 6.1) can be rearranged to determine power ($F_m v$) explicitly:

$$F_m v = \frac{v(bF_{m,o} - av)}{(v + b)} \quad (6.2)$$

On the basis of Equation 6.2 it is possible to demonstrate that power ($F_m v$) is maximum when F_m and v are one third and one quarter of their respective maximum values.

Because the force- and power-velocity relationships characterize the dynamic capability of muscle, they have considerable significance in the performance of movement (Fitts et al., 1991). Due to this effect, it is possible to alter muscle dynamics (force- and power-velocity relationships) and hence to influence performance capabilities. The following examples describe the estimation and measurement of these parameters, manipulation of the force-velocity relationship in cycling, and the effect of a warm-up on muscle dynamics.

EXAMPLE 1

The ability of muscle to generate power depends on its force capacity and shortening velocity. Because both cross-sectional area (an index of muscle force capacity) and contractile speed (as indicated by fast- and slow-twitch muscle fibers) vary between muscles, the ability to generate power differs across muscles. An example of this variation is shown in Figure 6.25 for the human soleus and medial gastrocnemius muscles. Although soleus has a much greater $F_{m,o}$, whereas medial gastrocnemius has a much greater v_{max}, the peak power produced by the two synergist muscles is similar but complementary in that each occurs at a different region of the power-velocity domain (Wickiewicz, Roy, Powell, Perrine, & Edgerton, 1984).

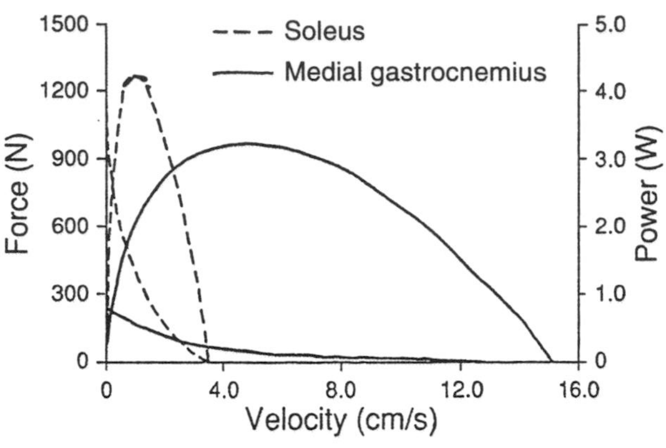

Figure 6.25 Predicted force-velocity and power-velocity relationships for the soleus and medial gastrocnemius muscles in humans. The force-velocity relationship has a peak force at velocity = 0, and force then declines as velocity increases. In contrast, the power-velocity relationship has a peak power at an intermediate velocity (1.0 cm/s for soleus and 5.0 cm/s for medial gastrocnemius).

Note. From "Morphological Basis of Skeletal Muscle Power Output" by V.R. Edgerton, R.R. Roy, R.J. Gregor, and S. Rugg. In *Human Muscle Power* (p. 55) by N.L. Jones, N. McCartney, and A.J. McComas (Eds.), 1986, Champaign, IL: Human Kinetics. Copyright 1986 by Human Kinetics Publishers, Inc. Adapted by permission.

EXAMPLE 2

One way to measure the force-velocity relationship for a muscle is to disconnect one end from bone and attach it to a load and then activate the muscle by supramaximally stimulating its nerve and measuring the force and velocity during the contraction. This procedure is referred to as **isotonic** loading. Different values for force and velocity are obtained when the load is varied (e.g., 3% to 75% $F_{m,o}$) between contractions; these measurements form the data set for the force-velocity curve (Figure 6.24). Subsequently, the power-velocity relationship can be obtained by determining power as the product of the force and velocity data and plotting it with the velocity values. This isotonic loading procedure enables us to determine the force-velocity relationship for concentric contractions and the power-velocity relationship for power production (Figure 6.24).

Gregor and colleagues (Figure 6.26) compared the force- and power-velocity relationships obtained in this way for soleus to values measured when a cat ran at about 2.2 m/s (cats can gallop at speeds up to 8 m/s) (Goslow, Reinking, & Stuart, 1973). This comparison of muscle dynamics required measures of both force and velocity for soleus while the cat was locomoting. As was done for the data shown in Figure 6.22, force was measured with a transducer that was placed on the soleus tendon, and muscle length was measured from a film record of the movement. The force- and power-velocity results for two cats are shown in Figure 6.26. Two features of the force-velocity measurement deserve emphasis: (a) the similarity of the curves for human (Figure 6.22) and cat locomotion (Figure 6.26a), which include eccentric (negative velocity) and concentric phases with the peak force occurring at about the transition from eccentric to concentric, and (b) the occurrence of greater forces during running than during isotonic loading, which can be explained by the use of an eccentric-concentric contraction during locomotion and only a concentric contraction during isotonic loading. Similarly, the power-velocity curves (Figure 6.26b) contain two phases, power absorption (negative velocity) and power production, and greater power was produced by soleus during running at 2.2 m/s than during isotonic loading.

EXAMPLE 3

In sprint *cycling*, the winner is the individual who generates the most power while encountering a minimum air resistance. Cyclists generate power by applying a force on the pedals that is transmitted to the ground. Because of friction, this force causes the wheel to rotate relative to the ground. As with isolated muscle, the legs of the cyclist exhibit a force-velocity relationship whereby the maximum force applied to the pedals decreases as crank velocity increases (Figure 6.27a). At a given crank velocity, a cyclist may apply a wide range of forces on the pedals, but the maximum is indicated by the line in Figure 6.27a; the line represents the upper limit. With these measured maximum force and velocity values, it is possible to calculate the power-velocity relationship

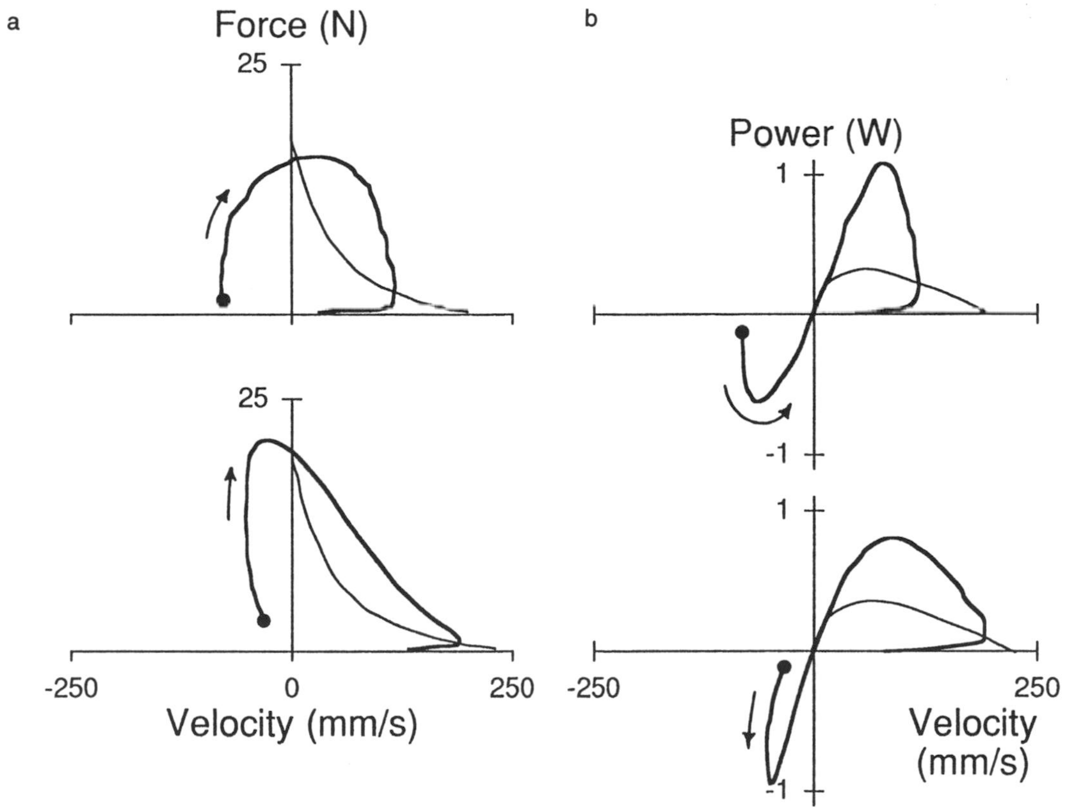

Figure 6.26 Force- and power-velocity measurements for the cat soleus muscle. The thin lines indicate values obtained with supramaximal stimulation of soleus and isotonic loading. The thick lines represent data acquired with a running speed of 2.2 m/s; footstrike occurs at the dot. The graphs are for two different cats (a and b).

Note. Adapted from "Mechanical Output of the Cat Soleus During Treadmill Locomotion: In Vivo vs. In Situ Characteristics" by R.J. Gregor, R.R. Roy, W.C. Whiting, R.G. Lovely, J.A. Hodgson, and V.R. Edgerton, 1988, *Journal of Biomechanics*, **21**, pp. 728-729, Copyright 1988, with kind permission from Pergamon Press, Headington Hill Hall, Oxford OX3 DBW, UK.

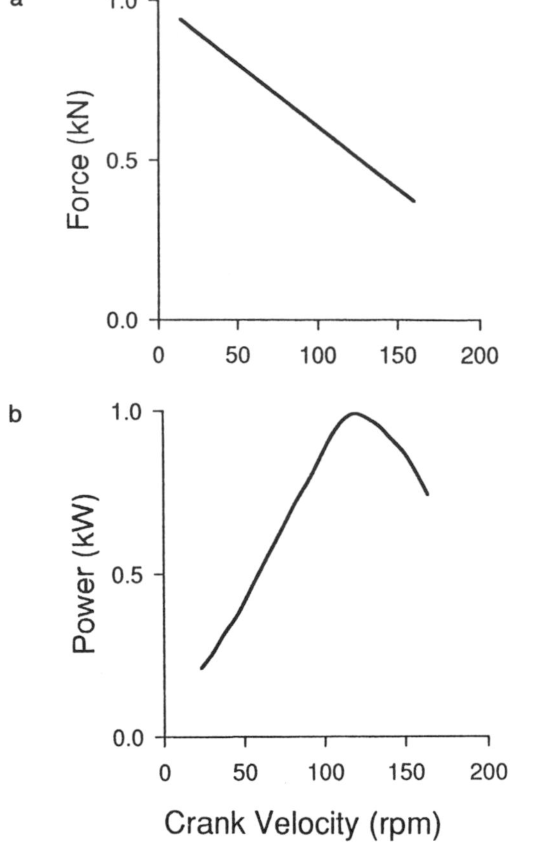

Figure 6.27 Effect of crank velocity (rpm, or revolutions per minute) on the (a) peak force and (b) power applied to the pedals during cycling. Peak force declines as crank velocity increases, but peak power occurs at an intermediate crank velocity.

Note. From "Measurement of Maximal Short-Term (Anaerobic) Power Output During Cycling" by A.J. Sargeant and A. Boreham. In *Women and Sport* (p. 122) by J. Borms, M. Hebbelinck, and A. Venerando (Eds.), 1981, Basel: S. Karger AG. Copyright 1981 by S. Karger AG, Basel. Adapted by permission.

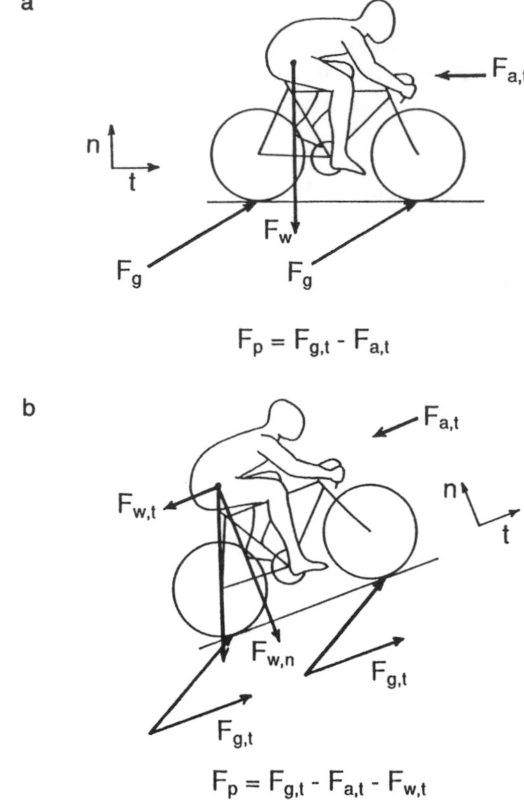

Figure 6.28 Free body diagram of a bicycle rider system. (a) The system traveling along level ground experiences a system weight ($\mathbf{F}_w$), air resistance ($\mathbf{F}_a$), and a ground reaction force ($\mathbf{F}_g$). The propulsive force ($\mathbf{F}_p$) acts in the direction the cyclist is progressing. (b) When the system travels up a hill, $\mathbf{F}_{w,t}$ and $\mathbf{F}_{a,t}$ oppose the forward motion of the system while $\mathbf{F}_{g,t}$ assists it.

that represents the maximum power that can be produced at each crank velocity (Figure 6.27b). As with isolated muscle, maximum power occurs at an intermediate velocity. For elite cyclists, optimum velocity appears to be about 110 rpm when the pedal force is about 500 N.

When a cyclist encounters a resistance, such as a gust of wind or a hill, the propulsive force experienced by the cyclist decreases. As shown in the free body diagram of Figure 6.28, the propulsive force ($\mathbf{F}_p$) is equal to the difference between the tangential components of the ground reaction force ($\mathbf{F}_{g,t}$), air resistance ($\mathbf{F}_{a,t}$), and weight vectors ($\mathbf{F}_{w,t}$). As a result of the increased resistance and decreased propulsive force, crank velocity decreases, and although pedal force may increase, the power generated by the cyclist decreases. To counteract this effect, the cyclist intuitively changes gears, which increases crank velocity, power production, and speed. Because of this effect, we can conclude that the reason for gears on a bicycle is to allow the cyclist to manipulate the force-velocity relationship and hence power production.

Muscle Architecture

Once a muscle has been activated by the nervous system, the force that it exerts depends not only on the mechanical properties of its contractile elements but also on the **muscle architecture**, *or physical arrangement of these contractile elements*. Similarly, neuromuscular function seems to be profoundly affected by the size and shape of muscle fibers and whole muscles (Edgerton & Roy, 1991). For example, both the maximum force that a muscle can exert and the maximum velocity of shortening are influenced by muscle architecture. To describe the effects of architecture on muscle force, we will first introduce a model of the sarcomere (three-component model) and then explain the effects of different sarcomere arrangements at the level of the muscle fiber, the whole muscle, and the single joint system.

Three-Element Model. The basic functional unit of muscle is the sarcomere; it contains all the elements necessary to exert force. The mechanical behavior of a sarcomere can, to a first approximation (see Winters & Woo, 1990, for a more detailed account), be represented by a **three-element model of muscle** (Figure 6.29). In this model, the three elements are a *parallel elastic element* (PE), a *series elastic element* (SE), and a *contractile element* (CE). These three elements are intended not to represent specific anatomical structures but *to account for the mechanical characteristics of the sarcomere.* The SE and the PE represent different types of elasticity and are indicated as springs. For example, the length-force relationship of muscle outlined in Figure 6.12 indicates that certain passive structures within muscle (e.g., connective tissue sheaths and cytoskeletal elements) will exert a passive force when unstimulated muscle is stretched; the PE represents this effect. As depicted in Figure 6.12, the contribution of the passive elements (PE) to total muscle force increases with muscle length and is greatest at the longest muscle lengths.

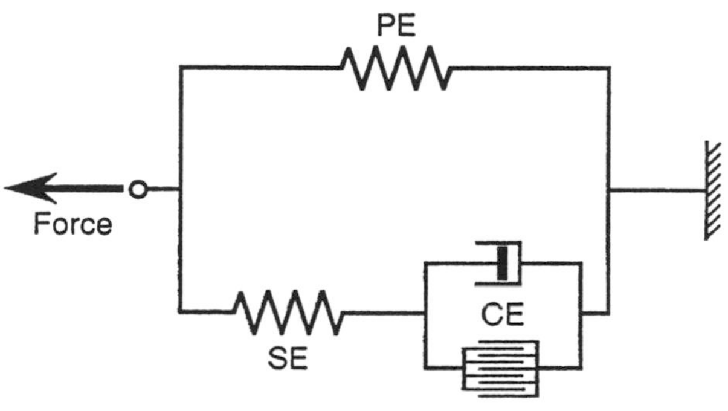

Figure 6.29 A three-element mechanical model of the sarcomere.

In addition to the PE elasticity, which lies in parallel with the force-exerting element, some of the structures that comprise the myofilaments and tendon possess an elasticity that is aligned in series with the force-generating units; this characteristic is identified as the SE. Within the sarcomere, the thick and thin filaments, crossbridges, and Z bands contribute to the SE, but the filaments provide the greatest contribution (Blangé, Karemaker, & Kramer, 1972; Suzuki & Sugi, 1983). The SE, which is apparent only when the muscle is contracting, is much less extensible than the PE and has a length of about 4% to 5% l_o (where l_o refers to the resting length of muscle).

The CE represents the force-velocity capability of the sarcomere. It contains two components, one that indicates the *force-generating capability* of the myofilaments and another that accounts for the *velocity-related effect of viscosity* on the force that is exerted by the sarcomere. The force-generating capability of the myofilaments is indicated by the interdigitated filaments, which represent the effect of myofilament overlap on force. The effect of viscosity is indicated by a dashpot—which is an element that can be described by the equation $F = Bv$, where F refers to force, B is a constant, and v corresponds to muscle velocity. The dashpot accounts for the av term in Equation 6.1 and indicates that more force (and hence energy) is needed to cause the myofilaments to slide past one another as the velocity of shortening increases; consequently, less force is exerted externally by the sarcomere. The dashpot is the component of the model that represents the decreased efficiency of fast contractions.

From these descriptions of the elements of the model, it should be apparent that the mechanical behavior of the active sarcomere is **viscoelastic**; that is, it has both *viscous* and *elastic* properties. As described by Winters (1990), viscoelasticity gives rise to (a) hysteresis in the force-length relationship during cyclic loading, whereby the force exerted at a particular sarcomere length during lengthening is greater than the force exerted at the same length during shortening (this means that the force-length relationship is not a single line but a loop); (b) force relaxation, in which there is a gradual decline in force when the sarcomere is held at a constant length; and (c) length creep, or the need to increase length in order to maintain a constant force (Figure 6.30). Another consequence of the in-series arrangement of the CE and SE is that in isometric contractions, even though whole-muscle length essentially remains constant, CE may shorten significantly when SE is stretched (Fellows & Rack, 1987; Griffiths, 1991).

The data illustrated in Figure 5.2 indicate that the peak force a muscle exerts in a twitch response is much less than that exerted in a tetanus. Why does the peak force differ for these two conditions? The difference in force is due to two factors, the quantity of calcium released by the sarcoplasmic reticulum and the mechanical behavior of muscle. A single action potential (twitch) does not release enough calcium to uncover sufficient binding sites for maximum force. But with a series of action potentials (tetanus), there is a progressive intracellular accumulation of calcium that eventually maximizes disinhibition of troponin and tropomyosin and permits the attachment of a sufficient number of crossbridges to exert the maximum force (Allen, Lee, & Westerblad, 1989). The effect of calcium is accompanied by a mechanical effect that can be explained by the three-element model. The state of activation of the contractile machinery reaches a maximum intensity within 4 ms after the action potential and is maintained at maximum for about 30 ms before it begins to decline (Hoyle, 1983). In response to this activation, the CE

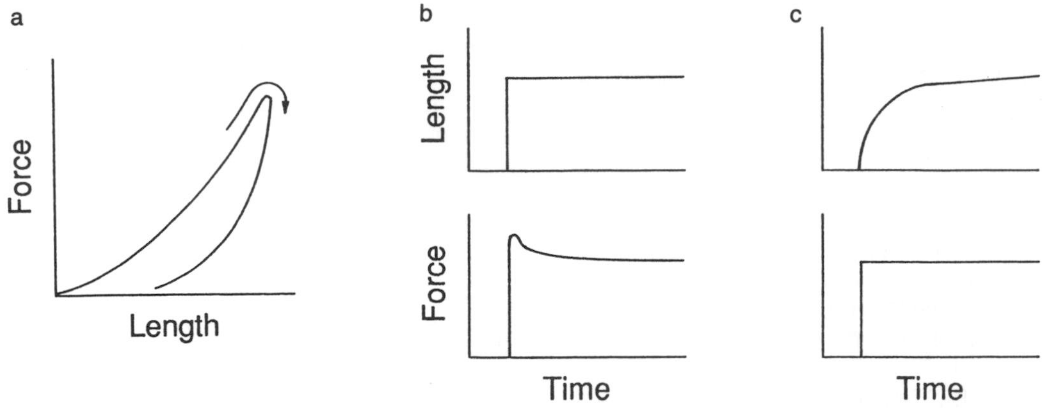

Figure 6.30 Viscoelastic properties: (a) Hysteresis in the force-length relationship; (b) force relaxation—force declines when length remains constant; (c) length creep—length increases when force is held constant.
Note. From "Hill-Based Muscle Models: A Systems Engineering Perspective" by J.M. Winters. In *Multiple Muscle Systems: Biomechanics and Organization* (p. 72) by J.M. Winters and S.L-Y. Woo (Eds.), 1990, New York: Springer-Verlag. Copyright 1990 by Springer-Verlag, Inc. Adapted by permission.

generates force, but this force is only registered externally by being transmitted through the SE. Because the SE may be represented as a spring, the slack must be first stretched out of the SE before it will transmit any force exerted by the CE. The characteristics of the SE are such that the state of activation of the CE has begun to decline in a twitch (muscle response to a single action potential) before the SE has been fully stretched. More simplistically, the CE generates a certain quantity of force in response to a single action potential. Some of this force is used to stretch the SE, and the remainder can be measured externally. In contrast, during a tetanus the SE becomes fully stretched after the first 5 to 10 action potentials, so all subsequent force exerted by the CE can be registered externally (Hoyle, 1983). In combination, the progressive intracellular accumulation of calcium and stretch of the SE results in a tetanic force that is greater than the twitch force. Tetanic force is about three to seven times greater than twitch force in motor units of human hand muscles (Thomas et al., 1991; Young & Mayer, 1981).

Muscle Fiber Level. The force-generating units of muscle, each of which can be represented by the three-element model, can be arranged in various ways at the levels of the muscle fiber, the whole muscle, and the single joint system. The two major architectural influences at the level of the muscle fiber are the arrangement of the force-generating units **in series** (i.e., end to end) and **in parallel** (i.e., side by side). Figure 6.31 shows the arrangement of three force-generating units in series (Figure 6.31a) and in parallel (Figure 6.31b).

A myofibril was previously described as a series of sarcomeres. Variations in myofibril length are generally due to differing numbers of sarcomeres in series; the length of each sarcomere within a myofibril does not change much. For example, sarcomere length in human wrist muscles averages about 2.2 μm and differs by only 2.1% within a given muscle (Lieber et al., 1990). When the leg is in an anatomical position, the average sarcomere length for various muscles has been estimated to range from 1.97 to 3.01 μm (Cutts, 1988). If each sarcomere or force-generating unit is represented by a three-element model, the resting myofibril length can be changed by altering the number of units in the chain (Figure 6.31a). When an action potential propagates along a muscle fiber, all of the sarcomeres within a myofibril experience about the same degree of excitation, and therefore each sarcomere undergoes about the same change in length (Δl). As a consequence, *the greater the number of sarcomeres (n) in series, the greater the change in length of the myofibril* (Δl) in response to the action potential; thus, $\Delta L = n(\Delta l)$. For example, a 100-sarcomere myofibril will experience a ΔL of 50 μm if each sarcomere produces a 0.5-μm change in length when activated by an action potential. In contrast, under the same conditions a 1,000-sarcomere myofibril will undergo a 500-μm length change. Furthermore, because a myofibril can change its length by about one third, the greater its length, the greater the absolute change in length.

In addition to this effect on displacement, another feature of the in-series arrangement is the effect on *shortening velocity*. The rate of change in length (velocity) of the myofibril is determined as the product of the number of sarcomeres and the average velocity of each sarcomere. The greater the number of sarcomeres in series, the greater the change in myofibril length and the greater the rate of change in length in response to a given stimulus; thus, $\Delta V = n(\Delta v)$.

In contrast to the in-series arrangement, if each myofibril comprised a single sarcomere, and three such myofibrils were arranged in parallel (Figure 6.31b), the change in myofibril length would be equivalent to that experi-

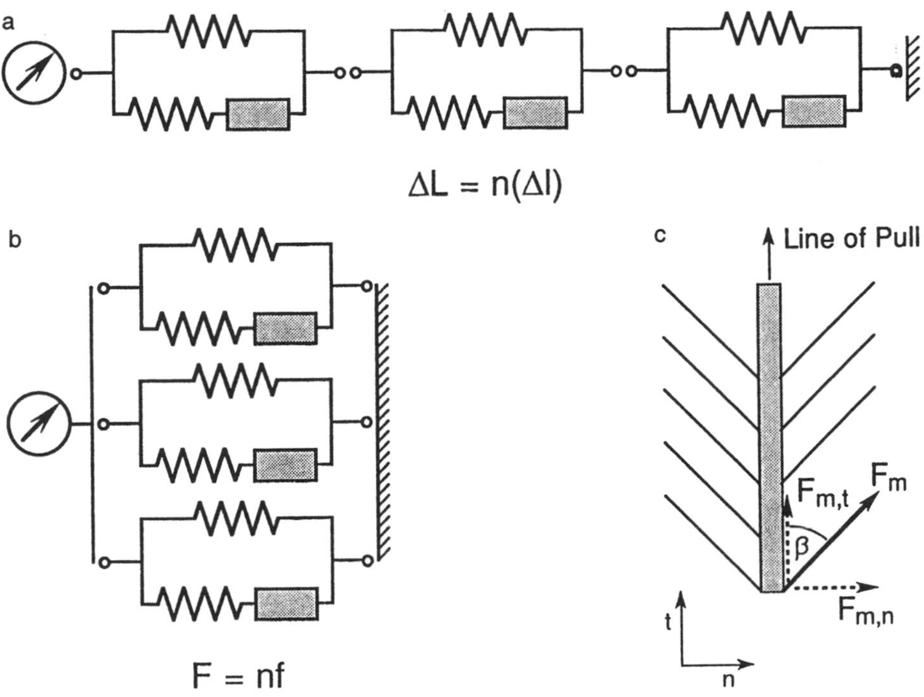

Figure 6.31 Architectural features of a muscle fiber: (a) an in-series arrangement of sarcomeres, with each sarcomere represented by a three-element model; (b) an in-parallel arrangement of sarcomeres; (c) the effect of pennation on the contribution of each muscle fiber to whole-muscle force and velocity of shortening. β = angle of pennation; F_m = muscle force or velocity vector; $F_{m,n}$ = normal component; $F_{m,t}$ = tangential component.

enced by the single sarcomere. In this configuration, however, the forces exerted by each sarcomere combine so that the muscle fiber force (F) is equal to the product of the number of myofibrils (n) and the average force exerted by a single myofibril (f); thus, $F = nf$. Consequently, the in-parallel arrangement of force-generating units maximizes force such that *muscle fiber force is proportional to the number of myofibrils in parallel.* This relationship was encountered earlier when the cross-sectional area of muscle was used as an index of the maximum force that muscle can exert (Roy & Edgerton, 1992).

A muscle fiber consists of myofibrils arranged in parallel, and a myofibril represents an in-series arrangement of sarcomeres; therefore, muscle fibers actually comprise an in-series and an in-parallel collection of force-generating units. Consequently, *a long, small-diameter muscle fiber represents a dominant in-series effect, and, conversely, a short, large-diameter fiber exhibits mainly in-parallel characteristics.* This distinction is apparent in measurements made on 25 leg muscles in five cadavers (Table 6.1) (see Roy & Edgerton, 1992, and Lieber, 1992, for similar data on arm muscles and Yamaguchi, Sawa, Moran, Fessler, & Winters, 1990, for data on lower extremity, trunk, upper extremity, hand, and head-neck musculature). Although these measurements were made on preserved cadavers and probably differ from values for living tissue, the data do allow comparison between muscles. The magnitude of the difference between muscle fibers in terms of in-series and in-parallel architecture can be substantial. If we assume a sarcomere length of 2.2 μm, then the average number of sarcomeres that are arranged in series in a muscle fiber ranges from 11,364 for soleus (25 mm) to 203,636 for sartorius (448 mm). Similarly, the cross-sectional area of muscle fibers can vary, as has been reported for the different muscle fiber types (Tables 5.1 and 5.2), and is influenced by such factors as age, gender, health, and training history. For example, the average cross-sectional area of Type II fibers in the deltoid of weight lifters was found to be 8,910 μm^2 compared with 5,060 μm^2 for Type I fibers. Also, in vastus lateralis the average cross-sectional area of Type II fibers was reported to be 4,814 μm^2 for males and 3,409 μm^2 for females. The variation in cross-sectional area (Table 6.1) is due to differences in the number of myofibrils that are in parallel.

Muscles are designed to capitalize on the distinction between these in-series and in-parallel effects. For example, muscles that support an upright posture (e.g., knee extensors, ankle plantarflexors), the **antigravity muscles**, are generally considered to be twice as strong as their antagonists. This implies that the cross-sectional area of the antigravity muscles should be twice as large. Indeed, the cross-sectional area of quadriceps femoris is about double that of the hamstrings (87 vs. 38 cm^2),

Table 6.1 Architectural Features of Human Leg Muscles

Muscle	Fiber length (mm)	Length ratio	Pennation angle (rad)	Functional CSA (cm^2)
Hip muscles				
Adductor brevis	103	0.70	0.00	4.7
Adductor longus	103	0.50	0.09	6.8
Adductor mangus	114	0.53	0.04	18.2
Pectineus	98	0.83	0.00	2.9
Hip and knee muscles				
Biceps femoris (long head)	80	0.26	0.03	12.8
Gracilis	264	0.83	0.04	1.8
Rectus femoris	64	0.20	0.13	12.7
Sartorius	448	0.88	0.00	1.7
Semimembranosus	63	0.27	0.27	16.9
Semitendinosus	155	0.46	0.09	5.4
Knee muscles				
Biceps femoris (short head)	130	0.52	0.37	—[a]
Vastus intermedius	72	0.23	0.05	22.3
Vastus lateralis	72	0.23	0.12	30.6
Vastus medialis	73	0.23	0.10	21.1
Knee and ankle muscles				
Medial gastrocnemius	37	0.16	0.25	32.4
Lateral gastrocnemius	55	0.25	0.19	—[b]
Ankle muscles				
Tibialis anterior	75	0.26	0.12	9.9
Extensor digitorum	80	0.25	0.17	5.6
Extensor hallucis	78	0.32	0.11	1.8
Flexor digitorum	31	0.13	0.13	5.1
Flexor hallucis	37	0.17	0.21	5.3
Peroneus brevis	39	0.17	0.12	5.7
Peroneus longus	41	0.15	0.15	12.3
Tibialis posterior	26	0.10	0.24	20.8
Soleus	25	0.08	0.48	58.0

Note. The data for fiber length, ratio of fiber length to muscle length, and pennation angle are mean values from five cadavers (Friederich & Brand, 1990; Wickiewicz, Roy, Powell, & Edgerton, 1983). The measures of functional cross-sectional area (CSA) are from three cadavers (Wickiewicz et al., 1983). Length ratio = ratio of fiber length to muscle length.
[a]This value is included in the measurement for the long head of biceps femoris.
[b]This value is included in the medial gastrocnemius measurement.

and the cross-sectional area of the plantarflexors is substantially larger than that of the dorsiflexors (139 vs. 17 cm^2). However, the average muscle fiber in hamstrings is longer than an average one in quadriceps femoris (43,000 vs. 31,200 sarcomeres in series), and the average fiber length in the dorsiflexors is greater than that in the plantarflexors (29,300 vs. 15,200 sarcomeres in series). Therefore, although the hamstrings and dorsiflexors are weaker than their antagonists, they have a greater capability for change in length and rate of change in length (shortening velocity).

Whole-Muscle Level. The organization of force-generating units at the level of the whole muscle also includes in-series and in-parallel effects on the force exerted by muscle. That is, variation in the arrangement of muscle fibers will influence the force, change in length, and shortening velocity of the whole muscle. In most muscles, the ratio of fiber length to muscle length is less than 1, ranging from 0.08 in soleus to 0.85 for some leg (sartorius, pectineus, gracilis) and wrist (extensor carpi radialis longus) muscles. Most muscles have ratios in the range of 0.2 to 0.5 (Table 6.1). Deviation of the ratio from 1, where a value of 1 indicates equal fiber and muscle lengths, is due to two in-series effects: pennation and staggered fibers.

In many muscles, the force-generating units are oriented at an angle to the line of pull of the muscle tendon;

that is, they are not completely in series (collinear) with the direction of the muscle force vector but instead attach to intramuscular connective tissue sheaths at an angle (Figure 6.31c). As a result, fiber length is shorter than muscle length because fibers do not traverse from one end of a muscle to the other. The angular deviation of the fibers from the line of pull of the muscle is referred to as the angle of **pennation** and usually varies from 0 to 0.4 rad (Table 6.1). Muscles can have fibers arranged with a common angle of pennation (unipennate), with two sets of fibers at different pennation angles (bipennate), or with many sets of fibers at a variety of angles (multipennate). Muscle fibers aligned with a zero angle of pennation are usually described as having a fusiform or parallel arrangement. The morphological types of muscle, based on their general form and fascicular architecture, are shown in Figure 6.32. This kind of diagram, however, serves only as a guide to muscle architecture, because it focuses on the organization of the fascicles rather than the muscle fibers.

The advantage of pennation is that a greater number of in-parallel fibers (and thus a greater cross-sectional area) can be packed into a given volume. This principle is exemplified by the soleus muscle, which has a small fiber-to-muscle-length ratio, a large pennation angle, and an extensive cross-sectional area (Table 6.1). The number of fibers that can be contained in a given volume is an important consideration because the volume available for various organs is often a limiting factor in biological design (Otten, 1988). For example, the muscles that control the fingers are located in either the hand or the forearm; those in the hand are small (because of restricted space), whereas those in the forearm are larger and stronger. If all the muscles that control the fingers were located in the hand, this would make the hand bulky and awkward for fine manipulation.

Although pennation increases the number of in-parallel fibers that can be accommodated in a given volume, pennation also reduces the contribution of fiber force and velocity of shortening to the whole-muscle values; the magnitude of the tangential component is less than the magnitude of the resultant and varies as the cosine of the angle of pennation (cos β). Even though the magnitude of the muscle fiber force (the length of F_m) remains constant as the angle of pennation varies, the magnitude of the tangential component ($F_{m,t}$) gets smaller. When the angle of pennation is 0, $F_{m,t}$ is equal to F_m. Because the volume of a muscle remains relatively constant during a contraction, the angle of pennation changes during the contraction. Consequently, the change in whole-muscle length is due to the change in length that the fiber experiences plus the displacement due to the change in pennation angle (Gans, 1982; Huijing, 1992a; Otten, 1988). The same is true for the velocity of contraction. At optimum muscle length, the changes in fiber length and pennation angle contribute 84% and 6%, respectively, to the shortening velocity of muscle. At shorter muscle lengths, however, the contributions of fiber length and angle are 35% and 31%, respectively (Zuurbier & Huijing, 1992).

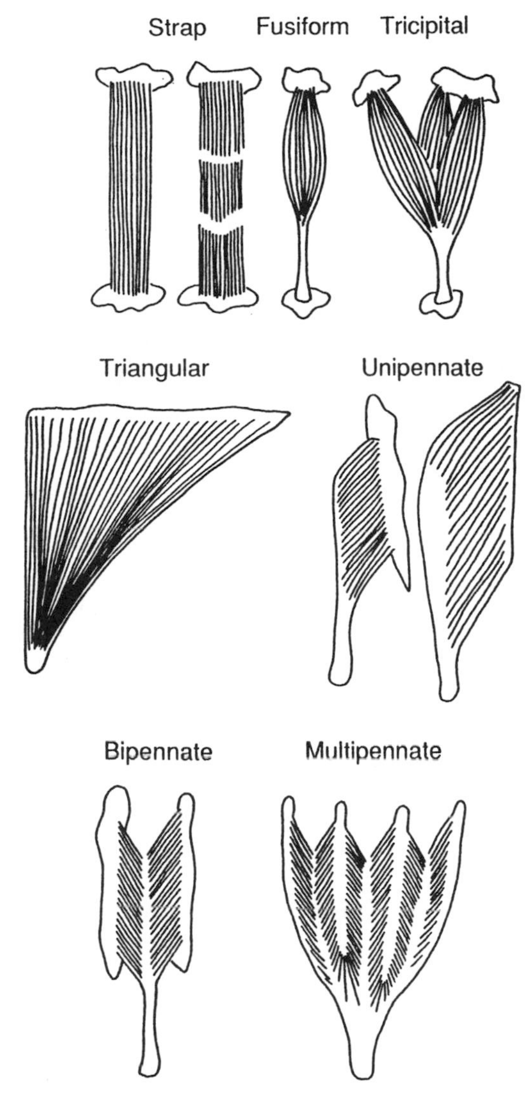

Figure 6.32 The traditional classification of muscles on the basis of general form and fascicular organization. *Note.* From *Gray's Anatomy* (36th ed.) (p. 524) by P.L. Williams and R. Warwick (Eds.), 1980, Edinburgh: Churchill Livingstone. Copyright 1980 by Longman Group Ltd. Adapted by permission.

The other in-series effect that can account for a fiber-to-muscle-length ratio of less than 1 is the serial attachment of short **staggered muscle fibers**. This effect can be demonstrated by plotting the longitudinal distribution of the muscle fibers belonging to a single motor unit (Figure 6.33). Visualization of these fibers is based on a technique that involves stimulating a single motor unit for a prolonged period so that its muscle fibers are depleted of glycogen and then performing serial sections of the excised muscle and reconstructing the architecture of the motor unit. With this procedure, it has been shown for the cat tibialis anterior muscle that the muscle fibers of a single motor unit can be distributed along the entire

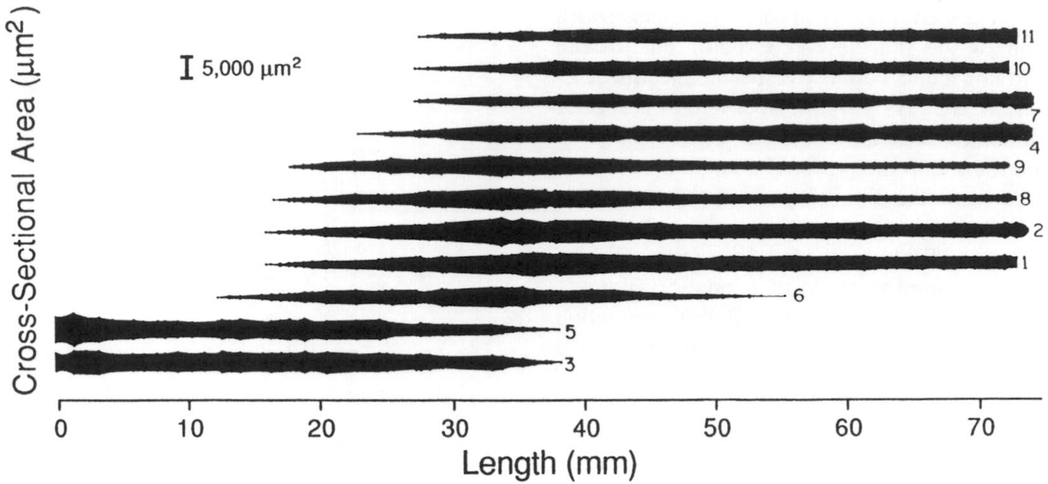

Figure 6.33 The cross-sectional area and relative longitudinal position of 11 fibers from a fast-twitch motor unit in the cat tibialis anterior muscle. On the length scale (x-axis), zero is the proximal end of the muscle.
Note. From "Physiological and Developmental Implications of Motor Unit Anatomy" by M. Ounjian, R.R. Roy, E. Eldred, A. Garfinkel, J.R. Payne, A. Armstrong, A.W. Toga, and V.R. Edgerton, 1991, *Journal of Neurobiology*, **22**, p. 555. Copyright 1991 by John Wiley & Sons, Inc. Adapted by permission.

length of the muscle and that at least one end terminates within the fascicle (Ounjian et al., 1991). The end that terminates within the fascicle usually, but not always, has a long taper, which appears to transmit force to the endomysium (Trotter, 1990; Trotter & Purslon, 1993). This means that the force exerted by one motor unit is not transmitted from one end of the muscle to the other by its own fibers but rather by cytoskeletal structures and connective tissue sheaths. For muscles that contain staggered fibers—and there seem to be many (Roy & Edgerton, 1992; Trotter, 1990; Trotter, 1993)—it is most appropriate to think of muscle fibers as networks of force generators that are embedded in connective tissue sheaths with force transmitted through the connective tissue to the tendon. This in-series arrangement of staggered fibers adds elasticity between the contractile element (CE) and the tendon (Figure 6.34).

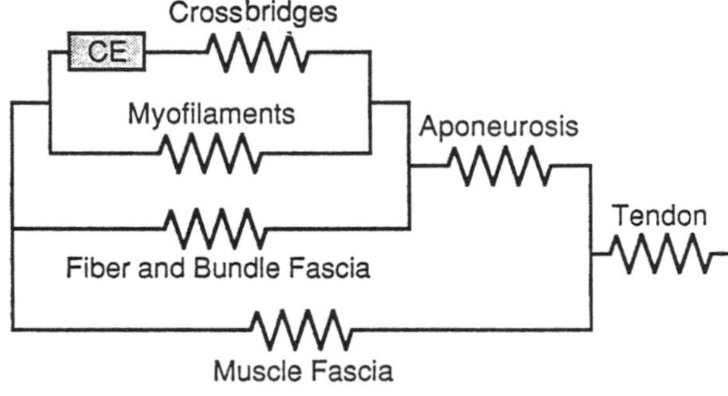

Figure 6.34 Three-element model of the sarcomere and its surrounding layers of elasticity.
Note. From "Mechanical Muscle Models" by P.A. Huijing. In *Strength and Power in Sport* (p. 145); by P.V. Komi (Ed.), 1992b, Oxford: Blackwell Scientific Publications. Copyright 1992 by International Olympic Committee. Adapted by permission.

As the data in Figure 6.33 show, the muscle fibers of a motor unit generally are not distributed throughout the entire muscle. In addition to the longitudinal effect, the territory of a motor unit in the cat tibialis anterior muscle may range from 8% to 22% of a cross-sectional area (see Figure 5.1 as an example of medial gastrocnemius); Type S units usually occupy the smallest areas. In contrast, motor unit territories range from 41% to 76% in the cat soleus muscle, which contains only Type S motor units (Bodine, Garfinkel, Roy, & Edgerton, 1988). Furthermore, motor unit territories appear to expand with age as surviving motor units reinnervate muscle fibers that are deprived of a nerve supply (Kanda & Hashizume, 1989). As a result of the in-series and in-parallel architecture of the muscle fibers belonging to a motor unit, the activation of a motor unit leads to a localized contraction within a muscle.

The regional activation of muscle also extends beyond the level of the motor unit to the level of the whole muscle. This effect, referred to as *compartmentalization*, involves the functional subdivision of a muscle into several units, or **compartments**, based on the innervation pattern of primary nerve branches (Fleckenstein et al., 1992; Weeks, 1989; Windhorst et al., 1989). A muscle is an anatomical concept and is defined by the fascial sheaths that distinguish bundles of muscle fibers. It appears, however, that some muscles are not activated as homogeneous units and that the activation can vary within different parts of the muscle; that is, the functional (physiological) definition of a muscle can differ from the anatomical definition. This variable activation within a muscle is based on the innervation pattern of primary nerve branches. For example, the biceps brachii muscle is innervated by three to six primary branches of the musculocutaneous nerve, and functionally the motor units in biceps brachii appear to comprise two distinct populations. One population, which is located in the lateral aspect of the long head, is active when a flexion torque is exerted about the elbow joint. The other population is active when the torque about the elbow joint includes both flexion and supination components (ter Haar Romeny, Denier van der Gon, & Gielen, 1982; van Zuylen, Gielen, & Denier van der Gon, 1988). Individual compartments can have different fiber type distributions and can be activated independently (English, 1984; English & Ledbetter, 1982). However, differences in innervation do not mean that individual compartments or synergists (e.g., vastus lateralis and vastus medialis oblique) are necessarily activated independently (Grabiner, Koh, & Mill, 1991).

The in-parallel effect at the whole-muscle level is the same as that at the muscle fiber level; the greater the cross-sectional area (in-parallel content), the greater the maximum force the muscle can exert. Because of pennation, however, measurement of cross-sectional area is more difficult for the whole muscle than for a single muscle fiber. Accurate measurement of cross-sectional area requires the measurement of the area of a section made perpendicular to the long axis of each fiber. Consequently, whole-muscle cross-sectional area cannot be determined by simply measuring at the thickest part of the muscle; the angle of pennation also must be taken into account. Such a measurement, referred to as the **functional** (or physiological) **cross-sectional area**, is a more accurate estimate of the maximum force that a muscle can exert (Fukunaga et al., 1992; Lieber, 1992). For the muscles listed in Table 6.1, functional cross-sectional area ranged from about 1.7 cm^2 (sartorius, gracilis, extensor hallicus) to 58.0 cm^2 (soleus).

At this point, a note of caution is needed about the use of the measure of cross-sectional area as an estimate of the force that a muscle fiber (or muscle) can exert. Until now we have included in this measure the total contents of the muscle that contribute to the cross-sectional area (e.g., ultrasound, CT, and MRI scans). Based on the earlier discussion of architectural features, however, the measurement of cross-sectional area is used as an index of the number of crossbridges (myofilaments) that are arranged in parallel. Such an index is valid if, and only if, the noncontractile protein content of the muscle is evenly distributed among the muscle fibers. It appears that this is not the case; for example, the proportion of sarcoplasmic reticulum in a muscle fiber can vary, accounting for up to 60% of the area, and mitochondria can account for up to 50% (Hoyle, 1983). Furthermore, these variations appear to be related to muscle fiber type. Consequently, the measurement of cross-sectional area, which includes all the intramuscular material, can be used only as an estimate of the force capabilities of muscle; in fact, this source of error probably contributes to the variation in the reported values for specific tension (discussed in Part I).

Single Joint System Level. In addition to the muscle fiber and whole-muscle effects, at least two other design features are known to influence the torque that a muscle can exert. These features, which involve the organization of muscles about a joint, are the points of attachment of the muscle relative to the joint and the proportion of whole-muscle length that contains contractile tissue. Figure 6.35 shows the geometry of a single-joint muscle. The attachment points determine the distance (c) from the proximal muscle attachment to the joint and the distance (q) from the distal muscle attachment to the joint and hence the angle of pull of the muscle (α) and the proportion of muscle force that contributes to rotation (L_c). Also shown in Figure 6.35 are the moment arm (d), the joint angle (E), the pulling force exerted by the muscle on the distal segment (F_m), and the normal and tangential components of the force (F_n and F_t). The major effects of varying the parameters c, q, and L_c on the torque exerted by a muscle include the following (van Mameren & Drukker, 1979):

1. Breadth—The angle-torque relationship has a sharper peak when c and q are equal. When either c or q is much greater than the other parameter, the peak is much broader; that is, the maximum torque values are attainable over a greater range of joint angles (Figure 6.36a).
2. Location—Peak torque is reached closer to full extension when c and q are equal but closer to the midrange position (e.g., elbow angle = 1.57 rad) when c and q are quite different (Figure 6.36b).

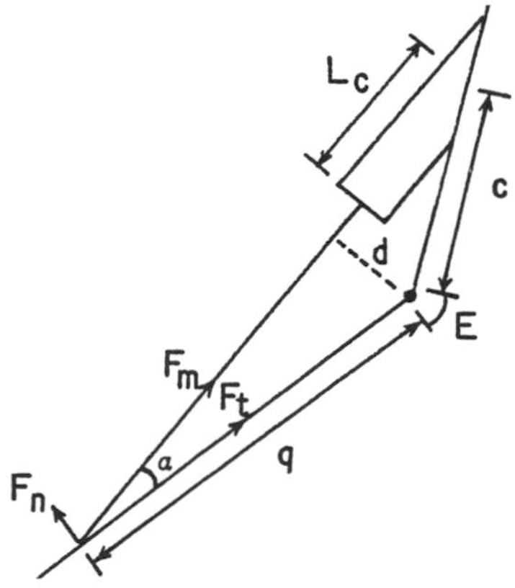

Figure 6.35 The geometry of a single-joint muscle.
Note. Adapted from "Attachment and Composition of Skeletal Muscles in Relation to Their Function" by H. van Mameren and J. Drukker, 1979, *Journal of Biomechanics*, **12**, p. 860, Copyright 1979, with kind permission from Pergamon Journals Ltd., Headington Hill Hall, Oxford OX3 DBW, UK.

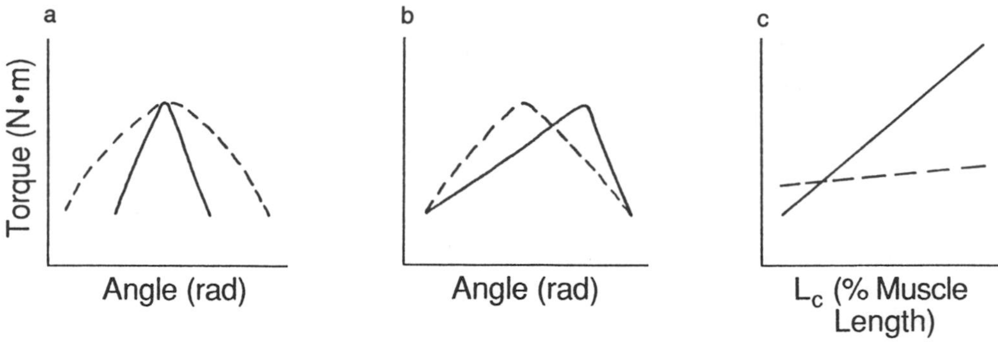

Figure 6.36 Effect of architecture on the angle-torque relationship (solid lines show relationships when c and q are equal, whereas the dashed line indicates the relationships for unequal values of c and q): (a) effect on the breadth of the peak; (b) effect on the location of the peak; (c) effect of L_c on the magnitude of the peak torque.

3. Magnitude—The maximum torque exerted by the muscle when c and q are quite different does not increase significantly when L_c is increased. When c and q are equal, the maximum torque is highly dependent on L_c. This interaction is readily apparent anatomically in that muscles with a c/q quotient near one always have a high L_c value, whereas muscles with a c/q quotient much less than 1 generally have long tendons (low L_c values).

The rotation of one body segment about another is typically controlled by a group of muscles. Often the parameters c, q, and L_c vary among the muscles in a single group; this arrangement provides the broadest range of properties for the muscle group. For example, consider the interactions of the major elbow flexor muscles—brachialis, biceps brachii, and brachioradialis. The brachialis muscle contains a substantial variation in c/q quotients and L_c values throughout the muscle; the deeper part has c/q quotients near 1 and high L_c values, whereas the superficial part has c/q values far from 1 and low L_c values. Because the different parts possess distinct features, the muscle contributes over the entire range of motion (Basmajian & Latif, 1957). In contrast, biceps brachii and brachioradialis have less variety in c/q quotients and L_c values, with the c/q quotient for biceps brachii being much greater than 1 and that for brachioradialis being much less than 1. Biceps brachii and brachioradialis, therefore, have angle-torque relationships with a sharp peak that occurs in the middle of the range of motion.

Variation in the c and q distances influences the magnitude of the moment arm of the muscle relative to the joint. Changes in the moment arm, however, interact with other features of muscle design, such as the number of sarcomeres in series, to influence muscle function. Lieber (1992) characterizes this effect by examining the ratio of muscle fiber length to moment arm length. The size of the ratio for a particular muscle indicates the extent to which the muscle contributes to the resultant muscle torque over the entire range of motion. When the ratio is high, the change in fiber length, and hence sarcomere length, is small relative to the angle-dependent change in moment arm. Recall from the length-tension relationship of sarcomeres (Figure 6.11) that the force exerted by a sarcomere remains greatest at intermediate sarcomere lengths. Consequently, for a high ratio of muscle fiber to moment arm, the muscle fiber remains at an intermediate length and capable of exerting maximal forces over a greater range of motion; this corresponds to a broad muscle length-tension relationship. Lieber reports that muscles with a high ratio include gluteus maximus (80), sartorius (11), and extensor carpi radialis longus (110), and those with a low ratio include soleus (0.9), vasti (1.8), hamstrings (1.8), and dorsiflexors (3.1). This relationship emphasizes that the entire architecture of a muscle, from sarcomere arrangement to joint organization, can influence the functional capabilities of the single joint system.

Summary

This chapter describes the factors that influence the torque that a muscle located in a single joint system can exert. These factors include the activation provided by the nervous system (neural factors) and the mechanical and architectural properties of the muscle (muscular factors). Mechanically, the length of the muscle and the rate at which length is changing both influence the force and power that the muscle produces. Furthermore, the length and size (cross-sectional area) of muscle fibers, the way muscle fibers are arranged within a muscle, and the location of the muscle around a joint influence the torque and shortening velocity of the muscle. Because of these effects, the nervous system must accommodate the musculoskeletal dynamics of the single joint system when it issues a command to achieve a desired muscle torque.

TRUE-FALSE QUESTIONS

1. There is usually a muscle fiber action potential every time there is a motor neuron action potential. **TRUE FALSE**
2. Motor unit recruitment and modulation of discharge rate can occur concurrently. **TRUE FALSE**
3. Orderly recruitment refers to the recruitment of small motor neurons before larger motor neurons. **TRUE FALSE**
4. When muscle force begins to decline, the first motor unit derecruited is the one that was recruited first. **TRUE FALSE**
5. Variation in muscle force is never due only to modulation of discharge rate. **TRUE FALSE**
6. The force at which a motor unit is activated is referred to as the recruitment threshold. **TRUE FALSE**
7. A motor neuron pool refers to the group of motor neurons that are activated during a particular task. **TRUE FALSE**
8. The number of dendrites, the innervation ratio, and the input resistance are examples of morphological characteristics of motor neurons. **TRUE FALSE**
9. Rheobase is an index of motor neuron excitability. **TRUE FALSE**
10. Sag is observed in a submaximal tetanic force profile. **TRUE FALSE**
11. Motor neuron excitability does not depend on size alone. **TRUE FALSE**
12. *Electrotonic* describes the spread of excitation along an excitable membrane. **TRUE FALSE**
13. Effective synaptic current is a measure of the net effect of the intrinsic factors involved in motor neuron activation. **TRUE FALSE**
14. Extrinsic factors can result in the activation of large motor units before small motor units. **TRUE FALSE**
15. Soleus is a slow-twitch muscle. **TRUE FALSE**
16. The brain does not determine the order of motor unit recruitment. **TRUE FALSE**
17. It is possible to selectively activate either slow- or fast-twitch motor units. **TRUE FALSE**
18. The force-frequency relationship for whole muscle is linear. **TRUE FALSE**
19. Tonic motor units increase discharge rate linearly over the entire range of muscle forces. **TRUE FALSE**
20. The unit of measurement for discharge rate is ms. **TRUE FALSE**
21. Muscle wisdom refers to the decline in discharge rate of motor units during fatiguing contractions. **TRUE FALSE**
22. A double discharge describes the appearance of two muscle action potentials in response to a single motor neuron action potential. **TRUE FALSE**
23. The regular discharge of action potentials by a motor unit is described as a synchronous discharge. **TRUE FALSE**
24. A muscle fiber action potential is an all-or-none event. **TRUE FALSE**
25. The crossbridge theory seeks to explain the sliding-filament hypothesis of muscle contraction. **TRUE FALSE**
26. Muscle force is greatest at the longest lengths. **TRUE FALSE**
27. At the longest muscle lengths, most of the force is due to crossbridge activity. **TRUE FALSE**
28. The magnitude of the force exerted by the passive elements varies between muscles. **TRUE FALSE**
29. Twitch force varies with muscle length. **TRUE FALSE**
30. Muscle torque is maximal at an intermediate joint angle for all muscles. **TRUE FALSE**
31. In general, the angle-torque relationship varies among the muscles that constitute a single muscle group. **TRUE FALSE**
32. The extensor torque about the knee joint is dominated by the contribution from the vasti muscles. **TRUE FALSE**
33. The muscle torque is greater than the load torque during a concentric contraction. **TRUE FALSE**
34. The magnitude of the torque exerted in eccentric contractions is greater than that exerted in concentric contractions. **TRUE FALSE**
35. The maximum knee extensor torque occurs at an intermediate angle. **TRUE FALSE**
36. Muscle fiber length can shorten during a concentric contraction. **TRUE FALSE**
37. Compliant tendons are stiff and resist stretching. **TRUE FALSE**
38. Efficiency is defined as the amount of work done for the energy consumed. **TRUE FALSE**
39. A muscle contraction becomes less efficient as speed of shortening increases. **TRUE FALSE**
40. The unit of measurement for eccentric force is N. **TRUE FALSE**

41. The stiffness of muscle is initially low as a contraction changes from isometric to eccentric. **TRUE FALSE**
42. The elbow flexor muscles are the prime movers for a chin-up. **TRUE FALSE**
43. The soleus and gastrocnemius muscles contribute to the Achilles tendon force. **TRUE FALSE**
44. A muscle produces power when it does positive work. **TRUE FALSE**
45. The rectus femoris muscle is one of the knee extensors. **TRUE FALSE**
46. The abbreviation $F_{m,o}$ stands for the maximum isometric force. **TRUE FALSE**
47. The Hill equation describes the decline in muscle force as a function of shortening velocity. **TRUE FALSE**
48. Viscosity describes the stiffness of muscle. **TRUE FALSE**
49. Power can be calculated as the rate of change in energy. **TRUE FALSE**
50. Soleus and gastrocnemius are not synergists. **TRUE FALSE**
51. Isotonic loading is an experimental procedure used to measure angle-torque relationships in muscle. **TRUE FALSE**
52. The unit of measurement for power is J. **TRUE FALSE**
53. A warm-up does not affect contraction speed. **TRUE FALSE**
54. Muscle architecture does not affect the maximum force that a muscle can exert. **TRUE FALSE**
55. Tendon does not influence the series elasticity of muscle. **TRUE FALSE**
56. The dashpot is an elastic element in the three-element model of muscle. **TRUE FALSE**
57. *In parallel* refers to an end-to-end arrangement. **TRUE FALSE**
58. Average sarcomere length is about 2.2 cm. **TRUE FALSE**
59. The greater the number of sarcomeres in series, the greater the maximum shortening velocity of muscle. **TRUE FALSE**
60. A muscle fiber consists of sarcomeres that are arranged in parallel and in series. **TRUE FALSE**
61. Antigravity muscles are generally stronger than their antagonists. **TRUE FALSE**
62. The angle of pennation usually varies from 0 to 0.4 rad. **TRUE FALSE**
63. The advantage of pennation is that a greater number of in-parallel fibers can be packed into a given volume. **TRUE FALSE**
64. Staggered muscle fibers represent an in-series effect. **TRUE FALSE**
65. The functional definition of muscle can be different to its anatomical one. **TRUE FALSE**
66. Muscle compartments are distinguished on the basis of the attachment points to connective tissue. **TRUE FALSE**
67. The measurement of functional cross-sectional area takes into account the length of the muscle fibers. **TRUE FALSE**
68. The location of the attachment points of a muscle to the skeleton can influence the angle-torque relationship of muscle. **TRUE FALSE**

MULTIPLE-CHOICE QUESTIONS

69. What does the size principle state?
    A. The recruitment of motor units occurs in a fixed order based on motor neuron size and begins with the smallest motor neuron.
    B. The gradual increase in the power demands of a task involves the progressive recruitment of more powerful motor units.
    C. The recruitment of motor neurons does not depend solely on motor neuron size but is influenced by other motor neuron characteristics (intrinsic factors) and by the organization of synaptic input (extrinsic factors) on the motor neurons.
    D. The performance of a particular movement is accomplished by the activation of motor units in a set sequence.
70. Which variables do not influence the recruitment order of motor units?
    A. Extrinsic factors
    B. Biophysical characteristics
    C. Twitch contraction time
    D. Motor neuron size
71. Which feature of a synapse is not an extrinsic factor that influences the excitability of motor neurons?
    A. Distribution of synapses
    B. Number of synaptic terminals
    C. Speed of action potential propagation
    D. Amount of neurotransmitter released at each synapse

72. Why is it not possible to selectively activate either slow-twitch or fast-twitch motor units?
   A. The extrinsic factors are not different for slow- and fast-twitch motor units.
   B. Slow-twitch units are smaller.
   C. Motor unit recruitment and modulation of discharge rate occur concurrently.
   D. There is an overlap in the size and excitability of Type S and Type FR units.

73. Which effect does not cause muscle force to change due to the pattern of action potential activity?
   A. Motor unit synchrony
   B. Muscle wisdom
   C. Force-frequency relationship
   D. Double discharge

74. Identify the *incorrect* statement or statements about motor unit activity.
   A. Recruitment continues up to 85% of maximum force for all muscles.
   B. Modulation of discharge rate and recruitment can occur concurrently.
   C. All slow-twitch units are recruited before any fast-twitch units are recruited.
   D. Synchronization is known to increase the force exerted by active units.

75. What does the crossbridge theory state?
   A. Muscle force varies with the amount of overlap between thick and thin filaments within a sarcomere.
   B. The thick and thin filaments slide past one another during a muscle contraction.
   C. The crossbridge cycle consists of an attach-rotate-detach sequence.
   D. Muscle force is proportional to the number of attached crossbridges.

76. Why does the force exerted by muscle vary with muscle length?
   A. It is due to changes in the moment arm.
   B. Passive force increases with muscle length.
   C. The amount of overlap of the thick and thin filaments varies.
   D. Most muscles have a pennate structure.

77. During an eccentric contraction, what is the relationship of muscle torque and load torque?
   A. The muscle torque is less than the load torque.
   B. The muscle and load torques are equal.
   C. The muscle torque is greater than the load torque.
   D. The load torque is less than the muscle torque.

78. What does tendon compliance refer to?
   A. Amount of passive force transmitted by tendon
   B. Stretch of tendon per unit of applied force
   C. Force transmitted by tendon per unit of cross-sectional area
   D. Stretch of tendon per unit of cross-sectional area

79. The decrease in muscle force with an increase in the velocity of a concentric contraction is not influenced by which of the following?
   A. A decline in the number of crossbridge attachments
   B. A decrease in the average force exerted by each crossbridge
   C. A reduction in the efficiency of the contraction
   D. An increased rate of crossbridge cycling

80. The force exerted by muscle is greatest during an eccentric contraction. This is probably because of which of the following?
   A. An increase in the number of attached crossbridges
   B. An increase in the quantity of Ca^{2+} that is released
   C. A change in the mechanism responsible for crossbridge detachment
   D. A stretch of incompletely activated sarcomeres

81. What does power absorption correspond to?
   A. A concentric contraction
   B. Negative work
   C. The rate of doing work on the surroundings
   D. A shortening contraction

82. When the shortening velocity of a contraction is equal to v_{max}, what is the value of the power produced by the muscle?
   A. Zero
   B. Maximum
   C. One third of maximum
   D. Determined by the force

83. Which relationship would not enable us to calculate power?
   A. $\frac{\Delta \text{ energy}}{\text{time}}$
   B. $\frac{v_{max}}{F_{m,o}}$
   C. Force × velocity
   D. $\frac{\text{work}}{\text{time}}$

84. Which is not one of the components of the three-element model of muscle?
   A. Passive element
   B. Series elastic element
   C. Contractile element
   D. Parallel elastic element

85. Which mechanism causes tetanic force to be greater than twitch force?
   A. Series elasticity
   B. Amount of calcium released from the sarcoplasmic reticulum
   C. Cross-sectional area
   D. Activation of the dashpot

86. How many sarcomeres are arranged in series (average length = 2.2 μm) in an 11.4-cm muscle fiber?
    A. 51,818
    B. 518,181
    C. 5,181
    D. 518
87. What do studies of motor unit architecture indicate about the muscle fibers of a motor unit?
    A. They are all the same length.
    B. They do not always extend from one end of a fascicle to another.
    C. They are not distributed across the entire cross section.
    D. They transmit force directly from one tendon to another.
88. The attachment points of a muscle to a skeleton and the proportion of the muscle that contains contractile protein do not influence which feature of the angle-torque relationship?
    A. Breadth of the peak
    B. Magnitude
    C. Range of motion
    D. Location of the peak torque in the range of motion

PROBLEMS

89. The size principle suggests that motor unit recruitment order is based upon differences in motor neuron size.
    A. What is meant by motor neuron size?
    B. What morphological features of the motor units seem related to motor neuron size?
    C. What types of synaptic input onto the motor neuron also seem to vary according to size?
90. The relationship between action potential rate and force is known to be sigmoidal.
    A. What is the qualitative difference between sigmoidal and linear relationships?
    B. How does muscle length affect this relationship?
    C. How does the relationship differ for slow- and fast-twitch motor units?
91. Angle-torque relationships for muscle typically have one of three forms: ascending, descending, or intermediate.
    A. What is the intermediate relationship?
    B. Suppose the following graph represents the most common form of the angle-torque relationship, such as for the elbow flexors. If the force exerted by the elbow flexor muscles remained constant at 3.5 kN over this range of motion, what would be the minimum and maximum values for the moment arm? 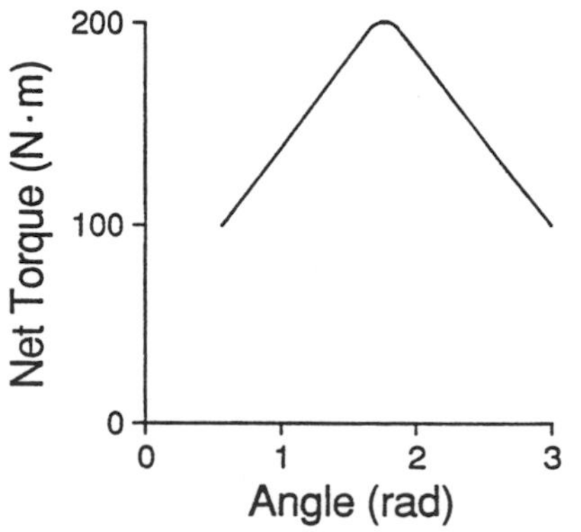
    C. Draw a free body diagram to show the moment arm referred to in B.
    D. If the graph shown in B was obtained under the conditions of a maximum voluntary contraction, is it reasonable to expect that the shape of the graph is due mainly to variation in the moment arm?
92. The terms *isometric*, *concentric*, and *eccentric* refer to differences in the mechanical conditions and not to different types of muscle contractions.
    A. What is the mechanical distinction between these terms?
    B. Can there be different types of muscle contractions? Explain your answer.
    C. The term *isotonic* means constant muscle tension. Why does this not occur under natural conditions, even when lifting a constant load?
93. The first law of thermodynamics states

$$\Delta C = \Delta H + W$$

where C = chemical energy, H = heat or thermal energy, and W = work. We know that, as the shortening velocity of a muscle contraction increases, the muscle becomes less efficient (work output/energy input). If the amount of ATP used remains the same, what happens to the heat generated by the muscles as velocity increases? What happens to the work done by the muscle?

94. From the work-energy relationship, we know that the work done is equal to the amount of energy used.
    A. What are the SI units of measurement for work and energy?
    B. What are the mechanical sources of energy for work?
    C. Physiologically, we distinguish positive and negative work. Mechanically, we describe positive work as involving the flow of energy from a system to its environment and negative work as involving flow in the reverse direction. What does this mean?
95. Imagine that you are holding a heavy briefcase by the handle while waiting for a bus. Although no mechanical work is done on the briefcase, why is it not possible to hold the briefcase indefinitely?
96. If a cyclist is trying to maintain a constant speed and encounters a hill, often the cyclist will change gears. Why would the cyclist choose this strategy?
97. The viscoelasticity of muscle causes a hysteresis in the force-length relationship of muscle (Figure 6.30a), which means that muscle loses energy during repeated cycles of lengthening and shortening. How does Figure 6.30a indicate that muscle loses energy during this activity? (Hint: work = Δ energy.)
98. A unipennate muscle has 1,167 fibers, which have an average pennation angle of 0.4 rad to the line of pull of the tendon. The muscle comprises 25% Type S motor units, 20% type FR motor units, and 55% Type FF motor units. The average cross-sectional area is 1,370 μm^2 for the Type S units, 2,503 μm^2 for the Type FR units, and 4,980 μm^2 for the Type FF units. Calculate the maximum force that the muscle can exert in the direction of the line of pull of the tendon.
99. If each sarcomere produces a 0.5-µm change in length and there are 31.2×10^3 sarcomeres in a muscle fiber, how much could the muscle fiber maximally change its length?
100. Based on the data in Table 6.1 and a value for specific tension of 30 N/cm^2, calculate the maximum force that can be exerted by the quadriceps femoris.
101. Why is the architecture of the soleus muscle so different from that of most other muscles in the leg (Table 6.1)?

Summary of Part II

The goal of Part II (chapters 4 through 6) has been to define a biological model (the single joint system), to describe the interactions among its elements, and to identify the neural and muscular factors that influence the force exerted by muscle. As a result of reading these chapters, you should be familiar with the following features and concepts related to the single joint system:

- Observe that the morphological and mechanical features of the components of the single joint system are largely determined by the functions they serve
- Note the structural features of the five components of the simple joint system
- Understand why the single joint system provides a reasonable model to study the control of movement
- Realize that the motor unit is the basic functional unit of the system and be able to describe its features
- Conceive of the means by which information is transmitted rapidly throughout the system
- Identify the processes measured by common techniques used to monitor the activity of the single joint system
- Understand the mechanism that links the neural signal with muscle, resulting in a muscle contraction
- Comprehend the basic features of afferent feedback, including simple reflex circuits, and the prominent role of interneurons in processing information
- Acknowledge the contribution of afferent feedback to sensation associated with movement
- Know that muscle force is controlled by varying the activity of motor units
- Realize that muscle does not exert the same force under all conditions for a constant excitation by the nervous system
- Differentiate the effects of muscle architecture on the force exerted by muscle

PART III

Adaptability of the Motor System

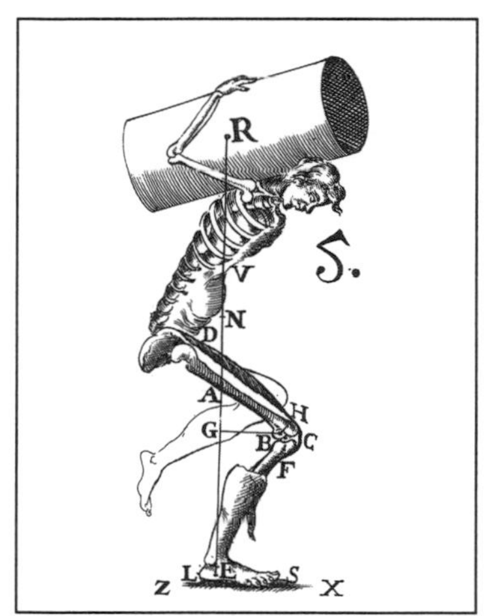

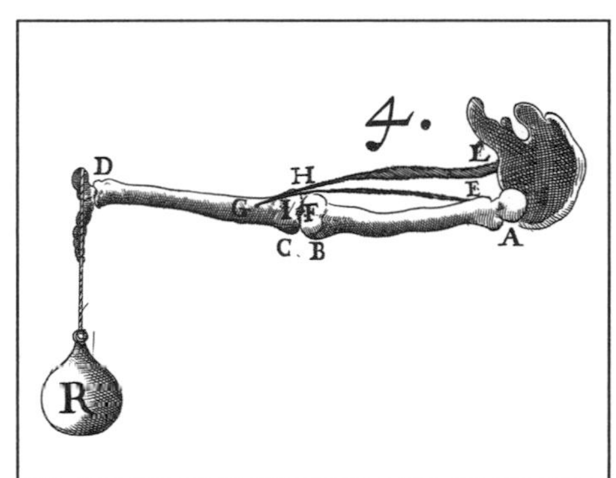

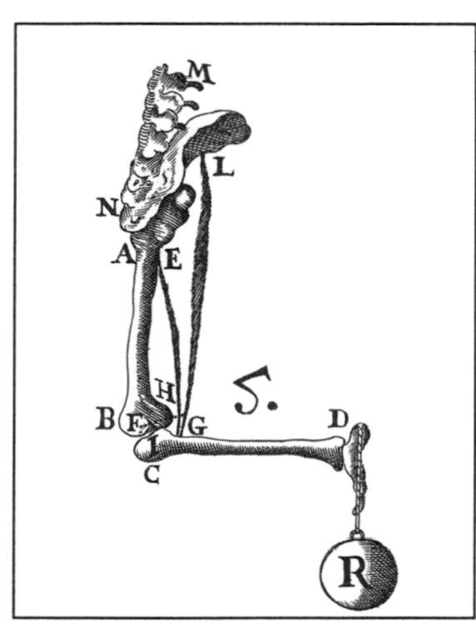

In this text, we have characterized human movement as an interaction between a biological system and its surroundings. In Part I, we identified the concepts and principles for the study of motion that have been derived from physics. In Part II, we developed a simple biological model (the single joint system) that enabled us to examine the biological processes involved in the performance of movement. However, the human body is more complex than a single joint system. Part III extends the single joint system model to provide a more complete account of those components of the human body that are directly involved in the performance of movement. Collectively, these components are referred to as the motor system and include both motor and sensory elements. Chapter 7 provides more detail on the neural and muscular organization of the motor system and describes some of the strategies that the motor system uses to perform movement. The last two chapters describe the acute (chapter 8) and chronic (chapter 9) adaptive capabilities of the system.

Objectives

To conclude our study of the neuromechanical basis of kinesiology, we examine the adaptive capabilities of the motor system. The goal of Part III is to provide a realistic description of the motor system, to the extent that the research literature permits, and to document the ways in which the motor system adapts to various types of physical stress. Specific objectives include the following:

- To describe the organization of the suprasegmental components of the nervous system and the flow of information associated with the control of movement
- To extend the description of the musculoskeletal system to consider the interaction of multiple single joint systems
- To characterize the movement strategies that have been identified
- To explain the effect of altering core temperature on performance capabilities
- To indicate the techniques that have been developed to alter flexibility
- To outline the multifactorial basis of muscle fatigue and to identify the processes that can be impaired in different tasks
- To examine the sensory adaptations that occur during fatiguing contractions
- To describe the potentiating capabilities of muscle
- To establish the principles of exercise prescription
- To define the performance characteristics of strength and power and the mechanisms that mediate changes in these parameters
- To document the adaptations that accompany periods of reduced activity
- To evaluate the changes that occur with aging and to identify the physiological basis of these changes

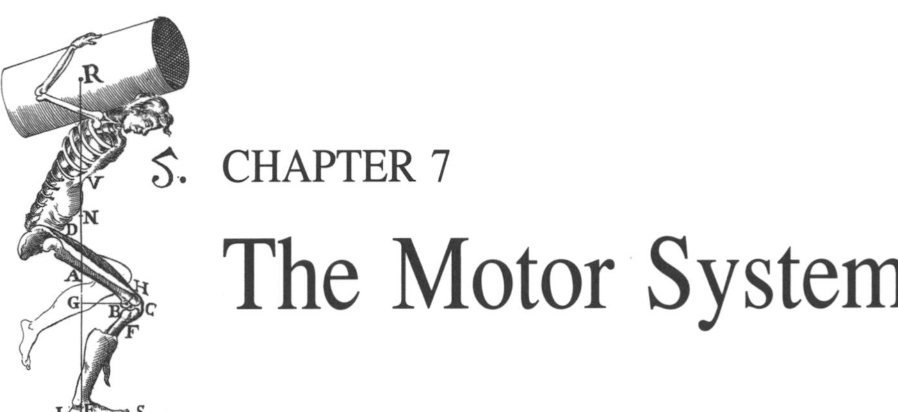

CHAPTER 7

The Motor System

In Part II we developed a model of the motor system that we called the single joint system. Using this model we focused on some of the fundamental physiological and biomechanical features of the human body that enable it to generate movement. The human body, however, is much more than a single joint system. We did not, for example, consider how the single joint system initiates movement. Chapter 7 expands on the single joint system by examining the role of other neural elements in the initiation and modulation of single joint system activity, by considering the consequences of musculoskeletal design for the control of movement, and by identifying some of the strategies used by the motor system to perform movement.

Suprasegmental Organization

Until now, we have considered only the motor neuron as the efferent element of the motor system and have paid scant attention to the origin of the neural commands that control muscle. Clearly, the idea to produce a movement does not begin and end with the motor neuron. Many central nervous system (CNS) structures contribute to the development of the motor neuron signals that activate muscle. The brain, of course, is critical and has three major functions related to the control of movement: motivation, perception, and activation of muscle. Motivation and perception are important in initiating and modifying movement. Figure 7.1 characterizes the flow of information through the nervous system that results in the excitation of motor neurons. There are four parts to this scheme: motivation, ideation, programming, and execution (Cheney, 1985). In the previous discussion of the single joint system, we focused on the execution phase.

The limbic system provides the *motivation* for many movements. It governs basic biological drives and emotional behavior, including such activities as hunger, thirst, reproduction, maternal behavior, and socialization. In addition, the limbic system is crucial to our ability to learn from experience. The limbic system has been described as the emotional motor system because it can elicit specific emotional behaviors and can alter the excitability of neurons directly involved in movement (Holstege, 1992). For example, many forms of stress (e.g., aggression, fear, sexual arousal) can induce analgesia while the excitability of motor neurons is increased. The limbic system can elicit behaviors through the sensorimotor system, which transforms the motivation into an *idea* and initiates the suprasegmental interactions that result in a command to execute a movement (see Brooks, 1986, and Rothwell, 1987, for more detail on these systems). The **limbic system** refers to a functional unit that includes a set of forebrain structures that are interconnected with the hypothalamus and parts of the midbrain. The hypothalamus is central to this system and regulates many vital factors, such as body temperature, heart rate, blood pressure, and water and food intake. The demands expressed by the limbic system are analyzed and integrated into ideas by the association cortex (e.g., prefrontal, parietal, and temporal lobes); these ideas project onto the sensorimotor cortex, cerebellum, part of the basal ganglia, and

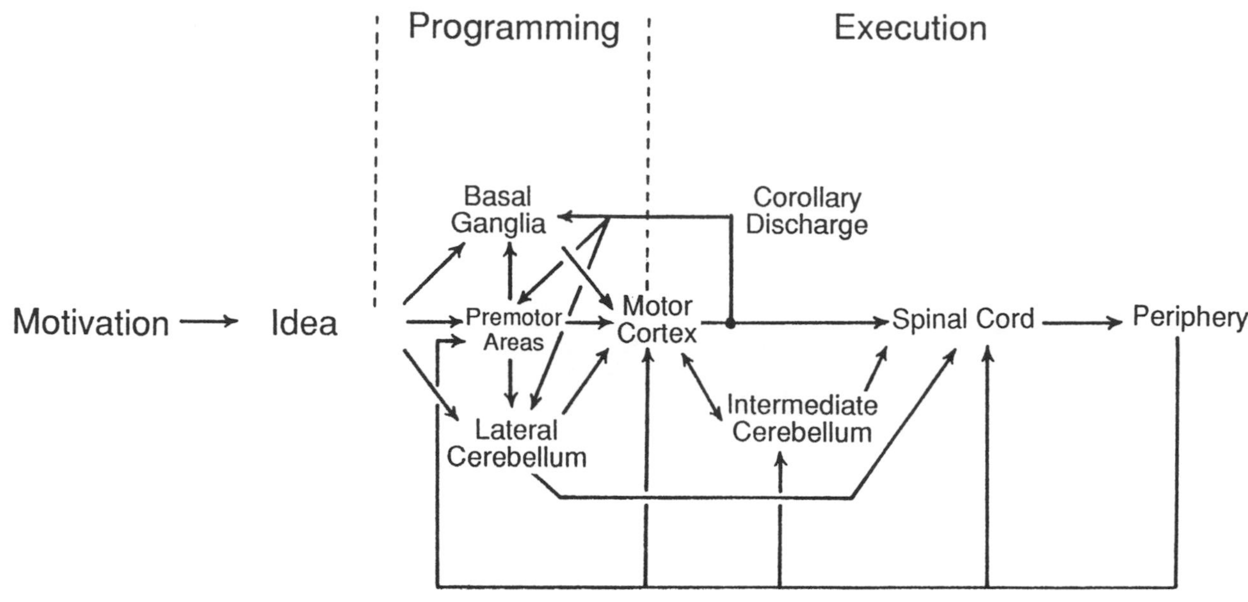

Figure 7.1 Flow chart of major neural interactions associated with the performance of a movement.
Note. From "Role of Cerebral Cortex in Voluntary Movements. A Review" by P.D. Cheney, 1985, *Physical Therapy*, **65**, p. 625. Copyright 1985 by the American Physical Therapy Association. Reprinted with permission.

associated subcortical nuclei for the development of a program to perform the movement.

Programming involves the conversion of an idea into the proper strength and pattern of muscle activity necessary for a desired movement. The major suprasegmental centers involved in programming include the association cortex (premotor), motor cortex, basal ganglia, and cerebellum. The motor cortex provides the brain with access to most motor neurons. In addition, the basal ganglia, cerebellum, and cerebral cortex can independently influence motor neurons via the brain stem. The neural output that emerges from the process of programming is known as the **central command** and is transmitted both to lower neural centers (brain stem and spinal cord) and back to the suprasegmental centers involved in the development of the program. The signal that is transmitted back to the suprasegmental centers is known as the **corollary discharge** and provides a reference that enables the system to interpret incoming afferent signals.

The central command activation of the lower neural centers initiates the *execution* phase of a movement. This phase involves the activation of motor neurons, both those in muscles directly associated with the movement and those in other muscles that must provide postural support to stabilize the body. Despite the many possible activation sequences of motor units that may be used for a given movement, it appears that learning results in the development of a specific activation sequence for each task. *A stereotyped sequence of commands sent from the spinal cord to the muscles to elicit a specific behavior* is known as a **motor program**. The motor program is the consequence of interactions between the programming activities of suprasegmental centers, spinal networks, and afferent feedback.

In addition to the efferent commands transmitted by the motor neurons, the execution phase involves the modification of movements by feedback from the sensory receptors (e.g., muscle spindle, tendon organ, joint receptor, cutaneous mechanoreceptor). The activation of these receptors results in afferent input that acts at the segmental level and also traverses in ascending pathways to suprasegmental centers (Figure 7.1). Although this input can be potent enough to initiate movement (e.g., reflexes), it is typically used to ensure consistency between the movement and the surroundings. This matching of input and output is accomplished by networks of interneurons that transform afferent input onto relevant motor neurons so that the output is appropriate for the conditions detected by the sensory receptors. If a particular afferent-efferent transformation occurs frequently enough, the neural network can learn to become more economical and can be activated with a minimum of input. *Neural networks that are capable of generating behaviorally relevant patterns of output in the absence of external timing cues* (by afferent input) are known as **central pattern generators**.

Given this motivation-ideation-programming-execution scheme of information flow through the nervous system, we next examine how the major components of the motor system interact in the control of movement. The suprasegmental centers interact via ascending and descending pathways with spinal networks and information received from the surroundings (Figure 7.2). The

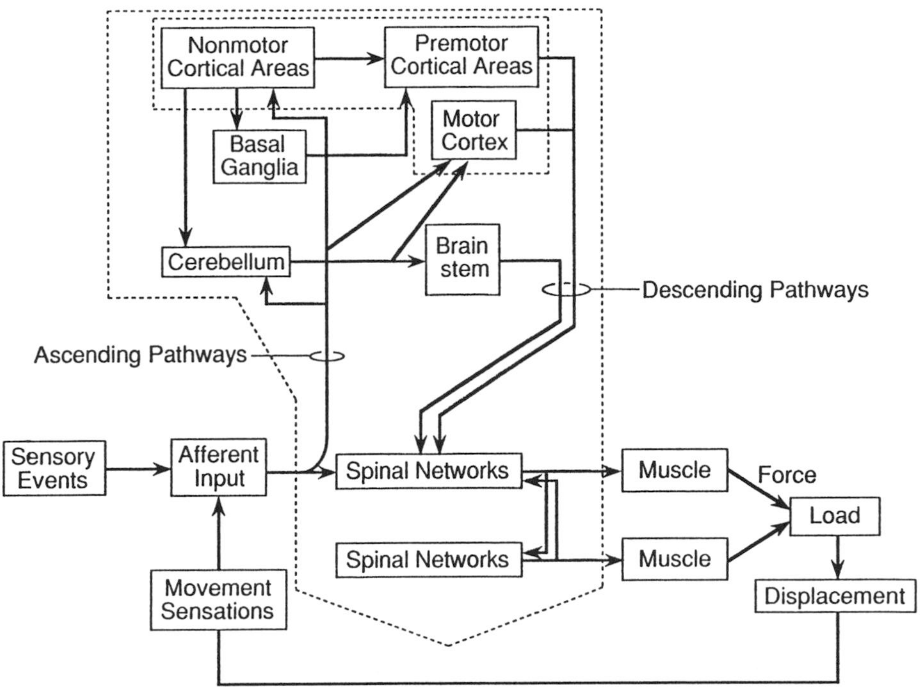

Figure 7.2 The major components of the motor system.
Note. From "Introduction to the Motor Systems" by C. Ghez. In *Principles of Neural Science* (2nd ed.) (p. 434) by E.R. Kandel and J.H. Schwartz (Eds.), 1985, New York: Elsevier Science Publishing Co., Inc. Copyright 1985 by Elsevier Science Publishing Co., Inc. Adapted by permission of Appleton & Lange, Norwalk, CT.

structures that contribute to the **descending pathways** include the sensorimotor cortex, brain stem, cerebellum, and basal ganglia (see Rothwell, 1987, and Kandel, Schwartz, & Jessell, 1991, for more details on this topic). We will focus on the role of the descending pathways, but because it is difficult for humans to perform movement without both efferent and afferent signals, we will also briefly discuss **ascending pathways** and the distribution of afferent input among the suprasegmental structures.

Sensorimotor Cortex

The cerebrum consists of left and right cerebral hemispheres that are connected by axon bundles called *commissures*. The outer shell of the cerebral hemisphere is a 3-mm layer of cortical neurons, known as the *cerebral cortex*, that represents a major integrating center of sensory input and motor output. The **sensorimotor cortex** contains the regions of the cerebral cortex that are located immediately anterior and posterior to the central sulcus (Figure 7.3). The three anterior components are the **primary motor area** (Brodmann's area 4), the **supplementary motor area** (area 6—between areas 4 and 8 on the medial surface), and the **premotor cortex** (area 6). The two posterior components are the **primary somatosensory cortex** (areas 1 to 3) and the **posterior parietal cortex** (areas 5 and 7). All of the sensorimotor cortex contributes to the descending pathway known as the *corticospinal tract*.

The input-output connections of the sensorimotor cortex are extensive, and the flow of information occurs over interacting parallel pathways (Georgopoulos, Ashe, Smyrnis, & Taira, 1992; Houk, Keifer, & Barto, 1993; Kalaska & Crammond, 1992). In general, the output from area 4 goes to the ventral spinal cord via the corticospinal tract and influences motor activity. The output from areas 1, 2, 3, and 5 goes to the dorsal spinal cord and modulates peripheral sensory input. In addition, area 4 receives afferent input from the cerebellum via the thalamus, and areas 1 to 3 receive afferent input from the periphery (Figure 7.4). Peripheral sensory input to the primary somatosensory cortex (areas 3a, 3b, 1, and 2) is discrete. Each area receives input from different types of sensory receptors and is arranged in a manner that provides a precise scheme corresponding to the dermatomes of the whole body. The input and output of the sensorimotor cortex has a **somatotopic** (body map) organization in which the cortical neurons influencing muscles and the sensory information from the different parts of the body are arranged in an orderly sequence (Figure 7.5). Two features of the motor map are prominent: The cortical cells influence muscles on the contralateral side of the body, and facial and hand muscles seem to have a larger representation. These topographic maps are also found in other regions of the sensorimotor cortex (posterior

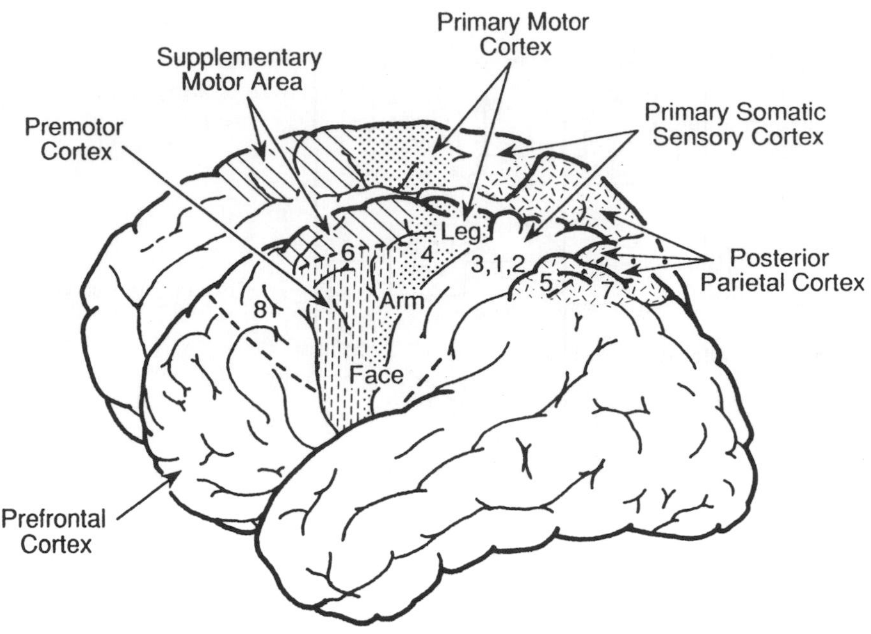

Figure 7.3 Components of the sensorimotor cortex. The figure shows the left and right halves of the cerebral cortex.
Note. From "Voluntary Movement" by C. Ghez. In *Principles of Neural Science* (3rd ed.) (p. 611) by E.R. Kandel, J.H. Schwartz, and T.M. Jessell (Eds.), 1991b, New York: Elsevier Science Publishing Co., Inc. Copyright 1991 by Elsevier Science Publishing Co., Inc. Adapted by permission of Appleton & Lange, Norwalk, CT.

parietal cortex, supplementary motor area, and premotor cortex) and in the cerebellum and thalamus.

Each area of the sensorimotor cortex has primary, but not exclusive, responsibility for certain aspects of movement generation. The *premotor cortex* is mainly concerned with the correct orientation of the body and limb prior to a movement and with the sensory guidance of limb movements. These effects are achieved by influencing axial and proximal muscles. The premotor cortex is also active when learning a new task. The *posterior parietal cortex* interprets and transforms sensory information so that the movement is consistent with surrounding conditions. It is involved in initiating movements that are directed to extrapersonal space. Neurons in area 5 are involved in the use of tactile sensory information to guide exploratory limb movements, whereas neurons in area 7 are more involved in the visual guidance of eye and limb movements (Cheney, 1985). Damage to either the premotor or posterior parietal cortex results in an inability to produce the correct strategy for movement (Jeannerod, 1988). The *supplementary motor area* is important in the planning of movements and ensures the correct sequence of muscle activation for such movements as playing a piano, writing, and speaking. The supplementary motor area is, for example, active during the performance of imagined movements (Roland, Larsen, Lassen, & Skinhøj, 1980). In addition, the supplementary motor area affects input-output coupling in the primary motor cortex. The main responsibility of the *primary somatosensory cortex* is to provide the sensory information required for specific planning and initiation of movement and for the modulation of ongoing movement. The *primary motor cortex* synthesizes input from other parts of the CNS and generates and transmits the central command to the brain stem and spinal cord neurons to initiate and modulate movement. In addition, the central command is sent to

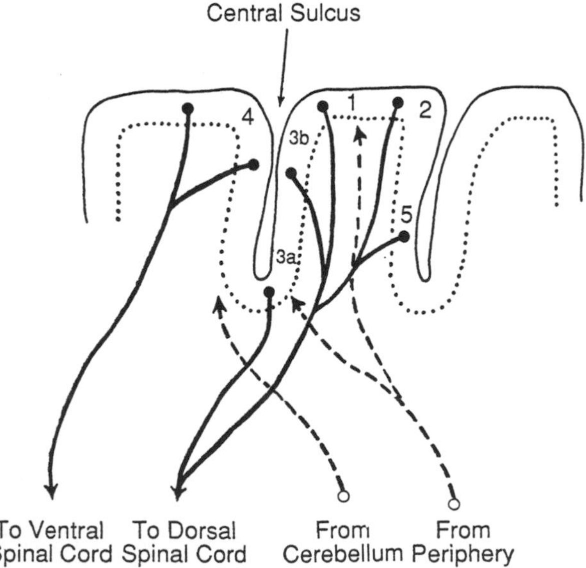

Figure 7.4 Input-output connections of the sensorimotor cortex. Input and output occur on both sides of the central sulcus.

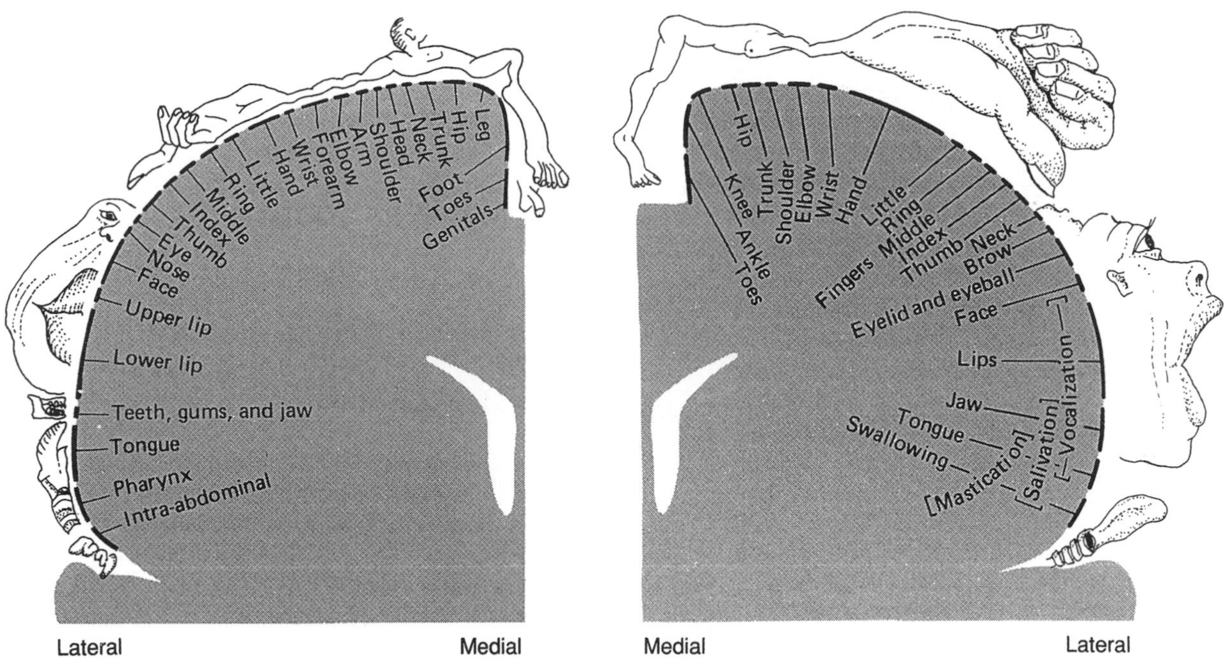

Figure 7.5 Somatotopic organization of the sensorimotor cortex: (a) sensory homunculus; (b) motor homunculus.
Note. From "Touch" by E.R. Kandel and T.M. Jessell. In *Principles of Neural Science* (3rd ed.) (p. 372) by E.R. Kandel, J.H. Schwartz, and T.M. Jessell (Eds.), 1991, New York: Elsevier Science Publishing Co., Inc. Copyright 1991 by Elsevier Science Publishing Co., Inc. Adapted by permission of Appleton & Lange, Norwalk, CT.

other brain areas, such as the basal ganglia and cerebellum, in the form of a corollary discharge. The central command appears to specify the muscles that need to be activated, the contraction force, and the timing of the contraction (Cheney, 1985). These five subcomponents of the sensorimotor cortex interact extensively to ensure that voluntary movements are based on prior knowledge of the effector system (relevant single joint systems), the surroundings of the system, and the expected consequences of the movement (Ghez, 1991a; Kalaska & Crammond, 1992).

It appears that the sensorimotor cortex can plan movements in terms of the participation of individual muscles, groups of muscles, or the movement itself. Analyses of neuronal activity indicate that the CNS can shift control from one neural system to another, that emotive needs can change motor output strategies, and that the motor cortex seems most needed for skilled motor performance. This flexibility seems to be possible due to the absence of a rigid hierarchical organization among the areas of the sensorimotor cortex. Thus different areas can take priority depending on the behavioral context of the movement (Houk et al., 1993; Kalaska & Crammond, 1992).

Brain Stem

The brain stem comprises the medulla oblongata, pons, and mesencephalon (Figure 7.6). The brain stem is largely responsible for the control of posture. It integrates information provided by the equilibrium organ (vestibular apparatus) and sensory receptors in the neck region along with input from the cerebral cortex and cerebellum. The **medulla oblongata** is the continuation of the brain into the spinal cord and is the site where most of the cranial nerves enter and exit the brain. Several vital autonomic nuclei concerned with respiration, heart action, and gastrointestinal function are located in the medulla oblongata. The **pons** lies between the medulla and the mesencephalon (midbrain). A large bundle of crossing fibers lying on the lower aspect of the pons interconnects the brain stem and the cerebellum. The **mesencephalon** merges anteriorly into the thalamus and hypothalamus.

Distributed in the brain stem are at least three motor centers that send efferent fibers to influence the motor neurons of the spinal cord (Figures 7.2 and 7.6). These motor centers include the red nucleus, lateral vestibular nucleus, and reticular formation. The **red nucleus** is located in the mesencephalon and gives rise to the **rubrospinal tract**, which crosses in the brain stem and influences contralateral segmental centers. The tract is arranged *somatotopically* (according to a body map), which means that the location of the axons in the tract depends on the motor neurons that they innervate. For example, axons of neurons in the dorsomedial part of the nucleus travel medially in the tract and innervate cervical segments. When stimulated alone electrically, the red nucleus mainly excites alpha and gamma motor neurons that innervate flexor muscles and inhibits motor

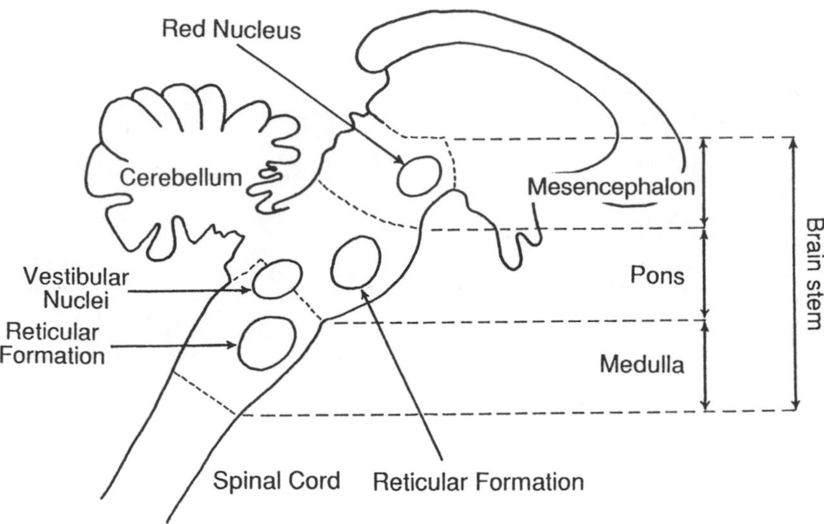

Figure 7.6 Components of the brain stem and location of the motor centers.
Note. From "Motor Systems" by R.F. Schmidt. In *Human Physiology* (p. 93) by R.F. Schmidt and G. Thews (Eds.), 1983, New York: Springer. Copyright 1983 by Springer-Verlag, Inc. Adapted by permission.

neurons that connect to extensor muscles. The red nucleus receives most of its input from the dentate nucleus (in the cerebellum) and the motor cortex. The **lateral vestibular nucleus** extends between the pons and medulla oblongata, with the neurons arranged somatotopically within the nucleus. The axons of cells located in this nucleus give rise to the **vestibulospinal tract**, which affects ipsilateral motor neurons. The effects of the lateral vestibular nucleus include excitation of alpha and gamma motor neurons that innervate extensor muscles and inhibition of motor neurons that connect to flexor motor neurons. The lateral vestibular nucleus receives input from the cerebellum and the equilibrium organ (labyrinth). The **reticular formation** produces two **reticulospinal tracts**, the pontine and medullary components. The reticulospinal tracts are diffuse—they have no somatotopic organization—and terminate along the length of the cord from the cervical to lumbar segments. The reticular formation receives input from the sensorimotor cortex, the fastigial nucleus (in the cerebellum), and ascending pathways of the spinal cord.

The descending pathways from suprasegmental structures can be classified into three pathways: the corticospinal tract and two groups (A and B) of brain stem tracts (Kuypers, 1985). The crossed, pre–central sulcus component of the **corticospinal tract** largely terminates laterally on or near motor neurons that innervate *distal muscles* (Figure 7.7). Some axons in the corticospinal tract terminate ventrally on or near motor neurons that innervate *axial* and *proximal muscles*. The corticospinal tract has a greater degree of topographic organization than either of the brain stem pathways (Groups A and B). The post–central sulcus component of the corticospinal tract modulates incoming sensory information from the periphery, including that required for the control of movement.

The **Group A pathway** lies in the ventromedial brain stem and consists of the vestibulospinal, reticulospinal, and tectospinal tracts; the **tectospinal tract** arises from neurons in the superior colliculus, which is a mesencephalic area concerned with orientation to visual stimuli. The tectospinal tract probably contributes to the head and neck orienting reactions to visual stimuli (Rothwell, 1987). Activation of the Group A pathway provides postural support for goal-directed movement involving synergistic muscles, especially axial and proximal muscles. The Group A axons travel in the ventral and ventromedial funiculi of the spinal cord and *terminate bilaterally onto motor neurons that innervate axial and proximal muscles* (Figure 7.7). The axons of the Group A neurons give off many collaterals, which seems necessary to coordinate the activity of many muscles.

The descending axons in the **Group B pathway** consist of the crossed rubrospinal and the crossed reticulospinal tracts. This pathway contributes to goal-directed motor activity by *influencing motor neurons that innervate distal flexor muscles*. The axons in this pathway have relatively few collaterals, and they end either directly on motor neurons or on short **propriospinal** axons of the spinal cord; propriospinal neurons are interneurons that convey information between spinal segments. The Group B axons travel in the contralateral dorsolateral funiculus of the spinal cord and terminate in the intermediate zone close to neurons that affect distal flexor muscles (Figure 7.7).

Cerebellum

The cerebellum is located inferior to the cerebral hemispheres and posterior to the brain stem. It is an unusual structure because it contains over half of the neurons in the brain, is highly ordered, and has no direct efferent

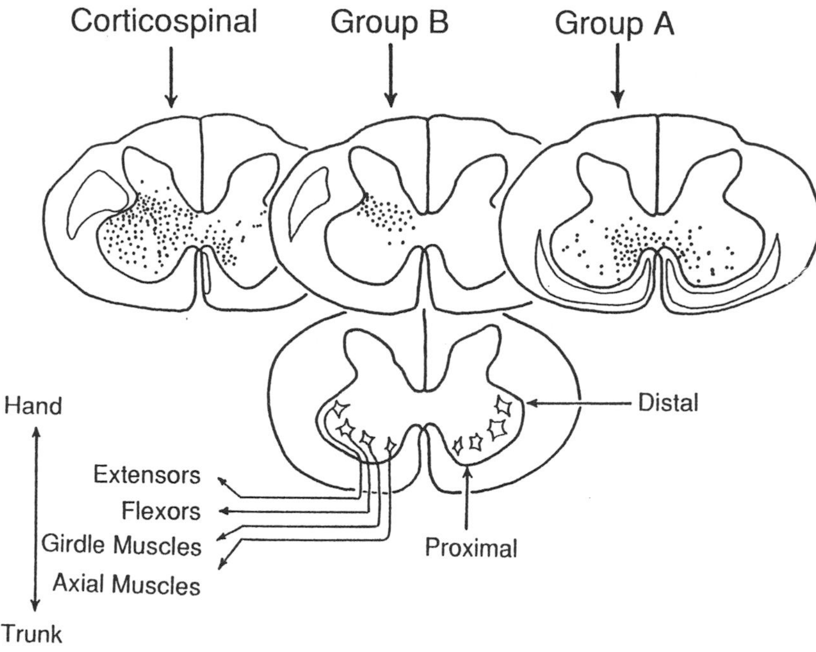

Figure 7.7 Termination zones of the three descending pathways. The upper three cross sections show the location of the terminations of the axons in the corticospinal tract, Group B pathway, and Group A pathway, respectively. The lower cross section indicates the topographic organization of the motor neurons in a cervical spinal cord segment.
Note. From "The Functional Organization of the Motor System in the Monkeys. II. The Effects of Lesions of the Descending Brain-Stem Pathways" by D.G. Lawrence and H.G.J.M. Kuypers, 1968, *Brain*, **XCI**, p. 26. Copyright 1968 by Oxford University Press. Adapted by permission.

connections to the spinal or brain stem motor nuclei despite its involvement in the control of movement. The cerebellum comprises an outer mantle of gray matter (cerebellar cortex), internal white matter, and three pairs of deep subcortical nuclei (fastigial, interpositus, and dentate). The cerebellum receives input from the periphery (somatosensory and visual), brain stem (vestibular), and cerebral cortex. The entire output of the cerebellum is transmitted through the deep subcortical nuclei and the vestibular nucleus in three symmetrical pairs of tracts that *connect the cerebellum and the brain stem* (Figure 7.8).

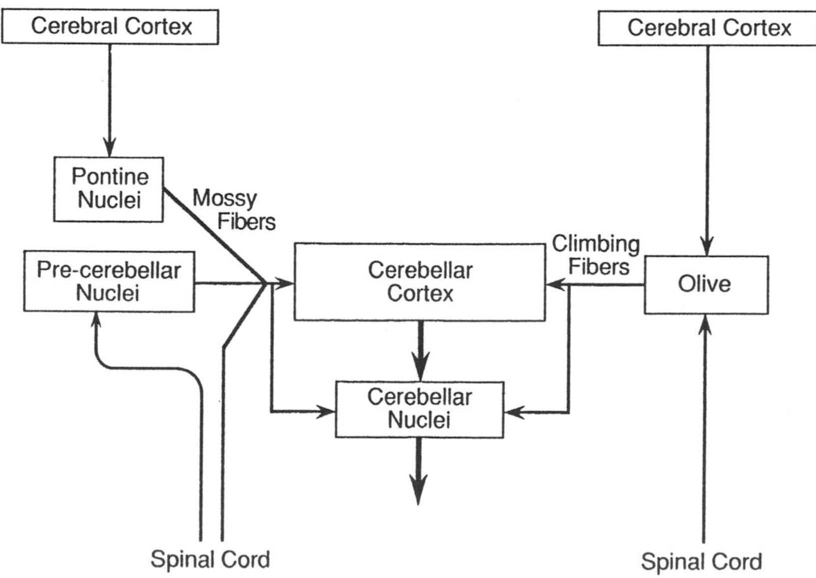

Figure 7.8 Afferent input to the cerebellum and output via the deep cerebellar nuclei.
Note. From *Control of Voluntary Human Movement* (p. 241) by J.C. Rothwell, 1987, Kent, United Kingdom: Croom Helm. Copyright 1987 by John C. Rothwell. Adapted by permission.

The cerebellum has three functional divisions, each of which receives input and sends output to distinct parts of the CNS (Ghez, 1991b). The three divisions are the spinocerebellum, cerebrocerebellum, and vestibulocerebellum (Figure 7.9). The **spinocerebellum**, which includes the vermis and intermediate hemispheres, receives most of its sensory input from the spinal cord. The output from the vermis region of the spinocerebellum goes to the *fastigial nucleus* whereas that from the intermediate hemispheres is directed to the *interposed* (or *interpositus*) *nucleus*. The output from these two deep nuclei is transmitted to the brain stem and spinal cord in the descending pathways and contributes to the control of limb movements. The **cerebrocerebellum**, which refers to the lateral hemispheres of the cerebellum, receives input from the cerebral cortex via the pons. The output goes initially to the *dentate nucleus* and then through the thalamus to the premotor and motor cortices. By this connection the cerebrocerebellum contributes to the planning of movements. The **vestibulocerebellum** occupies the flocculonodular lobe and receives input from the *vestibular nuclei* in the medulla and sends its output back to the same nuclei. The vestibulocerebellum contributes to the control of eye movements and balance during stance and gait.

The cerebellum seems to function as a **comparator**; it *compares the movements the CNS wants performed to those that actually take place*. This function is accomplished by the cerebellum receiving extensive afferent input and generating output that can modify motor commands. However, there is no consensus on the precise features of movement that are of concern to the cerebellum. Rothwell (1987) identifies three theories of cerebellar function: timing device, learning device, and coordinator. As a *timing device*, the cerebral cortex, it is thought, initiates the movement by activating the agonist muscle, and the cerebellum would stop the movement at the desired location. The cerebellum may function as a *learning device* by strengthening selected input-

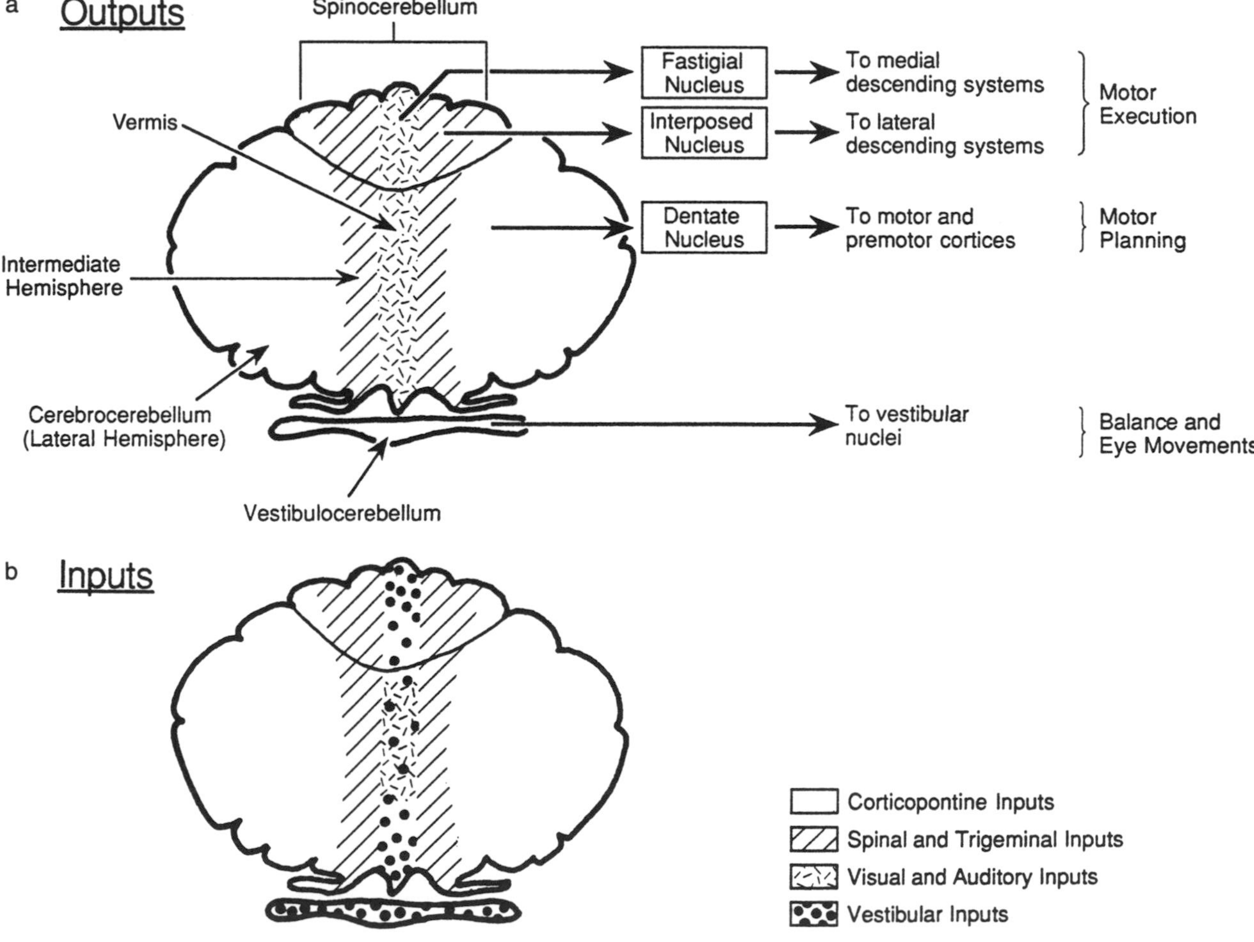

Figure 7.9 Three functional subdivisions of the cerebellum.
Note. From "The Cerebellum" by C. Ghez. In *Principles of Neural Science* (3rd ed.) (p. 633) by E.R. Kandel, J.H. Schwartz, and T.M. Jessell (Eds.), 1991a, New York: Elsevier Science Publishing Co., Inc. Copyright 1991 by Elsevier Science Publishing Co., Inc. Adapted by permission of Appleton & Lange, Norwalk, CT.

output connections so that, over time, less input is required to elicit the intended output. In this scheme, the cerebellar circuits are modified by experience so that recurring input strengthens certain synapses, and the cerebellum learns the appropriate movement for a given context. The cerebellum may also serve as a *coordinator* of muscle activity across several joints so that the activity at different joints is scaled to produce the intended limb movement. Clearly, the cerebellum is able to modulate ongoing movements; it also seems to be important in the learning of motor skills.

Basal Ganglia

The basal ganglia represent five closely related nuclei: caudate, putamen, globus pallidus, subthalamic nuclei, and substantia nigra. The caudate and putamen are indicated as a single nucleus called the striatum in Figure 7.10. The basal ganglia receive a major input from the cerebral cortex and send most of their output, via the thalamus, back to the cortex (Figure 7.10). The basal ganglia represent an *intermediate link between the nonmotor cerebral cortex and the primary motor cortex.* The basal ganglia are concerned with motor planning in that the output neurons code for various aspects of movement—such as direction, amplitude, and velocity—rather than the detailed activation of specific muscles and motor programs. The timing of neuronal discharge suggests that there are at least two types of cells in the basal ganglia: movement-related cells concerned with the ongoing movement rather than initiation of movement and cells that are active before movement begins and are related to the preparation for movement. In addition, the basal ganglia are involved in nonmovement behaviors, such as cognition and habitual learning.

Although the basal ganglia do not have direct connections with the spinal cord, a role for the basal ganglia in movement has been well documented by the study of patients with Parkinson's disease, Huntington's disease, and hemiballismus (Hallett, 1993). As described by Côté and Crutcher (1991), these diseases involve impairment of the basal ganglia and are characterized by (a) involuntary movements and tremor; (b) muscular rigidity—changes in posture and muscle tone; and (c) immobility—poverty and slowness of movement without paralysis.

Ascending Pathways

The human body contains a great diversity and number of sensory receptors. Those that have a more direct effect on movement were discussed in chapter 4 in the description of the single joint system. The afferent signals from these sensory receptors enter the spinal cord through the dorsal root and synapse onto neurons at both the segmental and suprasegmental levels. The

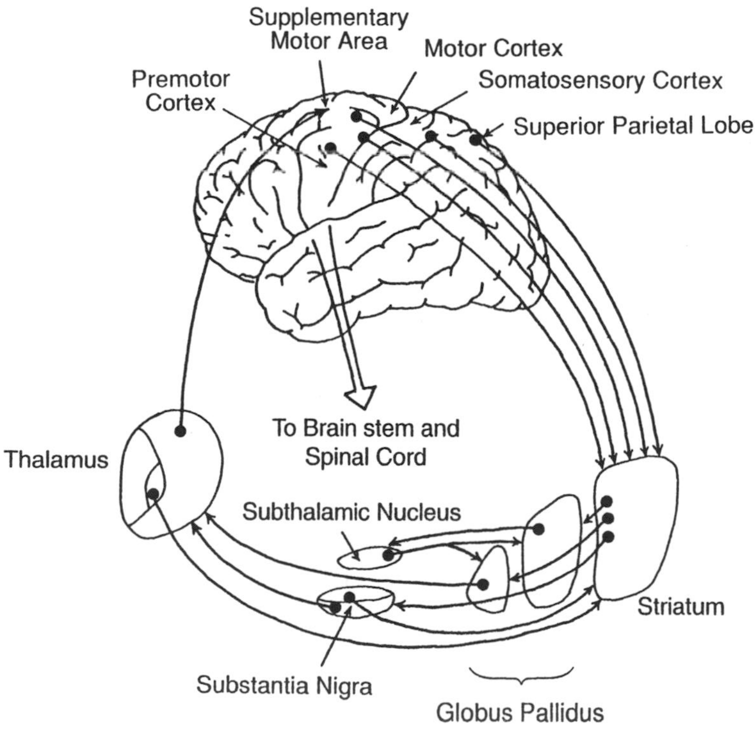

Figure 7.10 The motor circuit of the basal ganglia.
Note. From "The Basal Ganglia" by L. Côté, and M.D. Crutcher. In *Principles of Neural Science (3rd ed.)* (p. 651) by E.R. Kandel, J.H. Schwartz, and T.M. Jessell (Eds.), 1991, New York: Elsevier Science Publishing Co., Inc. Copyright 1991 by Elsevier Science Publishing Co., Inc. Adapted by permission of Appleton & Lange, Norwalk, CT.

groups of fibers that distribute the information to the suprasegmental centers are referred to as *ascending pathways*. Somatic sensory information from the arms, legs, and trunk ascends through the thalamus to the cerebral cortex in two major pathways: the dorsal column–medial lemniscus pathway and the anterolateral pathway. Somatic sensory information includes afferent signals from mechanoreceptors, thermoreceptors, nociceptors, and chemoreceptors. These two ascending pathways play a major role in **perception** whereby the afferent signals reach consciousness and are recognized (perceived).

The **dorsal column–medial lemniscus pathway** distributes information on limb proprioception and tactile sensation, including touch and vibration sense. Both ascending pathways send sensory information to the contralateral brain, but they cross the midline at different points (Figure 7.11). The dorsal column–medial lemniscus pathway crosses in the medulla and ascends through the brain stem as the medial lemniscus to the thalamus and the primary somatic sensory area of the somatosensory cortex. The **thalamus** integrates and distributes most of the sensory and motor information going to the cerebral cortex. The thalamus receives input about somatic sensation, audition, and vision and transmits information to the cerebellum, basal ganglia, and primary sensorimotor cortex. The afferent axons in the dorsal column are arranged somatotopically, with input from the sacrum most medial and that from the neck most lateral. A somatotopic organization of the afferent input is preserved all the way to the primary (SI) and secondary (SII) somatic sensory cortices and to the posterior parietal cortex. The information conveyed in the dorsal column–medial lemniscus pathway is not simply relayed from the sensory receptors in the periphery to the primary somatic sensory cortex but rather is processed to varying degrees along the pathway.

The **anterolateral pathway** conveys information mainly about pain and temperature and also some tactile and proprioceptive information. The axons in this pathway cross in the spinal cord and ascend in the lateral part of the spinal cord to the reticular formation in the brain stem and the thalamus. The anterolateral pathway actually comprises three parts: spinothalamic, spinoreticular, and spinocervical. The names describe the origin and termination of the parts. The spinothalamic and spinoreticular tracts carry noxious and thermal afferent signals. The spinocervical tract ends in an area of the midbrain that contains neurons involved in the descending control of pain. The information in these three tracts is also sent to the primary and secondary somatic sensory cortices and to the posterior parietal cortex.

Although the two pathways project to the thalamus, the axons synapse on separate populations of neurons. The information from the different classes of sensory receptors remains segregated in the spinal cord, brain stem, and thalamus and does not interact until reaching the somatic sensory areas of the cerebral cortex. The

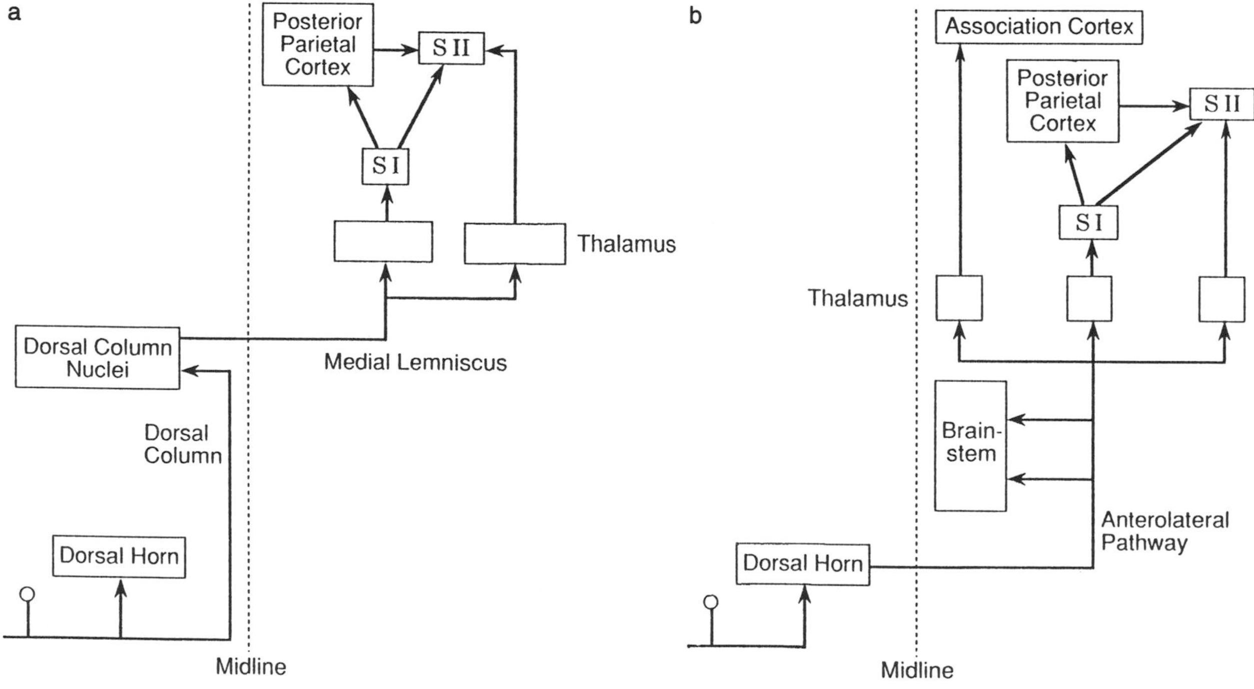

Figure 7.11 Ascending pathways: (a) dorsal column–medial lemniscus; (b) anterolateral.
Note. From "Anatomy of the Somatic Sensory System" by J.H. Martin and T.M. Jessell. In *Principles of Neural Science* (3rd ed.), (p. 359) by E.R. Kandel, J.H. Schwartz, and T.M. Jessell (Eds.), 1991, New York: Elsevier Science Publishing Co., Inc. Copyright 1991 by Elsevier Science Publishing Co., Inc. Adapted by permission of Appleton & Lange, Norwalk, CT.

sensory information is distributed among several areas of the cerebral cortex and to subcortical areas, where it influences the flow of information in descending pathways.

Musculoskeletal Organization

The musculoskeletal system consists of the peripheral parts of the motor system and comprises muscle and the connective tissue elements that form the skeleton. In addition to the musculoskeletal system, the *motor system* includes those components of the nervous system that are involved in the production of movement. Figure 7.12 characterizes the biomechanical features of the musculoskeletal system (Zajac & Gordon, 1989). In this scheme, the signals transmitted by motor units (EMG) represent the commands issued by the nervous system to activate muscle and to produce a force that is transmitted to the skeleton and appears as a torque about the joint. The force exerted by muscle depends on its mechanical properties (chapter 6), whereas the torque is affected by both the muscle force vector and its moment arm relative to the joint axis of rotation (chapter 2). The biomechanical effect of muscle activation, however, is not limited to the joint or joints that the muscle crosses but rather is distributed throughout the musculoskeletal system (see the Intersegmental Dynamics section in chapter 3). As a result of this musculoskeletal organization, the activation of a muscle by the nervous system can affect the kinematics (angle, velocity, and acceleration) of many joints throughout the body. Furthermore, as the kinematics change, this can also affect muscle dynamics (length-force and force-velocity relationships) and joint geometry (moment-arm length).

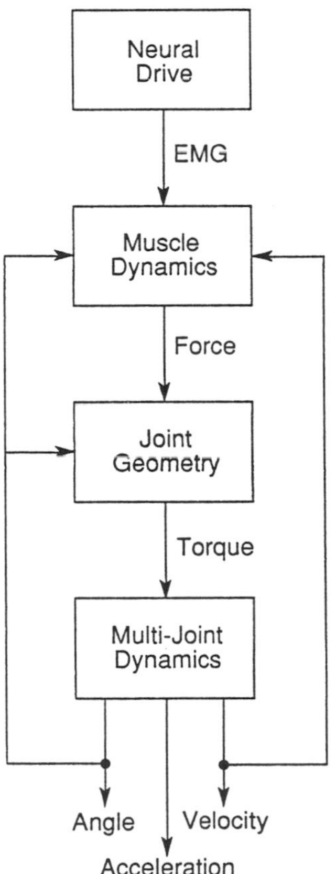

Figure 7.12 Biomechanical consequences (EMG, force, torque, angle, velocity, acceleration) of the activation of the musculoskeletal system by the nervous system.
Note. From "Determining Muscle's Force and Action in Multi-Articular Movement" by F.E. Zajac and M.E. Gordon. In *Exercise and Sport Science Reviews* (Vol. 17) (p. 188) by K.B. Pandolf (Ed.), 1989, Baltimore: Williams & Wilkins. Copyright 1989 by American College of Sports Medicine. Adapted by permission.

This scheme emphasizes an important point. A given command from the nervous system does not always elicit the same biomechanical effect from the motor system. Rather, *the outcome of the neural command depends on the state of the musculoskeletal system* (e.g., muscle length, muscle velocity, moment-arm length). These two factors, the quality and quantity of the neural drive and the state of the musculoskeletal system, are at least equally important in determining the ability of the motor system to perform movement (Bernstein, 1984). To produce a desired movement, the nervous system must know the state of the musculoskeletal system. Two significant constraints imposed by musculoskeletal design are that movement is caused by net muscle activity and that muscle attachment sites vary about a joint.

Net Muscle Activity

In chapter 2 we emphasized that human movement generally is due to an imbalance of the forces acting on the body, rather than to the application of a single, isolated force. Due to musculoskeletal organization, this generalization also applies to the effect of muscle activity about a joint. We do not typically activate a single muscle by itself; usually we activate a group of synergist muscles or an agonist-antagonist set of muscles. Because of this mode of operation, it is important to reemphasize that *resultant muscle force* and *resultant muscle torque* (chapter 2) are the net effect of muscle activity about a joint. We learned in chapter 3 how to use Newtonian mechanics to calculate the resultant muscle force for both static and dynamic conditions. In this approach, we defined a free body diagram, accounted for all the kinematic and kinetic effects acting on the system except the resultant muscle force (or resultant muscle torque), and then calculated the resultant muscle force as the residual moment (chapter 3). This provided an estimate of the net muscle activity and the contribution of the muscle activity to the movement.

Bone-on-Bone Force. One limitation to studying movement as the product of the resultant muscle force is that we are unable to determine the absolute force exerted by muscle. Although it is the net muscle force that influences a movement, the absolute force is an important consideration in such fields as prosthetic design. In chapter 2 we introduced the concept of a *joint reaction force* to account for the *net* forces generated by bone-on-bone contact between adjacent body segments. These forces can be exerted by tissues (e.g., capsule, ligament, muscle) and by external agents (e.g., ground reaction force). In contrast, the term *bone-on-bone force* represents the *absolute* force at the joint due to the contact of body segments. The difference in magnitude between the joint reaction force and the bone-on-bone force can be substantial.

An interesting example of this difference was provided by Galea and Norman (1985) in a study of the bone-on-bone forces at the ankle joint during a rapid ballet movement—a spring from flat feet onto the toes. In this analysis, both the joint reaction force and the bone-on-bone force due to the forces exerted by the muscles involved in the movement were determined. The model used to perform these calculations (Figure 7.13a) took into account the major muscles crossing the ankle joint and therefore those that contributed to the muscle-dependent component of the bone-on-bone force. The muscles included extensor hallucis longus (EHL), tibialis anterior (TA), flexor hallucis longus (FHL), peroneus longus (PL), and gastrocnemius/soleus (G/S). For the calculations, the force exerted by each of these muscles was estimated based on its EMG, length, and rate of change in length (Hof, 1984; Hof & van den Berg, 1981a, 1981b, 1981c, 1981d).

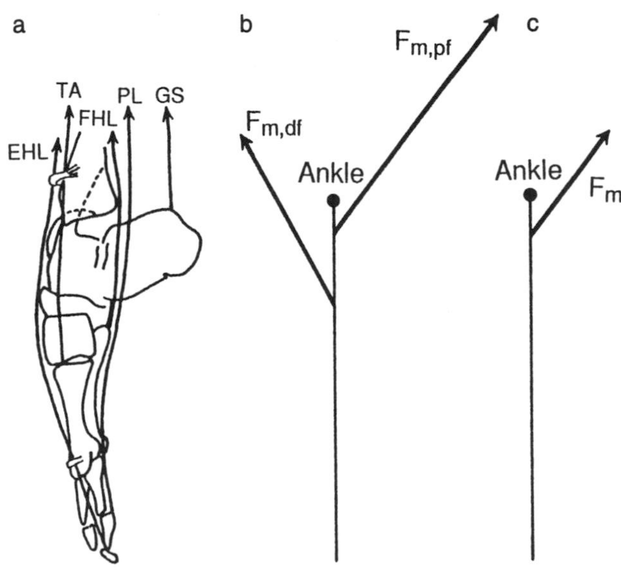

Figure 7.13 Bone-on-bone forces at the ankle joint during a full pointe: (a) schematic view of the foot showing the lines of action of the major muscles crossing the ankle joint; (b) separation of the muscular effects into those that exert dorsiflexor and plantarflexor torques; (c) the resultant muscle force.

Note. From "Bone-on-Bone Forces at the Ankle Joint During a Rapid Dynamic Movement" by V. Galea and R.W. Norman. In *Biomechanics IX-A* (p. 72) by D.A. Winter, et al. (Eds.), 1985, Champaign, IL: Human Kinetics. Copyright 1985 by Human Kinetics Publishers, Inc. Adapted by permission.

Once Galea and Norman (1985) had estimated a force for each muscle, they grouped the forces (Figure 7.13b) into those that contributed to a plantarflexor torque ($\mathbf{F}_{m,pf}$) and those that contributed to a dorsiflexor torque ($\mathbf{F}_{m,df}$). The bone-on-bone force was calculated as the tangential component of the $\mathbf{F}_{m,pf}$ vector *plus* the tangential component of the $\mathbf{F}_{m,df}$ vector. The joint reaction force, however, was calculated as the tangential component of the resultant muscle force ($\mathbf{F}_m$), where $\mathbf{F}_m$ is the *difference* between $\mathbf{F}_{m,pf}$ and $\mathbf{F}_{m,df}$ (Figure 7.13c). In terms of vector algebra, which considers both the magnitude and direction of vectors, $\mathbf{F}_m$ is always defined as the vector sum of $\mathbf{F}_{m,pf}$ and $\mathbf{F}_{m,df}$. For one subject, Galea and Norman calculated an average joint reaction force due to muscle activity of 732 N during the movement and an average bone-on-bone force of 6,068 N. This is a difference of approximately one order of magnitude.

Muscle Activity and Movement. Despite this difference in the absolute and net effects at a joint, *it is the net and not the absolute muscle force that influences a movement*. For this reason, the ability to determine the net muscle activity exerted by the musculoskeletal system can be useful in the prescription of exercise and rehabilitation activities. It is possible to determine net muscle activity by a qualitative analysis of a movement. For example, consider a person lying supine on a bench and performing forearm curls (elbow flexion-extension movements) with a 120-N weight in the right hand (Figure 7.14a). If the individual executes a *slow* movement throughout the full range of motion, what is the net muscle activity as the arm moves from Position 1 through Position 4? Which muscle group is primarily responsible for the movement? Is it active concentrically or eccentrically? What provides the load, and about what point is it acting?

The load against which the muscle group acts is the weight held in the person's hand $\mathbf{F}_{w,h}$ plus the weight of the forearm-hand segment $\mathbf{F}_{w,s}$. The appropriate free body diagram indicates that the axis of rotation is the elbow joint (Figure 7.14b). The free body diagram also includes the resultant muscle force ($\mathbf{F}_m$) and the joint reaction force ($\mathbf{F}_j$). The analysis concerns the torques that are acting about the elbow joint ($\mathbf{T}_m$ and $\mathbf{T}_l$), and the free body diagram can be simplified as shown in Figure 7.14c. In this simplification, the load torque represents the effects of the system weight ($\mathbf{F}_{w,s}$ + $\mathbf{F}_{w,h}$). The direction of the weight vectors is obtained by the right-hand-thumb rule; the fingers of the right hand are curled in the direction that each force (resultant muscle force and resultant load force) will cause the system to rotate, and the right-hand thumb is extended. If the

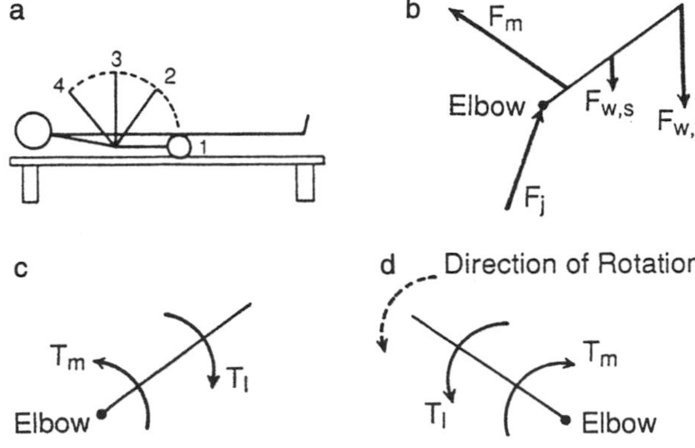

Figure 7.14 Determination of the pattern of net muscle activity for a *slow* movement. (a) A subject performs an elbow flexion-extension movement with a dumbbell in the right hand. (b) Free body diagram isolates the forearm-hand segment. (c) Resultant muscle and load torques are determined. (d) The net muscle activity is based on comparison of the directions of the torques and the movement.

thumb extends out from the page, the torque vector is drawn in a counterclockwise direction. And if the thumb extends into the page, the torque vector is drawn in a clockwise direction.

From these diagrams we can determine the muscle group mainly responsible for the movement and the direction of its change in length (concentric or eccentric). The appropriate steps involve drawing

1. the body segment in the approximate position,
2. the load torque vector ($\mathbf{T}_l$),
3. the resultant muscle torque vector ($\mathbf{T}_m$) in the opposite direction to the load torque vector, and
4. the direction of rotation of the body segment.

For example, consider the movement of the forearm-hand segment in the forearm curl from Position 3 to Position 4. Note that the body segment is drawn not at either Position 3 or Position 4 but rather midway between the two positions (Figure 7.14d). From this diagram, we can deduce the direction of the resultant muscle torque vector and determine whether the net muscle activity is concentric or eccentric. For *slow* movements, *the direction of the resultant muscle torque vector is opposite to that of the load torque vector*. To determine the muscle group represented by the resultant muscle torque vector, imagine that no other forces or torques are acting on the system and that the segment rotates in the same direction as the resultant muscle torque vector. *If this direction is one of extension, then the responsible muscle group is the extensors; if the resultant muscle torque vector produces flexion, then the vector represents the flexor muscles.* Alternatively, recall that muscles can only exert a pulling force. So when the resultant muscle force ($\mathbf{F}_m$) vector is drawn on the free body diagram, it is possible to determine which muscles (e.g., flexors or extensors, abductors or adductors) would exert such a force. In the example shown in Figure 7.14, a clockwise torque indicates the elbow extensors and counterclockwise represents the elbow flexors (i.e., $\mathbf{T}_m$ represents an elbow extensor torque).

By comparing the directions of the resultant muscle torque vector and the segment rotation, we can determine whether the net muscle activity is concentric or eccentric. If the directions of the resultant muscle torque vector and the segment rotation are the same, the muscle group is experiencing a concentric contraction. If the segment rotation and the resultant muscle torque are not acting in the same direction (as in the movement from Position 3 to Position 4), the net muscle activity is an eccentric contraction. Thus the movement from Position 3 to Position 4 is controlled by an eccentric contraction of the elbow extensors. In general, if an extensor muscle group produces extension or a flexor group causes flexion, the net muscle activity is concentric. In contrast, flexion controlled by an extensor group or extension controlled by a flexor group represents net eccentric activity. Using these procedures, we can determine qualitatively the net muscle activity for such slow movements as the forearm curl over the entire range of motion (Table 7.1).

Table 7.1 Net Muscle Activity for the Forearm-Curl Exercise Shown in Figure 7.14

Movement	1→2	2→3	3→4	4→3	3→2	2→1
Elbow flexors	C	C			E	E
Elbow extensors			E	C		

Note. C = concentric; E = eccentric.

The example of the forearm curl with a supine subject emphasizes an important feature of the way in which we use our muscles. Suppose that the individual moves from the supine position to an upright posture and performs a similar exercise of raising and lowering an extended arm between Positions 1 and 2 (Figure 7.15). What patterns of net muscle activity are needed to complete this movement? Using the procedure that has just been described, we can determine that the movement is controlled by the shoulder flexors—that is, both going from Position 1 to Position 2 and from Position 2 to Position 1. The muscle activity would be concentric going up and eccentric going down. The difference in the patterns of activity between the supine and upright postures has to do with *the orientation of the individual (and hence the movement) relative to the direction of gravity*. The gravitational environment has a major effect on the pattern of muscle activation. For example, imagine that, in both the upright standing and supine examples, a vertical line (gravity) passes through the axis of rotation (shoulder and elbow, respectively). The major qualitative difference is that the movement never crosses this line in the upright exercise, whereas it does in the supine exercise. For *slow* exercises, crossing this line means changing the muscle group that controls the movement.

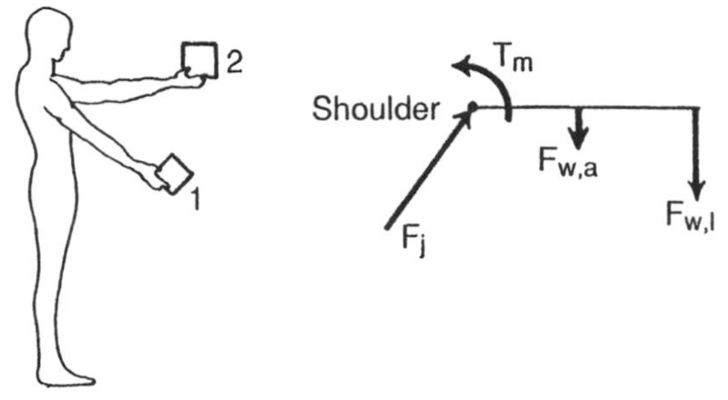

Figure 7.15 A shoulder flexion-extension movement. Because the load torque ($\mathbf{F}_{w,a}$ + $\mathbf{F}_{w,l}$) acts in the same direction throughout the movement, the direction of the resultant muscle torque ($\mathbf{T}_m$) is constant throughout the range of motion.

Note that the analysis just employed applies only to conditions that we call **quasistatic**—that is, those instances that are almost static and in which the acceleration of the system and its various parts is quite small (Miller, 1980; Toussaint et al., 1992). *Under quasistatic conditions we can approximate the mechanical state of the muscle by doing a static analysis* (i.e., $\Sigma\mathbf{F} = 0$). Researchers do not agree on the boundary that separates a quasistatic state from one that is not. Alexander and Vernon (1975), for example, have used a version of the quasistatic approach to calculate the resultant muscle torques about the ankle and knee joints during the stance phase of a jog. Others, however, think that such conditions require a complete dynamic analysis (i.e., $\Sigma\mathbf{F} = ma$) (Mann, 1981; Miller & Munro, 1985).

Certainly a dynamic analysis is necessary for more vigorous activities, such as kicking. For example (Miller, 1980), consider the pattern of muscle activity about the knee joint during the execution of a soccer toe kick (Figure 7.16). To demonstrate that a quasistatic analysis is not appropriate, let us perform a quasistatic analysis and then compare the results with those from a dynamic analysis. The four positions of the kicking leg (the one with the filled-in shoe) from Figure 7.16 are superimposed onto a stick figure, first relative to the hip joint and then with respect to the knee joint (Figure 7.17). Notice that the position of the shank relative to the knee in Figure 7.17 is the same as its original position in Figure 7.16. Next, let us determine the pattern of net muscle activity about the knee joint during the kick as we did for the forearm-curl example. Because the movement (Positions 1 through 4) never crosses the vertical line that passes through the axis of rotation (knee), the load-torque vector, and therefore the resultant muscle force vector, act in the same direction throughout the movement and the net muscle activity is shown in Table 7.2.

On the basis of this quasistatic rationale, the motion about the knee joint during the toe kick (Figure 7.16)

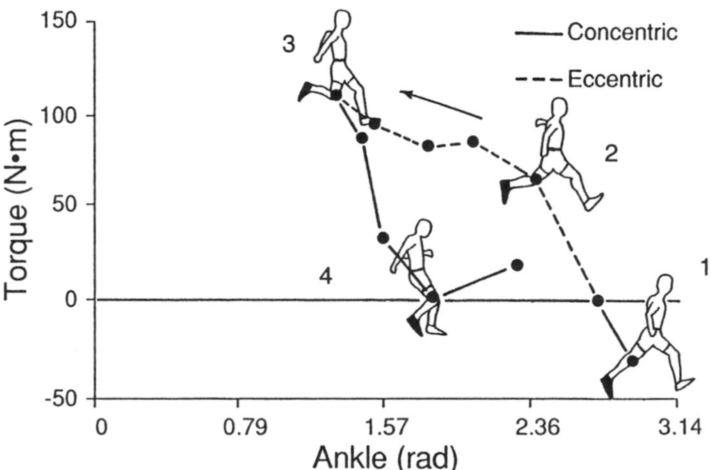

Figure 7.16 Resultant muscle torque about the knee joint of the kicking leg as a function of knee angle during the swing preceding ball contact in a fast soccer toe kick. The total time for the swing was 170 ms, with 17 ms between each of the large dots on the diagram. To interpret the graph, begin at the bottom right-hand corner and proceed in a counterclockwise direction from Position 1 to Position 4.

Note. From "Body Segment Contributions to Sport Skill Performance: Two Contrasting Approaches" by D.I. Miller, 1980, *Research Quarterly for Exercise and Sport*, **51**, p. 225, a publication of the American Alliance for Health, Physical Education, Recreation and Dance, 1900 Association Drive, Reston, VA 22091. Reprinted by permission.

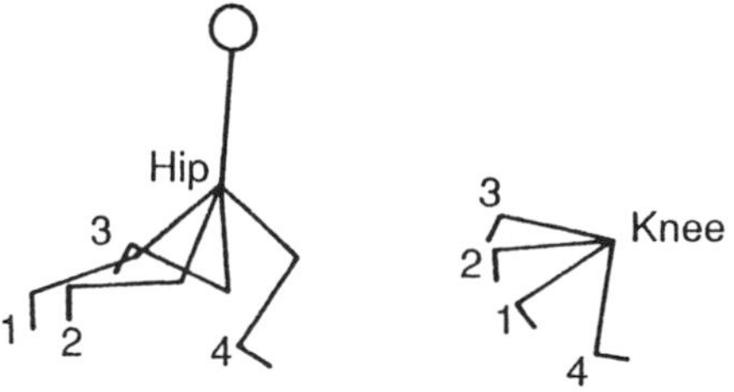

Figure 7.17 Angular displacement of the shank relative to the knee during the soccer toe kick shown in Figure 7.16.

Table 7.2 Net Muscle Activity for the Soccer Toe Kick Shown in Figure 7.16 Based on a Quasistatic Analysis

Movement	1→2	2→3	3→4
Knee flexors	C	C	E
Knee extensors			

Note. C = concentric; E = eccentric.

is supposedly controlled by the knee flexor muscles. Our intuition tells us that it would be difficult to convince anybody that the knee flexors were primarily responsible for this kicking movement. The obvious shortcoming of this analysis is the assumption that the movement is *slow*. Clearly, it is not slow! In chapter 2 we considered some of the forces that are commonly encountered in human movement. Recall that inertial forces were effects due to the motion of an object or segment. In fast movements, inertial forces become substantial and drastically alter the pattern of muscle activity that is necessary for the control of movement. Indeed, for an event such as a strong kick, it even becomes necessary to consider the way in which the motion of the shank will affect the thigh and, conversely, how thigh motion will affect the shank (Putnam, 1983, 1991).

The magnitude of the inertial effects can be seen by comparing the resultant muscle torque about the knee joint in Figure 7.16 with that deduced by the quasistatic approach. Interpretation of the angle-torque graph (Figure 7.16) begins in the lower right-hand corner and progresses in a counterclockwise direction. The graph indicates several interesting features: Except for the first part of the kick, the resultant muscle torque is due to the knee extensors and not the flexors, as suggested by the quasistatic analysis; most of the movement from Position 1 to Position 3, during which knee angle decreases, is controlled by eccentric extensor activity; from Position 3 to contact with the ball, the net activity is a concentric extensor effect; and the peak resultant muscle torque occurs at the change from eccentric to concentric activity. These observations indicate that a quasistatic analysis, which ignores the effects due to acceleration, is inappropriate for revealing the pattern of net muscle activity for tasks such as a rapid kick.

One technique that can be used to derive qualitatively the information shown in Figure 7.16 is to align the resultant muscle torque and angular velocity graphs (Figure 7.18). To illustrate this procedure, let us return to the example of horizontal extension-flexion at the elbow joint (Figure 3.28). Half of the movement involves extension (positive velocity), and half involves flexion. The change in direction of the movement from extension to flexion occurs when velocity is zero. The first part of the extension phase is produced by a net extensor torque; hence the extensors are active eccentrically. The rest of the extension phase, however, is accomplished by net flexor activity, which means that the flexors are being actively lengthened (i.e., an eccentric contraction). By similar reasoning, the flexion phase is controlled by concentric (C) flexor activity and then by eccentric (E) extensor activity (Figure 7.18). We arrived at this same pattern of net activity by other means in chapter 1 (Figure 1.9).

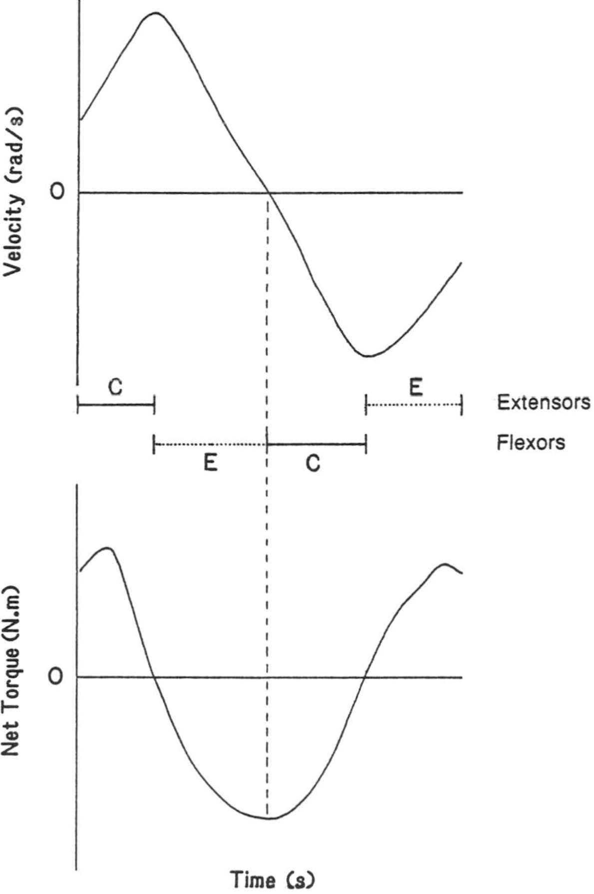

Figure 7.18 Deduction of the net muscle activity responsible for an elbow extension-flexion movement performed in a horizontal plane. Positive velocity represents extension at the elbow joint. Positive net torque indicates a resultant extensor muscle torque.

This discussion of net muscle activity emphasizes an important consequence of musculoskeletal design. Because of intersegmental dynamics (chapter 3), the nervous system controls fast and slow movements differently; that is, the pattern of net muscle activity changes with movement speed to accommodate inertial effects between segments. Also, fast movements typically involve an eccentric-concentric sequence in which the peak resultant muscle torque occurs at the transition from the eccentric to the concentric phase. Nonetheless, whether a movement is performed slowly or rapidly, the contribution of muscle is determined by the net activity among synergists and antagonists.

Muscle Attachments

One feature of musculoskeletal organization that has received considerable attention is the distribution of activity among the muscles involved in a movement. Frequently, groups of muscles are activated by the nervous system to perform a particular movement. The term **synergy** refers to *a group of muscles that are constrained to act as one unit.* Because muscles can only pull and not push, a minimum of one pair of muscles must cross a joint to control each degree of freedom (e.g., flexion-extension, abduction-adduction). For example, the elbow joint needs at least one muscle to control flexion and one to control extension. Often, however, the arrangement of muscles is not this simple, and usually several muscles contribute to the same action. The primary elbow flexors, for example, include biceps brachii, brachialis, and brachioradialis. Why do we have so many muscles? Do we have more muscles than we actually need?

Although each of these three muscles can contribute to an elbow flexor torque, they are not equivalent force generators, because they have different architectures and therefore different capabilities (Buchanan, Rovai, & Rymer, 1989; Funk, An, Morrey, & Daube, 1987). As we discussed in chapter 6, differences in muscle fiber length, muscle cross-sectional area, and muscle attachment points affect the range of motion, contraction velocity, maximum force, and shape of the torque-angle relationship. Table 7.3 outlines some of the architectural differences between the three main elbow flexor muscles; this diversity exists at many joints throughout the body (e.g., ankle, knee, hip, vertebral column, shoulder). Not only can the contributions of the three elbow flexors vary over the range of motion, but variation in the attachment points can enable a muscle to contribute to more than one mechanical action. For example, because of the location of its distal attachment, biceps brachii is able to contribute to both the elbow flexor and forearm-supinator torques. Buchanan et al. have shown that for a variety of flexion-extension, internal-external rotation (varus-valgus), and supination-pronation loads at the wrist, there is a strong association between the activity of brachialis and brachioradialis but not between biceps brachii and brachioradialis (Figure 7.19). Increasing supination loads were associated with increases in biceps brachii but not in brachioradialis EMG; conversely, increasing pronation loads were accommodated by increasing brachioradialis but not biceps brachii EMG. Biceps brachii was typically most active when brachioradialis and brachialis were least active. With this diversity in design and associated function, the nervous system can vary the activation of synergist muscles depending on the details of the task.

Table 7.3 Architectural Features of the Three Main Human Elbow Flexor Muscles

Muscle	Muscle length (cm)	Fiber length (cm)	Muscle CSA (cm^2)	Distance to joint	
				Proximal (cm)	Distal (cm)
Biceps brachii	24.0	17.8	5.8	30	4
Brachialis	20.1	14.4	7.4	10	3
Brachioradialis	23.7	14.7	2.0	6	25

Note. The distance to the joint indicates the location of the center of the attachment point (proximal and distal) to the axis of rotation of the elbow joint.
The distance data were kindly provided by Scott L. Delp, Ph.D.

Two-Joint Muscles. One common variation in the points of attachment for a muscle is the number of joints that the muscle spans. For example, a significant number of muscles span two joints and hence are referred to as two-joint muscles (e.g., biceps brachii, rectus femoris). Two-joint muscles provide at least three advantages in the control of the musculoskeletal system.

First, two-joint muscles couple the motion at the two joints that they cross (Arsenault & Chapman, 1974; Fujiwara & Basmajian, 1975; van Ingen Schenau, Bobbert, & van Soest, 1990). For example, biceps brachii crosses both the elbow and the shoulder joints and thus controls both elbow and shoulder flexion. Because these two movements occur concurrently in many daily activities, it is useful to have a muscle that contributes to both actions. In addition, this coupling can be achieved by a reduction in the EMG of a single-joint muscle and an increase in the activity of the two-joint muscle. Yamashita (1988) has reported a reduction in the EMG of gluteus maximus (a single-joint hip extensor) and an increase in the EMG of semimembranosus (a two-joint hip extensor and knee flexor) during concurrent hip extension and knee extension. Zajac (1993) suggests that single-joint muscles produce the propulsive energy for a vertical jump while the two-joint muscles refine the coordination.

Second, the shortening velocity of a two-joint muscle is less than that of its single-joint synergists (van Ingen Schenau et al., 1990). For example, the shortening velocity of rectus femoris during concurrent hip flexion

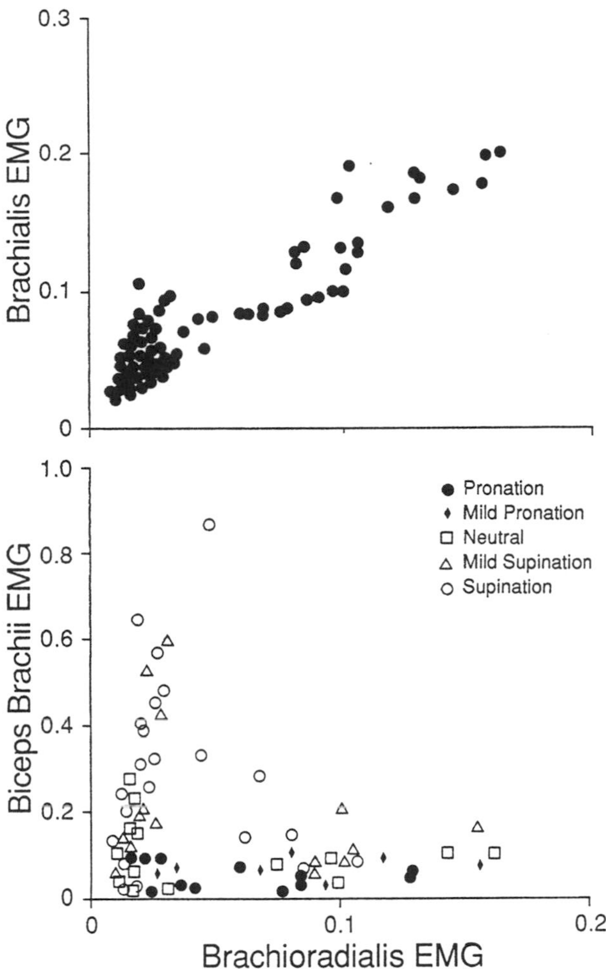

Figure 7.19 Comparison of changes in EMG activity (expressed in arbitrary units) among the three main elbow flexor muscles during the application of various loads at the wrist. Pronation loads result in the covariation of EMG in brachialis and brachioradialis. In contrast, there was not a general covariation of EMG activity in biceps brachii and brachioradialis with different loads.
Note. From "Strategies for Muscle Activation During Isometric Torque Generation at the Human Elbow" by T.S. Buchanan, G.P. Rovai, and W.Z. Rymer, 1989, *Journal of Neurophysiology*, **62**, p. 1209. Copyright 1989 by The American Physiological Society. Adapted by permission.

and knee extension is less than the shortening velocity of the vasti, and the shortening velocity of gastrocnemius is less than the shortening velocity of soleus during concurrent knee extension and plantar flexion. The advantage of the lesser shortening velocity for the two-joint muscles is that they are higher on the force-velocity relationship (Figure 6.19) compared with the one-joint muscle and are capable of exerting a force that is a greater proportion of the isometric maximum.

Third, two-joint muscles can redistribute muscle torque and joint power throughout a limb (Gielen, van Ingen Schenau, Tax, & Theeuwen, 1990; Toussaint et al., 1992; van Ingen Schenau et al., 1990). Figure 7.20 represents a model of the human leg comprising a pelvis, thigh, and shank with several single- and two-joint muscles. In this model, muscles 1 and 3 are single-joint hip and knee extensors, muscles 2 and 4 are single-joint hip and knee flexors, and muscles 5 and 6 are two-joint muscles. Concurrent hip and knee extension can be performed by activating the two single-joint extensors (muscles 1 and 3). Because muscle 5 (two-joint) exerts a flexor torque about the hip joint and an extensor torque about the knee, concurrent activation of muscle 5 with muscles 1 and 3 will result in a reduction in the net torque at the hip but an increase in the net torque at the knee. Based on this interaction, the two-joint muscle (muscle 5) is described as redistributing some of the muscle torque and joint power from the hip to the knee. Conversely, activation of muscle 6 along with muscles 1 and 3 will result in a redistribution from the knee to the hip.

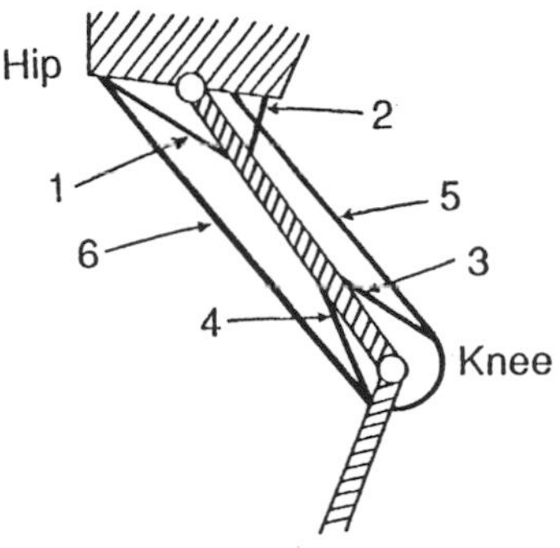

Figure 7.20 Model of the human leg with six muscles arranged around the hip and knee joints. Muscles 1 through 4 cross a single joint; muscles 5 and 6 cross both joints.
Note. From "The Unique Action of Bi-Articular Muscles in Leg Extensions" by G.J. van Ingen Schenau, M.F. Bobbert, and A.J. van Soest. In *Multiple Muscle Systems* (p. 647) by J.M. Winters and S.L.-Y. Woo (Eds.), 1990, New York: Springer-Verlag. Copyright 1990 by Springer-Verlag New York, Inc. Adapted by permission.

Moment-Arm Effects. In Part I, we discussed the notion that, as muscle moves through its range of motion, the *moment arm* from the muscle force vector to the axis of rotation (joint) changes. For the elbow flexors, the moment arm is greatest in a midrange position (elbow angle about 1.57 rad) and least at full extension and full flexion. Based on this relationship, we would expect *the muscle force necessary to support a constant-torque load to vary inversely with the changes in the moment arm.* That is, the resultant muscle torque must be constant over the range of motion to support a constant-torque load. However, recall that the resultant muscle torque is equal to the product of muscle force and moment arm. Thus, if the moment arm changes, muscle force must change in an inverse manner to maintain a constant product. This relationship is difficult to assess directly, because both moment arm and muscle force are difficult quantities to measure. Instead, Hasan and Enoka (1985) examined the EMG and found that it varied in the manner that would be expected of muscle force, given the change that occurs in the moment arm (Figure 7.21).

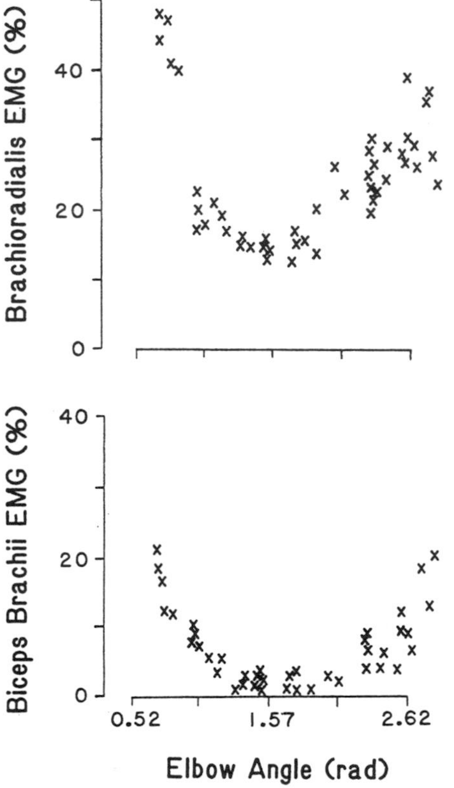

Figure 7.21 EMG activity necessary for two concurrently active elbow flexor muscles (biceps brachii and brachioradialis) to support a constant-torque load over a substantial range of motion of the elbow joint. EMG was measured as the average voltage level needed to support the load and is expressed as a percentage of the maximum voluntary contraction.

Note. From "Isometric Torque-Angle Relationship and Movement-Related Activity of Human Elbow Flexors: Implications for the Equilibrium-Point Hypothesis" by Z. Hasan and R.M. Enoka, 1985, *Experimental Brain Research*, **59**, p. 447. Copyright 1985 by Springer-Verlag, Inc. Adapted by permission.

Movement Strategies

Due to the complexity of both the interactions among the suprasegmental centers and the organization of the musculoskeletal system, many different activation sequences and combinations of muscles can be used to perform a desired movement. In fact, many more options are available than are needed for most movements (Bernstein, 1984). What rules does the nervous system use to choose from among these options? What strategies do we use to perform different movements? It seems desirable to have many degrees of freedom (skeletal, anatomical, and neural options) available when we are learning a new movement. But once we have learned the movement (e.g., breathing, walking, throwing), we use only a subset of the available options. The topic of *movement strategies concerns the study of the neural activation patterns that are associated with different types of movements.* A common approach to the study of movement strategies is to identify variables that are modulated to control movement (e.g., duration and intensity of EMG). Such analyses have been applied to movements that range from the perturbation of the thumb to the performance of the clean by an Olympic weight lifter. The selection of an appropriate strategy, however, is further complicated by the necessity to assure that the system remains stable during the movement. Accordingly, we must first consider the issue of posture before we examine the fundamental movement strategies that have been identified.

Posture

It has been known for some time that the execution of movement is accompanied by the need to maintain postural stability. This association was aptly described by Sherrington (1931): *Posture accompanies movement like a shadow* (Martin, 1977). Because of the rigid link structure of the musculoskeletal system, all movements involve postural activity that is designed to ensure the stability of the system. This postural activity may be as focal as the activation of an antagonist muscle or so widespread as to involve activation of the trunk muscles. Postural activity is automatic and specific to the task and does not require the voluntary (conscious) activation of muscle by the nervous system to maintain the stability of the system.

Whole-Body Stability. **Posture** is a neuromechanical response that concerns the maintenance of equilibrium. A system is in mechanical *equilibrium* when the forces acting on the system sum to zero ($\Sigma \mathbf{F} = 0$). However, this system has **stability** only if following a perturbation it returns to its position of equilibrium. The purpose of postural activity is to *maintain the stability of the musculoskeletal system.* This involves maintaining the position of the system relative to its base of support and ensuring the desired orientation of the body segments not involved in the movement. The ability to maintain an upright posture depends on the location of the line of action of the total-body weight vector relative to the base of support. In an upright posture, the base of support is determined by the position of the feet and includes the area underneath and between the feet. The farther apart the feet are, the larger the base of support and the more stable the individual. Also, stability is inversely related to the height of the center of gravity (Hayes, 1982). A person in an upright stance is in equilibrium as long as the line of action of the weight vector remains within the boundaries of the base of support, and the person is stable as long as the musculoskeletal system can accommodate perturbations and return to an equilibrium position.

When we stand in an upright position, our bodies sway backward and forward. The muscular activity that prevents us from losing balance and falling represents our automatic postural control activity. As an individual sways backward and forward, visual, somatosensory, and vestibular sensory receptors detect these fluctuations and elicit compensatory responses from the appropriate muscles (Dietz, 1992). Incidentally, one of the mechanisms attributed to the increased incidence of falling among the elderly is a decline in the ability to detect and control the forward-backward sway of the body (Horak et al., 1989).

A common method of studing these responses involves having subjects or patients stand on a platform that can be moved suddenly in several different directions (Nashner, 1971, 1972). With the use of this protocol, there appear to be different response strategies depending on the perturbation. For example, Keshner and Allum (1990) found that when the platform was rotated toes up (Figure 7.22a) subjects responded with an initial forward rotation of the legs and trunk and a backward rotation of the head followed later by a smaller backward rotation of the trunk. These segmental displacements resulted in a stiffening of the body with simultaneous activation of the muscles on the dorsal surface of the body for toes-up rotation and on the ventral surface for toes-down rotation. In contrast, when the platform was translated backward the subjects responded with large backward rotation of the legs and forward rotation of the trunk; that is, these two segments rotated in opposite directions (Figure 7.22b). This represented a coordinated, multisegmental response to the perturbation that comprised an ascending sequence of muscle activation (distal to proximal) on the ventral surface of the body for a backward translation and a descending sequence for a forward translation.

Similarly, Cordo and Nashner (1982) found that *postural adjustments were influenced by the body part that provided the contact with the surroundings.* When a subject was in an upright position and only the feet contacted the surroundings, the response to a perturbation was initiated in the leg muscles (Figure 7.23). However, when the hands were used to provide support with the surroundings, the response to a perturbation was initiated in the arm muscles.

Perhaps the most important body segment for whole-body stability is the trunk. Because many movements require postural activation of trunk muscles, the trunk is

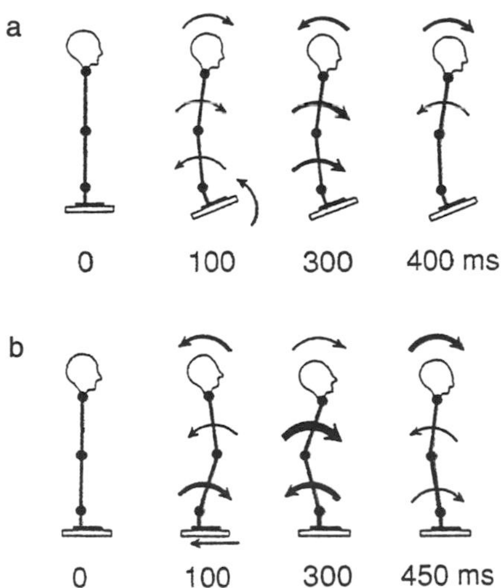

Figure 7.22 Generalized responses to (a) toes-up rotation and (b) backward translation of a support platform. Each perturbation elicited a three-phase response that consisted of an initial unstable phase (100 ms), a compensatory phase (300 ms), and a final correction phase (400 to 450 ms). *Note.* From "Muscle Activation Patterns Coordinating Postural Stability from Head to Foot" by E.A. Keshner and J.H.J. Allum. In *Multiple Muscle Systems* (p. 484) by J.M. Winters and S.L.-Y. Woo (Eds.), 1990, New York: Springer-Verlag. Copyright 1990 by Springer-Verlag, New York, Inc. Adapted by permission.

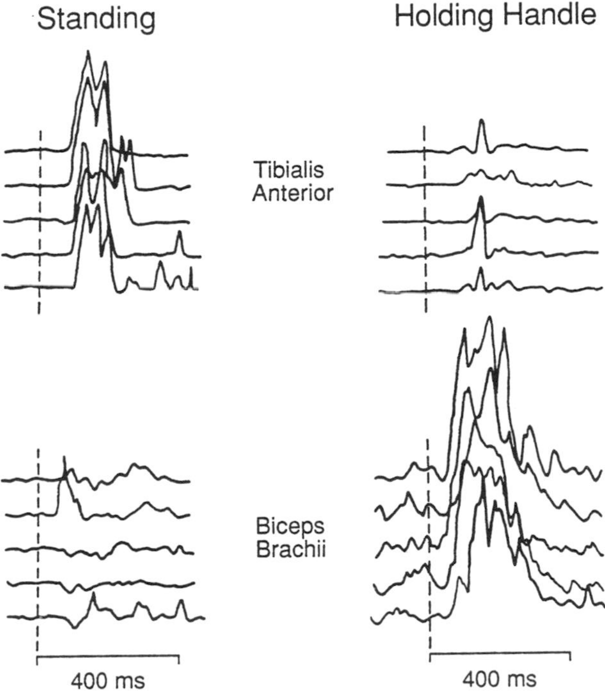

Figure 7.23 EMG responses to a postural disturbance. The vertical dashed line indicates the moment of forward translation of the platform. Each panel shows results from five trials. The responses in tibialis anterior were greatest for the upright stance condition, whereas those for the biceps brachii were greater when the subject held the handle. *Note.* From "Properties of Postural Adjustments Associated with Rapid Arm Movements" by P.J. Cordo and L.M. Nashner, 1982, *Journal of Neurophysiology*, **47**, p. 296. Copyright 1982 by The American Physiological Society. Adapted by permission.

often overused and sustains misuse injuries. Repetitive ergonomic tasks can place undue stress on the musculoskeletal elements of the trunk, especially the vertebral column. The high incidence of incapacitating low-back pain has attracted a lot of attention to the issue of vertebral column stability and postural activity of the trunk muscles (Crisco & Panjabi, 1990). To study this system, the trunk can be modeled as two rigid bodies, the thoracic rib cage superiorly and the pelvis inferiorly, connected by the lumbar vertebrae (Bergmark, 1989). The stability of the spine is controlled by global muscles that connect the two rigid bodies (rib cage and pelvis) and by local muscles and ligaments that attach to the lumbar vertebrae. The global muscles (rib cage–pelvis), however, seem most important in the distribution of external loads supported by the trunk. Crisco and Panjabi demonstrated that the spinal muscles that spanned several vertebrae and those that were most lateral to the vertebral column were most effective at stabilizing the spine.

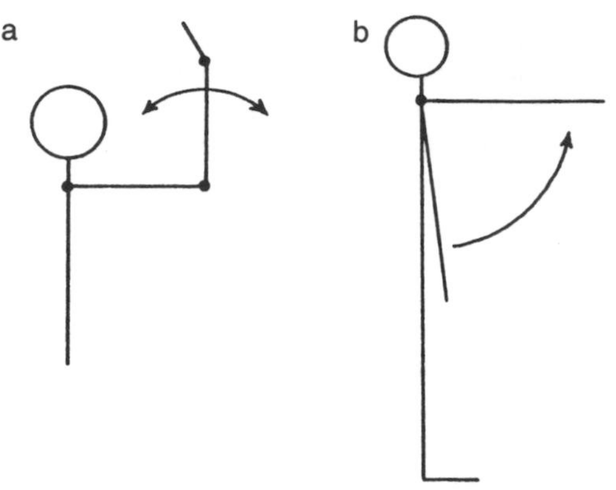

Figure 7.24 Distribution of postural activity with rapid movements. (a) Rapid alternating flexion-extension at the elbow joint requires postural activation of shoulder and wrist muscles to control the inertial effect due to forearm motion. (b) Rapid elevation of the arm to an extended horizontal position is preceded by postural activity in the legs.

Body Segment Orientation. In addition to maintaining a stable whole-body position, postural activity is also concerned with maintaining the orientation of body segments both within and between limbs (Hoy, Zernicke, & Smith, 1985). Consider, for example, the movement shown in Figure 7.24a. The subject is asked to shake the forearm rapidly in a forward-backward motion while keeping the upper arm horizontal. This is done by alternately activating the elbow flexor and extensor muscles. The task also requires substantial activation of the muscles that cross the shoulder joint. This is necessary to stabilize the upper arm and to minimize the inertial effects of the forearm motion on other body segments. Furthermore, the muscles that cross the wrist need to be activated to control the motion of the hand, which could vary from an uncontrolled flail, to a slow wave, to no relative motion between the hand and forearm.

Similarly, suppose an individual standing upright is asked to raise an arm as rapidly as possible to a horizontal position (Figure 7.24b). This is a reaction-time task, in which the movement is done as rapidly as possible following a signal to begin. The anterior deltoid muscle is the prime mover for the task. The fastest a muscle can be activated (the minimum reaction time) following a signal is about 120 ms, which is true for anterior deltoid activity in this task. However, about 50 ms before the anterior deltoid activity, the hamstrings on the same side of the body are activated (Belen'kii, Gurfinkel', & Pal'tsev, 1967). The activity of the leg muscles probably serves at least two purposes: It acts as an *anticipatory stabilization* against the inertial effects of the ensuing arm movement and it provides a rigid connection between the limb motion and the associated ground reaction force applied at the feet. By increasing the rigidity of the body segments not involved in the movement, the anticipatory postural activity can probably facilitate the subsequent movement (Bouisset & Zattara, 1990), even to the extent of transferring energy through intersegmental dynamics. Patients with Parkinson's disease have difficulty combining anticipatory postural adjustments and intentional (voluntary) movements (Rogers, 1991).

Interlimb postural responses are prominent when the two limbs are being used to support a posture. For example, single-leg displacements (perturbations) during stance, balancing, or gait evoke a bilateral postural response of similar latency in both legs (Dietz, 1992). The bilateral response probably provides a more stable base from which to compensate for the perturbation. Similarly, if a hand that is grasping something in the surroundings for support is perturbed, this can elicit an automatic postural response (e.g., increased handgrip) in the contralateral arm and hand (Marsden, Merton, & Morton, 1983). This interlimb coupling of the arms is thought to be one of the mechanisms underlying the accidental discharge of handguns.

Adaptability of Postural Responses. Automatic postural responses are *context-dependent*: The response is tailored to meet the needs of the system-surroundings interaction. For example, the response is focused in the limb (e.g., leg or arm) providing support and the type of response elicited during gait depends on the phase of the gait cycle during which the perturbation was applied. These observations suggest that *the control of posture is not simply based on a set of reflex responses, nor is it a preprogrammed response that is triggered by a disturbance*. Rather, the control of posture is an adaptable feature of the motor system that relies on the integration of afferent input and efferent output.

This adaptability is evident in the modulation of postural responses associated with the toes-up rotation and backward-translation protocol described in Figure 7.22.

In this experiment, the platform was translated backward for four trials (Figure 7.25a) and then it was rotated toes-up for four trials (Figure 7.25b). A common effect of both perturbations was the stretch experienced by the calf muscles of the subject standing on the platform. Whereas a stretch-evoked response stabilizes the subject as a result of the translation by reducing the forward body sway, it destabilizes the subject after the toes-up rotation by increasing body sway. Accordingly, the subject learns to facilitate the response over four translation trials and to adapt (diminish) the response over four rotation trials.

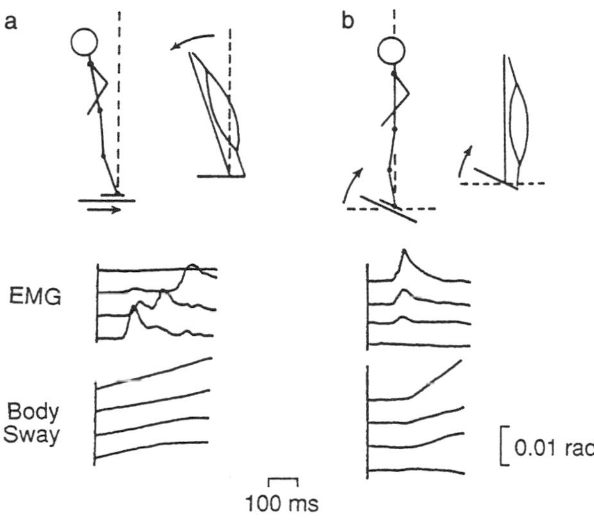

Figure 7.25 Postural responses of a subject when the base of support (platform) was (a) translated backward and (b) rotated toes-up. The stretch response in the calf muscles was measured as the integrated EMG, the effect of which is shown relative to the sway experienced by the subject. For both the EMG and body sway traces, Trial 1 is the top line and Trial 4 the bottom line.
Note. From "Adapting Reflexes Controlling the Human Posture" by L.M. Nashner, 1976, *Experimental Brain Research*, **26**, pp. 62, 65, 66. Copyright 1976 by Springer-Verlag, Inc. Adapted by permission.

A puzzling feature of the automatic postural responses is the appearance of *short-latency activity in muscles that are not stretched by the perturbation.* This distribution of automatic responses has been reported for leg muscles following displacement of the support surface (Nashner, Woollacott, & Tuma, 1979); for shoulder and elbow muscles after imposed disturbances of the arm (Lacquaniti & Soechting, 1986); for wrist muscles after perturbation of the elbow joint (Koshland, Hasan, & Gerilovsky, 1991); and bilaterally after a unilateral disturbance (Dietz & Berger, 1982). For example, Koshland et al. found that a perturbation of the elbow joint consistently elicited responses in wrist muscles that shortened as a result of the perturbation (Figure 7.26a). This activation pattern, as proposed by Gielen, Ramaekers, and van Zuylen (1988), was similar to a volitional elbow flexion movement, even though the mechanical effects at the wrist joint were not similar for the two conditions (Figure 7.26b). These observations suggest that these automatic postural responses are not mediated solely by local stretch reflexes but rather represent a coordinated and predetermined motor pattern.

Fundamental Movement Strategies

Previously we defined a motor program as a stereotyped sequence of commands sent from the spinal cord to the muscles to elicit a specific behavior. The identification of some of the fundamental units that are used in various movement strategies provides information on the commands encoded in motor programs.

Goal-Directed Movement. When an individual performs a goal-directed movement such as reaching for an object, the major agonist-antagonist muscles controlling the movement often exhibit a **three-burst pattern of EMG**. This sequence of EMG activity is evident in movements of moderate to fast speed. It comprises an initial agonist burst of EMG that is followed by a burst of antagonist EMG and a second burst of agonist EMG (Wachholder & Altenburger, 1926). These three bursts usually overlap, but the peaks occur in an agonist-antagonist-agonist order (Figure 7.27). This sequence has been referred to as an ABC set, where A denotes action burst, B indicates braking burst, and C represents clamping burst (Hannaford & Stark, 1985). The action burst accelerates the limb toward the target position, the braking burst slows down the limb as it approaches the target position, and the clamping burst fixes the limb at the target position.

Although the presence of a three-burst pattern of EMG for unidirectional movements has been known for several years, no unifying hypothesis has been formulated to account for the variety of experimental observations (reviewed in Hasan, Enoka, & Stuart, 1985). There is no doubt that the three-burst pattern concerns stopping a movement at a targeted position, but the control strategy adopted by the central nervous system has not yet been determined. For example, many investigators have attempted to determine which parts of the sequence are preprogrammed (under open-loop control) and which can be altered by feedback (closed-loop control). Two major limitations, however, have prevented a complete understanding of this EMG sequence: First, the three-burst pattern is drastically affected by the features of movement, including the instructions given to the subject (Gottlieb, Corcos, Agarwal, & Latash, 1990a), and it has proven difficult to control systematically all the variables that influence the three-burst pattern (Karst & Hasan, 1987). Second, not enough attention has been paid to the mechanical conditions, such as the orientation of the subject (plane in which the movement is

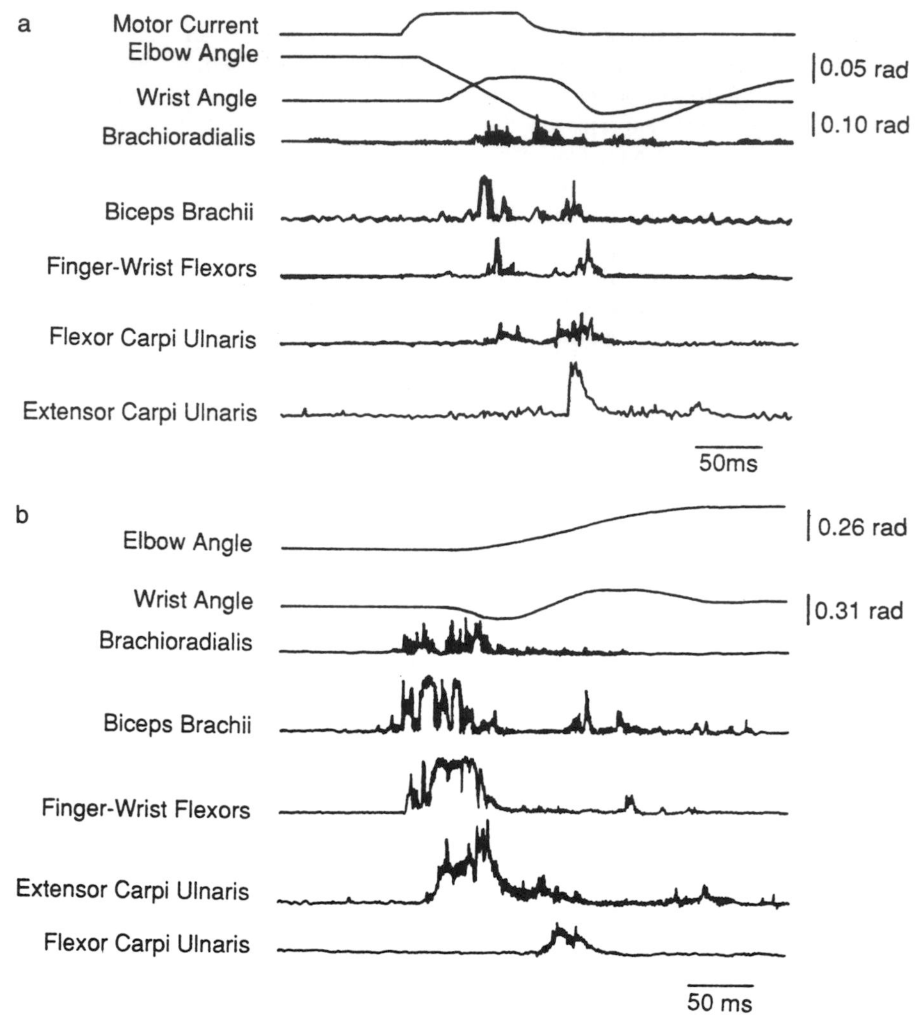

Figure 7.26 Joint angles and EMG records from wrist muscles during (a) an elbow joint perturbation and (b) a volitional elbow flexion movement. For the perturbation, the elbow joint was extended by a motor, which produced a flexion response at the wrist. The volitional elbow flexion, however, produced a slight initial wrist extension followed by wrist flexion. Nonetheless, both the finger flexor and wrist flexor muscles were active in both instances.

Note. From ''Activity of Wrist Muscles Elicited During Imposed or Voluntary Movements About the Elbow Joint'' by G.F. Koshland, Z. Hasan, and L. Gerilovsky, 1991, *Journal of Motor Behavior*, **23**, pp. 94, 95. Copyright 1991 by Heldref Publications. Adapted by permission.

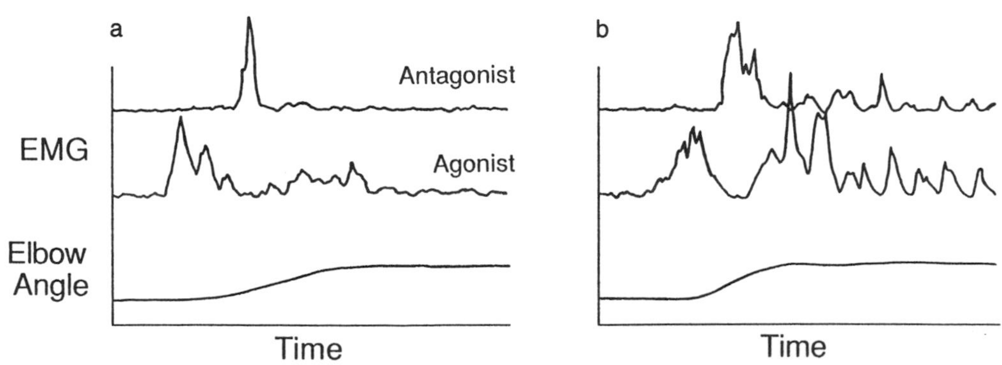

Figure 7.27 The three-burst pattern of EMG activity for an elbow flexion movement (0.50 rad) (the EMG data are rectified and filtered): (a) a distinct sequence; (b) a movement displaying coactivation at the termination of the movement.

Note. Data from Karst and Hasan (1987).

performed) and the fact that the agonist (or antagonist) for a particular movement often includes more than one muscle.

Nonetheless, based on both experimental measurements and computer simulations, Gottlieb and colleagues have proposed a **dual-strategy hypothesis** to account for the performance of goal-directed, elbow flexion movements (Corcos, Gottlieb, & Agarwal, 1989; Gottlieb, Corcos, & Agarwal, 1989). The hypothesis can explain variations in the EMG (agonist and antagonist) and resultant muscle torque that accompany movements involving different displacements, speeds, and loads. The hypothesis is based on a model of motor neuron excitation as a rectangular pulse that can be modulated by variation of its amplitude and duration. Modulation of excitation amplitude produces a *speed-dependent strategy*, which involves an increase in EMG and torque amplitude and a decrease in the latency to the onset of antagonist EMG (Figure 7.28a). In contrast, modulation of excitation duration produces a *speed-independent strategy*, which involves similar initial slopes for agonist EMG and resultant muscle torque, increasing EMG and torque durations, and an increase in the latency to the onset of the antagonist EMG (Figure 7.28b). When subjects are required to perform a rapid task within certain time constraints, they choose the speed-dependent strategy. However, when subjects are free to choose movement speed, they select the speed-independent strategy, which involves selecting an excitation intensity (amplitude) and then matching that with the appropriate duration to achieve the desired movement (Gottlieb et al., 1990a).

But movements are controlled not only by premovement CNS commands that impinge on motor neurons but also by afferent feedback that occurs during the movement. For this reason, Feldman (1986) has proposed that the controlled parameter in such goal-directed movements is the threshold (λ) of the tonic stretch reflex (Latash, 1993). In this model, which has been termed the **lambda (λ) model**, movements are controlled by variation in the threshold of the tonic stretch reflex. Latash and Gottlieb (1991a) have demonstrated that it is possible to simulate the EMG patterns of single-joint movements associated with the speed-dependent and speed-independent strategies by using the λ model. In this scheme, the speed-dependent strategy is equivalent to varying λ at a constant rate for different durations, and the speed-independent strategy is equivalent to varying λ at different rates. Latash and Gottlieb (1991a) concluded that some characteristics of the EMG patterns (e.g., onset of antagonist EMG) were defined exclusively by central commands and that others (e.g., EMG amplitude and duration) depended equally on the afferent signals generated during the movement. Furthermore, there appear to be no differences between the EMG patterns for such goal-directed movements and the initial part of a blocked (hence isometric) movement (Latash & Gottlieb, 1991b).

Similarly the activation patterns of seven to nine arm muscles involved in movements (shoulder and elbow

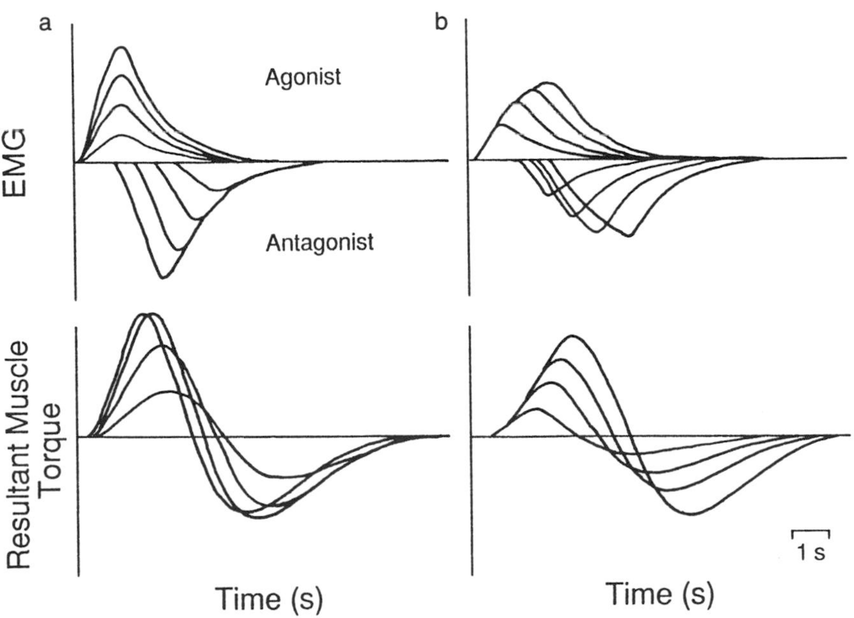

Figure 7.28 Simulated agonist and antagonist EMG and resultant muscle torque for a rapid elbow flexion movement over four trials. For the speed-dependent strategy (a), *amplitude* of excitation increased over the four trials, while for the speed-independent strategy (b) *duration* of excitation increased.

Note. From "Principles Underlying Single-Joint Movement Strategies" by G.L. Gottlieb, D.M. Corcos, G.C. Agarwal, and M.L. Latash. *Multiple Muscle Systems* (p. 241), by J.M. Winters and S.L.-Y. Woo (Eds.), 1990, New York: Springer-Verlag. Copyright 1990 by Springer-Verlag New York, Inc. Adapted by permission.

rotations) in a sagittal plane suggested simple relationships between EMG and the direction and speed of the movement (Flanders, 1991; Flanders & Herrmann, 1992). The directional kinematic requirements for the movements were met by EMG patterns that were scaled in amplitude and delayed in time, depending on movement direction. Furthermore, when these movements were performed at different speeds by varying movement time, the EMG patterns were characterized by two additive components: a tonic component that counteracted the effect of gravity and a phasic component that was associated with the change in movement speed. These observations suggest a relatively simple control strategy for mechanically complex arm movements.

Despite the elegance of these hypotheses, there remains some uncertainty as to whether these rules apply to *multijoint* movements. Movements that involve displacements about more than one joint raise the possibility of significant effects due to intersegmental dynamics. Consider, for example, a movement performed in a horizontal plane with the right arm that moves the hand from right to left about 35 cm. The movement involves flexion at the elbow and horizontal flexion at the shoulder. If the elbow flexors contract vigorously and the shoulder muscles are relaxed, there will be flexion at the elbow joint and, due to intersegmental dynamics, extension at the shoulder joint, despite the activity of biceps brachii. Because of this effect, the motor command must account for both these intersegmental interactions as well as generate a displacement in the desired direction. When Karst and Hasan (1991a, 1991b) studied the initial muscle activity for this type of movement, they found regions of the work space where a desired flexion movement was not always initiated with flexor muscle activity and conversely that a desired extension movement was not always initiated with extensor muscle activity. Instead, the initial EMG activity was predicted by simple rules related to the direction of the hand movement relative to the orientation of the forearm. Such results suggest that the control of multijoint movements may not simply comprise a summation of single-joint strategies (Hasan, 1991).

Coactivation. One of the observations made by Karst and Hasan (1987) concerning the three-burst pattern of EMG was the variability in the degree of coactivation at the end of the movement (Figure 7.27b). Coactivation refers to the *concurrent activity in the muscles comprising an agonist-antagonist set*. Surprisingly little information is available on coactivation. In particular, conditions under which we use coactivation have largely been unexplained. *Coactivation has the mechanical effect of making a joint stiffer and more difficult to perturb* (Baratta et al., 1988; Kornecki, 1992). Stiffness (N/m) is apparent as the *slope* of a force-length (or torque-angle) relationship. Figure 7.29 shows why coactivation

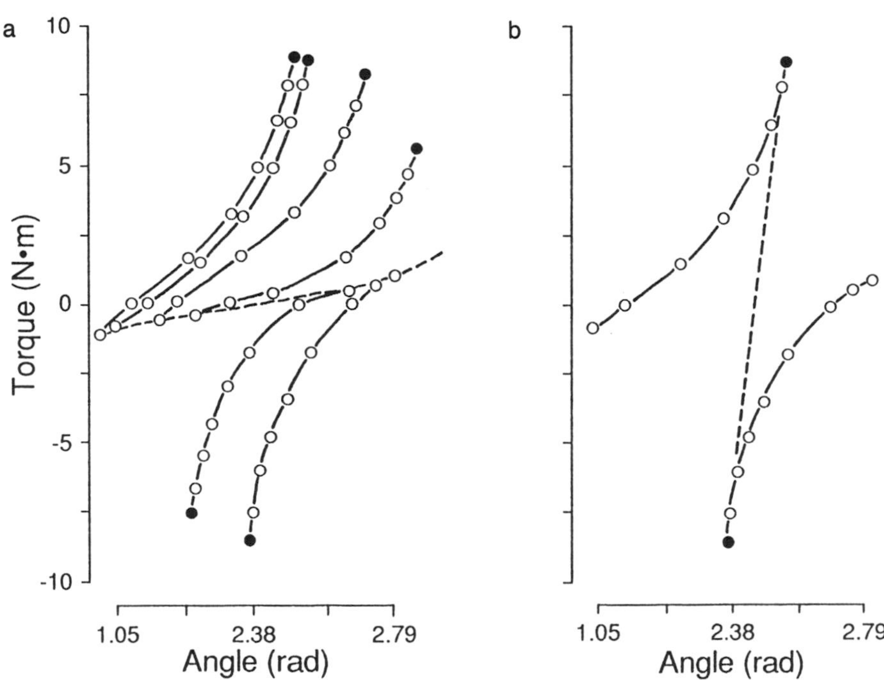

Figure 7.29 The stiffness at a joint depends on the net muscle activity. (a) Family of angle-torque relationships for the elbow flexors (positive torques) and elbow extensors (negative torques). The dashed line represents the passive angle-torque relationship. (b) The choice of the appropriate angle-torque relationship for the elbow flexors and extensors results in a net relationship (dashed line) that has a greater slope and therefore a higher stiffness.

Note. (a) is from "Once More on the Equilibrium-Point Hypothesis (λ Model) for Motor Control" by A.G. Feldman, 1986, *Journal of Motor Behavior*, **18**, p. 21. Copyright 1986 by Heldref Publications. Adapted by permission.

can increase the stiffness at a joint. Each muscle group (elbow flexors and extensors) is capable of producing a family of angle-torque relationships depending on the level of activation (Figure 7.29a). From these families of relationships, it is possible to choose a pair, one for each muscle group, such that the net angle-torque relationship has a greater slope and therefore a higher stiffness.

Because it increases the stiffness and hence the stability of a joint, coactivation would seem like a useful feature when we learn novel tasks or when we perform a movement that requires a high degree of accuracy. For example, Person (1958) monitored the EMG of biceps and triceps brachii over a 2-week period of training subjects to file and cut with a chisel. Over the course of the training, the EMG activity changed from a high level of coactivation to one of alternating activity with little coactivation. Similarly, Carolan and Cafarelli (1992) found that 8 weeks of training with a maximal isometric knee extension contraction produced an increase in strength and a decrease in the coactivation of hamstring muscles. The reduction in coactivation occurred in both the trained and untrained legs.

Skilled performance, however, is not characterized by the absence of coactivation. For example, Enoka (1983) found that skilled weight lifters performing the clean (Figure 1.7) used coactivation of vastus lateralis and biceps femoris throughout the movement (Figure 7.30). There are at least three reasons why individuals might use coactivation for such tasks. First, for movements that involve changes in direction (e.g., extension to flexion), it seems more economical to modulate the level of tonic activity in an agonist-antagonist set of muscles than to alternately turn them on and off. Hasan (1986) has argued that under certain conditions coactivation actually decreases the effort involved in performing the movement, and some empirical observations support this suggestion (Engelhorn, 1983). Second, because it increases the stiffness and hence the stability of a joint, coactivation may be a desirable strategy when individuals move heavy loads or loads that may be perturbed. Third, one of the capabilities of two-joint muscles is the transfer of power from one joint to another. As described previously, coactivation of a single-joint hip extensor (e.g., gluteus maximus) and a two-joint hip flexor (e.g., rectus femoris) has the net effect of increasing the extensor torque at the knee joint. Consequently, coactivation at the hip joint can result in an increase in the torque at the knee joint. This strategy may be useful for conditions requiring maximal force or in supplementing a fatigue-induced decline in muscle force (Dimitrijevic et al., 1992).

It appears that the neural strategy associated with coactivation is somewhat different from that in which an agonist muscle group alone is activated. When an unexpected change in the load is applied to the elbow flexors or extensors, the reflex response (motor unit discharge) depends on whether or not the load prior to the perturbation was supported by coactivation and on whether the subject was instructed to let go or resist (Gydikov, Kossev, Radicheva, & Tankov, 1981). There was one maximum in the reflex EMG when there was no coactivation and the subjects were instructed to let go. In contrast, there were two maxima in the reflex EMG when there was coactivation. This difference was probably due to the role of coactivation in increasing joint stability.

Individuals who experience a stroke exhibit varying degrees of residual motor function following the cerebrovascular accident. The remaining capabilities in the

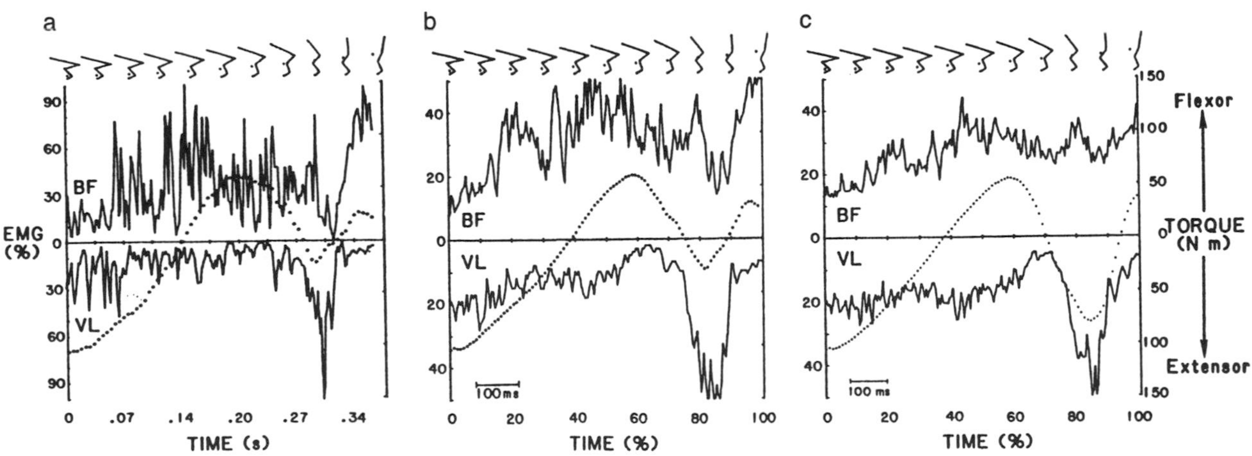

Figure 7.30 Resultant muscle torque (dotted line) and EMG records (solid line) about the knee joint during the clean in weight lifting (the EMG records for biceps femoris [BF] and vastus lateralis [VL] were rectified, filtered, and expressed as a percentage of the respective maximum values): (a) the profiles for one trial; (b) the mean profiles of 6 trials for 1 subject; (c) the mean profile of 15 trials for 3 subjects.

Note. From "Muscular Control of a Learned Movement: The Speed Control System Hypothesis" by R.M. Enoka, 1983, *Experimental Brain Research*, **51**, p. 140. Copyright 1983 by Springer-Verlag, Inc. Adapted by permission.

arm can be assessed by evaluating the ability of patients to perform such tasks as touching the opposite knee, placing the hand on the chin, and raising the arm overhead (Gowland, deBruin, Basmajian, Plews, & Burcea, 1992). The inability of a patient to perform one of these tasks is due to the generation of an inadequate net muscle torque. Neurodevelopmental theory (Bobath, 1978) suggests that this inability is due to an inappropriate coactivation of muscle because of a failure to inhibit antagonist activity. It appears, however, that the problem has more to do with muscle weakness and insufficient motor unit activity rather than with heightened coactivation (Burke, 1988; Gowland et al., 1992; Tang & Rymer, 1981).

Stretch-Shorten Cycle. A common pattern of muscle activation, particularly during high-performance tasks, is to use *an eccentric-concentric sequence in which the active muscle is first lengthened and then shortened.* The advantage of this strategy is that a muscle can perform more positive work if it is actively stretched before being allowed to shorten (Cavagna & Citterio, 1974; Fenn, 1924). The result of this **stretch-shorten cycle** is that *a greater quantity of work is done during the concentric contraction than would be done if the muscle simply performed a concentric contraction by itself.* The experimental evidence for this conclusion is based on the work done by an isolated muscle (Figure 7.31). The experiment had two parts: The muscle was first stretched and then stimulated before it was allowed to shorten and perform positive work (Figure 7.31a); next the muscle was first stimulated and then stretched before it performed positive work (Figure 7.31b). The results for each part of the experiment are shown as length-time, force-time, and force-length graphs. The critical comparison is contained in the force-length graphs. Phase c shows the change in force and length as the muscle performs work. Because work is defined as the product of force and displacement, the area under the force-length graph during the c phase represents the work done during each part of the experiment. Clearly, the area under the curve is greater for the second part of this experiment, which consisted of the stretch (lengthening) of an active muscle; this corresponds to the stretch-shorten cycle.

The work-energy relationship (chapter 3) states that an increase in the work done by a muscle requires an increased expenditure of energy. Where might this additional energy come from? The typical two-part rationale (e.g., Cavagna, 1977) is as follows. First, the eccentric contraction loads the series-elastic element by stretching it (three-element model of muscle in chapter 6), which can be envisaged as a transfer of energy from the load to the series elastic element; this represents the *storage* of elastic energy. For example, if an elastic band is held with one end in each hand and then stretched, some of the arm-hand muscle activity involved in stretching the band is stored in the band as elastic energy. And second, once released, the molecular structure of the elastic band will use this elastic energy to return to its original shape. Similarly, as the ratio of muscle force to load force changes and the muscle

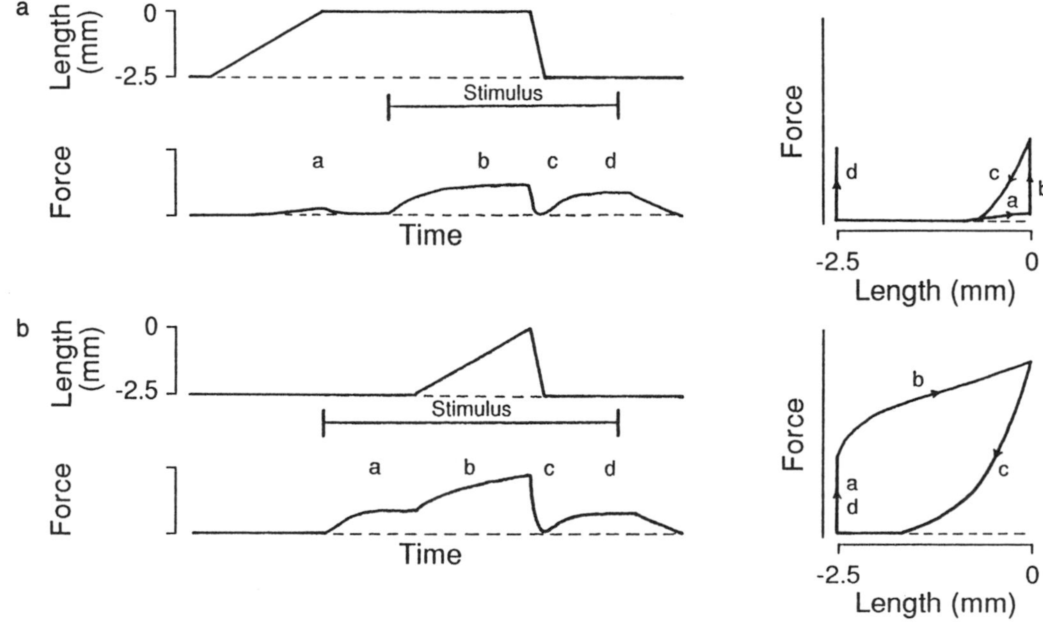

Figure 7.31 Positive work done by an isolated muscle during (a) isometric-concentric and (b) eccentric-concentric activation sequences. The work done by the muscle (area under phase c of the force-length curve) is greater for the eccentric-concentric (stretch-shorten) contraction.

Note. From "Effect of Stretching on the Elastic Characteristics and the Contractile Component of Frog Striated Muscle" by G.A. Cavagna and G. Citterio, 1974, *Journal of Physiology (London)*, **239**, p. 4. Copyright 1974 by Cambridge University Press. Adapted by permission.

undergoes a concentric contraction, the elastic energy stored in the series elastic element can be recovered and used to contribute to the shortening contraction (positive work).

This phenomenon can be stated algebraically by using the first law of thermodynamics (chapter 3):

$$U = \Delta E_h + \Delta E_m + \Delta E_{p,s}$$

where U = work, ΔE_h= change in heat or thermal energy, ΔE_m = change in chemical energy, and $\Delta E_{p,s}$ = change in elastic energy. The eccentric-concentric sequence in Figure 7.31 resulted in an increase in the work (U) done during the concentric activity. When muscle is activated, ATP is supplied by a variety of metabolic processes as the essential unit of chemical energy (E_m). In both the generation and use of ATP, some of the energy is degraded as heat (E_h). As the above equation states, if both the chemical energy used and the heat given off remain constant (let $E_{p,s}$ = 0 for now), the amount of work done will remain the same. But the point of Figure 7.31 is that the work done increased with an eccentric-concentric (stretch-shorten) sequence. One explanation for this is that either E_m or E_h changed. The explanation based on the phenomenon of storage and utilization of elastic energy, however, is that *additional energy ($E_{p,s}$) beyond that provided by chemical means is made available for the performance of work.* According to this point of view, E_m and E_h may vary little between the isometric-concentric and eccentric-concentric modes, but most of the extra work can be done because of the elastic energy ($E_{p,s}$) contribution.

This ability to use stored elastic energy is affected by three variables: time, magnitude of stretch, and velocity of stretch. Cavagna (1977) has demonstrated that there should be no time delay between the eccentric and concentric contractions; otherwise, some of the stored elastic energy is lost (dissipated). Presumably the energy loss is due to the detachment and reattachment of crossbridges during the delay such that, following reattachment, the myofilaments are under less stretch. Similarly, if the magnitude of the lengthening contraction is too great, fewer crossbridges remain attached following the stretch, and hence less elastic energy is stored (Edman, Elzinga, & Noble, 1978). Provided the crossbridges remain attached, however, the greater the velocity of stretch, the greater the storage of elastic energy (e.g., Rack & Westbury, 1974).

Despite the widespread use of the phenomenon of the storage and utilization of elastic energy to account for the increased positive work associated with eccentric-concentric contractions, the enhanced positive work is probably also due to a substantial increase in the amount of available chemical (E_m) energy (Jaric, Gavrilovic, & Ivancevic, 1985). This increase in the available chemical energy is called the **preload effect**. For example, note that the force at the beginning of phase c on the force-length graph in Figure 7.31 is greater during the eccentric-concentric condition than during the isometric-concentric condition; this corresponds to the right-most peak in the force-length graphs. Clearly, the force at the beginning of the concentric phase in the eccentric-concentric condition is greater.

An estimate of the relative contributions of the elastic energy and the preload effects can be obtained by considering the height that subjects can jump using two types of vertical jumps (Komi & Bosco, 1978). The *squat jump* begins from a squat position (knee angle about 2 rad) and simply involves an extension of the knee and ankle joints; the arms are kept stretched overhead to minimize their contribution to the jump. The *countermovement jump* begins from an upright posture and involves, in one continuous movement, squatting down to a knee angle of about 2 rad and then extending the knee and ankle joints as in the squat jump. The major difference between these two techniques is the manner in which the powerful knee extensors are used (they perform about 50% of the work during the maximum vertical jump) (Hubley & Wells, 1983); namely, the squat jump involves only an isometric-concentric contraction of the knee extensors, whereas a countermovement jump requires an eccentric-concentric sequence.

Whether these jumps are performed on one or two legs also alters the magnitude of the initial concentric knee extensor torque. The *one-legged jumps generate a preload effect* compared with the two-legged jumps. This point is illustrated in Figure 7.32, which shows the torque-angle relationship for a one-legged and a two-legged squat jump. The torque represents the resultant muscle torque about the knee joint for a single leg. The knee angle changes from 2.0 rad to complete extension (3.14 rad) during the takeoff phase. The main point shown in Figure 7.32 is that the initial torque (at 2.0 rad) during the one-legged jump is about twice that for the two-legged jump. In the one-legged jump, the single leg must support the whole body weight, whereas in the two-legged jump, a single leg need support only half of that weight. Because of this difference, the initial load supported by the single leg is much greater in the one-legged jump, and hence we say the limb (particularly the muscles about the knee joint) has been preloaded. Also, the greater initial torque exerted by the knee extensors requires a greater supply of chemical energy (E_m).

By comparing the height reached in the vertical jumps (Table 7.4), we can examine the contribution of elastic energy (storage and utilization) and chemical energy (preload effect) to the performance. Where does the energy come from (elastic or chemical) for the greater work done (height achieved) in the countermovement jumps? A comparison of the one- versus two-legged jumps illustrates the preload effect (chemical energy), whereas differences between squat and countermovement jumps include both effects (chemical and elastic). The differences in height jumped with the four combinations illustrate three features of mus-

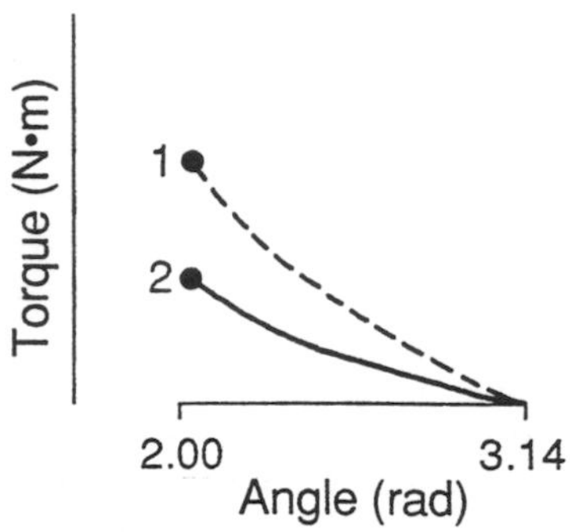

Figure 7.32 Idealized resultant muscle torque-angle relationship about one knee joint during one-legged (dashed line) and two-legged (solid line) squat jumps.

cle use in this type of task. First, for both the one- and two-legged jumps, the subjects reached a greater height with the countermovement jump than with the squat jump; this observation supports the superiority of the eccentric-concentric contraction over the isometric-concentric contraction. Second, the heights attained with the two-legged jumps were greater than those attained with the one-legged jumps due to the greater muscle mass (and hence chemical energy) available to the former. Third, the height jumped in the two-legged situations was not twice that achieved in the one-legged jumps despite the fact that twice the muscle mass was available. That is, the heights attained in the two-legged squat jump were on average 147% of those attained in the one-legged squat jump. If the difference between the two jumps were simply the quantity of muscle mass involved, we would expect the height reached in the two-legged jump to be 200% of the one-legged height. However, because the height reached in the one-legged jump was greater than 50% of the two-legged height, the increased performance of the single leg in the one-legged jump is due to the greater amount of chemical energy (preload effect). Such analyses suggest that the storage and utilization of elastic energy is of lesser importance than the preload effect (increased chemical energy) in such tasks as the vertical jump but may be more important in such activities as hopping (Fukashiro & Komi, 1987; van Ingen Schenau, 1984).

Table 7.4 The Vertical Heights Attained With One- and Two-Legged Squat and Countermovement Jumps

	Squat	Countermovement
One-legged	22.1 ± 5.9[a]	24.0 ± 6.6
Two-legged	32.4 ± 9.1	36.4 ± 8.5

Note. [a]The values are mean ± *SD* in centimeters for 44 subjects.

Many human movements have evolved to capitalize on the stretch-shorten effect. For example, the normal mode of performing the vertical jump (e.g., basketball, net ball, volleyball) is the countermovement style, which, as discussed above, maximizes the height that can be reached. The stretch-shorten cycle is also used in kicking and running. The foot is first placed on the ground in running, and the knee flexes and extends (Figures 1.14 and 1.15) as the total-body center of gravity passes over the foot. This pattern of knee flexion-extension is accomplished by an eccentric-concentric sequence of knee extensor activity. However, this mode of activation does not always enhance the ensuing positive work; Cavagna (1977) has estimated that the elastic energy recovered in running makes significant contributions to the power generated by the muscles only at running speeds greater than about 6.5 m/s. A similar observation has been reported for the takeoff in the long and triple jumps (Luthanen & Komi, 1980).

Our final example of the stretch-shorten cycle concerns the experimental situation in which a barbell is suspended from the ceiling so that it is horizontal and located at shoulder height. The subject is asked to stand under the barbell so that it touches the shoulders, but not to support the weight of the barbell, and to grasp the ends of the barbell, one with each hand. Once in this position, the subject is given a signal to rotate the barbell in the horizontal plane by approximately 1.57 rad as rapidly as possible. How might the subject accomplish this task? Grieve (1969) performed this experiment and found that subjects first rotated the hips in the desired direction while the shoulders and barbell were briefly rotated in the opposite direction. This action had the effect of actively stretching (eccentric contraction) the trunk muscles, which were subsequently used concentrically to rotate the shoulders and barbell. The peak torque exerted during this movement occurred during the eccentric phase. This observation provides support for the notion of leading with the hips in a golf drive; the torque applied to the golf club will be greater if the muscle activity comprises an eccentric-concentric sequence rather than just a concentric contraction.

Manipulation. In our discussion of movement strategies, we have focused on activities such as walking, running, throwing, kicking, reaching, and stabilizing. In the spectrum of movement capabilities, however, a qualitatively different type of movement involves the exploration of our environment. This distinction is typified by the different functions performed by the hands compared with those of the arms and legs. The hand and brain are close partners in the human's ability to explore the physical world and to reshape it (Lemon, 1993). Both exploring and reshaping an environment depend on accurate descriptions of mechanical events when objects are bought in close contact with the hand. Much of this information is provided by the mechanoreceptive afferent units that innervate the hairless skin of the hand. These sensory receptors participate in a behavior called **active touch** (Johansson & Cole, 1992; Johansson & Vallbo, 1983; Johansson & Westling, 1984, 1987; Phillips, 1986).

The role of cutaneous mechanoreceptors in controlling motor output has been explored in the study of grip-force responses to unexpected changes in load (Johansson, Häger, & Bäckström, 1992; Johansson, Häger, & Riso, 1992; Johansson, Riso, Häger, & Bäckström, 1992). When an object is squeezed between the index finger and the thumb in a pinch grip, an unexpected change in the pulling load elicits an *automatic adjustment* in the grip-force after a brief delay. The latency of the grip-force response decreases (174 to 80 ms) as the rate of change in pulling load increases. The automatic response consists of two parts: The initial phase involves a bell-shaped rate of change in force; it is followed by a second phase that includes a slow increase in the rate of change in force (Figure 7.33). Similar biphasic responses have been reported for isometric tasks, compensation for body sway elicited by translation of the support surface, and eye movements during smooth pursuit. The initial phase is a standard element that has a constant duration but an amplitude that varies with the rate of change in the pulling load. The second phase appears for longer duration increases in grip force and is abolished if the index finger and thumb are anesthetized. This suggests that the second phase is *dependent on afferent feedback from mechanoreceptors in the hand.*

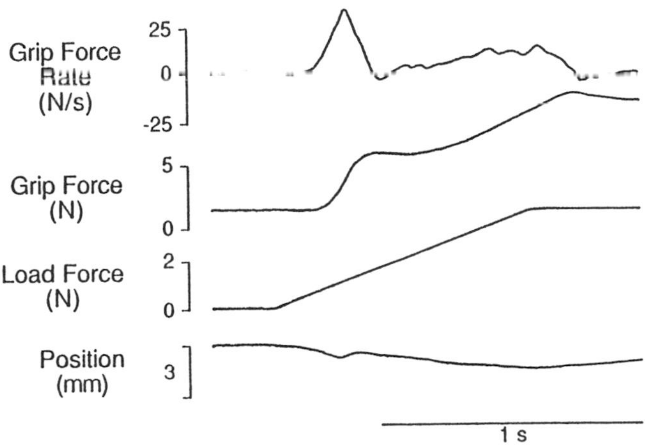

Figure 7.33 Automatic grip-force response to an unexpected change in the load force. A motor applied the load force, and the subject was required to prevent movement (or change in position). The rate of change in grip force (top trace) consists of two phases, an initial bell-shaped response followed by a more gradual change.
Note. From "Somatosensory Control of Precision Grip During Unpredictable Pulling Loads" by R.S. Johansson, R. Riso, C. Häger, and L. Bäckström, 1992, *Experimental Brain Research*, **89**, p. 185. Copyright 1992 by Springer-Verlag, Inc. Adapted by permission.

Motor Unit Activation Patterns. Based on studies that began in the 1930s and continue today a general rule has emerged concerning the sequence of motor unit activation. As we discussed in chapter 6, this rule asserts that motor units are recruited in an orderly sequence that proceeds from the smallest to the largest. However, there are three limitations to this rule. First, studies on motor unit recruitment have usually been limited to forces that are less than 30% of maximum. As a result, there is some uncertainty whether orderly recruitment occurs, or is even needed, at high forces (Stuart & Enoka, 1983).

Second, the force at which a motor unit is recruited (the recruitment threshold) differs for isometric and movement (concentric and eccentric) conditions, and the direction of the difference can vary among a group of synergist muscles (Tax, Denier van der Gon, & Erkelens, 1990). These data suggest that the intent of the subject, to move or not to move, can influence the recruitment threshold of the motor unit.

Third, recruitment order is not an invariant property of the motor unit pool, and there have been reports of alterations in this order. The boundary conditions that can alter recruitment order include eccentric contractions, fatigue, and cutaneous afferent feedback. In addition, aging is known to be associated with changes in motor unit size (due to an increased innervation ratio and decreased cross-sectional area of some muscle fibers), and it is unclear how well orderly recruitment is preserved in the presence of these peripheral changes (Galganski, Fuglevand, & Enoka, 1993).

Let us consider the boundary conditions that can alter the recruitment order of motor units. As we discussed in chapter 6, the force exerted by muscle is greater during an eccentric contraction than during an isometric or concentric contraction. Although our understanding of crossbridge mechanics is incomplete, it is clear that the mechanical behavior of muscle is different for concentric and eccentric contractions (Morgan, 1990). For example, changes in muscle performance during fatiguing contractions differ for eccentric and concentric contractions (Enoka & Stuart, 1992; Tesch, Dudley, Duvoisin, Hather, & Harris, 1990). In addition to these mechanical differences, recruitment order may also differ during eccentric contractions when larger (high-threshold) motor units can be recruited before smaller motor units (Nardone, Romanò, & Schieppati, 1989). Figure 7.34 shows motor unit activity in the lateral gastrocnemius (LG) muscle of a human subject during concentric and eccentric contractions. In the first trial (Figure 7.34a), the intramuscular electrode detected no local motor unit activity during the slow plantarflexion movement (concentric), but motor units were recruited during the eccentric contraction (negative slope on the position trace). In the second trial (Figure 7.34b), the plantarflexion movement was performed more rapidly, and the same motor unit was recruited during the con-

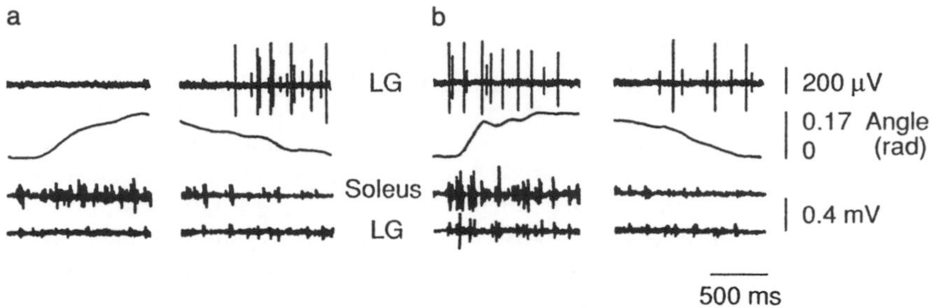

Figure 7.34 Motor unit activity in the lateral gastrocnemius (LG) muscle during concentric and eccentric contractions. The four traces (from top to bottom) represent intramuscular EMG, ankle position, soleus EMG, and lateral gastrocnemius EMG. When the ankle angle increased (upward), the muscles performed a concentric contraction.

Note. From "Selective Recruitment of High-Threshold Human Motor Units During Voluntary Isotonic Lengthening and Active Muscles" by A. Nardone, C. Romanò, and M. Schieppati, 1989, *Journal of Physiology (London)*, **409**, p. 456. Copyright 1989 by Cambridge University Press. Adapted by permission.

centric contraction. These differences in recruitment order are underscored by the whole-muscle EMG records (bottom two traces in Figure 7.34) during the concentric and eccentric contractions; note the shift in activity between soleus and gastrocnemius.

Recall that we previously distinguished concentric and eccentric contractions on the basis of the muscle torque–load torque ratio. The ratio was greater than 1 for concentric contractions but less than 1 for eccentric contractions; that is, the difference between the two was simply one of mechanics. However, if recruitment order does differ for concentric and eccentric contractions, then there is also a neural difference between the two. Perhaps the motor system has at least two schemes of motor unit recruitment, one for concentric contractions and another for eccentric contractions. Because we know before a movement is performed whether the contraction should be concentric or eccentric, it should not be difficult to select the appropriate recruitment sequence. If larger motor units are preferentially activated during eccentric contractions, this might explain why it is more difficult to control force precisely during eccentric contractions.

A second boundary condition that may alter recruitment order is fatigue. As we will discuss in chapter 8, fatigue is a general term used to describe a class of acute effects that impair motor performance. Many physiological processes, including motor unit activation, exhibit adaptive capabilities when an individual performs a fatiguing contraction. The fatigue-related adaptations that concern motor unit activity include a reduction in discharge rate (Bigland-Ritchie, Johansson, Lippold, Smith, et al., 1983; Dietz, 1978; Seyffarth, 1940), a cessation of activity (Person, 1974), and *flexibility in the motor units that contribute to a task* (Enoka, Robinson, & Kossev, 1989). Although the effect of fatigue on recruitment order has not been tested directly, Enoka and colleagues (1989) have found that when the same task is performed before and after a fatiguing contraction, the motor units involved in the task could vary (Figure 7.35). This finding suggests some fatigue-related flexibility in the recruitment order of motor units.

The third boundary condition that appears to alter recruitment order is changes in cutaneous feedback. Garnett and Stephens (1981) have found that *the muscle force at which a motor unit is recruited (recruitment*

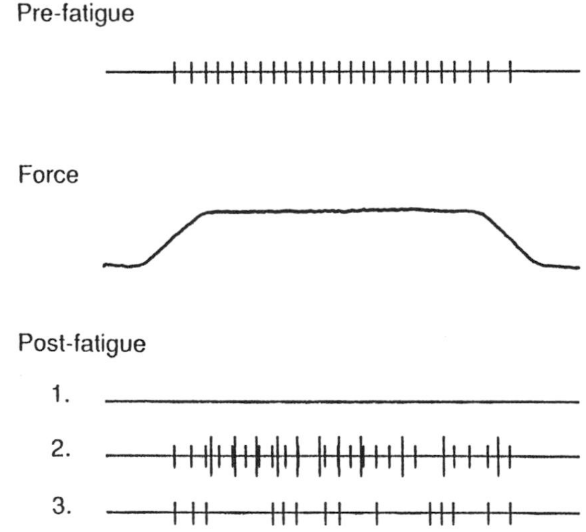

Figure 7.35 Motor unit activity in the first dorsal interosseus muscle during an isometric contraction that was performed before and after a fatiguing contraction. The force exerted during the task is shown in the middle trace. The pre-fatigue trace shows typical motor unit activity during the task. The post-fatigue traces illustrate three effects of the fatiguing contraction on motor unit activity: (1) The motor unit was not recruited during the same task; (2) additional motor units were recruited; and (3) motor unit discharge was more variable. Trace #1 indicates that the task was completed by activating a different group of motor units.

Note. Data from Enoka et al. (1989).

threshold) can be altered by manipulating feedback from cutaneous afferents (Figure 7.36). Furthermore, the effect depends on whether the motor unit has a low or high recruitment threshold. For low-threshold motor units, the recruitment threshold is increased with appropriate cutaneous stimulation (Figure 7.36a). For high-threshold motor units, the recruitment threshold is decreased (Figure 7.36b). In this experiment, the cutaneous stimulation consisted of electrical stimulation at 4× perception threshold of the digital nerves of the index finger, which felt like squeezing of the index finger. The advantage of a strategy whereby the recruitment threshold of high-threshold (presumably high-force) motor units can be lowered is that these units can be activated earlier when conditions (as indicated by the cutaneous stimulation) dictate that a rapid, powerful response is necessary.

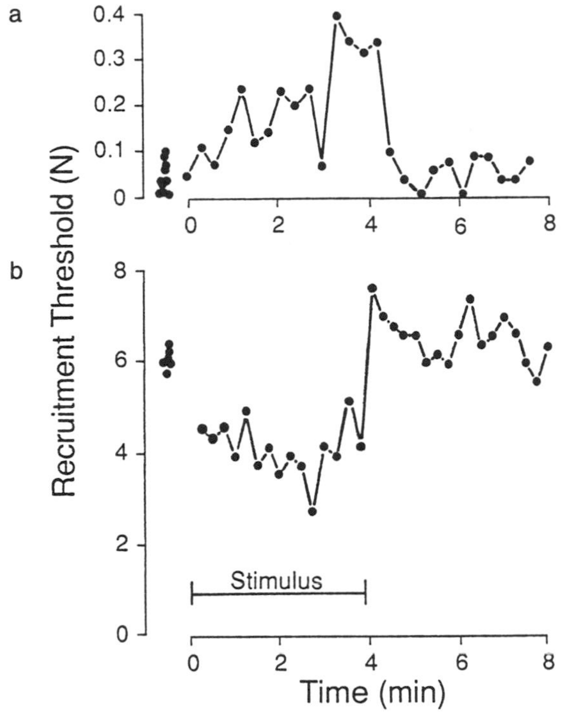

Figure 7.36 Changes in the recruitment threshold of (a) a low-threshold and (b) a high-threshold motor unit in a hand muscle with cutaneous stimulation of the index finger. Control values (prestimulation) for recruitment threshold are shown to the left of time zero.
Note. From "Changes in the Recruitment Threshold of Motor Units Produced by Cutaneous Stimulation in Man" by R. Garnett and J.A. Stephens, 1981, *Journal of Physiology (London)*, **311**, p. 469. Copyright 1981 by Cambridge University Press. Adapted by permission.

Summary

This chapter expands the model of the single joint system to examine the role of suprasegmental centers, to consider multijoint aspects of musculoskeletal design, and to identify movement strategies used by the nervous system in the control of movement. First, the chapter provided an overview of the contribution of the suprasegmental centers—which include the sensorimotor cortex, brain stem, cerebellum, basal ganglia, and ascending pathways—to the flow of information through the motor system. Second, we discussed the concept of net muscle activity as the primary determinant of movement and considered the effects of variation in muscle attachment on the output of the musculoskeletal system. Third, we identified fundamental strategies used by the nervous system in the control of posture and movement. The movement-related strategies included the activation patterns used for goal-directed movements, the utility of coactivation, the basis of the stretch-shorten cycle, the reliance on afferent feedback for such movements as manipulation, and the general rules underlying motor unit activation.

TRUE-FALSE QUESTIONS

1. Motivation is important for the initiation of movement. **TRUE FALSE**
2. The association cortex governs basic biological drives and and emotional behavior. **TRUE FALSE**
3. The limbic system transforms motivation into an idea. **TRUE FALSE**
4. The association cortex includes the prefrontal, parietal, and temporal lobes. **TRUE FALSE**
5. The central command is transmitted to lower neural centers from the motor cortex. **TRUE FALSE**
6. Corollary discharge is the motor command from the brain stem to the spinal cord. **TRUE FALSE**
7. A motor program represents a stereotyped sequence of afferent signals from peripheral sensory receptors. **TRUE FALSE**
8. A neural network is the connective tissue structure that supports neuronal organization in the spinal cord. **TRUE FALSE**
9. Central pattern generators generate commands that activate muscle. **TRUE FALSE**

10. Commissures connect the right and left cerebral hemispheres. **TRUE FALSE**
11. The primary motor area is part of the sensorimotor cortex. **TRUE FALSE**
12. The primary somatosensory cortex is anterior to the central sulcus. **TRUE FALSE**
13. The posterior parietal cortex is located in Brodmann's area 4. **TRUE FALSE**
14. The corticospinal tract is a descending pathway. **TRUE FALSE**
15. Output from the sensorimotor cortex goes to distinct, nonoverlapping regions of the spinal cord. **TRUE FALSE**
16. *Somatotopic* means body map. **TRUE FALSE**
17. The sensorimotor cortex contains several somatotopic schemes. **TRUE FALSE**
18. The primary somatosensory cortex provides sensory information for the planning and initiation of movement. **TRUE FALSE**
19. The premotor cortex is primarily concerned with the planning and sequencing of movements. **TRUE FALSE**
20. The posterior parietal cortex is active during the performance of imagined movements. **TRUE FALSE**
21. Neurons in area 5 are involved in the use of tactile sensory information to guide exploratory limb movements. **TRUE FALSE**
22. The sensorimotor cortex can plan movements in terms of the participation of groups of muscles. **TRUE FALSE**
23. The pons is located in the brain stem. **TRUE FALSE**
24. The medulla is located between the pons and the spinal cord. **TRUE FALSE**
25. The rubrospinal tract is the descending pathway from the red nucleus. **TRUE FALSE**
26. The reticular formation is diffuse, with nuclei in both the mesencephalon and the medulla. **TRUE FALSE**
27. The corticospinal tract terminates near motor neurons that innervate distal muscles. **TRUE FALSE**
28. The Group B brain stem pathway provides postural support for goal-directed movements. **TRUE FALSE**
29. Propriospinal neurons convey only proprioceptive information. **TRUE FALSE**
30. The cerebellum has no direct connections to the spinal cord. **TRUE FALSE**
31. The cerebellum receives input through the fastigial, interpositus, and dentate nuclei. **TRUE FALSE**
32. The spinocerebellum refers to the lateral hemispheres of the cerebellum. **TRUE FALSE**
33. The cerebrocerebellum receives input from the cerebral cortex. **TRUE FALSE**
34. A comparator, such as the cerebellum, compares desired movements to actual movements. **TRUE FALSE**
35. The cerebellum seems to be exclusively concerned with motor learning. **TRUE FALSE**
36. The basal ganglia represent an intermediate link between the motor cortex and the brain stem. **TRUE FALSE**
37. The basal ganglia do not affect movement. **TRUE FALSE**
38. Ascending pathways convey afferent information from peripheral receptors to suprasegmental centers. **TRUE FALSE**
39. Perception involves the conscious interpretation of sensory information. **TRUE FALSE**
40. The dorsal column–medial lemniscus pathway distributes information on touch and vibration. **TRUE FALSE**
41. The thalamus processes only sensory information. **TRUE FALSE**
42. The anterolateral pathway conveys signals to the primary somatosensory cortex. **TRUE FALSE**
43. The biomechanical effect of muscle activation is not limited to the joint or joints that the muscle crosses. **TRUE FALSE**
44. The state of the musculoskeletal system is just as important as the neural drive in movement performance. **TRUE FALSE**
45. A free body diagram is valid only for static conditions. **TRUE FALSE**
46. Resultant muscle force represents the sum, or addition, of all the muscle forces about a joint. **TRUE FALSE**
47. The joint reaction force accounts for the net joint forces. **TRUE FALSE**
48. The joint reaction force is measured in N/m^2. **TRUE FALSE**
49. The bone-on-bone force at a joint is never greater than body weight. **TRUE FALSE**
50. For a slow movement, the muscle torque opposes the load torque. **TRUE FALSE**
51. When the muscle force vector and the displacement act in the same direction, the muscle is contracting concentrically. **TRUE FALSE**
52. For rapid movements, the direction of the weight vector is perpendicular to the body segment. **TRUE FALSE**
53. The curved arrow indicates that the torque vector is perpendicular to the page. **TRUE FALSE**

54. If a flexor muscle group produces extension, the net muscle activity is eccentric. **TRUE FALSE**
55. When the movement crosses the line of gravity (through the axis of rotation) during a slow movement, the net muscle vector changes direction. **TRUE FALSE**
56. A quasistatic analysis requires the use of the equation $\Sigma \mathbf{F} = ma$. **TRUE FALSE**
57. The net muscle force about a joint is never zero. **TRUE FALSE**
58. Inertial forces exert significant effects during rapid movements. **TRUE FALSE**
59. Fast and slow movements are controlled differently by the nervous system. **TRUE FALSE**
60. Fast movements typically involve an eccentric-concentric sequence of muscle activity. **TRUE FALSE**
61. The three primary elbow flexor muscles have similar mechanical actions. **TRUE FALSE**
62. Activation levels of brachioradialis and brachialis are similar for many tasks. **TRUE FALSE**
63. Biceps brachii activity covaries with brachioradialis activity for many tasks. **TRUE FALSE**
64. Rectus femoris is a two-joint muscle. **TRUE FALSE**
65. The shortening velocity of a two-joint muscle is less than that of its single-joint synergists. **TRUE FALSE**
66. The net effect of activating a two-joint muscle can be focused at only one of the two joints that it crosses. **TRUE FALSE**
67. A moment arm can be measured in cm. **TRUE FALSE**
68. Muscle moment arms usually change throughout the range of motion. **TRUE FALSE**
69. Postural activity does not require the voluntary (conscious) activation of muscle. **TRUE FALSE**
70. Stability and equilibrium mean the same thing. **TRUE FALSE**
71. A system in equilibrium can have a nonzero acceleration. **TRUE FALSE**
72. Postural adjustments are initiated by afferent feedback. **TRUE FALSE**
73. Disturbance of the support surface elicits a distal-to-proximal response. **TRUE FALSE**
74. Postural adjustments always begin at the feet. **TRUE FALSE**
75. Postural activity controls the effects of inertial forces between body segments. **TRUE FALSE**
76. Postural adjustments consist of a set of reflex responses. **TRUE FALSE**
77. The three-burst pattern of EMG occurs in slow, bidirectional movements. **TRUE FALSE**
78. Movements that are performed under open-loop control are altered by afferent feedback. **TRUE FALSE**
79. Subjects choose the speed-dependent strategy when performing a rapid task within certain time constraints. **TRUE FALSE**
80. The speed-independent strategy involves varying EMG duration for a constant EMG amplitude. **TRUE FALSE**
81. The λ model involves modulation of the threshold for the H reflex. **TRUE FALSE**
82. The speed-dependent strategy is equivalent to varying λ at different rates. **TRUE FALSE**
83. The onset of EMG during a single-joint movement seems to depend exclusively on afferent signals generated during the movement. **TRUE FALSE**
84. Coactivation increases stiffness about a joint. **TRUE FALSE**
85. Stiffness refers to the slope of the torque-time graph. **TRUE FALSE**
86. Coactivation occurs only when an individual is learning a new skill. **TRUE FALSE**
87. The inability of stroke patients to perform many movements is due to increased coactivation. **TRUE FALSE**
88. A muscle can perform more negative work if it is actively stretched before being allowed to shorten. **TRUE FALSE**
89. Work is measured as the area under a force-displacement graph. **TRUE FALSE**
90. The increased work done during the concentric part of an eccentric-concentric contraction is due to the availability of an increased quantity of energy. **TRUE FALSE**
91. Elastic energy cannot be stored in the series elastic element of muscle. **TRUE FALSE**
92. The work-energy relationship can be stated as the first law of thermodynamics. **TRUE FALSE**
93. The amount of elastic energy that can be stored in muscle increases as the time between the eccentric and concentric contractions increases. **TRUE FALSE**
94. The preload effect is based on an increased availability of elastic energy. **TRUE FALSE**
95. The increased performance of the countermovement jump compared to the squat jump is due more to the storage and utilization of elastic energy than to the preload effect. **TRUE FALSE**
96. Active touch refers to the exploration of an object by the hand. **TRUE FALSE**

97. The hand is capable of automatic adjustments of the pinch grip. **TRUE FALSE**

98. Motor unit recruitment order can change during eccentric contractions. **TRUE FALSE**

99. Motor unit discharge rate increases during fatigue. **TRUE FALSE**

100. Cutaneous afferent feedback can alter recruitment order. **TRUE FALSE**

MULTIPLE-CHOICE QUESTIONS

101. Which of the following is not one of the parts of the scheme that describes information flow through the nervous system?
   A. Perception
   B. Programming
   C. Motivation
   D. Execution

102. What does the limbic system contribute to the control of movement?
   A. Ideation
   B. Programming
   C. Motivation
   D. Perception

103. Which suprasegmental center issues the central command?
   A. Basal ganglia
   B. Motor cortex
   C. Brain stem
   D. Cerebellum

104. Which components of the CNS receive the corollary discharge?
   A. Motor cortex
   B. Cerebellum
   C. Brain stem
   D. Spinal cord

105. What is a motor program?
   A. A neural network that can generate a behaviorally relevant pattern of output in the absence of external timing cues
   B. A copy of the central command that is sent from the motor cortex back to the suprasegmental centers
   C. A group of muscles constrained to act as one unit
   D. A stereotyped sequence of commands sent from the spinal cord to the muscles to elicit a specific behavior

106. Which structures do not contribute to descending pathways?
   A. Sensorimotor cortex
   B. Spinal cord
   C. Basal ganglia
   D. Peripheral receptors

107. Which regions are part of the sensorimotor cortex?
   A. Primary somatosensory cortex
   B. Supplementary motor area
   C. Thalamus
   D. Mesencephalon

108. Which Brodmann's area or areas defines the primary motor area?
   A. 1, 2, and 3
   B. 4
   C. 5 and 7
   D. 8

109. What is somatotopy?
   A. A body map in the CNS
   B. The region of the body innervated by a single nerve
   C. The separation of the motor and sensory functions in the sensorimotor cortex
   D. The hierarchical organization of the suprasegmental centers

110. Which part of the sensorimotor cortex is active during the performance of imagined movements?
   A. Premotor cortex
   B. Primary somatosensory cortex
   C. Supplementary motor area
   D. Posterior parietal cortex

111. Which structure is not in the brain stem?
   A. Medulla oblongata
   B. Pons
   C. Mesencephalon
   D. Thalamus

112. Which descending pathways do not arise from the brain stem?
   A. Dorsal column–medial lemniscus pathway
   B. Anterolateral pathway
   C. Rubrospinal tract
   D. Corticospinal tract

113. Which pathway is not part of the Group A pathway?
   A. Tectospinal tract
   B. Rubrospinal tract
   C. Vestibulospinal tract
   D. Reticulospinal tract

114. Which muscles do the descending axons in the Group B pathway influence?
   A. Distal flexor muscles
   B. Axial muscles
   C. Proximal muscles
   D. Distal extensor muscles

115. Which structure is not one of the deep subcortical nuclei in the cerebellum?
   A. Dentate
   B. Vestibular
   C. Interpositus
   D. Fastigial

116. Which statement best describes the cerebrocerebellum?
   A. Receives input from the cerebral cortex and sends its output to the dentate nucleus
   B. Receives most of its sensory input from the spinal cord and sends its output to both the fastigial and the interposed nuclei
   C. Receives input from the vestibular nuclei in the medulla and sends its output back to the same nuclei
   D. Receives input from the cerebral cortex and sends its output to the fastigial nucleus

117. Which five nuclei comprise the basal ganglia?
   A. Caudate, dentate, globus pallidus, putamen, substantia nigra
   B. Subthalamic nuclei, putamen, interpositus, globus pallidus, caudate
   C. Putamen, substantia nigra, subthalamic nuclei, dentate, caudate
   D. Caudate, putamen, globus pallidus, subthalamic nuclei, substantia nigra

118. What is the function of the basal ganglia?
   A. Activation of specific muscles and motor programs
   B. Timing of agonist-antagonist activation
   C. Motor planning such as direction and amplitude
   D. Providing motivation to the sensorimotor cortex

119. What is perception?
   A. The interpretation of corollary discharge
   B. Afferent signals that reach consciousness and contribute to sensation
   C. Assessment of the sense of effort
   D. Any afferent information that reaches the sensorimotor cortex

120. Which parameter does not influence the state of the musculoskeletal system?
   A. Central command
   B. Muscle length
   C. Moment-arm length
   D. Muscle velocity

121. What is the best synonym for the term *resultant muscle force*?
   A. Resultant muscle torque
   B. Net muscle activity
   C. Total muscle force
   D. Net muscle force

122. What feature of force does the term *joint reaction force* indicate?
   A. Net
   B. Absolute
   C. Total
   D. Average

123. What is the critical feature of a slow movement that enables us to determine the net muscle activity associated with the movement?
   A. The direction of the displacement is given
   B. The inability of muscle to perform an eccentric contraction during a slow movement
   C. The direction of the load torque
   D. Muscle can exert only a pulling force

124. For which conditions is a quasistatic analysis appropriate?
   A. Slow jog
   B. Football punt
   C. Baseball pitch
   D. Isometric contraction

125. What aspects of muscle function are not affected by variation in muscle attachment sites?
   A. Range of motion
   B. Torque-angle relationship
   C. Maximum force
   D. Contraction velocity

126. What are the advantages of having two-joint muscles?
   A. The shortening velocity is less than that of a single-joint muscle.
   B. Muscle fibers are longer than in a single-joint muscle.
   C. Muscle torque and joint power are redistributed throughout a limb.
   D. Common motions are coupled.

127. How would the neural drive to biceps brachii change over its range of motion in order to keep the torque exerted by the muscle constant?
   A. Increase continually
   B. Change with a U-shape
   C. Decrease continually
   D. Change with an inverted U-shape

128. What is the best definition of stability?
   A. Mechanical equilibrium
   B. The return to equilibrium following a perturbation
   C. A maximal base of support
   D. A stationary system that does not move

129. Which sensory information is important in the control of upright posture?
    A. Auditory
    B. Somatosensory
    C. Vestibular
    D. Visual
130. Which statement does not correctly describe the postural responses to disturbance of a support surface?
    A. Postural adjustments are influenced by the body part that contacts the surroundings.
    B. The most important body segment for whole-body stability is the trunk.
    C. Disturbances are accommodated by a proximal-to-distal sequence of muscle activation.
    D. Response strategies depend on the type of perturbation.
131. Which muscles does a subject activate first for a reaction time task that involves raising a straight arm from the side to the horizontal?
    A. Anterior deltoid
    B. Trunk
    C. Quadriceps femoris
    D. Hamstrings
132. For what types of movements is the three-burst pattern of EMG used?
    A. Slow and unidirectional
    B. Fast and bidirectional
    C. Isometric contractions
    D. Moderate-to-fast speed and unidirectional
133. What does the dual-strategy hypothesis attempt to explain?
    A. EMG patterns for movements of differing displacements, speeds, and loads
    B. Resultant muscle torques for movements confined to a horizontal plane
    C. The control of movement by variation in the threshold of the tonic stretch reflex
    D. The role of afferent feedback in shaping EMG patterns during movement
134. Why is the control of multijoint movements more difficult than single-joint movements?
    A. The properties of two-joint muscles differ markedly from those of single-joint muscles.
    B. There is more than one axis of rotation.
    C. The presence of intersegmental dynamics complicates control.
    D. Displacements of segment endpoints are much greater.
135. What are the advantages of coactivation when a person is lifting heavy loads?
    A. An increase in the stiffness and hence stability of a joint
    B. Decreased specificity of the neural command
    C. The ability to transfer power from one joint to another
    D. An increase in economy for movements that involve changes in direction
136. Which term is not part of the first law of thermodynamics?
    A. Change in chemical energy
    B. Change in heat or thermal energy
    C. Change in elastic energy
    D. Change in kinetic energy
137. What variables affect the ability of muscle to use stored elastic energy?
    A. The available chemical energy
    B. Time between the eccentric and concentric contractions
    C. Velocity of stretch
    D. Magnitude of stretch
138. Which activities do not include a movement that contains a stretch-shorten cycle?
    A. Swimming
    B. Baseball
    C. Running
    D. Weight lifting
139. Which sensory receptors are most important in active touch?
    A. Cutaneous mechanoreceptors
    B. Muscle spindles
    C. Joint receptors
    D. Nociceptors
140. What activities result in an alteration of motor unit recruitment order?
    A. Cutaneous stimulation
    B. Eccentric contractions
    C. Muscle fatigue
    D. Rapid movements

PROBLEMS

141. What parts of the nervous system can provide motivation and initiate the commands that subsequently result in the performance of a movement?
142. Why is the conversion of an idea into the proper strength and pattern of muscle activity necessary for a desired movement referred to as *programming*?

143. Why is the corollary discharge a necessary part of the control scheme used by the central nervous system?

144. A motor program is defined as a stereotyped sequence of commands sent from the spinal cord to the muscles that elicit a specific behavior. What reasons suggest that a given motor unit can participate in a variety of motor programs?

145. What is the difference between a motor program and a central pattern generator?

146. Which suprasegmental structures and pathways (ascending and descending) have a somatotopic organization?

147. Describe the role of the following regions of the sensorimotor cortex in the generation of movement:
   A. Premotor cortex
   B. Primary motor cortex
   C. Posterior parietal cortex
   D. Primary somatosensory cortex
   E. Supplementary motor area

148. What are the differences between the Group A and Group B descending pathways from the suprasegmental structures?

149. Although the cerebellum contains over half of the neurons in the brain, it has no direct efferent connections to the spinal or brain stem nuclei.
   A. By what pathways does the cerebellum influence the control of movement?
   B. What is a comparator?
   C. Describe the three theories on how the cerebellum might function as a comparator.

150. Describe some evidence that suggests that the basal ganglia are involved in the control of movement.

151. Why does a given command from the nervous system not always elicit the same biomechanical effect from the musculoskeletal system?

152. The bone-on-bone force represents the absolute force at a joint and can be affected by both muscle force and ground reaction force.
   A. Why is the bone-on-bone force so much larger than the joint reaction force?
   B. Draw a diagram to help explain how these two forces affect the bone-on-bone force at the ankle joint.
   C. One way to determine the bone-on-bone force is to determine the force exerted by each muscle that crosses the joint. How can we measure the force exerted by muscle in a human subject?

153. An individual (body weight = 586 N) performs an exercise that involves raising and lowering the straight legs between Positions 1 and 2.

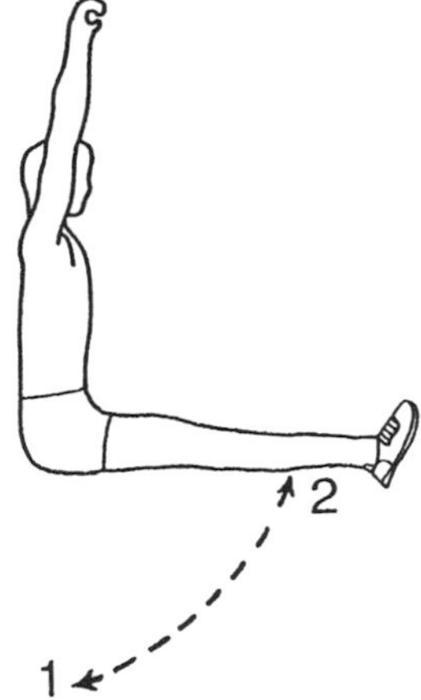

   A. Suppose you wanted to determine the torque exerted by the muscles at the hip joint during this exercise. Draw and label the appropriate free body diagram.
   B. Describe the pattern of net muscle activity (hip flexors and extensors) during one complete slow repetition (raising and lowering of legs) of this exercise.
   C. Given the following dimensions, calculate the resultant muscle torque required for the individual to maintain the legs in Position 2 (the legs are horizontal):

leg length (hip joint to ankle)	=	72 cm
weight of both legs	=	30% of body weight
CG location (hip to ankle)	=	43.4% of leg length from the hip

154. An athlete recovering from knee surgery performs a slow knee extension exercise with a light load.

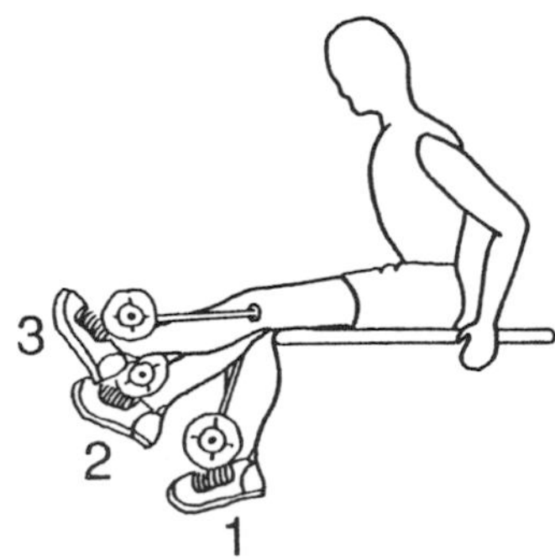

   A. Identify the net pattern of muscle activity for one slow repetition of the exercise.

Movement	1→2	2→3	3→2	2→1
Knee flexors				
Knee extensors				

B. If the limb is held stationary at Positions 1, 2, and then 3, which of these positions will require the following:

Greatest torque to maintain? ______________

Least torque to maintain? ______________

C. Calculate the resultant muscle torque about the knee joint that the athlete must exert to hold the leg in Position 3. In this position, the leg is 0.21 rad below the horizontal. The following system dimensions apply:

body weight (BW)	= 758 N
leg length (knee to ankle)	= 42 cm
load	= 172 N
load acts perpendicular to the leg	= 37 cm from the knee
limb (leg + foot) CG	= 41.3% of leg length from knee
limb (leg + foot) weight	= 0.045 BW − 1.75

Draw a free body diagram before you attempt to solve the problem.

155. The performance of a particular movement is associated with the following hip joint velocity-time relationship:

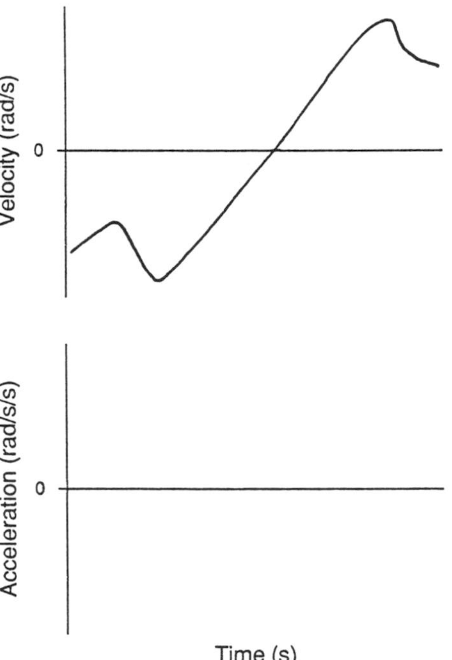

A. From the velocity-time graph, derive the acceleration-time graph as we did in chapter 1.

B. Because acceleration is proportional to force (**F** = *ma*), the acceleration-time graph also represents the profile of the resultant muscle torque about the hip joint. Assume that positive velocity indicates extension and that positive acceleration (torque) represents a net flexor torque. Derive the pattern of net muscle activity associated with the movement.

156. The neural control of a movement depends on the speed at which the movement is performed.

A. Draw velocity-time and torque-time graphs for a rapid performance of the movement shown in Figure 7.15. Assume that the individual used a light weight and did the movement as fast as possible.

B. Describe the net muscle activity associated with such a performance.

157. Two-joint muscles occur across many joints in the human body. They afford a number of advantages for the control of movement.

A. Why is the shortening velocity of a two-joint muscle less than that of its single-joint synergists? What is the functional consequence of this difference?

B. How could rectus femoris (a two-joint muscle that crosses the hip and knee joints) increase the torque exerted by the knee extensors?

158. Draw moment arm–angle and torque-angle graphs for the task shown in Figure 7.21.

159. The performance of a movement involves a close association between muscles concerned with the maintenance of posture and those that produce the movement.

A. Why are postural adjustments described as automatic?

B. Describe conditions when postural activity is concerned with maintaining the stability of the system and conditions when it needs to keep the system in equilibrium.

C. Why is there a greater incidence of falls among elderly individuals?

D. Provide an example that helps explain why postural activity is necessary to counteract the effects of intersegmental dynamics.

160. Describe a movement that would exhibit a three-burst pattern of EMG. Draw the velocity- and EMG-time graphs for the movement.

161. Most studies on the movement strategies have focused on goal-directed movements.

A. Why do scientists analyze EMG patterns when trying to understand the control of movement by the nervous system?

B. What parameters does the dual-strategy hypothesis propose are important in the control of movement? Why choose these parameters?

C. Explain the association between the dual-strategy hypothesis and the λ model of the control of movement.

162. Suppose that you were performing therapy on an injured athlete by doing passive manipulation and you concluded that the joint was stiffer than in healthy subjects. Draw a graph to show that the joint was indeed stiffer in the athlete. Label the graph with the correct variables and units of measurement. Draw one line for the athlete and one for a healthy subject.

163. A common form of muscle activation is to use an eccentric-concentric sequence.
   A. Describe the concepts of the storage and utilization of elastic energy and the preload effect.
   B. How could the stretch reflex possibly affect the increased positive work done during the stretch-shorten cycle? Why is the stretch reflex not a significant contributor to this effect?

164. A typical experimental strategy to examine the stretch-shorten cycle is to compare the performance of the squat jump and the countermovement jump.
   A. Draw force-time graphs for the vertical component of the ground reaction force for both the squat jump and the countermovement jump.
   B. For the countermovement jump, why must the propulsion impulse (area above body weight) be larger than the braking component (area below body weight) for the individual to actually leave the ground?
   C. Suppose that the impulses during the countermovement jump had the following magnitudes:

Propulsion	320 N·s
Braking	82 N·s

   Use the impulse-momentum relationship (chapter 3) to determine the change in the velocity of the jumper (mass = 64 kg) during the takeoff phase.

165. Motor unit recruitment order is reasonably consistent but appears to be somewhat adaptable given the appropriate stress.
   A. What are the advantages of having the motor units recruited in the spinal cord–determined order of smallest to largest?
   B. Why might the control of muscle force be less precise during eccentric contractions?
   C. What features of motor unit organization enable recruitment order to be altered by such events as fatigue and cutaneous stimulation?
   D. Cite one advantage to the musculoskeletal system for each of the three boundary conditions that appear to alter recruitment order.

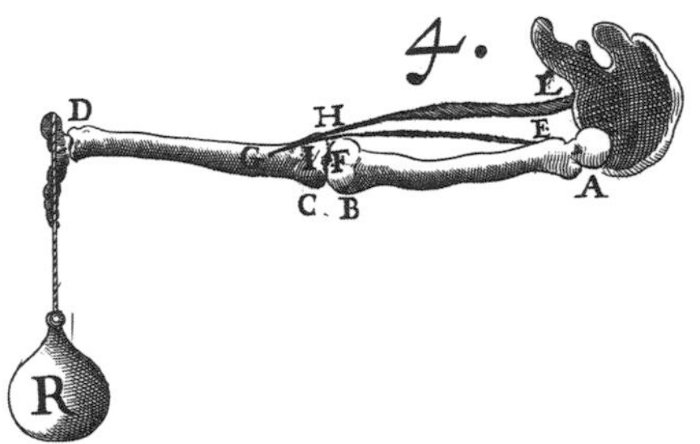

CHAPTER 8
Acute Adaptations

One prominent characteristic of the motor system is its adaptability. When subjected to an acute or chronic stress, the motor system can adapt to the altered demands of usage. These adaptations can be extensive and have been shown to affect most aspects of the system, both morphological and functional. In this chapter we consider the immediate (acute) response of the motor system to the stress associated with a single bout of physical activity. We will examine the effects of a warm-up, the techniques and mechanisms underlying changes in flexibility, muscle damage and soreness, the mechanisms that cause muscle fatigue, and the phenomenon of muscle potentiation.

Warm-Up Effects

Often when an individual undertakes a bout of physical activity the initial activity includes light exercises to prepare the body for the ensuing stress. The purpose of the light exercises is to elicit a warm-up effect, which includes increasing core temperature and disrupting transient connective tissue bonds. The increase in core temperature will improve the biomechanical performance of the motor system, and the stretch may reduce the possibility of a muscle strain (Garrett, 1990; Stanish & Hubley-Kozey, 1984). The effects of a warm-up should be distinguished from the effects of exercises designed to increase flexibility—that is, those intended to increase the range of motion about a joint.

Temperature

The warm-up has a significant effect on temperature-dependent physiological processes. The increase in core temperature as the result of a warm-up can produce an increase in the dissociation of oxygen from hemoglobin and myoglobin, an enhancement of metabolic reactions, a greater muscle blood flow, a decline in muscle viscosity, an increase in the extensibility of connective tissue, and an increase in the conduction velocity of action potentials (Shellock & Prentice, 1985). We are most interested in the effect of a warm-up on the mechanical output of muscle.

Generally the height that a person can reach in a vertical jump increases after a warm-up. This happens because a warm-up increases the maximum power that a muscle can produce (Figure 8.1) and because jump height depends on the quantity of power produced. The enhancement of jump performance following a warm-up is elicited by the effect of muscle temperature on contraction speed. For example, Davies and Young (1983) found that increasing muscle temperature by 3.1°C decreased contraction time and half relaxation time by 7% and 22%, respectively, but did not affect twitch or tetanic tension. In contrast, decreasing muscle temperature by 8.4°C (by immersing the leg in an ice bath) increased both contraction time (38%) and half relaxation time (93%). Observations such as these indicate that relaxation is more dependent on muscle temperature than force developments. In terms of the force-velocity relationship, changes in temperature within the physiological range affect the maximum velocity of shortening (12% · °C^{-1}) but not the maximum isometric force (Binkhorst, Hoofd, & Vissers, 1977). For a muscle group with a given force-velocity relationship, the effect of changing muscle temperature is not to alter $F_{m,o}$ but to shift v_{max} to the right by 12% for each degree (°C) increase in temperature (Figure 8.1). The net effect of the change in contraction speed is an increase in peak power.

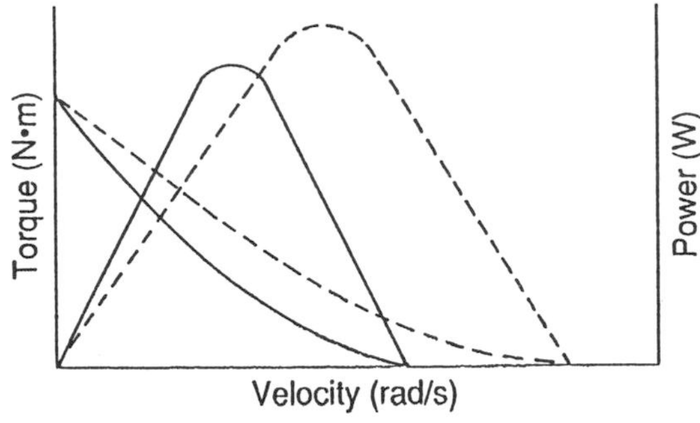

Figure 8.1 Idealized effect of a change in temperature due to a warm-up on the torque- and power-velocity relationships. The power-velocity relationships are bell-shaped. An increase in temperature shifts v_{max} and the peak power to the right.

When the change in temperature is substantial, there can be an effect on the maximum isometric force. For example, Ranatunga, Sharpe, and Turnbull (1987) found that the maximum isometric force of a hand muscle remained relatively constant on cooling to 25°C but decreased by about 30% when cooled to 12°C to 15°C. Similarly, Bergh and Ekblom (1979) reported a change in maximum isometric torque of the knee extensors from 262 N·m at 30.4°C to 312 N·m at 38.5°C (2.4%·°C^{-1}), with the temperature measured in the vastus lateralis muscle. These changes did affect performance; the subjects increased vertical jump height by 44% (17 cm) and maximum power production in cycling by 32% (316 W) when temperature was increased from 30.4°C to 38.5°C.

Ingjer and Strømme (1979) report that the best strategy for inducing changes in muscle temperature, as measured by performance in a maximum-effort 4-min treadmill run, is by doing an *active-related warm-up* (see also Shellock, 1986; Shellock & Prentice, 1985). An active warm-up is one in which the changes in temperature result from muscle activity rather than from a passive external heat source, such as a warm bath. Similarly, in a related warm-up, the muscles activated during the warm-up are subsequently used in the event. Ingjer and Strømme were able to raise the intramuscular temperature of the lateral part of quadriceps femoris from 35.9°C to about 38.4°C with both active and passive techniques, but they argued in favor of the active warm-up because of the superior performance by the subjects during a 4-min maximum aerobic treadmill run. In short, a greater part of the total energy expenditure is provided by aerobic processes when the work is preceded by an active warm-up. These authors further suggest that the duration of the warm-up should be greater than 5 min and that the warm-up should be done at an intensity equivalent to a 7.5-min/mile pace for a trained athlete or sufficient to cause perspiration and an increase in heart rate in an untrained individual. The increase in muscle temperature from such activities will be lost by about 15 min after the warm-up; therefore, the time between the warm-up and the event should be no longer than 15 min. In addition, a warm-up done to achieve the benefits of an increase in core temperature is probably necessary only if the speed of the task requires it and the environmental temperature is low.

Stiffness

The structural elements of muscle resist an increase in length (stretch). The greater the stretch, the greater the passive resistance. As we noted in chapter 6 (Figure 6.12), there is a curvilinear relationship between muscle force and muscle length when a passive muscle is stretched. Because **stiffness** (N/m or N/mm) is defined as the slope of a force-length relationship, the curvilinear relationship indicates that the stiffness of a stretched passive muscle increases as a function of length. Figure 8.2 shows the force-length relationships for three muscles from the cat hindlimb; the three components of each graph are total, passive, and active force. In each graph, the passive force-length component is shown by the line that begins to increase at a normalized length of 1.0, the active force-length component is shown by the bell-shaped curve, and the total force-length relationship is shown by the upper line (sum of the passive and active components). The shape of the total force-length component is different for each muscle because of differences in the active and passive components. Muscles that have narrow active and steep passive length-force curves (e.g., soleus, lateral gastrocnemius) tend to have greater cross-sectional areas, slower shortening velocity, and large pennation angles (Gareis, Solomonow, Baratta, Best, & D'Ambrasia, 1992).

In human subjects, the passive muscle resistance can be characterized as a torque-angle relationship (Figure 8.3c). For example, Hufschmidt and Mauritz (1985) measured the passive resistance of the plantarflexor muscles in control subjects and in patients with varying degrees of spasticity. They used a torque motor to displace the ankle joint through a 0.35-rad range of motion (Figure 8.3a)—from 0.175 rad (10°) of plantarflexion to 0.175 rad of dorsiflexion—and measured the resistance exerted by the plantarflexor muscles through the foot (Figure 8.3b). The area enclosed by the torque-angle

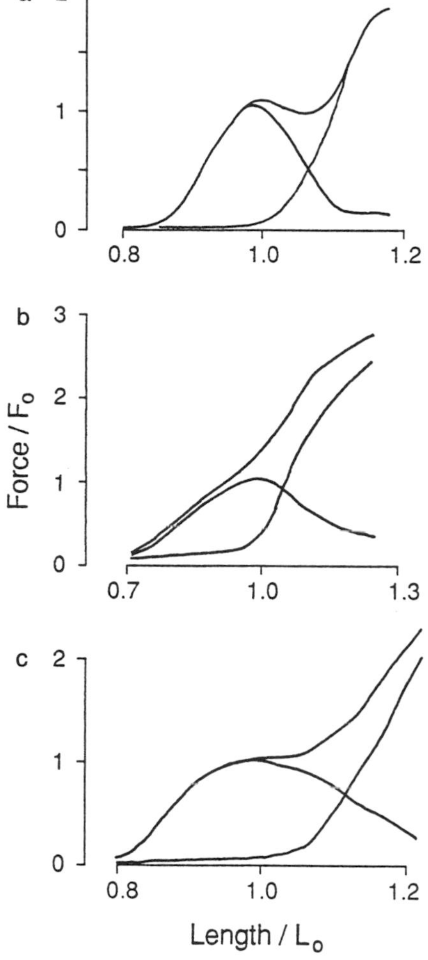

Figure 8.2 Force-length graphs for three muscles from the cat hindlimb: (a) lateral gastrocnemius, (b) soleus, and (c) extensor digitorum longus. Force was normalized to the peak force (F_o) at resting length (L_o). Muscle length was normalized to resting length (L_o), where L_o is the muscle length at which passive force was first evident.

Note. Adapted from "The Isometric Length-Force Models of Nine Different Skeletal Muscles" by H. Gareis, M. Solomonow, R. Baratta, R. Best, and R. D'Ambrosia, 1992, *Journal of Biomechanics*, **25**, pp. 908, 909, Copyright 1992, with kind permission from Pergamon Press Ltd, Headington Hill Hall, Oxford 0X3 OBW, UK.

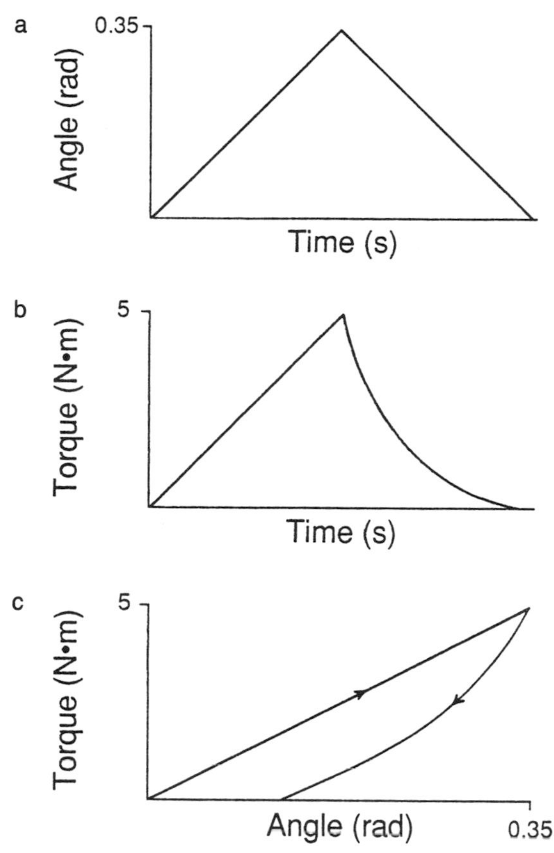

Figure 8.3 Torque-angle relationship associated with the passive stretch and release of the plantarflexor muscles: (a) the increase (stretch) and decrease (release) in angle; (b) the passive torque exerted by the muscles; (c) the resulting torque-angle relationship, which does not overlap during the stretch and release phases.

Note. From "Chronic Transformation of Muscle in Spasticity: A Peripheral Contribution to Increased Tone" by A. Hufschmidt and K.H. Mauritz, 1985, *Journal of Neurology, Neurosurgery, and Psychiatry*, **48**, p. 678. Copyright 1985 by Journal of Neurology, Neurosurgery, and Psychiatry. Adapted by permission.

relationship indicates the energy absorbed by the plantarflexor muscles during the passive stretch-shorten cycle. Because spasticity involves changes in muscle that result in an increased passive resistance to angular displacement about a joint, patients who had exhibited spastic symptoms for more than 1 year had stiffer muscles than healthy subjects (Hufschmidt & Mauritz, 1985). This effect is greater for the extensor (antigravity) muscles (Dietz, 1992).

An interesting property of passive muscle is that its stiffness (slope of the torque-angle relationship) increases as the time between the stretches increases (Figure 8.4). This effect—which has been observed in single muscle fibers, finger muscles, and the plantarflexor muscles—seems to involve an increase in muscle stiffness that occurs over a 30-min rest interval following muscle activation (Hufschmidt & Mauritz, 1985; Kilgore & Mobley, 1991; Lakie & Robson, 1988a). The increase in stiffness is most rapid immediately after the activity and then becomes more gradual. The increased stiffness can be eliminated by active or passive movements but not by isometric contractions (Lakie & Robson, 1988a).

This effect has been attributed to the thixotropic property of muscle. **Thixotropy** is a property exhibited by various gels, such as muscle. The gel becomes a fluid when shaken, stirred, or otherwise disturbed, and it sets

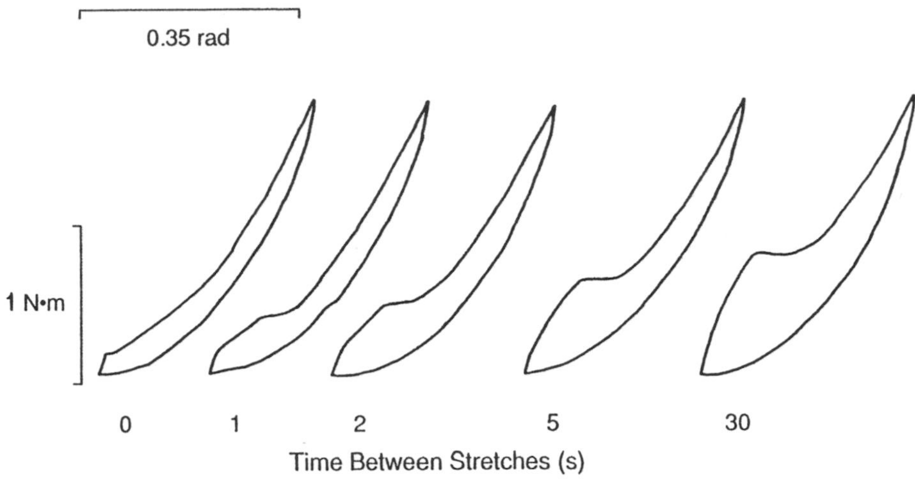

Figure 8.4 Effect of variable rest intervals on the torque-angle relationship associated with the passive stretch and release of the plantarflexor muscles.
Note. From "Chronic Transformation of Muscle in Spasticity: A Peripheral Contribution to Increased Tone" by A. Hufschmidt and K.H. Mauritz, 1985, *Journal of Neurology, Neurosurgery, and Psychiatry*, **48**, p. 678. Copyright 1985 by Journal of Neurology, Neurosurgery, and Psychiatry. Adapted by permission.

again when allowed to stand. A simple example is the behavior of ketchup. If a bottle of ketchup has been unused for a few hours or longer, it is often difficult to pour the contents out of the bottle. If, however, the bottle is first shaken, the contents experience a change of state and are much easier to pour. The increased stiffness of resting muscle is thought to occur by a qualitatively similar mechanism. The molecular rearrangement in muscle may involve the development of stable bonds between actin and myosin filaments. With inactivity, the number of bonds increases and hence the stiffness of muscle increases. However, with a brief stretch or period of physical activity, many, if not all, of the bonds are broken, and muscle stiffness decreases (Hagbarth, Hagglund, Nordin, & Wallin, 1985; Lakie & Robson, 1988b; Proske, Morgan, & Gregory, 1993; Wiegner, 1987).

The functional consequences of muscle thixotropy are widespread. One purpose of a warm-up is probably to minimize muscle stiffness by moving most major muscle groups through a complete range of motion (Wiktorsson-Möller, Öberg, Ekstrand, & Gillquist, 1983). The warm-up disturbs the actin-myosin bonds that have developed and thereby reduces the passive stiffness of muscle. Because we prefer to begin activity or movement with a warm-up, this implies that the nervous system prefers to control muscle when it is in a state of minimum passive stiffness. In contrast, an increase in stiffness due to inactivity makes muscle less responsive to perturbations and may make the control of posture easier. One aspect of muscle thixotropy that is largely unexplored is the associated afferent effects. For example, is increased stiffness due to muscle thixotropy the main reason why we stretch after a period of inactivity (e.g., sitting for a long time)? If so, is there an afferent signal that triggers this response? Also, what is the role of muscle thixotropy as a stimulus for activity during sleep? Furthermore, the time course of both muscle and joint (connective tissue) thixotropy must determine the need for passive range-of-motion activities in patients who are immobile. However, the effects of passive motion on joint tissues and properties (e.g., stiffness) is complex. For example, 16 hours of passive motion per day is necessary to prevent joint stiffness, but this amount of activity does not maintain bone density (Gebhard, Kabo, & Meals, 1993). This type of interaction complicates the selection of a treatment to prevent joint stiffness.

Muscle Tone. The thixotropic property of muscle endows it with a passive stiffness that resists changes in its length. Clinicians refer to this resistance to stretch by a relaxed muscle as **muscle tone**. The resistance to stretch can, of course, be supplemented by the stretch reflex, but this does not occur in a relaxed subject when the rate of stretch has a low to moderate amplitude. In a relaxed subject, the fusimotor system is not active enough to sensitize the muscle spindles to the stretch (unexpected change in length), and the motor neurons are in a low state of excitability. Apparently, muscle tone is similar in relaxed control subjects and in totally anesthetized patients (Rothwell, 1987).

Alterations in muscle tone can be used by clinicians to identify pathological conditions. Decreases in muscle tone, known as **hypotonus**, appear in patients with lesions in the cerebellar hemispheres and in patients who have had spinal transections. Rothwell (1987) suggests that hypotonus is probably due to a decreased level of excitability of the stretch reflex.

An increase in muscle tone, referred to as **hypertonus**, is probably caused by a steady state of motor neuron activity despite the intent to relax completely.

The two most common forms of hypertonus are spasticity and rigidity. **Spasticity** describes a pathologically induced state of heightened excitability of the stretch reflex. When a spastic muscle is stretched, it responds with a more vigorous stretch reflex than normal muscle would. Furthermore, the exaggerated stretch reflex increases with the velocity of the stretch. Many mechanisms underlie spasticity, including changes in the excitability of motor neurons, postsynaptic hypersensitivity to a neurotransmitter, enlargement of motor unit territories by sprouting, and increases in the passive thixotropic properties of muscle. Spasticity can be induced by a lesion in the central nervous system (primarily in the corticospinal tract) and by spinal transection. The symptoms associated with spasticity include increased passive resistance to movement in one direction, a hyperactive tendon tap reflex, the adoption of a characteristic posture by the involved limb, an apparent inability to relax the involved muscle, and an inability to move the involved joint quickly or in alternating directions. One misconception associated with spasticity is that the changes in muscle tone impair movement capabilities. This is not correct. Spasticity in an antagonist muscle is not the primary factor that impairs the ability of an agonist muscle to perform a movement. The impairment is due to an inability of the agonist muscle to recruit a sufficient number of motor units (McComas, Sica, Upton, & Aguilera, 1973; Sahrmann & Norton, 1977; Tang & Rymer, 1981). Consequently, the appropriate clinical protocol is to improve the activation patterns in the agonist rather than attempting to reduce the spasticity in the antagonist. For example, the long-term application of transcutaneous electrical nerve stimulation (TENS) to the common peroneal nerve significantly increased the dorsiflexion force (agonist) but not the plantarflexion force and thus reduced clinical spasticity (Levin & Hui-Chan, 1992).

The other form of hypertonia is **rigidity**. Rigidity and spasticity have markedly different symptoms. The symptoms associated with rigidity include a bidirectional resistance to passive movement that is independent of the movement velocity and occurs in the absence of an exaggerated tendon tap reflex. The most common example of rigidity occurs in Parkinson's disease and involves a persistent muscle contraction that can appear in passive manipulation as a series of interrupted jerks (cogwheel rigidity).

Flexibility

Frequently no distinction is made between warm-up exercises and those designed to increase flexibility. One function of warm-up exercises is to reduce thixotropic-associated muscle stiffness, which is different from increasing the range of motion, or flexibility, about a joint. The main difference is the duration of the effect. The benefits of the warm-up are designed to last for the subsequent bout of physical activity, whereas the goal of flexibility exercises is to induce a more long-term change in the range of motion.

In general, flexibility exercises are designed to increase the range of motion either with the limb muscles passive (static) or with one or more of the muscles attempting to assist the stretch (dynamic). A typical example of a **static stretch** is an attempt to increase the hip flexion–knee extension range of motion by bending forward from an upright erect posture, keeping the knees extended, and attempting to touch the toes with the fingers. The individual is instructed to keep the leg muscles passive and to maintain the stretch position for 15 to 30 s. The individual will feel that the range of motion is restricted by the tension that develops in the hamstring muscles. One variation of this exercise is to bounce up and down while attempting to touch the toes rather than sustaining one continuous stretch; the bouncing version is called a **ballistic stretch**. The same toe-touch exercise also can be performed with the individual seated on the floor. When the exercise is done in this position, gravity does not assist in the stretch of the hamstring muscles.

Although the static stretch is the most commonly used exercise to improve flexibility among athletes and recreational participants, clinicians tend to favor exercises that combine stretching with activation of either the agonist or the antagonist muscles. Three of these exercises have been derived from a rehabilitation technique known as *proprioceptive neuromuscular facilitation* (PNF) (Knott & Voss, 1968). The **hold-relax (HR) stretch** involves an initial maximum isometric contraction of the muscle to be stretched (the antagonist), followed by relaxation and stretch of the muscle to the limit of the range of motion. This type of exercise can be performed with the assistance of a partner or a therapist (Figure 8.5). The **agonist-contract (AC)** stretch requires the assistance of a partner. The partner moves the participant's limb so that the joint is at the limit of rotation. The participant contracts the agonist (e.g., quadriceps femoris) while the partner applies a force to the limb to stretch the antagonist muscle (e.g., hamstrings). The **hold relax–agonist contraction** (HR-AC) technique is a combination of the HR and AC techniques. For the example shown in Figure 8.5, the HR-AC technique would involve an initial maximum isometric contraction of the hamstrings followed by a relaxation and stretch of the hamstrings; the hamstring stretch would be accomplished by manual assistance from the partner and by contraction of the quadriceps femoris.

The PNF stretches were designed on the basis of known connections and effects within the nervous system. The *HR technique* is intended to stretch the muscle while the alpha motor neurons are least excitable so that afferent input from the length detectors (muscle spindles) is least likely to elicit a stretch-evoked activation of muscle. To examine this possibility, both Hoffmann (H) and tendon tap (T) reflexes have been measured immediately

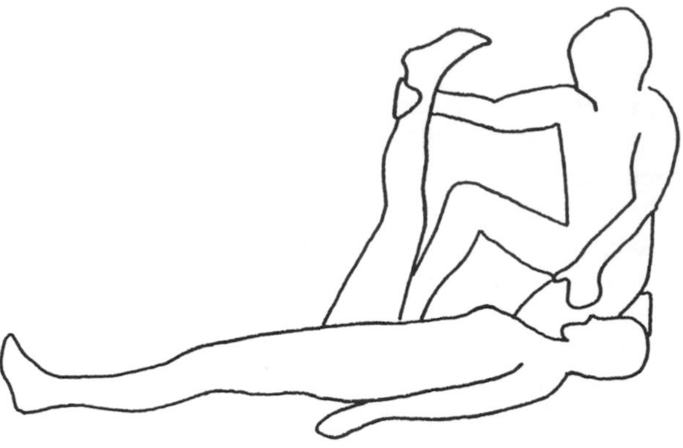

Figure 8.5 An example of the HR-AC technique in which the partner stretches the hamstrings maximally.

after isometric contractions. Recall from chapter 5 that the H reflex is used to test the level of excitability of the motor neuron pool, whereas the T reflex tests the combined effect of muscle spindle responsiveness, motor neuron excitability, and the level of presynaptic inhibition. Studies have shown that the amplitudes of both the H and T reflexes are depressed after an isometric contraction and that the depression of the T reflexes is greater than that of the H reflexes (Enoka, Hutton, & Eldred, 1980; Guissard, Duchateau, & Hainaut, 1988; Moore & Kukulka, 1991). These findings suggest that the excitability of both the muscle spindle and the motor neurons is decreased immediately after an isometric contraction and that this depression lasts about 10 s (Crone & Nielsen, 1989). Furthermore, the depression of the reflexes is similar for contractions that vary in duration from 1 to 30 s.

The rationale for the *AC technique* is to activate the reciprocal inhibition reflex (chapter 5) onto the antagonist (muscle to be stretched) by contraction of the agonist. In this scheme, voluntary activation of the agonist involves activation of the alpha (α) and gamma (γ) motor neurons and the interneuron (I) that mediates the reciprocal inhibition reflex (Figure 8.6) (Nielsen, Kagamihara, Crone, & Hultborn, 1992). This interneuron is known as the Ia inhibitory interneuron (chapter 5). Activation of this interneuron causes action potentials to be transmitted to motor neurons that innervate the antagonist and subsequently causes a reduction in the excitability of these motor neurons. Consequently, the AC technique is presumed to involve an activation of the agonist and, through reciprocal inhibition, a relaxation of the antagonist. Etnyre and Abraham (1988) have shown that the plantarflexor muscles (e.g., soleus) are electrically silent (EMG), as would be expected with reciprocal inhibition, during performance of the AC stretch and contraction of the dorsiflexor muscles.

Several studies have compared the effectiveness of stretching techniques (Condon & Hutton, 1987; Etnyre & Abraham, 1986; Guissard et al., 1988; Moore &

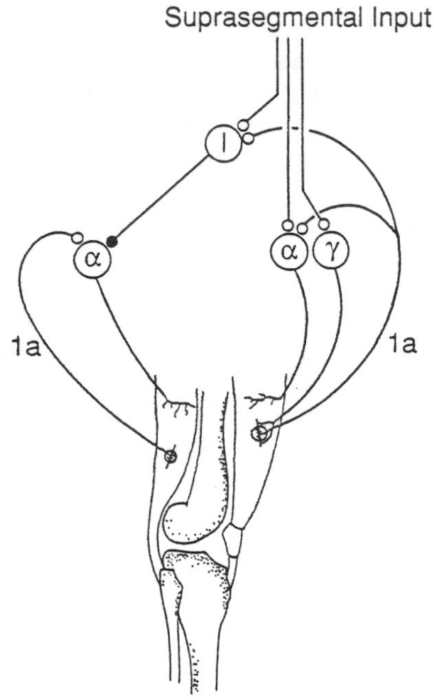

Figure 8.6 Neural pathways activated during the AC stretch of the hamstrings. Activation of the quadriceps femoris by suprasegmental centers results in the concurrent activation of motor neurons innervating quadriceps femoris and the Ia inhibitory interneuron that mediates the reciprocal inhibition reflex. This connection supposedly causes the antagonist (hamstrings) to relax during activation of the agonist (quadriceps femoris).

Hutton, 1980; Wallin, Ekblom, Grahn, & Nordenborg, 1985). In general, the findings indicate that the improvement in flexibility with static stretching is similar to that achieved with ballistic stretching, but ballistic stretches are more likely to produce muscle soreness (Shellock & Prentice, 1985). When the tests of flexibility involve stretching to the physiological limit rather than stretching with a constant torque, the PNF stretches (HR, AC, and HR-AC techniques) provide greater improvements in flexibility than do static and ballistic stretches. There does not, however, seem to be a consistent difference between the three PNF techniques.

The studies on the PNF stretching techniques have not determined the relative significance of the neural mechanisms in the improvement of flexibility. It has been suggested that the range of motion about most joints is limited by one or more connective tissue structures, including those in muscle (Sapega, Quedenfeld, Moyer, & Butler, 1981). The contribution of the various joint structures to the stiffness of the joint varies among joints; for the metacarpophalangeal joint it has been found to be joint capsule, 47%; muscle, 41%; tendons, 10%; and skin, 2% (Johns & Wright, 1962). This means that the range of motion of the metacarpophalangeal joint was determined by the passive properties of these tissues in this proportion. It is unclear how muscle relaxation influences these proportions. Although the PNF techniques can minimize the stretch-induced activation of muscle, the functional significance of this reduction for improved flexibility has not been identified. Alternatively, because the PNF techniques involve muscle contractions, the improvements in range of motion due to these techniques may be related to temperature effects.

In the absence of a neural input, successive stretches can increase the length of a muscle-tendon unit (Taylor, Dalton, Seaber, & Garrett, 1990). In these experiments, a hindlimb muscle (extensor digitorum longus) was stretched from an initial force of about 2 N to a length that produced a passive force of 78 N and was held at this length for 30 s. This stretch cycle was repeated 10 times. Figure 8.7a indicates that the muscles had to be stretched to longer lengths in subsequent stretches in order to exert a force of 78 N; the total increase in length over the 10 stretches was about 3.5%. The converse protocol was also applied in which the muscle was stretched by 10% and the force exerted in successive stretches was measured (Figure 8.7b). These measurements indicate that in the absence of neural input, the load on a muscle-tendon unit due to its passive properties (connective tissue) will decrease with stretches to a prescribed length.

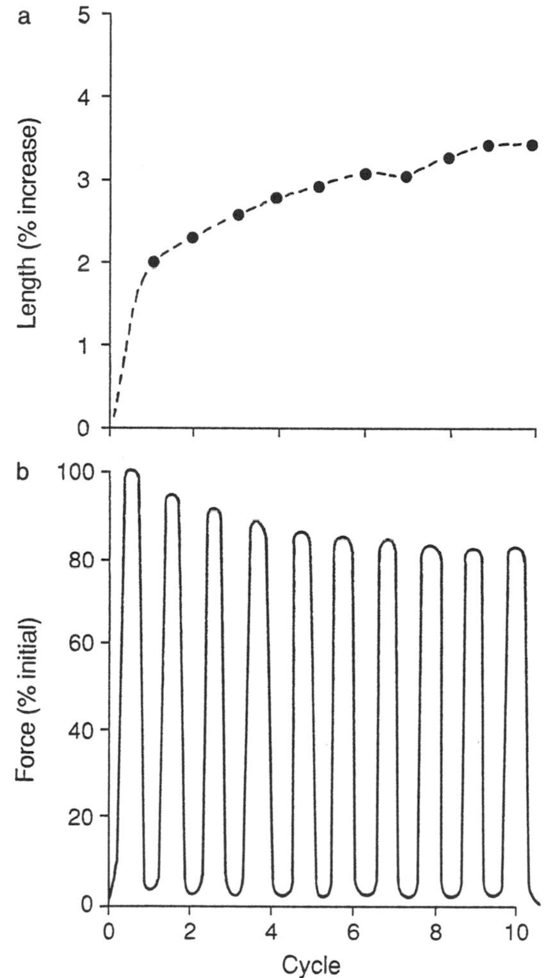

Figure 8.7 Stretch-induced changes in (a) length and (b) force of a muscle-tendon unit.
Note. From "The Viscoelastic Properties of Muscle-Tendon Units" by D.C. Taylor, J. Dalton, A.V. Seaber, and W.E. Garrett, 1990, *American Journal of Sports Medicine*, **18**, pp. 303, 304. Copyright 1990 by the American Orthopaedic Society for Sports Medicine. Adapted by permission.

Despite the uncertain contribution of neural quiescence to changes in flexibility, connective tissue seems to play a significant role in limiting range of motion (Gebhard et al., 1993; Sapega et al., 1981; Shellock & Prentice, 1985). For this reason, flexibility exercises should be directed at altering the length of connective tissue structures. To achieve this goal, flexibility exercises should cause plastic rather than elastic changes in connective tissue, because plastic changes produce more permanent changes in tissue length. The relative proportions of elasticity and plasticity within a stretch are determined by the amount and duration of the applied force and the tissue temperature. The low-force, long-duration stretch optimizes the plastic changes. There is a trade-off between tissue elongation and weakening: *Whenever plastic changes are induced in a tissue, this involves a molecular reorganization and weakening of the tissue.* However, this stress stimulates the tissue to adapt, and the weakening is only short-term, provided that the stress is not too great. In reality, flexibility training merely involves the stressing of tissue so that it adapts to a new length without any long-term loss of strength. Tissue is most extensible when the temperature is higher, such as at the end of a workout, and the amount of long-term elongation is greatest if the stretch is applied while the tissue cools (Sapega et al., 1981). Some individuals even suggest applying ice packs during stretching to promote the cooling effect. Hence, *effective stretching for flexibility should employ low-force, long-duration stretches and should be performed at the end of a workout.*

Muscle Soreness and Damage

Strenuous physical activity can have diverse effects on muscle, ranging from subcellular damage of muscle fibers to stretch-induced muscle injuries (strains). The subcellular damage, which most active individuals have experienced, frequently produces an inflammatory response and is associated with muscle soreness that begins hours after the exercise has been completed. In contrast, strain injuries typically occur as an acute painful injury during high-power tasks and require clinical intervention.

Muscle Soreness

Because the soreness associated with subcellular damage is not evident until 24 to 48 hr after the exercise, it has been described as **delayed-onset muscle soreness**. This term distinguishes the postexercise soreness from the exertional pain that can occur during exercise (Asmussen, 1953, 1956). The clinical symptoms associated with delayed-onset muscle soreness include an increase in plasma enzymes (e.g., creatine kinase), myoglobin, and protein metabolites from injured muscles; structural damage to subcellular components of muscle fibers, as seen with light and electron microscopy; and temporary increases in muscle weakness (Armstrong, 1990).

Cause of Muscle Soreness. Although there is no consensus on the actual mechanism responsible for delayed-onset muscle soreness, there is agreement that it occurs more frequently with eccentric contractions than with other modes of muscle activation (Fridén, 1984; McCully & Faulkner, 1985; Newham, 1988; Stauber, 1989). Why does muscle soreness occur with eccentric contractions? Recall that an eccentric contraction is one in which an active muscle is lengthened because the torque due to the load was greater than that exerted by the muscle. In this scheme, crossbridge cycling may involve an altered detachment phase in which the actomyosin bond is broken mechanically (Morgan, 1990), and there may be selective activation of the larger motor units (Nardone et al., 1989). These differences result in eccentric exercise being characterized by the recruitment of fewer motor units, a lower oxygen cost for cycling (Figure 8.8a), a gradual increase in oxygen consumption (Figure 8.8b), and the gradual increase in EMG (Figure 8.8c) during constant-level downhill running, greater postexercise muscle weakness, and higher peak intramuscular pressure due to an increase in the water content of muscle (Figure 8.8) (Dick & Cavanagh, 1987; Evans & Cannon, 1991; Fridén, Sfakianos, Hargens, & Akeson, 1988; Newham, Jones, & Edwards, 1983; Stauber, 1989).

Physical activity has been proposed to activate both metabolic and mechanical events that may damage muscle and lead to soreness (Armstrong, 1990). The metabolic factors include (a) high temperature, which may disrupt protein structures; (b) insufficient mitochondrial respiration, which may reduce ATP levels and hence the energy available to remove calcium from the cytoplasm; (c) lowered pH due to an increase in lactic acid; and (d) free O_2 radical production and lipid peroxidation, which may initiate muscle injury. Because eccentric contractions are associated with high forces, the mechanical factor commonly thought to elicit muscle soreness is high *stress* (Warren, Hayes, Lowe, & Armstrong, 1993; Warren, Hayes, Lowe, Prior, & Armstrong, 1993; McCully & Faulkner, 1985). It has been suggested that a high stress (force per unit area) can result in the disruption of sarcolemmal, sarcoplasmic reticular, and myofibrillar structures. However, Fridén and Lieber have shown that the amount of *strain* rather than stress is the more significant mechanical factor (Fridén & Leiber, 1992; Leiber & Fridén, 1993). Although these studies suggest that several events may be responsible for muscle soreness, the single mechanism that appears to be common to these metabolic and mechanical events is the *loss of cellular calcium homeostasis* (Clarkson, Cyrnes, McCarmick, Turcotte, & White, 1986; Jackson, Jones, & Edwards, 1984). A high intracellular calcium concentration activates proteolytic and lipolytic systems that may trigger **autogenesis**, the process of degrading cellular structures (Armstrong, 1990).

Time Course of the Response. Structural abnormalities following exercise include sarcolemmal dis-

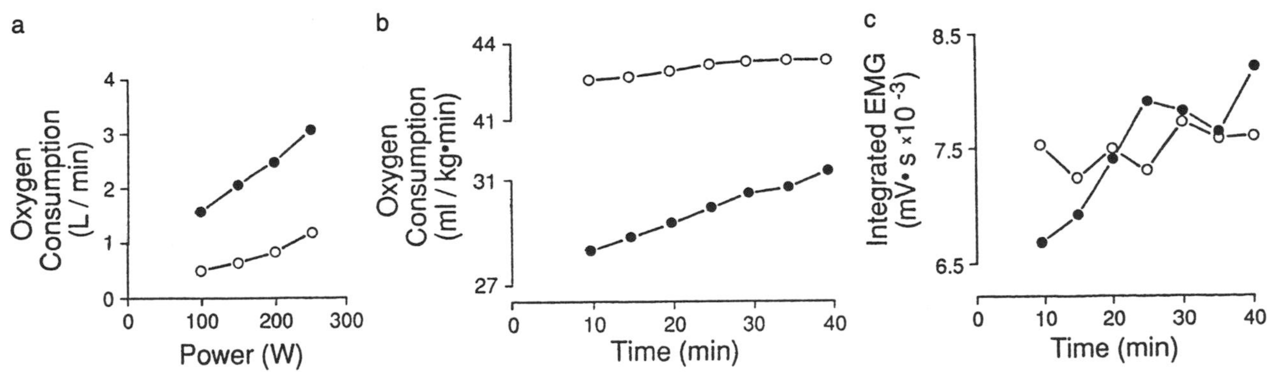

Figure 8.8 Oxygen consumption and EMG activity associated with eccentric contractions. (a) Oxygen costs of performing concentric (○) and eccentric (●) contractions on a cycle ergometer.
Note. From "The Metabolic Effects of Exercise-Induced Muscle Damage" by W.J. Evans and J.G. Cannon, 1991, in J.O. Holloszy (Ed.), *Exercise and Sport Sciences Reviews*, Vol. 19 (p. 101), Baltimore: Williams & Wilkins. Copyright 1991 by Williams & Wilkins. Adapted by permission.
(b) Oxygen consumption associated with running on a level grade and downhill on a 10% grade. (c) Integrated EMG from vastus lateralis during running.
Note. From "An Explanation of the Upward Drift in Oxygen Uptake During Prolonged Sub-Maximal Downhill Running" (Panels Band C) by R.W. Dick and P.R. Cavanaugh, 1987, *Medicine and Science in Sports and Exercise*, **19**, pp. 313, 314. Copyright 1987 by the American College of Sports Medicine. Adapted by permission.

ruption, dilated transverse tubule system, distortion of myofibrillar components, fragmented sarcoplasmic reticulum, lesions of the plasma membrane, cytoskeletal damage, change in the extracellular myofiber matrix, and swollen mitochondria (Fridén & Lieber, 1992; Stauber, 1989). Many of the structural effects of eccentric contractions involve the Z band. Micrographs have shown a clear connection between Z-band streaming and possible damage of the cytoskeletal intermediate filaments with high-tension contractions (Fridén, Kjörell, & Thornell, 1984).

Although this damage is apparent immediately after the exercise, changes continue during the postexercise period; these include an increase in the release of muscle-specific enzymes and an increase in cell permeability. A common marker of muscle damage is the enzyme creatine kinase, which indicates the turnover of proteins rendered dysfunctional by physical or oxidative stress (Evans & Cannon, 1991). The efflux of enzymes occurs by diffusion through exercise-induced holes in the plasma membrane of the muscle cell. The postexercise increase in the skeletal muscle enzymes in the circulation is related to the type and intensity of the exercise. It is greater with eccentric contractions than with concentric ones, and it increases with exercise intensity. Interestingly, estrogen may directly affect the muscle cell membrane and reduce the efflux of skeletal muscle enzymes. Because of this effect, women tend to exhibit lower postexercise levels of circulating creatine kinase (Evans & Cannon, 1991).

Delayed-onset muscle soreness is most pronounced at 24 to 48 hr after exercise (Figure 8.9). Although there is evidence of structural damage immediately after strenuous exercise, significant changes also occur in the days after exercise (Howell, Chleboun, & Conatser, 1993). From 4 hr to 4 days after the exercise, there is an increase in phagocytic activity, which marks the presence of an inflammatory response. Because of this association in time, it has been suggested that *delayed-onset muscle soreness is a consequence of the inflammatory response* (Stauber, 1989). However, the consequences of strenuous exercise, especially eccentric contractions, are not confined to a few days after exercise; MRI studies have indicated marked swelling of injured muscles for up to 10 days and increased signal intensity for up to 60 days after exercise (Fleckenstein & Shellock, 1991; Shellock, Fukunaga, Mink, & Edgerton, 1991b).

Fiber-Type Effect. Type IIb (FG) muscle fibers seem to be preferentially damaged by exercise that is associated with delayed-onset muscle soreness (Fridén, Seger, & Ekblom, 1988). An analysis of muscle ultrastructure indicated that 2 hr after repetitive bouts of sprint running, 36% of the muscle fibers examined in vastus lateralis exhibited significant changes that

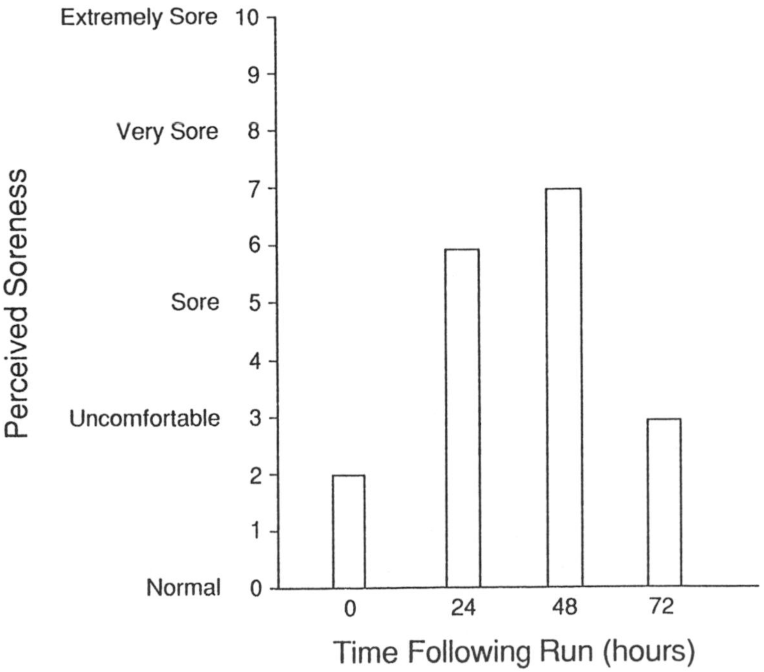

Figure 8.9 Changes in perceived muscle soreness for 72 hr after a 40-min downhill run.
Note. From "An Explanation of the Upward Drift in Oxygen Uptake During Prolonged Sub-Maximal Downhill Running" by R.W. Dick and P.R. Cavanaugh, 1987, *Medicine and Science in Sports and Exercise*, **19**, p. 316. Copyright 1987 by the American College of Sports Medicine. Adapted by permission.

affected mainly the Z band. Eighty percent of these abnormalities occurred in the Type IIb fibers. Recall that Type IIb fibers have narrow Z bands and a low oxidative capacity. These observations have led Fridén and Lieber (1992) to propose the following muscle-damage scheme: Fatigue of the Type IIb (FG) fibers results in an inability to regenerate ATP and the development of a state of rigor (high stiffness) so that a subsequent stretch produces mechanical damage of the myofibrillar and cytoskeletal structures.

The damage to muscle fibers is not limited to the contractile proteins and the intramuscular cytoskeleton. One of the most plastic tissues is connective tissue, and it may also be affected by exercise-induced damage. The effect of exercise on connective tissue is often assessed by measuring the level of urinary excretion of hydroxyproline, a component of mature collagen. Hydroxyproline excretion increases after exercise involving eccentric contractions (Stauber, 1989). In addition, muscle subjected to forced lengthening demonstrates histological changes in the extracellular connective tissue matrix. Because Type II muscle fibers have a less developed extracellular matrix, they are more susceptible to stretch-induced damage.

Regeneration. Armstrong (1990) has identified four stages associated with exercise-induced muscle injury: the initial, autogenetic, phagocytic, and regenerative stages. The initial stage is the exercise bout that produces the damage. The autogenetic stage involves the activation of the proteolytic and lipolytic systems to begin degrading cellular structures. The phagocytic stage occurs from about 4 hr to 4 days after the exercise bout and contains the inflammatory response. The final stage involves the regeneration of the muscle fibers and appears to begin after about 4 days.

The functional consequences of these responses are initial decreases in muscle strength and in the range of motion due to swelling and an increase in the pressure on pain receptors (Stauber, 1989). These effects are short-term, however, and the muscle fibers and connective tissue in the damaged muscles do regenerate. Regeneration requires an intact basal lamina, an uninterrupted vascular supply, and a functional nerve, and it probably involves satellite cells and processes that are similar to those present during development. Muscle fibers can be repaired and can resume normal function after damage due to exercise. Although the repair process seems to alter the fine structure of sarcomeres (Fridén, 1984), the fibers are not different physiologically. Nonetheless, a single bout of exercise seems to prevent muscle damage from subsequent exercise for about 6 weeks (Byrnes et al., 1985; Newham, Jones, & Clarkson, 1987). The adaptation may involve a strengthening of the connective tissue (Clarkson & Tremblay, 1988). However, neither stretching nor warm-up prior to exercise seems to prevent the delayed-onset muscle soreness after an intense exercise bout (High, Howley, & Franks, 1989).

Muscle Strain

In contrast to the microinjuries that often accompany eccentric contractions, muscle can experience acute and painful events, such as cramps and strains. **Muscle cramp** is a painful, involuntary shortening of muscle that appears to be triggered by peripheral stimuli (Bertolasi, De-Grandis, Bongiovanni, Zanette, & Gasperini, 1993). **Muscle strain** is a substantial strain that is immediately recognized as an injury. Muscle strains are also referred to as *pulls* and *tears*. Clinical reports indicate that these strains invariably occur at the muscle-tendon junction; muscle strains have been reported for medial gastrocnemius, rectus femoris, triceps brachii, adductor longus, pectoralis major, and semimembranosus (Garrett, 1990). Muscles most prone to such an injury are the two-joint muscles (because they can be stretched more), muscles that limit the range of motion about a joint, and muscles that have a high proportion of Type II muscle fibers (Garrett, Califf, & Bassett, 1984). Furthermore, the injury most often occurs during powerful eccentric contractions, when the force is several times greater than the maximal isometric force. The injury frequently involves bleeding and subsequent accumulation of the blood in the subcutaneous spaces. The most appropriate immediate treatment is to rest the muscle and to apply ice and compression. Rehabilitation should include physical therapy to improve the range of motion and the prescription of *functional* strengthening exercises.

Experiments on the characteristics of the muscle strain injury have indicated that the disruption occurs near the muscle-tendon junction despite differences in the architectural features of the muscles tested and the direction of the strain (Garrett, 1990). When different muscle-tendon units were stretched passively (muscle not contracting), the strain injury did not occur after a constant fiber strain. However, there was no difference in the total strain to failure when a given muscle-tendon unit was stretched while it was passive versus when it was active (muscle stimulated electrically to contract). Nonetheless, muscle contraction can double the strain energy that the muscle-tendon unit can absorb during the stretch (eccentric contraction) (Garrett, Safran, Seaber, Glisson, & Ribbeck, 1987). Figure 8.10a shows strain energy as the area under the force-length graph, with stretch-induced failure occurring at the peak force. In Figure 8.10b, the strain energy (area under the curve) absorbed by the active muscle is about twice that absorbed by the passive muscle, although the peak force at failure is only about 15% greater in the active muscle. Because of the effect of muscle contraction, any factor, such as fatigue or weakness, that tends to reduce the contractile capability of muscle may predispose the muscle-tendon unit to a muscle strain.

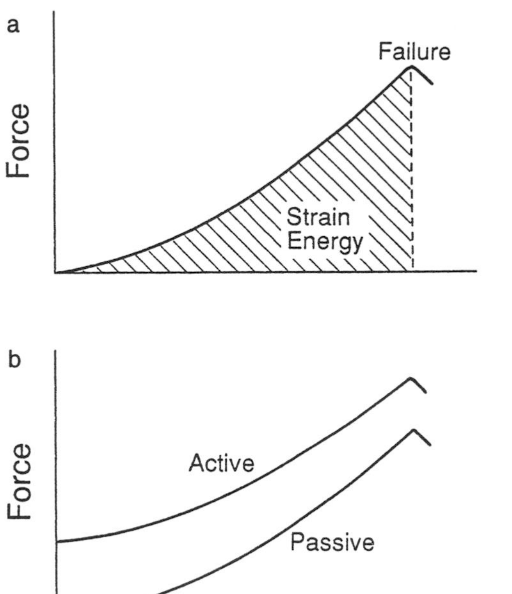

Figure 8.10 Strain energy absorbed by a muscle-tendon unit during a stretch to failure.
Note. From "Biomechanical Comparison of Stimulated and Nonstimulated Skeletal Muscle Pulled to Failure" by W.E. Garrett, M.R. Safran, A.V. Seaber, R.R. Glisson, and B.M. Ribbeck, 1987, *American Journal of Sports Medicine*, **15**, p. 452. Copyright 1987 by the American Orthopaedic Society for Sports Medicine. Adapted by permission.

Muscle Fatigue

Muscle fatigue refers to *a class of acute effects that impair performance*. These effects involve both motor and sensory processes. Because the processes involved in performance extend from the limbic system to the crossbridge, considerable attention has been focused on identifying the process that fails and therefore produces fatigue. The last 50 years of study, however, have clearly demonstrated that fatigue is not caused by the impairment of a single process but rather that the mechanisms responsible for fatigue vary from one condition to another. This effect has been termed the **task dependency** of muscle fatigue (Enoka & Stuart, 1992). To understand muscle fatigue and its causes, we need to identify the conditions that can impair the different processes and contribute to muscle fatigue. This section outlines what is known about these conditions and discusses some of the sensory adaptations that occur during fatiguing contractions.

Task Dependency

When an individual performs a task, the requirements of the task (e.g., amount of force, muscles involved, duration of the activity) stress various physiological processes associated with the motor performance. As the requirements change, so do the processes that experience the greatest stress. The task variables that appear to influence the stressed processes include the level of subject motivation, the neural strategy (pattern of muscle activation and motor command), the intensity and duration of the activity, the speed of a contraction, and the extent to which an activity is sustained continuously. The physiological processes that can be impaired by these variables include the central nervous system drive to motor neurons, the muscles and motor units that are activated, neuromuscular propagation, excitation-contraction coupling, the availability of metabolic substrates, the intracellular milieu, the contractile apparatus, and muscle blood flow.

Central Drive. Typically the maximality of the excitation provided by the nervous system is tested by comparing the force exerted during a maximum voluntary contraction (MVC) to the force that can be elicited artificially with electrical stimulation. This approach involves applying single shocks or a brief train of shocks to the nerve during a fatiguing contraction. This test has been applied to both sustained and intermittent contractions and to maximal and submaximal contractions. When subjects performed a sustained 60-s maximum voluntary contraction with a thumb muscle (adductor pollicis), the force declined by 30% to 50% but this decrease in force could not be supplemented by an electric shock (Bigland-Ritchie et al., 1982). Similarly, when subjects performed an intermittent (6-s contraction, 4-s rest) submaximal contraction (target force was 50% of maximum) with the quadriceps femoris, the maximal voluntary and electrically elicited forces declined in parallel (Figure 8.11) (Bigland-Ritchie, Furbush, & Woods, 1986). In this experiment, the maximum voluntary and electrically elicited forces were elicited periodically during the submaximal contraction. The parallel decline in the voluntary and artificial forces suggests that the central drive remained maximal during these tasks.

Two tests of artificial stimuli were applied to the muscle in the experiment shown in Figure 8.11. One test consisted of a train of eight stimuli that were delivered at a rate of 50 Hz. The proportion of the muscle activated by this stimulus depended on the amount of electric current that was passed through the muscle by the stimulating device. For the muscle fibers that were activated, a stimulus rate of 50 Hz elicited close to the maximum force. The second test used was the **twitch superimposition** (Tw_s) test, which is also known as the twitch interpolation and twitch occlusion test. The twitch superimposition test involves applying one to three supramaximal, electric shocks to the nerve during a maximum voluntary contraction and observing whether force increases. If the maximum voluntary contraction force increases, the central drive during the contraction is not

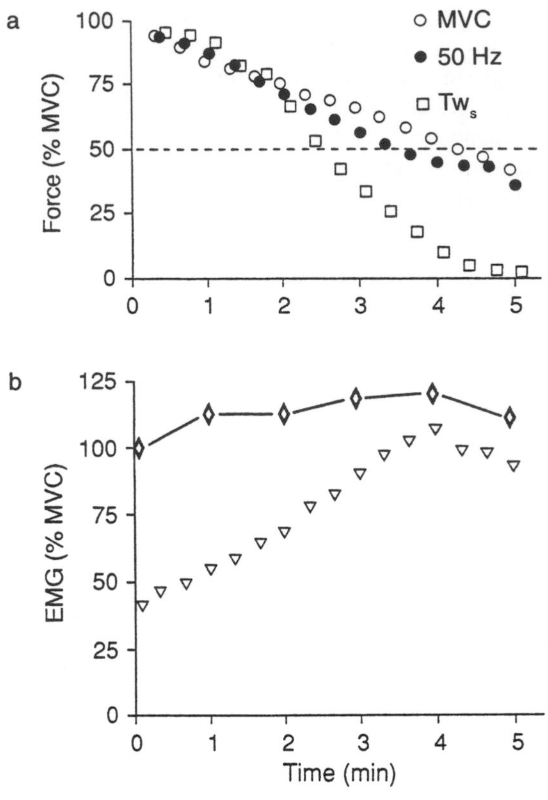

Figure 8.11 (a) Decline in maximum voluntary contraction (MVC) and in artificially elicited forces during a fatiguing contraction with the quadriceps femoris muscle; (b) change in the integrated EMG for quadriceps femoris during MVC and intermittent submaximal contractions over the course of the task.

Note. From "Fatigue of Intermittent Submaximal Voluntary Contractions: Central and Peripheral Factors" by B. Bigland-Ritchie, F. Furbush, and J.J. Woods, 1986, *Journal of Applied Physiology*, **61**, p. 424. Copyright 1986 by the American Physiological Society. Adapted by permission.

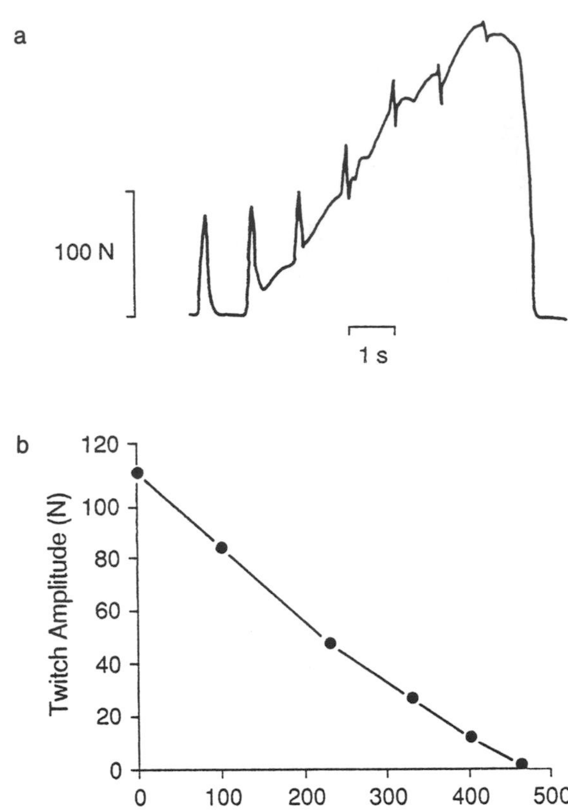

Figure 8.12 The twitch superimposition method: (a) decrease in the amplitude of the twitch force elicited by an electric shock as the voluntary force exerted by quadriceps femoris increased; (b) decline in twitch force as a function of voluntary force. The twitch was elicited at various levels of force during voluntary contractions. The maximum voluntary contraction (MVC) force was about 470 N.

Note. From "Fatigue of Submaximal Static Contractions" by B. Bigland-Ritchie, E. Cafarelli, and N.K. Vøllestad, 1986, *Acta Physiologica Scandinavica*, Suppl. **556**, p. 138. Copyright 1986 by the Blackwell Scientific Publications. Adapted by permission.

maximal. In the experiment depicted in Figure 8.12, as the voluntary force increased, the twitch force elicited by a superimposed electric shock decreased. During the intermittent fatiguing contraction (Figure 8.11), the amplitude of the superimposed twitch force decreased to zero by the time the maximum voluntary contraction force reached the target force (50% MVC); therefore, the central drive was maximal at the end of the fatigue test.

The tests that have been conducted on fatiguing contractions indicate that the neural drive to muscle provided by the central nervous system is not always maximal and that the reduction in central drive can be a factor that contributes to the decline in force. There are at least three examples in the research literature of the inability of humans to generate an adequate central drive during fatiguing activity. First, subjects who are not motivated maximally do not exhibit a parallel decline in the voluntary and artificial (electrical stimulus) forces (Bigland-Ritchie et al., 1986; Gandevia & McKenzie, 1985; Grimby, Hannerz, & Hedman, 1981). This inability appears as a deviation of the two force lines (MVC and 50 Hz) in Figure 8.11a. A lack of motivation probably results in an inadequate central drive to the appropriate motor neurons. Second, it seems more difficult to maximally activate some muscles than others (Bigland-Ritchie et al., 1986; Milner-Brown & Miller, 1989; Thomas, Woods, & Bigland-Ritchie, 1989). For example, in one study, subjects could maximally activate the tibialis anterior muscle, as assessed by twitch superimposition, but 10 of 17 male and 4 of 11 female subjects could not maximally activate the plantarflexors (Belanger & McComas, 1981). Third, the fatigability of muscle differs for concentric and eccentric contractions. This difference is shown in Figure 8.13. Subjects performed maximal contractions with the quadriceps femoris on an isokinetic device that was set with a 1.2-rad range of motion and a speed of

3.14 rad/s. The subjects performed three bouts of 32 contractions each for both the concentric and eccentric conditions. The torque exerted during the concentric contractions decreased, and that for the eccentric contractions increased slightly (Figure 8.13a). Furthermore, the integrated EMG recordings for both vastus lateralis and rectus femoris were less during the eccentric contractions but increased for both the concentric and eccentric contractions (Figure 8.13b). The dissociation between torque and EMG for the two contractions (concentric and eccentric) suggests a fundamental difference in muscle activation.

Another approach that uses electrical stimuli to assess the role of central drive in limiting performance in sustained activities involves manipulating humoral factors that circulate in the cerebrospinal fluid. In these experiments, pharmacological agents are infused into the cerebrospinal fluid of experimental animals (typically rats), and the effect on endurance time is assessed. The results indicate that some agents (e.g., amphetamine, a central nervous system stimulant) can increase the time to reach exhaustion, whereas other agents (e.g., 6-hydroxydopamine, a neurotoxin that destroys catecholaminergic fibers) decrease the time it takes to reach exhaustion (Bhagat & Wheeler, 1973; Heyes, Garnett, & Coates, 1985). These observations suggest that humoral factors such as epinephrine and the neurotransmitter 5-hydroxytryptamine can play a significant role in sustaining the central drive during fatiguing tasks (Blomstrand, Celsing, & Newsholme, 1988; Chaouloff, Kennett, Serrurier, Merino, & Curzon, 1986; Heyes, Garnett, & Coates, 1988). It appears, therefore, that the levels of hormones circulating in the cerebrospinal fluid are an important factor in the ability to sustain the central drive.

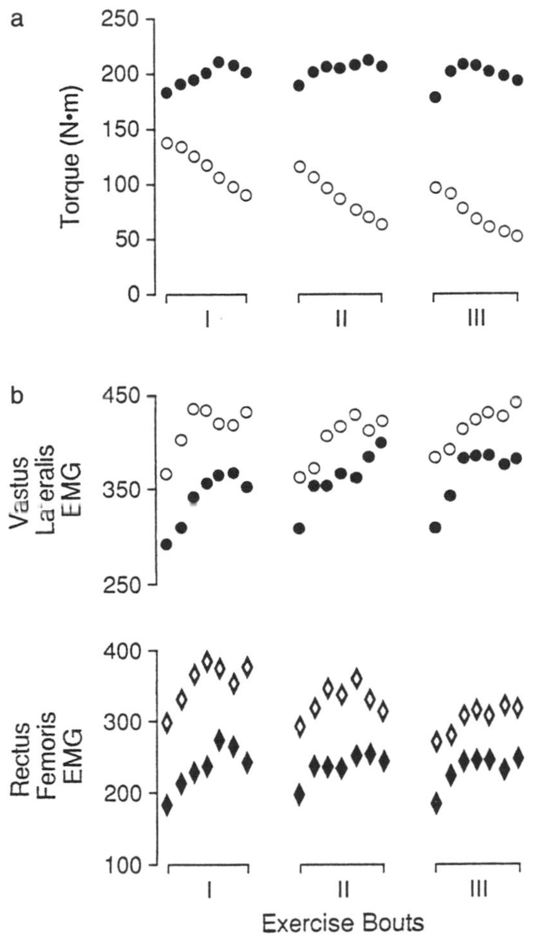

Figure 8.13 Changes in (a) torque and (b) EMG during concentric (open shapes) and eccentric (solid shapes) fatiguing contractions with the quadriceps femoris muscle. *Note*. From "Force and EMG Signal Patterns During Repeated Bouts of Concentric and Eccentric Muscle Actions" by P.A. Tesch, G.A. Dudley, M.R. Duvoisin, B.M. Hather, and R.T. Harris, 1990, *Acta Physiologica Scandinavica*, **138**, p. 266. Copyright 1990 by the Blackwell Scientific Publications. Adapted by permission.

Neural Strategy. A resultant muscle force about a joint can be achieved by a variety of muscle activation patterns. This flexibility exists among a group of synergist muscles and perhaps among the motor units within individual muscles. Because of this possibility, one way the motor system can delay the onset of force decline (fatigue) is to vary the contribution of synergist muscles to the resultant muscle force. This possibility is available only when the task requires submaximal forces, but most activities of daily living are in this category. A good example of this distribution of activity among synergist muscles was obtained by Sjøgaard and colleagues, who had subjects sustain an isometric knee extensor force at 5% of the maximum voluntary contraction force for 1 hr (Sjøgaard, Kiens, Jørgensen, & Saltin, 1986; Sjøgaard, Savard, & Juel, 1988). Despite the low intensity of the task, the subjects did experience fatigue, as indicated by a 12% decline in the maximum voluntary contraction force and an increase in the associated effort at the end of the hour. During the fatiguing contraction, the subjects were able to maintain a constant force (5% MVC) while the activity switched from rectus femoris to vastus lateralis (Figure 8.14). This observation indicates that the neural drive to the muscles changed while the net output remained constant. Similarly, when the knee extensor muscles perform low intensity, intermittent contractions to the endurnace limit, there is a progressive increase in the coactivation of the knee flexor muscles (Psek & Cafarelli, 1993).

Even though the motor performance of a single limb may remain constant, the rate of decline in force depends on whether one or two limbs are active during the task. For example, Rube and Secher (1990) examined the ability of subjects to perform an isometric leg extension task (concurrent knee and hip extension) with either one or two legs. Subjects were required to perform 150 maximal voluntary contractions for both conditions (one and two legs) before and after 5 weeks of strength training. The subjects were separated into three groups:

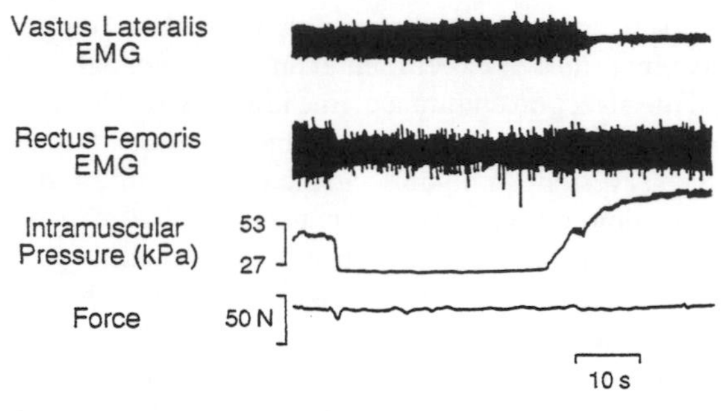

Figure 8.14 Variable contributions of vastus lateralis and rectus femoris to a sustained knee extensor force. Intramuscular pressure was measured in the rectus femoris muscle. *Note.* From "Intramuscular Pressure, EMG and Blood Flow During Low-Level Prolonged Static Contraction in Man" by G. Sjøgaard, B. Kiens, K. Jørgensen, and B. Saltin, 1986, *Acta Physiologica Scandinavica*, **128**, p. 479. Copyright 1986 by the Blackwell Scientific Publications. Adapted by permission.

control group, one-legged training group, and two-legged training group. Both the one- and two-legged groups increased strength after 5 weeks of training while the control group stayed the same. Furthermore, the rate of decline in force during the 150 contractions was less after training (Figure 8.15). But the effect was specific to the training mode. The one-legged training group was only less fatigable during the one-legged task, and the two-legged training group was only less fatigable during the two-legged task. These data suggest that the neural strategy adopted during the one-legged task differed from that used during the two-legged task.

As exercise training can alter the biomechanical and physiological properties of muscle, so too can chronic activity patterns associated with pathology (Grabiner, Koh, & Andrish, 1992). For example, the fatigability of the quadriceps femoris is altered in patients who have undergone unilateral, anterior cruciate ligament reconstruction. In these patients, the quadriceps femoris of the involved leg was *less* fatigable than that of the uninvolved leg (Snyder-Mackler, Binder-Macleod, & Williams, 1993).

At the motor unit level, two observations suggest that motor unit activation (neural strategy) may vary during fatiguing contractions. First, the motor units recruited during a submaximal, isometric contraction can vary before and after a fatiguing contraction. When subjects performed a ramp-and-hold task (ramp increase in force to a target followed by a ramp decrease back to baseline) with a hand muscle, some motor units were not recruited again after a fatiguing contraction, although the task was identical to that performed before

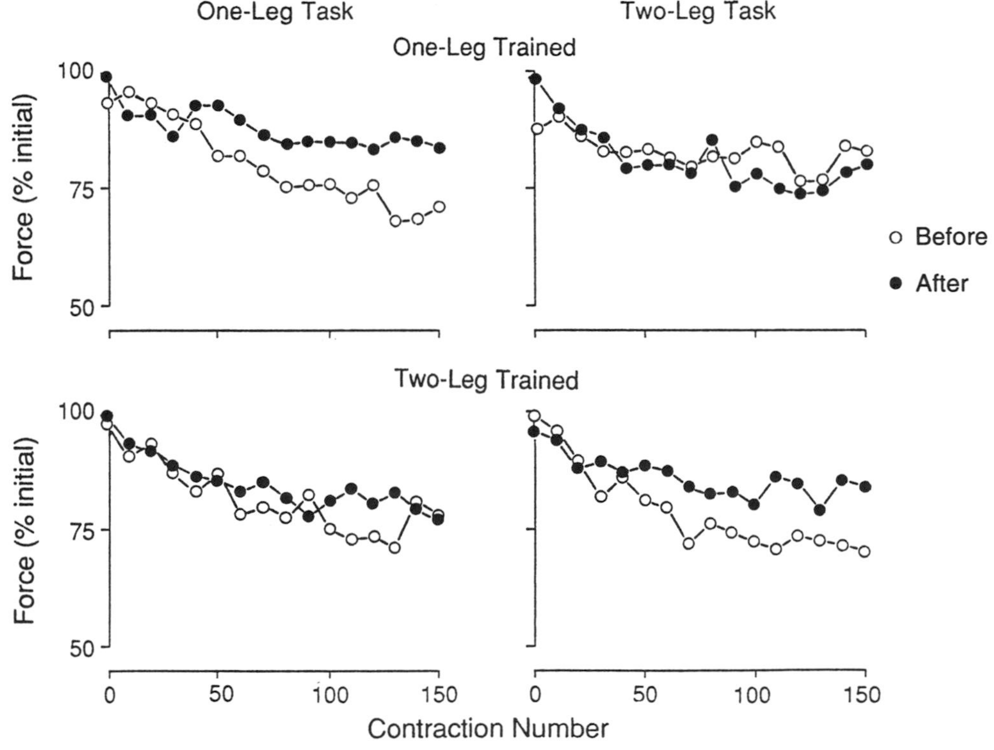

Figure 8.15 Effects of training on the decline in force during one- and two-legged performances of a leg extension task. *Note.* From "Effect of Training on Central Factors in Fatigue Following Two- and One-Leg Static Exercise in Man" by N. Rube and N.H. Secher, 1990, *Acta Physiologica Scandinavica*, **141**, p. 91. Copyright 1990 by Blackwell Scientific Publications. Adapted by permission.

the fatigue task (Figure 7.35) (Enoka, Robinson, & Kossev, 1989). This finding suggests that there is some flexibility in motor unit recruitment based on the activation history of the muscle (Person, 1974). Second, the recruitment order of motor units may differ for concentric and eccentric contractions (Figure 7.34) (Nardone, Romanò, & Schieppati, 1989). Neither the range of forces over which this difference may exist nor its significance for fatigue is known.

In addition to these differences in motor unit recruitment, the ability of motor units to sustain an activity also depends on whether the contraction is isometric. For example, slow-twitch motor units can sustain an isometric force longer than fast-twitch motor units (Burke et al., 1973), and slow-twitch muscle (soleus) can sustain a greater force during isovelocity shortening contractions than fast-twitch (extensor digitorum longus) muscle (Brooks & Faulkner, 1991). However, fast-twitch muscle can sustain a greater power production for 5 min compared with slow-twitch muscle (Brooks & Faulkner, 1991). This difference in power output is due to biochemical differences in the optimum velocity of shortening for slow- and fast-twitch muscle. Fast-twitch muscle has a higher optimum velocity of shortening, which combines with a lower sustained force to produce a greater power production. Undoubtedly this difference in the ability to exert force and to produce power has functional significance because many movements rely on the ability of muscle to produce power rather than simply to exert a force.

Neuromuscular Propagation. Several processes are involved in converting an axonal action potential into a sarcolemmal action potential. Collectively, these processes are referred to as *neuromuscular propagation*. Sustained activity can impair some of the processes involved in neuromuscular propagation, and this can contribute to the decline in force associated with fatigue. Potential impairments include a failure of the axonal action potential to invade all the branches of the axon (branch-point failure), a depletion of neurotransmitter, a reduction in exocytosis, and a decrease in the sensitivity of the postsynaptic membrane (Krnjevic & Miledi, 1958; Kugelberg & Lindegren, 1979; Spira, Yarom, & Parnas, 1976).

The most common way to test for impairment of neuromuscular propagation in human subjects is to elicit M waves before, during, and after a fatiguing contraction. Recall that an M wave is measured by applying an electric shock to a nerve to generate action potentials in the axons of alpha motor neurons and measuring the EMG response in muscle (Figure 5.28). A decline in M-wave amplitude is interpreted as an impairment of one or more of the processes involved in converting the axonal action potential (initiated by the electric shock) into a muscle (sarcolemmal) action potential. Figure 8.16 shows an example of a decline in M wave immediately after a fatiguing contraction and the eventual recovery of the wave form after 10 min of rest. This type of decline in M-wave amplitude tends to occur in long-duration, low-intensity contractions and less frequently in short-duration, high-intensity contractions (Bellemare & Garzaniti, 1988; Bigland-Ritchie et al., 1982; Fuglevand, Zackowski, Huey, & Enoka, 1993; Kranz, Williams, Cassell, Caddy, & Silberstein, 1983; Milner-Brown & Miller, 1986). Modeling studies suggest that impairment of neuromuscular propagation is one of several mechanisms that can contribute to the decline in force during these types of tasks (Fuglevand et al., 1993).

Excitation-Contraction Coupling. Under normal conditions, excitation by the nervous system results in the activation of muscle and the associated cycling of crossbridges. Seven processes (Fitts & Metzger, 1988) are involved in the conversion of the excitation (action potential) into a muscle fiber force (Figure 5.22): (1) propagation of the action potential along the sarcolemma; (2) propagation of the action potential down the transverse tubule; (3) change in the Ca^{2+} conductance of the sarcoplasmic reticulum; (4) movement of Ca^{2+} down its concentration gradient into the sarcoplasm;

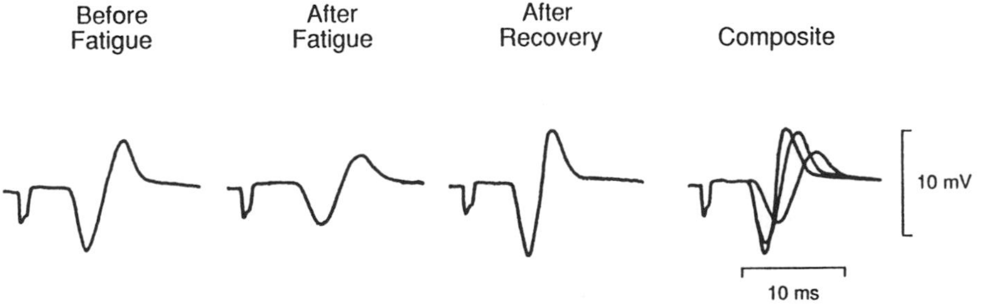

Figure 8.16 M waves elicited in a hand muscle (first dorsal interosseus) by stimulation of the ulnar nerve. M waves were elicited before and after a fatiguing contraction and after 10 min of recovery. The fatiguing contraction involved sustaining an isometric force at 35% of the maximum voluntary contraction force for as long as possible. M-wave amplitude declined immediately after the fatigue task but recovered quickly.

Note. From "Impairment of Neuromuscular Propagation During Human Fatiguing Contractions at Submaximal Forces" by A.J. Fuglevand, K.M. Zackowski, K.A. Huey, and R.M. Enoka, 1993, *Journal of Physiology*, **460**, p. 556. Copyright 1993 by the Physiological Society. Adapted by permission.

(5) reuptake of Ca^{2+} by the sarcoplasmic reticulum; (6) binding of Ca^{2+} to troponin; and (7) interaction of myosin and actin and the performance of work by the cross-bridge.

Some of these processes are known to be affected by intracellular changes associated with sustained activity. For example, changes in the intracellular milieu reduce the amount of Ca^{2+} released (Process 4) and the amount of Ca^{2+} returned (Process 5) to the sarcoplasmic reticulum (Allen et al., 1989; Blinks, Rüdel, & Taylor, 1978; Dawson, Gadian, & Wilkie, 1980; Westerblad & Lännergren, 1990). The decline in Process 4 (the quantity of Ca^{2+} released) is most probably due to an impairment of the release process, perhaps due to an inactivation of Ca^{2+} release channels and a depletion of Ca^{2+} stores in the sarcoplasmic reticulum (Blinks et al., 1978; Webb, Hibberd, Golman, & Trentham, 1986). The decline in Process 5 may be due to an increase in the intracellular binding of Ca^{2+}, which would reduce the concentration gradient and hence the flux of Ca^{2+} across the sarcoplasmic reticulum. Despite these possibilities, the relative importance of these mechanisms to the in vivo decline of force remains to be determined.

One way to demonstrate a role for the impairment of excitation-contraction coupling in fatigue is to show that the decline in force cannot be ascribed to neural or metabolic factors. This has been shown (Bigland-Ritchie, Cafarelli, & Vøllestad, 1986) for a task in which subjects were asked to perform a series of 6-s contractions with the quadriceps femoris muscle to a force level that was 30% of maximum. There was a 4-s rest between each 6-s contraction. The subjects did this sequence for 30 min. The force exerted during a maximum voluntary contraction (MVC) declined in parallel with the electrically elicited force (Figure 8.11a), which suggests that the subjects voluntarily exerted as much force as the muscle was capable of producing. There were no significant changes in muscle lactate, ATP, or phosphocreatine, and the depletion of glycogen was minimal and confined to the Type I and Type IIa muscle fibers. Consequently, the decline in MVC force could not be explained by an inadequate neural drive (the M waves were not reduced), acidosis, or lack of metabolic substrates. Furthermore, the twitch response decreased more than the tetanus or MVC force, which is regarded as evidence of an impairment of excitation-contraction coupling. For these reasons, Bigland-Ritchie et al. (1986) concluded that the process of excitation-contraction coupling was impaired during this task and that this impairment contributed to the decline in force.

Another experimental approach that has been used to demonstrate a role for the impairment of excitation-contraction coupling in muscle fatigue has been to monitor the recovery from fatigue (Figure 8.17). These experiments involve comparing the recovery of tetanic force, M waves, and twitch responses: Tetanic force provides a measure of the excitation that can be delivered to the muscle independent of the central nervous system, M-wave amplitude indicates the stability of neuromuscular propagation, and twitch force is an index of the excitation-contraction response to a single pulse of excitation. For fatiguing contractions that are sustained for a long duration at a low force, tetanic force recovers rapidly (2 min), recovery from impaired neuromuscular propagation takes a little longer (4 to 6 min), and the impairment of excitation-contraction coupling can last for hours (Edwards, Hill, Jones, & Merton, 1977; Fuglevand et al., 1993; Miller et al., 1987). Excitation-contraction coupling seems to be impaired by long-duration contractions in human subjects but by as little as 2 min of stimulation for Type FF motor units (Burke et al., 1973; Enoka et al., 1992).

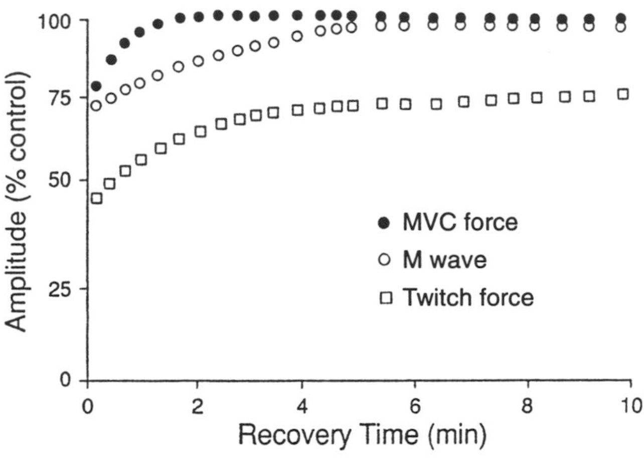

Figure 8.17 Recovery of muscle force from a fatiguing contraction. The fatiguing contraction concluded at time zero.

Metabolic Substrates. Two critical factors in a muscle contraction are excitation of the muscle by the nervous system and the supply of metabolic energy to the muscle. Although the ability to sustain a muscle contraction does not always depend on the availability of metabolic energy, for some tasks the depletion of metabolic substrate is associated with a decline in performance. For example, when subjects exercised on a cycle ergometer at a rate of 70% to 80% of maximal aerobic power, exhaustion and the inability to sustain the forces necessary for the task coincided with the depletion of glycogen from the muscle fibers of vastus lateralis (Hermansen, Hultman, & Saltin, 1967). Similarly, when exercising subjects are fed glucose or have it infused intravenously, they are able to exercise longer (Coggan & Coyle, 1987; Coyle, Coggan, Hemmert, & Ivy, 1986). Consequently, the availability of carbohydrates determines how long motivated subjects can ride a cycle ergometer at 65% to 85% of maximal aerobic power (Broberg & Sahlin, 1989; Costill & Hargreaves, 1992; Hultman, Bergström, Spriet, & Söderlund, 1990).

The products of energy metabolism are also known to influence muscle force. One popular candidate is H^+

concentration. Although H^+ can inhibit glycolysis, this interaction does not appear to be a major mechanism that causes the decline in force. For example, reducing intracellular pH (7.0 to 6.6) in single, intact muscle fibers by increasing the CO_2 in the extracellular medium, causes only a moderate decline in the number of attached crossbridges, a decrease in the force exerted by each crossbridge, and a reduction in the speed of crossbridge cycling during shortening contractions (Edman & Lou, 1990; Lännergren & Westerblad, 1989). It is probable that other products of ATP hydrolysis (e.g., Mg-ADP, P_i) can contribute to the decline in force. An increase in the concentration of P_i, for example, can reduce the maximum isometric force but does not affect the maximum speed of shortening. Conversely, an increase in the concentration of Mg-ADP can cause a small increase in the maximum isometric force and a modest decline in the maximum speed of shortening (Chase & Kushmerick, 1988; Cooke, Franks, Luciani, & Pate, 1988). The task conditions under which these products might be important in muscle fatigue are not yet known.

Blood Flow. The impairment of blood flow to active muscle was one of the first mechanisms identified that could contribute to muscle fatigue. An increase in muscle blood flow with motor activity is necessary for the supply of substrate, the removal of metabolites, and the dissipation of heat. When a muscle is active, however, there is an increase in intramuscular pressure that compresses blood vessels and can eventually occlude blood flow. For example, when the knee extensors sustain an isometric contraction for as long as possible at a submaximal force (5% to 50% of the maximum voluntary contraction), blood flow decreases as the level of the sustained force increases (Sjøgaard et al., 1988). It is not possible, however, to identify a specific force level at which blood flow is occluded, because intramuscular pressure can vary markedly within a group of synergist muscles and even within a single muscle for a given contraction intensity. Nonetheless, blood flow is probably not significantly impaired for tasks that involve less than 15% of the maximum voluntary contraction force (Gaffney, Sjøgaard, & Saltin, 1990; Sjøgaard et al., 1988). When blood pressure is artificially increased above the intramuscular pressure, there is a significant increase in endurance time for contractions in which the force was less than 60% of maximum (Petrofsky & Hendershot, 1984). At higher forces, there is no effect.

In addition to this obvious effect of intramuscular pressure on muscle blood flow, an elevated blood flow (hyperperfusion) can decrease fatigue by a mechanism that is independent of O_2 or substrate delivery. A reduction in blood flow (ischemia) exerts an effect on mechanical performance that is independent of reduced oxygen content in the blood (hypoxemia). The increase in blood flow associated with hyperperfusion may improve the removal of metabolites and thereby diminish the inhibitory effect of metabolite accumulation (Barclay, 1986; Stainsby, Brechue, O'Drobinak, & Barclay, 1990).

Sensory Adaptations

Much less is known about the changes that occur in sensory processes than about those that occur in motor processes during fatiguing contractions. There have been some studies, albeit with conflicting observations, on the effects of fatigue on the sensitivity of large-diameter proprioceptive afferents. In addition, some attention has been directed to the sensory-based phenomena referred to as muscle wisdom and the sense of effort.

Proprioceptive Afferents and Reflexes. Experiments performed on anesthetized animals have generally shown that fatigue enhances the sensitivity of muscle spindle afferents (Groups Ia and II) in response to single motor unit contractions (Christakos & Windhorst, 1986; Zytnicki, Lafleur, Horcholle-Bossavit, Lamy, & Jami, 1990). However, results from conscious humans indicate that fatigue produces a decline in fusimotor drive to muscle spindles (Bongiovanni & Hagbarth, 1990; Duchateau & Hainaut, 1993) and a reduction in spindle discharge during sustained isometric contractions (Macefield, Hagbarth, Gorman, Gandevia, & Burke, 1991). The effects of fatigue on the Group Ib afferent of the tendon organ are also uncertain. In experimental animals, the sensitivity of tendon organs to motor unit contractions does not change during fatiguing contractions; tendon organ discharge varies linearly with motor unit force (Gregory, 1990). In contrast, the response of tendon organs to whole-muscle stretch has been reported to decline with fatigue (Hutton & Nelson, 1986), and there is a reduction in the inhibitory effect of Group Ib feedback on motor neurons (Zytnicki et al., 1990).

Despite the uncertain effects of fatigue on the proprioceptive afferents, some evidence suggests that fatigue enhances the reflex EMG and motor neuron responses to brief perturbations (Darling & Hayes, 1983; Windhorst et al., 1986). For example, in one experiment the torque and EMG response of the elbow flexor muscles to a stretch were examined when subjects exerted the same torque before and after a fatiguing contraction (Kirsch & Rymer, 1987). The fatigue task consisted of 20 repetitions of a 25-s isometric contraction to a force of 50% of maximum. Although the fatiguing contraction substantially reduced the force that the elbow flexor muscles could exert, the torque response to the muscle stretch was reasonably similar to that measured before the fatigue task. The ability of the subjects to respond with a comparable torque was due to the increase in the EMG response after the exercise (Figure 8.18). Because the stretch was applied 10 min after the end of the fatiguing contraction, the increase in EMG was interpreted as an increase in the stretch-elicited neural drive after the fatiguing contraction (see also Kirsch & Rymer,

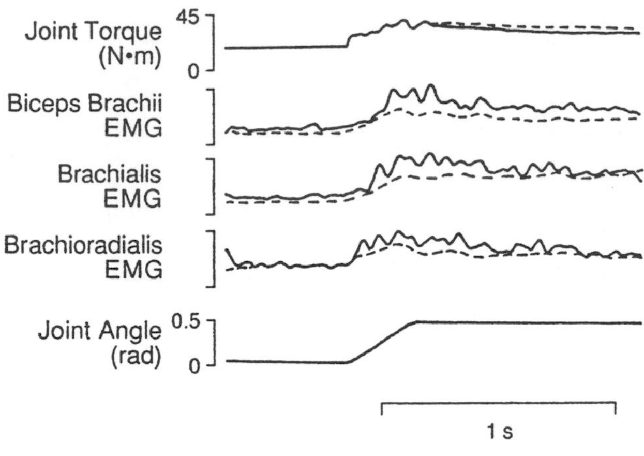

Figure 8.18 Increase in the EMG response to muscle stretch after (solid line) a fatiguing contraction.
Note. From "Neural Compensation for Muscular Fatigue: Evidence for Significant Force Regulation in Man" by R.F. Kirsch and W.Z. Rymer, 1987, *Journal of Neurophysiology*, **57**, p. 1901. Copyright 1987 by The American Physiological Society. Adapted by permission.

1992). Also in that experiment, the joint was rapidly displaced by 0.5 rad before and after the fatigue task. The displacement response was similar for the before and after trials.

Muscle Wisdom. The term *muscle wisdom* is used to describe an effect that has been observed in three types of experiments: A muscle is better able to sustain a force that is elicited by electrical stimulation if the stimulus rate declines over time rather than remaining constant (Figure 6.9) (Binder-Macleod & Guerin, 1990; Marsden, Meadows, & Merton, 1983); motor neuron discharge rate declines during a fatiguing contraction (Bigland-Ritchie, Johansson, Lippold, Smith, et al., 1983; Dietz, 1978; Garland, Enoka, Serrano, & Robinson, 1994); and fatigue is associated with a progressive decrease in the rate of relaxation in force (Bigland-Ritchie, Johansson, Lippold, & Woods, 1983; Gordon, Enoka, Karst, & Stuart, 1990; Hultman & Sjøholm, 1983; Lännergren & Westerblad, 1991). Because the relaxation rate decreases, the twitch duration increases, and therefore the same degree of fusion in a tetanus (force) can be achieved with a lower rate of activation (Figure 8.19). The change in relaxation rate occurs because of biochemical changes that are associated with excitation-contraction coupling. *Muscle wisdom, therefore, describes the ability of muscle to reduce the discharge of its motor neurons to match the change in the biochemically mediated reduction in relaxation rate.*

There are at least three mechanisms that the system can use to control the decline in discharge rate during a fatiguing contraction (Enoka & Stuart, 1992). First, some combination of afferent feedback from peripheral sources probably inhibits the discharge of interneurons and motor neurons (Bigland-Ritchie, Dawson, Johansson, & Lippold, 1986; Garland, 1991; Seyffarth, 1940). This feedback may involve reflex inhibition by Group III and Group IV afferents and disfacilitation of Group Ia afferents (i.e., a reduction in fusimotor-controlled feedback from muscle spindles). Second, there is probably an adaptation in the excitability of motor neurons so that the same synaptic input elicits fewer action potentials (Kernell & Monster, 1982a, 1982b). **Adaptation** occurs due to an intrinsic biophysical property of the motor neuron membrane. Third, the discharge of a motor neuron may decline during a fatiguing contraction due to a reduction in the central drive. Maton (1991) recorded from cells in area 4 of the motor cortex of monkeys as they performed repetitive isometric torques and found that the output was modulated in concert with the EMG. This suggests that the discharge of motor

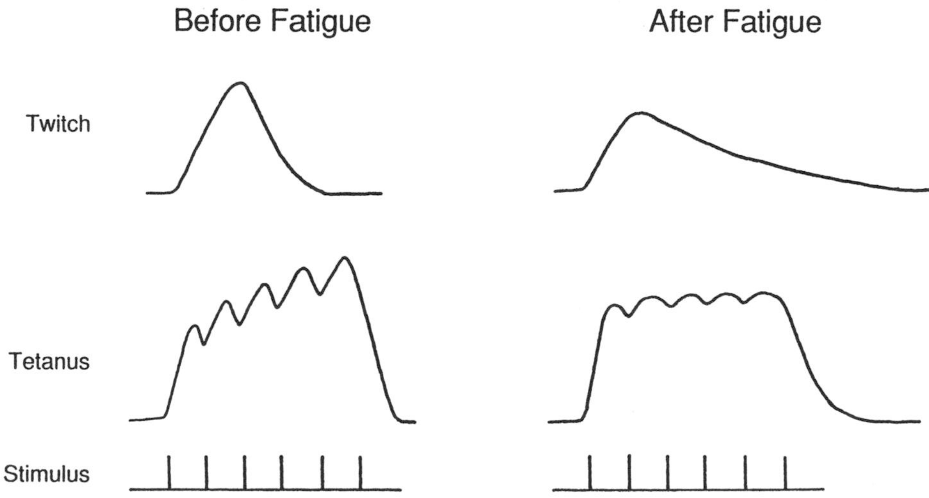

Figure 8.19 A fatiguing contraction is usually associated with an increase in the relaxation time of a twitch and an increase in the degree of fusion of an unfused tetanus.

neurons may be modulated by descending signals during fatiguing contractions. It is probable that all three mechanisms (sensory feedback, spinal neuron adaptations, and descending modulation) can contribute to the fatigue-related decline in motor neuron discharge rate but that the relative contributions depend on the details of the task, especially duration (Enoka & Stuart, 1992).

Sense of Effort. The process of motor programming that is performed by the suprasegmental centers results in a descending motor command that is known as the central command (chapter 7). In the performance of a movement, this signal is sent to lower neural centers (brain stem and spinal cord) and also back to the suprasegmental centers as a corollary discharge to aid in the interpretation of incoming afferent information. The corollary discharge that projects to the primary somatosensory cortex provides the basis for the *sense of effort* (see chapter 5). The sense of effort is a sensation that indicates the effort required to generate a specific muscle force. This sensation is independent of the mechanisms that impair the ability of muscle to exert a force. For example, imagine standing stationary in an upright position with a heavy load in one hand. After a few minutes, although you will still be able to support the load, you will notice that the task requires an increase in the effort necessary to support the load. To support the load, you will need to increase the descending command, which will lead to a greater corollary discharge and the perception of an increase in the effort associated with the task. For sustained submaximal contractions that are performed by a motivated subject, the effort will always increase before the force begins to decline.

The typical way to study effort during a fatiguing contraction is to have a subject perform a contralateral matching experiment. In this protocol, the subject exerts a sustained force with one (test) limb and periodically indicates with the other (matching) limb the effort associated with the activation of the test limb. The matching limb typically exerts a force until the sensations, or effort, are equal for the two limbs. The results of such an experiment are shown in Figure 8.20. The filled circles represent the load supported by the test arm. The line with open circles represents the estimated load by the matching arm when the test arm was not supporting the load continuously; the load was estimated once every 15 to 45 s for the 10-min duration. The line with open circles indicates the estimated load by the matching arm when the test arm supported the load continuously and was fatigued. The increase in the matching force when the test arm supported the load is interpreted as an increase in the effort during the fatiguing contraction.

Despite the prevalence of muscle fatigue as a prominent symptom in a wide variety of diseases of the CNS, minimal attention has been paid to the role of CNS pathology in fatigue. Some of this inattention is due to the difficulty associated with designing and interpreting appropriate psychophysical experiments. For example,

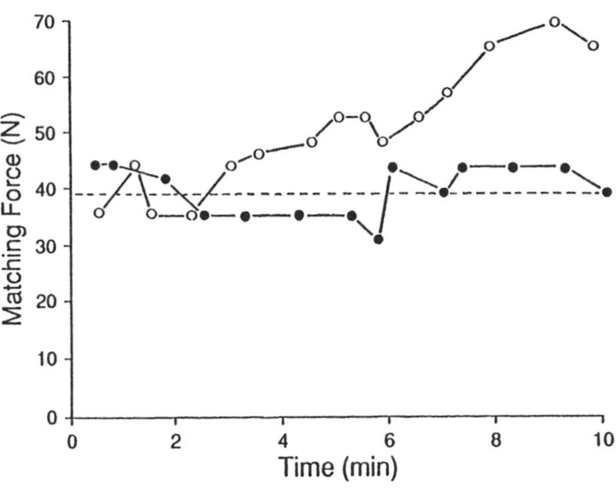

Figure 8.20 Increase in the effort during a fatiguing contraction.
Note. From "Estimation of Weights and Tensions and Apparent Involvement of a Sense of Effort" by D.I. McCloskey, P. Ebeling, and G.M. Goodwin, 1974, *Experimental Neurology*, **42**, p. 226. Copyright 1974 by the Academic Press, Inc. Adapted by permission.

which afferent signals are used by an intact human to assess the effort associated with performance? Most evidence suggests that the sense of effort during a fatiguing contraction seems to rely predominantly on the corollary discharge (Cafarelli, 1988; Jones & Hunter, 1983), with feedback from peripheral receptors contributing to other sensations (e.g., sense of heaviness or tension). Studies on patients with effort syndromes and chronic fatigue syndrome suggest that the sense of effort may entail at least two components: one associated with the impairment of performance and another above the level of the motor cortex, wherein the perceived effort does not relate directly to motor performance (Lloyd, Gandevia, & Hales, 1991). These studies suggest the need for advanced psychophysical testing and therapy for such patients.

Muscle Potentiation

In contrast to the performance-reducing effects of fatigue, several acute mechanisms can enhance the output of the neuromuscular system (Hutton, 1984; McComas, 1977). After a brief period of activity, these mechanisms can increase both the electrical and mechanical output above resting values. Examples of these capabilities include the potentiation of monosynaptic responses, miniature end-plate potentials, M waves, twitch force, and the discharge of muscle spindle receptors.

Monosynaptic Responses

It seems reasonable to assume that most processes in the motor system can be augmented by brief periods of activity. This assumption is true for input-output relationships of the spinal cord. Lloyd (1949) delivered single electric shocks to the muscle nerve of an experimental animal and measured the output in the ventral root. The experimental preparation is schematized in Figure 8.21a. Because the ventral root was cut, the electric shock generated action potentials that were transmitted along the afferent axons and into the spinal cord. The synaptic input by the afferent axons activated motor neurons, and the monosynaptic response was measured (Figure 8.21a). The input-output relationship involved populations of afferent and efferent axons. The monosynaptic responses were measured before and after tetanic stimulation of the nerve (12-s duration at 555 Hz), shown at time zero in Figure 8.21b. The tetanic stimulation increased the amplitude of the monosynaptic response by seven times, compared with the control value (before the tetanus). The potentiation of the monosynaptic response decayed over a 3-min interval (Figure 8.21b). A similar effect has been reported for potentiation of the H and tendon tap reflexes after a brief period of high-frequency electrical stimulation (Hagbarth, 1962).

When Lloyd (1949) activated different afferent pathways after the tetanic stimulation, the potentiation was limited to the afferent pathway that received the tetanic stimulation. Therefore, the mechanism underlying the potentiation was presynaptic; that is, it was located before the synaptic contact with the motor neurons. Given the type of stimuli that elicit potentiation of the monosynaptic response, the mechanism is thought to involve Group Ia afferents (Hutton, 1984). The effects may include an increase in the quantity of neurotransmitter released, an increase in the efficacy of the neurotransmitter, or a reduction in axonal branch-point failure along the Group Ia afferents (Kuno, 1964; Lüscher, Ruenzel, & Henneman, 1979).

Miniature End-Plate Potentials

At the neuromuscular junction, the spontaneous release of neurotransmitter (ACh) elicits miniature end-plate potentials in the postsynaptic membrane (see chapter 4). The miniature end-plate potentials are not constant events; the amplitude is greater for fast-twitch compared with slow-twitch muscle fibers, and the frequency can decline as a function of age (Alshuaib & Fahim, 1990; Lømo & Waerhaug, 1985). Similarly, a brief period of high-frequency stimulation can increase the amplitude and frequency of the spontaneously released miniature end-plate potentials (Pawson & Grinnell, 1990; Vrbová & Wareham, 1976). The increase in frequency lasts for a few minutes, whereas the increase in amplitude has been reported to last for several hours. Furthermore, the effect is greater at stronger neuromuscular junctions (in this context, *strength* is defined as the quantity of neurotransmitter released per unit length of the junction).

The mechanisms seem to involve an activity-dependent increase in the sensitivity of the postsynaptic membrane and in the influx of Ca^{2+} into the presynaptic terminal. The increase in postsynaptic membrane sensitivity means that a quantum of neurotransmitter will elicit a greater response (amplitude of the synaptic potential) in the postsynaptic membrane. In addition, because Ca^{2+} is necessary for the fusion of vesicles to the

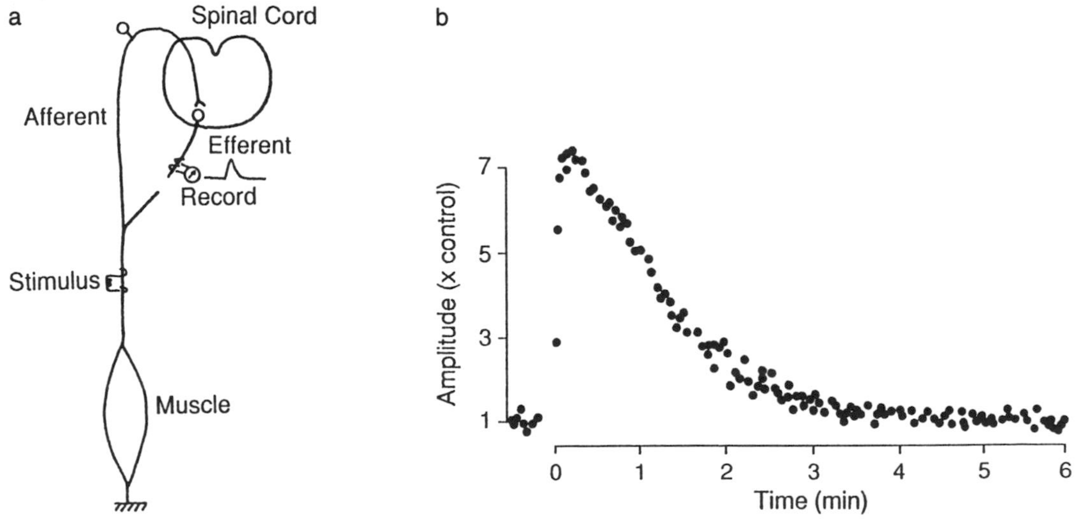

Figure 8.21 Potentiation of monosynaptic responses: (a) the experimental model; (b) amplitude of the monosynaptic response.
Note. From "Post-Tetanic Potentiation of Response in Monosynaptic Reflex Pathways of the Spinal Cord" by D.P.C. Lloyd, 1949, *Journal of General Physiology*, **33**, p. 149. Copyright 1949 by The Rockefeller University Press. Adapted by permission.

presynaptic membrane and the subsequent release of neurotransmitter, a greater influx of Ca^{2+} will lead to an enhanced frequency of spontaneous release of neurotransmitter.

M-Wave Amplitude

The processes involved in converting an axonal action potential into a sarcolemmal action potential are collectively referred to as neuromuscular propagation. One way to test the stability of neuromuscular propagation is to elicit M waves (see chapter 5). This test involves stimulating the muscle nerve and measuring the subsequent compound muscle action potential. Although M-wave amplitude can be reduced by presynaptic factors (e.g., branch-point failure, neurotransmitter depletion), only postsynaptic factors can increase M-wave amplitude. These factors include a reduction in the temporal dispersion of the muscle fiber action potentials and an increase in the amplitude of individual muscle fiber action potentials.

When a human subject performs a voluntary contraction or when a muscle is activated with electrical stimulation, there is often an initial transient increase in M-wave amplitude (Figure 8.22). For example, when subjects performed 20 maximal contractions (3-s duration for each contraction, 1.5 s between contractions) with the thenar muscles, M-wave amplitude increased immediately after the first contraction and then reached a plateau at an average increase of about 24% for the 100-s task (Hicks, Fenton, Garner, & McComas, 1989). Studies of single muscle fibers that were subjected to repetitive stimulation suggest that the increase in M-wave amplitude was at least partially due to an activity-dependent increase in the amplitude of muscle fiber action potentials (Hicks & McComas, 1989). It seems that activation increases the activity of the Na^+-K^+ pump, which lowers (hyperpolarizes) the resting membrane potential and produces a greater change in voltage (amplitude) across the membrane during an action potential. Therefore, at least some of the initial increase in M-wave amplitude at the onset of activity is due to an increase in the activity of the Na^+-K^+ pumps, and this seems to cause a modest increase in force (Enoka et al., 1992).

Posttetanic Potentiation

Perhaps the best known of the acute potentiation responses is the effect on twitch force. The magnitude of the twitch force is extremely variable and depends on the activation history of the muscle. A twitch elicited in a resting muscle does not represent the maximal twitch. Rather, twitch force is maximal following a brief tetanus; this effect is known as **posttetanic potentiation** of twitch force (Belanger, McComas, & Elder, 1983; Brown & von Euler, 1938; Schiff, 1858). The posttetanic potentiation of twitch force can be substantial and can be elicited by both voluntary contractions and electrical stimulation. For example, when the ankle dorsiflexor muscles of human volunteers were intermittently stimulated with electric shocks, the potentiation of twitch force ranged from 29% to 150% after 20 to 40 s of stimulation (Garner, Hicks, & McComas, 1989). Furthermore, twitch force can be potentiated by both maximal and submaximal voluntary contractions (Vandervoort, Quinlan, & McComas, 1983). A special case of potentiation achieved with submaximal activation involves the progressive increase in twitch force when a series of twitches are elicited in close succession; this is known as the **staircase effect**, or **treppe** (Bowditch, 1871; Krarup, 1981).

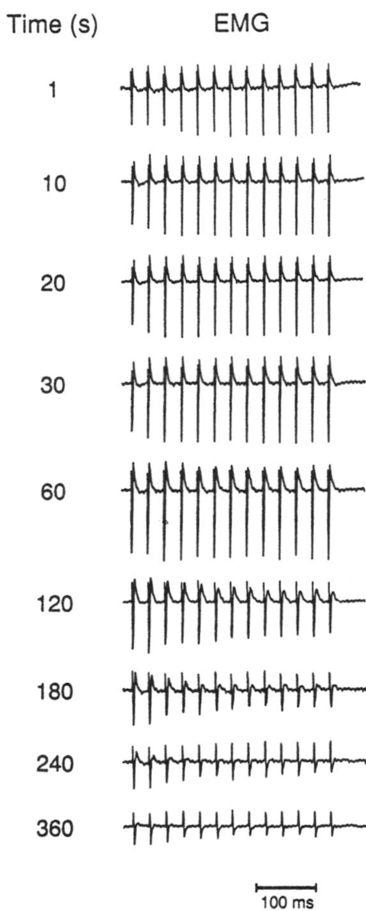

Figure 8.22 Changes in EMG during a 360-s fatiguing contraction involving electrical stimulation of a cat hindlimb muscle. The stimulus regimen consisted of 330-ms trains of 13 stimuli given once each second. The EMG represents the summed muscle action potentials (similar to M waves) that were elicited by the electric shocks. The EMG amplitude increased during the first 60 s of the test and then declined.

Note. From "Fatigue-Related Changes in Motor Unit Action Potentials of Adult Cats" by R.M. Enoka, N. Trayanova, Y. Laouris, L. Bevan, R.M. Reinking, and D.G. Stuart, 1992, *Muscle & Nerve*, **14**, p. 143. Copyright 1992 by John Wiley & Sons, Inc. Adapted by permission.

At least two processes may be involved in posttetanic twitch potentiation (Grange & Houston, 1991; Vandervoort et al., 1983). An early potentiation occurs after brief contractions and decays relatively quickly. After a delay of about 60 s, a late potentiating process emerges, which reaches a peak at about 200 s and then decays to control levels after about 8 to 12 min of recovery. The mechanisms underlying these potentiation processes probably involve an alteration in calcium kinetics (Duchateau & Hainaut, 1986) and the phosphorylation of myosin light chains (Grange, Vandenboom, & Houston, 1993; Sweeney, Bowman, & Stull, 1993; Sweeney & Stull, 1990).

Potentiation of the submaximal force occurs in all three types of motor units (Types S, FR, and FF). When motor units were activated with a stimulus that elicited a submaximal tetanic force, the average potentiation (increase in peak force) was greater among the fast-twitch motor units (50% to 60% of control in Types FR and FF) than the average increase in the slow-twitch motor units (20% of control in Type S). However, the incidence of potentiation among the motor units was greater for the fatigue-resistant motor units (60% to 75% for Types S and FR) than for the fatigable motor units (40% for Type FF). Because the occurrence of potentiation was distributed across all three types of motor units, the mechanisms underlying potentiation are different from those that define motor unit type (Gordon, Enoka, & Stuart, 1990).

The study of posttetanic twitch potentiation has emphasized that the processes of potentiation and fatigue occur concurrently, beginning from the onset of activation. For example, when the extensor digitorum muscle of rats was stimulated with a protocol that reduced the submaximal tetanic force to an average peak force of 36% of the control value, 50% of the muscles exhibited posttetanic twitch potentiation (Rankin, Enoka, Volz, & Stuart, 1988). This effect was not observed for soleus. The coexistence of potentiation and fatigue has also been observed in the human quadriceps femoris muscle after a 60-s maximal voluntary contraction (Grange & Houston, 1991).

A scheme for the interaction of these two processes and the net effect on twitch force during a specific protocol is shown in Figure 8.23. In the experiment from which the scheme was devised, the ankle dorsiflexor muscles of human volunteers were electrically stimulated to elicit a 3-s submaximal tetanus once every 5 s for a duration of 180 s. Between the tetani, a twitch was elicited by a single electrical shock. For this particular protocol, twitch force increased and then decreased during the stimulation period and then subsequently increased and decreased during the recovery period. Garner et al. (1989) proposed that the time course of the change in twitch force was due to the interaction of the potentiating and fatigue processes.

Postcontraction Sensory Discharge

In addition to the effects on motor processes, a brief period of intense activity can influence the behavior of sensory processes. One example of this effect is the

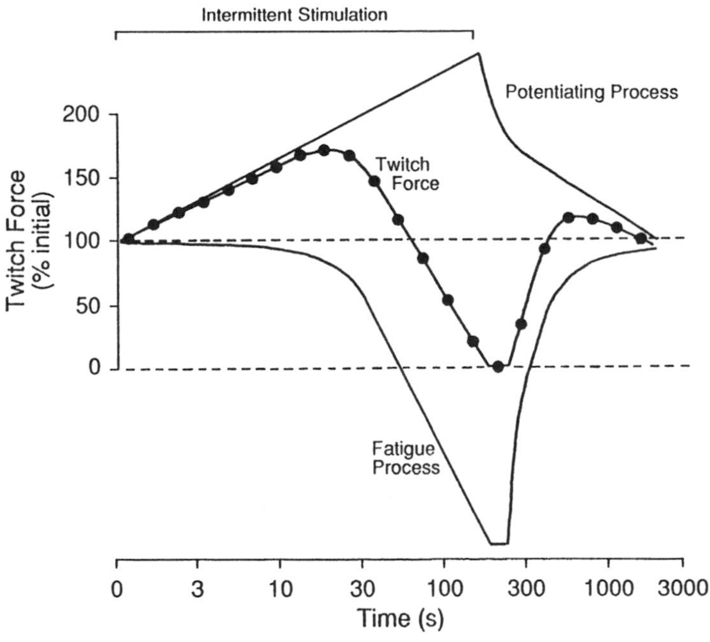

Figure 8.23 Coexistence of potentiation and fatigue and the net effect on the twitch force. When a muscle is stimulated electrically, the contraction elicits processes that both diminish (fatigue) and enhance (potentiate) the muscle force. The result is a nonlinear change in the amplitude of the twitch force.

Note. From "Prolongation of Twitch Potentiating Mechanism Throughout Muscle Fatigue and Recovery" by S.H. Garner, A.L. Hicks, and A.J. McComas, 1989, *Experimental Neurology*, **103**, p. 280. Copyright 1989 by Academic Press, Inc. Adapted by permission.

increased neural activity that has been recorded in the dorsal roots of experimental animals after a contraction; this phenomenon has been termed the **postcontraction sensory discharge** (Hutton, Smith, & Eldred, 1973). The increase in the dorsal root activity is primarily due to an increase in muscle spindle discharge, mainly the Group Ia afferents. Furthermore, the postcontraction sensory discharge is abolished if the muscle is stretched immediately after the contraction. The mechanism responsible for this phenomenon is the development of stable crossbridges in the intrafusal fibers. These crossbridges develop during the muscle contraction (α-γ coactivation) and persist after the extrafusal fibers relax so that the muscle spindle is in a state of increased tension and the resting discharge is higher than before the contraction (Gregory, Morgan, & Proske, 1986; Hutton et al., 1973). Consequently, when the muscle is stretched, these crossbridge bonds are broken and the postcontraction sensory discharge is abolished.

Because postcontraction sensory discharge increases the excitatory input to the motor neuron pool, it influences subsequent activity for up to 15 min with a peak effect at 5 to 20 s after the contraction. For example, it enables the system to respond more rapidly and forcefully to subsequent perturbations of muscle length, it influences kinesthetic sensations, and it opposes the relaxation of muscle that may be desired for stretching maneuvers. The strength of postcontraction sensory discharge can be sufficient to increase the resting discharge of motor neurons (Suzuki & Hutton, 1976).

Summary

This is the first of two chapters to examine the effects of physical activity on the neuromuscular system. This chapter focuses on the immediate (acute) response of the system to the stress associated with a single bout of physical activity and addresses five topics that characterize the adaptive capabilities of the system: warm-up effects, flexibility, muscle soreness and damage, muscle fatigue, and muscle potentiation. First, we discussed warm-up effects and described the influence of changes in temperature on the mechanical output of the system and on the passive stiffness of muscle and connective tissue. Second, we distinguished warm-up activities from those related to flexibility and described techniques used to alter flexibility and the factors that limit joint range of motion. Third, we examined the consequences of strenuous activity on the integrity of muscle and connective tissue. This included a consideration of the cause and time course of muscle soreness, the effect on different muscle fiber types, the regeneration of tissue following muscle damage, and the result of excessive strains. Fourth, we considered the impairment effects of sustained activity on the ability to exert force (muscle fatigue) and the perceived effort associated with the activity. Because the mechanisms that cause muscle fatigue vary with the details of the task, we attempted to identify the conditions under which the different mechanisms can be impaired and hence contribute to muscle fatigue. Fifth, we characterized the potentiating effects of acute activity on monosynaptic responses, miniature end-plate potentials, M-wave amplitude, posttetanic twitch potentiation, and postcontraction sensory discharge. These examples are not intended to represent a complete summary of the acute adaptive responses, but rather to illustrate the capabilities of the system.

TRUE-FALSE QUESTIONS

1. The sole purpose of a warm-up is to increase core temperature. **TRUE FALSE**
2. A warm-up increases the maximum force a muscle can exert but not the power that it can produce. **TRUE FALSE**
3. An increase in muscle temperature due to a warm-up decreases both contraction time and half relaxation time. **TRUE FALSE**
4. Half relaxation time is influenced more by temperature than contraction time. **TRUE FALSE**
5. An 8°C change in muscle temperature is a big change. **TRUE FALSE**
6. It is not possible to raise muscle temperature by more than 1°C by using a passive external heat source. **TRUE FALSE**
7. The unit of measurement for stiffness is N·m. **TRUE FALSE**
8. Stiffness can be examined graphically as the slope of a force-length graph. **TRUE FALSE**
9. The passive force-length graph is identical for all muscles. **TRUE FALSE**
10. It requires more energy to stretch a stiff muscle than one that is less stiff. **TRUE FALSE**
11. A thixotropic material is one that changes its state due to increases in core temperature. **TRUE FALSE**
12. Thixotropy is measured with the same unit of measurement as stiffness. **TRUE FALSE**
13. Muscle tone is the resistance that a relaxed muscle offers to stretch. **TRUE FALSE**

14. Increases in muscle tone are known as *hypertonus*. TRUE FALSE
15. The two most common forms of hypotonus are spasticity and rigidity. TRUE FALSE
16. Spasticity involves an increase in the passive thixotropic property of muscle. TRUE FALSE
17. The impairment of movement in spasticity is due to an inability to recruit a sufficient number of motor units in the agonist muscle. TRUE FALSE
18. Rigidity is a velocity-dependent resistance to movement. TRUE FALSE
19. The goal of flexibility exercises is to induce long-term changes in the range of motion about a joint. TRUE FALSE
20. A ballistic stretch involves bouncing movements. TRUE FALSE
21. In the agonist-contract stretch technique, the antagonist muscle is stretched. TRUE FALSE
22. Motor neuron excitability is depressed for about 10 min immediately after a maximal isometric contraction. TRUE FALSE
23. The rationale for the AC stretching technique is based on the recurrent inhibition reflex. TRUE FALSE
24. The antagonist muscle is relaxed during the AC stretch. TRUE FALSE
25. The improvement in flexibility is similar for static and ballistic stretches. TRUE FALSE
26. The HR-AC technique produces the greatest improvement in flexibility of all the stretching techniques. TRUE FALSE
27. Flexibility exercises involve altering the length of connective tissue. TRUE FALSE
28. Flexibility exercises should be performed at the beginning of a workout. TRUE FALSE
29. Connective tissue is most extensible when the temperature is high. TRUE FALSE
30. When tissue is stretched to cause plastic changes, this weakens the tissue. TRUE FALSE
31. Delayed-onset muscle soreness begins immediately after the exercise. TRUE FALSE
32. Delayed-onset muscle soreness causes structural damage of muscle fibers but no loss of strength. TRUE FALSE
33. Eccentric contractions are more likely to cause delayed-onset muscle soreness than are other types of muscle contractions. TRUE FALSE
34. Eccentric exercise produces a higher peak intramuscular pressure but involves a lower oxygen cost. TRUE FALSE
35. The amount of stress, rather than strain, is the more significant mechanical factor in muscle soreness. TRUE FALSE
36. The loss of extracellular calcium may trigger processes that degrade cellular structures. TRUE FALSE
37. Z-band damage frequently accompanies muscle soreness. TRUE FALSE
38. There is a continued release of muscle-specific enzymes by damaged muscle fibers after the exercise has stopped. TRUE FALSE
39. One way to assess the extent of muscle damage is to measure the quantity of skeletal muscle enzymes in the circulation. TRUE FALSE
40. Plasma membranes are damaged by exercise. TRUE FALSE
41. Delayed-onset muscle soreness may be a consequence of the inflammatory response. TRUE FALSE
42. Type IIb muscle fibers, which belong to Type S motor units, are preferentially damaged by exercise. TRUE FALSE
43. Both the intramuscular cytoskeleton and the extracellular connective tissue matrix can be damaged by exercise. TRUE FALSE
44. The regeneration of muscle fibers begins about 4 days after the exercise. TRUE FALSE
45. Muscle fibers never recover normal function after damage due to exercise. TRUE FALSE
46. Stretching and warm-up exercises can prevent delayed-onset muscle soreness. TRUE FALSE
47. Muscle strains typically occur at muscle-tendon junctions. TRUE FALSE
48. Muscles with a high proportion of Type II muscle fibers are more susceptible to muscle strains. TRUE FALSE
49. Strains occur most often during concentric contractions. TRUE FALSE
50. The amount of strain that a muscle-tendon unit can withstand is no different whether the muscle is active or passive. TRUE FALSE
51. An active muscle can absorb a greater amount of strain energy. TRUE FALSE
52. Due to task dependency, the mechanisms that cause fatigue vary. TRUE FALSE
53. Muscle fatigue is a class of chronic effects that impair performance. TRUE FALSE
54. Neuromuscular propagation is a task variable that causes fatigue. TRUE FALSE
55. Neural strategy does not include subject motivation. TRUE FALSE

56. When the voluntary and electrically elicited forces are equal, central drive is considered to be maximal. **TRUE FALSE**
57. The twitch superimposition technique involves a single shock applied to a muscle during a voluntary contraction. **TRUE FALSE**
58. The SI unit of measurement for twitch force is kg. **TRUE FALSE**
59. The electrically elicited force declines more rapidly than voluntary force in an unmotivated subject. **TRUE FALSE**
60. Subjects find it difficult to maximally activate the plantarflexor muscles. **TRUE FALSE**
61. Concentric contractions are more fatigable than eccentric contractions. **TRUE FALSE**
62. Amphetamine can decrease fatigability. **TRUE FALSE**
63. Hormones in the cerebrospinal fluid influence the ability to sustain the central drive. **TRUE FALSE**
64. Synergist muscles are always activated by equal amounts during a sustained activity. **TRUE FALSE**
65. Muscles do not fatigue when the force is less than 10% of maximum. **TRUE FALSE**
66. The recruitment order of motor units may differ for concentric and eccentric contractions. **TRUE FALSE**
67. Recruitment order is not affected by fatigue. **TRUE FALSE**
68. Fast-twitch muscle can sustain a greater power output than slow-twitch muscle. **TRUE FALSE**
69. The term *neuromuscular propagation* refers to several processes. **TRUE FALSE**
70. At the neuromuscular junction, the sarcolemma is the postsynaptic membrane. **TRUE FALSE**
71. M waves are elicited by applying an electric shock to the muscle nerve. **TRUE FALSE**
72. The M wave is generally measured as an EMG response in the muscle. **TRUE FALSE**
73. Neuromuscular propagation is typically impaired in short-duration maximal voluntary contractions. **TRUE FALSE**
74. Ca^{2+} conductance is normally low in the sarcoplasmic reticulum during resting conditions. **TRUE FALSE**
75. Ca^{2+} binds to troponin when released into the sarcoplasm. **TRUE FALSE**
76. The amount of Ca^{2+} released from the sarcoplasmic reticulum can decline during fatigue. **TRUE FALSE**
77. The twitch-tetanus ratio is used an index of the stability of neuromuscular propagation. **TRUE FALSE**
78. Recovery from impairment of excitation-contraction coupling takes several hours. **TRUE FALSE**
79. Excitation-contraction coupling and neuromuscular propagation are impaired by long-duration contractions. **TRUE FALSE**
80. Exhaustion can coincide with the depletion of glycogen from muscle fibers. **TRUE FALSE**
81. A reduction in intracellular pH does not affect the force exerted by individual crossbridges. **TRUE FALSE**
82. An inadequate supply of ATP, rather than the products of hydrolysis, impairs force during fatiguing contractions. **TRUE FALSE**
83. Muscle blood flow is not occluded until muscle force exceeds 64% of maximum. **TRUE FALSE**
84. The only effects of blood flow on fatigue are the delivery of O_2 and substrates. **TRUE FALSE**
85. Muscle spindle discharge in humans declines during fatiguing contractions. **TRUE FALSE**
86. A decline in muscle spindle discharge could be caused by an increase in the fusimotor drive to muscle spindles. **TRUE FALSE**
87. The response of tendon organs to whole-muscle stretch declines with fatigue. **TRUE FALSE**
88. The stretch reflex EMG is increased with fatigue. **TRUE FALSE**
89. The discharge rate of motor neurons declines during fatigue. **TRUE FALSE**
90. Relaxation time increases with fatigue. **TRUE FALSE**
91. According to muscle wisdom, a change in relaxation rate causes a change in the discharge rate of motor neurons. **TRUE FALSE**
92. Disfacilitation refers to the removal of excitatory input. **TRUE FALSE**
93. Motor neuron discharge during fatigue is affected by changes in synaptic input (sensory feedback and descending drive). **TRUE FALSE**
94. Motor neuron adaptation refers to the decline in discharge rate due to inhibitory sensory feedback. **TRUE FALSE**
95. The sense of effort is derived from muscle spindle feedback. **TRUE FALSE**
96. The sense of effort is a perceptual capability of the suprasegmental centers. **TRUE FALSE**
97. Potentiation of monosynaptic responses involves Group Ia afferents. **TRUE FALSE**

98. Miniature end-plate potentials occur spontaneously. TRUE FALSE
99. The frequency but not the amplitude of the miniature end-plate potentials changes after brief activity. TRUE FALSE
100. The quantity of neurotransmitter released can vary between neuromuscular junctions. TRUE FALSE
101. The spontaneous release of neurotransmitter increases with an increase in the extracellular concentration of Ca^{2+}. TRUE FALSE
102. An increase in M-wave amplitude can be caused by an increase in the quantity of neurotransmitter that is released. TRUE FALSE
103. The amplitude of a muscle fiber action potential can increase at the onset of a period of activity. TRUE FALSE
104. Twitch force is not maximal in a resting muscle. TRUE FALSE
105. Twitch force cannot be potentiated by electrically stimulated contractions. TRUE FALSE
106. *Treppe* refers to the irregularities in the force during an unfused tetanus. TRUE FALSE
107. Potentiation of twitch force only occurs in Type FF motor units. TRUE FALSE
108. Twitch force potentiation is one of the characteristics that can be used to distinguish the three types of motor units. TRUE FALSE
109. Muscles can be fatigued yet exhibit potentiation of twitch force. TRUE FALSE
110. Postcontraction sensory discharge mainly involves an increase in the resting discharge of muscle spindles. TRUE FALSE
111. The increase in muscle spindle discharge occurs because of the increase in tension of the intrafusal muscle fibers. TRUE FALSE
112. Postcontraction sensory discharge can cause motor neurons to discharge action potentials. TRUE FALSE

MULTIPLE-CHOICE QUESTIONS

113. What effect is not produced by the increase in core temperature due to a warm-up?
    A. A decline in muscle viscosity
    B. An increase in thixotropy
    C. An increase in the extensibility of connective tissue
    D. A greater muscle blood flow
114. A warm-up has been shown to increase muscle temperature and subsequently to enhance power production. Why is this?
    A. Stretching exercises increase flexibility.
    B. The v_{max} value increases.
    C. The half relaxation time becomes longer.
    D. Contraction time decreases.
115. Stiffness is defined as the slope of which relationship?
    A. Force-length
    B. Force-time
    C. Deformation-time
    D. Torque-time
116. Which of the following are not characteristic of a muscle that has narrow active and steep passive force-length relationships?
    A. Large pennation angles
    B. Greater cross-sectional areas
    C. Faster shortening velocity
    D. A pennation angle of zero
117. What is the cause of the thixotropic property of muscle?
    A. Changes in viscosity due to changes in temperature
    B. Coactivation of the muscles about a joint
    C. An increase in the feedback from sensory receptors that is sufficient to sustain a low level of muscle activity
    D. An increase in the crossbridge connections in resting muscle
118. What units of measurement could be used to measure the mechanical consequences of thixotropy?
    A. N·m
    B. N/mm^2
    C. N/m
    D. N·m/rad
119. What is muscle tone?
    A. The response of motor neurons to muscle stretch
    B. The resistance to stretch by a relaxed muscle
    C. A ratio of muscle volume to the quantity of contractile proteins
    D. The level of EMG activity in a resting muscle
120. Which of the following is not a symptom associated with spasticity?
    A. An apparent inability to relax muscle
    B. A hyperactive tendon tap reflex
    C. Bidirectional resistance to passive movement
    D. An inability to perform a movement in alternating directions

121. Which expression best describes the effectiveness of the different stretching techniques?
A. static = ballistic < HR = AC = HR-AC
B. ballistic < static = HR = AC = HR-AC
C. ballistic < static = HR = AC < HR-AC
D. ballistic = static < HR = AC < HR-AC

122. Which reflexes have been used to document the decrease in excitability of the motor neuron pool with the HR stretching technique?
A. H reflex
B. Recurrent inhibition reflex
C. Reciprocal inhibition reflex
D. T reflex

123. Contracting the agonist muscle while stretching the antagonist (AC technique) causes relaxation of the antagonist by which mechanisms?
A. Postactivation depression of motor neuron excitability
B. Reciprocal inhibition reflex
C. Increase in thixotropy due to an increase in temperature
D. Stretch reflex

124. What will happen to a passive muscle-tendon unit when it is stretched repetitively to a constant force?
A. The length of the muscle-tendon unit will increase.
B. Thixotropy of the muscle-tendon unit will increase.
C. The force will decrease.
D. The force and length will increase linearly.

125. What is the best way to induce plastic changes in connective tissue?
A. Low-force, long-duration stretches at the end of a workout
B. Intermediate-force, rapid stretches at the beginning of a workout
C. Short-duration, low-force contractions at the end of a workout
D. Long-duration, high-force contractions at the end of a workout

126. Which of the following is not a clinical symptom associated with delayed-onset muscle soreness?
A. An increase in protein metabolites from injured muscles
B. Structural damage to subcellular components
C. Immediate increase in pain associated with the injured muscles
D. An increase in plasma enzymes

127. In which ways is eccentric exercise different from concentric exercise?
A. Greater postexercise muscle soreness and weakness
B. A lower oxygen cost
C. Lower peak intramuscular pressure
D. Greater EMG for the same force

128. What metabolic factors might lead to muscle soreness?
A. An inability to remove intracellular Ca^{2+}
B. Disruption of protein structures due to the decrease in temperature
C. High strain experienced by the intracellular structures
D. Free O_2 radical production

129. What is the most prominent structural abnormality that is evident following exercise?
A. Z-band streaming
B. Lesions of the plasma membrane
C. Change in the extracellular myofiber matrix
D. Swollen mitochondria

130. Which muscle fiber types seem to be most affected by exercise that induces delayed-onset muscle soreness?
A. IIb
B. I
C. Ia
D. IIa

131. Which substance reveals the involvement of connective tissue damage in delayed-onset muscle soreness?
A. Creatine kinase
B. Ca^{2+}
C. Hydroxyproline
D. Titin

132. Which stage of exercise-induced muscle injury is associated with the inflammatory response?
A. Regeneration
B. Initial
C. Phagocytic
D. Autogenetic

133. What is not needed for the regeneration of muscle fibers to occur?
A. Functional nerve
B. Existing myofilaments
C. An intact basal lamina
D. An uninterrupted vascular supply

134. Which of the following is not a task variable that affects the mechanisms associated with muscle fatigue?
A. Contraction intensity
B. Muscle blood flow
C. Subject motivation
D. Central drive

135. Which fatigue mechanism is the twitch superimposition technique used to test?
A. Central drive
B. Neuromuscular propagation
C. Excitation-contraction coupling
D. Neural strategy

136. Which examples demonstrate an inadequate central drive during fatiguing activity?
   A. Reduction of M-wave amplitude
   B. Differences in the fatigability of concentric and eccentric contractions
   C. Unmotivated subjects
   D. Reduction in the twitch-tetanus ratio
137. What happens to the EMG during a fatiguing eccentric contraction?
   A. EMG declines in parallel with force.
   B. EMG increases and becomes greater than that associated with concentric contractions.
   C. EMG remains constant.
   D. EMG increases but remains less than that for concentric contractions.
138. What effect can hormones circulating in the cerebrospinal fluid have on the time to exhaustion?
   A. Cause an increase in the time to exhaustion
   B. Have no effect
   C. Decrease the time to exhaustion
   D. Increase the central drive but not influence the time to exhaustion
139. Which of the following are examples of neural strategy effects associated with fatiguing contractions?
   A. Variable activity among synergist muscles
   B. Difference in fatigability for isometric and anisometric contractions
   C. Specificity of training effects
   D. Flexibility in the recruitment order of motor units
140. Why can fast-twitch muscle sustain a greater power production compared with slow-twitch muscle?
   A. Slow-twitch muscle exerts less force.
   B. Fast-twitch muscle has a higher optimum velocity for shortening.
   C. Fast-twitch muscle fatigues more rapidly.
   D. Slow-twitch muscle has a smaller cross-sectional area.
141. Which processes can impair neuromuscular propagation?
   A. Axonal branch-point failure
   B. A decrease in the sensitivity of the presynaptic membrane
   C. A reduction in exocytosis
   D. Depletion of neurotransmitter
142. Which of the following is not a process of excitation-contraction coupling?
   A. Reuptake of Ca^{2+} by the sarcoplasmic reticulum
   B. Propagation of the action potential along the sarcolemma
   C. Release of neurotransmitter at the neuromuscular junction
   D. Disinhibition of troponin
143. Which excitation-contraction processes are known to be impaired during fatiguing contractions?
   A. Ca^{2+} reuptake
   B. M-wave amplitude
   C. Ca^{2+} release
   D. Action potential propagation down the transverse tubule
144. At what intensity of maximal aerobic power does the depletion of glycogen coincide with the time to exhaustion?
   A. <35%
   B. 100%
   C. 50% to 60%
   D. 65% to 85%
145. What effects can a reduction in intracellular pH have on crossbridge activity?
   A. Reduce the speed of cycling during shortening contractions
   B. Increase the number of attached crossbridges
   C. Inhibit the uncovering of binding sites
   D. Reduce the force exerted by individual crossbridges
146. Why does the occlusion of blood flow cause a reduction in muscle force?
   A. Reduces the availability of O_2
   B. Slows the removal of metabolites
   C. Decreases muscle temperature
   D. Impairs action potential propagation
147. Why might the spindle discharge decrease during a sustained contraction?
   A. An enhancement of the sensitivity of muscle spindle afferents
   B. A decline in fusimotor drive
   C. A reduction in motor neuron excitability
   D. An increase in the relaxation time of intrafusal fibers
148. Why is a muscle better able to sustain a force if the stimulus rate declines over time?
   A. Motor neuron discharge declines.
   B. The relaxation rate of muscle decreases.
   C. Sensory feedback inhibits motor neuron activity.
   D. Adaptation of motor neuron excitability.
149. What signal is used to derive the sense of effort?
   A. Motor program
   B. Proprioceptive feedback
   C. Corollary discharge
   D. Descending command
150. What mechanisms could cause a potentiation of monosynaptic responses?
   A. An increase in afferent input
   B. A greater release of neurotransmitter at the neuromuscular junction
   C. A reduction in axonal branch-point failure
   D. An increase in neurotransmitter efficacy

151. Why will an increase in the influx of Ca^{2+} into the presynaptic terminal of the neuromuscular junction cause an increase in the frequency of miniature end-plate potentials?
   A. It increases the neurotransmitter released in each quantum.
   B. It enhances the manufacture of neurotransmitter.
   C. It makes the membrane more excitable for the action potential.
   D. It increases the rate at which vesicles fuse with the presynaptic membrane.

152. What causes M waves to potentiate at the onset of activity?
   A. An increase in Na^+-K^+ pump activity
   B. A reduction in branch-point failure
   C. An increase in the quantity of neurotransmitter released
   D. Depolarization of the resting membrane potential

153. What are the mechanisms that cause posttetanic potentiation of twitch force?
   A. Decrease in muscle stiffness
   B. An increase in sarcolemmal excitability
   C. Enhanced Ca^{2+} kinetics
   D. Improved biochemical reactions involving ATP

154. What causes the postcontraction sensory discharge of Group Ia afferents?
   A. An increase in fusimotor drive
   B. A decrease in motor neuron inhibition
   C. An increase in the resting tension of intrafusal fibers
   D. Enhanced sensitivity of the muscle spindle receptors

PROBLEMS

155. Suppose that during a warm-up a muscle experienced a change in temperature of 7°C from an initial value of 31°C. The maximum isometric torque ($F_{m,o}$) was 238 N·m, and the maximum angular velocity (ω_{max}) was 6.8 rad/s, prior to the change in temperature. The value of $F_{m,o}$ changed at a rate of 2.5%·°C^{-1} and ω_{max} changed at a rate of 12.3%·°C^{-1}.
   A. What are the values for $F_{m,o}$ and ω_{max} after the warm-up?
   B. On graph paper, draw a graph of the force-velocity relationships before and after the warm-up. Show the relationships as straight lines between $F_{m,o}$ and ω_{max}.
   C. From the graph of the linear force-velocity relationships, estimate the maximum power produced by the muscle before and after the warm-up. Recall that power can be calculated as the product of force and velocity.

156. The following data set represents the force exerted by muscle when it was stretched and then released (allowed to shorten). In one trial the muscle was passive during the stretch-shorten cycle, whereas in the other trial the muscle was activated with electrical stimulation.
   A. Plot the data for both trials on a force-length graph.
   B. What was the average stiffness of the muscle during the steepest part of the stretch phase of each trial?
   C. The area enclosed by the force-length graph for the complete stretch-shorten cycle indicates the energy absorbed by the muscle. Approximately how much energy was absorbed by the muscle during each trial?

Length (mm)	Passive force (N)	Active force (N)
0.00	0	50
0.30	0	220
0.60	0	270
0.90	0	290
1.20	0	275
1.50	8	273
1.80	15	265
2.10	27	260
2.40	42	262
2.70	59	258
3.00	90	260
2.70	59	195
2.40	42	135
2.10	27	83
1.80	15	55
1.50	8	45
1.20	0	35
0.90	0	25
0.60	0	23
0.30	0	20
0.00	0	20

157. Calculate the stiffness for the initial part of the stretch phase of the torque-angle graphs shown in Figure 8.4. Why does the initial stiffness increase with an increase in the rest interval between activity and the stretch?

158. The HR technique supposedly permits the stretching of muscle while the motor neurons are in a relative state of lowered excitability.
   A. What is the H reflex (draw the appropriate neural circuit), and why does it provide an assessment of the excitability of the motor neuron pool?
   B. What is the T reflex (draw the appropriate neural circuit)? Describe what it assesses.
   C. What does a comparison of the H and T reflexes following a maximal voluntary contraction indicate?

159. The AC technique supposedly permits the stretching of muscle because activating the reciprocal inhibition reflex lowers the excitability of the muscle's motor neurons.
   A. Draw the neural circuit for the reciprocal inhibition reflex.
   B. Why is the reciprocal inhibition reflex activated when the agonist muscle is not stretched?
   C. How is it possible to shut down the reciprocal inhibition reflex?

160. Delayed-onset muscle soreness is more frequently associated with eccentric contractions.
   A. Why are eccentric contractions more likely to induce soreness?
   B. What is the difference between stress and strain? Why does strain induce muscle soreness?
   C. According to the Fridén and Lieber (1992) scheme, why are Type IIb muscle fibers more susceptible to exercise-induced damage?

161. It is not uncommon for an elite sprinter to experience a muscle strain of the hamstrings during an event such as the 100-m sprint.
   A. Draw a torque-angle graph for the knee joint during one complete stride for such an event. The torque should be the resultant muscle torque about the knee joint (See Figures 7.16 and 7.18).
   B. Indicate on the graph where muscle strain is most likely to occur and explain the reason for choosing that point.

162. Consider the fatigue experiment shown in Figure 8.11.
   A. What does it mean when the MVC force and the 50-Hz force decline in parallel?
   B. Why did the EMG for the submaximal contractions increase during the fatiguing contraction?
   C. What is the significance of the twitch superimposition force reaching zero at about 5 min?

163. The data shown in Figure 8.13 were obtained from an experimental protocol that involved three bouts of 32 maximal contractions. Why do these data for the eccentric and concentric contractions suggest that the central drive is not always maximal during voluntary contractions?

164. One set of processes associated with muscle fatigue has been termed *neural strategy*.
   A. What is a neural strategy?
   B. How can the neural strategy influence muscle fatigue? Give one example at the whole-muscle level and another at the motor unit level.
   C. Propose an experiment that will examine the role of a neural strategy during a fatiguing contraction.

165. The M wave is a probe that can be used to test the efficacy of neuromuscular propagation.
   A. What is an M wave and how is it measured?
   B. What processes can increase M-wave amplitude?
   C. What processes can reduce M-wave amplitude?
   D. During a sustained contraction, when does M-wave amplitude increase and when does it decrease?

166. When a Type FF motor unit was stimulated with a 360-s regimen of electrical stimulation, the force and EMG declined as shown in Figure 8.24. The stimulation consisted of 330-ms trains of 13 shocks that were given once each second. The figure shows the decline in force (filled circles) and EMG (open circles) for a Type FF motor unit during a 360-s fatiguing contraction (data from Enoka et al., 1992). Why can such data be used as indirect evidence that the muscle fatigue (decline in force) was due to an impairment of excitation-contraction coupling?

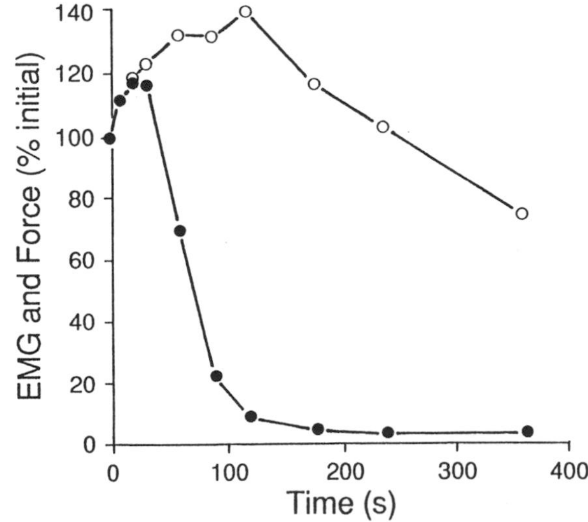

167. Some fatigue processes (e.g., central drive, blood flow, neuromuscular propagation) are known to be significant during normal activities performed by human subjects. Which processes have not yet been demonstrated to be significant?

168. Some evidence suggests that fatigue enhances the reflex EMG and motor neuron responses to brief perturbations. Why is this a reasonable acute adaptation by the system?

169. Muscle wisdom describes the ability of muscle to reduce the discharge of its motor neurons to match the change in the biochemically mediated reduction in relaxation rate.
   A. Why is this adaptation regarded as *wisdom*?
   B. Why does the decline in motor neuron discharge rate not cause a reduction in muscle force?
   C. What synaptic inputs might contribute to the decline in motor neuron discharge?
   D. How could adaptation (the intrinsic biophysical property of motor neurons) contribute to the decline in discharge rate?

170. The assessment of the sense of effort is described as a psychophysical experiment.
   A. What is psychophysics?
   B. The sense of effort has been assessed by a contralateral matching protocol, by ratings of perceived exertion, and by the measurement of blood pressure and EMG. The measurements of blood pressure and EMG are used as an index of the central (descending) command. Why is this not appropriate?

171. How can a reduction in axonal branch-point failure cause a potentiation of monosynaptic responses?

172. What effect might an increase in the frequency and amplitude of miniature end-plate potentials have on the postsynaptic membrane?

173. Posttetanic potentiation of twitch force indicates that twitch force is maximal after a brief tetanus.
   A. Because this type of potentiation can be elicited by both voluntary contractions and electrical stimulation, where must the mechanism reside?
   B. What factors determine the amplitude of twitch force?
   C. How can the alteration of Ca^{2+} kinetics influence twitch force?
   D. Explain the amplitude and incidence of posttetanic twitch potentiation among the three motor unit types based on ability to release Ca^{2+} in response to a single action potential.
   E. What enables a fatigued muscle to exhibit posttetanic twitch potentiation?

174. Why does a brief maximal voluntary contraction increase the resting discharge of Group Ia afferents?

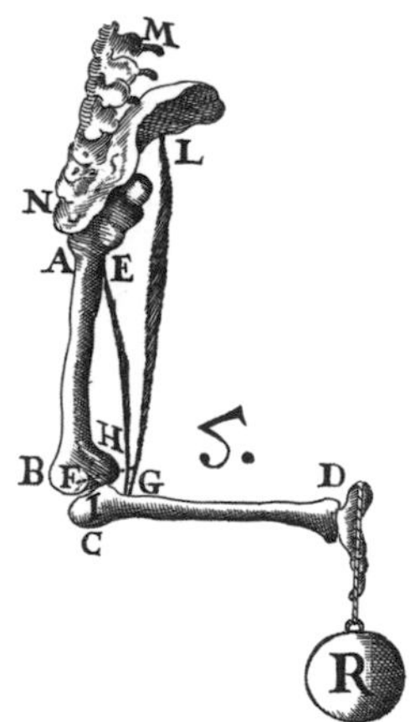

CHAPTER 9
Chronic Adaptations

In chapter 8, we examined the adaptive capabilities of the motor system in response to a single bout of physical activity. We learned that the adaptations can be extensive and that the response involves the processes and components of the system that are stressed by the activity. Chapter 9 addresses the cumulative (chronic) response of the motor system to the stress associated with long-term physical activity. We will examine the adaptations associated with strength and power training, reduced activity, recovery from injury, and aging.

Principles of Training

Substantial effort has been focused on determining the neuromechanical basis of muscle strength. As a result of these efforts, several rules for the prescription of exercise have been elaborated. These rules are often referred to as the *principles of training*. One such rule is the **overload principle** (DeLorme, 1946), which may be stated as follows:

> To increase their size or functional ability, muscle fibers must be taxed toward their present capacity to respond.

This principle implies that *there is a threshold point that must be exceeded before an adaptive response will occur*. Normally the threshold point is expressed as a percentage of maximum. For example, it has been suggested that the threshold for isometric exercises is about 40% of maximum; that is, adaptations will occur only if the force exceeds 40% of maximum. Because the maximum torque that a muscle can exert changes with time due to variations in the level of activity (i.e., training and detraining), so too does the absolute load that will exceed threshold changes. This is apparent in individuals who train regularly and then experience a period of inactivity (e.g., having a limb in a cast, being confined to bed for illness); upon recovery, they find that they cannot resume training at their preinactivity levels.

In addition to manipulating the training load relative to maximum capabilities, exercise prescription must match the mode of training to the desired effect. This constraint is embodied in the **specificity principle** (McCafferty & Horvath, 1977):

> Training adaptations are specific to the cells and their structural and functional elements that are overloaded.

This principle states *that the induced change is specific to the exercise stress*. If an individual performs a strength-training program, then only this characteristic (strength) and not others (e.g., endurance) will exhibit an adaptation (Boulay et al., 1985; Dudley & Djamil, 1985; Sale, MacDougall, Jacobs, & Garner, 1990). However, there appears to be some uncertainty regarding the boundaries associated with the phenomena of specificity. For example, some investigators have reported that strength training appears to elicit adaptations that are specific to the task, muscle length, and muscle velocity used in the training (Jones, Rutherford, & Parker, 1989; Sale & MacDougall, 1981). In contrast, others have found an absence of specificity for concentric and eccentric contractions (Peterson et al., 1990) and between rapid and isometric contractions (Behm & Sale, 1993). Although the specificity principle emphasizes the need to carefully match the training or rehabilitation program with the desired outcome, the degree of association remains to be determined.

An important corollary of the specificity principle is that the effects produced by training also depend on the pretraining capabilities of the individual. Figure 9.1 illustrates this point by showing the change in performance (squat lift) of two groups of subjects (strength athletes vs. nonathletes) over the course of a training program. The strength athletes were power lifters and bodybuilders. The nonathletes were not novices; they were trained individuals who used weight training as part of their programs, but they did not compete in strength events (power lifting, body building). The strength athletes had a pretraining maximum squat-lift load of 1,400 N, and the nonathletes could manage a load of 446 N. As Figure 9.1 shows, the strength athletes experienced a smaller absolute change and rate of change (slope of the graph) in performance than the nonathletes. This difference was due to a greater initial strength and familiarity with the movement by the strength athletes. The initial rapid increase in strength exhibited by the nonathletes reflects neural adaptations that are elicited by the training program.

Finally, the **reversibility principle** formalizes an aspect of exercise-induced adaptation that was mentioned in the discussion of the overload principle:

Training-induced adaptations are transient.

As we emphasized in chapter 8, the functional and structural status of the motor system adapts to accommodate the level (overload principle) and type (specificity principle) of stress that the motor system experiences. One corollary of this interaction is that the adaptations acquired as a result of a training program are retained only as long as the physical demands warrant that level of performance. When the training ceases, the system adapts to the new (lower) requirements; this down-regulation is often referred to as *detraining* (Chi et al., 1983; Thorstensson, 1977). One extensively examined example of detraining is the adaptations experienced by astronauts in a microgravity (weightless) environment; we will consider some of these adaptations later in the chapter.

Strength

The dynamic capabilities of a single joint system can be characterized by the force-velocity relationship (Figure 6.19). Qualitatively, the relationship comprises four distinct regions: (a) velocity = 0 (isometric or P_o), (b) velocity > 0 (concentric), (c) velocity < 0 (eccentric), and (d) maximum velocity (v_{max}). The measurements of muscle strength and muscle power correspond to the maximum outputs for regions (a) and (b), respectively. Muscle strength is defined as *the magnitude of the torque exerted by a muscle or muscles in a single maximal isometric contraction of unrestricted duration* (Atha, 1981; Enoka, 1988b). By this definition there is only one type of strength—that measured under isometric conditions. Therefore, it is inappropriate to talk of isotonic, dynamic, isokinetic, or static strength. Traditionally, people interested in human performance have used *static strength* to refer to P_o and *dynamic strength* for measures of torque when velocity was nonzero. This cumbersome terminology is responsible for much of the confusion in the strength-training literature. We shall use a narrow definition of strength, one confined to the

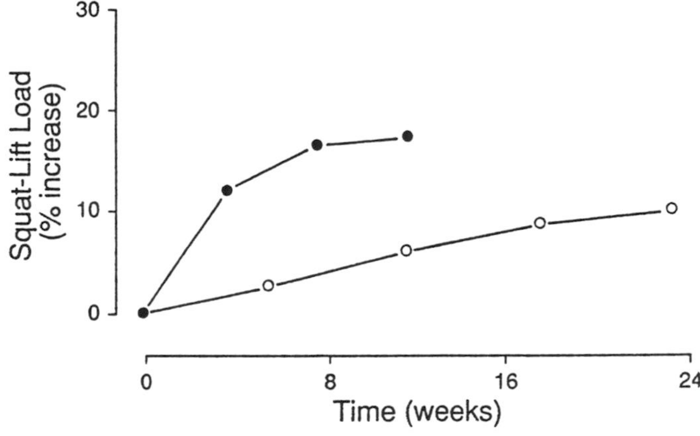

Figure 9.1 Percentage change in maximum performance in the squat lift after a strength-training program. Training involved concentric knee extension exercises in which the load varied from 80% to 100% of the concentric maximum.
Note. From "Research Overview: Factors Influencing Trainability of Muscular Strength During Short Term and Prolonged Training" by K. Häkkinen, 1985, *National Strength & Conditioning Association Journal*, **7**(2), pp. 32-37. Copyright 1985 by the National Strength & Conditioning Association. Adapted by permission.

measure of peak torque under isometric conditions. We shall characterize the nonzero-velocity region by the measure of power.

Before considering the adaptations that occur with strength training, we will first discuss the training and loading techniques used by weight lifters. This is necessary to define the terminology and to describe the changes that have been produced with the different training procedures. Then we will describe the adaptations associated with strength training. Because strength is defined as the maximum isometric muscle torque and torque is the consequence of *neural* (motor unit recruitment and discharge rate), *mechanical* (moment arm), and *muscular* (length and cross-sectional area) interactions, the adaptations produced by strength training are distributed throughout the motor system. However, we will focus on the neural and muscular adaptations.

Training Techniques

According to the crossbridge theory of muscle contraction, once Ca^{2+} has inhibited the inhibitory effect of the troponin-tropomyosin complex, myosin crossbridges attach to binding sites on actin and undergo a conformational change that results in the generation of force. Despite this single-mechanism account of muscle activation, the strength-training literature is largely organized around the different types of muscle contractions. It seems more appropriate for a discussion of strength training to emphasize variations in the load–muscle torque relationship by examining the isometric, concentric, and eccentric conditions. Such a focus is employed here in considering the effectiveness of strength-training techniques for achieving increases in muscle strength and work.

Isometric Training. An isometric contraction (*iso* = constant, *metric* = whole-muscle length) was previously defined as a condition in which the torque due to the load is matched by a torque of equal magnitude but opposite direction exerted by the muscle. Although there is no change in whole-muscle length with an isometric contraction, the muscle fibers shorten (chapter 6). In the 1950s, Hettinger and Muller popularized isometric exercises as a fitness activity (Hettinger, 1961). These exercises supposedly produced good hypertrophic responses, took little time, were convenient, and were economical. The Hettinger and Muller scheme was characterized by *consideration of threshold points, intensity levels, and the systematization of an exercise program*. For example, it was suggested for the elbow flexors that daily exercises at 40% to 50% of maximum would produce strength gains, intensity levels of 20% to 30% of maximum would maintain the status quo, and intensities of 20% of maximum would result in a loss of strength. The isometric scheme also proposed a trade-off between exercise duration and intensity; for durations of 1 to 2 s, the exercise had to be done at 100% of maximum to provide sufficient stress, whereas at a duration of 4 to 6 s the intensity could be lowered to 66% of maximum. Hettinger and Muller suggested that one 4- to 6-s contraction per day at 40% to 50% of maximum would produce the maximum gains in strength (e.g., 2% a week for the elbow flexors). Finally, Hettinger and Muller recognized that threshold values had to be reset periodically, and they encouraged rechecking the maximum value.

Although many have challenged their ideas, Hettinger and Muller made a significant contribution in that they emphasized a quantitative approach to exercise prescription. The popularity of isometric exercises has subsided in recent years, but such exercises are still advocated as part of a strength-training program. For example, the following isometric exercise regimens have been proposed: (a) for untrained or trained individuals, daily workouts of five to eight repetitions per muscle group in which the active muscles exert a force of 40% to 50% of maximum, and (b) for elite athletes, performance of exercises at several joint angles with forces of 80% to 100% of maximum, each exercise lasting 5 to 10 s, and two to five repetitions per muscle group with 30 s to several minutes between repetitions (Harre, 1982).

There are, however, at least two major concerns associated with isometric exercises. First, because movements are dominated by concentric and eccentric muscle activity in which muscle length changes, isometric exercises may not be specific to the training goals of the program, thus violating the specificity principle. For example, Kitai and Sale (1989) found that the increase in strength with isometric training was confined to a 0.17-rad (10°) range of motion about the ankle joint angle at which the exercises were performed. Second, when the force that a muscle exerts exceeds approximately 15% of maximum (see chapter 8), blood vessels begin to become occluded, and peripheral resistance to blood flow increases substantially. The net effect of this occlusion is an increase in heart rate and blood pressure (Seals, Washburn, Hanson, Painter, & Nagle, 1983), a dangerous consequence for some individuals.

An example of the adaptations that can be achieved with isometric training was demonstrated in a study by Kitai and Sale (1989). They had 6 subjects perform 6 weeks of training with isometric exercises. The subjects trained three times a week, performing two sets of 10 repetitions. Each repetition was a 5-s maximal (100%) isometric contraction of the plantarflexor muscles with the ankle joint fixed at a right angle between the shank and the foot. Two contractions were performed each minute, and there was a 2-min rest between sets. The training resulted in an increase in strength (maximal isometric torque) at the training angle of 18% (25.5 N·m) and increases of 17% and 14% at adjacent (0.17 rad) plantarflexion and dorsiflexion angles. These increases in strength with isometric training are similar to values reported by other groups (Davies, Parker,

Rutherford, & Jones, 1988; Garfinkel & Cafarelli, 1992). With a similar protocol but for a duration of 16 weeks, Alway, MacDougall, and Sale (1989) found that isometric training produced a 44% increase in the maximum plantarflexor torque.

Dynamic Training. When the muscle torque does not equal the load torque, the muscle activity is often referred to as *dynamic contraction* because the muscle length changes. This term is typically used to refer to concentric and eccentric conditions. In contrast, an isometric contraction may be described as a *static contraction* because the system is in equilibrium; that is, the muscle and load torques balance each other. The notion of system equilibrium, however, is not confined to isometric conditions but also includes nonaccelerating (constant-velocity) conditions. The term *isokinetic* refers to contractions that produce a constant angular velocity of the limb; this constraint is typically imposed by an external device (i.e., an isokinetic machine).

In the modern era of strength training, DeLorme (1945) was perhaps the first to use dynamic exercises systematically. The DeLorme technique, which uses **progressive resistance exercises**, involves performing three sets of 10 repetitions in which the load is increased for each set. The maximum load is determined by trial and error; the lifter experiments with different loads to identify the load that can be lifted exactly 10 times. This load is referred to as the 10-RM load (RM = repetition maximum). The first set of 10 repetitions is done with one-half the 10-RM load, the second set with three-fourths the 10-RM load, and the final set with the full 10-RM load. The rationale behind this approach is that the first two sets serve as a warm-up for the maximum effort of the third set. The reverse protocol was the basis for the Oxford technique (Zinovieff, 1951), the rationale being that the second (75% 10-RM) and third (50% 10-RM) sets loaded the muscle while it was fatigued after performing the first, maximum-effort set. The distinction between the DeLorme and Oxford techniques has been referred to as *ascending-* and *descending-pyramid loading*, respectively (McDonagh & Davies, 1984). The many systems for varying the load in a weight-training program (Fleck & Kraemer, 1987) include the single set, multiple set, bulk, light-to-heavy, heavy-to-light, triangle programs, super set, circuit program, peripheral heart action, tri-set, multipoundage, cheat, split routine, blitz program, isolated exercise, exhaustion set, burn, forced repetition, super pump, functional isometrics, and double progressive.

Because dynamic exercises include both concentric and eccentric muscle activity, it is misleading to suggest that a 10-RM load represents the maximum that can be handled for all phases of each repetition. From the torque-velocity relationship (Figure 6.19), we know that the torque a muscle can exert is greater for an eccentric contraction than for a concentric one. Consequently, in a movement that involves alternating concentric and eccentric contractions, the capabilities of the concentric contraction limit performance and thus determine the 10-RM load. In addition, because the amount of exercise stress depends on the magnitude of the load relative to maximum capabilities (overload principle), it is probably the concentric component of a concentric-eccentric sequence that experiences the greater stress and subsequent adaptation (Hortobágyi & Katch, 1990).

Because muscle can exert a greater force during an eccentric contraction, a common belief is that eccentric exercises should provide a more intense training stimulus and hence greater strength gains. While the overload principle suggests that *it is not the absolute force that determines the quantity of the stimulus but rather the size of the force relative to maximum*, eccentric contractions seem to elicit more significant adaptions than concentric contractions. Indeed, most studies have found that concentric-only and eccentric-only exercise programs produce similar strength gains and enhanced work capacities. But training programs that use both concentric and eccentric contractions produce strength gains superior to those of either mode by itself (Colliander & Tesch, 1990; Dudley, Tesch, Miller, & Buchanan, 1991; Häkkinen, 1985). For example, 12 weeks of a knee extensor strength-training program (three times a week) produced an 18% increase in the peak torque for subjects who trained with concentric contractions and a 36% increase in peak torque for subjects who trained with concentric-eccentric contractions (Colliander & Tesch, 1990).

As previously discussed, experimental evidence suggests that there are significant differences between concentric and eccentric contractions:

1. Submaximal concentric and eccentric contractions may involve the activation of different motor units (Nardone et al., 1989).
2. The high stresses associated with eccentric contractions probably influence the behavior of individual crossbridges and sarcomeres. Detachment of the crossbridge during an eccentric contraction probably involves a mechanical disruption of the chemical bond in contrast to the more orderly binding of ATP and crossbridge detachment that occurs during the normal crossbridge cycle. Because of differences in the quantity of contractile proteins and amount of filament overlap among sarcomeres, the maximum force that each sarcomere can exert varies along the length of a muscle. Consequently, the lengthening of contracting muscle (an eccentric contraction) involves the stretching and popping of individual sarcomeres as each reaches its yields stress (Morgan, 1990).
3. Although the force is greater during an eccentric contraction, the EMG is substantially less than during a concentric contraction. This suggests that individuals are incapable of maximally activating a muscle during an eccentric contraction (Naka-

zawa, Kawakami, Fukunaga, Yano, Miyashita, 1993; Tesch et al., 1990; Westing, Cresswell, & Thorstensson, 1991).

4. Delayed-onset muscle soreness is preferentially elicited by activity that involves eccentric contractions (Fridén, 1984; Newham, 1988; Stauber, 1989).
5. Steady-state running at a submaximal intensity on a treadmill at a level grade is associated with a constant oxygen consumption. In contrast, steady-state downhill (−10% grade) running at the same intensity is accomplished by a gradual increase in oxygen consumption and in the EMG of leg muscles (Dick & Cavanagh, 1987). Downhill running involves a significant reliance on eccentric contractions of the leg muscles.
6. Level running at a submaximal intensity on a treadmill for 20 min produced a 9% reduction in the amplitude of the soleus H reflex. Downhill running (−10% grade) at the same intensity reduced the H reflex amplitude by 25% (Bulbulian & Bowles, 1992).
7. Eccentric exercises provide a more effective stimulus for hypertrophy (Dudley et al., 1991). This effect may be mediated by a differential control (transcription vs. translation) of protein synthesis compared with concentric contractions (Wong & Booth, 1990a, 1990b).

Although training with concentric-only or eccentric-only exercises does not appear to produce differential effects on strength, there are probably differences at submaximal levels of activation, ranging from such effects as connective tissue morphology to the coordination and control of muscle activity.

Accommodation Devices. Exercise machines in which the load is controlled by gear or friction systems (e.g., Cybex, Biodex), by hydraulic cylinders (e.g., Kin-com, Lido, Omnitron), or by pneumatic systems (e.g., Keiser) provide an *accommodating resistance*, which can generate a load equal in magnitude but opposite in direction to the force exerted by the subject. One consequence of the gear systems and of some hydraulic devices is *a movement in which the angular velocity of the displaced body segment is constant*, and hence **isokinetic**, for constant speed. An isokinetic contraction represents the dynamic condition (because muscle length changes) in which the quotient of muscle torque to load torque equals 1. When no net torque acts on the system, the acceleration of the system is 0 and velocity is constant. Obviously the two torques (muscle and load) are not equal at the beginning or at the end of an isokinetic contraction; otherwise the movement would never start or stop, because both starting and stopping require a nonzero acceleration. Isokinetic devices can be used with concentric and eccentric contractions. For a concentric test, the individual pushes against the device and power flows from the individual as positive work is done on the surroundings. For an eccentric test, however, the individual must resist the load imposed by the device; power flows from the device to the individual and negative work is done by the surroundings on the individual. Although these machines have some safety features, it is necessary that the operator of the machine provide adequate supervision and instruction.

Many researchers have used isokinetic devices (Figure 9.2a) to quantify the torque, work, and power output of muscle. Typically, this is accomplished by measuring the resistance provided by the machine ($\mathbf{F}_l$ in Figure 9.2b) and expressing the effort in the selected units. An appropriate free body diagram of the system (Figure 9.2b), however, indicates that the machine load is not the only factor affecting the load torque; the weight of the limb should also be considered. Winter, Wells, and Orr (1981) have demonstrated that neglecting the acceleration components (acceleration due to gravity and the acceleration of the limb to reach the speed of the machine) can lead to an error of 500% at the highest speeds in determining the amount of work performed by the muscles. However, these acceleration effects can be determined and incorporated into the

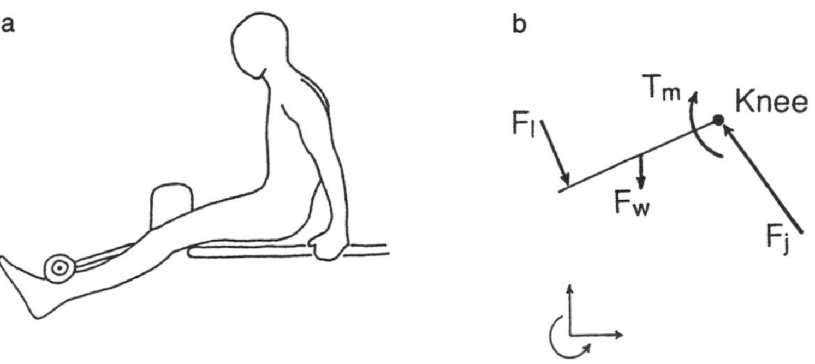

Figure 9.2 A subject performing an isokinetic knee extension exercise: (a) the Cybex isokinetic dynamometer; (b) free body diagram of the forces acting on the shank. $\mathbf{F}_j$ = joint reaction force; $\mathbf{F}_w$ = limb weight; $\mathbf{F}_l$ = machine load; $\mathbf{T}_m$ = resultant muscle torque.

appropriate calculations to allow more accurate measurements (Gransberg & Knutsson, 1983). Without this correction, the torque exerted by the machine ($\mathbf{F}_l$) is not a measure of the resultant muscle torque.

One *disadvantage* of isokinetic devices is that they violate the principle of specificity. Natural movements do not involve a constant angular velocity of a limb and hence are not isokinetic. Moreover, the maximum speed on many isokinetic devices (5 rad/s) is much less than the maximum angular velocity achieved during movements such as running, jumping, and throwing. Despite this limitation, isokinetic training can produce adaptations comparable to those obtained with less constrained techniques, although these changes are specific to the training speed (Osternig, 1986). For example, when subjects trained the knee extensors for 7 weeks on an isokinetic device at 3.14 rad/s, there were significant increases in peak torque (about 14%) for contractions from 0 to 3.14 rad/s but not at 4.2 and 5.2 rad/s (Lesmes, Costill, Coyle, & Fink, 1978). Similarly, subjects who trained for 10 weeks at 1.1 rad/s experienced a 9% increase (25 N·m) in peak torque, whereas those who trained at 4.2 rad/s had a 20% increase (30 N·m) in peak torque; each of these increases was specific to the training speed (Ewing, Wolfe, Rogers, Amundson, & Stull, 1990). Furthermore, strength-trained athletes have a greater muscle cross-sectional area and a greater isokinetic torque than control subjects have (Alway, Stray-Gundersen, Gumbt, & Gonyea, 1990; Sale & MacDougall, 1984).

Interestingly, the specificity of isokinetic training seems to depend on the intended rather than the actual speed of movement. Behm and Sale (1993) trained subjects for 16 weeks to perform rapid (5.23 rad/s) isokinetic contractions. The subjects trained one leg at a time with an ankle dorsiflexion task. Although the intent was to perform the rapid isokinetic contraction with each leg, one leg was prevented from moving and hence performed an isometric contraction. The training produced an increase in the peak torque for both limbs with the 5.23 rad/s isokinetic contractions. The increase in peak torque was greater than the increase at lower velocities. This observation suggests that the attempt to perform the rapid contraction was more important for the specific adaptation than whether the contraction was isometric or concentric.

Isokinetic devices offer several *advantages*. One significant advantage is that a muscle group may be stressed differently throughout its range of motion. This accommodation is useful in rehabilitation settings. When one specific locus in a range of motion is painful, the patient can reduce the effort at this point yet exercise the joint system in the other nonpainful regions. Furthermore, the patient can simply stop in the middle of an exercise without having to worry about controlling the load. Because of the accommodation property, the resistance provided by the device varies in proportion to the capabilities of the user over the range of motion. Isokinetic devices also provide substantial support for the user, removing the need to provide some of the stabilizing support that must be generated with other forms of exercise.

In contrast to isokinetic devices, hydraulic and pneumatic systems provide an accommodating resistance but do not control movement speed. Hydraulic devices provide resistance as individuals exert force against a leaky, fluid-filled chamber. The muscles perform concentric contractions against this resistance. The resistance is manipulated by varying the diameter of the valve through which the fluid passes from one chamber to another; the resistance is high when the diameter of the valve is small. The hydraulic system offers an accommodating resistance. The greater the force exerted by the individual, the greater the resistance provided by the device. However, movement speed increases as the applied force increases. Nonetheless, hydraulic devices can elicit strength gains comparable to those achieved with free weights. For example, Hortobágyi and Katch (1990) trained one group of subjects with free weights and another group with a hydraulic device for 12 weeks. The average improvements in the maximum bench press and squat loads were 23% for the free-weights subjects and 20% for the hydraulic-device subjects. Furthermore, changes in force, velocity, and power evaluated by isokinetic and hydraulic dynamometers did not differ (~ 8%) for the two groups. Similarly, pneumatic systems provide a variable resistance by using a compressor to alter the pressure and hence the resistance against which the individual works. The pressure can be changed rapidly, simply by depressing a button. One useful feature of pneumatic systems is that the pressure can be increased sufficiently to enable the performance of eccentric contractions; however, the user must increase the pressure cautiously, because he or she has little control of the machine under these conditions.

Plyometric Training. In contrast to the isokinetic device, plyometric exercises were designed to train a specific movement pattern, the eccentric-concentric sequence of muscle activity. Recall that such a sequence involves a change in the muscle torque–load torque ratio from less than 1 to greater than 1. Such exercises include starting at the top of a long flight of stairs and jumping down the stairs, one or several at a time (a vertical drop of 50 to 80 cm), with both feet together.

Most human movements involve an eccentric-concentric sequence of muscle activity. The advantage of this strategy is that the initial eccentric contraction enables the muscle to perform more positive work during the subsequent concentric contraction. Based on the work-energy relationship (chapter 3), we know that the muscle must have an additional supply of energy for it to perform more work. In chapter 7, we discussed how the storage and utilization of elastic energy and the preload effect serve as mechanisms that provide the additional energy. Elastic energy is stored in the series elastic component during the eccentric contraction and then used during the concentric contraction as the contractile component and the series elastic component both shorten. With the preload effect, the eccentric contraction causes an increase in the muscle force at the onset of the concentric contraction so that the area under the force-distance graph (work) is greater. The force is greater with the eccentric contraction due to such effects as an alteration in the crossbridge detachment mechanism, an enhancement of contractile activity (e.g., increased release of Ca^{2+}, stretch of less completely activated sarcomeres), and a possible activation of bigger and stronger motor units.

Apparently plyometric exercises induce a training effect. For example, Blattner and Noble (1979) trained two groups of subjects for 8 weeks, one group on an isokinetic device and the other with plyometric exercises. The two groups increased their vertical jump heights by about the same amount (5 cm) as a result of the training. Similarly, Häkkinen, Komi, and Alén (1985) found that jump training caused a minor increase in the maximal knee extensor isometric force but a substantial increase in the rate of force development and an increase in the area of fast-twitch fibers. Although it is difficult to train experimental animals to perform jump training, Dooley, Bach, and Luff (1990) managed to train rats to perform 30 jumps a day, 5 days a week, for at least 8 weeks. The jumps were estimated to be between 30% to 67% of the maximum height that could be achieved. The training produced some changes in the medial gastrocnemius muscle but not in the soleus. There was a 15% increase in the maximum tetanic force, a 3% increase in the maximum rate of force development, a 15% increase in fatigability, and a 4% decrease in the percentage of Type IIa muscle fibers. These observations indicate that regular vertical jump exercises performed at a moderate intensity can induce adaptations in the involved muscles.

Neuromuscular Electrical Stimulation. Reports from the former Soviet Union in the early 1970s (Elson, 1974) suggested yet another means of strengthening muscle, that of applying electric shocks to the muscle. Such procedures, which are referred to as **neuromuscular electrical stimulation**, involve stimulating the muscle with a protocol designed to minimize the pain and discomfort associated with this procedure (Moreno-Aranda & Seireg, 1981a, 1981b, 1981c). Neuromuscular electrical stimulation has been used since the 18th century as a rehabilitation tool (Hainaut & Duchateau, 1992; Liberson, Holmquest, Scot, & Dow, 1961), but only since the 1970s has it been applied to noninjured active athletes as a supplement to conventional training.

The rationale for neuromuscular electrical stimulation is that individuals are unable to activate muscle maximally and that electric shocks can elicit the difference between the maximum voluntary contraction force and the maximum capability of the muscle. Two lines of experimental evidence document the inability of human subjects to exert the maximum muscle force with a voluntary contraction (Enoka & Fuglevand, 1993):

1. Neural insufficiency—the force exerted during a maximum voluntary contraction can be increased with single electric shocks (Belanger & McComas, 1981; Hales & Gandevia, 1988) and with trains of electric shocks (Davies, Dooley, McDonagh, & White, 1985; Duchateau & Hainaut, 1987, 1988; McDonagh, Hayward, & Davies, 1983; Young, McDonagh, & Davies, 1985). This means that the neural drive is *insufficient* to elicit the maximum force that the muscle can exert.

2. Neural supplementation—the maximum voluntary contraction force can be *supplemented* by manipulation of afferent feedback. For example, Howard and Enoka (1991) examined the effect of right-leg neuromuscular electrical stimulation on the maximum voluntary contraction force of the left-leg knee extensors. Neuromuscular electrical stimulation of the quadriceps femoris of the left leg elicited a strong cutaneous sensation and a right-leg force that was 40% of maximum. In addition, the right-leg stimulation caused an 11% increase in the left-leg maximum voluntary contraction force. Similarly, the therapeutic procedure of rapid brushing has been shown to cause an increase in the quadriceps femoris and biceps femoris EMG during a maximum voluntary contraction (Matyas, Galea, & Spicer, 1986). Furthermore, stretching of the contralateral muscles can enhance the maximum voluntary contraction force (Lagasse, 1974; Morris, 1974).

Neuromuscular electrical stimulation can be applied with a variety of protocols. The parameters that can be varied include stimulus frequency, stimulus waveform, stimulus intensity, electrode size, and electrode type. The simplest stimulus protocol is to apply a train of rectangular pulses (Figure 9.3a). The problem with this scheme is that it requires a frequency of about 100 Hz to elicit the maximum force in a muscle, but this frequency also produces a significant pain response (Barr, Nielsen, & Soderberg, 1986). This problem can be circumvented with a protocol of high-frequency stimulation (10 kHz) that is modulated (i.e., turned on and off) at a lower frequency (50 to 100 Hz); this scheme is attributed to Kots (1971) and is shown in Figure 9.3b. Moreno-Aranda and Seireg (1981a, 1981b, 1981c) tested such a protocol and found that the optimum regimen involved the application of the stimulus for 1.5 s once every 6 s for 60 s, followed by a 60 s rest. The specific details of the optimum protocol probably differ among muscle groups. Nonetheless, this type of a protocol minimizes the pain associated with the procedure and can elicit a force

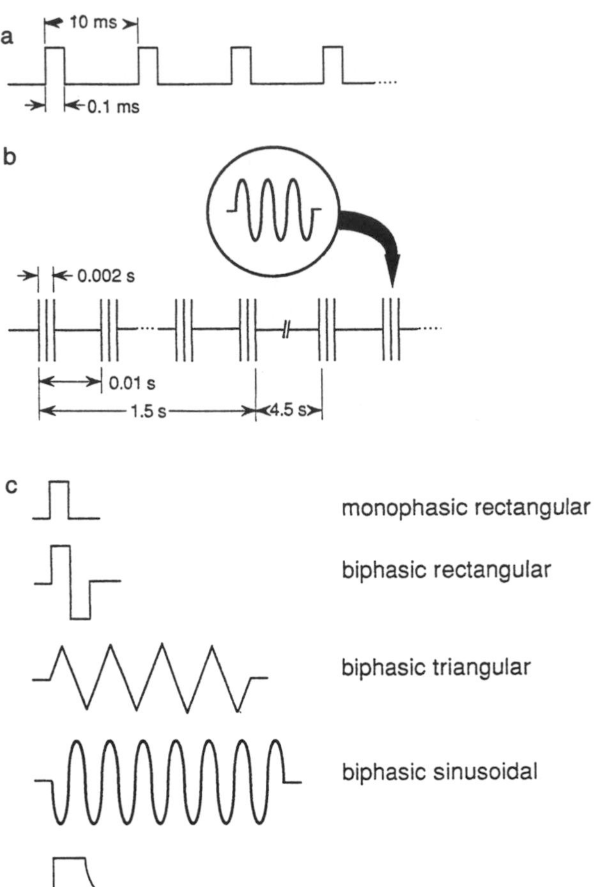

Figure 9.3 Selected stimulus regimens used in neuromuscular stimulation: (a) a conventional train of low-frequency (100-Hz) stimuli with a pulse width of 0.1 ms; (b) a pattern of high-frequency stimulation (10-kHz sine wave) that is modulated at a low frequency (100 Hz) with 0.01 s between trains of stimuli; (c) waveform shapes commonly used in neuromuscular stimulation. The biphasic waveforms vary about a zero line.

Note. From "Muscle Strength and Its Development: New Perspectives" by R.M. Enoka, 1988b, *Sports Medicine*, **6**, p. 149. Copyright 1988 by ADIS Press Limited. Adapted by permission.

equivalent to the maximum voluntary contraction force (Delitto, Brown, Strube, Rose, & Lehman, 1989).

In addition to varying stimulus frequency, it is also possible to use stimulus waveforms that have different shapes (Figure 9.3c). There are two reasons for changing waveform shape. First, the shape of the stimulus waveform can affect the *comfort* associated with neuromuscular electrical stimulation. Commercially available clinical stimulators provide a variety of waveform shapes (e.g., rectangular, triangular, sinusoidal) that can deliver electric current in either a positive (monophasic) or positive-negative (biphasic) pulse. Although no waveform shape is universally preferred, subjects and patients do have individual preferences (Baker, Bowman, & McNeal, 1988; Delitto & Rose, 1986). This may be an important consideration in patient compliance with a prescribed protocol. Second, conventional stimulus waveforms (e.g., biphasic rectangular pulses) are known to *preferentially activate large-diameter motor units* in contrast to the small-to-large recruitment order that occurs with voluntary activation (Fang & Mortimer, 1991c; Trimble & Enoka, 1991). Although this recruitment order may be advantageous in strength training and recovering from injury, it causes a problem in the functional electrical stimulation of paralyzed muscle. Because large-diameter motor units tend to be more fatigable, conventional stimulus waveforms fatigue paralyzed muscle too rapidly, which limits the usefulness of this technique for restoring function (Fang & Mortimer, 1991a). With the use of a quasitrapezoidal waveform and a tripolar stimulating electrode, Fang and Mortimer (1991b, 1991c) devised a technique to preferentially recruit fatigue-resistant motor units. The tripolar-cuff electrode causes a differential block (action potentials cannot be propagated) by hyperpolarizing the membrane of large-diameter axons and enabling the small-diameter axons to be activated selectively. This technique may be useful for the functional electrical stimulation of paralyzed muscle.

Because the axolemma is more easily excited than the sarcolemma, the electric current that is passed through a muscle to cause a muscle contraction *generates action potentials in the intramuscular nerve branches rather than directly exciting muscle fibers* (Hultman, Sjöholm, Jäderholm-Ek, & Krynicki, 1983). Therefore, electrodes that have a large area enable the current to be more readily disbursed throughout the muscle. This is especially true for large muscles such as quadriceps femoris. The electrode also should have a low *impedance*—that is, a low resistance to the flow of electric current. With a large electrode, the current density (nA/cm^2) is less, and it is possible to pass a greater amount of current without damaging the underlying tissue. Large carbonized-rubber electrodes seem particularly effective at meeting these needs (Lieber & Kelly, 1991).

Neuromuscular electrical stimulation involves the artificial generation of action potentials in the peripheral motor system. Action potentials also can be generated with **magnetic stimulation** (Lotz, Dunne, & Daube, 1989). This is done by using a magnetic stimulator to generate a magnetic field that varies over time and induces an electric field and the flow of current. When the current is strong enough and flows for a sufficient duration, it depolarizes excitable membranes and generates action potentials. As with neuromuscular electrical stimulation, magnetic stimulation activates intramuscular nerve branches. In contrast, the threshold for a motor response is lower than that for a sensory response with magnetic stimulation. Therefore, magnetic stimulation is less painful than electrical stimulation. An additional advantage of magnetic stimulation is that there is less decline in the stimulus over distance compared with electrical stimulation, and the stimulus is more widespread.

Although researchers disagree on the maximum capabilities of neuromuscular electrical stimulation, there is no doubt that it is possible to increase strength in both healthy and injured muscle with this technique (Enoka, 1988b; Hainaut & Duchateau, 1992; Miller & Thépaut-Mathieu, 1993). A common approach is to use the Kots protocol (stimulus frequency of 2,500 Hz modulated in 50-Hz bursts), with 10 repetitions in each of 15 to 25 training sessions. A single repetition lasts 60 s and comprises 10 s of stimulation and 50 s of rest. The stimulus intensity is set at the maximum tolerable level. The increase in strength (isometric contraction) ranges from 0.6% to 3.6% per session, with an average of about 1.6% per session. This is comparable to the strength gain that can be achieved with other training modalities (Hainaut & Duchateau, 1992).

Three aspects related to the application of neuromuscular electrical stimulation to healthy muscle are interesting. First, a comparable increase in strength can be achieved at a much lower force with stimulation than with voluntary activation of muscle. For example, Laughman, Youdas, Garrett, and Chao (1983) trained the quadriceps femoris muscles of two groups of subjects, one group with isometric exercises and the other group with neuromuscular electrical stimulation. Although both groups exhibited similar increases in strength (18% and 22%, respectively) after 5 weeks of training, these were accomplished with average training intensities of 78% (isometric) and 33% (electrical stimulation) of maximum. This effect may be due to the preferential activation of the large-diameter motor units (Type II) with neuromuscular electrical stimulation. The activation of muscle fibers is clearly different for a voluntary contraction compared with neuromuscular electrical stimulation (Adams, Harris, Woodard, & Dudley, 1993). Furthermore, this difference varies among subjects and throughout the volume of the muscle (quadriceps femoris), as detected by MRI.

Second, activation of the muscles in one limb with neuromuscular electrical stimulation can elicit a modest increase in strength in the contralateral limb. The rationale for the use of neuromuscular electrical stimulation is to supplant an inadequate voluntary activation of muscle with artificial activation of the nerve in the muscle. However, the artificial stimulation increases not only the strength of the activated muscle but also the strength of the contralateral nonstimulated muscle. For example, Laughman et al. (1983) reported an increase of about 15% of the maximum voluntary contraction force for the quadriceps femoris of the left leg after 5 weeks of training (neuromuscular electrical stimulation) for the quadriceps femoris of the right leg. This adaptation must be due to changes that occur in the neural circuitry as a consequence of the action potentials that are sent back to the spinal cord from the artificial activation of the muscle nerve.

Third, despite the conclusion that strength gains are comparable with voluntary activation and neuromuscular electrical stimulation (Hainaut & Duchateau, 1992), a convincing case has been made for the substantial enhancement of voluntary performance with the use of electrical stimulation (Delitto et al., 1989). A 140-day study (which was divided into four periods, each about a month in duration) was performed on an experienced (11 years) and highly motivated weight lifter. The dependent variable was the maximum load that could be lifted in the clean and jerk, snatch, and squat lifts. In the second and fourth periods of training, the voluntary training regimen (3 hr/day) was supplemented with high-intensity (200-mA) neuromuscular electrical stimulation of the quadriceps femoris. The results (Figure 9.4) indicate that performance of all three lifts improved substantially with the application of the neuromuscular electrical stimulation. Furthermore, others have shown that the maximum eccentric torque can be increased with neuromuscular electrical stimulation (Westing et al., 1990).

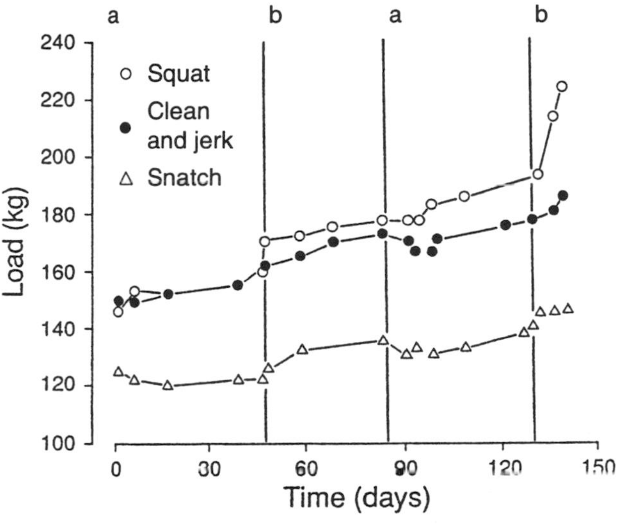

Figure 9.4 Changes in the maximum loads lifted by an experienced weight lifter for three lifts. During the b periods, the regular training was supplemented with neuromuscular electrical stimulation of the quadriceps femoris.
Note. From "Electrical Stimulation of Quadriceps Femoris in an Elite Weight Lifter: A Single Subject Experiment" by A. Delitto, M. Brown, M.J. Strube, S.J. Rose, and R.C. Lehman, 1989, *International Journal of Sports Medicine*, **10**, p. 189. Copyright 1989 by Georg Thieme Verlag Stuttgart. Adapted by permission.

Neuromuscular electrical stimulation is a useful modality in rehabilitation. Some investigators have suggested that it is easier to strengthen hypotrophic muscle with electrical stimulation than with voluntary activation (Godfrey, Jayawardena, & Welsh, 1986; Williams, Morrissey, & Brewster, 1986). In the absence of a neural drive (patients with spinal cord injury), neuromuscular electrical stimulation can preserve muscle properties. For example, stimulation (20 Hz) of tibialis anterior for 1 to 2 hr over a 6-week interval was sufficient to reduce the fatigability of the muscle (increased proportion of Type I muscle fibers) and to increase contraction and relaxation times but not to increase muscle fiber size or strength (Stein et al., 1992). Neuromuscular electrical stimulation can induce a substantial cardiorespiratory stress in paraplegic patients, and with several months of training there is a marked improvement in endurance capabilities (Petrofsky & Stacy, 1992). Neuromuscular electrical stimulation is more effective than voluntary exercise at minimizing the atrophy of quadriceps femoris and in improving the restoration of gait in patients following reconstruction of the anterior cruciate ligament (Synder-Mackler, Ladin, Schepsis, & Young, 1991).

Loading Techniques

One should consider several issues when deciding how to vary the load in order to induce strength gains; these include progressive resistance exercises, the magnitude of the load, and the way in which the load varies over the range of motion.

Progressive Resistance Exercises. The load used in strength training can be manipulated in a number of ways. DeLorme (1945), one of the first to address this issue, proposed varying the load systematically from one set of repetitions to another; this technique has become known as **progressive resistance exercise**. As indicated previously, DeLorme's technique involves three sets of repetitions of an exercise in which the load increases with each set. In conventional weight-training programs, this typically means increasing the weight of the barbell from one set to another. However, the torque exerted by muscle depends on *the size of the moment arm* and *the speed of the movement* in addition to the magnitude of the external load (the amount of the barbell weight). Thus progressive resistance exercises can be accomplished without changing the size of the load (barbell weight) by simply varying moment-arm length or the speed of the movement. For example, Figure 9.5a shows an individual in the middle of a bent-knee sit-up. In the four positions illustrated (Figure 9.5b), the arms move from being at the person's side to being stretched overhead. The change in the arm position does not alter the weight of the upper body, but it shifts the center of gravity toward the head and thus increases the moment arm of the system weight relative to the hip joint (dotted line in Figure 9.5b). The net effect is an increase in the torque that the hip flexors must exert to accomplish the movement.

Magnitude of the Load. When defining the concept of progressive resistance exercises, DeLorme (1945) proposed that each of the three sets include 10 repetitions. However, most strength-training programs currently advocate 1 to 8 repetitions in a set. *When an individual lifts a heavier barbell, the resultant muscle torque will increase in direct relation to the increase in load if, and only if, the kinematics of the movement remain the same.* For example, if an individual performs two squats, one with a barbell weight of 500 N and the other with a weight of 1,000 N, the torque exerted by the knee extensors with the heavier load will be about twice that for the lighter load if the movement is performed in exactly the same way each time. When subjects lifted loads of 40%, 60%, and 80% of the 4-RM load, the kinematics of the movement were altered (e.g., by flexing more at the hip during a squat) so that the resultant muscle torque did not increase in proportion

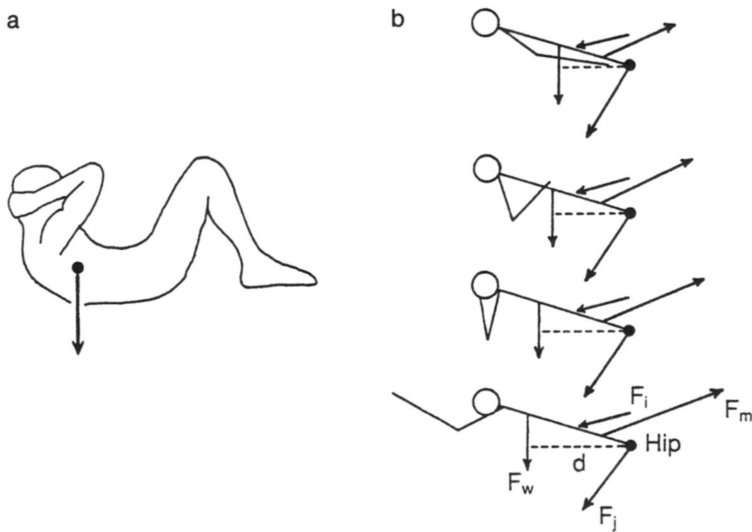

Figure 9.5 Subject performing a bent-knee sit-up. (a) Whole-body figure. (b) Changing arm position on the moment arm (*d*) of the upper-body weight vector ($\mathbf{F}_w$) with respect to the hip joint increases the required resultant hip flexion mucle force ($\mathbf{F}_m$) to perform a slow sit-up. $\mathbf{F}_i$ and $\mathbf{F}_j$ represent the intraabdominal pressure and the joint reaction forces, respectively.

to the load (Hay, Andrews, & Vaughan, 1980; Hay, Andrews, Vaughan, & Ueya, 1983). In weight lifting and in rehabilitation, it is necessary to focus on the form of the movement to ensure that the load experienced by the muscle increases in proportion to the load and that the exercise stresses the intended muscles. Because the heaviest loads tend to alter the kinematics of a movement, the most favored loads for strength training appear to be about 4- to 6-RM with four sets (three to six) of each exercise (Atha, 1981; Fleck & Kraemer, 1987; McDonagh & Davies, 1984; Sale & MacDougall, 1981). According to the scheme proposed by Sale and MacDougall (1981), a 5- to 6-RM load would be about 85% to 90% of the maximum load (Figure 9.6). In contrast, body builders use 8- to 12-RM loads and do three to six sets over several different parts of the range of motion about a joint to increase muscle mass and definition.

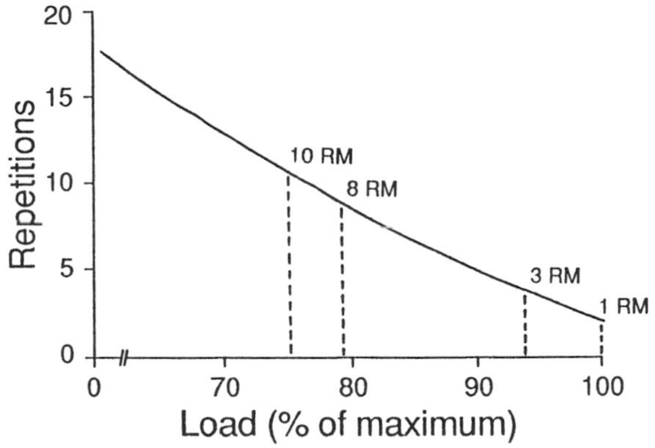

Figure 9.6 The relationship between the number of repetitions in an exercise set (RM = repetition maximum) and the magnitude of the load as a percentage of maximum. *Note.* From "Specificity in Strength Training: A Review for the Coach and Athlete" by D.G. Sale and D. MacDougall, 1981, *Journal of Applied Sport Sciences*, **6**, p. 90. Copyright 1981 by the Canadian Association of Sport Sciences. Adapted by permission.

McDonagh and Davies (1984) summarized the effect of load with the following statements, which are based on studies that involved *untrained* subjects: Loads less than 66% of maximum do not increase strength even with 150 contractions per day; loads greater than 66% of maximum increase strength from 0.5% to 1.0% each training session; and with loads of greater than 66% of maximum, as few as 10 repetitions per session can increase strength. Harre (1982) further suggested that beginners use loads of 60% to 80% of maximum with 8 to 10 repetitions in each set and that elite athletes use loads of 80% to 100% of maximum with two to five repetitions per set.

An alternative to manipulating the magnitude of the load during weight lifting is to vary the speed of the movement. Some investigators have compared the strength gains achieved with conventional weight-lifting exercises to those that result from rapid movements (so-called explosive exercises). Because the maximum force that a muscle can exert declines as the speed of shortening increases (force-velocity relationship), there is no obvious reason to suppose that rapid movements might provide a more effective training stimulus. Accordingly, Häkkinen and Komi (1986) found that knee extensor strength increased in subjects following 20 weeks of weight-lifting training (squat exercise), but strength did not increase in another group of subjects who trained for a similar duration with rapid vertical jumps. In contrast, training of the elbow flexor muscles with heavy loads (90% of 1 RM) and with rapid movements (10% of 1 RM) produced comparable increases in muscle strength and the cross-sectional area of muscle fibers (Dahl, Aaserud, & Jensen, 1992).

Constant and Variable Loads. With the exception of isometric exercises, strength-training activities involve changing the length of an active muscle over a prescribed range of motion. Because the torque a muscle can exert changes with joint angle, it is necessary to indicate not only the size of the load (as discussed previously) but also the way in which the load changes over the range of motion. Billions of dollars have been spent by manufacturers of weight-lifting equipment in addressing this issue. The controversy concerns **constant-** and **variable-load** training. This is essentially a disagreement over free weights (i.e., barbells and dumbbells) versus machines (e.g., Nautilus, Universal); the load applied to the body segment remains constant with free weights and varies over the range of motion with machines.

For example, consider the forearm-curl exercise, which involves alternate flexion-extension of the elbow joint against a load. With free weights (Figure 9.7a), the load held by the individual remains *constant* throughout the exercise and acts vertically downward. The load ($\mathbf{F}_l$) in the barbell exercise is the weight of the barbell. The length of the load vector remains constant over the range of motion (Figure 9.7a)—recall that the length of a weight vector indicates the magnitude of the load. With some machines, however, the load moved by the individual *varies* over the range of motion (Figure 9.7b). The load ($\mathbf{F}_l$) provided by the machine is a torque due to the product of the weight stack and the moment arm (*d*) of the cam. Although the weight remains constant for a particular exercise, the length of the moment arm (*d*) from the axis of rotation (*O*) to the point at which the chain (that connects to the weight stack) leaves the cam changes throughout the range of motion (Figure 9.7b). This is shown in the free body diagram for the machine by the change in the length of the load vector over the range of motion.

Another difference between constant and variable loads is that with free weights the system operates on a force (barbell weight), whereas the machine acts

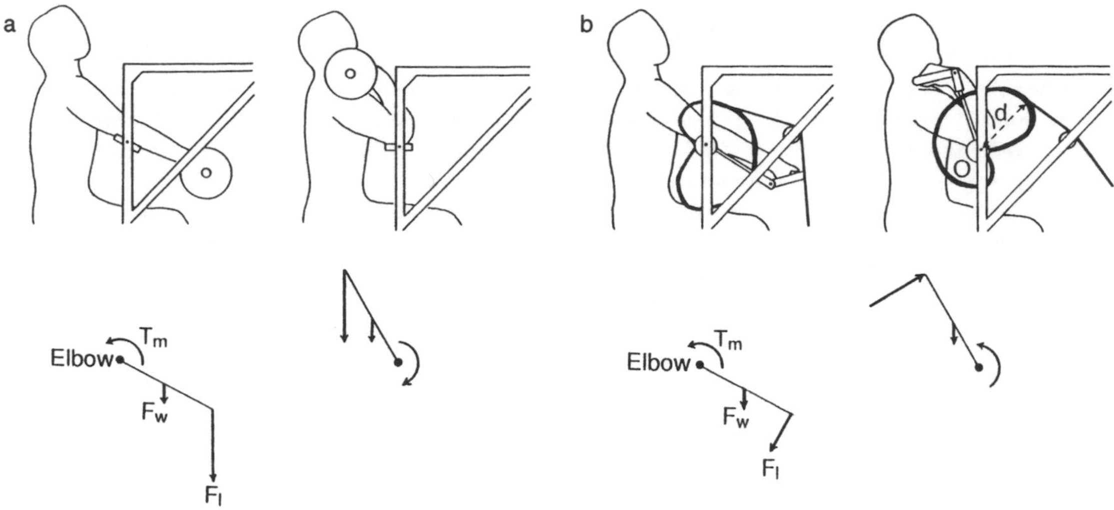

Figure 9.7 Loading conditions associated with (a) constant- and (b) variable-load training for the forearm curl.

against a torque (weight stack × moment arm). For the free weights, the load (barbell weight) remains constant over the range of motion, whereas for the machine the load (torque) varies. If we ignore the weight of the limb, the load experienced by the muscles is the product of the external load (barbell weight or machine torque) and the moment-arm distance (*s* in Figure 9.8). For the barbell load (Figure 9.8a), *s* varies substantially over the range of motion, much more than for the machine load (Figure 9.8b). The concept of variable load arose from consideration of the torque-angle relationship of muscle; recall the ascending, descending, and intermediate relationships (Figure 6.14). The most common relationship (intermediate) depicts the maximum torque as occurring in the middle of the range of motion. Accordingly, *variable-load devices are designed so that the load changes over the range of motion to match the torque-angle relationship of muscle.*

The differences in the stress applied to muscle with constant and variable loads are shown in Figure 9.9. The solid line represents the maximum isometric torque-angle relationship for the elbow flexor muscle group. The other two curves indicate the elbow flexion torque required to move a moderate load over the indicated range of motion with constant (barbell—dotted line) and variable (Nautilus machine—dashed line) loads. The load employed with both the barbell and the Nautilus machine was 60% of 4 RM. The shape of the curve for the variable load more closely matches that of the maximum capability. Thus some people claim that this loading technique is superior because it stresses the muscle more evenly over its entire range of motion. In contrast, free weights (constant loads) produce a more variable load over the range of motion, and the peak torque for one part of the curve exceeds that associated with the variable load. Despite the rationale provided for variable loads, differences in body size (limb dimensions) among users probably negate or substantially diminish this benefit (Garhammer, 1989).

Neural Adaptations

One common approach to examining the mechanisms underlying training-induced increases in strength is to

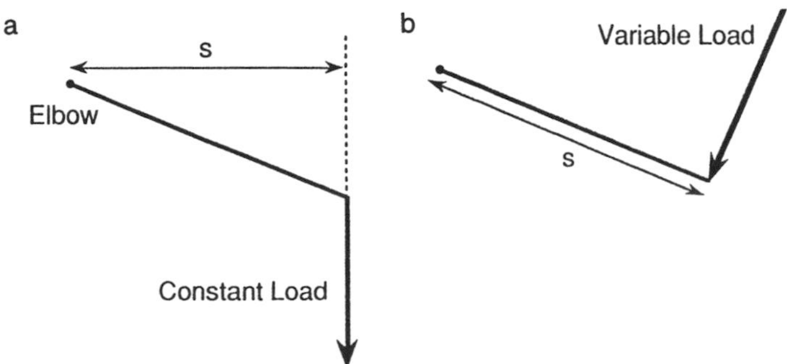

Figure 9.8 The load experienced by the muscles about the elbow joint in response to (a) free weights (constant load) or (b) machine load (variable load). For simplicity, the variable load is assumed to act perpendicular to the forearm throughout the range of motion.

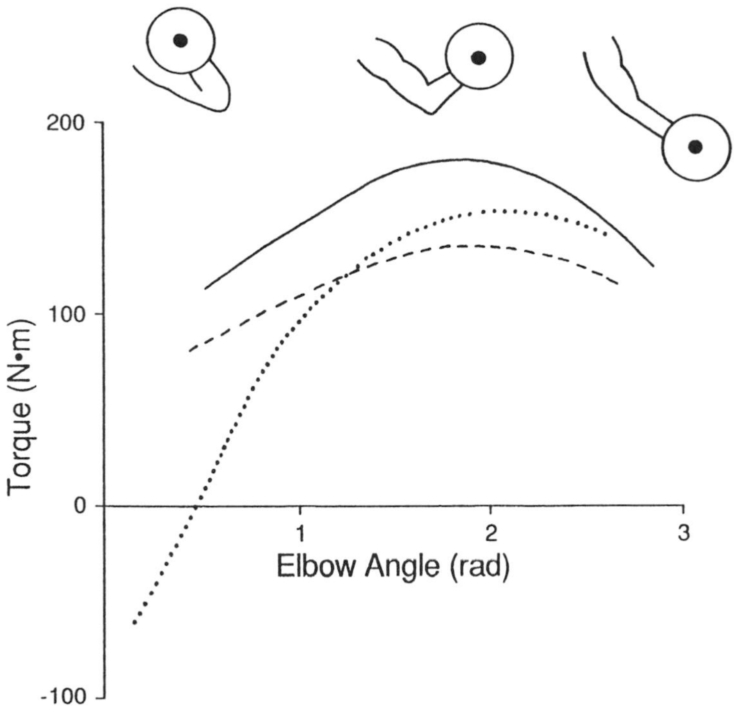

Figure 9.9 Resultant muscle torque about the elbow joint during a forearm-curl exercise with a barbell and a Nautilus machine in comparison with the maximum torque that can be exerted over the range of motion.
Note. From "Dynamic Variable Resistance and the Universal System" by F. Smith, 1982, *National Strength & Conditioning Association Journal*, **4**, pp. 14-19. Copyright 1982 by the National Strength & Conditioning Association. Adapted by permission.

compare the time course of changes in EMG and muscle size (e.g., whole-muscle or muscle fiber cross-sectional area). These studies have found that the change in EMG begins before the change in muscle size, that the change in EMG is generally more substantial, and that there can be a change in EMG and strength with training

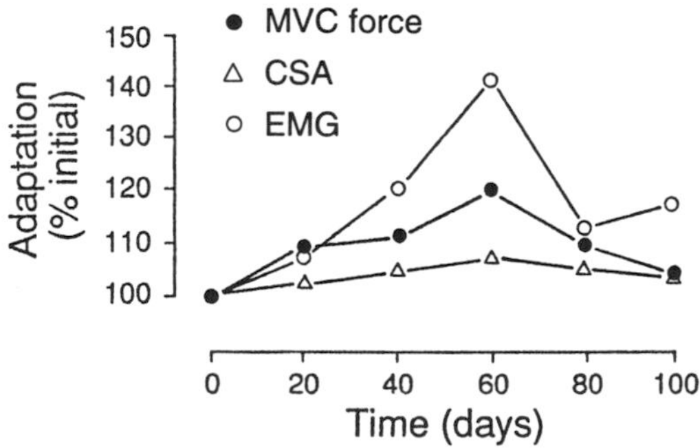

Figure 9.10 Changes in cross-sectional area of the quadriceps femoris, integrated EMG of vastus lateralis during a maximum contraction, and the maximum voluntary contraction force during isokinetic training and detraining.
Note. From "Changes in Force, Cross-Sectional Area, and Neural Activation During Strength Training and Detraining of the Human Quadriceps" by M.V. Narici, G.S. Roi, L. Landoni, A.E. Minetti, and P. Ceretelli, 1989, *European Journal of Applied Physiology*, **59**, p. 313. Copyright 1989 by Springer-Verlag. Adapted by permission.

programs that are too brief to induce morphological changes (Häkkinen, Alén, & Komi, 1985; Häkkinen & Komi, 1983; Rutherford & Jones, 1986). For example, Figure 9.10 shows the time course of changes in the cross-sectional area of quadriceps femoris (as determined by MRI), integrated EMG, and the maximum isometric force in response to 60 days of training and 40 days of detraining (Narici, Roi, Landoni, Minetti, & Ceretelli, 1989). In this study, the increase in EMG was disproportionately larger than the increase in muscle size. Such observations support the conclusion that *it is possible to obtain an increase in strength without an adaptation in the muscle but not without an adaptation in the nervous system.*

Despite the evidence that neural adaptations contribute to changes in strength, the difficult task has been to identify the types of neural adaptations that occur. Adaptations seem to occur at all levels of the nervous system, from the high-level controller to motor neurons. The following examples indicate the types of adaptations that have been shown to occur in the nervous system:

1. Imagined contractions—Perhaps the most convincing evidence for neural adaptations was obtained in a study on the strength gains achieved with isometric contractions. In this study, one group of subjects actually performed the contractions, and another group simply

imagined performing the contractions (Yue & Cole, 1992). The study was performed on a hand muscle and involved a 4-week training program that consisted of five training sessions per week. The subjects who imagined performing maximum isometric contractions did not activate the muscle (there was no EMG). The results are shown in Figure 9.11. The maximum isometric force increased in the trained hand for both groups of subjects (imaginary group, 22%; contraction group, 30%) but not for a control group of subjects who did not train. Furthermore, the maximum isometric force of the untrained hand also increased for both groups of subjects (imaginary group, 10%; contraction group, 14%) but to a lesser extent than for the trained hand. Because the muscle was not actually activated, the increase in strength experienced by the subjects who imagined the contractions must have been due to adaptations in the nervous system.

2. Coordination—Rutherford and Jones (1986) trained a group of subjects for 12 weeks with a bilateral knee extension task. The subjects used a 6-RM load, which was about 80% of maximum. Both the maximum isometric voluntary contraction force (strength) and the training load were recorded over the course of the training program. The change in the strength-training load relationship from before training to after training is shown in Figure 9.12. The increase in strength, which is evident as a shift along the x-axis, was 20% for the male subjects and 4% for the female subjects. In contrast, the training load increased (a shift along the y-axis) by 200% for the male subjects and 240% for the female subjects. Because strength was tested with an isometric contraction, whereas the training involved dynamic contractions, the greater increase in training load must have been due to an improvement in the coordination associated with the knee extension task; this adaptation would be mediated by the nervous system.

3. Coactivation—When subjects perform a maximum voluntary contraction, there is typically significant EMG activity in the antagonist muscle (Dimitrijevic et al., 1992). For example, Carolan and Cafarelli (1992) recorded EMG activity in the biceps femoris muscle that was as much as 22% of maximum during a maximum isometric contraction of the knee extensor muscles. Coactivation has a negative effect on the measurement of strength because the force measured during a strength test represents the net output (the difference) due to the activation of an agonist-antagonist group of muscles. Consequently, an apparent increase in strength (net force) may be due simply to a reduced amount of coactivation. Carolan and Cafarelli observed such an adaptation when subjects trained the knee extensors of one leg with maximum isometric contractions for 8 weeks. The coactivation of biceps femoris during the knee extensor task decreased by about 20% after 1 week of training, and this contributed to an increase in the measured knee extensor force (strength). Furthermore, the reduction in coactivation of biceps femoris was also observed in the contralateral (untrained) leg. Some of the increase in strength, therefore, was due to an alteration in the activation pattern of the involved muscles.

4. Cross education—A common observation in rehabilitation and in strength training is that the chronic manipulation of one side of the body can cause changes in the other side. The adaptation in motor capabilities that occurs in one limb as a consequence of training the contralateral limb is referred to as **cross education**. For example, cross education is evident in Figure 9.11,

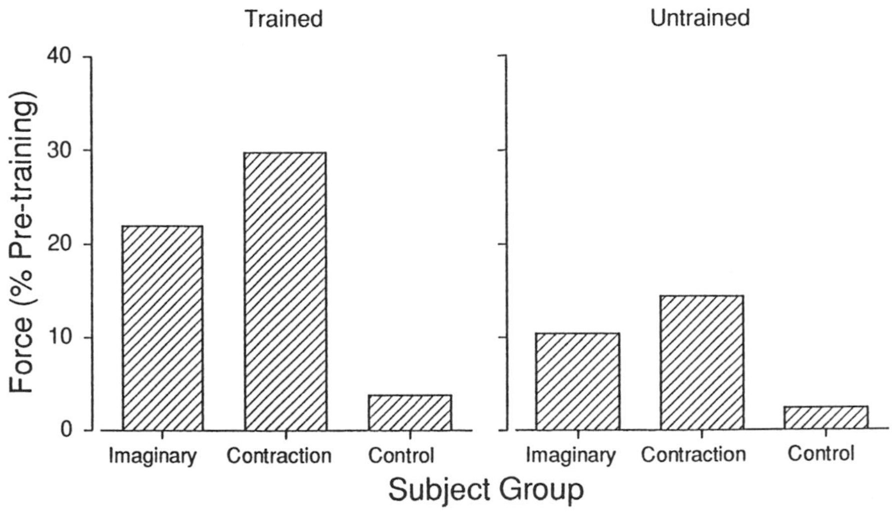

Figure 9.11 Average training-induced increases in the maximum isometric force of the trained (left) and untrained (right) fifth finger for three groups of subjects: One group imagined performing contractions, another group performed voluntary contractions, and the third (control) group did not train.
Note. Data from Yue and Cole (1992).

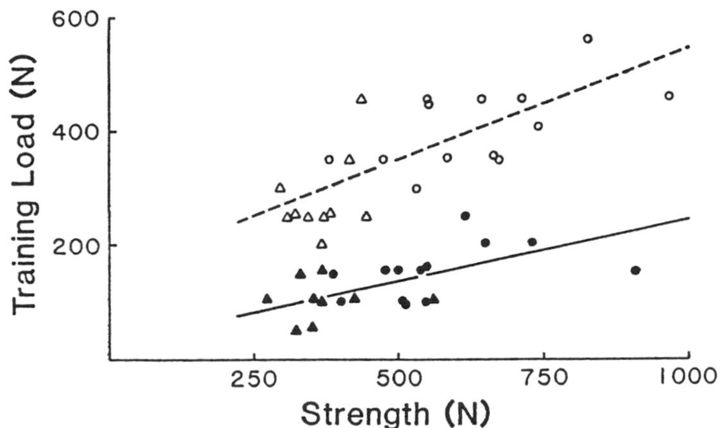

Figure 9.12 The relationship between strength and training load before (solid) and after (open) a 12-week training program by female (triangles) and male (circles) subjects.
Note. From "The Role of Learning and Coordination in Strength Training" by O.M. Rutherford and D.A. Jones, 1986, *European Journal of Applied Physiology*, **55**, p. 105. Copyright 1986 by Springer-Verlag, Inc. Adapted by permission.

which shows an increase in the strength of the right (untrained) hand as a result of training the left hand. Cross education is a robust effect that has been reported for a number of training programs (Figure 9.13), including imaginary contractions and the activation of muscle with both voluntary contractions and neuromuscular electrical stimulation. Because cross education occurs without activation (no EMG) of the muscles in the contralateral limb (Devine, LeVeau, & Yack, 1981; Panin, Lindenauer, Weiss, & Ebel, 1961) and without alterations in muscle fiber areas or enzyme activities (Houston, Froese, Valeriote, Green, & Ranney, 1983), the increase in strength of the untrained limb must be due to adaptations that occur in the nervous system. Furthermore, the amplitude of the EMG (normalized to the maximum evoked motor response, the M wave) associated with the maximum voluntary contraction increases in the untrained limb (Butler & Darling, 1990).

5. Bilateral deficit—In contrast to the facilitative interlimb effects observed with cross education, the concurrent activation of both limbs typically causes a reduction in strength. The decline in force that occurs during a maximum voluntary contraction under these conditions is referred to as a **bilateral deficit** (Howard & Enoka, 1991; Schantz, Moritani, Karlson, Johansson, & Lindh, 1989; Secher, Rube, & Elers, 1988). However,

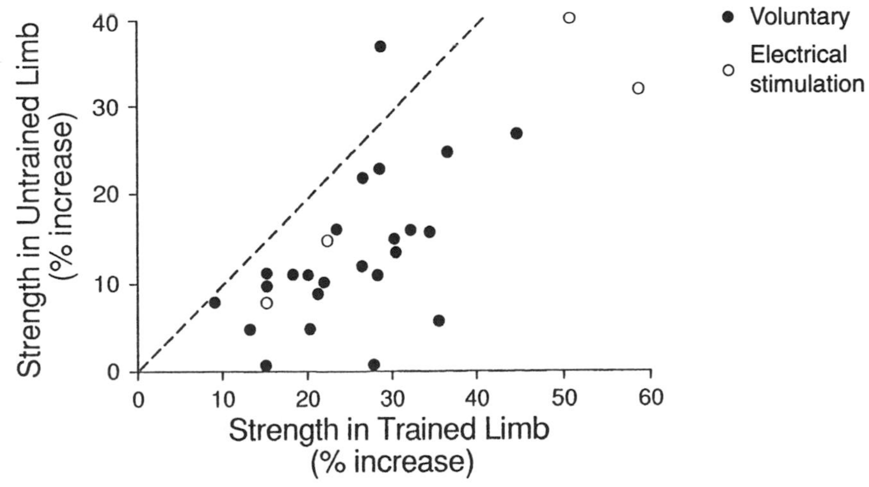

Figure 9.13 Results of 28 studies of cross education show the range of associations between increases in strength for the trained and the untrained limbs. Strength, measured as the maximum isometric force, was increased by voluntary contractions or neuromuscular electrical stimulation in several different muscles. Data points below the line of identity indicate that the strength increase in the trained limb was greater than that in the untrained limb.
Note. From "Muscle Strength and Its Development: New Perspectives" by R.M. Enoka, 1988b, *Sports Medicine*, **6**, p. 155. Copyright 1988 by ADIS Press Limited. Adapted by permission.

the bilateral deficit is limited to homologous muscles. An example of a bilateral deficit is shown in Figure 9.14 for a task that involved the isometric contraction of the elbow extensor muscles (triceps brachii) with the elbow at about a right angle. Each column in Figure 9.14 corresponds to one trial. In Trial 1 (left column), the subject performed a maximum isometric contraction with the right arm. In Trial 3 (right column), the subject performed a maximum isometric contraction with the left arm. In Trial 2 (middle column), the subject concurrently activated the elbow extensors in both arms maximally. Figure 9.14 shows the EMG for triceps brachii (the agonist) and biceps brachii (the antagonist). The critical feature of the graph is that both the maximum force exerted by each arm and the quantity of EMG were *less* during the bilateral contraction than the force and EMG during the single-limb trial. The reduction in force with a bilateral deficit averages about –5% to −10% of maximum. Because of the decline in agonist EMG and the increased amount of coactivation during the bilateral contraction, the bilateral deficit must include a contribution from the nervous system (Enoka, 1988b; Howard & Enoka, 1991; Koh, Grabiner, and Clough, 1993; Vandervoort, Sale, & Moroz, 1984).

Although most studies on bilateral interactions have reported a bilateral deficit, it appears that bilateral interactions are modifiable with training. For example, Howard and Enoka (1991) compared the maximum isometric force exerted during one- and two-limb knee extensor contractions and found a bilateral deficit for untrained subjects (–9.5%) and elite cyclists (-6.6%) but a bilateral facilitation for weight lifters (+6.2%). A **bilateral facilitation** means that the maximum isometric force occurred during the two-limb contraction. Furthermore, Howard (1987) found that 3 weeks of bilateral training by untrained subjects was sufficient to change a bilateral deficit (–3.7%) to a bilateral facilitation (+4.2%).

6. Reflex potentiation—When a maximal M wave is elicited in a muscle at rest, there is usually no H-reflex response. But when the same stimulus is applied to a muscle that is performing a voluntary contraction, there are two reflex responses (V_1 and V_2). V_1 occurs at the same latency as an H reflex, while V_2 has a longer latency (Upton, McComas, & Sica, 1971). The effect of strength training on these responses has been evaluated by calculating a potentiation ratio that consists of the amplitude of the response (V_1 and V_2) relative to the amplitude of the M wave in the resting muscle. Reflex potentiation does not occur in all muscles, is more pronounced in weight lifters and elite sprinters, and increases with strength training (Sale, 1988). The magnitude of reflex potentiation is presumed to provide an index of the motor unit activation achieved during a voluntary contraction.

7. Synchronization—One of the most frequently cited examples of neural adaptations that accompany strength training is a change in the amount of motor unit synchronization. This property describes the temporal relationship between the discharge of motor units. It is a measure of the temporal coincidence of action potentials from different motor units. A high degree of synchrony means that motor units tend to discharge action potentials at about the same time. The most appropriate way to test synchrony is to perform a cross-correlation analysis on the discharge of a pair of motor units (Datta & Stephens, 1990; Moore, Perkel, & Segundo, 1966; Nordstrom et al., 1992). This type of analysis has indicated that there is a modest degree of synchrony between motor units.

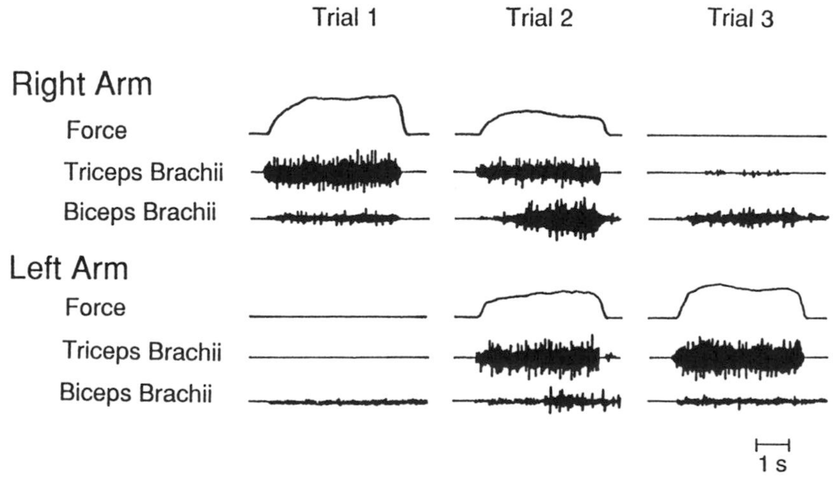

Figure 9.14 Maximum force and EMG of the elbow extensor muscles during maximum isometric contractions of the right arm, both arms, and the left arm.
Note. From "Decrease in Human Voluntary Isometric Arm Strength Induced by Simultaneous Bilateral Exertion" by T. Ohtsuki, 1983, *Behavioural Brain Research*, **7**, p. 169. Copyright 1983 by Elsevier Biomedical Press. Adapted by permission.

Another way to measure synchronization has been to compare the discharge of a single motor unit to a population of units (Milner-Brown, Stein, & Yemm, 1973). Because the discharge of the population of motor units is measured as the EMG, the technique has been referred to as the surface-EMG estimate of motor unit synchronization. With this technique, Milner-Brown et al. (1975) found that the amount of synchronization in a hand muscle increased after a 6-week program of strength training. The validity of this technique, however, has been questioned (Yue, Fuglevand, Nordstrom, & Enoka, in press). Therefore, the possible role of changes in motor unit synchronization with strength training remains to be determined.

The neural adaptations described in this section are important during the initial phase of a training program (Sale, 1988), especially for novice weight lifters, and contribute significantly to the changes that occur in patients with selected neuromuscular disorders (McCartney, Moroz, Garner, & McComas, 1988).

Muscular Adaptations

In contrast to the neural adaptations associated with changes in strength, it is readily apparent that changes in strength can be caused by mechanisms that increase muscle size (e.g., body builders, weight lifters, strength athletes). The increase in muscle size (cross-sectional area) can be caused by **hypertrophy**, an increase in the cross-sectional area of individual muscle fibers, or by **hyperplasia**, an increase in the number of muscle fibers. Most experimental evidence suggests that the typical response of human subjects to strength training involves hypertrophy but that hyperplasia may occur under some conditions (Antonio & Gonyea, 1993; Gonyea, Sale, Gonyea, & Mikesky, 1986; MacDougall, Sale, Alway, & Sutton, 1984; Sjöstrom, Lexell, Eriksson, & Taylor, 1991; Tamaki, Uchiyama, & Nakano, 1992).

The magnitude of the increase in *cross-sectional area* depends on several factors, including the initial strength of the individual, the duration of the training, and the training technique used (e.g., isometric, dynamic, eccentric). In novice subjects, 6 weeks of isometric training increased the cross-sectional area of the elbow flexors (biceps brachii, brachioradialis) by about 5% (Davies et al., 1988), whereas 8 weeks of isometric training increased the cross-sectional area of the quadriceps femoris by 15% (Garfinkel & Cafarelli, 1992). Similarly, 60 days of isokinetic training (2.1 rad/s) with the knee extensors increased the cross-sectional area of the quadriceps femoris by 9% (Narici et al., 1989). Furthermore, 19 weeks of eccentric-concentric knee extensor exercises produced a greater increase in cross-sectional area than did exercises that involved only concentric contractions (Hather, Tesch, Buchanan, & Dudley, 1991). In contrast, 24 weeks of dynamic training by experienced body builders failed to elicit an increase in the cross-sectional area of muscle fibers in biceps brachii (Alway, Grumbt, Stray-Gundersen, & Gonyea, 1992).

The effect of strength training on the different muscle fiber types can also be diverse (Figure 9.15) (Alway et al., 1989; Häkkinen, 1985; Hather et al., 1991). For example, 16 weeks of isometric training with the triceps surae did not alter the fiber-type proportions in either soleus or lateral gastrocnemius. In contrast, 19 weeks of a knee extension exercise produced an increase in the proportion of Type IIa muscle fibers and a decrease in the proportion of Type IIb muscle fibers in vastus lateralis. This included a reduction in the Type IIb myosin heavy chains, which presumably reflects a change in gene expression. Similarly, the 16-week isometric program increased the cross-sectional area of Type I (20%) and Type II (27%) muscle fibers in soleus and of the Type II fibers (50%), but not the Type I fibers, in lateral gastrocnemius. These observations show that not all the muscle fibers in an active synergist group of muscles receive the same training stimulus. Furthermore, the increase in cross-sectional area of muscle fibers is greater for an eccentric-concentric program (Type I, 14% increase; Type II, 32% increase) than for a concentric program (Type II, 27% increase).

Although strength training can induce hypertrophy, differences in muscle size account for only about 50% of the differences in strength between individuals. For example, Figure 9.16 shows the scatter of the data points about the regression line between the maximum isometric strength and cross-sectional area for quadriceps femoris. This raises the question of whether other changes, besides neural adaptations, can occur in muscle to account for differences in strength. One possibility is differences in **specific tension**. Recall that specific tension refers to the intrinsic strength of muscle and is measured as the force that muscle can exert per unit of cross-sectional area (N/cm^2). Because the force that a muscle can exert is equal to the product of cross-sectional area and specific tension, differences in specific tension can contribute to differences in strength between individuals.

Some investigators have suggested that specific tension may vary among the muscle fiber types, with Type II fibers having a greater specific tension than Type I fibers (Jones et al., 1989). Indeed, specific tension has been found to be greatest for Type FF motor units and least for Type S motor units (Bodine et al., 1987; Kanda & Hashizume, 1992; McDonagh et al., 1980). This finding means that the force exerted by the contractile apparatus and measured at the muscle tendon is greatest for Type FF motor units. Such an effect might be caused by an increase in the density of myofilaments in muscle fibers, but this does not appear likely (Claasen et al., 1989). Furthermore, the specific tension of isolated slow- and fast-twitch muscle fibers does not differ (24.5 vs. 24.3 N/cm^2) (Lucas et al., 1987). Therefore, the difference in specific tension for the motor unit types must be related to the way that force is transmitted from the muscle fiber to the muscle tendon. Recall the complex cytoskeleton that surrounds single muscle fibers (Figure 4.9) and is responsible for transmitting the force generated by the contractile apparatus to the intramuscular connective tissue. It appears that the characteristics of this tissue vary with fiber type (Kovanen, Suominen, & Heikkinen, 1984a, 1984b). The concentration and hence the tensile strength of endomysial colla-

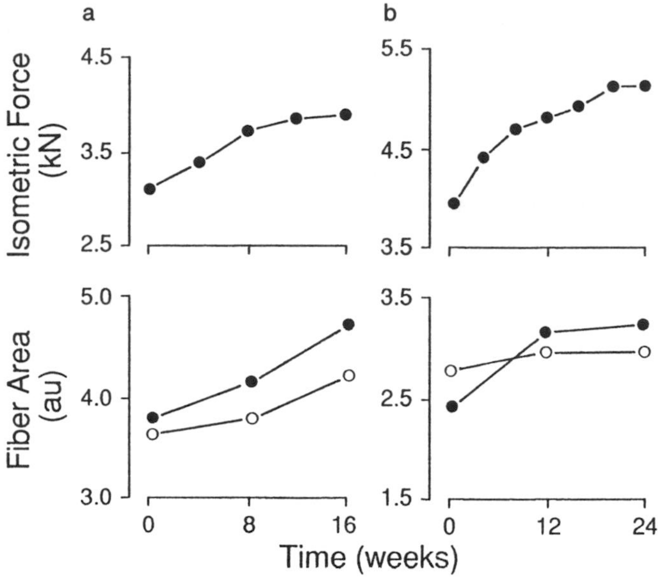

Figure 9.15 Effects on strength and muscle fiber area associated with (a) a continuous concentric-eccentric training program and (b) an intermittent concentric and concentric-eccentric program. Strength is measured as the maximum isometric force exerted by the knee extensors. The muscle fiber areas were measured for the Type I (open circles) and Type II (filled circles) fibers. The top row indicates the time course of the increase in strength, and the bottom row shows the change in muscle fiber area.

Note. From "Research Overview: Factors Influencing Trainability of Muscular Strength During Short Term and Prolonged Training" by K. Häkkinen, 1985, *National Strength & Conditioning Association Journal*, **7**, pp. 32-37. Copyright 1985 by the National Strength & Conditioning Association. Adapted by permission.

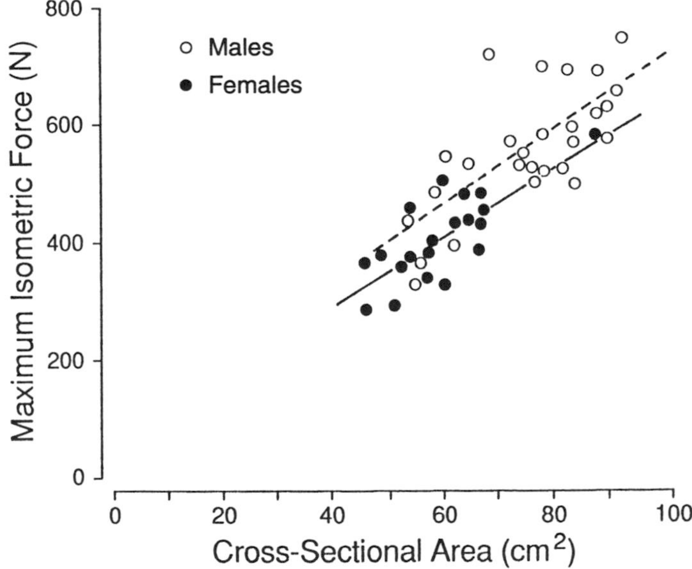

Figure 9.16 Relationship between the cross-sectional area of quadriceps femoris and the maximum isometric force (strength).

Note. From "Physiological Changes in Skeletal Muscle as a Result of Strength Training" by D.A. Jones, O.M. Rutherford, and D.F. Parker, 1989, *Quarterly Journal of Experimental Physiology*, **74**, p. 245. Copyright 1989 by The Physiological Society. Adapted by permission.

gen are significantly greater for slow-twitch muscle fibers than for fast-twitch fibers. Furthermore, these properties are adaptable with training. These observations suggest that specific tension varies with muscle fiber type, and that this effect may be due to differences in connective tissue rather than the contractile apparatus.

Although it is widely accepted that muscle hypertrophy requires a change in the ratio of protein synthesis to degradation, the mechanisms controlling the ratio appear complex and have not been clearly defined. The potential stimuli for muscle hypertrophy include hormonal, metabolic, and mechanical factors (Jones et al., 1989). *Hormonal stimuli* (e.g., insulin, growth hormone, testosterone) are unlikely to provide the key stimulus for hypertrophy, because unlike the distribution of a hormone, the change in cross-sectional area can be limited to one muscle fiber type or to one muscle. In addition, the increases in muscle size, strength, and muscle protein that occur with strength training are not further enhanced by the administration of growth hormone (Yarasheski et al., 1992). Nonetheless, hormones may have a permissive effect for other factors. The thyroid hormone (T3), for example, seems to have a significant effect on the expression of Type I and Type IIa myosin heavy chains (Caiozzo, Herrick, & Baldwin, 1991, 1992; Fitzsimons, Herrick, & Baldwin, 1990).

Activity that stresses *metabolic factors* tends to result in improvements in endurance rather than in strength. Furthermore, the metabolic cost of eccentric contractions is less than that of concentric contractions, yet the eccentric activity provides a greater stimulus for muscle hypertrophy.

In contrast, evidence suggests that *mechanical stimuli* are important for muscle hypertrophy (Vandenburgh, 1992). The mechanism by which this occurs involves the application of a mechanical stimulus (e.g., stretch, contraction) that leads to the release of a second messenger and the subsequent modulation of the rates of protein synthesis and degradation. For example, intermittent stretch of skeletal muscle cells in culture results in an increase in the synthesis of different prostaglandins that can modulate protein synthesis and protein degradation (Vandenburgh, Hatfaludy, Sohar, & Shansky, 1990). There are two classes of second messengers: (a) extracellular matrix molecules (e.g., proteoglycans, collagen, laminin, fibronectin) that surround cells, transmit the mechanical stimulus to the cell surface, and influence nuclear events and cell growth and (b) stretch-induced alterations in plasma membrane–associated molecules (e.g., Na^+-K^+ ATPase, ion channels, phospholipases, G proteins) and associated cytoplasmic second messengers (e.g., prostaglandins, cAMP, inositol phosphates, intracellular Ca^{2+}, protein kinase C).

Mechanical stimuli influence not only the quantity but also the quality of muscle tissue (protein) that is synthesized. This is accomplished by regulating the expression of genes that determine muscle fiber phenotype, which depends on the protein-isoform genes that are transcribed. This effect has been examined by determining the consequences of mechanical stimuli on the myosin heavy chain genes (Goldspink et al., 1992). These genes encode the myosin crossbridge and are different for slow- and fast-twitch muscle (Table 4.3). Goldspink and colleagues concluded that the fast myosin heavy chain gene is the default gene and that the expression of the slow myosin heavy chain depends on the mechanical environment. For example, they found that stimulation regimes eliciting maximal force, especially when the protocol involved muscle stretch, were effective at resulting in the expression of the slow myosin heavy chain and repressing the expression of the fast myosin heavy chain. However, much remains to be learned about the connection between mechanical stimuli and the activation or repression of different muscle genes.

In addition to potential hormonal, metabolic, and mechanical factors in the hypertrophic response, muscular adaptations are also determined by developmental factors. Cells undergoing myogenesis in the absence of neural input develop distinct subpopulations of myotubes (which subsequently become mature muscle fibers) that have a limited capacity to change phenotype (Hoh, 1991; Hoh & Hughes, 1988). Furthermore, this fiber-type differentiation produces muscle fibers that can respond to certain physiological perturbations (e.g., weight training, unloading, denervation) and those that do not respond: *responders* and *nonresponders*. The exact proportions of these fiber types in different muscles and the variability between individuals remains to be determined. Consequently, a given training stimulus is unlikely to elicit the same hypertrophic response in all muscles or to be similar for different individuals.

Muscle Power

At the beginning of this chapter, we noted that strength and power are measures of the output of the motor system. One useful way of distinguishing between these two parameters is to consider the force-velocity relationship of muscle (Figure 6.19). We characterized the force-velocity relationship as comprising four distinct regions: velocity = 0, force = 0, velocity > 0, and velocity < 0. We defined strength as the point on the force-velocity curve where velocity = 0, whereas the power produced by the motor system corresponds to the region of velocity ≠ 0.

Many investigators have traditionally quantified performance capabilities by *static strength* and *dynamic strength*. These terms both represent measures of resultant muscle torque, the difference being whether or not muscle length changes. We know from the torque-velocity relationship that the maximum torque a muscle can exert decreases as the velocity of shortening increases. This means that measures of dynamic strength depend to a great extent on the rate at which muscle length changes. Furthermore, a thorough characterization of dynamic strength requires that the maximum torque be measured for a number of different speeds. These procedures can be simplified, however, by measuring the peak power produced by the system, which represents the combination of force and velocity and produces the maximum mechanical effect.

Let us review the material that we covered earlier and ask, What features of the motor system influence its ability to produce power? Recall from chapter 3 that power can be determined as the product of force and velocity. As a result, those factors that affect either muscle force or the velocity of shortening will determine the power that can be produced. Given an adequate neural input to muscle, *the principal determinants of power production are the number of in-parallel activated muscle fibers and the rate at which the myofilaments can convert energy into mechanical work.* The force that muscle can exert is proportional to the number of force-generating units in parallel (chapter 6); muscle force increases with cross-sectional area. Although power production is maximal when muscle force is about one third of maximum (Figure 6.24), power production increases as muscle becomes stronger (cross-sectional area increases) and hence the one-third value increases. A similar rationale applies to the effect of muscle velocity on power production. The maximum speed at which a muscle can shorten (v_{max}) is determined by the enzyme myosin ATPase. This enzyme controls the speed of the interaction between actin and myosin and hence the rate of crossbridge cycling. The quantity of myosin ATPase activity can change with alterations in physical activity levels. Power production is maximal when the velocity of shortening is about one quarter of v_{max}.

Power Production and Movement

Success in many athletic endeavors depends critically on the performer's ability to sustain the maximum power production possible for the duration of the event. Of course, the maximum sustainable power is inversely related to the duration of the event (Figure 3.27). Despite this significant role for power production, it is one of the least examined biomechanical parameters in the analysis of human movement. This lack of attention can be attributed partly to the difficulty associated with measuring power and the abstract nature of the parameter. The next few sections provide examples that address these two issues.

Power Production and Whole-Body Tasks.

The power produced by the motor system can be determined by a task performance (e.g., vertical jump, weight lifting), with the use of an ergometer, or by an isolated-muscle experiment. The evaluation of a task performance provides an index of whole-body power. The vertical jump is commonly used for this purpose. In this approach, a subject performs a vertical jump on a force platform so that we can measure the ground reaction force (Figure 9.17). From the ground reaction force we can determine the average force ($\bar{\mathbf{F}}$) applied to the ground and the average velocity ($\bar{\mathbf{v}}$) of the center of gravity and hence determine the average power produced during the jump (Bosco & Komi, 1980). To focus on only the ability of the subject to produce power, we confined the subject to a squat jump that involves only a concentric contraction and not an eccentric-concentric contraction, which is characteristic of the countermovement jump. We obtain ($\bar{\mathbf{F}}$) by dividing the magnitude of the impulse (shaded area in Figure 9.17) by the duration of the impulse. From Figure 9.17, these values are approximately as follows:

$$\text{impulse} = 179 \text{ N·s}$$
$$\text{time} = 0.29 \text{ s}$$
$$\bar{\mathbf{F}} = 617 \text{ N}$$

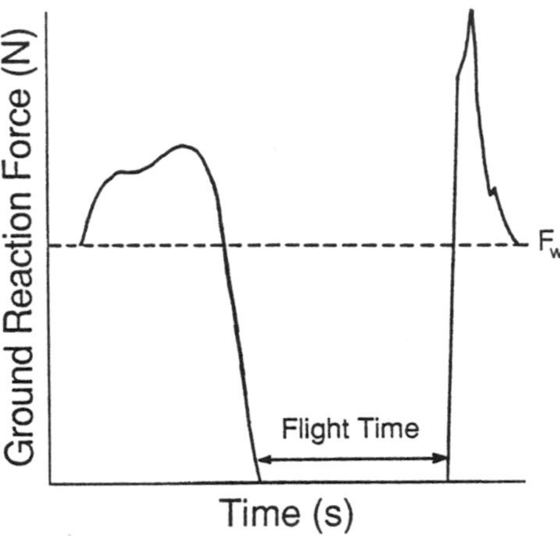

Figure 9.17 Vertical component of the ground reaction force during a squat jump. The subject performed a vertical jump for maximum height.

To get an approximate value of the average velocity ($\bar{\mathbf{v}}$) during the takeoff phase of the jump, we can use the equations of chapter 1. If we assume that the jumper left the ground and landed again in approximately the same position, then one half of the flight time was spent going up, and one half was spent coming down. Consider only the second half of the flight phase, the coming-down phase. From Equation 1.3, we can determine the landing velocity ($\mathbf{v}_f$) because we know the acceleration due to gravity (–9.81 m/s²), the initial velocity was 0 (velocity at the peak), and we can measure from Figure 9.17 that one half of the flight time was 0.20 s:

$$\mathbf{v}_f = \mathbf{v}_i + \mathbf{a}t$$
$$= 0 + (9.81 \times 0.20)$$
$$\mathbf{v}_f = 1.96 \text{ m/s}$$

The average velocity ($\bar{\mathbf{v}}$) during the takeoff phase is then determined as the average of the initial and final velocities:

$$\bar{\mathbf{v}} = \frac{\mathbf{v}_f + \mathbf{v}_i}{2}$$

$$= \frac{1.96 + 0}{2}$$

$$= 0.98 \text{ m/s}$$

Finally, the average power produced by the jumper during the takeoff phase is determined as the product of the average force and the average velocity.

$$\text{average power} = \bar{\mathbf{F}} \times \bar{\mathbf{v}}$$

$$= 617 \times 0.98$$

$$= 605 \text{ W}$$

Note: 1 W = 1 J/s = 1 N·m/s

This approach indicates that the average power produced by the jumper during the squat jump shown in Figure 9.17 was 605 W. Because the arms were not used in the squat jump, the value mainly represents the effect of the lower extremity muscles. For comparison, van Ingen Schenau, Bobbert, Huijing, and Woittiez (1985) measured the torque-velocity relationship and determined the maximum power produced by the plantarflexor muscles during the vertical jump to be 2,499 W. Similarly, Josephson (1993) estimated the maximum power for muscle to range from 7 to 500 W/kg, with the sustainable power ranging from 5 to 150 W/kg.

The power produced during weight-lifting events can be calculated by determining the rate of change in mechanical energy (Garhammer, 1980). Because the quantity of work done can be determined from the change in energy, and power equals the rate of doing work, power can be calculated from the rate of change in energy:

$$\text{power} = \frac{\text{change in energy}}{\text{time}}$$

$$= \frac{E_{k,t} + E_{p,g}}{\text{time}}$$

$$= \frac{0.5 \text{ mv}^2 + \text{mgh}}{\text{time}} \quad (9.1)$$

where $E_{k,t}$ refers to the maximum kinetic energy, $E_{p,g}$ represents potential energy, and time indicates the time to maximum kinetic energy.

An example of this approach is provided in Table 9.1. This table indicates the power delivered to a barbell by a weight lifter during three lifts: (a) the clean, a movement that requires the displacement of the barbell from the floor to the lifter's chest in one continuous rapid motion; (b) the squat, a movement in which a lifter supports a barbell on the shoulders and goes from a standing erect position to a knee-flexed position (thighs parallel to the floor) and then returns to standing; and (c) the bench press, a movement in which the lifter, in a supine position, lowers the barbell from a straight-arm location above the chest down to touch the chest and then raises it back to the initial position. Data have been taken from the literature (clean—Enoka, 1979; squat—McLaughlin, Dillman, & Lardner, 1977; bench press—Madsen & McLaughlin, 1984) to illustrate the differences in power production for the three lifts (Table 9.1).

Table 9.1 Power Delivered to a Barbell During the Three Types of Lifts

	Clean	Squat	Bench press
Barbell weight (N)	1,226	3,694	1,815
Peak barbell velocity (m/s)	2	0.30	1.54
Height (m)	0.30	0.30	0.06
Time to $E_{k,t}$ (s)	0.35	1.30	0.70
$E_{k,t}$ (J)	250	17	219
$E_{p,g}$ (J)	368	1,108	109
Power (W)	1,766	865	469

These data suggest that the peak power delivered to the barbell is greatest during the clean and least for the bench press. This is interesting because the squat and bench press are two of the three lifts that make up the sport of power lifting. Indeed, these data suggest that success in the squat and bench press lifts is not determined solely by power production. Furthermore, we could speculate that success in power lifting is related more to strength, as we have defined it.

Production of Power About a Single Joint. In contrast to the methods of estimating whole-body or whole-limb average power, an ergometer allows us to focus more specifically on one muscle group (Grabiner & Jeziorowski, 1992). Such an ergometer permits the measurement of the torque-angular velocity relationship about a joint and enables the calculation of the associated power. For example, DeKoning, Binkhorst, Vos, and van't Hof (1985) measured the torque-angular velocity relationship about the elbow joint for three groups of subjects: untrained females, untrained males, and arm-trained males (i.e., track-and-field athletes, rowers, weight lifters, body builders, handball players, karate exponents, and tug-of-war competitors). Figure 9.18 shows that the torque-velocity curves are different for the three groups of subjects. The strength ($\mathbf{P}_o$ = isometric torque) for the groups varies in the order that would be expected: arm-trained males, untrained males, and untrained females. There was, however, only a small difference in the maximum angular velocity (ω_{max}) among the groups (Table 9.2). Nonetheless, there was a significant difference among the groups in the peak power that the elbow-flexor muscles could produce. The values for power were calculated from the torque-velocity data shown in Figure 9.18.

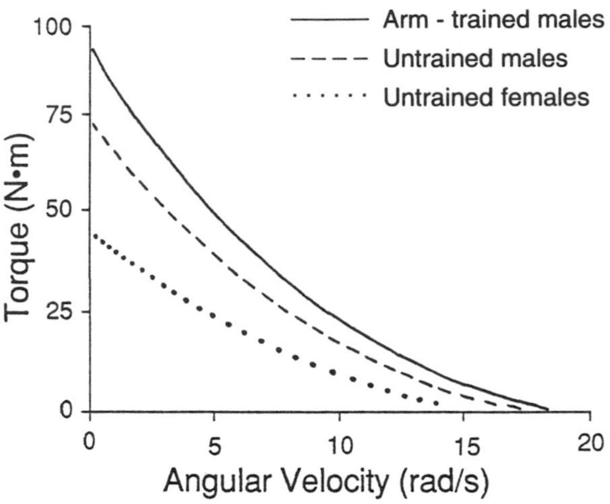

Figure 9.18 Relationships between torque and angular velocity for the elbow flexor muscles of three groups of subjects.
Note. From "The Force-Velocity Relationship of Arm Flexion in Untrained Males and Females and Arm-Trained Athletes" by F.L. DeKoning, R.A. Binkhorst, J.A. Vos, and M.A. van't Hof, 1985, *European Journal of Applied Physiology*, **54**, p. 92. Copyright 1985 by Springer-Verlag, Inc. Adapted by permission.

Another approach that can be used to determine power production about a single joint is to perform a biomechanical analysis and to calculate the product of torque and angular velocity about a joint (Enoka, 1988a; Winter, 1983). This approach requires a description of limb kinematics and data on a contact force (e.g., ground reaction force, pedal force). For example, the power that a subject can apply to the pedal of a cycle ergometer can be determined by calculating the product of pedal force and pedal velocity or by summing the joint power (torque × angular velocity) for the hip, knee, and ankle joints (van Ingen Schenau, van Woensel, Boots, Snackers, & de Groot, 1990). These two methods produce similar results and indicate the power produced by the leg throughout one complete pedal revolution (Figure 9.19a). However, the individual joint powers provide the additional information on the contribution of the three joints to the total leg power (Figure 9.19b).

Power Production for Isolated Muscle. Although there are few reports on the topic in the literature due to the technical demands of the experimental protocol, the power production capabilities of the motor system can also be measured at the level of the single muscle or muscle fiber. Brooks and Faulkner (1991) compared the ability of slow-twitch muscle (soleus) and fast-twitch muscle (extensor digitorum longus) of mice to sustain isometric force and to sustain power production. The peak isometric force that extensor digitorum longus could exert was greater than that for soleus (363 vs. 273 mN). However, when force was expressed relative to muscle size (cross-sectional area), peak force was similar for the two muscles (24.5 vs. 23.7 N/cm^2 for extensor digitorum longus and soleus, respectively). Nonetheless, extensor digitorum longus was more fatigable and experienced a more rapid decline in isometric force with sustained activation; maximum sustainable force was 1.38 N/cm^2 for extensor digitorum longus and 4.58 N/cm^2 for soleus. In contrast, when power production was expressed relative to muscle mass, the peak sustainable power was greater for extensor digitorum longus (9.1 W/kg) compared with soleus (7.4 W/kg). Although force declined more for extensor digitorum longus with sustained activation, the greater velocity of shortening resulted in a greater power. This observation suggests that Type S and Type FR motor units are important for the ability to sustain isometric force, but Type FR and Type FF motor units are more important for sustaining power production (see also, Rome et al., 1988).

Eccentric-Concentric Contractions. Power is defined as the rate of doing work—that is, the amount of work done per unit time. Because work is equal to the change in energy, power can also be considered in terms of the *rate of change in energy*. The power that

Table 9.2 Elbow Flexor Strength (P_o), Maximum Angular Velocity (ω_{max}), and Peak Power for Arm-Trained Males and Untrained Males and Females

	Arm-trained males	Untrained males	Untrained females
P_o (N·m)	90.9 ±15.6	68.5 ±11.0	42.7 ±6.9
ω_{max} (rad/s)	17.0 ±1.6	16.6 ±1.5	14.9 ±1.3
Power (W)	253.0 ±58.0	195.0 ±46.0	111.0 ±24.0

Note. Values are given as mean ± *SD*.

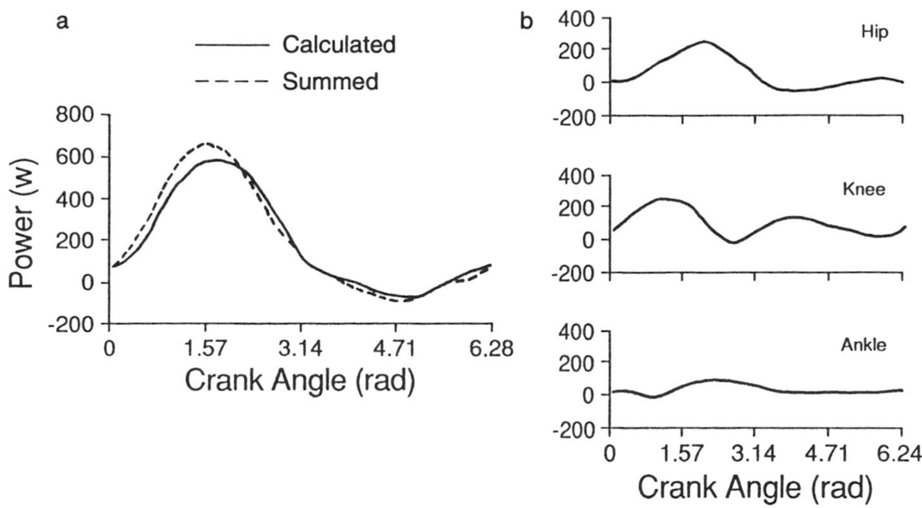

Figure 9.19 Power production during a single pedal cycle on a cycle ergometer: (a) power produced as a function of crank angle; (b) joint power (product of torque and angular velocity) for the hip, knee, and ankle as a function of crank angle. *Note.* From "Determination and Interpretation of Mechanical Power in Human Movement: Application to Ergometer Cycling" by G.J. van Ingen Schenau, W.W.L.M. van Woensel, P.J.M. Boots, R.W. Snackers, and G. deGroot, 1990, *European Journal of Applied Physiology*, **61**, p. 13. Copyright 1990 by Springer-Verlag, Inc. Adapted by permission.

a muscle can produce depends on how quickly the energy is used to perform work. The principal source of energy for muscle is chemical energy (ATP). However, we have also discussed how the muscle can use mechanical (elastic) energy to perform positive work and produce power; this is known as the storage and utilization of elastic energy and occurs during an eccentric-concentric contraction. When a muscle performs an eccentric contraction before a concentric contraction, it is able to perform more positive work during the shortening contraction. This enhanced performance is attributed, at least partly, to the ability of muscle to store energy during the stretch (eccentric contraction) and then use this energy during the concentric contraction.

The height that an individual can reach in a vertical jump depends on how much power the muscles can produce. The most effective vertical jump technique is the countermovement jump, which involves an eccentric-concentric contraction of the quadriceps femoris and triceps surae muscles. This association indicates that the additional energy provided by the eccentric-concentric contraction represents a significant contribution to the power produced by these muscles. This is probably why we intuitively use eccentric-concentric contractions for most movements.

Power Production and Training

Because of the technical demands associated with measuring joint power, few studies have considered the effects of training on power production. Duchateau and Hainaut (1984) performed one such study on a hand muscle (adductor pollicis) of human volunteers. Two groups of subjects trained the muscle at a moderate intensity (10 repetitions) every day for 3 months. One group used maximum isometric contractions, and the other group did rapid dynamic contractions against a load that was 30% to 40% of maximum. The moderate load for the dynamic contractions was chosen because we know that muscle produces maximum power when the contraction force is about one third of maximum.

The effects of the training protocols on the force-velocity relationship were consistent with the principle of specificity. The group that trained with the greater loads experienced a significant increase (dashed lines) in the maximum isometric force, whereas the other group did not (Figure 9.20). Surprisingly, however, the isometric group who trained with maximum contractions produced a 51% increase in maximum power compared with a 19% increase for the dynamic group (Figure 9.20). This finding suggests that the greatest increase in power is obtained with high-force rather than high-velocity training. In contrast, other studies have found that training loads at 30% of maximum produce greater increases in power production than do loads of either 0% or 100% (Moritani, 1992). Furthermore, Moritani has reported that power training is accompanied by significant changes in the EMG, including an increase in the quantity of EMG and a decrease in the mean power frequency. Taken together, these observations suggest that a training-induced increase in power production involves a neural adaptation but that the most effective load for increasing power production remains uncertain.

Adaptation to Reduced Use

One popular technique used by experimentalists to identify the fundamental properties of the motor system is to perturb the system and measure the response. A typical

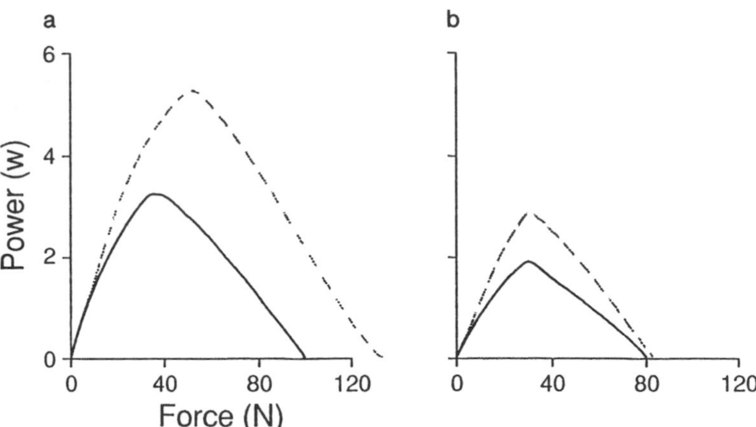

Figure 9.20 The effects of (a) isometric and (b) dynamic training on muscle power production. Isometric training involved a 5-s maximum isometric contraction each minute for 10 min, every day for 3 months. Dynamic training involved the same number of repetitions and length of training, but the exercise comprised a rapid concentric movement.
Note. From "Isometric or Dynamic Training: Differential Effects on Mechanical Properties of a Human Muscle" by J. Duchateau and K. Hainaut, 1984, *Journal of Applied Physiology*, **56**, p. 298. Copyright 1984 by the American Physiological Society. Adapted by permission.

perturbation involves altering the amount of activity performed by the system; this can involve either an increase or a decrease in the amount of activity. An assumption frequently expressed in the research literature is that the excitation of muscle by the nervous system is a critical factor in determining the properties of the muscle. Consequently, several experimental models have been developed to alter the connection between the nervous system and muscle and, by subtraction, to determine the role of the nervous system in defining muscle properties (Table 9.3). In this section, we consider, as examples, three commonly used models of reduced use and examine the types of adaptations that have been observed with each model. (For information on the other models, see Gordon & Pattullo, 1993; Lieber, 1992; Pette & Vrbová, 1992; Roy, Baldwin, & Edgerton, 1991.) The profound effects of reduced use on the structural components of the neuromuscular system (e.g., bone, articular cartilage, tendon, ligament) are not considered here, but there is also substantial literature on these adaptations (Snow-Harter & Marcus, 1991; Zernicke et al., 1990).

Table 9.3 Experimental Models Used to Study the Effects of Changes in Activity

	Neural	Muscle
Increased use	Electrical stimulation	Stretch
	Exercise	Exercise
		Compensatory hypertrophy
Reduced use	Denervation	Tenotomy
	Tetrodotoxin	Immobilization
	Curare	Hindlimb suspension
	Colchicine	Barbiturate sleep
	Rhizotomy	Bed rest
	Spinal transection	Water immersion
	Spinal isolation	

Limb Immobilization

Individuals who have an injured limb immobilized in a cast for a few weeks often experience loss of muscle mass and function, which is apparent upon removal of the cast. This adaptation is of obvious concern to the clinician, who would like to know how to minimize the loss of mass and function (Gardiner & Lapointe, 1982; Videman, 1987). However, the scientist regards this adaptation as an opportunity to characterize the changes that occur in the motor system with this type of stress (reduction in activity) and to identify the mechanisms that mediate these changes.

Reduction in Activity. Several studies conducted on both animals and humans have examined the adaptations that occur with limb immobilization. Before we can compare these adaptations to those reported for other models, we need to characterize the effect of immobilization on the daily activation patterns of muscles. As would be expected, when a rat hindlimb was immobilized for 4 weeks with an external brace, there was a substantial decline the amount of activity (EMG) recorded in the soleus and medial gastrocnemius muscles over a 24-hr period (Figure 9.21a) (Fournier, Roy, Perham, Simard, & Edgerton, 1983). However, the amount of decline in the EMG depended on the length at which the muscle was immobilized; the decline was greatest for short muscle lengths and negligible for long lengths. The integrated EMG decreased by 77% for soleus and

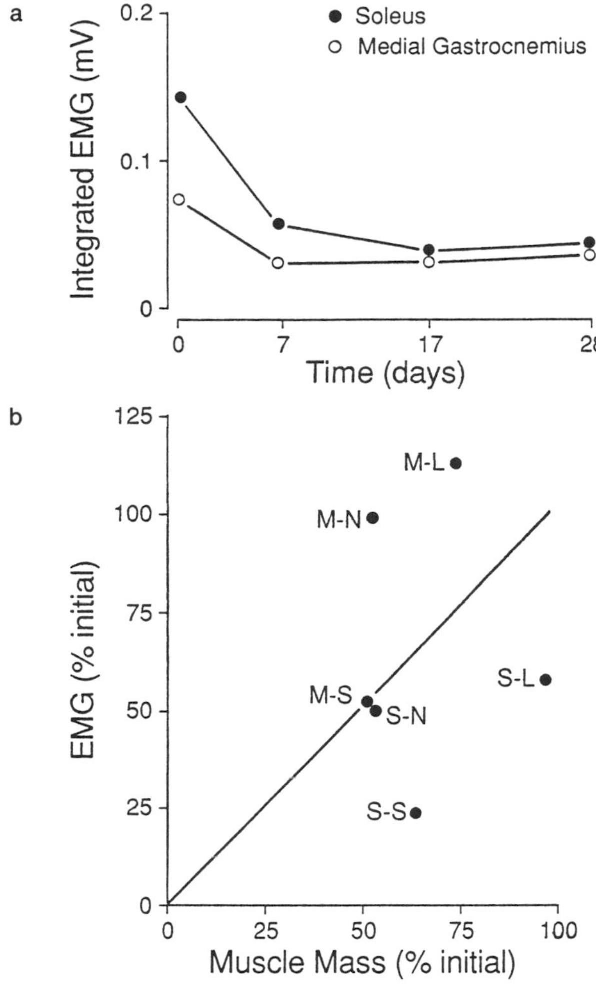

Figure 9.21 Reduction in activity and muscle mass produced by 4 weeks of limb immobilization of the rat hindlimb: (a) decline in integrated EMG for the soleus and medial gastrocnemius muscles constrained to be at a short length; (b) relationship between decline in EMG and reduction in muscle mass. S-S = soleus, short length; S-N = soleus, neutral length; S-L = soleus, long length; M-S = medial gastrocnemius, short length; M-N = medial gastrocnemius, neutral length; M-L = medial gastrocnemius, long length.

Note. From "Is Limb Immobilization a Model of Muscle Disuse?" by M. Fournier, R.R. Roy, H. Perham, C.P. Simard, and V.R. Edgerton, 1983, *Experimental Neurology*, **80**, p. 153. Copyright 1990 by Academic Press, Inc. Adapted by permission.

50% for medial gastrocnemius at the shortest lengths. These reductions in EMG were accompanied by a 36% loss of muscle mass for soleus and a 47% loss of muscle mass for medial gastrocnemius at the shortest lengths.

The amount of atrophy (loss of muscle mass) reported in the literature is quite variable for limb immobilization. The reduction in muscle mass (wet weight) has been recorded as high as 50% (Fournier et al., 1983) and the decrease in muscle fiber cross-sectional area as much as 42% (Nicks, Beneke, Key, & Timson, 1989) with a few weeks of limb immobilization. Others, however, have reported a modest reduction of 17% in wet weight (Robinson, Enoka, & Stuart, 1991) and a 14% decrease in the mean diameter of Type I but not Type II muscle fibers (Gibson et al., 1987) for a similar duration of immobilization. The atrophy is probably due both to a decline in the rate of protein synthesis (Gibson et al., 1987) and a loss of muscle fibers (Oishi, Ishihara, & Katsuta, 1992).

Despite the reduction in EMG and the muscle atrophy that has been reported to occur with limb immobilization, these results are difficult to interpret because of the dissociation that has been observed between the decline in EMG and the muscle atrophy and the dissociation between the muscle atrophy and loss of function. These effects are apparent in Figure 9.21b, in which several data points deviate from the line of identity. Fournier et al. (1983) found no reduction in EMG for the medial gastrocnemius muscle when it was immobilized at a neutral length, but the muscle experienced a 54% decline in mass. Similarly, rat soleus and medial gastrocnemius muscles experienced similar declines in wet weight (43% to 52%) after 4 weeks of limb immobilization at short and neutral lengths. However, the decline in maximum isometric force when both muscles were immobilized at a short length was 72% to 77% compared to 45% at the neutral length (Simard, Spector, & Edgerton, 1982; Spector, Simard, Fournier, Sternlicht, & Edgerton, 1982). The decrease in maximum isometric force did not parallel the reduction in wet weight.

Some of the dissociation between the muscle atrophy (loss of mass) and the decline in force can be explained by the use of loss of mass as an index of atrophy. As we discussed previously, the maximum force a muscle can exert is closely related to its cross-sectional area, not to the quantity of muscle mass (weight). When muscle atrophy is expressed in terms of cross-sectional area, there is a much tighter correlation with the decline in force (Lieber, 1992). However, a dissociation has also been reported for Type S and Type FR motor units in a cat hindlimb muscle (Nordstrom, Enoka, Callister, Reinking, & Stuart, 1993). There was a decline in cross-sectional area of about 25% for both the Type I and IIa muscle fibers but a reduction in the maximum isometric force of 40% for Type FR and 52% for Type S motor units. Clearly, the remodeling that occurs in the neuromuscular apparatus during short-term immobilization is more complex than can be predicted by a linear relationship between the decline in EMG, loss of muscle mass, and impairment of performance. Probably the altered neuromechanical conditions (e.g., fixed muscle length, possibility of frequent isometric contractions, altered sensory feedback) are just as important as the reduction in neuromuscular activity in determining the nature of the adaptations.

Neuromuscular Adaptations. Given the absence of a simple relationship between decreased activity and impaired performance with limb immobilization, scientists have examined the adaptive processes in more detail. Most elements of the system will adapt, given the appropriate stress. For example, 7 days of immobilization of the rat soleus muscle at a short length reduced muscle mass by 37%, depolarized muscle fiber membranes by 5 mV, decreased the frequency of miniature end-plate potentials by 60%, and reduced Na^+-K^+ transport across the membrane by 25% (Zemková et al., 1990). Similarly, 3 weeks of immobilization of the rat plantaris muscle at a long length increased the postsynaptic areas of junctional folds and clefts at the neuromuscular junctions of Type I and Type II muscle fibers (Pachter & Eberstein, 1986).

One observation that has been reported in numerous limb-immobilization studies is the *slow-to-fast conversion* of muscle fiber types. There is a decline in the proportion of Type SO muscle fibers and an increase in the proportion of Type FOG fibers that seems to consist of a conversion of the fiber enzyme pattern of the Type FOG fibers (Fitts, Brimmer, Heywood-Cooksey, & Timmerman, 1989; Lieber, Fridén, Hargens, Danzig, & Gershuni, 1988; Oishi et al., 1992). The typical explanation for this adaptation is that the muscle fibers most affected by the immobilization are those whose activity is reduced the most, namely, the Type SO muscle fibers. Despite the appeal of this rationale and its observation in muscles of the rat, dog, and human, motor unit studies on the cat hindlimb muscle have not found a similar change in motor unit proportions or a differential reduction in cross-sectional area after several weeks of immobilization (Mayer et al., 1981; Nordstrom et al., 1993; Robinson et al., 1991). This effect, however, may be due to maintained EMG activity in the immobilized cat hindlimb (Mayer et al., 1981). Nonetheless, the decline in force seems to be greatest in Type S and Type FR motor units of the cat hindlimb muscle, which is consistent with the slow-to-fast hypothesis.

Limb immobilization has a profound effect on performance. The muscle atrophy associated with limb immobilization results in a substantial loss of strength and an impairment of most activities of daily living (Imms, Hackett, Prestidge, & Fox, 1977; Imms & MacDonald, 1978). For example, 6 weeks of immobilization due to a fracture produces a 55% decline in the maximum voluntary contraction (MVC) force and a 45% reduction in the MVC EMG in a hand muscle (Duchateau & Hainaut, 1991). This duration of immobilization appeared to inhibit the ability of subjects to generate a central drive that was sufficient to maximally activate the hand muscle. Nonetheless, these changes did not impair the ability of the subjects to sustain their maximal force for 60 s. Similarly, Davies, Rutherford, and Thomas (1987) found that 3 weeks of immobilization in a plaster cast produced a reduction in the electrically elicited (10%) and the MVC force (23%) but did not affect the fatigability of the triceps surae muscle in human volunteers.

The consequences of limb immobilization are also apparent at the level of the motor unit. Six to 8 weeks of immobilization alters both the properties and behavior of motor units in human hand muscles (Duchateau & Hainaut, 1990). When recruitment threshold was expressed relative to maximum force, there was an increase in the number of high-threshold motor units in the immobilized muscle. However, the average force

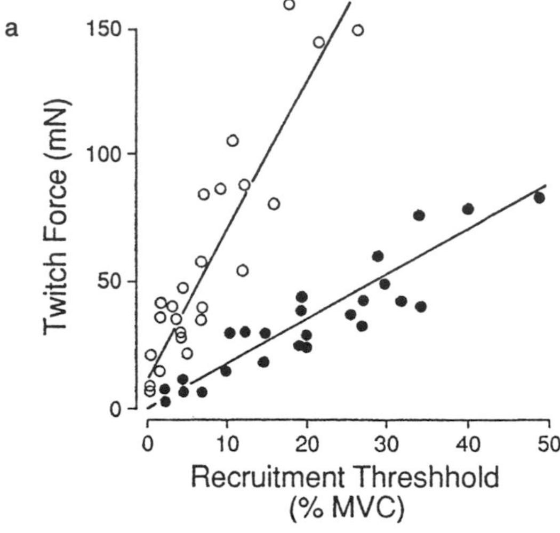

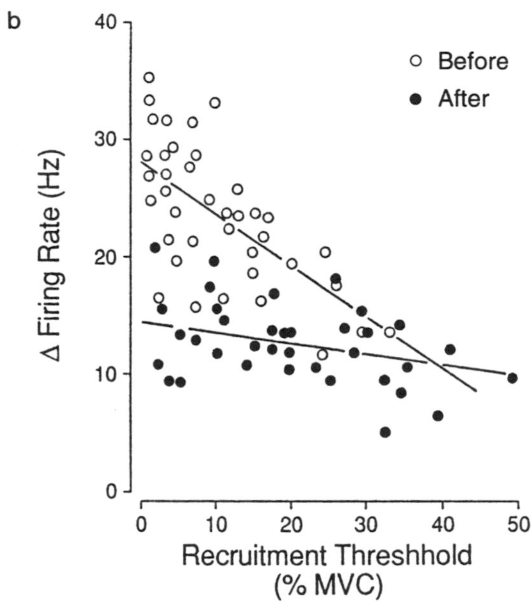

Figure 9.22 Immobilization-induced change in motor unit twitch force and range of discharge rate modulation as a function of recruitment threshold. Each data point corresponds to a single motor unit.

Note. From "Effects of Immobilization on Contractile Properties, Recruitment, and Firing Rates of Human Motor Units" by J. Duchateau and K. Hainaut, 1990, *Journal of Physiology (London)*, **422**, pp. 60, 62. Copyright 1990 by Cambridge University Press. Adapted by permission.

exerted by these units was less (Figure 9.22a), and there was a decline in the peak-to-peak amplitude of the motor unit action potentials. Although recruitment order was not affected by the period of immobilization, there was an increase in the range of recruitment (Figure 9.22b) and a decrease in the range of discharge rate modulation. This indicates a change in the activation strategy used to grade muscle force.

Hindlimb Suspension

When astronauts were first sent into space two principal concerns were the extent to which astronauts could maneuver outside the space vehicle and the type of physiological adaptations that would occur in a microgravity environment. Both questions were explored by experiments performed in space and on Earth. On the maneuverability question, biomechanical studies were performed on the dynamics of movement in a weightless environment. These involved some of the first segmental analysis studies (chapters 2 and 3). To examine the physiological adaptations, scientists developed an animal model that mimicked many of the changes that were known to occur during spaceflight. The model involves suspending the hindlimbs of an animal, usually a rat, off the ground for a few weeks so that the limbs are free to move but cannot touch the ground or any support surface (Figure 9.23a). This experimental technique is known as **hindlimb suspension**. The animal can perform many of its daily functions and experiences only minimal levels of stress, which are transient and vary between animals (Thomason & Booth, 1990). However, the animal is prevented from participating in group social activities, which are important for such physiological functions as temperature regulation.

Reduction in Activity. Four important ways in which the hindlimb-suspension model mimics spaceflight are a cephalic shift in fluid, loss of bone mineral content, decreased growth, and removal of the need for postural activity in the leg muscles. Despite the reduced need for hindlimb postural support, the EMG activity recorded in ankle muscles is not substantially depressed for the duration of the suspension period (Figure 9.24a). The activity of soleus and medial gastrocnemius are depressed for about the first 3 days of suspension but then recover to control levels (Alford, Roy, Hodgson, & Edgerton, 1987; Riley et al., 1990). Paradoxically, the muscles atrophy during the period when EMG levels seem to be normal (Figure 9.24b), which means that there is a dissociation between the levels of activity and the loss of muscle mass (Figure 9.25). This is probably due to the detrimental effects of unresisted concentric contractions (Gordon & Pattullo, 1993). Numerous studies have found that the soleus muscle (slow-twitch ankle extensor) experiences greater atrophy than either its fast-twitch synergist (medial gastrocnemius, plantaris) or its antagonist (tibialis anterior, extensor digitorum longus).

Perhaps the dissociation between EMG and muscle atrophy can be explained by the behavior of the animal (Figure 9.23b). In a load-bearing stance, the range of motion at the ankle joint extends from about 0.5 rad to

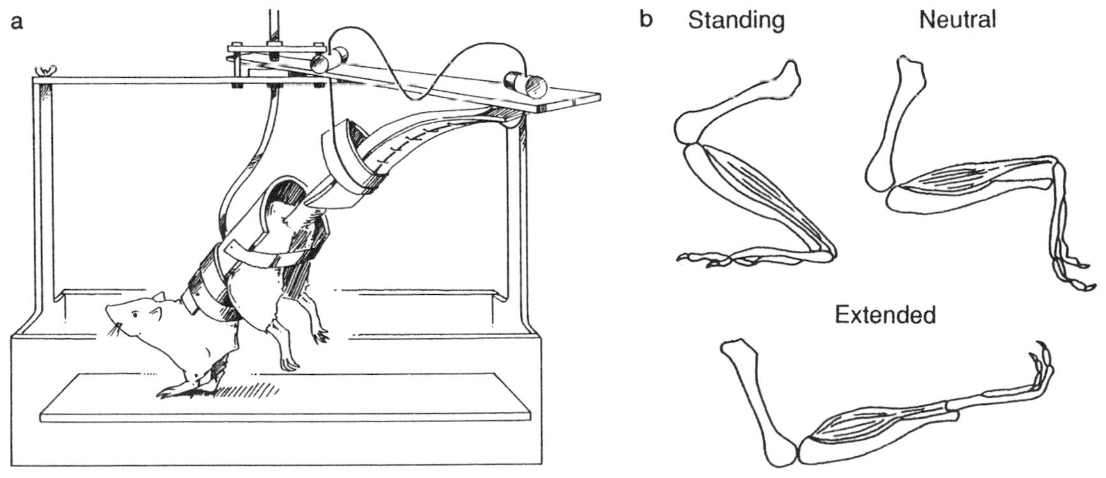

Figure 9.23 The hindlimb-suspension model. (a) A rat with its hindlimbs suspended off the ground. The rat can move around its cage by using its front legs.
Note. Figure provided by Dr. Charles M. Tipton.
(b) Change in the range of motion for the soleus muscle during hindlimb suspension. The left figure shows the minimum angle at the ankle joint during normal standing, the middle figure indicates the neutral position during hindlimb suspension, and the right figure represents the maximum angle for both conditions.
Note. From "Rat Hindlimb Unloading: Soleus Histochemistry, Ultrastructure, and Electromyography" by D.A. Riley, G.R. Slocum, J.L.W. Bain, F.R. Sedlack, T.E. Sowa, and J.W. Mellender, 1990, *Journal of Applied Physiology*, **69**, p. 63. Copyright 1990 by the American Physiological Society. Adapted by permission.

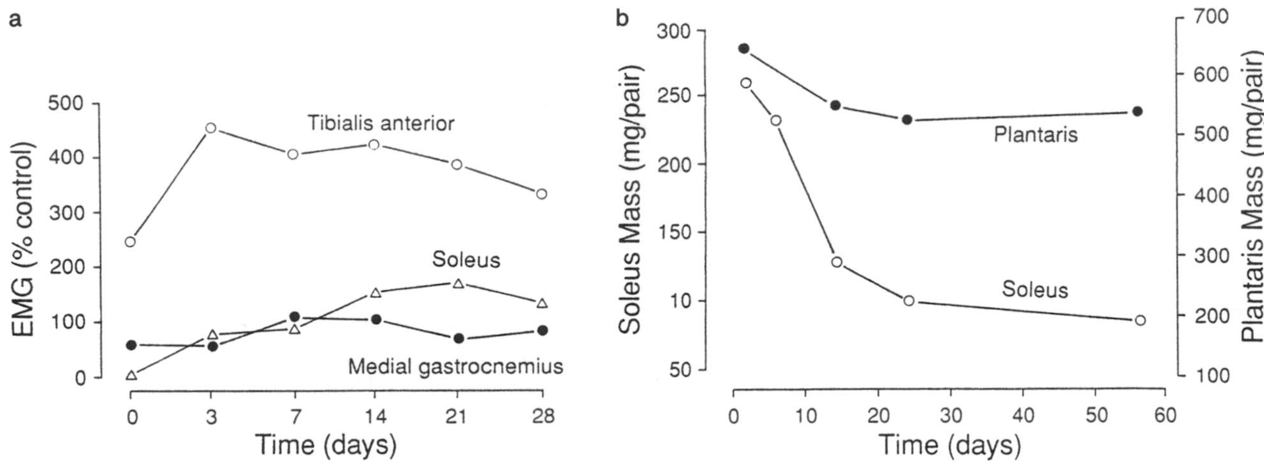

Figure 9.24 Results of 28 days of hindlimb suspension: (a) integrated EMG activity in three rat hindlimb muscles; (b) loss of mass in two muscles. Muscle mass is expressed in milligrams (mg) for the two hindlimb muscles of each animal.
Note. From "Atrophy of the Soleus Muscle by Hindlimb Unweighting" by D.B. Thomason and F.W. Booth, 1990, *Journal of Applied Physiology*, **68**, pp. 2, 4. Copyright 1990 by the American Physiological Society. Adapted by permission.

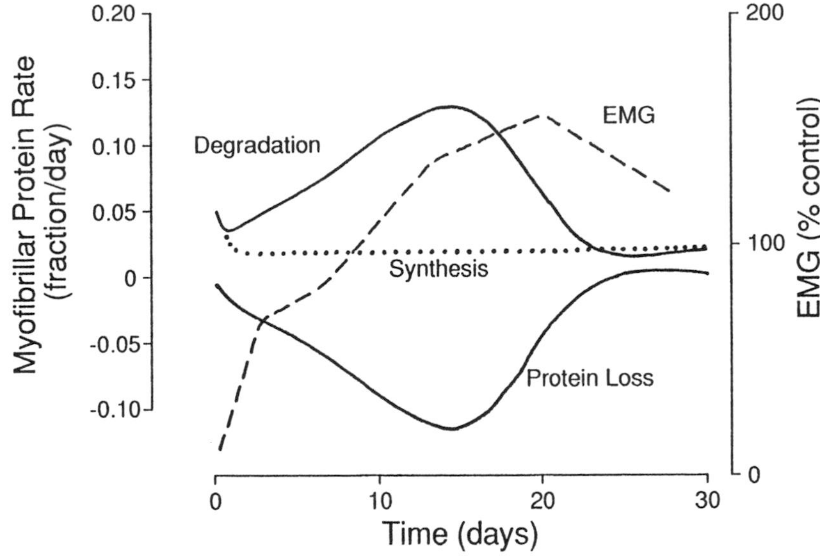

Figure 9.25 Time course of protein degradation, protein synthesis, protein loss, and integrated EMG activity in the rat soleus muscle during 28 days of hindlimb suspension. The net loss of protein is largely due to an increase in protein degradation despite an increase in EMG.
Note. From "Atrophy of the Soleus Muscle by Hindlimb Unweighting" by D.B. Thomason and F.W. Booth, 1990, *Journal of Applied Physiology*, **68**, p. 8. Copyright 1990 by the American Physiological Society. Adapted by permssion.

3.14 rad. However, after a few days of suspension the ankle adopts a neutral angle of 1.57 rad, which reduces its range of motion substantially. Also, recall that the EMG is affected by muscle length, and for the same muscle force the EMG is greater at shorter muscle lengths. Presumably, with the hindlimbs suspended, the forces exerted by the muscles are substantially reduced despite the maintained levels of EMG.

Another suspension model has been developed recently, which involves human subjects (Dudley et al., 1992; Hather, Adams, Tesch, & Dudley, 1992). In this model, the subjects wear a shoe on the right foot that has a raised sole (10-cm height) and use crutches for ambulation. This discourages the subjects from using the left leg for load bearing and thus represents a human model of one-legged suspension. When subjects performed this task for 6 weeks, they experienced substantial reductions in whole-muscle and muscle fiber cross-sectional area. For example, the cross-sectional areas of the vasti were reduced by 16%, the soleus by 17%, and gastrocnemius by 26%. Biopsies of the vastus lateralis muscle indicated reductions in cross-sectional area of 12% for the Type I fibers and 15% for the Type II muscle fibers. Because of the effect on the Type II fibers, these observations suggest some differences exist between the response of rat and human muscle to hindlimb suspension.

Neuromuscular Adaptations. Many animal studies have found a preferential effect of hindlimb suspension on slow-twitch muscle (Fitts, Metzger, Riley, & Unsworth, 1986; Roy et al., 1991; Thomason, Herrick, & Baldwin, 1987). Consequently, many studies have focused on the effects elicited in soleus (Thomason & Booth, 1990). A few weeks of suspension produces a decline in the proportion of Type I fibers in soleus with little effect on synergist (medial gastrocnemius) or antagonist (tibialis anterior) muscles. The suspension produces a decrease in the concentration of myofibrillar and myosin proteins, with an increased appearance of the fast myosin isoforms. However, this effect occurs in some but not all Type I muscle fibers of soleus (Gardetto, Schluter, & Fitts, 1989).

Accompanying these changes in protein content and myosin isoforms are some changes in the metabolic capabilities of the muscle fibers (Roy et al., 1991). The concentrations of succinate dehydrogenase and citrate synthase (enzymes involved in oxidative metabolism) increase in soleus with hindlimb suspension. But after several weeks of suspension, predominantly fast-twitch muscles (medial gastrocnemius, tibialis anterior) exhibit lower levels of some enzymes (succinate dehydrogenase) that characterize oxidative metabolism but not of others (citrate synthase). In contrast, both Type I and Type II muscle fibers in extensor and flexor muscles either maintain or elevate the levels of enzymes (e.g., alpha glycerophosphate dehydrogenase) associated with glycolytic metabolism.

These quantitative and qualitative changes in protein and enzyme content with hindlimb suspension produce a range of adaptations in the mechanical properties of muscle. The maximum force capacity of the soleus muscle is reduced, but the decline is greater than would be anticipated based on the loss of muscle mass. This means that there is a change in the specific tension (Herbert, Roy, & Edgerton, 1988; Thomason & Booth, 1990). However, specific tension (N/cm^2) is not altered in the predominantly fast-twitch muscles, which are less affected by hindlimb suspension. At the muscle-fiber level, Type I muscle fibers from both soleus and medial gastrocnemius show a significant reduction in diameter and peak force, whereas the Type IIa fibers from medial gastrocnemius exhibit a decline in diameter but no change in peak force (Gardetto et al., 1989). In contrast to these results obtained with rat muscle, Dudley and colleagues found a 21% decrease in the strength of the knee extensor muscles in human subjects after 6 weeks of one-legged suspension (Dudley et al., 1992; Hather et al., 1992). This decline in strength did not appear to involve a preferential effect on Type I muscle fibers.

In addition to the changes in force capacity, there are alterations in contraction speed (Roy et al., 1991; Thomason & Booth, 1990). The soleus muscle becomes faster, which should be expected due to the shift in myosin isoforms and the reduction in the proportion of Type I muscle fibers. This is evident by an increase in the maximum velocity of shortening (v_{max}) for the soleus muscle and some of its Type I muscle fibers; not all of the muscle fibers indicate an increase in v_{max}, which is consistent with the change in myosin isoforms limited to a subpopulation of these muscle fibers (Gardetto et al., 1989). With this increase in v_{max}, there is a shift in the force-velocity relationship to the right (Figure 6.24). Other effects on contraction speed include a reduction in both contraction time and half relaxation time, which are probably related to changes in Ca^{2+} kinetics. Hindlimb suspension lowers the Ca^{2+} concentration necessary to activate the contractile apparatus and increases Ca^{2+} cooperatively (Gardetto et al., 1989).

Similar to the observation reported for limb immobilization is the finding that a few weeks of hindlimb suspension does not alter the fatigability of the soleus muscle. The rat medial gastrocnemius muscle experiences a more complicated adaptation that involves no affect on fatigability after 7 days of suspension, but apparently the muscle becomes much more fatigable after 28 days of suspension (Winiarski, Roy, Alford, Chiang, & Edgerton, 1987).

Spinal Transection

In contrast to limb immobilization and hindlimb suspension, which involve constraint, **spinal transection** imposes a reduction in use by disconnecting parts of the nervous system from muscle. The separation is imposed in experimental animals at the level of the spinal cord, usually T12-T13. This disruption is referred to as an *upper motor neuron lesion* because it eliminates supraspinal control of the hindlimbs. This lesion is distinct from a *lower motor neuron lesion* in which the muscle is separated from all components of the nervous system (i.e., denervation).

Reduction in Activity. As described by Lieber (1992), spinal cord transection (spinalization) produces an immediate flaccid paralysis in which the hindlimbs are dragged by the experimental animal. About 3 to 4 weeks after the spinalization, the muscles develop spasticity, which eventually leads to sustained extensor activity with no apparent voluntary activation of the muscles. In this ''paralyzed'' state, however, the neuromuscular system can still be electrically activated and, with appropriate support and afferent feedback, the animals can be trained to perform hindlimb locomotion on a treadmill for which the force exerted by soleus is essentially similar to that of normal animals (Gregor et al., 1988; Lovely, Gregor, Roy, & Edgerton, 1990). Furthermore, spinalization at 2 weeks of age eventually (after 5 to 6 months) produces a 75% reduction in the quantity of integrated EMG and a 66% decrease in the duration of activity in the soleus muscle. The effects are less pronounced for gastrocnemius and include no change in the quantity of integrated EMG but a 66% reduction in the duration of activity (Roy et al., 1991).

Neuromuscular Adaptations. As with limb immobilization and hindlimb suspension, spinal transection produces muscle atrophy with a preferential effect on slow-twitch muscle and a slow-to-fast conversion of muscle fiber types (West, Roy, & Edgerton, 1986). The conversion of fiber types and loss of muscle mass is most pronounced in single-joint muscles involved in weight bearing. Soleus and medial gastrocnemius atrophy by about 45% and 30%, respectively, by 2 weeks after spinal transection. There is a decrease in the proportion of Type I muscle fibers and an increase in Type II fibers, which result in an increase in the proportion of Type FR motor units in soleus and an increase of Type FF and Type FI (intermediate fast-twitch) motor units in medial gastrocnemius (Munson, Foehring, Lofton, Zengel, & Sypert, 1986). These changes include an increase in the expression of fast myosin isoforms, 50% of the fibers in soleus reacting with a slow myosin heavy chain antibody, and an increase in myosin ATPase of 50% for soleus and 30% for medial gastrocnemius (Jiang, Roy, & Edgerton, 1990; Roy, Sacks, Baldwin, Short, & Edgerton, 1984).

The slow-to-fast conversion of muscle fiber types also influences the activities of enzymes associated with oxidative and glycolytic metabolism (Roy et al., 1991). These effects, however, are complex because the magnitude of the metabolic adaptations at the single-fiber level depend on both the age at spinalization and the muscle fiber type as defined by myosin properties. For example, there appear to be different effects on citrate synthase and succinate dehydrogenase, two enzymes involved in the citric acid cycle. In contrast, spinalization does not seem to alter the relationship between the glycolytic potential of muscle fiber and its type as defined by myosin ATPase.

The physiological adaptations exhibited by spinalized muscles are consistent with the changes in contractile and metabolic proteins (Figure 9.26). The slow-to-fast conversion in the cat soleus muscle is associated with a decrease in both time to peak force and half relaxation time and an increase in v_{max}. The increase in the maximum velocity of shortening (v_{max}) is to be expected because of the increase in myosin ATPase, the increase in proportion of Type II muscle fibers, and

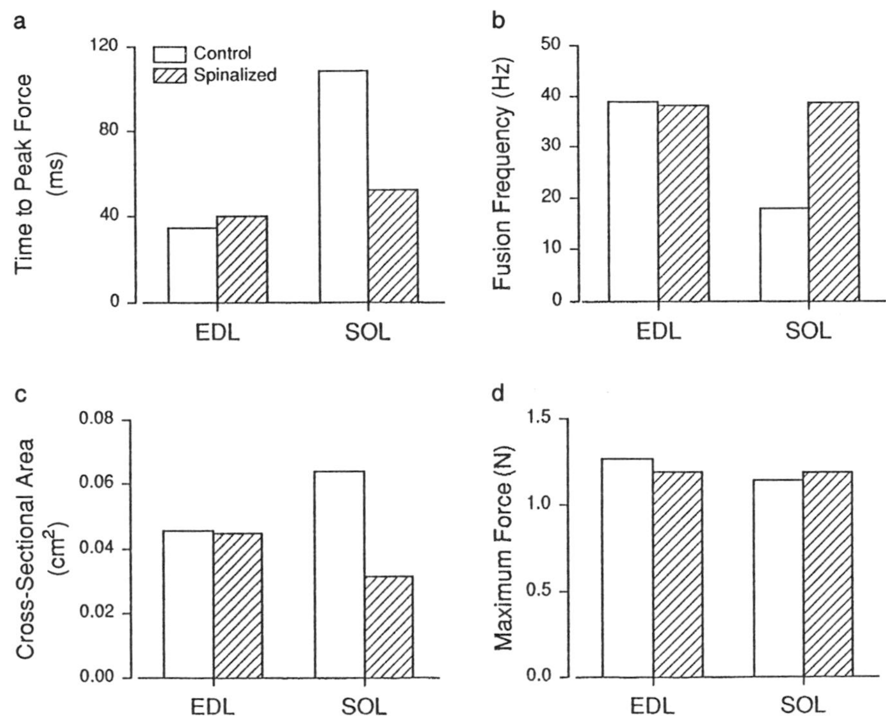

Figure 9.26 Contractile properties of soleus (SOL) and extensor digitorum longus (EDL) muscles in normal and spinalized rat hindlimbs: (a) time to reach the peak twitch force; (b) stimulus frequency at which the tetanic force became fused (smooth); (c) cross-sectional area; (d) maximum tetanic force. The greater effects occur in the soleus muscle.
Note. From *Skeletal Muscle Structure and Function* (p. 229) by R.L. Lieber, 1992, Baltimore: Williams & Wilkins. Copyright 1992 by Williams & Wilkins. Adapted by permission.

the increased incidence of fast myosin isoforms (Roy et al., 1984). The change in the time course of the twitch (time to peak force and half relaxation time) probably reflects changes in Ca^{2+} kinetics. In the cat medial gastrocnemius muscle, there is no change in the time course of the twitch, but v_{max} increases due to the increase in myosin ATPase. In both slow- and fast-twitch muscle, there is an increase in glycolytic enzyme activity after spinal cord transection.

The fatigability of the soleus muscle is not affected by spinalization (Gordon & Pattullo, 1993). For medial gastrocnemius, however, fatigability can change depending on the age at spinal transection. For young animals there is no effect on the fatigability of medial gastrocnemius, but when an adult animal experiences a spinal transection, the medial gastrocnemius muscle becomes more fatigable. Because these tests of fatigability include standard stimulus regimens, it would be interesting to use a variety of experimental protocols (recall the task dependency of fatigue) to determine how well each of the different processes tolerate spinalization. This information would be useful in the clinical management of human paraplegia.

Because the spinal transection model involves the removal of supraspinal control over spinal networks, there are changes in the properties and connectivity of spinal cord neurons. For example, motor neurons that innervate the soleus muscle become less excitable and resemble motor neurons that innervate fast-twitch muscle. These changes include a decrease in the afterhyperpolarization and an increase in rheobase (Cope, Bodine, Fournier, & Edgerton, 1986). There are, however, mixed reports on changes in the motor neurons that innervate the medial gastrocnemius muscle (Czéh et al., 1978; Foehring & Munson, 1990; Foehring, Sypert, & Munson, 1987a, 1987b). The changes in motor neuron properties are accompanied by changes in segmental reflexes. In humans with cervical spinal cord lesions, it is possible to elicit interlimb responses that are not normally present in intact individuals. For example, brief electric shocks to the lower extremities are sufficient to elicit unusual ipsilateral and contralateral upper extremity responses (Calancie, 1991). Because these responses are only present in patients with no signs of motor recovery, Calancie and colleagues are examining the usefulness of this test for predicting the prognosis of an acute spinal cord injury.

Motor Recovery From Injury

As the discussion in this text suggests, the neuromuscular system has remarkable adaptive capabilities. For example, the neuromuscular system can alter its contractile and metabolic proteins as a result of reductions in activity, and muscle can recover from the damage induced by eccentric contractions. We now consider the capabilities of the system to recover motor function after an injury to the peripheral nerve or a lesion in the CNS.

Peripheral Nervous System

It has been known for some time that the neuromuscular system is capable of some recovery of function after an injury to a peripheral nerve. Because neurons in the adult CNS are not capable of cell division, this recovery depends on the ability of the neurons to reinnervate appropriate targets. When an axon experiences a lesion (**axotomy**), degenerative changes occur in both the axon distal to the lesion and in the neuron (Figure 9.27). About 2 to 3 days after the axotomy, the soma begins to swell and may double in size, and the rough endoplasmic reticulum (stained as Nissl substance) breaks apart and moves to the periphery of the soma (Figure 9.27, b and c). The dissolution of the Nissl substance is referred to as **chromatolysis**. The process of chromatolysis lasts 1 to 3 weeks and seems to involve the massive resynthesis of the proteins necessary for regeneration of the axon and the formation of sprouts. If a sprout is able to invade the remaining myelin fragments (Figure 9.27c), then it is likely that reinnervation of the original target will occur. Invasion of the appropriate myelin fragment appears to be the most critical step in the recovery of function.

Because recovery depends on the ability of sprouts to reinnervate appropriate targets, the nature of the injury to a peripheral nerve is an important determinant in the extent of the recovery of function. In general, a complete transection of the nerve offers the worst prognosis, a partial denervation (sparing of some axons) is less severe, and a crush injury has only minimal long-term effects. Injuries that result in a transection of peripheral nerve trunks are usually accompanied by lasting motor and sensory deficits, which represent poor functional recovery. There are surgical techniques for resuturing a severed nerve with the aim of permitting regenerating axons to develop sprouts and reinnervate original targets. These procedures, however, typically result in a significant number of misdirected reinnervations (Table 9.4). When this occurs, motor neurons that originally innervated one muscle now innervate another muscle. Consider the number of misdirected reinnervations for the first dorsal interosseus muscle (abducts the index finger) shown in Table 9.4: Thirty-nine percent of the motor units reinnervated the correct muscle, 11% of the motor units recorded in this muscle previously innervated abductor digiti minimi (abducts the fifth finger), 22% previously innervated adductor pollicis (adducts the thumb), and 28% previously innervated other muscles (Thomas, Stein, Gordon, Lee, & Elleker, 1987).

An obvious consequence of these misdirected reinnervations is that the activation of a motor neuron pool does not lead to the selective activation of one muscle. For example, Table 9.4 indicates that activating the

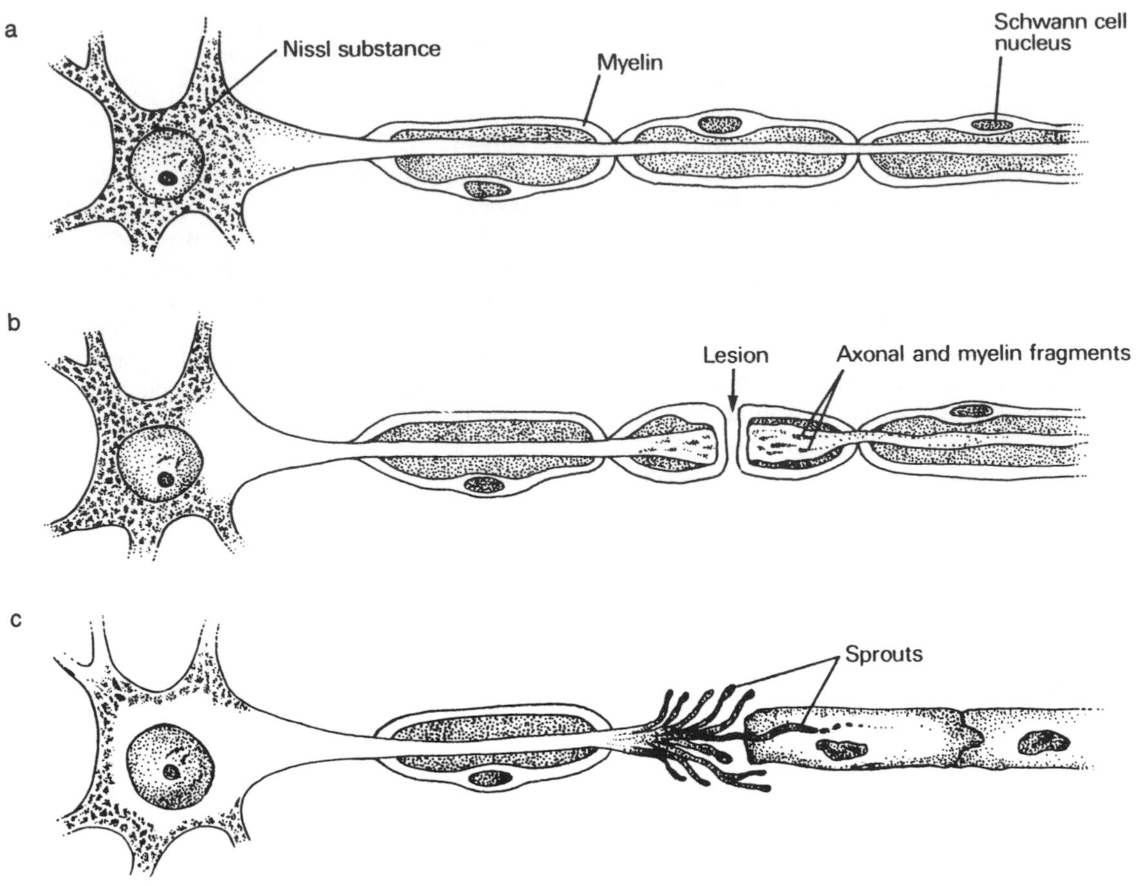

Figure 9.27 Changes in the distal axon and in the neuron following axotomy: (a) a normal soma and axon of a neuron; (b) 2 to 3 days after axotomy; (c) 1 to 3 weeks after axotomy.
Note. From "Reactions of Neurons to Injury" by J.P. Kelly. In *Principles of Neural Science* (2nd ed.) (p. 191) by E.R. Kandel and J.H. Schwartz (Eds.), 1985, New York: Elsevier Science Publishing Co., Inc. Copyright 1985 by Elsevier Science Publishing Co., Inc. Reprinted by permission of Appleton & Lange, Norwalk, CT.

Table 9.4 Origin of Motor Axons in Reinnervated Muscles After Complete Transection of the Ulnar Nerve at the Level of the Wrist

After Transection	First dorsal interosseus	Abductor digiti minimi	Adductor pollicis	Others
Abductor digiti minimi	15 (31%)	16 (34%)	9 (18%)	8 (17%)
First dorsal interosseus	22 (39%)	6 (11%)	12 (22%)	16 (28%)

Note. Data are from Thomas et al. (1987).

motor neuron pool for first dorsal interosseus will result in activation of at least the first dorsal interosseus (index finger abduction) and abductor digiti minimi (fifth finger abduction) muscles. Similarly, activating the motor neuron pool for adductor pollicis (thumb adduction) will also activate both first dorsal interosseus and abductor digiti minimi. This reorganization obviously impairs motor coordination for fine movements. These misdirections also occur with sensory axons and make it difficult for patients to localize a sensory stimulus, although it can be detected. Furthermore, the intermingling of motor neurons and muscle fibers disrupts the relationship between the force at which motor units are recruited (recruitment threshold) and the amplitude of the twitch force. The relationship is frequently used as an index of orderly recruitment based on the size principle, whereby small units (low-twitch forces) are recruited at low forces. The absence of a relationship between recruitment threshold and twitch amplitude suggests a loss of the ability to recruit motor units according to size and to finely grade the force exerted by a muscle. Taken together, the misdirected reinnervations and loss of size-related recruitment indicate a poor coordination for fine movements. However, the recovery of function is sufficient for power movements.

Compared with the recovery associated with a complete nerve transection, there is less impairment following recovery from a partial transection of the nerve close to the muscle. The principal recovery mechanism following a partial lesion is somewhat different from that associated with a complete transection; it still involves axonal growth and reinnervation, but the axons of surviving motor units develop sprouts to reinnervate the muscle fibers that have been denervated. This type of sprouting, referred to as **collateral sprouting**, by surviving axons appears to be confined to the distal region of the motor axon and to occur close to the target. Apparently all the motor units within a pool compensate for these types of nerve injuries by collateral sprouting (Rafuse, Gordon, & Orozco, 1992). The number of additional fibers reinnervated depends on the size of the motor unit; larger motor units are capable of supporting a larger number of muscle fibers. Furthermore, the extent of the sprouting depends on the number of fibers that were denervated (Figure 9.28). The greater the extent of the partial lesion and hence the number of muscle fibers that are denervated, the greater the amount of collateral sprouting and the greater the size of the surviving motor units.

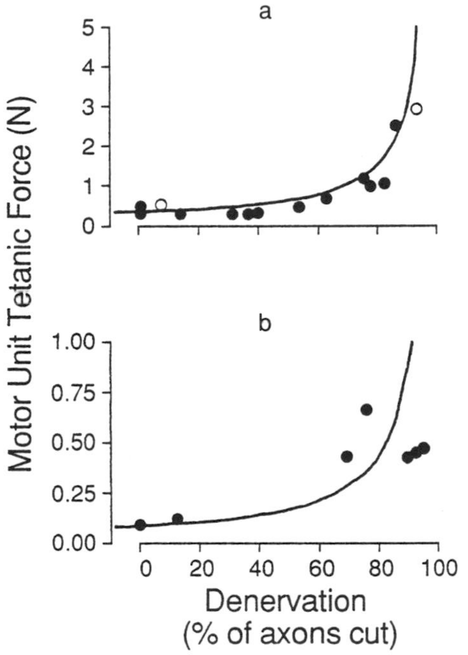

Figure 9.28 Tetanic force of motor units as a function of the degree of partial denervation (percentage of the nerve that was lesioned): (a) motor units in medial (filled circles) and lateral (open circles) gastrocnemius; (b) motor units in soleus.

Note. From "Proportional Enlargement of Motor Units After Partial Denervation of Cat Triceps Surae Muscles" by V.F. Rafuse, T. Gordon, and R. Orozco, 1992, *Journal of Neurophysiology*, **68**, p. 1266. Copyright 1992 by The American Physiological Society. Adapted by permission.

Collateral sprouting can account for up to about 80% of the motor neuron loss (Figure 9.29), with motor units capable of enlarging to about five times the original size (Gordon, Yang, Ayer, Stein, & Tyreman, 1993). These observations indicate that when as much as 80% of the peripheral nerve is cut, the neuromuscular system can recover the maximum force capability. This recovery could result from three mechanisms: an increase in the innervation ratio, an increase in the average cross-sectional area of muscle fibers, or an increase in the specific tension. The most important mechanism appears to be an increase in the innervation ratio, which must be due to collateral sprouting (Tötösy de Zepetnek, Zung, Erdebil, & Gordon, 1992). The potential for collateral sprouting does not depend on descending drive, because it occurs to the same extent when the spinal cord is transected above the level of the motor neuron pool (Figure 9.29).

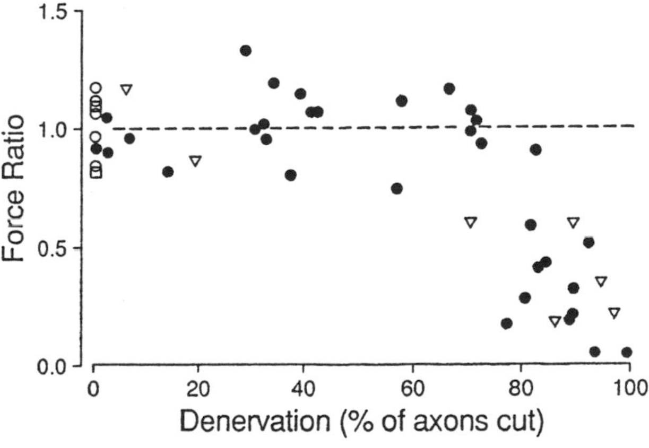

Figure 9.29 Maximum tetanic force exerted by the tibialis anterior muscle of the rat after different amounts of denervation by partially sectioning the muscle nerve. The force is expressed as a ratio of the force exerted by the partially denervated muscle relative to that of the intact contralateral muscle. The forces are for muscles from control animals (open circles), animals with a partially denervated muscle (filled circles), animals with a spinal cord transection (open squares), and animals with both a spinal cord transection and a partial nerve transection (open diamonds).

Note. Adapted from "Recovery Potential of Muscle After Partial Denervation: A Comparison Between Rats and Humans" by T. Gordon, J.F. Yang, K. Ayer, R.B. Stein, and N. Tyreman, 1993, *Brain Research Bulletin*, **30**, p. 480, Copyright 1990, with kind permission from Pergamon Press Ltd., Headington Hill Hall, Oxford OX3 DBW, UK.

Central Nervous System

Researchers disagree on the recuperative abilities of the CNS. Although most believe that the CNS is capable of reorganization, the evidence is scarce that the system can recover from injury. One example that the CNS is capable of reorganization is the **synaptogenesis** (formation of new synapses) that occurs with enhanced activity. When electrodes were implanted in the thalamus of the cat and corticocortical connections were subjected to long-term stimulation (4 days), the results were an increased density of specific classes of synapses in layers II and III in the motor cortex, an increase in certain structural features of synapses, and alterations in the patterns of synaptic activity (Keller, Arissian, & Asanuma, 1992). This reorganization has been suggested to underlie processes associated with motor learning and memory.

Other attempts to examine sprouting in the CNS have focused on the ability of the spinal cord to compensate for removal of selected inputs or pathways. Three classical preparations are used to examine CNS sprouting: the spared-root preparation, a hemisection model, and a deafferentation preparation. The **spared-root preparation** involves transection of all the dorsal roots supplying a hindlimb, except one (usually L6). The animal recovers from the surgery and learns to reuse the limb. Once the motor recovery has been measured (weeks to months later), experiments are performed to determine the extent of the sprouting in the spinal cord by the spared dorsal root. The *hemisection model* involves removing half (hemi) of the spinal cord at some appropriate level, and then monitoring both the recovery of function and subsequently the extent of sprouting in the spinal cord. The *deafferentation preparation* follows the same principle and measures the recovery of function and sprouting after transection of the dorsal (afferent) roots.

These experiments are difficult to perform, and the data analysis is tedious. There is no consensus on the results; some scientists conclude the sprouting does occur, and others suggest that is does not (Goldberger, Murray, & Tessler, 1992). Some of the difficulties encountered include distinguishing between sprouting and unmasking and accounting for interanimal variability. As is the case for the peripheral nervous system, there is some evidence that collateral sprouting occurs in the spinal cord with the spared-root preparation. There are also clear examples in which the recovery of function can be explained by unmasking (O'Hara & Goshgarian, 1991). **Unmasking** refers to the activation of a dormant neural pathway (or set of synapses) that is not needed until another (primary) pathway has been interrupted. Furthermore, because the evidence for sprouting relies on the measurement of synaptic density, and there is considerable variability in this parameter between animals, it is possible that evidence of sprouting is missed when data from different animals are combined rather than examined individually (Goldberger et al., 1992). Nevertheless, it appears that sprouting occurs continuously in the normal CNS, with synpatic connections degenerating and being renewed. This provides a potential mechanism for plasticity in the damaged nervous system.

One of the critical factors in CNS regeneration is the presence or absence of scar tissue. When the spinal cord is crushed, there is massive glial and connective tissue scarring from which there is minimal recovery (Freed, deMedinaceli, & Wyatt, 1985). Sprouts cannot cross physical barriers such as scars, and this severely limits its regeneration.

Adaptations With Age

Senescence, or aging, is generally accompanied by a marked decline in the capabilities of the motor system. Although these age-related changes can often be attributed to pathological processes, even healthy and vigorous elderly individuals experience reductions in performance capabilities that seem to represent the natural consequences of aging. In this section we describe some of the changes that occur in the motor system with age and then examine mechanisms that seem to mediate these adaptations.

Movement Capabilities in Elderly Subjects

Throughout this text, we have emphasized the concept that the motor unit represents the functional unit of the motor system; it represents the link between the nervous system and muscle. Because aging causes significant changes in motor unit properties, it is accompanied by marked declines in most aspects of movement. These changes include a decline in strength, a reduction in the magnitude of reflex responses, a slowing of rapid reactions, an increased postural instability and diminished postural control, decreased control of submaximal force, and reduced manipulative capabilities.

Strength. One of the most prominent effects of age on the motor system is the unavoidable decline in muscle mass and strength (Figure 9.30) (Doherty et al., 1993a; Schultz, 1992). The decline in strength has been observed in many different muscles and species (Frolkis, Martynenko, & Zamostyan, 1976; Grimby & Saltin, 1983; Kanda & Hashizume, 1989; Narici, Bordini, & Ceretelli, 1991). At least two aspects of this adaptation are interesting: Several different studies have found that the decline in strength among humans seems to begin at about age 60, and the decline in strength varies greatly among elderly individuals. The latter point is evident in Figure 9.30 by the increased scatter of the data points after age 60. The variability in strength suggests that the mechanism, or mechanisms, controlling the decline

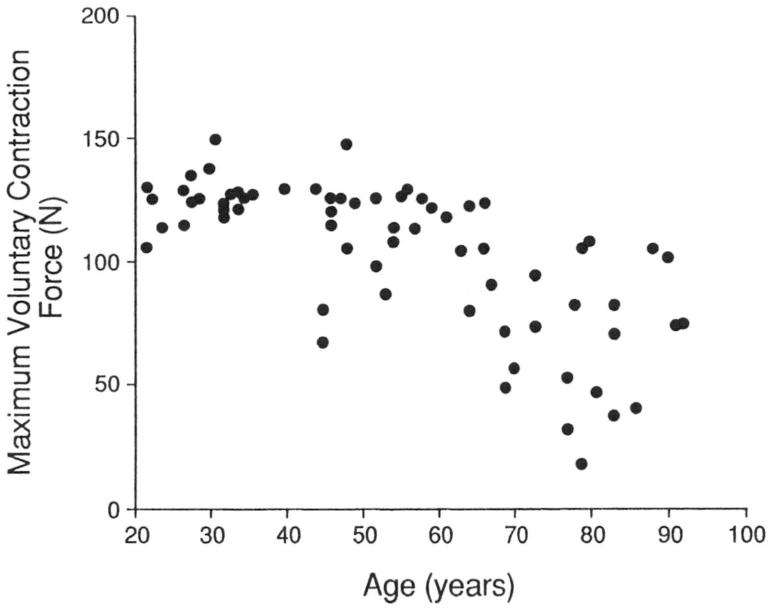

Figure 9.30 Maximum voluntary contraction force for a hand muscle (adductor pollicis) as a function of age.
Note. From "Effect of Aging on Human Adductor Pollicis Muscle Function" by M.V. Narici, M. Bordini, and P. Ceretelli, 1991, *Journal of Applied Physiology*, **71**, p. 1278. Copyright 1991 by The American Physiological Society. Adapted by permission.

in strength are activated to different degrees among individuals and can perhaps be attenuated by appropriate interventions. Older individuals are able to increase strength with appropriate training, and this increase in strength is due partially to hypertrophy of Type II muscle fibers (Doherty et al., 1993b; Keen, Yue, & Enoka, 1994; Roman et al., 1993).

Fatigability. In contrast to the decline in strength with age, there is less certainty about the changes in fatigability with age. A number of fatigue protocols have been used to examine the effect of age on the fatigability of muscle. The studies that have used voluntary activation of muscle have all found no change in fatigability with age (Hicks et al., 1992; Laforest, St-Pierre, Cyr, & Guyton, 1990; Larsson, Sjödin, & Karlsson, 1978). These studies have involved submaximal and maximal voluntary contractions, isometric and dynamic contractions, and continuous and intermittent contractions performed by hand and leg muscles. Other studies that have activated muscle with electrical stimulation have found mixed results. There have been reports of no change, an increase, and a decrease in fatigability. For example, 30 s of ulnar-nerve stimulation at 20 Hz (Lennmarken, Bergman, J. Larsson, & L.-E. Larsson, 1985) resulted in an increased fatigability in elderly men (67 years) but not in elderly women (63 years), whereas 30-Hz stimulation (Narici et al., 1991) resulted in a decrease in fatigability with age (70 male subjects, ages 20 to 91). These inconsistencies suggest a need to determine how fatigue mechanisms affect performance by elderly subjects in these different tasks.

Reflex Responses and Rapid Reactions. The peak-to-peak amplitude of the H-reflex EMG decreases with age. Vandervoort and Hayes (1989) found that the H-reflex amplitude in soleus for a group of elderly women (82 years) was only 43% of that for younger women (26 years); the peak-to-peak amplitudes were 2.4 and 5.6 mV, respectively. This difference in H-reflex amplitude, however, could be explained by the reduction in muscle fiber excitability with age. As with the H reflex, the peak-to-peak amplitude of the M wave was similarly different for the two groups of subjects: 4.4 vs. 9.3 mV. Recall that the M wave represents the direct activation of muscle and provides an index of muscle fiber excitability (Hicks et al., 1992), whereas the H reflex involves the reflex activation of the motor neurons in the spinal cord. For the elderly group of subjects, the H-reflex stimulus resulted in the activation of 55% of the muscle fibers that contributed to the M wave, and for the younger subjects the proportion was 60%. Consequently, the proportion of the motor unit pool that was activated by the H-reflex stimulus did not differ with age.

In contrast to the artificially elicited H reflex, differences have been found between young (27 years) and elderly (75 years) subjects in the tendon tap reflex elicited in the quadriceps femoris muscle (Burke, Kamen, & Koceja, 1989). In this study, a rubber-tipped implement struck the patellar tendon and elicited a tendon tap reflex that was measured as the force exerted at the ankle. The amplitude of the reflex response was greater in the younger subjects (30 vs. 18 N) and occurred with a briefer latency (60 vs. 79 ms). The difference in the amplitude of the response might be related to the difference in strength between the two groups, but the difference in latency suggests an age-related difference in the detection, transmission, and processing of the stimulus. Recall that the tendon tap reflex involves excitation of muscle spindles, propagation of afferent signals back to the spinal cord, activation of the alpha motor neurons, and excitation of the associated muscle fibers. It is unclear which of these processes are altered and contribute to the change in latency with age.

The study of rapid reactions to visual cues or sudden movements provides some further clues on the segmental adaptations that occur with aging. Warabi, Noda, and Kato (1986) had subjects perform rapid responses to displacements of a target. Performance was characterized by measured **reaction time**, which is the time from target displacement to the beginning of the response by the subject. The target was displaced by either 0.17, 0.34, or 0.70 rad, and the subjects were instructed to respond with either the eyes or the hand. The response of the eyes was measured with an electrooculogram (EOG), and that of the hand was measured as the movement of a joystick. Figure 9.31 shows that the reaction time increased with age, that it was greater for the larger target displacements, and that it had a similar magnitude for the two different motor systems (eyes and hands). Because the motor components of these movements (e.g., peak velocity, duration, amplitude) were not altered with age, Warabi et al (1986) concluded that the principal effect of age on the rapid reaction task was an impairment of sensory processing. This conclusion is consistent with other reports of a decline in sensory capabilities with age (Dyck, Karnes, O'Brien, & Zimmerman, 1984; Verillo, 1979).

Maintenance of Posture. Aging is typically accompanied by a reduced ability to control both posture and gait (Rogers, Kukulka, & Soderberg, 1992; Schultz, 1992; Winter, Patla, & Frank, 1990). These effects are most readily documented as a decline in the amount of walking and in the performance of the daily activities of living among elderly individuals. The adaptations that underlie these reductions are complex and involve both motor and sensory processes. Alterations in the control of posture are significant because they influence the stability of an individual and affect the likelihood of accidental falls and injuries.

Maintenance of an upright posture requires that an individual keep the direction of the weight vector within the base of support. This requires the involvement of several different processes: (a) sensory information to detect the orientation and motion of the individual; (b) selection of an appropriate response strategy to maintain stability; and (c) activation of the muscles that can overcome the postural imbalance. The sensory information can be derived from visual, somatosensory, and vestibular sources. Not all of this information is necessary, and an individual can choose from among these sensory signals. Furthermore, patients with deficits in one of these sensory systems can readily learn to rely on the other two. However, age seems to diminish the ability of an individual to select the appropriate sensory information (Horak et al., 1989). This decline is compounded by a reduced ability to control postural sway (forward-backward displacement of the center of gravity), even though the amount of sway increases with age (Hasselkus & Shambes, 1975). Although we may regard the maintenance of a stable upright posture as a simple task, there are a number of processes that if impaired can reduce an individual's stability. Horak et al. concluded that the age-related decline in the ability to control posture can be caused by several different factors, including abnormal selection of sensory information, poor detection of disequilibrium, abnormal selection of postural adjustment response, prolonged latencies for rapid responses, reduced perception of stability limits, weakness of involved muscles, and impaired ability to coordinate activity among synergist muscles.

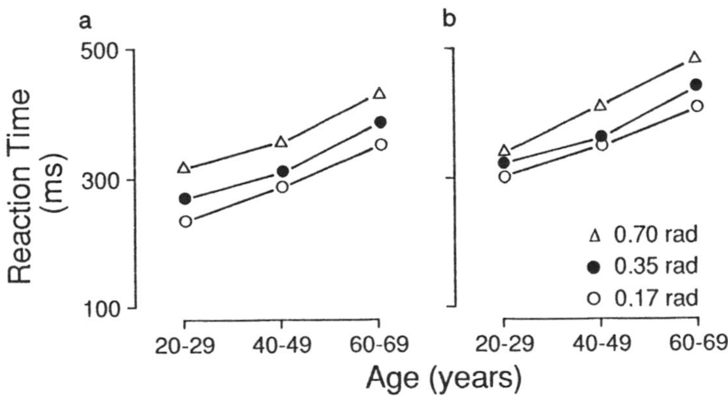

Figure 9.31 Changes in reaction times for (a) eye and (b) hand movements as a function of age. The reaction time tasks involved target displacements with amplitudes of 0.17 rad, 0.34 rad, and 0.70 rad.
Note. From "Effect of Aging on Sensorimotor Functions of Eye and Hand Movements" by T. Warabi, H. Noda, and T. Kato, 1986, *Experimental Neurology*, **92**, p. 690. Copyright 1986 by Academic Press, Inc. Adapted by permission.

Control of Submaximal Force. In contrast to the extensive literature on posture and gait in elderly subjects, much less is known about the effects of age on simpler movements and the behavior of motor units in the gradation of force (Kamen & DeLuca, 1989). Although maximum capabilities decline with age (e.g., strength, flexibility, sensory perceptions), the consequences of these changes for the control of submaximal force have not been well described. Some submaximal tasks have been examined, such as the control of isometric force, the kinematics of arm movements, and the control of grip force (Bennett & Castiello, 1994). For example, when subjects were asked to sustain a force at a constant, submaximal value for about 20 s (Figure 9.32), the force fluctuated about the target value (Galganski et al., 1993). The fluctuation in force (standard deviation) increased with both the target level (5%, 20%, 35%, and 50% of maximum) and the age of the subject; the averages ages for the two groups of subjects were 28 and 67. However, when the force fluctuations were expressed relative to the value of the target force (coefficient of variation), the fluctuations decreased with the increase in target force, although the elderly subjects still had greater fluctuations. The force fluctuations are unrelated to the changes in motor unit size and decline significantly with training (Keen et al., in press). These data indicate that the elderly subjects were less able to maintain constant force at submaximal values.

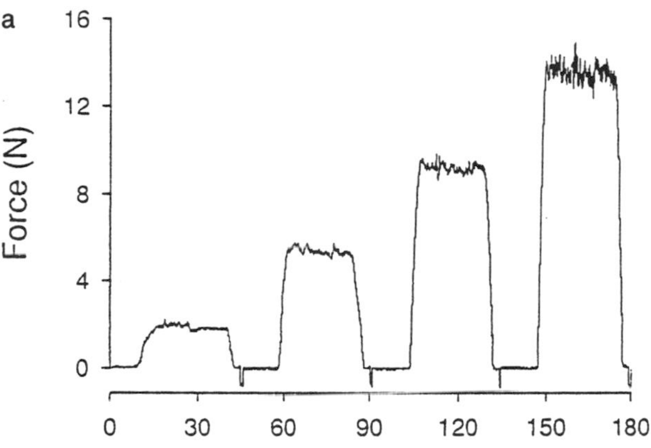

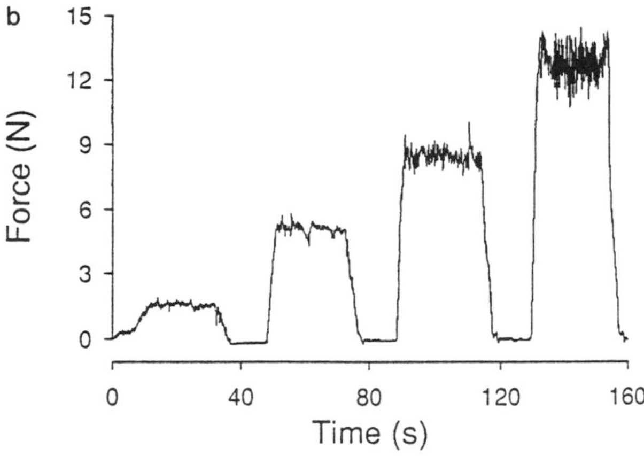

Figure 9.32 Force exerted by a young (a) and an elderly (b) subject when instructed to keep the force constant at each of the four target forces for 20 s. The isometric force was exerted by the index finger.

Note. From "Reduced Control of Motor Output in a Human Hand Muscle of Elderly Subjects During Submaximal Contractions" by M.E. Galganski, A.J. Fuglevand, and R.M. Enoka, 1993, *Journal of Neurophysiology*, **69**, p. 2110. Copyright 1993 by The American Physiological Society. Adapted by permission.

A similar effect was observed by Cole (1991) in the control of grip force during the manipulation of small objects. Subjects were instructed to lift a small object (1.7-cm width, 1.6-N weight) in a vertical direction about 5 cm, and the pinch force exerted by the thumb and index finger was measured. The slipperiness of the sides of the object could be altered. The elderly subjects (81 years) exerted a force that was about 2 times greater than that exerted by the younger subjects (22 years) and 2.5 times greater than the force needed to prevent the object from slipping. Cole attributed the use of excessive grip forces by the older subjects to an adaptation in tactile afferents that reduces hand sensibilities and to alterations in the glabrous skin that increase hand slipperiness. Clearly, performance capabilities for such tasks as manipulation depend on the maintained integrity of both sensory and motor systems.

Physiological Adaptations With Age

As we have learned throughout this text, movement is a complex event. The control of movement is difficult to understand because it involves the integration of numerous sensory and motor processes. In addition, many of these processes can change with age. For example, motor neurons die and those that survive can reinnervate additional muscle fibers. Other age-related changes include a decline in the rate of axonal transport, an alteration of the biophysical properties of motor neurons, an increase in the peak-to-peak amplitude of muscle fiber action potentials, a reduction in the frequency of miniature end-plate potentials, a decrease in the amplitude of M waves, changes in the myosin heavy chain isoform, and a reduction in tactile sensibility. This list does not include all the age-related changes that have been reported in the research literature; rather the list provides an overview of the types of changes that are known to occur.

Of these changes, there is considerable interest in motor neuron death because of the functional consequences for movement. There are at least two significant consequences: The decline in muscle mass with age is primarily due to the death of motor neurons and the progressive atrophy and disintegration of denervated muscle fibers, and surviving motor units are enticed by denervated muscle fibers to develop collateral sprouts from the intramuscular axon and reinnervate the muscle fibers (Campbell et al., 1973; Doherty, Vandervoort, Taylor, & Brown, 1993; Kanda & Hashizume, 1989; Rosenheimer & Smith, 1990; Stålberg & Fawcett, 1982; Tsujimoto & Kuno, 1988). The result of these processes is a decline in the number of motor units with age but an increase in the size (innervation ratio) of surviving motor units. The decline in motor unit number with age parallels the decrease in muscle strength (Figure 9.30).

The reinnervation of denervated muscle fibers by surviving motor units means that the innervation ratio of these units increases. Recall from chapter 5 that the force exerted by a motor unit depends on its innervation ratio, the average cross-sectional area of the muscle fibers, and the specific tension. Because of the increase in innervation ratio, the peak-to-peak force exerted by single motor units increases as a function of age (Galganski et al., 1993). This affects the ability of an individual to smoothly grade muscle force in two ways. First, compared with younger subjects, older subjects experience a substantial reduction in the number of low-force motor units. Typically a motor unit pool consists of many low-force units with an exponentially declining number of high-force units. The low-force units are important for finely controlled movements that involve submaximal forces. Second, the smoothness (absence of fluctuations) of the force exerted by a muscle depends on the force contributed by the most recently recruited motor unit (Christakos, 1982). Because motor units are recruited at low discharge rates that produce unfused tetani, the recruitment of a motor unit contributes fluctuations to the net force, and the amount of fluctuation increases with later-recruited (larger) motor units (Figure 9.33). The increase in motor unit force with age results in larger fluctuations with each motor unit recruited.

It appears that the death of motor neurons with aging is not a random process but that larger motor neurons seem to be preferentially affected. This conclusion is based on the changes that occur in the distribution and proportion of muscle fiber and motor unit types. For example, Larsson (1983) reported a significant decline in the proportion and cross-sectional area of Type II muscle fibers in vastus lateralis with age (Figure 9.34) and a grouping, rather than a random distribution, of the different fiber types. Similarly, Lexell and Taylor (1991) found a significant decline (35%) in the cross-sectional area of Type II muscle fibers in human vastus lateralis. Using an animal model, Kanda and Hashizume (1989) also found a decrease in the proportion of Type II muscle fibers in older medial gastrocnemius muscles and a reduction in the tetanic force exerted by Type FF and Type FR motor units. Furthermore, the Type S motor units in these muscles exerted a greater tetanic force, which was due to an increase in the innervation ratio with no change in the average cross-sectional area of the muscle fibers or the specific tension. As would be expected from a denervation-reinnervation process,

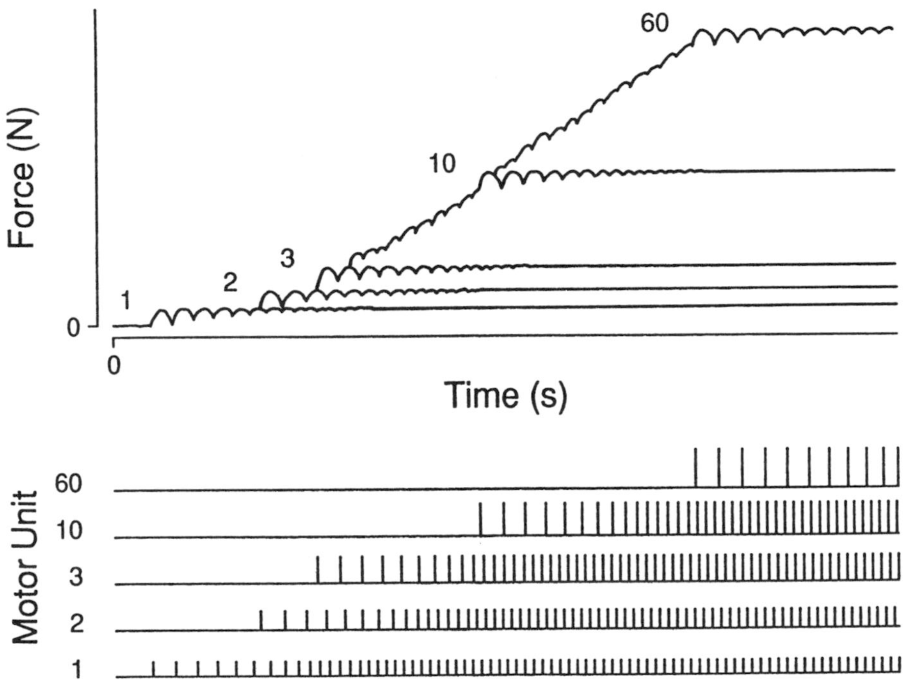

Figure 9.33 Motor unit contributions to the force exerted by a muscle. The hypothetical scheme shows an increase in muscle force over time (outer line in force-time graph) and shows the force contributed by five motor units (from among the many that were active). The lower traces show the action potentials of those five motor units. When each motor unit is recruited, its discharge rate is low but then increases.

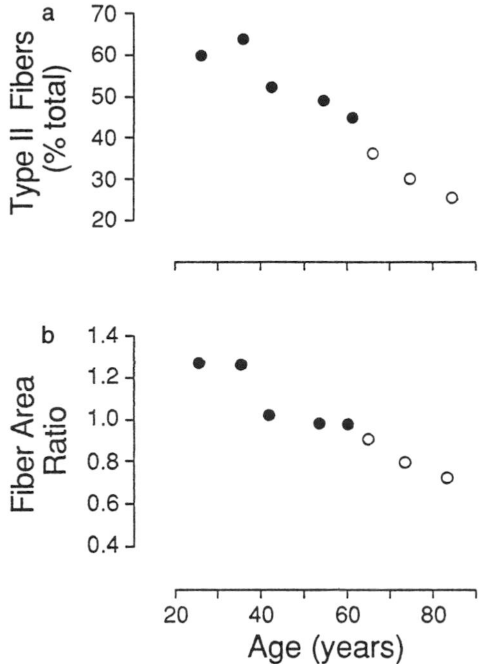

Figure 9.34 Fiber type proportions (a) and relative cross-sectional area (b) in the human vastus lateralis as a function of age. The data were obtained from 100 subjects aged 22 to 89 years. There is a substantial decline in the proportion of Type II muscle fibers and in the ratio of the cross-sectional area of Type II fibers relative to Type I fibers. *Note*. From "Histochemical Characteristics of Human Skeletal Muscle During Aging" by L. Larsson, 1983, *Acta Physiologica Scandinavica*, **117**, p. 470. Copyright 1983 by Scandinavian Physiological Society. Adapted by permission.

there is an increase in the density of muscle fibers for the Type S motor units with age (Kanda & Hashizume, 1989). These observations suggest that aging is associated with a preferential degeneration of Type FF and Type FR motor units and a reinnervation of adjacent muscle fibers by Type S motor units.

Given the dramatic loss of muscle mass that occurs with aging and the role of motor neuron death in this process, there is considerable clinical interest in identifying strategies that might minimize these reductions. One possibility is to determine the role of exercise and maintained levels of physical activity. Strength-training programs have been shown to increase the strength (quadriceps femoris) and mobility of frail, 90-year-old volunteers (Fiatarone et al., 1990). These adaptations include substantial increases in total muscle area and the cross-sectional area of Type I and Type II muscle fibers of arm (biceps brachii) and leg (vastus lateralis) muscles (Brown, McCartney, & Sale, 1990; Frontera, Meredith, O'Reilly, Knuttgen, & Evans, 1988; Keen et al., 1994). Such studies indicate that older individuals retain the ability to hypertrophy muscle, to increase the turnover of myofibrillar protein, and to activate satellite cells for the repair of damaged muscle fibers.

Summary

This is the final of three chapters examining the effects of physical activity on the neuromuscular system. This chapter focuses on the long-term (chronic) response of the system to the stress associated with a single bout of physical activity. After an introductory discussion on chronic activity and principles of training, this chapter addresses five topics that characterize the chronic adaptive capabilities of the system: strength, muscle power, adaptation to reduced use, motor recovery from injury, and adaptations with age. First, we defined the concept of strength, discussed training and loading techniques used to increase strength, and then examined the neural and muscular adaptations that accompany strength training. Second, we reexamined the concept of power production and its importance for movement and then considered training techniques that can be used to increase power production. Third, we reviewed the neuromuscular adaptations that occur when the level of physical activity is reduced. For examples of these adaptations, we considered the effects of limb immobilization, hindlimb suspension, and spinal transection. Fourth, we discussed the ability of the neuromuscular system to recover motor function after an injury to the nervous system. We distinguished between the capabilities of the peripheral and central nervous systems. Fifth, we characterized the adaptations that accompany an increase in age. We considered the consequences for the movement capabilities of elderly individuals and examined the mechanisms that mediate these adaptations. These examples do not represent a complete summary of the chronic adaptive responses but illustrate the capabilities of the system.

TRUE-FALSE QUESTIONS

1. The overload principle states that there is a threshold point that must be exceeded before an adaptive response will occur. TRUE FALSE
2. According to the reversibility principle, the type of change that occurs is specific to the exercise stress. TRUE FALSE

3. Muscle strength is measured when velocity equals zero. **TRUE FALSE**
4. Muscle length does not influence the measurement of strength. **TRUE FALSE**
5. Muscle fiber length does not change during an isometric contraction. **TRUE FALSE**
6. Isometric exercises produce an increase in strength. **TRUE FALSE**
7. In an isokinetic contraction, muscle length shortens at a constant velocity. **TRUE FALSE**
8. A 1-RM load is a weight that can be lifted only once. **TRUE FALSE**
9. Progressive resistance exercises involve an increase in the number of repetitions across sets. **TRUE FALSE**
10. The Oxford technique is an example of ascending-pyramid loading. **TRUE FALSE**
11. The 10-RM load is limited by the maximum force that can be exerted during a concentric contraction. **TRUE FALSE**
12. Eccentric exercises produce greater increases in strength than do concentric exercises. **TRUE FALSE**
13. Eccentric exercises provide a more effective stimulus for hypertrophy. **TRUE FALSE**
14. An accommodating device is one whose resistance varies to match the force exerted by the subject. **TRUE FALSE**
15. Other than at the beginning and end of the movement, an accommodation device has a zero acceleration during an isokinetic contraction. **TRUE FALSE**
16. Eccentric contractions cannot be performed on accommodation devices. **TRUE FALSE**
17. Hydraulic devices do not control movement speed. **TRUE FALSE**
18. Plyometric exercises are based on eccentric-concentric contractions. **TRUE FALSE**
19. Vertical jump height can be increased with plyometric training. **TRUE FALSE**
20. The neural insufficiency hypothesis suggests that the MVC force can be increased by manipulation of afferent feedback. **TRUE FALSE**
21. A stimulus rate of 10 Hz is sufficient to elicit the MVC force. **TRUE FALSE**
22. Biphasic rectangular electric shocks preferentially activate large-diameter motor units. **TRUE FALSE**
23. Neuromuscular electrical stimulation generates action potentials directly in muscle fibers. **TRUE FALSE**
24. Impedance is a measure of the resistance to the flow of electric current. **TRUE FALSE**
25. The unit of measurement for impedance is the mho. **TRUE FALSE**
26. Magnetic stimulation can elicit action potentials in intramuscular nerve branches. **TRUE FALSE**
27. Neuromuscular electrical stimulation protocols do not increase strength. **TRUE FALSE**
28. Neuromuscular electrical stimulation can minimize the atrophy in the quadriceps femoris of patients with spinal cord injuries. **TRUE FALSE**
29. Progressive resistance exercises can be performed only with barbells. **TRUE FALSE**
30. The most favored loads for strength training are 8- to 12-RM loads. **TRUE FALSE**
31. The torque due to the load does not vary about a joint during constant-load training. **TRUE FALSE**
32. Free weights (barbells) are used for variable-load training. **TRUE FALSE**
33. With strength training, the change in EMG occurs before a change in muscle size. **TRUE FALSE**
34. A program of imagined contractions can produce an increase in strength. **TRUE FALSE**
35. Coactivation increases the MVC force measured about a joint. **TRUE FALSE**
36. The amount of coactivation declines with strength training. **TRUE FALSE**
37. Cross education refers to the reduction in MVC force that occurs during bilateral contractions. **TRUE FALSE**
38. Cross education causes an increase in the MVC EMG of the untrained limb. **TRUE FALSE**
39. The bilateral deficit has a magnitude of about 20%. **TRUE FALSE**
40. Weight lifters exhibit a bilateral facilitation. **TRUE FALSE**
41. Synchronization is measured by comparing the amplitude of the M wave to the H response. **TRUE FALSE**
42. The increase in muscle size can be caused by hypertrophy or hyperplasia. **TRUE FALSE**
43. Hypertrophy occurs more frequently than hyperplasia with strength training. **TRUE FALSE**
44. Six months of strength training does not increase the cross-sectional area of muscle fibers in experienced body builders. **TRUE FALSE**
45. Type I muscle fibers do not hypertrophy with strength training. **TRUE FALSE**
46. Specific tension is greatest for Type S motor units. **TRUE FALSE**

47. Variation in the specific tension of motor units is due to differences in the density of the myofilaments. **TRUE FALSE**

48. Mechanical stimuli are able to modulate the rates of protein synthesis and degradation in muscle. **TRUE FALSE**

49. Mechanical stimuli influence the types of myosin heavy chains that are transcribed. **TRUE FALSE**

50. Power production represents the output of muscle during isometric contractions. **TRUE FALSE**

51. Power production is maximal when muscle force is about one third of maximum. **TRUE FALSE**

52. Power can be calculated as the product of force and work. **TRUE FALSE**

53. The average power produced during a vertical jump is about 600 W. **TRUE FALSE**

54. A weight lifter produces more power during the bench press than during the squat lift. **TRUE FALSE**

55. $\mathbf{P}_o$ = isometric torque. **TRUE FALSE**

56. The power produced by the muscles about a joint can be calculated as the product of torque and angular velocity. **TRUE FALSE**

57. Slow-twitch muscle is able to sustain power production at a greater value compared with fast-twitch muscle. **TRUE FALSE**

58. The most effective training loads for increasing power production are 100% MVC. **TRUE FALSE**

59. The decline in EMG during limb immobilization is influenced by the length at which the muscle is immobilized. **TRUE FALSE**

60. There is a parallel relationship between the decline in EMG and the loss of muscle mass with limb immobilization. **TRUE FALSE**

61. The medial gastrocnemius muscle experiences the most changes with limb immobilization. **TRUE FALSE**

62. Limb immobilization is associated with a fast-to-slow conversion of muscle fiber types. **TRUE FALSE**

63. The fatigability of muscle increases with limb immobilization. **TRUE FALSE**

64. There is an increase in the number of high-threshold motor units in immobilized muscle. **TRUE FALSE**

65. EMG activity in soleus is not substantially depressed for the entire period of hindlimb suspension. **TRUE FALSE**

66. Because the EMG activity does not decrease with hindlimb suspension, there is no muscle atrophy. **TRUE FALSE**

67. Hindlimb muscles can perform unresisted concentric contractions during hindlimb suspension. **TRUE FALSE**

68. The cross-sectional area of both Type I and Type II muscle fibers decreases with one-legged suspension in humans. **TRUE FALSE**

69. Hindlimb suspension produces a decline in the proportion of Type I muscle fibers in soleus. **TRUE FALSE**

70. Hindlimb suspension causes reductions in all enzymes associated with oxidative metabolism. **TRUE FALSE**

71. With hindlimb suspension the decline in tetanic force is greater than would be expected based on the loss of muscle mass in soleus. **TRUE FALSE**

72. Specific tension declines in slow-twitch muscle with hindlimb suspension. **TRUE FALSE**

73. Not all of the Type I fibers in soleus increase v_{max} after a few weeks of hindlimb suspension. **TRUE FALSE**

74. The spinal cord neurons receive no afferent feedback in the spinal transection model. **TRUE FALSE**

75. Spinal cord transection produces an immediate spasticity in muscle. **TRUE FALSE**

76. Muscle in a spinal transection preparation can be activated electrically. **TRUE FALSE**

77. Spinal transection produces muscle atrophy and a fast-to-slow conversion of muscle fiber types. **TRUE FALSE**

78. The metabolic adaptations at the muscle fiber level depend on the age at spinalization. **TRUE FALSE**

79. The maximum velocity of shortening decreases in a spinalized soleus muscle. **TRUE FALSE**

80. Changes in the time course of the twitch are due to changes in Ca^{2+} kinetics. **TRUE FALSE**

81. Spinalization causes changes in motor neuron properties. **TRUE FALSE**

82. Unusual reflexes can be elicited in patients with spinal cord injuries. **TRUE FALSE**

83. A cut axon can develop sprouts. **TRUE FALSE**

84. The process of chromatolysis occurs in muscle fibers deprived of a neural connection. **TRUE FALSE**

85. Cut axons can grow and reinnervate muscle fibers. **TRUE FALSE**

86. Regenerating axons can only reinnervate the original muscle fibers that were innervated prior to the axotomy. **TRUE FALSE**

87. Fine motor control is impaired after recovery from a complete nerve transection. **TRUE FALSE**

88. Collateral sprouting occurs at the distal ends of axons. **TRUE FALSE**
89. All motor units are able to develop collateral sprouts. **TRUE FALSE**
90. Collateral sprouting can replace only about 30% of the lost nerve supply with partial denervation. **TRUE FALSE**
91. The innervation ratio of motor units increases with collateral sprouting. **TRUE FALSE**
92. Collateral sprouting requires an intact connection with supraspinal centers. **TRUE FALSE**
93. The CNS is capable of synaptogenesis. **TRUE FALSE**
94. The spared-root preparation involves the removal of one half of the spinal cord. **TRUE FALSE**
95. Deafferentation occurs with a transection of the spinal cord. **TRUE FALSE**
96. The number of motor units in a muscle declines with age. **TRUE FALSE**
97. The average size of a motor unit decreases with age. **TRUE FALSE**
98. The fatigability of muscle during voluntary contractions is greater in older individuals. **TRUE FALSE**
99. The decrease in the peak-to-peak amplitude of the H reflex with age is due to a decline in muscle mass. **TRUE FALSE**
100. The latency of reflex responses does not change with age. **TRUE FALSE**
101. Sensory capabilities are impaired in older persons. **TRUE FALSE**
102. Elderly subjects have difficulty selecting an appropriate strategy to maintain an upright posture. **TRUE FALSE**
103. Older individuals have greater difficulty maintaining a constant submaximal isometric force. **TRUE FALSE**
104. Because of the loss of strength, grip forces are frequently too low in older persons. **TRUE FALSE**
105. The rate of axonal transport declines with age. **TRUE FALSE**
106. The decline in muscle mass with age is mainly due to the atrophy and disintegration of damaged muscle fibers. **TRUE FALSE**
107. The innervation ratio of motor units declines with age. **TRUE FALSE**
108. Type II muscle fibers seem to change the most with age. **TRUE FALSE**

MULTIPLE-CHOICE QUESTIONS

109. Which of the following is not one of the principles of training?
   A. Reversibility
   B. Transmissibility
   C. Overload
   D. Specificity
110. Which principle of training invokes the concept of a threshold point?
   A. Reversibility
   B. Transmissibility
   C. Overload
   D. Specificity
111. The measurement of muscle strength corresponds to which region of the force-velocity relationship?
   A. Velocity = 0
   B. Maximum velocity
   C. Velocity > 0
   D. Velocity < 0
112. Which parameter does not influence muscle strength?
   A. Rate of change in muscle length
   B. Moment arm
   C. Cross-sectional area
   D. Motor unit recruitment
113. What were the advantages ascribed to isometric exercises?
   A. Convenience
   B. Increased strength over the entire range of motion
   C. Good hypertrophic response
   D. Absence of risk factors
114. At what percentage of the maximum muscle force does blood flow begin to be occluded?
   A. 5%
   B. 15%
   C. 35%
   D. 55%
115. How much does muscle strength increase after 6 weeks of isometric exercises?
   A. 10%
   B. 20%
   C. 40%
   D. 80%
116. For what conditions does the muscle torque not equal the load torque?
   A. Isokinetic contraction
   B. Isometric contraction
   C. Dynamic contraction
   D. Static contraction

117. Which statements are correct?
   A. The magnitude of the 1-RM load is limited by the eccentric contraction force.
   B. Eccentric exercises produce greater increases in strength than concentric exercises.
   C. The amount of exercise stress depends on the magnitude of the load relative to maximum capabilities.
   D. Eccentric exercises provide a more effective stimulus for hypertrophy.

118. Why is the EMG during an eccentric MVC less than that during a concentric MVC?
   A. Submaximal concentric and eccentric contractions involve the activation of different motor units.
   B. There is a greater decline in the amplitude of H reflex during downhill running.
   C. The central drive is less than maximal.
   D. The EMG in leg muscles gradually increases during steady-state downhill running.

119. Which systems cannot provide an accommodating resistance?
   A. Gear systems
   B. Free weights
   C. Pneumatic systems
   D. Hydraulic cylinders

120. What are the advantages of an isokinetic device?
   A. Fulfills the requirements of the principle of specificity
   B. Provides increases in strength greater than those achieved with free weights
   C. Provides support for the user
   D. Enables the user to exercise over a limited part of the range of motion

121. Which phenomenon best describes the physiological basis of plyometric exercises?
   A. Work-energy relationship
   B. Stretch reflex
   C. Impulse-momentum relationship
   D. Eccentric-concentric contraction sequence

122. Which concepts account for the inability of an individual to exert the maximum muscle force with a voluntary contraction?
   A. Neural supplementation
   B. Cross education
   C. Bilateral facilitation
   D. Neural insufficiency

123. Which features are not influenced by the shape of the stimulus waveform in neuromuscular electrical stimulation?
   A. Comfort experienced by the subject
   B. Recruitment order of motor units
   C. The force-frequency relationship
   D. Electrode impedance

124. Which characteristics are not correct for magnetic stimulation?
   A. Generates action potentials in intramuscular nerve branches
   B. Is less painful than electrical stimulation
   C. Has a higher threshold for a motor response compared with a sensory response
   D. Has less of a decline in stimulus intensity over distance compared with electrical stimulation

125. Which observation concerning the use of neuromuscular electrical stimulation is not correct?
   A. The strength gains achieved with stimulation can improve voluntary performance.
   B. Stimulation can elicit a training response in the contralateral (nonstimulated) limb.
   C. A comparable increase in strength can be achieved at a much lower force with stimulation compared with voluntary exercise.
   D. It is more difficult to strengthen hypotrophic muscle.

126. What is the optimal load for strength training?
   A. 1 RM
   B. 2 to 3 RM
   C. 4 to 6 RM
   D. 8 to 12 RM

127. What percentage of maximum does a 5- to 6-RM load represent?
   A. 50% to 60%
   B. 70% to 80%
   C. 80% to 90%
   D. 90% to 100%

128. Why are variable-resistance devices supposed to be superior to free weights?
   A. The load remains constant over the range of motion.
   B. The variation in load closely approximates the maximum torque-angle relationship.
   C. The load torque about the joint varies more than that for free weights.
   D. The maximum load torque occurs in the middle of the range of motion.

129. Which observation does not indicate a role for neural adaptations in strength training?
   A. Strength can increase with training programs that are too brief to induce morphological changes.
   B. The change in EMG begins before a change in muscle size.
   C. The change in EMG is more substantial than the change in muscle size.
   D. There is no change in strength with a program of imagined contractions.

130. In which training protocols has cross education been observed?
   A. Voluntary contractions
   B. Endurance training
   C. Neuromuscular electrical stimulation
   D. Imagined contractions

131. What does reflex potentiation indicate?
   A. The synchronized discharge of motor units during an H reflex.
   B. The decline in M-wave amplitude in one limb while the other limb performs a voluntary contraction
   C. An index of motor unit activation that is used during a voluntary contraction
   D. The change from a bilateral deficit to a facilitation with training

132. What is hyperplasia?
   A. An increase in specific tension
   B. An increase in the number of muscle fibers
   C. An increase in the cross-sectional area of muscle fibers
   D. An increase in the cytoskeletal matrix

133. What could explain the difference in specific tension between motor units?
   A. Differences in the concentration of endomysial collagen
   B. Variation in the length of the muscle fibers
   C. Differences in the density of the myofilaments
   D. Variable cross-sectional area

134. Why are hormonal influences not the key factors that control muscle hypertrophy?
   A. They act too slowly.
   B. Hormones act systemically and influence many muscles.
   C. The metabolic cost of eccentric contractions is less than that of concentric contractions.
   D. The administration of growth hormone enhances the changes that occur with strength training.

135. How could mechanical stimuli modulate protein synthesis and degradation?
   A. Plasma membrane-associated molecules could act as second messengers.
   B. Through a stretch-induced increase in intracellular Ca^{2+}.
   C. An increase in the spontaneous generation of miniature end-plate potentials.
   D. Through enhanced transmission of the mechanical stimulus by the extracellular matrix.

136. Which two principal variables determine the quantity of power that can be exerted by a muscle?
   A. Cross-sectional area of muscle
   B. Myosin ATPase
   C. Ca^{2+} kinetics
   D. Muscle fiber length

137. How much power can the plantarflexor muscles produce during a vertical jump?
   A. 25 W
   B. 250 W
   C. 2.5 kW
   D. 25 kW

138. According to the principle of specificity, which training load (% MVC) should produce the greatest increase in power production?
   A. 10%
   B. 30%
   C. 60%
   D. 90%

139. Which experimental models are used to examine the adaptations in muscle with reduced use?
   A. Immobilization
   B. Spinal transection
   C. Tetrodotoxin
   D. Stretch

140. With limb immobilization, which conditions are associated with the greatest decrease in EMG?
   A. Soleus at a short length
   B. Medial gastrocnemius at a short length
   C. Soleus at a long length
   D. Medial gastrocnemius at a neutral length

141. What could explain the greater decline in motor unit force compared with cross-sectional area in a hindlimb muscle (of a cat) with limb immobilization?
   A. Decrease in the density of the myofilaments
   B. Decline in the innervation ratio
   C. Insufficient central drive
   D. Reduction in the concentration of endomysial collagen

142. Which of the following does not happen with limb immobilization?
   A. Fast-to-slow conversion of muscle fiber types
   B. Muscle atrophy
   C. Reduction in the MVC EMG
   D. Alterations in the recruitment order of motor units

143. In which experimental model does the EMG not remain depressed for the duration of the perturbation?
   A. Spinal transection
   B. Hindlimb suspension
   C. Limb immobilization
   D. One-legged suspension

144. Which of the following adaptations occur with hindlimb suspension?
   A. An increase in the concentrations of oxidative enzymes in medial gastrocnemius
   B. A preferential atrophy of Type I muscle fibers in human muscle
   C. An increase in the maximum velocity of shortening for some, but not all, Type I muscle fibers in soleus
   D. A reduction in twitch contraction time and half relaxation time
145. Which experimental models do not involve a slow-to-fast conversion of muscle fiber types?
   A. Spinal transection
   B. Hindlimb suspension
   C. Limb immobilization
   D. One-legged suspension
146. Which adaptations do not occur with a spinal transection?
   A. Muscles become incapable of being activated by electrical stimulation.
   B. The proportion of Type I muscle fibers decrease.
   C. There is no change in fatigability for the soleus muscle.
   D. The properties of some motor neurons change.
147. Which process involves the resynthesis of proteins by a neuron that has had its axon cut?
   A. Synaptogenesis
   B. Axotomy
   C. Chromatolysis
   D. Reinnervation
148. When does collateral sprouting occur?
   A. When a transected nerve is resutured and reinnervates muscle
   B. After partial section (denervation) of a nerve
   C. With spinal transection
   D. As part of the aging process
149. Which adaptation could not contribute to an increase in motor unit force after a partial section of a nerve?
   A. Increase in the innervation ratio
   B. Decrease in specific tension
   C. Increase in muscle fiber cross-sectional area
   D. Increase in the density of myofilaments
150. Which experimental model is not used to study CNS sprouting?
   A. Hemisection
   B. Spared-root preparation
   C. Deafferentation
   D. Unmasking
151. Why does muscle mass decline with age?
   A. Motor neurons die and deprive muscle fibers of neural innervation.
   B. Disease produces muscle atrophy.
   C. Older individuals become sedentary.
   D. Muscle is not stressed (exercised) enough to maintain high levels of protein synthesis.
152. Which adaptation is probably the most important reason for the delayed motor responses in older persons?
   A. Fewer motor neurons
   B. Decline in the number of muscle fibers
   C. Insufficient central drive from suprasegmental centers
   D. Reduction in sensory capabilities
153. Which factors contribute to the age-related decline in the ability to control posture?
   A. Reduced perception of stability limits
   B. Impaired ability to coordinate the activity of synergist muscles
   C. Poor detection of disequilibrium
   D. Abnormal selection of sensory information
154. Why do older individuals have greater difficulty maintaining a constant submaximal force?
   A. Decline in muscle mass
   B. Increased motor unit twitch force
   C. Fewer motor neurons
   D. Loss of Type II muscle fibers

PROBLEMS

155. An individual can lift a load of 1,050 N during a leg-lift exercise.
   A. What does the overload principle state about the load that this individual must use to gain strength with this exercise?
   B. Suppose the individual decided to use a load of 4 to 5 RM. What would be the magnitude of the load?
   C. According to McDonagh and Davies (1984), what is the minimum load that must be used in order to increase strength?
156. According to the principle of specificity, training adaptations are specific to the exercise stress. This suggests that gains in strength are achieved by performing strength exercises and improvements in endurance are best achieved by performing fatiguing exercises. Does strength training affect (increase or decrease) endurance? Similarly, does endurance training influence strength? Consult the research literature to answer this question; you may begin by checking the references cited at the back of this book.

157. Strength and maximum power production measure different regions of the force-velocity curve. Suppose you measured these two parameters in a group of 20 subjects. Would you expect to find a correlation between the two parameters? Use the force-velocity relationship to explain why you would or would not expect strength and maximum power production to be correlated.

158. Describe the advantages and disadvantages of isometric exercises.

159. The force that a muscle can exert is greater during an eccentric contraction than during a concentric contraction.
   A. Why do eccentric exercises not produce greater increases in strength?
   B. Why is it desirable to include eccentric exercises in a training program in order to obtain a good hypertrophic response?
   C. List five ways in which eccentric contractions differ from concentric contractions.

160. Why does an accommodation device produce an isokinetic contraction?

161. Why does the recruitment order of motor units with neuromuscular electrical stimulation differ from that of voluntary activation of the motor unit pool?

162. Neuromuscular electrical stimulation has been used as an artifical way to activate muscle.
   A. Where are the action potentials generated?
   B. The neural insufficiency and neural supplementation concepts are two reasons given for using electrical stimulation as a supplement. Explain these two concepts and distinguish between them.
   C. List the three research findings on neuromuscular electrical stimulation that are described as interesting in the chapter.

163. Consider the individual performing a bent-knee sit-up in Figure 9.5. Suppose the parameters have the following dimensions:

$\mathbf{F}_i$ = 1,200 N
$\mathbf{F}_j$ = 1,570 N
$\mathbf{F}_w$ = 487 N
Angle of push for $\mathbf{F}_i$ = 0.67 rad
Distance from hips to point of application for $\mathbf{F}_i$ = 27 cm
Angle of pull for $\mathbf{F}_m$ = 0.45 rad
Distance from hips to point of application for $\mathbf{F}_m$ = 8 cm

164. Machines that provide a variable load are supposedly designed to match the capabilities of selected joint systems.
   A. For the elbow joint, draw a diagram to show how the moment arm on the cam should change to match the resultant muscle torque capabilities.
   B. Draw a diagram to show how the load applied with a circular cam would change over the range of motion.

165. The statement is made that it is possible to obtain an increase in strength without an adaptation in the muscle but not without an adaptation in the nervous system.
   A. What does this statement mean?
   B. What research evidence supports this statement?
   C. Which populations of subjects would demonstrate this effect?

166. Some of the neural adaptations that occur with strength training involve the interactions between limbs. Typically an individual exerts a lower maximal force with a given limb when both limbs are active compared with the maximal force associated with a single-limb contraction. Howard (1987) measured this effect for the knee extensor muscle before and after a 3-week strength-training program. MVC force and EMG (vastus lateralis) were measured for one-legged (right and left separately) and two-legged (both legs together) contractions.

	One-legged		Two-legged	
	Right	Left	Right	Left
Force (N)				
Before	782	776	762	726
After	950	897	981	966
EMG (mV)				
Before	0.55	0.51	0.53	0.51
After	0.67	0.65	0.69	0.84

Use these data and the following equation to determine the bilateral index (BI) for EMG and force before and after the training. *Two-legged* refers to the sum of the right and left during the two-legged contraction, and *right leg + left leg* indicates the sum of the one-legged contractions. Explain what the results suggest about the types of adaptations that occurred.

$$\text{BI (\%)} = \left[100 \times \left(\frac{\text{two legged}}{\text{right leg + left leg}}\right)\right] - 100$$

167. Why are Type II muscle fibers more likely than Type I fibers to increase cross-sectional area with strength training?

168. Several studies have found a strong correlation between muscle size (e.g., cross-sectional area) and the maximum force that a muscle can exert. Figure 9.16, however, shows considerable variability between the cross-sectional area of quadriceps femoris and the maximum isometric force. List as many reasons as possible that could account for this variability.

169. Assume a linear force-velocity relationship, and estimate the maximum power that can be produced by the knee flexor muscles of an individual. The knee flexor muscles have a strength of 237 N·m and a maximum angular velocity of 17.3 rad/s.

170. Brooks and Faulkner (1991) expressed power with the unit of measurement W/kg. Why did they do this?

171. The drugs tetrodotoxin, curare, and colchicine can be used to induce models of reduced use in order to study the adaptive capabilities of the neuromuscular system. Consult the literature (e.g., Gordon & Pattulo, 1993; Lieber, 1992; Pette & Vrbová, 1992; Roy et al., 1991) to determine the effect of each of these three drugs.

172. Most models of reduced use have indicated a conversion of muscle fiber types from slow-twitch to fast-twitch.
    A. What happens to the metabolic enzymes with these models? How are these effects tested?
    B. What happens to the contractile proteins with reduced use?
    C. Why does this conversion not happen in all Type I muscle fibers?

173. Contrast the decline in activity that is achieved with the three models of reduced use (limb immobilization, hindlimb suspension, and spinal transection).

174. Based on the data of Tötösy de Zepetnek et al. (1992), motor units in normal and reinnervated muscle have the following characteristics. Calculate the average maximum force exerted by each motor unit type (S, FR, FF) in normal and reinnervated muscle.

175. Which two age-related adaptations seem to have the greatest impact on the impairment of motor capabilities?

176. The peak-to-peak amplitude of both the H reflex and the M wave decrease with age. What does this tell us about changes in the motor system with age?

177. Aging is associated with a decline in Type IIb muscle fibers. In contrast, reduced use produces a slow-to-fast conversion of muscle fiber types. In each case, however, most studies have found no change in fatigability. Why is there no effect on fatigability in both paradigms?

		Normal			Reinnervated	
	S	FR	FF	S	FR	FF
Muscle fiber CSA* (μm^2)	1,317	2,048	4,134	2,252	2,424	3,897
Innervation ratio	91	91	152	110	97	469
Specific tension (N/cm^2)	27.3	24.5	23.5	24.4	25.6	25.6

*CSA = cross-sectional area

Summary of Part III

The goal of Part III (chapters 7 to 9) has been to extend the concept of the single joint system to provide a more complete account of those components of the human body that are involved in the performance of movement. This description has included more detail on the neural and muscular organization and operation of the motor system (chapter 7), as well as a discussion of the acute (chapter 8) and chronic (chapter 9) adaptive capabilities of the system. As a result of reading these chapters, you should be familiar with the following features and concepts related to the motor system:

- Observe the organization of the suprasegmental components of the nervous system and the flow of information associated with the control of movement
- Note the structure of the musculoskeletal system as a set of multiple single joint systems
- Understand the movement strategies used by the motor system
- Realize the effect of altering core temperature on performance capabilities
- Conceive of the rationale for the techniques that have been developed to alter flexibility
- Identify the multifactorial basis of muscle fatigue and acknowledge that performance can be impaired by different processes depending on the details of the task
- Understand the sensory adaptations that occur during fatiguing contractions
- Comprehend the potentiating capabilities of muscle
- Know the performance characteristics of strength and power and the mechanisms that mediate changes in these capabilities
- Realize the extent of the adaptations that occur in the neuromuscular system with a reduction in the level of physical activity
- Acknowledge the ability of the motor system to recover some function following an injury to the nervous system
- Identify the age-related adaptations in motor function and the mechanisms that mediate these changes

APPENDIX A

SI Units

The abbreviation SI is derived from *Le Système Internationale d'Unités*, which represents the modern metric system. There are seven base units in this measurement system from which the other units of measurement are derived.

Base	*Derived*
1. **length**—meter (m) Defined as 1,650,763.73 wavelengths of an isotope of the element krypton (Kr-86). 1 in. = 2.54 cm 1 ft = 30.48 cm = 0.3048 m 1 yd = 0.9144 m 1 mile = 1,609 m = 1.609 km	**area**—square meters (m^2) Two-dimensional measure of length. 1 ft^2 = 0.0929 m^2 1 acre = 0.4047 hectares (ha) **volume**—cubic meters (m^3) Three-dimensional measure of length. Although the liter is not an SI unit of measurement, volume is often measured in liters (L), where 1 mL of H_2O = 1 cm^3 1 L = 0.001 m^3
2. **mass**—kilogram (kg) Defined as the mass of a platinum iridium cylinder preserved in Sevres, France. 1 lb = 0.454 kg	**density**—kilogram/cubic meters (kg/m^3) Mass per unit volume. **energy, work**—joule (J) Energy denotes the capacity to perform work, and work refers to the application of a force over a distance. 1 J = 1 N·m 1 kcal = 4.183 kJ 1 kpm = 9.807 J **force**—newton (N) One newton is a force that accelerates a 1-kg mass at a rate of 1 m/s^2. 1 N = 1 $kg \cdot m/s^2$ 1 kg-force = 9.81 N 1 lb-force = 4.45 N

Base	*Derived*
	impulse—newton·second (N·s) The application of a force over an interval of time; the area under a force-time curve.
	moment of inertia (kg·m^2) The resistance that an object offers to a change in its state of angular motion; a measure of the proximity of the mass of an object to an axis of rotation.
	power—watt (W) The rate of performing work. 1 W = 1 J/s 1 horsepower = 736 W
	pressure—pascal (Pa) The force applied per unit area. 1 Pa = 1 N/m^2 1 mmHg = 133.3 Pa
	torque—moment of force (N·m) The rotary effect of a force. 1 ft-lb = 1.356 N·m
3. **time**—second (s) Defined in terms of one characteristic frequency of a cesium clock (9,192,631,770 cycles of radiation associated with a specified transition of the cesium-133 atom).	**acceleration** (m/s^2) Time rate of change in velocity. Gravity produces an acceleration of 9.807 m/s^2. 1 ft/s^2 = 0.3048 m/s^2
	frequency—hertz (Hz) The number of cycles per second. 1 Hz = 1 cycle/s
	momentum (kg·m/s) Quantity of motion. 1 slug · ft/s = 4.447 kg·m/s
	speed—velocity (m/s) Time rate of change in position, where speed refers to the size of the change and velocity indicates its size and direction. 1 ft/s = 0.3048 m/s 1 mph = 0.447 m/s = 1.609 km/hr
4. **electric current**—ampere (A) Rate of flow of charged particles.	**capacitance**—farad (F) The property of an electrical system of conductors and insulators that enables it to store electric charge when a potential difference exists between the conductors.
	conductance—mho The reciprocal of resistance, thus the ease with which charged particles move through an object.
	resistance—ohm (Ω) The difficulty with which charged particles flow through an object.
	voltage—volt (V) The difference in net distribution of charged particles between two locations.

Base	*Derived*
5. **temperature**—kelvin (K) A measure of the velocity of vibration of the molecules of a body. There is 100 K from ice point to steam point. 0 K = absolute zero	**Celsius** (°C) 0°C = 273.15 K **Fahrenheit** (°F) 32°F = 0°C = 273.15 K °F = 1.8°C + 32
6. **amount of substance**—mole (mol) The amount of substance containing the same number of particles as are in 12 g (1 mol) of the nuclide ^{12}C.	**concentration** (mol/m^3) Amount of substance per unit volume.
7. **luminous intensity**—candela (cd) The luminous intensity of 1/600,000 of 1 m^2 of a black body at the temperature of freezing platinum (2,045 K).	**lumen** (1m) Measure of light flux
Supplementary Unit	
8. **angle**—radian (rad) Measurement of an angle in a plane (i.e., two-dimensional angle). 1 rad = 57.3° Theta (θ) equals one radian when the radius (*r*) and arc (*a*) are equal (see Figure A.1).	**steradian** (sr) Three-dimensional angle, also known as a solid angle.

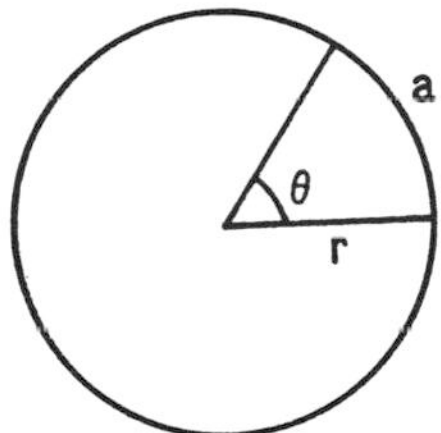

Figure A.1 Definition of a radian.

It is preferable for numbers to range from 0.1 to 9,999. Prefixes (Table A.1) can be attached to the units of measurement to represent a smaller or larger amount of the unit. For example, as indicated previously, the SI unit of length is the meter (m). The average height of U.S. males is about 1.73 m. To express larger distances (e.g., marathon — 26 miles, 385 yards), it is more convenient to refer to thousands of meters. The prefix kilo (k) represents 1,000; thus, 1 km = 1,000 m. A marathon race, therefore, covers a distance of approximately 42.4 km. Similarly, prefixes referring to parts of a meter can be used for small distances. For example, there are 1,000 millimeters (mm) in one meter (1 mm = 0.001 m).

Table A.1 Common Prefixes Used With SI Units of Measurement

Prefix	Abbreviation	Multiplication factor		
Tera	T	1,000,000,000,000	=	10^{12}
Giga	G	1,000,000,000	=	10^{9}
Mega	M	1,000,000	=	10^{6}
Kilo	k	1,000	=	10^{3}
Hecto	h	100	=	10^{2}
Deci	d	0.1	=	10^{-1}
Centi	c	0.01	=	10^{-2}
Milli	m	0.001	=	10^{-3}
Micro	µ	0.00 001	=	10^{-6}
Nano	n	0.000 000 001	=	10^{-9}
Pico	p	0.000 000 000 001	=	10^{-12}

APPENDIX B

Basic Mathematics

Algebra

Fractions

A ratio between two quantities, such as 3/4, in which the numerator (3) is divided by the denominator (4), is referred to as a **fraction**. A similar result is obtained by mulitplying the numerator by the inverse of the denominator;

$$\frac{3}{4} = 3 \times 4^{-1} = 0.75$$

Multiplication. Multiply both the number and units of measurement of all numerators together, and place the product over the result obtained with the denominators.

$$60 \text{ kg} \times \frac{16 \text{ m}}{2 \text{ s}} = \frac{60 \times 16 \text{ kg·m}}{2 \text{ s}}$$

$$= 480 \frac{\text{kg·m}}{\text{s}}$$

$$= 480 \text{ kg·m·s}^{-1} \text{ (or kg·m/s)}$$

Division. Invert the dividing fraction and multiply as done in the previous example.

$$480 \frac{\text{kg·m}}{\text{s}} \div \frac{16 \text{ m}}{2 \text{ s}} = 480 \frac{\text{kg·m}}{\text{s}} \times \frac{2 \text{ s}}{16 \text{ m}}$$

$$= \frac{480 \times 2 \text{ kg·m·s}}{16 \text{ m·s}}$$

$$= 60 \text{ kg}$$

Addition and Subtraction. Establish a common denominator into which the denominator of all fractions will divide, multiply the numerator to convert each fraction to the common denominator, and then perform the addition and subtraction.

$$\frac{76\text{ N}}{1\text{ s}} - \frac{12\text{ N}}{0.5\text{ s}} + \frac{3\text{ N}}{3\text{ s}} = \frac{[(3 \times 76) - (12 \times 6) + (3 \times 1)]\text{ N}}{3\text{ s}}$$

$$= \frac{[228 - 72 + 3]\text{ N}}{3\text{ s}}$$

$$= \frac{159\text{ N}}{3\text{ s}}$$

$$= 53\ \frac{\text{N}}{\text{s}}\ (\text{N}\cdot\text{s}^{-1}\text{, or N/s})$$

Percent. To convert fractions to percentages, divide the numerator by the denominator and multiply by 100.

$$\frac{0.4\text{ m}}{1.3\text{ m}} = 0.4 \times 1.3^{-1} \times 100$$

$$= 31\%$$

In this procedure, the units of measurement cancel, and the result (%) is described as dimensionless (i.e., unitless).

Equations

An expression of the equality of two sets of relationships is known as an equation. Often the relations include one or several quantities whose values are unknown.

Solving for One Unknown Quantity. To determine the value of one unknown quantity, we need to rearrange the equation to isolate the unknown quantity on one side of the equation.

$$(100 \times d) - \frac{50}{2.5} + 37 = (16 \times 9)$$

$$(100 \times d) - \frac{50}{2.5} + \frac{50}{2.5} + 37 = (16 \times 9) + \frac{50}{2.5}$$

$$(100 \times d) + 37 = (16 \times 9) + \frac{50}{2.5}$$

$$(100 \times d) + 37 - 37 = (16 \times 9) + \frac{50}{2.5} - 37$$

$$(100 \times d) = (16 \times 9) + \frac{50}{2.5} - 37$$

$$(100 \times d)/100 = [(16 \times 9) + \frac{50}{2.5} - 37]/100$$

$$d = [(16 \times 9) + \frac{50}{2.5} - 37]/100$$

$$d = 1.27$$

Solving for Two Unknown Quantities. Suppose an individual is going to perform a weight-lifting workout and wants to lift a total load of only 3,900 N for a particular exercise and wants to do so with several repetitions with 500-N and 200-N loads. Find all possible combinations of 500- and 200-N loads that the individual could lift to achieve the total of 3,900 N. Let f = the number of repetitions with the 500-N load and t = the number of repetitions with the 200-N load.

$$500f + 200t = 3{,}900$$

$$5f + 2t = 39$$

$$t = \frac{39 - 5f}{2}$$

Because the number of repetitions performed with each load must be a *whole* number, f is confined to odd numbers. Furthermore, if $f > 7$, t will be negative, so f can have values of 1, 3, 5, and 7.

f	t	(f, t)
1	17	(1, 17)
3	12	(3, 12)
5	7	(5, 7)
7	2	(7, 2)

The possible combinations of repetitions to achieve the 3,900 N total is given by the set under the (f, t) column.

Finding the Equation of a Line. Two methods can be used to determine the equation for a line. These are the **point-slope equation of a line,**

$$y - y_1 = m(x - x_1) \tag{B.1}$$

and the **slope-intercept equation of a line,**

$$y = mx + b \tag{B.2}$$

If a point and a slope (e.g., 4, –3 and –2/5) are given, the equation of the line can be determined as follows:

$$y - y_1 = m(x - x_1)$$

$$y - (-3) = -\frac{2}{5}(x - 4)$$

$$5(y + 3) = -2(x - 4)$$

$$5y + 15 = -2x + 8$$

$$2x + 5y = -7$$

If two points are given [e.g., (1, –2) and (5, 1)], the equation of the line can be determined by first finding the slope of the line and then using the point-slope equation:

$$m = \frac{1 - (-2)}{5 - 1}$$

$$m = \frac{3}{4}$$

$$y - y_1 = m(x - x_1)$$

$$y - (-2) = \frac{3}{4}(x - 1)$$

$$4y + 8 = 3x - 3$$

$$3x - 4y = 11$$

If a slope and a y-intercept are given (e.g., -2 and $\frac{4}{3}$), the equation of the line can be determined by using the slope-intercept equation:

$$y = mx + b$$

$$y = -2x + \frac{4}{3}$$

$$3y = -6x + 4$$

Inverse Relations. A function defined by the equation

$$xy = k, \qquad \text{or} \qquad y = \frac{k}{x}$$

is called an **inverse relation**. Typically x and y are positive, so they are inversely related: As x gets larger, y gets smaller (Figure B.1). If the two variables (x and y) are related by a constant (k), then the inverse relationship is linear and the slope of the line is given by k.

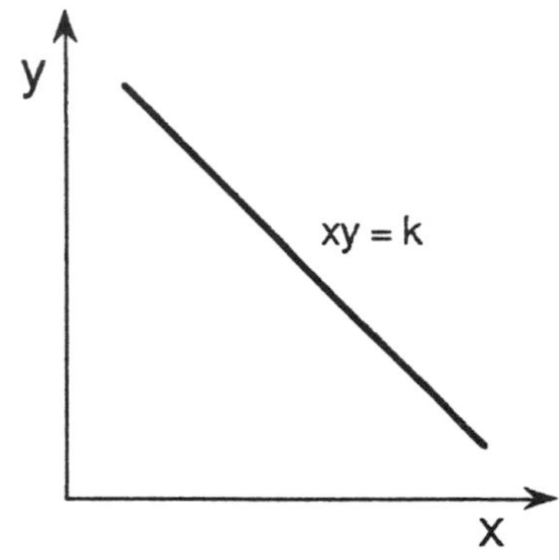

Figure B.1 An inverse function.

Examples of inverse relationships include the algebraic definitions of centripetal force (Equation 2.1) and Newton's law of gravitation (Equation 2.4). According to the law of gravitation, the force of attraction between two masses ($\mathbf{F} \propto \frac{m_1 m_2}{\mathbf{r}^2}$) depends on the product of the masses and the inverse of the distance squared between the centers of the two masses.

Polynomials. A sum of terms that are powers of a variable is termed a **polynomial**. For example, $3x^3 + 4x - 5$ is a third-degree polynomial of the independent variable x. The degree of the polynomial is defined by the highest power in the expression. Dependent variables (e.g., y) can be defined by polynomials:

$$y = 3x^3 + 4x - 5$$

Polynomials are sometimes used to represent the change in a kinematic variable (i.e., position, velocity, acceleration) as a function of time. For example, a position-time relationship may be represented by a fifth- or seventh-degree polynomial (Wood, 1982). The equations of motion (Equations 1.3, 1.4, and 1.5) are polynomial equations. Second-degree polynomials are known as **quadratic** relationships, and third-degree polynomials are referred to as **cubic** functions.

Two or more polynomials can be added by simply summing like terms. Similarly, polynomials can be subtracted by adding the opposite of like terms. To multiply polynomials, multiply each term of one polynomial by each term of the other polynomial, and add the resulting like terms. For example,

$$
\begin{aligned}
(2x + 3)(x^3 - 5) &= 2x(x^3 - 5) + 3(x^3 - 5) \\
&= 2x^4 - 10x + 3x^3 - 15 \\
&= 2x^4 + 3x^3 - 10x - 15
\end{aligned}
$$

Recall that the multiplication of polynomials involves the laws of exponents:

$$x^m \cdot x^n = x^{m+n} \tag{B.3}$$

$$(xy)^m = x^m y^m \tag{B.4}$$

$$(x^m)^n = x^{mn} \tag{B.5}$$

$$\frac{x^m}{x^n} = x^{m-n} \text{ for } m > n \tag{B.6}$$

$$\frac{x^m}{x^n} = \frac{1}{x^{n-m}} \text{ for } m < n \tag{B.7}$$

$$\left(\frac{x}{y}\right)^m = \frac{x^m}{y^m} \tag{B.8}$$

Circles

A circle is a set of points confined to a single plane and equidistant from a single point. A circle with center (h, k) and radius r is described by the equation

$$(x - h)^2 + (y - k)^2 = r^2 \tag{B.9}$$

The properties of a circle include the following:

a = arc length

r = radius

D = diameter = $2r$

π = pi = ratio of circumference to diameter = 3.1415926 . . .

C = circumference = $2\pi r$

$A = \text{area} = \frac{C^2}{4}\pi = \pi r^2$

θ = angle

When $a = r$ (as in Figure B.2), $\theta = 1$ rad (57.3°); 2π rad = 360°. Because a radian is defined as a ratio of distances, it is dimensionless.

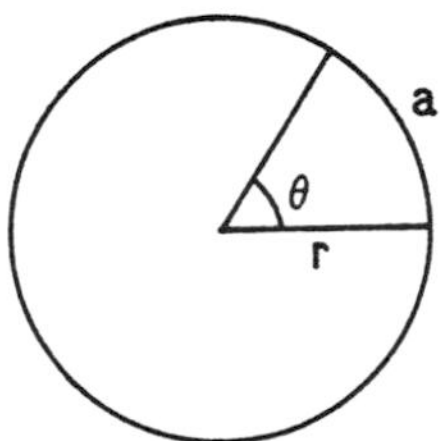

Figure B.2 Characteristics of a circle.

Parabolas

A **parabola** is a set of points that are equidistant from a fixed line (the **directrix**) and a fixed point not on the line (the **focus**) (Figure B.3). The **vertex** is midway between the focus and the directrix, and the graph is symmetrical about the *axis*, a line that connects the focus and the vertex.

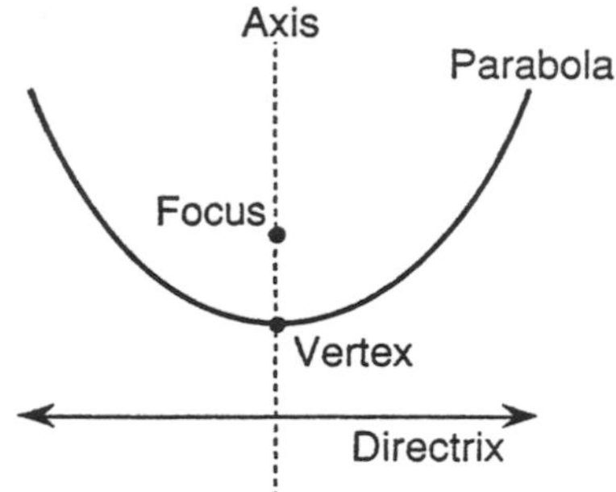

Figure B.3 Characteristics of a parabola.

The equation for a parabola can have the following form:

$$y - k = a(x - h)^2 \tag{B.10}$$

If the distance between the focus and the vertex is c, then $a = \frac{1}{4c}$. If $a > 0$, the parabola opens upward, if a < 0 the parabola opens downward (Figure B.4a). Projectiles that experience a minimal effect due to air resistance have a parabolic trajectory that opens downward. The upward-downward parabolas have the following features:

Vertex, $V(h, k)$
Focus, $F(h, k + c)$
Directrix, $y = k + c$
Axis of symmetry, $x = h$

A parabola that opens horizontally has the equation

$$x - h = a(y - k)^2 \tag{B.11}$$

The parabola opens to the right if $a > 0$ (Figure B.4b) and to the left if $a < 0$. Furthermore, the parabola has the following features:

Vertex, $V(h, k)$
Focus, $F(h + c, k)$
Directrix, $x = h - c$
Axis of symmetry, $y = k$

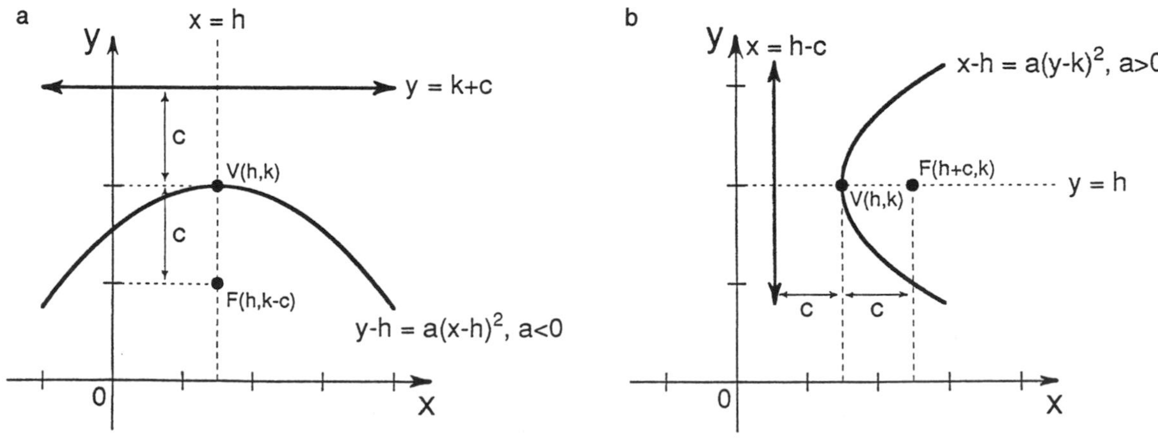

Figure B.4 Characteristics of parabolas that open downward (a) and to the right (b).

Exponentials and Logarithms

An **exponent** is a superscript placed to the right of a variable to indicate its power. For example, y^4 has an exponent of 4 and means y multiplied by itself four times ($y \times y \times y \times y$). An **exponential function** has the general form $y = b^x$, where x is the independent variable, b is a constant, and $b > 0$ and $b \neq 1$. If $b > 1$, then the function curves upward to the right (Figure B.5). If $0 < b < 1$, the graph curves upward to the left (Figure B.5). The graph of every exponential function has a y-intercept of 1 because $b^0 = 1$.

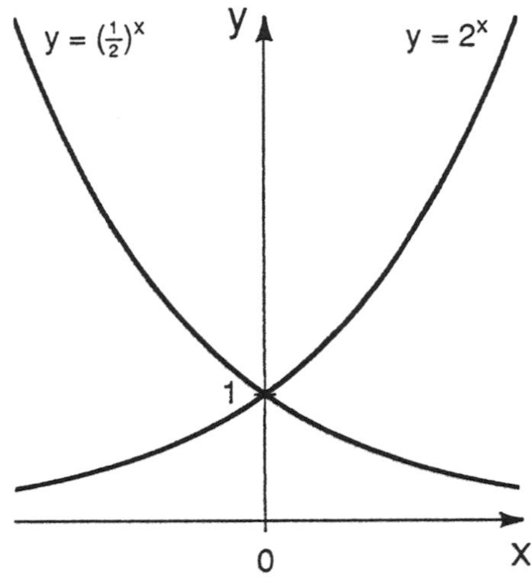

Figure B.5 Characteristics of exponential functions.

The most important exponential function is $y = e^x$ (sometimes written as $y = \exp x$), where e is the base of the natural logarithms (2.7182818). The constant e is the base of the natural system of logarithms (ln). The relationship between $y = x$, $y = e^x$, and $y = \ln x$ is shown in Figure B.6. Many biological processes occur as exponential functions. For example, when a bacteria population (N_o) doubles every d days, the number of bacteria in the population (N) at time t can be determined by

$$N = N_o \cdot 2^{t/d} \tag{B.12}$$

Similarly, when a chemical compound (N) decays with a half-life of h, the amount remaining (N) at time t can be determined by

$$N = N_o\ (0.5)^{t/h} \tag{B.13}$$

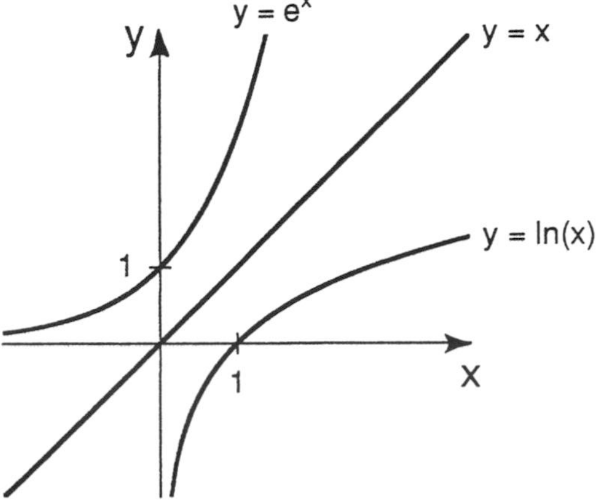

Figure B.6 Graphs of exponential and logarithmic functions.

Logarithms were devised as computational aids. Numbers are converted to logarithms, and such operations as multiplication, division, and square root are performed by addition, subtraction, and division. Once the operation has been performed, the result is converted with an antilogarithm to yield the answer. The logarithm of a number is the value (x) to which the number (b) must be raised to yield the specified number (n). If $b^x = n$, then $\log_b n = x$; thus, the graph of a logarithmic function is simply the inverse of the exponential function (Figure B.6). The number b is the logarithm base; common logarithms have a base of 10, whereas natural logarithms have a base of e (2.71828). A logarithm is written as the sum of an integer (the characteristic) and a decimal (the mantissa). The *characteristic* indicates the location of the decimal point and is positive for numbers greater than 1 and negative for numbers less than 1. The *mantissa* is the logarithm of the digits in the number. For example, $\log_{10} 25 = 1.3979$, where 1 is the characteristic ($1 = 10^1$) and 0.3979 is the mantissa; and $\ln 25 = 3.2189$, where 3 is the characteristic ($3 = e^3$) and 0.2189 is the mantissa. The laws of exponents can be used to derive the laws of logarithms:

$$\log_{10} XY = \log_{10} X + \log_{10} Y \tag{B.14}$$

$$\log_{10} \frac{X}{Y} = \log_{10} X - \log_{10} Y \tag{B.15}$$

$$\log_{10} X^k = k \log_{10} X \tag{B.16}$$

Logarithms are useful in biology for plotting graphs and interpreting statistics. When the values on an axis are expressed as logarithms, it is possible to plot data over a wide range of values on a single graph. Such graphs can have the values on only one axis expressed as logarithms (a semi-log graph) or the values on both axes can be transformed to logarithms (a log-log graph). Sometimes biologists find it necessary to transform a data set to logarithms so that parametric statistical procedures can be performed (e.g., analysis of variance). This transformation modifies the variance within the data set so that the different groups have equal variances and the assumption of variance homogeneity is not violated.

Trigonometry

Trigonometry is a branch of mathematics that concerns relations among the sides and angles of a triangle (*tri* = three, *gonia* = angle, and *metria* = measure).

Right Triangle

Most problems in motion analysis involve the use of right triangles, which are triangles containing internal right angles (1.57 rad, or 90°). The characteristics of a right triangle include the following (Figure B.7):

A = right angle

a = hypotenuse (longest side, which is opposite the right angle)

$B + C = A = 1.57$ rad

$a^2 = b^2 + c^2$ (**Pythagorean relationship**) (B.17)

area = $0.5\ bc$

$$h = \frac{bc}{a}$$

$$m = \frac{c^2}{a}$$

$$n = \frac{b^2}{a}$$

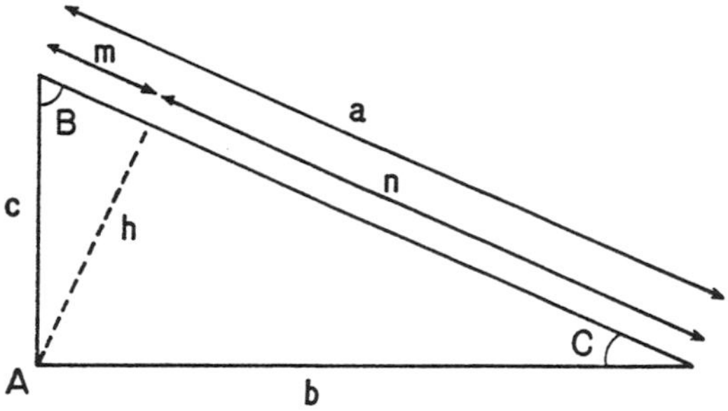

Figure B.7 Features of a right triangle.

Trigonometric Functions

The angles within a right triangle are defined by the ratios of the sides of the triangle. Given that B and C represent the acute angles of a right triangle and that a, b, and c are the sides opposite the respective angles (Figure B.7), the following relations are true:

$$\textbf{sine} = \frac{\text{opposite side}}{\text{hypotenuse}} \qquad \sin B = \frac{b}{a} \qquad \sin C = \frac{c}{a} \tag{B.18}$$

$$\textbf{cosine} = \frac{\text{adjacent side}}{\text{hypotenuse}} \qquad \cos B = \frac{c}{a} \qquad \cos C = \frac{b}{a} \tag{B.19}$$

$$\textbf{tangent} = \frac{\text{opposite side}}{\text{adjacent side}} \qquad \tan B = \frac{b}{c} \qquad \tan C = \frac{c}{b} \tag{B.20}$$

$$\textbf{cotangent} = \frac{\text{adjacent side}}{\text{opposite side}} \qquad \cot B = \frac{c}{b} \qquad \cot C = \frac{b}{c} \tag{B.21}$$

$$\textbf{secant} = \frac{\text{hypotenuse}}{\text{adjacent side}} \qquad \sec B = \frac{a}{c} \qquad \sec C = \frac{b}{c} \tag{B.22}$$

$$\textbf{cosecant} = \frac{\text{hypotenuse}}{\text{opposite side}} \qquad \csc B = \frac{a}{b} \qquad \csc C = \frac{a}{c} \tag{B.23}$$

As the angle in a triangle varies from 0 to 6.28 rad (2π), the trigonometric functions sine and cosine vary between +1 and −1 (Figure B.8). The sine curve is symmetric with respect to the origin, whereas the cosine term is symmetric about the y-axis. If the cosine function were shifted to the left by $\pi/2$, it would coincide with the sine function. When the sine or cosine term is multiplied by a coefficient (e.g., $y = a \sin x$), the amplitude will vary between $\pm a$. When the sine or cosine term is added to a constant (e.g., $b + a \sin x$), the function is shifted up ($+b$) or down ($-b$) the y-axis. When the variable (x) is multiplied by a constant (e.g., $\sin ax$), there are a repetitions of the function within 6.28 rad. In general, the functions $y = b \sin ax$ and $y = b \cos ax$ have an amplitude of a and a period of $2\pi/b$.

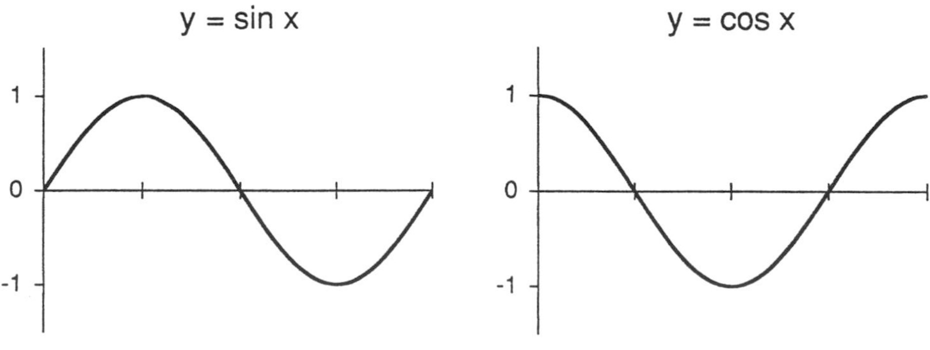

Figure B.8 Sine and cosine functions.

For the triangle shown in Figure B.9a, find θ, b, and a.

Find θ:

$$\theta = 3.14 - (1.57 + 0.90)$$
$$\theta = 0.67 \text{ rad}$$

Find b:

$$\sin 0.9 = \frac{b}{5}$$
$$b = 5 \sin 0.9$$
$$b = 3.92 \text{ cm}$$

or

$$\cos \theta = \frac{b}{5}$$
$$b = 5 \cos 0.67$$
$$b = 3.92 \text{ cm}$$

Find a:

$$\cos 0.9 = \frac{a}{5}$$
$$a = 5 \cos 0.9$$
$$a = 3.11 \text{ cm}$$

or

$$\sin \theta = \frac{a}{5}$$
$$a = 5 \sin 0.67$$
$$a = 3.11 \text{ cm}$$

Check:

$$\text{hypotenuse} = \sqrt{a^2 + b^2}$$
$$= \sqrt{3.92^2 + 3.11^2}$$
$$= \sqrt{15.37^2 + 9.67^2}$$
$$= \sqrt{25.04}$$
$$= 5 \text{ cm}$$

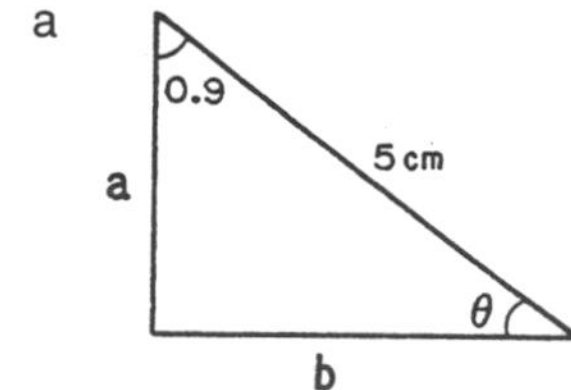

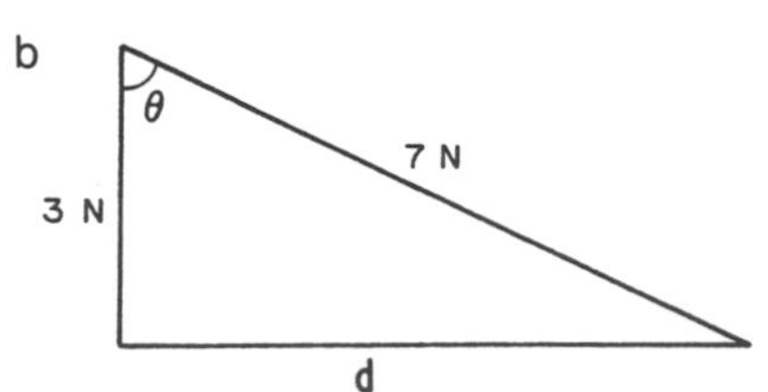

Figure B.9 Right triangles with some unknown functions.

For the triangle shown in Figure B.9b, find θ and *d*.
Find θ:

$$\cos\theta = \frac{3}{7}$$

$$\theta = \cos^{-1}\frac{3}{7}\ (\cos^{-1} = \text{arc cosine})$$

$$\theta = \cos^{-1} 0.4286$$

$$\theta = 1.13 \text{ rad}$$

Find *d*:

$$d^2 = 7^2 - 3^2$$

$$d^2 = 49 - 9$$

$$d^2 = 40$$

$$d = 6.32 \text{ N}$$

or

$$\sin\theta = \frac{d}{7}$$

$$d = 7 \sin 1.13$$

$$d = 6.32 \text{ N}$$

Any Triangle

1. Law of cosines—For any triangle *ABC* (Figure B.10), the following relationships apply:

$$c^2 = a^2 + b^2 - 2ab \cos C \tag{B.24}$$

$$b^2 = a^2 + c^2 - 2ac \cos B \tag{B.25}$$

$$a^2 = b^2 + c^2 - 2bc \cos A \tag{B.26}$$

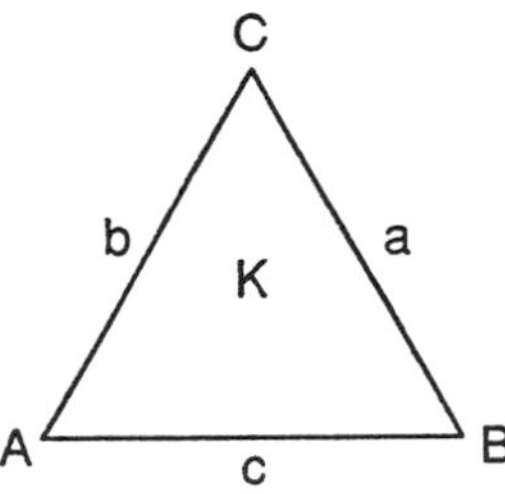

Figure B.10 Triangle *ABC*.

2. Law of sines—For any triangle *ABC* (Figure B.10), the following relationships apply:

$$\frac{\sin A}{a} = \frac{\sin B}{b} = \frac{\sin C}{c} \tag{B.27}$$

3. Area of a triangle—The area (K) of any triangle ABC (Figure B.10) can be determined by the following:

$$K = 0.5bc \sin A \tag{B.28}$$

$$K = 0.5ac \sin B \tag{B.29}$$

$$K = 0.5ab \sin C \tag{B.30}$$

$$K = 0.5a^2 \frac{\sin B \sin C}{\sin A} \tag{B.31}$$

$$K = 0.5b^2 \frac{\sin A \sin C}{\sin B} \tag{B.32}$$

$$K = 0.5c^2 \frac{\sin A \sin B}{\sin C} \tag{B.33}$$

Vectors

Characteristics

Quantities that convey magnitude and direction are called **vectors** (e.g., velocity, acceleration, force, momentum). Those variables that are defined by a magnitude only are called **scalars** (e.g., mass, length, speed, time, temperature). Vectors can be represented graphically (Figure B.11) as a directed line segment (i.e., as arrows). The length of the arrow specifies the magnitude of the vector, and its direction is indicated by a line of action and a sense (direction of arrowhead).

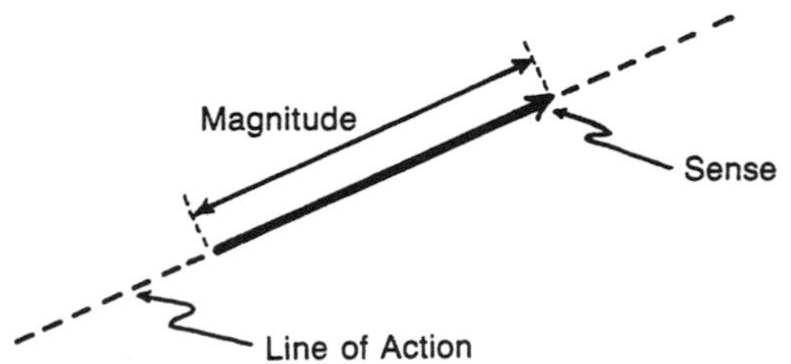

Figure B.11 Features of a vector.

Force Parallelogram Law

Two forces ($\mathbf{F}_p$ and $\mathbf{F}_q$) acting on an object can be replaced by their resultant ($\mathbf{F}_r$), which is the diagonal on the parallelogram having $\mathbf{F}_p$ and $\mathbf{F}_q$ as sides (Figure B.12). By definition, the opposite sides of a parallelogram are parallel to one another and therefore have the same length. The replacement of a force system (two or more forces) by a single resultant force is called the **composition** of a force system. The term **resolution** refers to the substitution of a given force by two or more components.

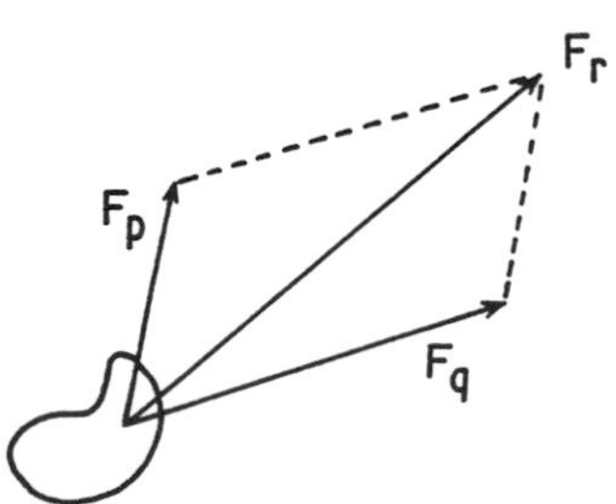

Figure B.12 A force parallelogram.

For example, suppose the runner depicted in Figure 2.6c experienced the following ground reaction force components:

Vertical ($F_{g,z}$)	812 N	(positive = upward)
Forward-backward ($F_{g,y}$)	−286 N	(positive = forward)
Side-to-side ($F_{g,x}$)	61 N	(positive = lateral)

A. Draw a diagram of the two components and the resultant force in each of the *sagittal* (divides body into left and right), *frontal* (divides body into front and back), and *horizontal* (divides body into upper and lower) planes.
B. Calculate the magnitude and direction for each of these resultants. (Answer: Sagittal = 861 N; frontal = 814 N; horizontal = 292 N.)
C. Use the Pythagorean relationship (Equation B.17) to determine the magnitude of the resultant ($\mathbf{F}_g$) ground reaction force ($\mathbf{F}_g = \sqrt{F_{g,z} + F_{g,y} + F_{g,x}}$). (Answer: $\mathbf{F}_g$ = 863 N.)

Vector Algebra

Analysis of biomechanical systems usually involves many vector quantities. Such analysis requires the manipulation of vectors with rigorous techniques. This section on vector algebra introduces some basic elements of these procedures.

Right-Hand Coordinate System. Throughout the text, it has been emphasized that movement and its associated mechanics are relative phenomena. For example, changes in position are noted as a shift in an object's location relative to some reference. In the simplest case, the reference has been a set of *x*- and *y*-axes. However, because movement can occur in three dimensions, the standard reference is a set of *xyz*-axes. The usual orientation is shown in Figure B.13.

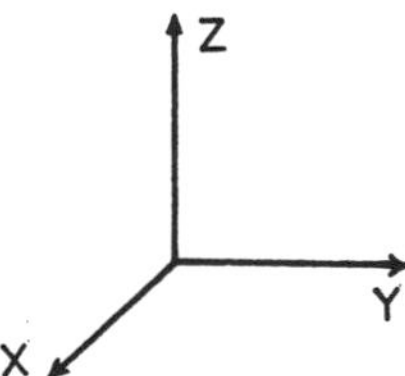

Figure B.13 Right-hand coordinate system.

If you curl the fingers of your right hand in the direction of rotating the positive *x*-axis onto the positive *y*-axis, then the direction of your extended thumb indicates the positive *z*-axis. For this reason, this configuration of the *xyz*-axes is known as the *right-hand coordinate system.*

Vector quantities can be related to the right-hand coordinate system. The direction of a specific vector quantity can be said to be so many units in the *x*-direction, so many in the *y*-direction, and so many in the *z*-direction. This idea is abbreviated with the use of the terms **i**, **j**, and **k**, which are known as unit vectors. The symbol **i** represents the *x*-direction, **j** refers to the *y*-direction, and **k** indicates the *z*-direction (Figure B.14).

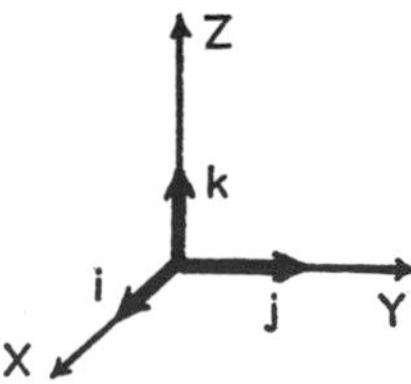

Figure B.14 Relationship between unit vectors and the right-hand coordinate system.

A vector described by the expression 61**i** – 286**j** + 812**k** has a direction of 61 units in the *x*-direction, 286 units in the negative *y*-direction, and 812 units in the *z*-direction. The magnitude of the vector could be determined by the Pythagorean theorem (Equation B.17):

$$\mathbf{F}_g = \sqrt{61^2 + (-286)^2 + 812^2)}$$

Scalar Product. When describing physical systems and the interaction of their components, we often manipulate vector quantities in algebraic expressions. The addition and subtraction of vectors is a simple procedure in that we add or subtract the **i**, **j**, and **k** terms separately. For example, consider the addition of $\mathbf{F}_p$ and $\mathbf{F}_q$:

$$\mathbf{F}_p = 10\mathbf{i} + 28\mathbf{j} + 92\mathbf{k} \qquad \text{and} \qquad \mathbf{F}_q = 3\mathbf{i} - 11\mathbf{j} + 46\mathbf{k}$$

$$\mathbf{F}_p + \mathbf{F}_q = (10 + 3)\mathbf{i} + (28 - 11)\mathbf{j} + (92 + 46)\mathbf{k}$$

$$= 13\mathbf{i} + 17\mathbf{j} + 138\mathbf{k}$$

Multiplication, however, is a more involved procedure. In fact, there are two procedures, scalar products and vector products. (Actually the procedures are more commonly known as dot and cross products, but the terms *scalar* and *vector* seem more appropriate for an overview of these techniques.) The distinction between the two procedures, scalar (dot) and vector (cross) products, has to do with the character of the result; that is, whether it is a scalar or a vector quantity.

The definition of a **scalar product** is given by the expression

$$\mathbf{d} \cdot \mathbf{F} = dF \cos \theta \tag{B.34}$$

The dot product (notice the dot between **d** and **F**) of the two vectors **d** and **F** is calculated as the magnitude of **d** (*d*) times the magnitude of **F** (*F*) times the cosine of the angle (θ) between the two vectors (Figure B.15a). This procedure gives the magnitude of **F** that is directed along **d** multiplied by the magnitude of *d*. The magnitude of **F** directed along **d** is shown in Figure B.15b as the base of the right triangle (i.e., the side that equals **F** cos θ). Thus the scalar product is just **d** times **F** cos θ.

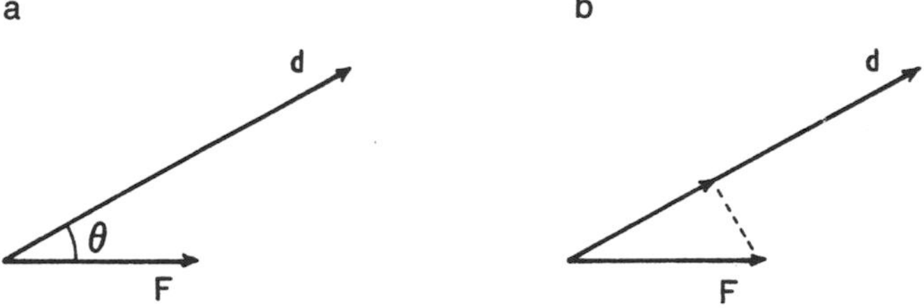

Figure B.15 The scalar product. The product involves determining how much of (a) **F** acts in the same direction as (b) **d**.

The multiplication of vectors with the scalar product is appropriate for calculating scalar quantities. One example is the calculation of work, where work is defined as the product of force and displacement (distance). Both force and displacement are vectors. The rigorous definition of work (a scalar quantity derived by the scalar product) is that it equals the component of force in the direction of the displacement times the displacement.

Vector Product. Alternatively, when the result of vector multiplication is a vector quantity, it is appropriate to use the technique for determining the vector (cross) product. In other words, when we are interested in multiplying two vectors together and we want to consider both magnitude and direction, we use the vector product. This is given by the expression

$$\mathbf{r} \times \mathbf{F} = rF \sin \theta \tag{B.35}$$

which reads, "**r** cross **F** is equal to the magnitude of **r** (r) times the magnitude of **F** (F) times the sine of the angle (θ) between the two vectors." In this instance, the direction of the product is perpendicular to the plane that contains the other two vectors. This relationship is illustrated in Figure B.16.

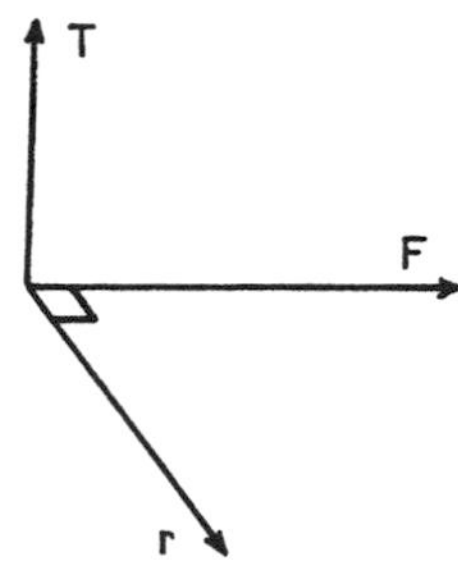

Figure B.16 The vector product of **r** and **F** is the vector **T**.

Consider the example in which a force (**F**) acts on a system that is constrained to rotate about an axis. The distance from this axis to the line of action of **F** is given by the vector **r**—we have called this perpendicular component of the distance the moment arm (chapter 2). The product of **r** and **F** is given by the vector product and is shown in Figure B.16 as **T**. We recognize this product as the torque. That is, torque is equal to the product of force and moment arm. What is important here, however, is that the product (torque, **T**) is a vector whose direction is perpendicular to the plane in which **r** and **F** lie. For example, if a net muscle force and the moment arm forming the force vector to the joint were located in the plane of this page, then the direction of the net muscle torque would be perpendicular to the page. The actual sense of the vector (Figure B.11) could be either toward or away from you and would depend on the directions of **r** and **F**.

There are several ways to determine the direction, including the sense, of a vector product. The most rigorous approach is to manipulate vectors in terms of their **i**, **j**, and **k** components. These procedures are beyond the scope of this text but are explained simply and well by Miller and Nelson (1973; Appendix B). This text uses a more qualitative procedure that is commonly known as the right-hand-thumb rule. As described in chapter 2 for determining the direction of torque vectors, the fingers of the right hand are curled in the direction of the rotation that the force produces about the axis, and the extended thumb indicates the direction of the associated torque vector.

Principle of Transmissibility. A force acting on a rigid body does not have to be restricted to a specific point of application. It is possible to slide the force vector along its line of action without altering the mechanical effect on the rigid body (Meriam & Kraige, 1987). This property is referred to as the **principle of transmissibility**. Because of this principle, force can be treated as a sliding vector, and it is sufficient to specify the magnitude, direction, and line of action of the force vector.

Curve Fitting

In biomechanics, we often measure one variable as a function of another. For example, we measure the ground reaction force as a function of time, knee angle as a function of thigh angle, and stride length as a function of running speed. Sometimes, however, it is useful to derive a mathematical expression that describes a particular function. Such a process is usually called *curve fitting*. Common curve-fitting techniques include finite differences, polynomial functions, cubic spline functions, digital filtering, and Fourier analysis (Wood, 1982). These techniques have the following forms:

Finite differences $$\dot{x}(t) = \frac{x_{i+1} - x_{i-1}}{2(\Delta t)} \quad (B.36)$$

$$\ddot{x}(t) = \frac{x_{i+1} - 2x_i + x_{i-1}}{(\Delta t)^2} \quad (B.37)$$

Polynomial $$x(t) = a_0 + a_1t + a_2t^2 + \cdots + a_nt^n \quad (B.38)$$

Digital filter $$X^1(nT) = a_0X(nT) + a_1X(nT-T) + a_2X(nT-2T) + b_1X^1(nT-T) + b_2X^1(nT-2T) \quad (B.39)$$

Fourier series $$x(t) = a_o + \sum_{i=1}^{n} [a_i \cos(2\pi\, it/T) + b_i \sin(2\pi\, it/T)] \quad (B.40)$$

One advantage of curve fitting is that it enables the biomechanist to manipulate the functions more readily. For example, it is easier to conduct statistical analysis on a ground reaction force record that is represented as a Fourier series than on the original data. Similarly, position-time data that are subjected to curve fitting can be readily differentiated to yield the associated velocity- and acceleration-time functions.

Fourier Analysis

An alternating signal (ac) is one that changes with time. It may be periodic or completely random, or a combination of both (Figure B.17). Also, any signal may have a direct current (dc) component about which the ac component fluctuates. These signals can be described in terms of frequency content, which provides a measure of the smoothness of the signal. The smoother a signal, the lower its frequency content, and vice versa. The sine wave in Figure B.17 is a low-frequency signal compared with the random signal, which contains many sharp peaks. Furthermore, a sine (or cosine) waveform is a single frequency, whereas any other signal, especially the random signal in Figure B.17, contains many frequencies (sine and cosine waves). These features are shown in Figure B.18, where these signals are shown in both the time domain (amplitude-time graph) and the frequency domain (amplitude-frequency graph). In the frequency domain, the sine wave is represented as a single frequency of a given amplitude. In contrast, the random signal contains many frequencies from some lower (f_1) to some upper (f_2) limit.

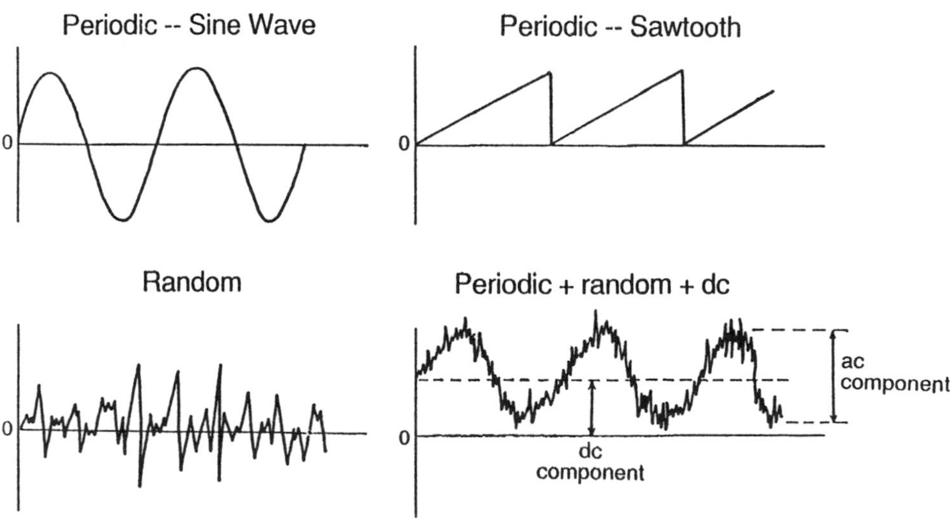

Figure B.17 Periodic and random signals that vary over time.
Note. From *Biomechanics and Motor Control of Human Movement* (p. 27) by D.A. Winter, 1990. Copyright 1990 by John Wiley & Sons, Inc. Adapted with permission.

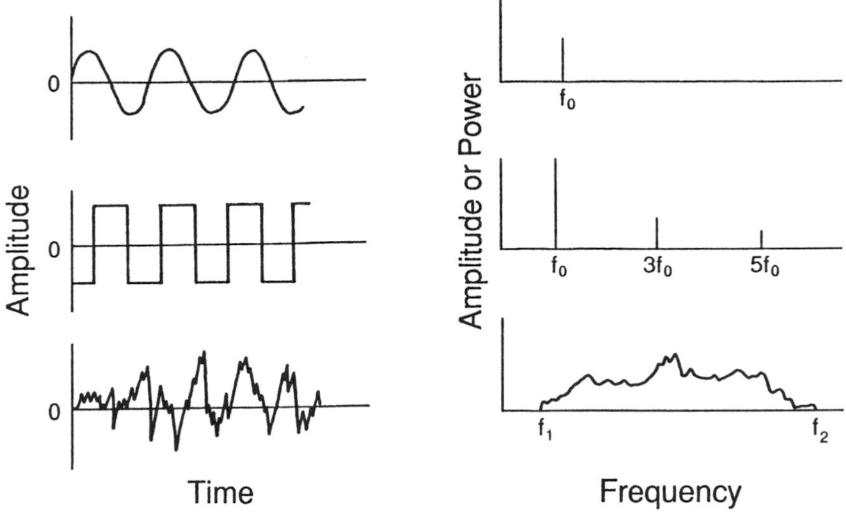

Figure B.18 Transformations (Fourier) of signals from the time domain to the frequency domain. *Note*. From *Biomechanics and Motor Control of Human Movement* (p. 28) by D.A. Winter, 1990. Copyright 1990 by John Wiley & Sons, Inc. Adapted with permission.

The technique of **Fourier analysis** (or Fourier transformation), which is only valid for periodic (cyclic) functions, involves *the derivation of a series of sine and cosine terms to represent the frequency content of a signal*. A nonperiodic signal, however, can be numerically converted to a periodic signal and subjected to a Fourier analysis (Wood, 1982). To analyze a periodic signal, we express the frequency content in terms of the **fundamental frequency** (f_0) and its multiples. The fundamental frequency represents the *single* sine + cosine term that best describes how the signal varies during one cycle (Figure B.19). Multiples of the fundamental frequency are referred to as **harmonics**. For example, the third harmonic is $3f_0$, and the ninth harmonic is $9f_0$. Once the fundamental frequency has been determined, then curve-fitting procedures are used to determine the size of the respective harmonics that are needed to approximate the signal. This scaling is accomplished by multiplying each harmonic with an appropriate **weighting coefficient**. These terms are collectively referred to as a Fourier series, which can be stated algebraically as

$$y(t) = a_o + \sum_{i=1}^{n} [a_i \cos (2\pi\ it/T) + b_i \sin (2\pi\ it/T)] \tag{B.41}$$

where

$y(t)$ = position as a function of time

a_o = constant (mean; accounts for the dc component)

a_i, b_i = Fourier series coefficients

i = harmonic number

n = maximum harmonic number

t = time, an instantaneous value (the independent variable) during the cycle

T = total time of one cycle

If Equation B.41 is utilized to provide a function for position-time data, it then can be differentiated with respect to time to produce Equation B.42, the velocity-time function:

$$y(t) = \sum_{i=1}^{n} \{[-a_i \sin(2\pi\, it/T)(2\pi\, i/T)] + [b_i \cos(2\pi\, it/T)(2\pi\, i/T)]\} \qquad \text{(B.42)}$$

Similarly, differentiation of Equation B.42 with respect to time produces an acceleration-time function:

$$y(t) = \sum_{i=1}^{n} \{[-a_i \cos(2\pi\, it/T)(2\pi\, i/T^2)] + [-b_i \sin(2\pi\, it/T)(2\pi\, i/T)^2]\} \qquad \text{(B.43)}$$

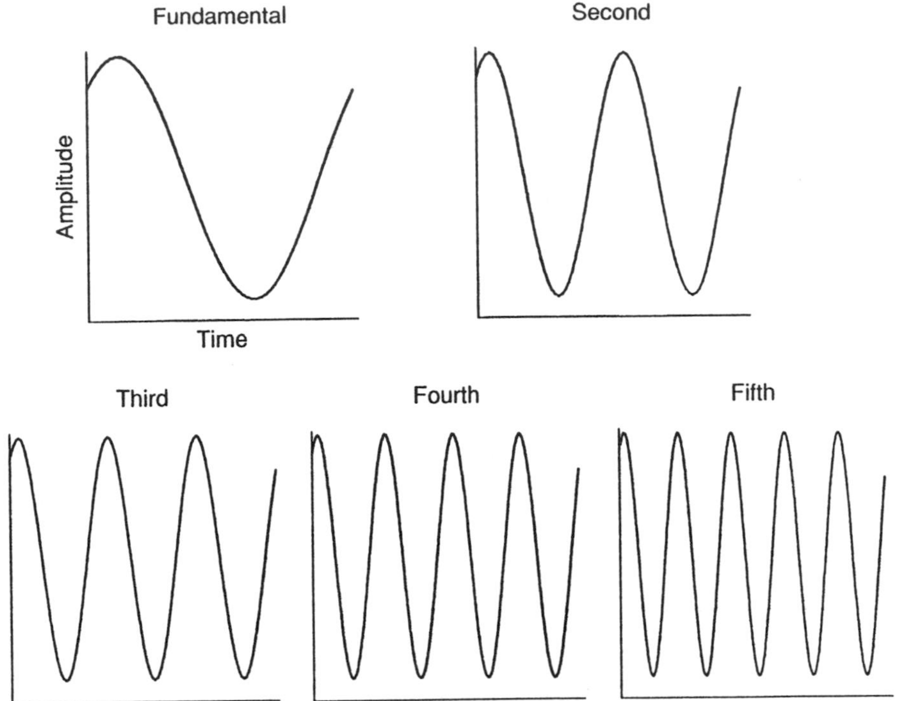

Figure B.19 The first five harmonics as multiples of the fundamental frequency.

The process of determining the components of equation B.41 consists of two stages, the first of which is to calculate the *fundamental frequency*. The term *frequency* refers to the number of cycles per second and is measured in hertz (Hz). A frequency of 1 Hz describes a cycle that takes 1 s to complete. One cycle corresponds to one complete revolution, which is represented as 2π rad (360°). The fundamental frequency is determined by summing weighted sine and cosine terms (i.e., cos $2\pi\, t/T$ and sin $2\pi\, t/T$), where T is the duration of the cycle and t is incremented from 0 to T. With this procedure it is possible to derive a best-fitting sine-cosine term for a data set (Figure B.20); this is the fundamental frequency. To improve the fit of the sine-cosine term to the data set, we add multiples of the fundamental frequency (i.e., harmonics) to the fundamental frequency. For example, the expression $2\pi\, it/T$, where i varies from 1 to 5, accounts for the generation of five of these harmonics (Figure B.19). Note that the fourth harmonic has a frequency (number of cycles) four times the fundamental.

Second, it is necessary to scale the contribution of each harmonic to the function. In general, the contribution decreases as harmonic number increases for human movement. The contribution of a harmonic to the Fourier series expression is weighted by means of a coefficient (i.e., a_o, a_i, b_i). The magnitude of these coefficients is determined

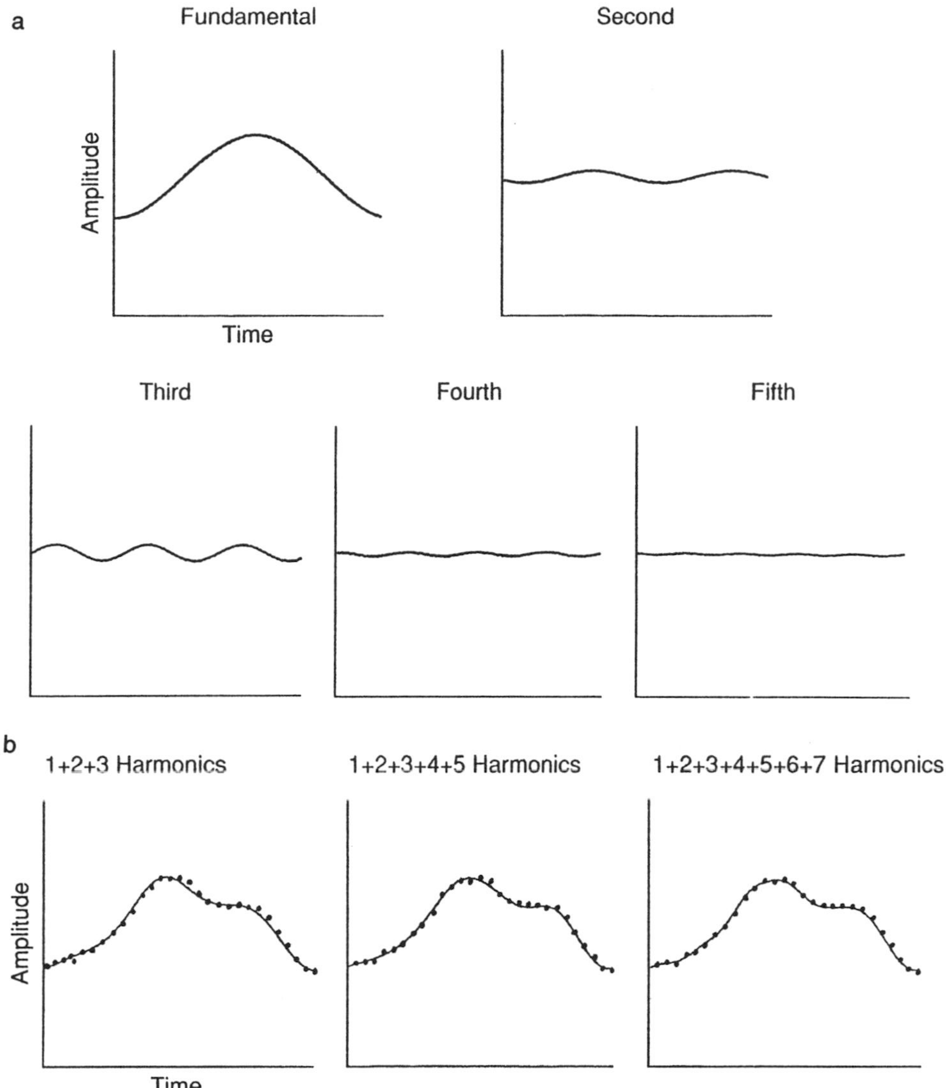

Figure B.20 Influence of the weighting coefficients on the contribution of the first five harmonics: (a) magnitude of the first five harmonics, as determined by least squares; (b) the match between the data and the trigonometric function comprising either 3, 5, or 7 harmonics.

by curve-fitting with a least squares regression technique. Explanation of this procedure is available in numerical analysis texts. Figure B.20 presents the contribution of the first five harmonics, as influenced by their respective coefficients. The cumulative effect of summing the first five harmonics is illustrated in Figure B.21. The coefficients and angles (radians) responsible for producing these diagrams are as follows:

$a_0 = 4.8332$

$a_1 = -0.4936$	$b_1 = -0.0478$	$\text{angle}_1 = 12.414$
$a_2 = -0.0285$	$b_2 = -0.0650$	$\text{angle}_2 = 24.828$
$a_3 = 0.0178$	$b_3 = 0.0942$	$\text{angle}_3 = 37.241$
$a_4 = 0.0195$	$b_4 = 0.0105$	$\text{angle}_4 = 49.655$
$a_5 = 0.0074$	$b_5 = 0.0046$	$\text{angle}_5 = 62.069$

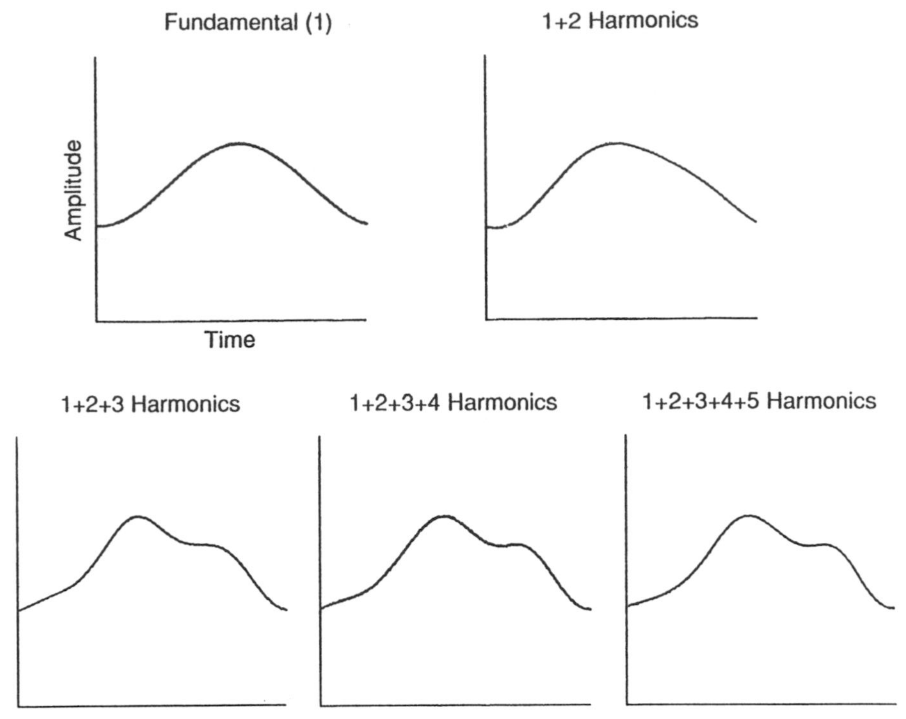

Figure B.21 Cumulative effect of the weighted contributions of the first five harmonics to the Fourier series.

Digital Filter

Another numerical analysis technique that is available for curve fitting is the **digital filter** (see chapter 2 in Winter, 1990). Unlike the Fourier analysis, the digital filter does not allow us to derive an equation to represent a signal. But like the Fourier analysis, the digital filter allows us to use a numerical technique to smooth a signal and to get rid of unwanted components. Typically, these components result from measurement errors and severely contaminate the estimates of velocity and acceleration from position-time data (Winter, Sidwall, & Hobson, 1974).

Previously we discussed the frequency content of signals (Fourier analysis) when we described how every signal can be represented by a fundamental frequency and its harmonics (multiples of the fundamental frequency). Higher order harmonics account for the high-frequency content (sharp edges and peaks) of a signal. When we digitize a movement from film or video, the measurement includes a true estimate of the position of the landmark in addition to errors due to such effects as camera vibration, distortions in the film and projection system, and inaccurate placement of the cursors. Because these error terms (referred to as noise) are generally random, they represent the high frequency component of a signal. Accordingly, one purpose of smoothing procedures (e.g., digital filter, Fourier analysis) is to remove the noise from a signal. The capability of these smoothing procedures is illustrated in Figure B.22.

The key step in the digital filter procedure is identifying the frequency that separates the desired signal from the noise; this is referred to as the **cutoff frequency** (f_c). Unfortunately this step is not a simple matter, because the frequencies of the desired signal and noise overlap to some extent (Figure B.23a). *The digital filter is a numerical procedure that manipulates the frequency spectrum of an input signal to produce an output signal that contains a substantially attenuated frequency content above* f_c. The numerical effect of a digital filter is shown in Figure B.23b, where the frequencies are either left untouched or are reduced. The cutoff frequency is defined as the frequency where the power of the signal is reduced to one half of its original value. The hypothetical effect of this procedure is shown in Figure B.23c (compare the amplitude-frequency graph of the input [$X_i(f)$] in Figure B.23a to the output [$X_o(f)$] in Figure B.23c).

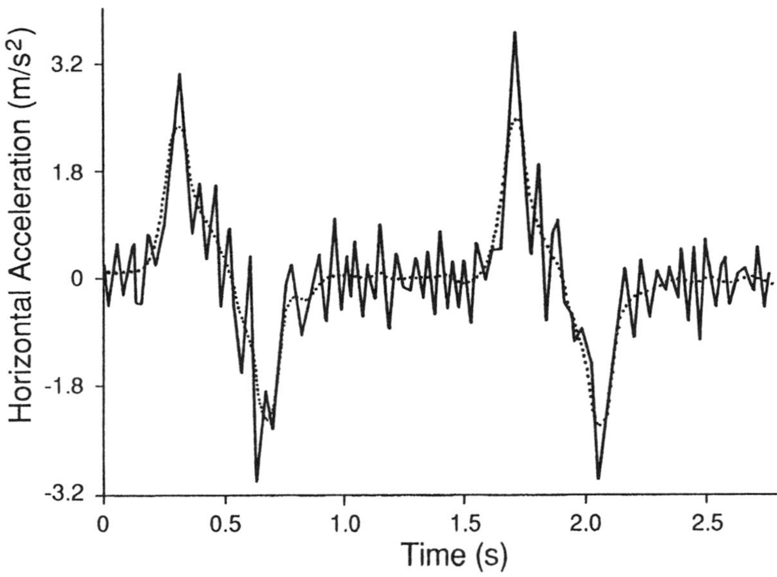

Figure B.22 Acceleration of a marker on the toe during walking. The fluctuating line was obtained from the original position-time data. The dotted, smooth line was derived from the position-time data after it was passed through a digital filter.

Note. Adapted from "Measurement and Reduction of Noise in Kinematics of Locomotion" by D.A. Winter, H.G. Sidwall, and D.A. Hobson, 1974, *Journal of Biomechanics*, **7**, p. 158. Copyright 1974 with kind permission from Pergamon Press Ltd., Headington Hill Hall, Oxford OX3 DBW, UK.

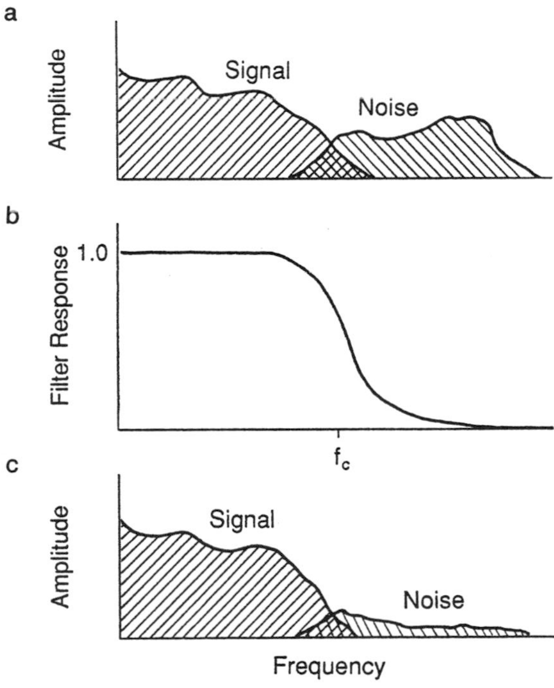

Figure B.23 The effect of a digital filter on the frequency content of a signal. (a) An original signal contains both desired information (signal) and noise. The noise is generally located in the higher frequency region of the frequency-amplitude graph. (b) The effect of a digital filter on the amplitude of different frequencies. At the cutoff frequency (f_c), the power of the original signal is reduced by one half. (c) Amplitude-frequency graph of the signal once it has been passed through the digital filter.

A commonly used digital filter (Butterworth, second-order) has the following form:

$$X^1(nT) = a_0X(nT) + a_1X(nT-T) + a_2X(nT-2T) + b_1X^1(nT-T) + b_2X^1(nT-2T)$$

where

X = unfiltered (input) original data

X^1 = filtered output data

T = time between data points

nT = n^{th} frame of data

$nT - T$ = one frame before the nth frame of data

a_0, a_1, a_2, b_1, b_2 = filter coefficients

From this expression, it should be apparent that the output $[X^1(nT)]$ is a weighted average of the immediate and past unfiltered data plus a weighted contribution from the past filtered output.

To implement a digital filter, we need to determine the values of the coefficients (a_0, a_1, a_2, b_1, b_2). However, to do this we must first specify f_c. Perhaps the most comprehensive procedure available to determine f_c is a **residual analysis** (Winter, 1990). This involves comparing the difference (residual) between filtered and unfiltered signals over a wide range of cutoff frequencies and choosing an f_c that minimizes both signal distortion and the amount of noise that passes through the filter. For many human movements, an f_c of 6 Hz is generally adequate. Once this has been determined, the coefficients can be obtained by calculating the ratio of the sampling frequency (f_s frames per second) to f_c. For an f_c of 6 Hz using video data obtained at 60 Hz, the ratio is 10 and the coefficients are

$$a_0 = 0.06746$$

$$a_1 = 0.13491$$

$$a_2 = 0.06746$$

$$b_1 = 1.14298$$

$$b_2 = -0.41280$$

The digital filter coefficients for different f_s/f_c ratios are available in Winter (1990).

The final consideration is to correct an unwanted distortion that the digital filter introduces in the filtered data. This effect is a phase distortion that can be observed as a shift of a sine wave along the horizontal axis; if one cycle equals 2π rad, then a phase distortion of $\pi/2$ rad is equal to a shift of the sine wave of one quarter of the cycle. This is the magnitude of the phase distortion that is introduced by a second-order Butterworth filter. To remove the phase distortion, we need to pass the data through the digital filter twice—once in the forward direction and once in the reverse direction. The forward direction should begin with the real first data point, and the reverse direction should begin with the last data point.

APPENDIX C

Conversion Factors

Acceleration

1 centimeter/second/second (cm/s^2)	=	0.036 kilometer/hour/second (km/hr/s)
1 centimeter/second/second (cm/s^2)	=	0.01 meter/second/second (m/s^2)
1 foot/second/second (ft/s^2)	=	30.48 centimeters/second/second (cm/s^2)
1 foot/second/second (ft/s^2)	=	1.097 kilometers/hour/second (km/hr/s)
1 foot/second/second (ft/s^2)	=	0.3048 meter/second/second (m/s^2)
1 kilometer/hour/second (km/hr/s)	=	27.78 centimeters/second/second (cm/s^2)
1 kilometer/hour/second (km/hr/s)	=	0.2778 meter/second/second (m/s^2)
1 meter/second/second (m/s^2)	=	100 centimeters/second/second (cm/s^2)
1 mile/hour/second	=	0.447 meter/second/second (m/s^2)
1 revolution/minute/minute	=	0.001745 radian/second/second (rad/s^2)
1 revolution/second/second	=	6.283 radians/second/second (rad/s^2)

Area

1 acre	=	4,047 square meters (m^2)
1 acre	=	0.4047 hectare (ha)
1 ares	=	100 square meters (m^2)
1 centare (ca)	=	1 square meter (m^2)
1 hectare (ha)	=	10,000 square meters (m^2)
1 square centimeter (cm^2)	=	0.0001 square meter (m^2)
1 square centimeter (cm^2)	=	100 square millimeters (mm^2)
1 square degree	=	0.00030462 steradian (sr)
1 square inch ($in.^2$)	=	6.4516 square centimeters (cm^2)
1 square inch ($in.^2$)	=	645.16 square millimeters (mm^2)
1 square foot (ft^2)	=	929 square centimeters (cm^2)
1 square foot (ft^2)	=	0.092903 square meter (m^2)
1 square foot (ft^2)	=	92,900 square millimeters (mm^2)
1 square inch ($in.^2$)	=	6.452 square centimeters (cm^2)
1 square kilometer (km^2)	=	1,000,000 square meters (m^2)

Area *(continued)*

1 square meter (m)	=	0.0001 hectare (ha)
1 square meter (m^2)	=	10,000 square centimeters (cm^2)
1 square mile ($mile^2$)	=	2.590 square kilometers (km^2)
1 square mile ($mile^2$)	=	2,590,000 square meters (m^2)
1 square millimeter (mm^2)	=	0.01 square centimeter (cm^2)
1 square yard (yd^2)	=	8,361 square centimeters (cm^2)
1 square yard (yd^2)	=	0.836127 square meter (m^2)

Density

1 pound/cubic foot (lb/ft^3)	=	16.01846 kilograms/cubic meter (kg/m^3)
1 slug/cubic foot ($slug/ft^3$)	=	515.3788 kilograms/cubic meter (kg/m^3)
1 pound/gallon (UK)	=	99.77633 kilograms/cubic meter (kg/m^3)
1 pound/gallon (USA)	=	119.8264 kilograms/cubic meter (kg/m^3)

Electricity

1 biot (Bi)	=	10 ampere (A)
1 ampere/square inch ($A/in.^2$)	=	1,550 amperes/square meter (A/m^2)
1 faraday/second	=	96,490 amperes (A)
1 faraday	=	96,490 coulombs (coulomb = ampere-second)
1 mho	=	1 siemen (S)

Energy and work

1 Btu (British thermal unit)	=	1,055 joules (J)
1 Btu	=	1.0548 kilojoules (kJ)
1 Btu	=	0.0002928 kilowatt-hour (kW-hr)
1 erg	=	0.0001 millijoule (mJ)
1 foot-poundal	=	0.04214 joule (J)
1 foot-pound force	=	1.355818 joules (J)
1 gram-centimeter	=	0.09807 millijoule (mJ)
1 horsepower-hour	=	2,684 kilojoules (kJ)
1 horsepower-hour	=	0.7457 kilowatt-hour (kW-hr)
1 kilocalorie (International)	=	4.1868 kilojoules (kJ)
1 kilocalorie	=	4,183 joules (J)
1 kilopond-meter (kp·m)	=	9.807 joules (J)
1 kilowatt-hour	=	3,600 kilojoules (kJ)

Force

1 dyne	=	0.01 millinewton (mN)
1 foot-pound (ft-lb)	=	1.356 joules (J)
1 foot-pound/second (ft-lb/s)	=	0.001356 kilowatt (kW)
1 gram (g)	=	9.807 millinewtons (mN)
1 kilogram-force (kg-f)	=	9.807 newtons (N)
1 kilopond (kp)	=	9.807 newtons (N)
1 poundal	=	0.138255 newton (N)
1 pound-force (lb-f)	=	4.448222 newtons (N)
1 stone (weight)	=	62.275 newtons (N)
1 ton (long)	=	9,964 newtons (N)
1 ton (metric)	=	9,807 newtons (N)

Length (angle)

1 angstrom (Å)	=	0.0001 micrometer (μm)
1 bolt	=	36.576 meters (m)
1 centimeter (cm)	=	0.00001 kilometer (km)
1 centimeter (cm)	=	0.01 meter (m)
1 centimeter (cm)	=	10 millimeters (mm)
1 chain	=	20.12 meters (m)
1 circumference	=	6.283 radians (rad)
1 degree	=	0.001745 radian (rad)
1 fathom	=	1.8288 meters (m)
1 foot (ft)	=	30.48 centimeters (cm)
1 foot (ft)	=	0.3048 meter (m)
1 foot (ft)	=	304.8 millimeters (mm)
1 furlong	=	201.17 meters (m)
1 hand	=	10.16 centimeters (cm)
1 inch (in.)	=	2.54 centimeters (cm)
1 inch (in.)	=	0.0254 meter (m)
1 inch (in.)	=	25.4 millimeters (mm)
1 light-year	=	9,460,910,000,000 kilometers (9.46×10^{12} km)
1 mile (nautical)	=	1.853 kilometers (km)
1 mile (nautical)	=	1,853 meters (m)
1 mile (statute)	=	1.609 kilometers (km)
1 mile (statute)	=	1,609 meters (m)
1 minute (angle)	=	0.0002909 radian (rad)
1 revolution	=	6.283 radians (rad)
1 rod	=	5.029 meters (m)
1 sphere (solid angle)	=	12.57 steradians (sr)
1 yard (yd)	=	91.44 centimeters (cm)
1 yard (yd)	=	0.9144 meter (m)

Luminous intensity

1 candela/square centimeter	=	10,000 candela/square meter (cd/m^2)
1 candela/square foot	=	10.76 candela/squqare meter (cd/m^2)
1 foot-candle	=	10.764 lumen/square meter (lm/m^2)
1 foot-lambert (fL)	=	3.426 candela/square meter (cd/m^2)
1 lambert (L)	=	3,183 candela/square meter (cd/m^2)
1 phot	=	10,000 lumen/square meter (lm/m^2)

Moment of inertia

1 slug-foot squared (slug-ft^2)	=	1.35582 kilogram-square meters ($kg \cdot m^2$)
1 pound-foot squared (lb-ft^2)	=	0.04214 kilogram-square meter ($kg \cdot m^2$)

Mass

1 hundredweight	=	50.8 kilograms (kg)
1 ounce (oz)	=	28.3495 grams (g)
1 pound (lb)	=	0.453592 kilogram (kg)
1 slug	=	14.59 kilograms (kg)
1 ton	=	1,016 kilograms (kg)

Power

1 Btu/hour	=	0.2931 watt (W)
1 Btu/minute	=	17.57 watts (W)
1 calorie/second	=	4.187 watts (W)
1 erg/second	=	10,000,000 megawatts (MW)
1 foot-pound/second	=	1.356 watts (W)
1 foot-poundal/second	=	0.04214 watt (W)
1 gram-calorie	=	0.001162 watt-hour
1 horsepower (UK)	=	745.7 watts (W)
1 horsepower (metric)	=	735.5 watts (W)
1 kilocalorie/minute (kcal/min)	=	69.767 watts (W)
1 kilogram-meter/second	=	9.807 watts (W)
1 kilopond-meter/minute (kpm/min)	=	0.1634 watt (W)

Pressure

1 atmosphere	=	760 millimeters of mercury (mmHg) at 0°C
1 atmosphere	=	101,340 pascals (Pa)
1 bar	=	100,031 pascals (Pa)
1 centimeter of mercury	=	1,333.224 pascals (Pa)
1 centimeter of water	=	0.738 millimeter of mercury (mmHg)
1 centimeter of water	=	98.3919 pascals (Pa)
1 dyne/square centimeter	=	0.10 pascal (Pa)
1 foot of water	=	2,989 pascals (Pa)
1 inch of mercury	=	3,386 pascals (Pa)
1 inch of water (4°C)	=	249 pascals (Pa)
1 kilogram-force/square meter	=	9.807 pascals (Pa)
1 millibar	=	100 pascals (Pa)
1 millimeter of mercury (1 torr)	=	133.322387 pascals (Pa)
1 pound-force/square foot (lb/ft^2)	=	47.88026 pascals (Pa)
1 pound-force/square inch (lb/in.^{2})	=	6.8948 kilopascals (kPa)
1 poundal/square foot	=	1.488 pascal (Pa)

Temperature

1 degree centigrade (°C)	=	(°C × 9/5) + 32 degrees Fahrenheit (°F)
1 degree centigrade (°C)	=	°C + 273.18 Kelvin

Torque

1 foot-pound (ft-lb)	=	1.356 newton-meters (N·m)
1 kilopond-meter (kp·m)	=	9.807 newton-meters (N·m)

Velocity

1 centimeter/second (cm/s)	=	0.036 kilometer/hour (km/hr, or kph)
1 centimeter/second (cm/s)	=	0.6 meter/minute (m/min)
1 foot/minute (ft/min)	=	0.508 centimeter/second (cm/s)
1 foot/minute (ft/min)	=	0.01829 kilometer/hour (km/hr, or kph)
1 foot/minute (ft/min)	=	0.3048 meter/minute (m/min)
1 foot/second (ft/s)	=	30.48 centimeters/second (cm/s)
1 foot/second (ft/s)	=	1.097 kilometers/hour (km/hr, or kph)
1 foot/second (ft/s)	=	18.29 meters/minute (m/min)

1 foot/second (ft/s)	=	0.3048 meter/second (m/s)
1 kilometer/hour (km/hr, or kph)	=	16.67 meters/minute (m/min)
1 kilometer/hour (km/hr, or kph)	=	0.2778 meter/second (m/s)
1 knot (kn)	=	1.8532 kilometers/hour (km/hr, or kph)
1 knot (kn)	=	51.48 centimeters/second (cm/s)
1 knot (kn)	=	0.5155 meter/second (m/s)
1 meter/minute (m/min)	=	1.667 centimeters/second (cm/s)
1 meter/minute (m/min)	=	0.06 kilometer/hour (km/hr, or kph)
1 meter/second (m/s)	=	3.6 kilometers/hour (km/hr, or kph)
1 meter/second (m/s)	=	0.06 kilometer/minute (km/min)
1 mile/hour (mph)	=	44.7 centimeters/second (cm/s)
1 mile/hour (mph)	=	1.6093 kilometers/hour (km/hr, or kph)
1 mile/hour (mph)	=	0.447 meter/second (m/s)
1 mile/minute	=	2,682 centimeters/second (cm/s)
1 mile/minute	=	1.6093 kilometers/minute (km/min)
1 revolution/minute	=	0.1047 radian/second (rad/s)
1 revolution/second	=	6.283 radians/second (rad/s)

Volume

1 barrel (UK)	=	0.1637 cubic meter (m^3)
1 barrel (USA)	=	0.11921 cubic meter (m^3)
1 bushel (UK)	=	0.03637 cubic meter (m^3)
1 bushel (USA)	=	0.03524 cubic meter (m^3)
1 bushel (USA)	=	35.24 liters (L)
1 cubic centimeter (cm^3)	=	0.000001 cubic meter (m^3)
1 cubic centimeter (cm^3)	=	0.001 liter (L)
1 cubic foot (ft^3)	=	0.02832 cubic meter (m^3)
1 cubic foot (ft^3)	=	28.32 liters (L)
1 cubic foot/minute (ft^3/min)	=	472 cubic centimeters/second (cm^3/s)
1 cubic foot/minute (ft^3/min)	=	0.472 liter/second (L/s)
1 cubic inch ($in.^3$)	=	16.387 cubic centimeters (cm^3)
1 cubic inch ($in.^3$)	=	0.0000164 cubic meter (m^3)
1 cubic inch ($in.^3$)	=	0.0164 liter (L)
1 cubic meter (m^3)	=	1,000,000 cubic centimeters (cm^3)
1 cubic meter (m^3)	=	1,000 liters (L)
1 cubic yard (yd^3)	=	0.7646 cubic meter (m^3)
1 cubic yard (yd^3)	=	764.6 liters (L)
1 cubic yard/minute (yd/min)	=	12.74 liters/second (L/s)
1 dram	=	3.6967 cubic centimeters (cm^3)
1 gallon (USA)	=	3,785 cubic centimeters (cm^3)
1 gallon (USA)	=	0.003785 cubic meter (m^3)
1 gallon	=	3.785 liters (L)
1 gallon/minute	=	0.06308 liter/second (L/s)
1 gill (UK)	=	142.07 cubic centimeters (cm^3)
1 gill (USA)	=	118.295 cubic centimeters (cm^3)
1 gill (USA)	=	0.1183 liter (L)
1 liter (L)	=	1,000 cubic centimeters (cm^3)
1 liter (L)	=	0.001 cubic meter (m^3)
1 ounce (fluid, USA)	=	0.02957 liter (L)
1 ounce (fluid, UK)	=	28.413 cubic centimeters (cm^3)

Volume *(continued)*

1 ounce (fluid, USA)	=	29.573 cubic centimeters (cm^3)
1 peck (UK)	=	9.0919 liters (L)
1 peck (USA)	=	8.8096 liters (L)
1 pint	=	473.2 cubic centimeters (cm^3)
1 pint	=	0.4732 liter (L)
1 quart	=	946.4 cubic centimeters (cm^3)
1 quart	=	0.9463 liter (L)

APPENDIX D

Body Segment Parameters

Body segment organization is used in many biomechanical studies. Chandler et al. (1975) used 14 body segments: head, trunk, upper arms, forearms, hands, thighs, shanks (or legs), and feet (see Figure D.1). In contrast, Zatsiorsky and Seluyanov (1983) separated the body into 16 segments: head, upper torso, middle torso, lower torso, upper arms, forearms, hands, thighs, shanks, and feet (see Figure D.2).

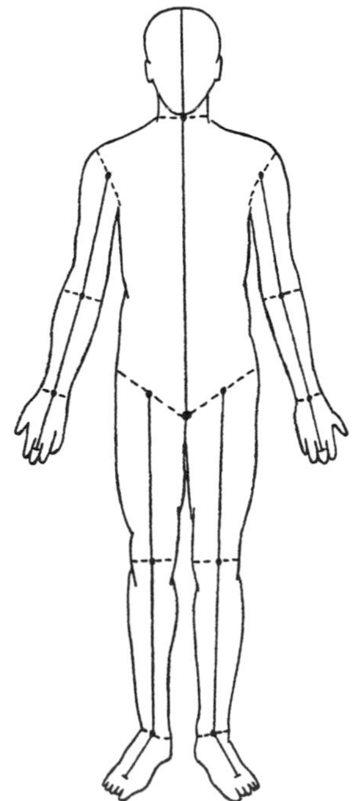

Figure D.1 The body segment organization used by Chandler et al. (1975).

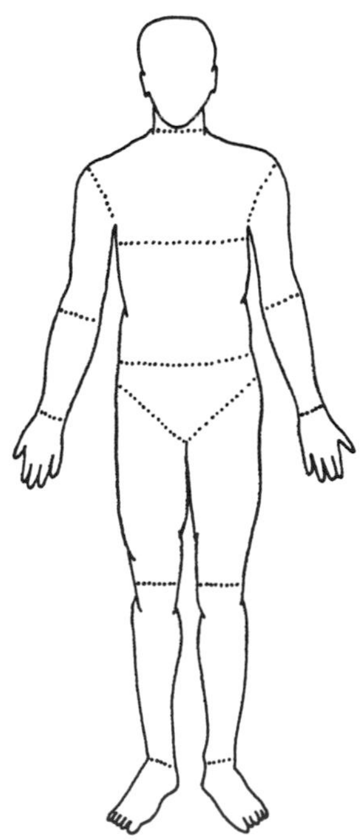

Figure D.2 The body segment organization used by Zatsiorsky and Seluyanov (1983).

APPENDIX E

Kinematic Data

This appendix contains three sets of kinematic data that can be used for student exercises:

1. Table E.1—Position of selected landmarks during the clean in weight lifting
2. Table E.2—Angle of the arms relative to a horizontal reference during a running forward somersault in gymnastics
3. Table E.3—Measurements of knee and elbow angle of a runner during one stride.

Table E.1 Position of Selected Landmarks During the Clean in Weight Lifting

Time	Metatarsal		Malleolus		Knee		Hip		Head		Shoulder		Elbow		Barbell	
(ms)	*x**	*y*	*x*	*y*	*x*	*y*	*x*	*y*	*x*	*y*	*x*	*y*	*x*	*y*	*x*	*y*
10	0.79	–0.56	0.69	–0.47	0.90	–0.13	0.45	–0.06	1.07	0.34	0.92	0.25	0.89	0.01	0.90	–0.27
20	0.79	–0.56	0.69	–0.47	0.90	–0.13	0.45	–0.05	1.07	0.34	0.92	0.25	0.89	0.01	0.90	–0.27
30	0.79	–0.56	0.69	–0.47	0.90	–0.13	0.44	–0.05	1.06	0.34	0.92	0.25	0.89	0.01	0.90	–0.27
40	0.79	–0.56	0.69	–0.47	0.89	–0.13	0.44	–0.04	1.06	0.35	0.92	0.25	0.89	0.01	0.90	–0.27
50	0.79	–0.56	0.69	–0.47	0.89	–0.13	0.44	–0.04	1.06	0.35	0.92	0.26	0.89	0.01	0.90	–0.27
60	0.79	–0.56	0.69	–0.47	0.89	–0.13	0.44	–0.03	1.06	0.35	0.92	0.26	0.89	0.02	0.90	–0.26
70	0.79	–0.56	0.69	–0.47	0.88	–0.13	0.44	–0.03	1.06	0.35	0.92	0.26	0.89	0.02	0.90	–0.26
80	0.79	–0.56	0.69	–0.47	0.88	–0.12	0.44	–0.02	1.06	0.36	0.92	0.26	0.90	0.02	0.90	–0.26
90	0.79	–0.56	0.69	–0.47	0.88	–0.12	0.44	–0.01	1.07	0.36	0.92	0.26	0.90	0.02	0.89	–0.26
100	0.79	–0.56	0.69	–0.47	0.87	–0.12	0.43	–0.00	1.07	0.36	0.92	0.27	0.90	0.02	0.89	–0.26
110	0.80	–0.56	0.69	–0.47	0.87	–0.12	0.43	0.00	1.07	0.37	0.92	0.27	0.90	0.03	0.89	–0.25
120	0.80	–0.56	0.69	–0.47	0.87	–0.12	0.43	0.01	1.07	0.37	0.92	0.27	0.90	0.03	0.89	–0.25
130	0.80	–0.56	0.69	–0.47	0.86	–0.12	0.43	0.02	1.07	0.37	0.92	0.28	0.90	0.03	0.89	–0.25
140	0.80	–0.56	0.69	–0.47	0.86	–0.11	0.43	0.03	1.07	0.38	0.92	0.28	0.90	0.04	0.89	–0.24
150	0.80	–0.56	0.69	–0.47	0.86	–0.11	0.43	0.04	1.07	0.38	0.92	0.28	0.90	0.04	0.89	–0.24
160	0.80	–0.56	0.69	–0.47	0.86	–0.11	0.43	0.04	1.07	0.39	0.92	0.29	0.90	0.04	0.89	–0.23
170	0.80	–0.56	0.69	–0.47	0.85	–0.11	0.43	0.05	1.07	0.39	0.92	0.29	0.90	0.05	0.89	–0.23
180	0.80	–0.56	0.69	–0.47	0.85	–0.11	0.43	0.06	1.07	0.39	0.92	0.30	0.90	0.05	0.88	–0.22
190	0.80	–0.56	0.69	–0.47	0.85	–0.11	0.43	0.07	1.07	0.40	0.93	0.30	0.89	0.06	0.88	–0.22
200	0.80	–0.56	0.69	–0.47	0.84	–0.10	0.42	0.08	1.07	0.40	0.93	0.30	0.89	0.06	0.88	–0.22
210	0.80	–0.56	0.69	–0.47	0.84	–0.10	0.42	0.09	1.08	0.41	0.93	0.31	0.89	0.07	0.88	–0.21
220	0.80	–0.56	0.69	–0.47	0.84	–0.10	0.42	0.10	1.08	0.41	0.93	0.31	0.89	0.07	0.88	–0.21
230	0.80	–0.56	0.69	–0.47	0.83	–0.10	0.42	0.11	1.08	0.42	0.93	0.32	0.89	0.08	0.88	–0.20
240	0.80	–0.56	0.69	–0.47	0.83	–0.10	0.42	0.11	1.08	0.42	0.93	0.32	0.89	0.08	0.88	–0.20
250	0.80	–0.56	0.69	–0.47	0.82	–0.09	0.42	0.12	1.08	0.43	0.93	0.33	0.89	0.09	0.88	–0.19
260	0.80	–0.56	0.69	–0.47	0.82	–0.09	0.42	0.13	1.08	0.43	0.93	0.34	0.89	0.10	0.88	–0.18
270	0.80	–0.56	0.69	–0.47	0.81	–0.09	0.42	0.14	1.08	0.44	0.93	0.34	0.89	0.10	0.87	–0.18
280	0.80	–0.56	0.69	–0.47	0.81	–0.09	0.42	0.15	1.08	0.45	0.93	0.35	0.89	0.11	0.87	–0.17
290	0.80	–0.56	0.69	–0.47	0.80	–0.09	0.42	0.16	1.08	0.45	0.93	0.35	0.89	0.11	0.87	–0.16
300	0.80	–0.56	0.69	–0.47	0.80	–0.09	0.42	0.17	1.08	0.46	0.93	0.36	0.89	0.12	0.87	–0.16
310	0.80	–0.56	0.69	–0.47	0.79	–0.09	0.42	0.17	1.08	0.47	0.93	0.37	0.89	0.13	0.87	–0.13
320	0.80	–0.56	0.69	–0.47	0.79	–0.08	0.42	0.18	1.08	0.47	0.93	0.37	0.89	0.13	0.87	–0.13
330	0.80	–0.56	0.69	–0.47	0.78	–0.08	0.42	0.19	1.08	0.48	0.93	0.38	0.89	0.14	0.87	–0.12
340	0.79	–0.56	0.69	–0.47	0.78	–0.08	0.42	0.20	1.08	0.49	0.93	0.39	0.89	0.15	0.87	–0.11
350	0.79	–0.56	0.69	–0.47	0.77	–0.08	0.42	0.20	1.08	0.49	0.93	0.39	0.89	0.16	0.87	–0.10
360	0.79	–0.56	0.69	–0.47	0.77	–0.08	0.42	0.21	1.08	0.50	0.93	0.40	0.89	0.17	0.86	–0.09
370	0.79	–0.56	0.69	–0.47	0.77	–0.08	0.42	0.21	1.08	0.51	0.93	0.41	0.89	0.17	0.86	–0.09
380	0.79	–0.56	0.69	–0.47	0.76	–0.08	0.42	0.22	1.08	0.52	0.93	0.41	0.89	0.18	0.86	–0.08
390	0.79	–0.56	0.69	–0.47	0.76	–0.08	0.42	0.22	1.08	0.53	0.93	0.42	0.89	0.19	0.86	–0.07
400	0.79	–0.56	0.69	–0.47	0.76	–0.08	0.42	0.23	1.08	0.54	0.93	0.43	0.89	0.20	0.86	–0.06
410	0.79	–0.56	0.69	–0.47	0.75	–0.08	0.43	0.23	1.08	0.55	0.93	0.44	0.88	0.21	0.86	–0.05
420	0.79	–0.56	0.69	–0.47	0.75	–0.08	0.43	0.24	1.08	0.56	0.93	0.45	0.88	0.22	0.86	–0.04
430	0.79	–0.56	0.69	–0.47	0.75	–0.08	0.43	0.24	1.08	0.57	0.93	0.46	0.88	0.23	0.86	–0.02
440	0.79	–0.56	0.69	–0.47	0.74	–0.08	0.44	0.24	1.07	0.58	0.93	0.47	0.88	0.24	0.86	–0.01
450	0.79	–0.56	0.69	–0.47	0.74	–0.08	0.44	0.24	1.07	0.60	0.93	0.48	0.88	0.25	0.85	–0.00
460	0.79	–0.56	0.69	–0.47	0.74	–0.08	0.44	0.24	1.07	0.61	0.92	0.49	0.88	0.26	0.85	0.01
470	0.79	–0.56	0.69	–0.47	0.73	–0.07	0.45	0.24	1.07	0.62	0.92	0.50	0.87	0.27	0.85	0.02
480	0.79	–0.56	0.69	–0.47	0.73	–0.07	0.46	0.24	1.06	0.64	0.92	0.51	0.87	0.29	0.85	0.03
490	0.79	–0.56	0.69	–0.47	0.73	–0.07	0.46	0.24	1.06	0.65	0.92	0.53	0.87	0.30	0.85	0.04
500	0.79	–0.56	0.69	–0.47	0.74	–0.08	0.47	0.24	1.05	0.67	0.91	0.54	0.87	0.31	0.85	0.06
510	0.79	–0.56	0.69	–0.47	0.74	–0.08	0.47	0.23	1.04	0.68	0.91	0.55	0.87	0.32	0.85	0.07
520	0.79	–0.56	0.69	–0.47	0.74	–0.08	0.48	0.23	1.03	0.70	0.90	0.56	0.86	0.33	0.85	0.08
530	0.79	–0.56	0.69	–0.47	0.75	–0.08	0.49	0.23	1.02	0.71	0.90	0.58	0.86	0.34	0.85	0.09

Time	Metatarsal		Malleolus		Knee		Hip		Head		Shoulder		Elbow		Barbell	
(ms)	*x**	*y*	*x*	*y*	*x*	*y*	*x*	*y*	*x*	*y*	*x*	*y*	*x*	*y*	*x*	*y*
540	0.79	−0.56	0.69	−0.46	0.76	−0.08	0.50	0.22	1.01	0.73	0.89	0.59	0.86	0.35	0.85	0.10
550	0.79	−0.56	0.69	−0.46	0.76	−0.08	0.50	0.22	1.00	0.75	0.89	0.60	0.86	0.37	0.85	0.11
560	0.79	−0.56	0.69	−0.46	0.78	−0.08	0.53	0.21	0.98	0.76	0.88	0.61	0.85	0.38	0.85	0.13
570	0.79	−0.56	0.69	−0.46	0.78	−0.08	0.54	0.20	0.96	0.80	0.87	0.63	0.85	0.39	0.85	0.14
580	0.79	−0.56	0.69	−0.46	0.78	−0.08	0.54	0.20	0.95	0.81	0.86	0.64	0.85	0.40	0.85	0.15
590	0.79	−0.56	0.69	−0.46	0.79	−0.08	0.56	0.19	0.93	0.83	0.85	0.66	0.84	0.41	0.85	0.16
600	0.79	−0.56	0.70	−0.46	0.80	−0.07	0.57	0.19	0.90	0.84	0.84	0.67	0.84	0.42	0.85	0.18
610	0.79	−0.56	0.70	−0.46	0.80	−0.07	0.59	0.19	0.88	0.86	0.83	0.68	0.83	0.44	0.86	0.19
620	0.79	−0.56	0.70	−0.46	0.80	−0.07	0.60	0.18	0.86	0.87	0.82	0.70	0.82	0.45	0.86	0.20
630	0.79	−0.56	0.70	−0.46	0.81	−0.07	0.62	0.18	0.84	0.88	0.81	0.71	0.82	0.46	0.86	0.22
640	0.79	−0.56	0.70	−0.46	0.81	−0.06	0.64	0.19	0.81	0.90	0.79	0.72	0.81	0.47	0.86	0.23
650	0.79	−0.56	0.70	−0.46	0.81	−0.06	0.65	0.19	0.79	0.91	0.78	0.74	0.81	0.49	0.86	0.24
660	0.79	−0.56	0.70	−0.46	0.81	−0.06	0.66	0.20	0.77	0.92	0.76	0.75	0.80	0.50	0.86	0.26
670	0.79	−0.56	0.70	−0.46	0.81	−0.06	0.67	0.21	0.74	0.93	0.75	0.76	0.79	0.51	0.86	0.27
680	0.79	−0.56	0.70	−0.46	0.80	−0.05	0.68	0.22	0.72	0.94	0.74	0.77	0.79	0.53	0.86	0.29
690	0.79	−0.56	0.70	−0.46	0.80	−0.05	0.69	0.23	0.71	0.94	0.73	0.78	0.78	0.54	0.87	0.29
700	0.79	−0.56	0.71	−0.45	0.79	−0.04	0.70	0.24	0.69	0.95	0.71	0.79	0.78	0.55	0.87	0.30
710	0.79	−0.56	0.71	−0.45	0.79	−0.04	0.70	0.24	0.67	0.96	0.70	0.80	0.77	0.56	0.87	0.30
720	0.79	−0.56	0.71	−0.45	0.79	−0.03	0.70	0.25	0.65	0.96	0.69	0.81	0.77	0.57	0.87	0.32

Data correspond to the forward-backward (*x*) and vertical (*y*) location of selected landmarks during the displacement of the barbell from the floor to about hip height.

**Note*. All values for *x* and *y* are in meters.

Data are from Enoka (1981).

Table E.2 Angle of the Arms Relative to a Horizontal Reference During a Running Forward Somersault in Gymnastics

Time	Subject									
(ms)	10	5	3	11	2	7	1	8	6	4
10	0.94	1.08	1.01	0.40	0.47	–0.80	1.01	0.87	0.63	0.89
20	1.03	1.13	1.08	0.49	0.54	–0.84	1.07	0.94	0.72	0.94
30	1.12	1.20	1.15	0.56	0.61	–0.86	1.13	1.00	0.79	1.01
40	1.19	1.26	1.22	0.65	0.68	–0.89	1.19	1.07	0.86	1.07
50	1.27	1.33	1.29	0.72	0.75	–0.93	1.26	1.13	0.93	1.12
60	1.36	1.40	1.36	0.80	0.82	–0.96	1.33	1.20	1.00	1.19
70	1.43	1.45	1.43	0.87	0.89	–1.00	1.38	1.27	1.07	1.24
80	1.52	1.52	1.50	0.96	0.96	–1.01	1.45	1.33	1.13	1.39
90	1.59	1.59	1.57	1.03	1.03	–1.05	1.52	1.40	1.20	1.36
100	1.68	1.66	1.64	1.12	1.10	–1.07	1.59	1.47	1.27	1.43
110	1.75	1.73	1.71	1.19	1.19	–1.10	1.66	1.54	1.34	1.50
120	1.82	1.78	1.78	1.27	1.26	–1.12	1.73	1.59	1.41	1.57
130	1.89	1.85	1.85	1.36	1.33	–1.12	1.80	1.66	1.48	1.62
140	1.96	1.92	1.92	1.43	1.41	–1.13	1.87	1.73	1.55	1.69
150	2.02	1.99	1.99	1.52	1.50	–1.13	1.94	1.80	1.62	1.76
160	2.08	2.04	2.06	1.59	1.57	–1.13	2.01	1.85	1.69	1.83
170	2.15	2.11	2.13	1.68	1.66	–1.13	2.06	1.92	1.76	1.90
180	2.20	2.16	2.18	1.75	1.75	–1.13	2.13	1.97	1.82	1.97
190	2.25	2.22	2.25	1.83	1.82	–1.12	2.18	2.04	1.89	2.02
200	2.29	2.27	2.30	1.90	1.90	–1.10	2.25	2.09	1.94	2.09
210	2.34	2.32	2.36	1.97	1.97	–1.08	2.30	2.16	1.99	2.16
220	2.37	2.36	2.43	2.04	2.06	–1.07	2.36	2.22	2.06	2.23
230	2.43	2.41	2.48	2.11	2.13	–1.03	2.41	2.27	2.11	2.29
240	2.46	2.44	2.53	2.18	2.20	–1.01	2.46	2.32	2.16	2.36
250	2.50	2.48	2.58	2.25	2.27	–0.98	2.51	2.37	2.22	2.41
260	2.53	2.53	2.65	2.30	2.36	–0.94	2.57	2.43	2.27	2.48
270	2.57	2.57	2.71	2.37	2.43	–0.91	2.60	2.48	2.30	2.53
280	2.60	2.60	2.76	2.43	2.50	–0.87	2.65	2.53	2.36	2.60
290	2.64	2.64	2.83	2.48	2.55	–0.86	2.71	2.58	2.41	2.65
300	2.67	2.67	2.88	2.53	2.62	–0.82	2.74	2.64	2.44	2.71
310	2.69	2.71	2.95	2.57	2.69	–0.79	2.79	2.69	2.50	2.78
320	2.72	2.74	3.02	2.62	2.76	–0.75	2.85	2.74	2.55	2.83
330	2.76	2.78		2.67	2.81	–0.73		2.79	2.58	
340	2.79	2.81		2.71	2.88	–0.70		2.85	2.64	
350	2.83	2.85			2.95			2.90	2.67	
360					3.02			2.95	2.71	
								3.00		

Data correspond to the shoulder-joint angle of the arms relative to the right horizontal. Each column contains two markers (underlines), one at the beginning of each data set and one at the end. The initial marker indicates the moment of contact with the force platform (footstrike), and the latter marker indicates the moment of takeoff.

Data are from Nissinen (1978).

Table E.3 Measurements of Knee and Elbow Angle of a Runner During One Stride

Time (ms)	Knee angle (rad)	Elbow angle (rad)	Event
0	2.48	1.97	
13	2.54	1.91	lFS
25	2.70	1.92	
38	2.49	1.87	
50	2.46	1.84	
63	2.36	1.86	
76	2.34	1.80	
88	2.39	1.77	
101	2.46	1.54	
114	2.66	1.39	
126	2.72	1.40	lTO
139	2.57	1.40	
152	2.33	1.40	
164	2.02	1.51	
176	1.80	1.75	
189	1.50	1.96	
202	1.33	2.16	
214	1.17	2.36	
227	1.00	2.44	
239	0.78	2.54	rFS
252	0.68	2.43	
265	0.61	2.40	
277	0.68	2.32	
290	0.75	2.21	
302	0.75	2.17	
315	0.82	2.20	
328	0.95	2.08	
340	1.11	1.97	
353	1.29	1.86	rTO
365	1.55	1.82	
378	1.76	1.75	
391	2.04	1.62	
403	2.18	1.63	
416	2.38	1.79	
428	2.40	1.80	
441	2.47	1.92	
454	2.54	1.82	lFS
466	2.53	1.97	
479	2.54	1.94	

Note. The data are for the left side of the runner. lFS = left footstrike; lTO = left takeoff; rFS = right footstrike; rTO = right takeoff.

Data provided by A.E. Atwater.

APPENDIX F

Abbreviations

$\mathbf{a}$	Acceleration
$\bar{\mathbf{a}}$	Average acceleration
A	Projected area
A	Ampere
A band	Anisotropic
ACh	Acetylcholine
AHP	Afterhyperpolarization
Alpha GPD	Alpha-glycerophosphate dehydrogenase
AMG	Acoustic myograph
AP	Action potential
ATP	Adenosine triphosphate
a_x	Acceleration in the x-direction
a_y	Acceleration in the y-direction
a_z	Acceleration in the z-direction
BB	Biceps brachii
BR	Brachioradialis
BW	Body weight
c	centi
C	Chemical energy
Ca^{2+}	Calcium ion
cd	Candela
C_D	Drag coefficient
CE	Contractile element
CG	Center of gravity
Cl^-	Chloride ion
C_L	Lift coefficient
cm	Centimeter
CNS	Central nervous system
CSA	Cross-sectional area
CT	Contraction time; computed tomography
d	deci
Da	Dalton

e	Coefficient of restitution
E	Modulus of elasticity; energy
EDL	Extensor digitorum longus
E_h	Heat energy
EHL	Extensor hallucis longus
E_k	Kinetic energy
$E_{k,r}$	Rotational kinetic energy
$E_{k,t}$	Translational kinetic energy
E_m	Metabolic energy
EMG	Electromyogram
EOG	Electroculogram
E_p	Potential energy
$E_{p,g}$	Potential energy due to gravity
$E_{p,s}$	Potential energy due to strain
EPP	End-plate potential
EPSP	Excitatory postsynaptic potential
F	Farad
$\mathbf{F}$	Force
$\bar{\mathbf{F}}$	Average force
$\mathbf{F}_a$	Air resistance force
F-actin	Fibrous actin
FBD	Free body diagram
$\mathbf{F}_c$	Centripetal force; compressive force
$\mathbf{F}_d$	Drag force
FDI	First dorsal interosseous
$\mathbf{F}_f$	Fluid resistance force
$\mathbf{F}_g$	Ground reaction force
FHL	Flexor hallucis longus
$\mathbf{F}_i$	Intraabdominal pressure
$\mathbf{F}_j$	Joint reaction force
$\mathbf{F}_l$	Lift force
$\mathbf{F}_m$	Muscle force
$F_{m,c}$	Clavicular component of muscle force
$F_{m,s}$	Sternal component of muscle force
F_n	Normal component of force
$\mathbf{F}_p$	Propulsive force
$\mathbf{F}_r$	Reaction force
$\mathbf{F}_s$	Shear force
F_t	Tangential component of force
$\mathbf{F}_w$	Weight
F_x	Force in the x-direction
F_y	Force in the y-direction
g	Conductance
g	gram
g_{Ca}	Calcium conductance
g_{K}	Potassium conductance
g_{Na}	Sodium conductance
G	Giga
$\mathbf{G}$	Linear momentum
G-actin	Globular actin
Group Ia	Muscle spindle afferent
Group Ib	Tendon organ afferent
GS	Gastrocnemius soleus
h	hecto
H	Heat or thermal energy
$\mathbf{H}$	Angular momentum

H^+	Hydrogen ion
$H^{B1/C1}$	Angular momentum of body segment 1 with respect to its center of gravity
H band	Hellerscheibe
$H^{C1/CS}$	Angular momentum of the center of gravity of body segment 1 with respect to the system center of gravity
HMM	Heavy meromyosin
HR	Hold-relax technique
HR-AC	Hold relax–agonist contraction
H reflex	Hoffmann reflex
HRT	Half relaxation time
Hz	Hertz
I	Moment of inertia
I band	Isotropic
I_{xx}	Moment of inertia about the somersault axis
I_{yy}	Moment of inertia about the cartwheel axis
I_{zz}	Moment of inertia about the twist axis
J	Joule
k	kilo
k	Proportionality constant; spring stiffness
K	Kelvin
K^+	Potassium ion
kg	Kilogram
L	Liter
L_c	Proportion of whole-muscle length containing contractile protein
lFS	Left footstrike
LG	Lateral gastrocnemius
lm	Lumen
l_o	Resting muscle length
LMM	Light meromyosin
LS	Lumbosacral joint
lTO	Left takeoff
m	Mass
m	Meter; milli
M	Mega
M	Moment of force
MAD	Mass acceleration diagram
M band	Mittelscheibe
MG	Medial gastrocnemius
mm	Millimeter
mol	Mole
MRI	Magnetic resonance imaging
MVC	Maximum voluntary contraction (usually isometric)
M wave	Motor response to electrical stimulation of the nerve
n	nano
N	Newton
Na^+	Sodium ion
NADH-TR	Nicotinamide neucleotide dehydrogenase tetrazolium reductase
NMJ	Neuromuscular junction
p	pico
Pa	Pascal
PE	Parallel elastic element
pH	A measure of the acidity or alkalinity of a substance; pH = 7 is neutral.
P_i	Inorganic phosphate

PL	Peroneus longus
PMG	Phonomyograph
PNF	Proprioceptive neuromuscular facilitation
P_o	Maximum isometric force
QF	Quadriceps femoris
$\mathbf{r}$	Position
$\mathbf{r}_f$	Final position
$\mathbf{r}_i$	Initial position
R	Renshaw cell
R	Resultant
rad	Radian
RCP	Renshaw cell pool
rFS	Right footstrike
Rh	Rheobase
RM	Repetition maximum
RMP	Resting membrane potential
R_N	Input resistance
RT	Relaxation time
rTO	Right takeoff
R_x	x-component of the resultant
R_y	y-component of the resultant
R_z	z-component of the resultant
s	Displacement
s	Second
S1	Subfragment 1
S2	Subfragment 2
$s_{A/K}$	Displacement of point A with respect to point K
SD	Standard deviation
SE	Standard error
SDH	Succinic dehydrogenase
SE	Series elastic component
S-I	Primary somatosensory cortex
S-II	Secondary somatosensory cortex
SI units	Le Système Internationale d'Unités
SMG	Sound myograph
sr	Steradian
t	Time
T	Tera
$\mathbf{T}$	Torque
TA	Tibialis anterior
TB	Triceps brachii
$\mathbf{T}_l$	Load torque
TM	Tropomyosin
$\mathbf{T}_m$	Muscle torque
TN	Troponin
TN-C	Ca^{2+}-binding component of troponin
TN-I	Inhibitory component of troponin
TN-T	Tropomyosin-binding component of troponin
T tubule	Transverse tubule
Tw_s	Twitch superimposition
Type FF	Fast-twitch, fatigable motor unit
Type FG	Fast-twitch, glycolytic muscle fiber
Type FOG	Fast-twitch, oxidative-glycolytic muscle fiber
Type FR	Fast-twitch, fatigue-resistant motor unit
Type I	Slow-twitch muscle fiber

Type II	Fast-twitch muscle fiber
Type S	Slow-twitch motor unit
Type SO	Slow-twitch, oxidative muscle fiber
U	Work
$\mathbf{v}$	Velocity
$\bar{\mathbf{v}}$	Average velocity
V	Volt
$v_{A/K}$	Velocity of point *A* with respect to point *K*
$\mathbf{v}_f$	Final velocity
$\mathbf{v}_i$	Initial velocity
VI	Vastus intermedius
VL	Vastus lateralis
V_m	Membrane voltage
VM	Vastus medialis
v_{max}	Maximum velocity of shortening
VO_2	Oxygen consumption
V_1	Short-latency reflex response to electrical stimulation during a voluntary contraction
V_2	Longer latency reflex response to electrical stimulation during a voluntary contraction
W	Watt
W	Weight
x	Side-to-side direction
XB	Crossbridge
y	Forward-backward direction
z	Vertical direction
Z band	Zwischenscheibe
α (alpha)	Angular acceleration; motor neuron
β (beta)	Angle of pennation; motor neuron
γ (gamma)	Fusimotor neuron
Δ (delta)	Change in
ε (epsilon)	Strain
θ (theta)	Angular position
μ (mu)	Micro; friction coefficient
π (pi)	3.14 rad
ρ (rho)	Fluid density
σ (sigma)	Stress
Σ	Sum of
ω (omega)	Angular velocity
$\times$	Cross product

Problem Answers

To assist students with the problems, the answers to odd-numbered problems are listed here.

Chapter 1

1. True
3. False
5. True
7. True
9. False
11. False
13. True
15. True
17. True
19. True
21. True
23. False
25. True
27. False
29. False
31. False
33. True
35. True
37. False
39. False
41. False
43. True
45. True
47. False
49. True
51. True
53. False
55. True
57. True
59. True
61. True
63. False
65. True
67. False
69. True
71. True
73. False
75. True
77. True
79. True
81. False
83. True
85. False
87. True
89. D
91. C
93. B
95. D
97. D
99. B
101. B
103. B,C
105. A
107. D
109. A,B
111. B
113. B,C
115. 1.73 m
117. Acceleration J
    Area G
    Mass F
    Height L
    Weight H
    Velocity B

Volume	E
Direction	M

119. A. 1.43 s

B.

Time	Position
0.0	0.00
0.1	0.05
0.2	0.20
0.3	0.44
0.4	0.78
0.5	1.28
0.6	1.77
0.7	2.40
0.8	3.14
0.9	3.97
1.0	4.91
1.1	5.94
1.2	7.06
1.3	8.29
1.4	9.61

C. −14.0 m/s

D. −9.81 m/s^2

121. 20 m/s, 2 m/s^2

123.

Acceleration (rad/s^2)

0

Time (s)

125.

Time (s)	Position (rad)	Velocity (rad/s)	Acceleration (rad/s^2)
0.115	2.57		
0.123		1.33	
0.130	2.59		45
0.138		2.00	
0.145	2.62		575
0.153		10.63	
0.161	2.79		669
0.169		21.33	
0.176	3.11		133
0.184		23.33	
0.191	3.46		356
0.199		28.67	
0.206	3.89		578
0.214		37.33	
0.221	4.45		222
0.229		40.67	
0.236	5.06		−920
0.244		26.88	
0.252	5.49		−263
0.260		22.67	
0.267	5.83		−1,200
0.275		4.67	
0.282	5.90		−178
0.290		2.00	
0.297	5.93		−133
0.305		0.00	
0.312	5.93		400
0.320		6.00	
0.327	6.02		

127. A. 6.1 m/s

B. −2.1 m/s^2

C. 2.9 m/s^2

129. A. −1.77 m/s

B. −0.88 m/s

C. 0.16 m

D. −9.81 m/s^2

131.

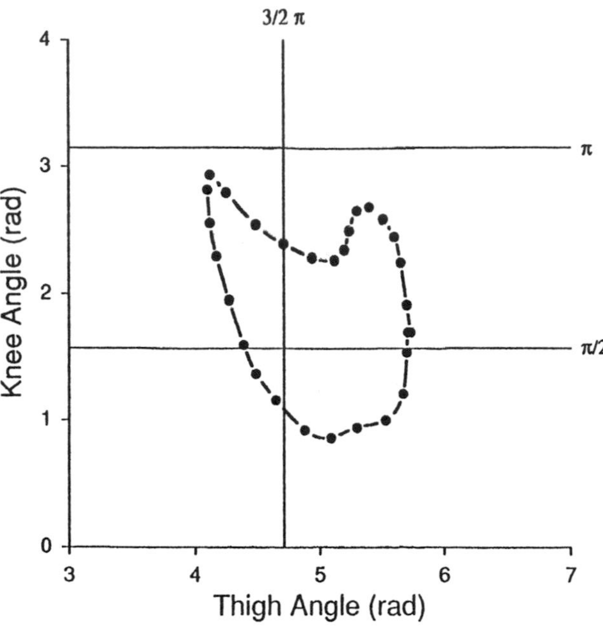

133. −21.2 m/s

135. A. 12.01 m

B. 54 m

137. 15.8 m

139. 192 m/s

Chapter 2

1. True
3. False
5. True
7. True
9. False
11. True
13. True
15. True
17. False
19. True
21. True
23. True
25. False
27. True
29. False
31. True
33. False
35. False

37. False
39. True
41. False
43. False
45. True
47. True
49. False
51. False
53. False
55. True
57. True
59. False
61. True
63. True
65. True
67. True
69. True
71. False
73. True
75. True
77. True
79. False
81. False
83. True
85. False
87. True
89. False
91. False
93. B
95. A
97. D
99. D
101. A,C,D
103. A,C,D
105. C
107. A
109. C
111. A,B,C
113. A
115. B
117. D
119. A
121. B
123. B
125. 678 N; up the midline
127. 0.39 rad to the left of vertical; 803 N
129. Normal = 165 N; tangential = 451 N.
131.

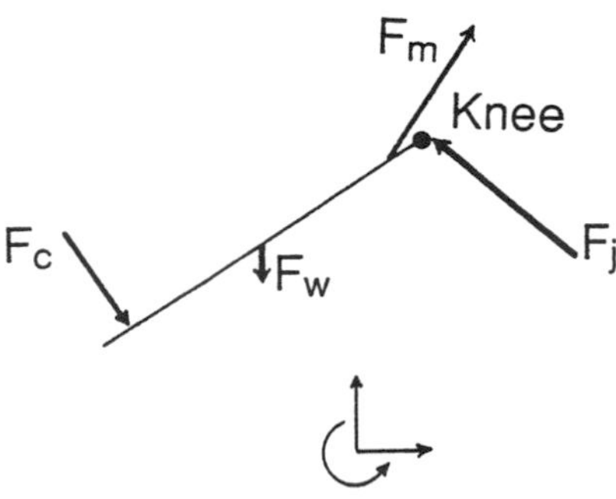

133. 271 N; 52%
135. A. Foot = 1.11 kg; shank = 3.45 kg.
B. 1.631 kg·m^2
137. 2,282 N
139. $\mathbf{F}_{s,max}$ = 698 N; $\mathbf{F}_s$ = 680 N. The shoe will not slide.
141. 1,825 N
143.

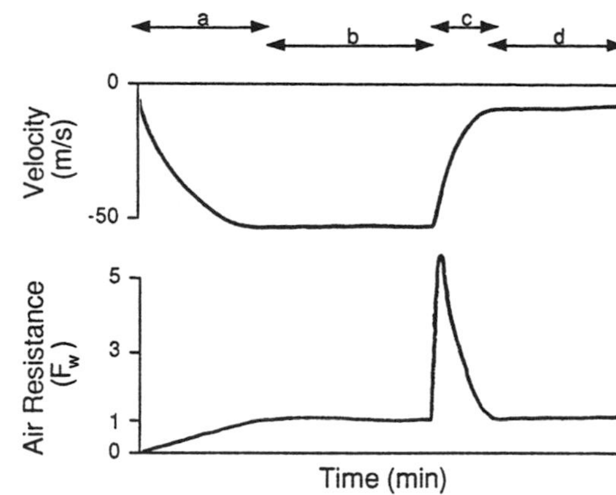

145. **F** = N; *k* = N/m; *x* = *m*.
147.

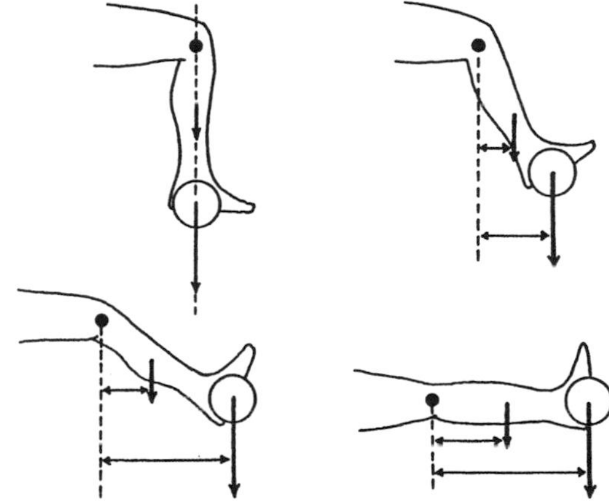

149. A, 36.6 N·m; B, 19.8 N·m.
151. Failure would occur at the initial position when the moment arm (*d*) is the smallest.

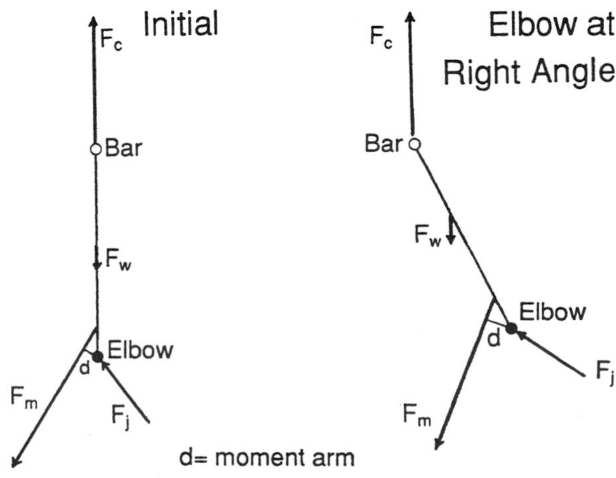

Extended = 4.38 N·m; flexed = 10.91 N·m.

Chapter 3

1. True
3. False
5. False
7. True
9. False
11. True
13. False
15. True
17. True
19. True
21. False
23. False
25. True
27. False
29. True
31. True
33. False
35. True
37. False
39. False
41. False
43. True
45. True
47. True
49. True
51. False
53. True
55. False
57. False
59. True
61. False
63. True
65. False
67. False
69. True
71. False
73. False
75. C
77. A
79. D
81. A,B,D
83. A
85. C
87. D
89. D
91. C,D
93. 1. O
    2. H
    3. C
    4. M
    5. H
    6. A
    7. O
    8. E
95. $\mathbf{F}_m$ = 156 N; normal = 88 N; tangential = 129 N.
97. 197 N
99. A. –4.39 N·m
    B. It is not possible to know without additional information.
101. 0.95 m; 55% of height
103. A. 10.3 6 N·m
    B. The knee extensors would control the movement, but the muscle force must be less than 103.6 N·m.
    C. 105 N·m
105. 15.1 m/s
107. Circle the arms backward so that they assume the backward momentum of the system and the performer can rotate forward to regain balance.
109. 5,059 N
111. 84 kg·m²/s
113. A. 42 m/s
    B. 456 N
115. A. 792 J
    B. Yes, $\mathbf{F}_{s,max}$ = 125 N.
117. The graph contains four alternating periods of power production and power absorption. When the person was in contact with the trampoline bed, there was first a period of power absorption followed by a period of power production. These two phases can be explained as a transfer of energy to the trampoline followed by the transfer of energy to the person.

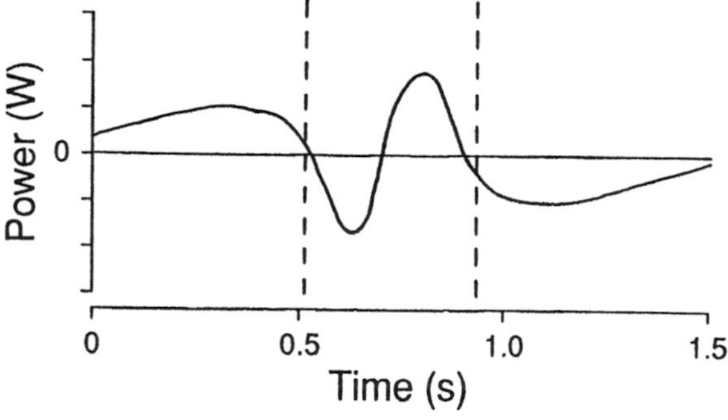

119.

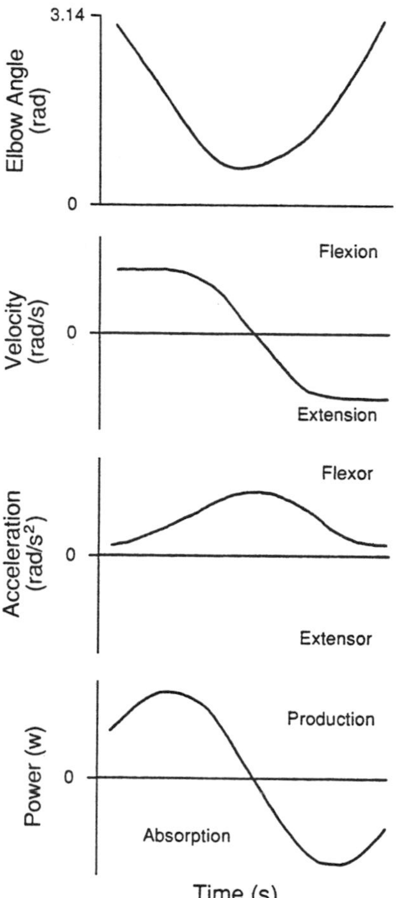

Chapter 4

1. True
3. False
5. False
7. True
9. True
11. False
13. True
15. True
17. True
19. False
21. True
23. False
25. False
27. False
29. False
31. True
33. False
35. False
37. False
39. True
41. True
43. True
45. True
47. True
49. False
51. True
53. True
55. False
57. True
59. True
61. True
63. False
65. False
67. True
69. False
71. True
73. True
75. False
77. False
79. False
81. True
83. True
85. False
87. False
89. True
91. False
93. False
95. True
97. True
99. True
101. False
103. True
105. False
107. False
109. True
111. True
113. False
115. B
117. B
119. A
121. C
123. A
125. D
127. D
129. C
131. B
133. C
135. C
137. B
139. A
141. D
143. A
145. C
147. C
149. C
151. B
153. A
155. A. To simplify the system and to focus on the elements that are important to the control of movement
     B. Rigid link, synovial joint, muscle, neuron, sensory receptor

C. These elements are not directly involved in the control of movement by the motor system. They serve support roles.

157.

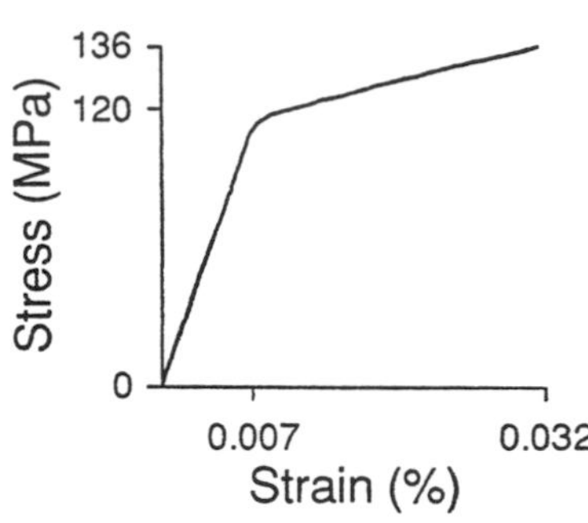

159. Exercises that impose cyclic impact loads on the long bones of the limbs. However, the intensity of the loads should progress slowly from low up to moderate values.

161. Cross-links bind microfibrils together to form fibrils, which represent the load-bearing unit of tendon and ligament. The greater the number of cross-links, the more secure the microfibrils and the stronger the fibrils. Similarly, the strength of the biochemical bonds (state of the cross-links) influences the strength of the fibrils and hence of tendon and ligament.

163. A. Bone has a greater elastic modulus, which indicates the slope of the stress-strain (load-deformation) relationship.
B. Bone has a greater strength, which is indicated by the parameter σ_{ult}.
C. Bone is stronger, but tendon and ligament can withstand a greater strain before failure.

165. A. Lubrication protects the articular cartilage.
B. To allow relative motion of opposing bones with minimal friction and disruption of tissue
C. Friction causes the articulating surfaces to lose tissue and to become worn down.

167. Flexion-extension about the somersault axis, abduction-adduction about the cartwheel axis, and rotation about the twist axis

169. A. 2,008
B. 1,890 fibrils and 4,309,200 sarcomeres
C. Sarcomeres are arranged both in series (in each myofibril) and in parallel (myofibrils arranged side by side).

171. A. A protein that is synthesized with a slightly different amino acid composition
B. There can be different isoforms for the heavy chain and the light chain of the myosin molecule. These isoforms are the product of different multigene families.
C. There is a strong relationship between the maximum velocity of shortening and the ATPase activity of the heavy chain that predominates in the muscle fiber. The functional significance of the light chains is unknown.

173. A. Neurons receive input, evaluate input, and transmit output.
B. Input arrives on the dendrites and soma, it is evaluated in the soma, and the output is transmitted from the axon hillock along the axon.
C. Direction of action potential propagation, the root in which the axons connect to the spinal cord, location of the soma, the sensory and motor functions served by each type of neuron, and the direction of information flow relative to the brain

175. Because the intrafusal fibers are about 36 times smaller than the extrafusal fibers

177. A. Excitability is changed by stretching the intrafusal fibers to make them taut and better able to detect very small changes in muscle length.
B.

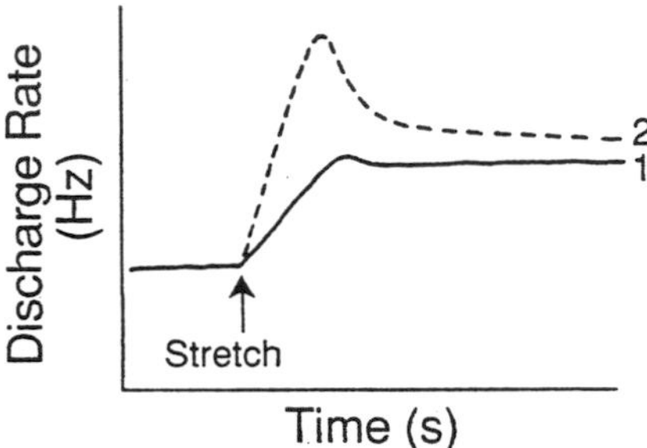

C. Either by descending commands from the brain or by afferent feedback from sensory receptors

Chapter 5

1. False
3. True
5. True
7. True
9. True
11. True
13. False
15. True
17. True
19. False
21. False
23. True
25. True
27. False
29. False
31. False
33. True
35. True
37. False
39. True
41. False
43. False
45. False
47. True
49. True

51. False
53. False
55. True
57. False
59. False
61. False
63. False
65. True
67. True
69. False
71. True
73. False
75. False
77. True
79. True
81. True
83. True
85. False
87. True
89. True
91. False
93. False
95. True
97. False
99. False
101. True
103. True
105. B
107. C
109. C,D
111. A,B
113. C
115. C
117. A
119. C
121. A
123. D
125. B
127. A
129. B
131. B
133. A,B,C,D
135. C
137. B
139. D
141. D
143. C
145. A,B,C
147. B
149. The motor unit represents the single nerve cell and the muscle fibers that it activates. Variation in muscle force is accomplished by controlling the activity of the motor units.
151. 1:2,353
153. Type FF motor units have the greatest innervation ratio and cross-sectional area of muscle fibers.
155. Morphologic—Diameter of soma, surface area of cell body, number of dendrites arising from the soma, diameter of axon, muscle fiber cross-sectional area

    Physiological—Input resistance, rheobase, afterhyperpolarization, conduction velocity, innervation ratio, tetanic force
157. A. Type FF
    B. 389
    C. 108 mN
    D. 19 mN
159. 3.2×10^{-15}
161. Motor neurons—Neuronal death is accompanied by reinnervation of some denervated muscle fibers. The result is a decline in the number of motor units but an increase in the average size of surviving motor units.

    Neurotrophism—The rate of axonal transport declines with age, which probably represents a decline in the sustaining neurotrophic interactions between neurons and muscle fibers.

    Miniature end-plate potentials—There is a decreased frequency in soleus (slow-twitch) but not in extensor digitorum longus (fast-twitch), which indicates a decline in spontaneous activity at the neuromuscular junctions of the soleus.

    Muscle action potentials—The peak-to-peak amplitude increases because of reinnervation.

    M waves—The decline in amplitude is due to a decrease in the excitability of muscle fibers.
163. Resting membrane potential—The membrane is most permeable to K^+, and hence the steady-state potential depends on the distribution of K^+ outside and inside the cell. The ratio of the concentration of an ion outside the cell relative to the inside of the cell is known as the equilibrium potential for that ion. For K^+, the equilibrium potential is negative and close to the resting membrane potential.

    Action potential—The equilibrium potential for Na^+ is positive, which means that it has a higher concentration outside the cell. A cell cannot generate an action potential without a positive equilibrium potential for Na^+, which provides the initial thrust for the action potential.
165. A. Differences—Conduction versus propagation, variable amplitude versus all-or-none amplitude, passive versus active processes. Synaptic potentials can be excitatory or inhibitory, whereas action potentials are excitatory; synaptic potentials are initiated by transmitter-gated events, whereas action potentials are activated by voltage-gated changes.

    Similarities—Ionic events underlying the change in membrane potential.

    B. Depolarization spreads from the active region along the membrane ahead of the action potential and leads to voltage-gated changes in the adjacent membrane.

167. A. Phonomyography measures the sounds associated with muscle contraction. These sounds are probably due to the lateral movement of muscle fibers.

B.

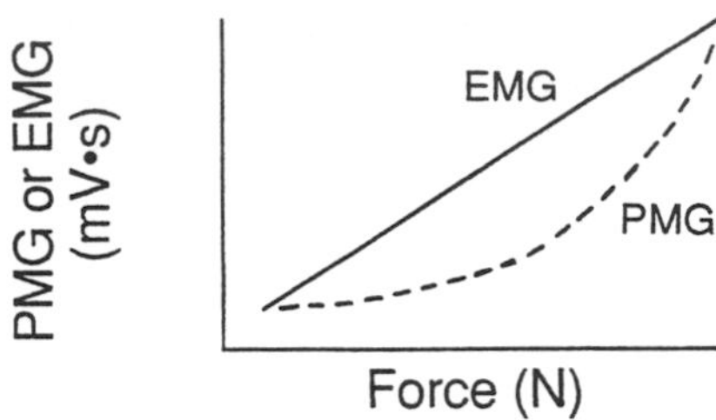

C.

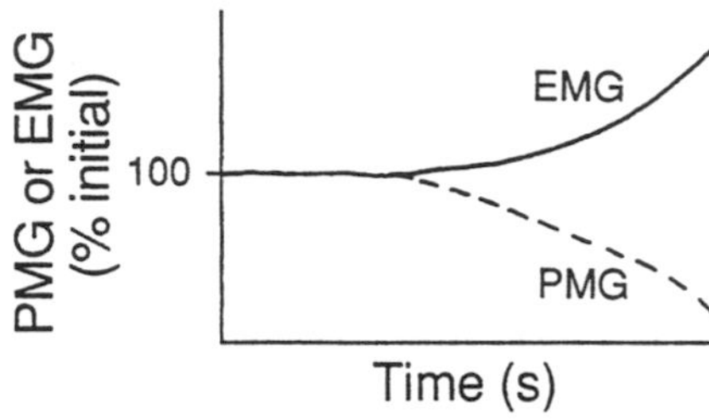

169. The seven steps of excitation-contraction coupling: (a) propagation of the sarcolemmal action potential; (b) propagation of the action potential down the T tubule; (c) coupling of the action potential to the change in Ca^{2+} conductance of the sarcoplasmic reticulum; (d) release of Ca^{2+} from the sarcoplasmic reticulum; (e) reuptake of Ca^{2+} by the sarcoplasmic reticulum; (f) Ca^{2+} binding to troponin; and (g) interaction of myosin and actin

171. Exteroception, proprioception, and consequences of action.

173. Under standardized conditions (e.g., examination room), these reflexes can provide a global measure of the excitability of the pathway from a sensory receptor back to the muscle.

175. A. The Renshaw cell can inhibit the Ia inhibitory interneuron and therefore disinhibit reciprocal inhibition.

B. It can modulate the reciprocal inhibition reflex so that it is not obligatory and is expressed only under appropriate conditions.

177. A. The interneuron receives input in the form of inhibitory and excitatory synaptic potentials from many different sources. This input is combined (spatially and temporally) to determine if it is necessary to discharge an action potential.

B. The final common pathway refers to the direct connection between a motor neuron and muscle fibers. The output of the interneuron can be directed to motor neurons and hence influence the commands that are transmitted over the final common pathway.

C. The interneuron serves as an integration site with convergence of inputs from afferent and supraspinal sources. This capability enables interneurons to alter input before it is transmitted to motor neurons and hence permits greater flexibility in input-output relationships.

179. A. Resistance reflex—A response that resists an unexpected disturbance

Assistance response—Postural and premovement activity that assists a movement

B. The involvement of proprioceptors enables the system to select a synergy that is appropriate for a given set of conditions.

C. Because muscle force (torque) is not constant and depends on the length of the muscle and its rate of change, it is necessary for the nervous system to be informed of these conditions so that it can more precisely send the appropriate command.

181. The sense of effort is a perception related to the intensity of the motor command required to perform a task. It is based on the magnitude of the motor command generated by supraspinal centers. The experimental evidence for a sense of effort consists of evaluations of the effort associated with a constant-force task when the muscle is weakened.

Chapter 6

1. True
3. False
5. False
7. False
9. True
11. True
13. False
15. True
17. True
19. False
21. True
23. False
25. True
27. False
29. True
31. True
33. True
35. True
37. False
39. True
41. False
43. True
45. True
47. True
49. True
51. False
53. False

55. False
57. False
59. True
61. True
63. True
65. True
67. False
69. A
71. C
73. C
75. C
77. A
79. C
81. B
83. B
85. A
87. B
89. A. Surface area of the cell body
    B. Number of dendrites, axon diameter, innervation ratio, muscle viber size
    C. Renshaw cell, Group Ia afferent, cutaneous afferent
91. A. Peak torque occurs at an intermediate joint angle.
    B. Minimum = 2.9 cm; maximum = 5.7 cm
    C.

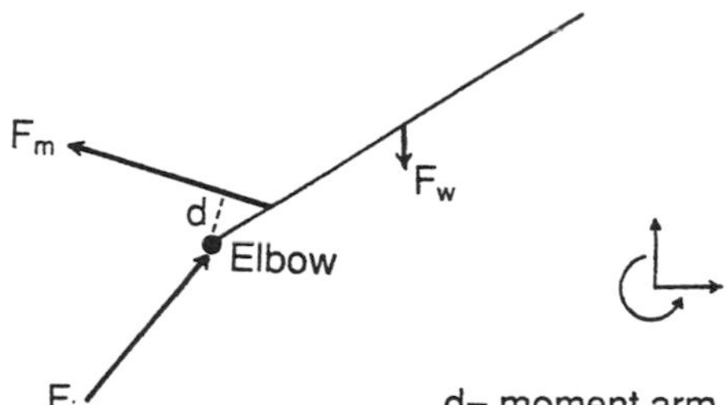

    D. Yes, although there will be some contribution from variations in the amount of overlap of the thick and thin filaments.
93. Heat production increases, and the amount of work performed decreases.
95. The muscles involved in supporting the briefcase fatigue and become unable to exert an adquate force to oppose the weight of the briefcase. When a person performs a submaximal contraction to exhaustion, the number of motor units recruited increases as the motor units already activated become unable to sustain the necessary force. Eventually an insufficient number of motor units are available to exert the required force, and the strategy for performing the task must change.
97. Both work and Δ energy are indicated as the area under a force-length graph. Because the area (Figure 6.30) was greater during the lengthening phase than during the shortening phase, there was a greater Δ energy during the lengthening contraction. This indicates that the muscle lost energy during the lengthening-shortening cycle.
99. 1.56 cm
101. Soleus has the largest functional cross-sectional area, which is due to short fiber lengths and large pennation angles.

Chapter 7

1. True
3. False
5. False
7. False
9. True
11. True
13. False
15. False
17. True
19. False
21. True
23. True
25. True
27. True
29. False
31. False
33. True
35. False
37. False
39. True
41. False
43. True
45. False
47. True
49. False
51. True
53. True
55. True
57. False
59. True
61. False
63. False
65. True
67. True
69. True
71. True
73. True
75. True
77. False
79. True
81. False
83. False
85. True
87. False
89. True
91. False
93. False
95. False
97. True
99. False
101. A

103. B
105. D
107. A,B
109. A
111. D
113. B
115. B
117. D
119. B
121. D
123. C
125. C
127. B
129. B,C,D
131. D
133. A
135. A,C,D
137. B,C,D
139. A
141. The limbic system and afferent feedback
143. To distinguish between a movement generated by the system and one imposed by an agent in the surroundings.
145. Central pattern generators are neural networks that do not require afferent input and typically generate commands for repetitive events (e.g., breathing, chewing, walking). A motor program refers to the output that is produced to perform a learned task. Afferent input probably contributes to the activation of the motor program.
147. A. Premotor cortex—Ensures correct orientation of the body and limb prior to a movement and the sensory guidance of limb movements
B. Primary motor cortex—Generates and transmits the central command to premotor, brain stem, and spinal cord neurons to initiate and modulate movement
C. Posterior parietal cortex—Interprets and transforms sensory information so that movement is consistent with surrounding conditions
D. Primary somatosensory cortex—Provides sensory information required for the specific planning and initiation of movement and for the modulation of ongoing movement
E. Supplementary motor area—Develops the plan and sequence for a movement
149. A. The output is distributed through three pairs of deep subcortical nuclei (fastigial, interpositus, dentate).
B. A system or device that compares a desired function (signal) to an actual function
C. Timing device—Compares the initial position to the desired termination and stops the movement at the desired location
Learning device—Compares input-output relationships and learns to minimize the input needed for a specific output
Coordinator—Compares and coordinates the muscle activity across several joints
151. Because the force exerted by muscle depends on muscle dynamics (length and velocity), muscle torque is influenced by joint geometry (moment arm), and inertial effects can be transmitted between segments depending on the speed of the movement.
153. A.

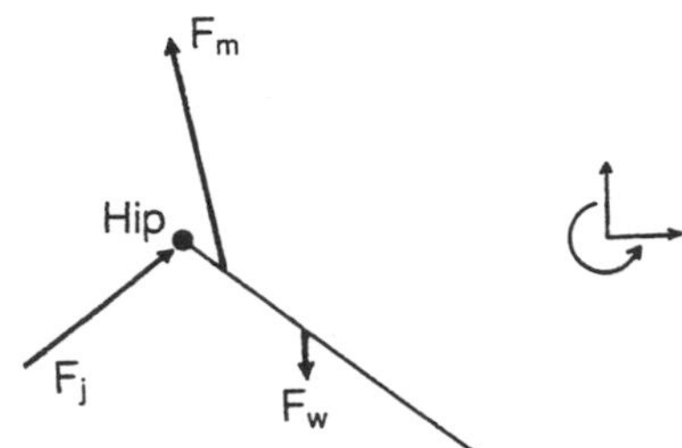

B. Up, concentric hip flexors; down, eccentric hip flexors
C. 54.9 N·m
155. A.

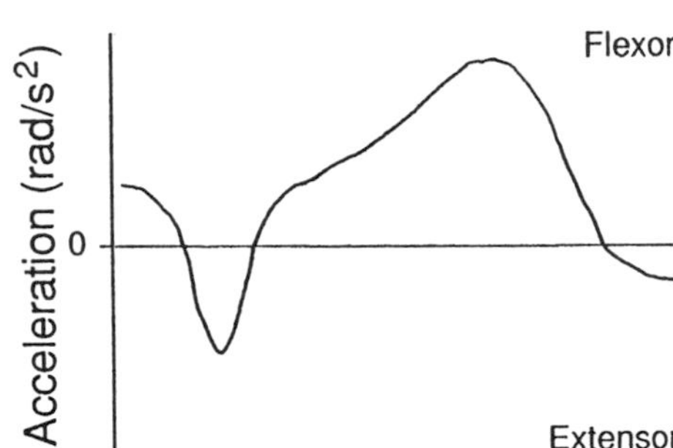

B.

Concentric
Flexors
Extensors
Eccentric

157. A. To achieve a given angular displacement about a joint, the amount of muscle shortening necessary decreases as the moment arm increases. Because two-joint muscles have longer moment arms, the shortening velocity associated with a given angular displacement is less than for the single-joint synergists. This lesser shortening velocity keeps the muscle higher on the force-velocity curve and enables it to exert a greater force.
B. As shown in Figure 7.20, rectus femoris (#5) could redistribute muscle torque from the hip to the knee by concurrent activation of gluteus maximus (#1), the vasti (#3), and rectus femoris.
159. A. Because they do not require conscious (voluntary) activation by the brain
B. Stability involves maintaining equilibrium and being able to accommodate perturbations without the loss of equilibrium. For example, equi-

librium involves keeping the vertical projection of the total-body weight vector within the boundaries of the base of support.

C. Elderly individuals have a decreased ability to detect perturbations due to impaired sensory processes and reduced motor abilities to counter the loss of equilibrium.

D. Two examples are shown in Figure 7.24. Describe another example.

161. A. It provides an index of the commands issued by the nervous system to muscle.

B. The important parameters are the amplitude and duration of motor neuron excitation. These parameters were chosen because this is the way in which the EMG changes for movements with various constraints.

C. The λ model includes a role for afferent feedback in the control of movement. In this model, movements are controlled by varying the threshold of the stretch reflex. The speed-dependent strategy (dual-strategy hypothesis) is equivalent to varying λ at a constant rate, whereas the speed-independent strategy corresponds to varying λ at different rates.

163. A. Storage and utilization of elastic energy—An agent stretches an active muscle, which can store energy and subsequently use some of the energy delivered to the muscle during the stretch. The energy is stored as $E_{p,s}$ and is used during a shortening contraction.

Preload effect—Refers to the use of an eccentric contraction to use the available E_m to develop the maximum muscle force prior to the shortening contraction

B. Supposedly the stretch of the active muscle during the eccentric contraction excites the muscle spindles and elicits the stretch reflex. However, the stretch reflex is unlikely to contribute significantly to the work done during the concentric contraction, because the force associated with a stretch reflex is too small.

165. A. The small motor units exert low forces, which is best for fine control, and they are the least fatigable. In addition, the specification of recruitment order by spinal cord mechanisms is much more economical than control by the brain.

B. Eccentric contractions seem to involve high-threshold motor units, which exert large forces.

C. The distribution of synaptic input and the intrinsic properties of motor neurons

D. Eccentric—Use of high-threshold motor units to control the large forces caused by external agents

Fatigue—Variable activation patterns enable motor units to recover from periods of sustained activity.

Cutaneous feedback—The activation of high-threshold (large-force) motor units produces a rapid response.

Chapter 8

1. False
3. True
5. True
7. False
9. False
11. False
13. True
15. False
17. True
19. True
21. True
23. False
25. True
27. True
29. True
31. False
33. True
35. False
37. True
39. True
41. True
43. True
45. False
47. True
49. False
51. True
53. False
55. True
57. True
59. False
61. True
63. True
65. False
67. False
69. True
71. True
73. False
75. True
77. False
79. True
81. False
83. False
85. True
87. True
89. True
91. False
93. True
95. False
97. True
99. False
101. False
103. True
105. False
107. False
109. True
111. True
113. B
115. A
117. D

119. B
121. A
123. B
125. A
127. A,B
129. A
131. C
133. B
135. A
137. D
139. A,C,D
141. A,C,D
143. A,C
145. A,D
147. B
149. C
151. D
153. C,D
155. A. $F_{m,o} = 381$ N·m
$\omega_{max} = 20.1$ rad/s
B.

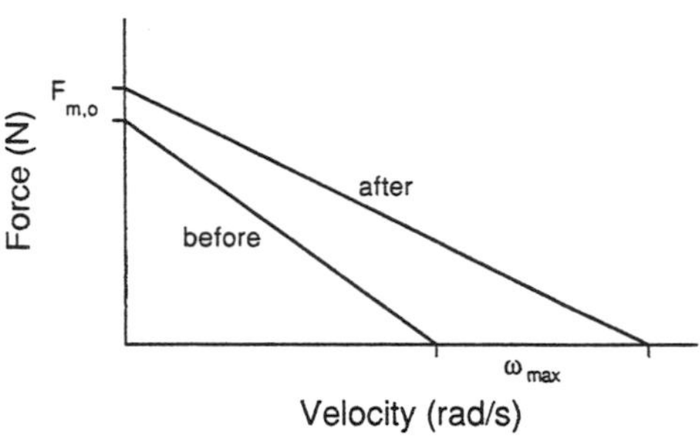

C. Before ~ 134 W; after ~ 631 W.
157. For the graph with 30 s between stretches, the stiffness is about 9.2 N·m/rad. The stiffness increases with rest interval because of an increase in the number of crossbridge bonds in the resting muscle.
159. A. See Figure 5.27.
B. Because descending input from suprasegmental centers activates the Ia inhibitory interneuron along with the alpha and gamma motor neurons
C. With disinhibition through the Renshaw cell
161. A.

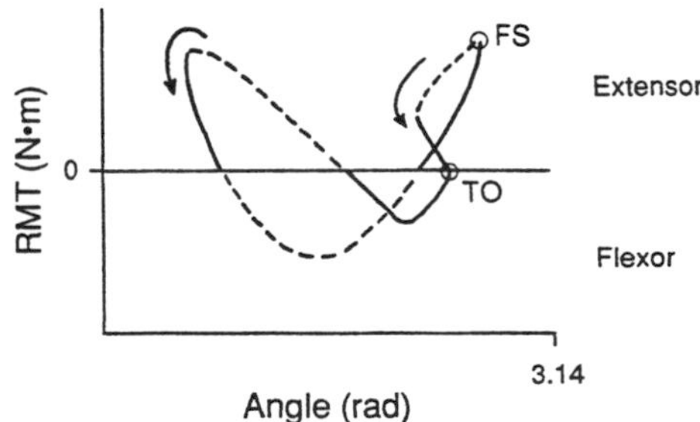

B. The strain is most likely to occur in the region where the peak flexor torque is greatest as the hamstrings change from an eccentric to a concentric contraction.
163. The force is greater but the EMG is less for the eccentric contractions than for the concentric contractions. This raises the question of whether the eccentric contraction force would be greater if the EMG could be increased to the level associated with the concentric contractions.
165. A. An M wave represents the EMG response to the activation of the alpha motor axons by an electrical stimulus.
B. M-wave amplitude is increased by postsynaptic factors that include a decrease in the temporal dispersion of muscle fiber action potentials and an increase in the amplitude of individual muscle fiber action potentials.
C. M-wave amplitude can be reduced by axonal branch-point failure, depletion of neurotransmitter, reduction in exocytosis, and a decrease in the sensitivity of the postsynaptic membrane.
D. The amplitude increases at the beginning of a sustained contraction and then declines with a low-force, long-duration contraction.
167. Metabolic effects on the crossbridge cycle
169. A. Because motor neuron discharge is altered so that it is no greater than necessary
B. The twitch relaxation rate decreases so that a lower discharge rate can achieve the same degree of fusion in tetanic force.
C. Reflex inhibition by Group III and Group IV afferents and by disfacilitation of Group Ia afferents
D. By lowering the excitability of motor neurons so that the same input elicits a lower discharge rate
171. The reduction will enable action potentials to invade more terminal collaterals and elicit end-plate potentials at a greater number of neuromuscular junctions—that is, activate more muscle fibers.
173. A. Distal to the motor neuron
B. Quantity of Ca^{2+} released and the series elasticity of muscle
C. Ca^{2+} kinetics can influence the time course of the twitch.
D. The increase in average posttetanic twitch potentiation was greater for Type FR and Type FF motor units, but the incidence was greater for the fatigue-resistant motor units (Types FR and S). The fast-twitch motor units release more Ca^{2+} per action potential but are probably more susceptible to impairment of neuromuscular propagation (e.g., branch-point failure).
E. Differences in the time courses of fatigue (force decline) and potentiation (Figure 8.24)

Chapter 9

1. True
3. True
5. False
7. False

9. False
11. True
13. True
15. True
17. True
19. True
21. False
23. False
25. False
27. False
29. False
31. False
33. True
35. False
37. False
39. False
41. False
43. True
45. False
47. False
49. True
51. True
53. True
55. True
57. False
59. True
61. False
63. False
65. False
67. True
69. True
71. True
73. True
75. False
77. False
79. False
81. True
83. True
85. True
87. True
89. True
91. True
93. True
95. False
97. False
99. True
101. True
103. True
105. True
107. False
109. B
111. A
113. A,C
115. B
117. C,D
119. B
121. D
123. C,D
125. D
127. C
129. D
131. C
133. A,C
135. A,D
137. C
139. A
141. A,D
143. B
145. D
147. C
149. B
151. A
153. A,B,C,D
155. A. The load should exceed some threshold value, which is expressed relative to the maximum capability.
     B. 893 N
     C. 693 N
157. Yes, because the force-velocity relationship moves proportionally along the x- and y-axes as a function of training (e.g., Figure 9.18).
159. A. Because the exercise stress is expressed as a percentage of maximum, not the absolute load.
     B. Eccentric exercises produce muscle damage that promotes protein synthesis and the reorganization of muscle.
     C. Eccentric contractions differ in at least the following ways:
     - Submaximal eccentric contractions involve the activation of different motor units.
     - Detachment of the crossbridge is different.
     - Some sarcomores are stretched too far and yield.
     - The maximal EMG is less.
     - Delayed-onset muscle soreness occurs most often with eccentric contractions.
     - Both oxygen consumption and EMG increase during steady-state exercise involving eccentric contractions.
     - There is a greater reduction in the amplitude of the H reflex when the activity involves eccentric contractions.
     - Eccentric contractions provide a greater stimulus for hypertrophy.
161. Axons of high-threshold (high-force) motor units have a greater diameter, which offers less resistance to the flow of current across the membrane that results from neuromuscular electrical stimulation.
163. Arms by the side (12 cm), $\mathbf{F}_m = 7.42$ kN.
     Arms overhead (18 cm), $\mathbf{F}_m = 8.25$ kN.
165. A. Strength increases can be caused by adaptations in either the nervous system by itself or by adaptation in both the nervous system and muscle.
     B. Strength increases can occur with training programs that are too brief to induce morphological changes; the increase in EMG is greater

than the increase in cross-sectional area; and the increase in strength occurs before an increase in cross-sectional area.

C. Populations most unfamiliar with the movements (e.g., novice weight lifters, older persons).

167. Type II muscle fibers are less active than Type I muscle fibers; therefore, a program of increased activity causes the Type II fibers to experience the greater change in activity.

169. 338 W

171. Tetrodotoxin—Blocks action potential propagation
Curare—Blocks transmission across the neuromuscular junction
Colchicine—Blocks axonal transport

173. Limb immobilization—There is a length and muscle (antigravity, slow-twitch) dependence with the greatest reduction in EMG (75%) occurring in soleus immobilized at short lengths and minimal reductions occurring in other muscles and neutral and long lengths.
Hindlimb suspension—There is a transient (3-day) depression in EMG that returns to normal for both soleus and medial gastrocnemius
Spinal transection—The muscles immediately become flaccid, then spastic, and finally assume a sustained extensor posture.

175. The two most influential adaptations are probably the death of motor neurons and the decline in sensory capabilities.

177. Aging—With a decrease in the proportion of Type IIb fibers, there is an increase in the fatigue-resistant (Types I and IIa) muscle fibers.
Reduced use—The slow-to-fast conversion involves a decline in the proportion of Type I fibers but an increase in the fatigue-resistant Type IIa fibers.

Glossary

A band—Anisotropic or dark striation in skeletal muscle due to the double refraction of light rays from a single source.

abduction—Movement away from the midline of the body or the body part.

acceleration—Rate of change in velocity with respect to time (m/s^2); the derivative of velocity with respect to time; the slope of a velocity-time graph.

accommodation device—A training device in which the resistance varies to match the force exerted by the user.

acetylcholine—A neurotransmitter that is used at, among other locations, the neuromuscular junction.

acetylcholinesterase—An enzyme found at the neuromuscular junction that serves the role of terminating the activity of the neurotransmitter acetylcholine.

actin—The major protein of the thin filament of the muscle fiber.

action potential—An all-or-none, propagated signal of neurons and muscle fibers. Physically, it represents a patch of membrane that contains a transient reversal of the Na^+ and K^+ distribution.

action potential threshold—The membrane potential at which the cell (neuron or muscle fiber) will generate an action potential.

active component—The part of a length-force relationship of muscle due to crossbridge activity.

active touch—The exploration of the surroundings by relying on the sensory receptors associated with the hand and its control.

adaptation—A decline in the discharge rate of a neuron despite a constant excitatory input. The decline is due to the biophysical properties of the neuron.

adaptive controller—A control system that modifies the elements of the control system rather than causing immediate changes in the output.

adduction—Movement toward the midline of the body or the body part.

afferent fiber—An axon that transmits signals from sensory receptors.

afterhyperpolarization—The trailing part of an action potential when the membrane is less excitable than in steady-state conditions. A period of hyperpolarization on the final phase of the action potential.

agonist-contract stretch—A PNF-based stretch for which an assistant places the joint at the limit of rotation and then has the individual contract the agonist (e.g., quadriceps femoris) to assist in further increasing the range of motion by stretching the antagonist (e.g., hamstrings).

alpha motor neuron—A motor neuron that innervates extrafusal (non–muscle spindle) muscle fibers.

angle-angle diagram—A graph that shows the changes in one angle (y-axis) as a function of another angle (x-axis).

angular—Motion in which not all parts of the object experience the same displacement.

anion—A negatively charged ion.

anisometric—A contraction that is not isometric.

anterolateral pathway—An ascending pathway that conveys information mainly about pain and temperature with some tactile and proprioceptive information.

anthropometric—Measurements of the human body.

anticipatory stabilization—An automatic postural adjustment that produces a movement and is intended to stabilize the system (e.g., limb, body) in anticipation of the mechanical disturbances associated with the movement.

antigravity muscles—The muscles that exert a torque to oppose the action of gravity when a human stands in an upright posture.

articular cartilage—A tissue that comprises a viscoelastic gel reinforced with collagen.

ascending pathways—Neural connections that distribute information from sensory receptors to the suprasegmental centers.

assistance responses—Adjustments that stabilize the system prior to a movement.

ATP hydrolysis—The breakdown of ATP to ADP, P_i, and energy.

atrophy—Loss of muscle mass.

autogenesis—The process of degrading cellular structures. This occurs with the exercise-induced damage of muscle.

axon—A tubular process that arises from the soma of a neuron and functions as a cable for transmitting the electrical signals generated by the neuron.

axonal transport—The movement of material on the inside of an axon. This material can be moved in both the orthograde (soma to synapse) and retrograde (synapse to soma) directions.

axotomy—Transection of an axon.

ballistic stretch—A technique to increase the range of motion about a joint by performing a series of rapid, bouncing stretches of the muscle.

basal ganglia—Five closely related nuclei (caudate, putamen, globus pallidus, subthalamic nuclei, and substantia nigra) that receive input from the cerebral cortex and send most of their output, via the thalamus, back to the cerebral cortex.

Bernoulli's principle—Fluid pressure is inversely related to fluid velocity.

beta motor neuron—A motor neuron that innervates both intrafusal and extrafusal muscle fibers.

bilateral deficit—The concurrent activation of two limbs that causes a reduction in the maximum voluntary contraction force of each limb.

bilateral facilitation—The concurrent activation of two limbs that causes an increase in the maximum voluntary contraction force of each limb.

biomechanics—The use of engineering mechanics to study biological systems.

biophysics—The application of physical principles and methods to biological systems.

boundary layer—The streamline closest to the object.

brain stem—The components of the central nervous system between the brain and the spinal cord, including the medulla oblongata, pons, and mesencephalon.

cartwheel axis—An axis that passes through the human body from front to back.

caudate nucleus—One of the basal ganglia nuclei.

center of gravity—The point about which the mass of the system is evenly distributed.

center of percussion—The location along a bat where no reaction force is felt at the hands when the bat strikes a ball.

center of pressure—the point of application of the ground reaction force.

central command—The neural signal transmitted by the motor cortex to lower motor centers (brain stem, spinal cord) for the execution of a movement.

central pattern generator—Neural network that is capable of generating behaviorally relevant patterns of motor output in the absence of afferent input.

centripetal force—A force directed toward the axis of rotation that is responsible for changing linear motion into angular motion.

cerebellum—A suprasegmental structure involved in the control of movement. It is located inferior to the cerebral hemispheres and posterior to the brain stem.

cerebrocerebellum—The laterial hemispheres of the cerebellum. Receives input from the cerebral cortex and sends output to the dentate nucleus.

characteristic—A typical or distinctive feature. The part of a logarithm that indicates the location of the decimal point.

chromatolysis—Dissolution of the Nissl substance (rough endoplasmic reticulum) and resynthesis of proteins following axotomy.

coactivation—Concurrent activation of muscles around a joint; usually describes activity in agonist-antagonist muscles.

coefficient of restitution—A ratio that describes the speed of an object after a collision compared with its speed before the collision.

collateral—A branch of the axon.

collateral sprouting—The development of sprouts (neurites) by distal segments of an axon for the purpose of reinnervating a denervated target. This usually occurs after damage or injury to a target cell (e.g., muscle fiber, neuron).

comparator—A device or structure that compares inputs provided by different sources.

compartment—A subvolume of muscle innervated by a primary branch of the muscle nerve.

compliance—The amount that a material can be forcibly stretched. This property is expressed as the amount of change in length per unit of force used to stretch the material (mm/N).

composition—The process of determining the resultant of several forces.

compressive force—A pushing force.

concentric—The mechanical condition in which the muscle torque is greater than the load torque and as a consequence the active muscle shortens.

conductance—A property of a membrane that depends on its permeability to an ion and the availability of the ion. Conductance is the inverse of electrical resistance.

conduction velocity—The speed at which an action potential is propagated along an excitable membrane.

conductivity—The ability to propagate a wave of excitation.

conservation of momentum—A concept stating that when a system is not subjected to an impulse, its momentum remains constant.

constant load—A training load that remains constant (e.g., barbell).

contractile element—One of the components in Hill-type models of muscle. It represents that ability of contractile proteins to exert a force.

contractility—The ability of an element to modify its length.

contraction—The muscle-activation state in which the crossbridges are cycling in response to an action potential. Muscle length may shorten, stay the same, or lengthen during this state of activation.

contraction time—The time from force onset to the peak force during a twitch.

control system—A set of subsystems and interconnecting communication channels that process information. A subsystem receives incoming messages and performs computations on them before distributing the information throughout the system. There are three types of automatic control systems: feedback controller, feedforward controller, and adaptive controller (Houk, 1988). A control system that is mainly used to compensate for disturbances is called a *regulator*.

corollary discharge—Internal signals that arise from motor commands and influence perception (Sperry, 1950).

corticospinal tract—A descending pathway with input from all of the sensorimotor cortex.

cosecant—A trigonometric function equal to the hypotenuse divided by the opposite side.

cosine—A trigonometric function equal to the adjacent side divided by the hypotenuse.

cotangent—A trigonometric function equal to the adjacent side divided by the opposite side.

crossbridge—The subfragment 1 extension of the myosin molecule. Subfragment 1 represents the globular component of the heavy meromyosin fragment.

crossbridge theory of muscle contraction—A theory that explains the force exerted during a muscle contraction as due to the action of crossbridges.

crossed-extensor reflex—An extension response in the contralateral limb to a noxious stimulus that elicits the flexor-withdrawal reflex.

cross education—An adaptation in motor capabilities that occurs in one limb as a consequence of training the contralateral limb.

cross-links—Biochemical bonds that hold the collagen molecules together in connective tissue.

cross product—The procedure used to multiply vectors that produces a *vector product*.

cross-sectional area—The area of the end-on view of an object (e.g., muscle) when it has been sectioned (cut) at right angles to its long axis.

cubic—third-degree polynomial

cutoff frequency—A frequency specified for a filter (digital or electronic) that marks the point at which the frequency content of a signal is altered. This alteration may involve the progressive reduction of frequencies above the cutoff frequency (see Figure B.23).

cytoskeleton—The set of structures that provides the physical framework for the interaction of the contractile proteins.

degree of freedom (of a joint)—An axis of rotation. A joint with 3 degrees of freedom permits rotation about three different axes.

delayed-onset muscle soreness—The perception of muscle soreness that is associated with subcellular damage and that occurs 24 to 48 hr after the exercise. The soreness may be due to the inflammatory response to the damage.

demineralization—The excessive loss of salts from bone.

dendrite—A process, other than the axon, that extends from the soma of a neuron.

denervation—A state in which the nerve has been cut; the muscle is without neural input.

dentate nucleus—One of the deep subcortical nuclei of the cerebellum.

depolarization—A reduction in the polarity (potential) across the membrance.

derecruitment—The inactivation of a motor unit.

descending pathways—Neural pathways that transmit information from suprasegmental centers to the brain stem and spinal cord.

digital filter—A numerical procedure that manipulates the frequency spectrum of an input signal to produce an output signal that contains a substantially attenuated frequency above the cutoff frequency.

directrix—A fixed line from which all the points in a parabola are equidistant.

discharge rate—The rate at which neurons discharge action potentials; the frequency of action potentials. The unit of measurement is hertz (Hz).

disinhibition—Inhibition of the inhibitor. For example, the Renshaw cell disinhibits the Ia inhibitory interneuron, and Ca^{2+} disinhibits the inhibitory effect of the regulatory proteins (troponin and tropomyosin).

displacement—A change in position. It is measured in meters (m).

dorsal column–medial lemniscus pathway—An ascending pathway that distributes information on limb proprioception and tactile sensation.

dorsal root—The entry zone on the back side of the spinal cord. Afferent axons enter the spinal cord in the dorsal root.

double discharge—The occurrence of two action potentials discharged by a single motor unit within an interval of about 10 ms.

drag—The component of the fluid resistance vector that acts parallel to the direction of fluid flow.

dual-strategy hypothesis—An explanation of the agonist-antagonist EMG patterns and resultant muscle torque that underlie movements of different displacements, speeds, and loads.

dynamic analysis—Mechanical analysis in which the forces acting on the system are not balanced and hence the system is accelerating. The right-hand side of Newton's law of acceleration is nonzero.

eccentric—The mechanical condition in which the load torque is greater than the muscle torque and as a consequence the active muscle is lengthened.

economy—The minimal amount of energy necessary to perform a prescribed quantity of work.

effective synaptic current—The net effect of input onto a neuron as measured by an intracellular microelectrode; presumably this represents the input that reaches the axon hillock.

efference copy—Internal, command-related signals that cancel sensory discharges due to motor commands (reafferent signals) and leave unaffected the sensory signals (exafferent signals) caused by external influences (von Holst, 1954).

efferent fiber—An axon that transmits action potentials from neurons.

efficiency—Ratio of the work done to the energy used. The more work performed per unit of energy expenditure, the more efficient the system.

elastic force—The passive property of a stretched material that tends to return the material toward its original length.

electrical potential—Voltage.

electrode—A probe that can be used to measure a physical quantity, such as the concentration of chemical (ion, metabolite, pH) or the flow of electricity (current).

electromyography—Measurement of the electrical activity in muscle (muscle action potentials) that is caused by axonal action potentials.

endomysium—A connective tissue matrix that surrounds individual muscle fibers.

endosarcomeric—The cytoskeletal components that maintain the orientation of the thick and thin filaments within the sarcomere.

end-plate potential—The synaptic potential that is generated in a muscle fiber in response to the release of a neurotransmitter.

epimysium—The outer layer of connective tissue that ensheathes an entire muscle.

excitation-contraction coupling—The electrochemical processes involved in converting a muscle action potential into mechanical work performed by the crossbridges.

excitatory postsynaptic potential—A depolarizing (excitatory) synaptic potential that is elicited in the postsynaptic membrane (distal to the synapse).

exocytosis—The process of fusion by a vesicle to the presynaptic membrance and the subsequent release of neurotransmitter.

exosarcomeric—The cytoskeletal components that maintain the lateral alignment of the sarcomeres.

exponent—The power of a variable. The exponent indicates the number of times the variable should be multiplied by itself; e.g., y^4 has an exponent of 4 and indicates that y should be multiplied by itself four times.

exponential function—Has the general form $y = b^x$, where x is the independent variable.

extension—An increase in the angle between two segments.

exteroceptor—A sensory receptor that detects selected external stimuli due to interactions between the system and its surroundings.

extracellular—Outside the cell.

extrafusal fiber—A skeletal muscle fiber that is not part of the muscle spindle.

F-actin—Fibrous actin; a strand of a few hundred G-actin molecules.

failure region—The part of a stress-strain relationship where the strain is so great that the material is physically disrupted.

fastigial nucleus—One of the deep subcortical nuclei of the cerebellum.

fatigue—A class of acute effects that impair motor performance.

feedback—Signals arising from various peripheral receptors that report to the nervous system the mechanical events in the neuromuscular system.

feedback controller—A system that generates forcing functions by comparing a desired performance (as dictated by command signals) with actual performance (sensed by feedback sensors).

feedforward controller—A system that generates commands without continuous negative feedback.

fibril—The basic load-bearing unit of tendon and ligament that consists of bundles of microfibrils held together by cross-links. The number and state of the cross-links determine the strength of the connective tissue.

filtering—A signal-processing term that describes the attenuation of some of the frequencies in the original signal. For example, an EMG signal can be filtered to remove the high frequencies, resulting in a much smoother EMG signal.

final common pathway—An expression characterizing the function of motor neurons as the route by which the nervous system controls muscle activity.

first law of thermodynamics—The performance of work requires the expenditure of energy; work = Δ energy.

flexion—A decrease in the angle between two segments.

flexor-withdrawal reflex—A flexion response of the leg to a noxious stimulus that is applied to the back of the leg.

fluid resistance—The resistance that a fluid offers to any object that passes through it. The magnitude of the resistance depends on the physical characteristics of the fluid and the extent to which the motion of the object disturbs the fluid.

focus—A point about which all the points in a parabola are equidistant.

force—A mechanical interaction between an object and its surroundings. The SI unit of measurement for force is the newton (N).

force-frequency relationship—The way in which an increase in the frequency of activation (e.g., discharge rate, electrical stimulation) causes muscle force to increase.

force-length relationship—The maximum force that a muscle can exert varies as a function of muscle length. Also referred to as the length-tension relationship.

force-velocity relationship—The effect that the rate of change in muscle length has on the maximum force that muscle can exert.

forward dynamics—A dynamic analysis to determine the kinematics that will be exhibited by a system based on the forces and torques that it experiences.

Fourier analysis—The derivation of a series of sine and cosine terms to represent the frequency content of a signal.

fraction—A ratio of two quantities.

free body diagram—A graphic-analysis technique that defines a system and indicates how the system interacts with its surroundings. A free body diagram represents a graphic version of the left-hand side of Newton's law of acceleration.

free nerve ending—Small-diameter axon that senses abnormal mechanical stress and chemical agents.

friction—Resistance to the relative motion of one body sliding, rolling, or flowing over another body with which it is in contact. In movements involving contacts with a support surface, friction is the resultant of the two horizontal components of the ground reaction force.

functional cross-sectional area—The measure of muscle cross-sectional area that takes into account muscle fiber pennation. The measurement is made perpendicular to the long axis of the muscle fibers.

fundamental frequency—The single sine and cosine term that best describes how a signal varies during one cycle.

fused tetanus—A smooth force response elicited by a high frequency of activation.

fusimotor—Because the muscle spindle has a fusiform shape, gamma motor neurons are sometimes called fusimotor neurons and their axons are referred to as fusimotor fibers.

G-actin—Globular actin molecule.

gamma motor neuron—A motor neuron that innervates intrafusal (muscle spindle) muscle fibers.

generator potential—The electrical potential that a sensory receptor generates in response to the stimulus to which it is sensitive.

globus pallidus—One of the basal ganglia nuclei.

Golgi ending—Thinly encapsulated, fusiform corpuscle that is similar to a tendon organ and may sense tension in ligament.

gravitational potential energy—The energy possessed by a system due to its location in a gravitational field above a baseline.

gravity—The force of attraction between an object and a planet (an object with large mass). The force of gravity causes an acceleration of 9.81 m/s^2 on Earth.

ground reaction force—The reaction force provided by the horizontal supporting surface.

Group A pathway—Descending pathways that include the vestibulospinal, reticulospinal, and tectospinal tracts.

Group B pathway—Descending pathways that include the crossed rubrospinal and crossed reticulospinal tracts.

Group Ia afferent—A muscle spindle afferent.

Group Ib afferent—The tendon organ afferent.

Group II afferent—Afferents with a diameter one class lower than the Group I afferents. The muscle spindle has a Group II afferent.

half relaxation time—The time of decay of a twitch response from the peak force to one half of the peak force.

harmonic—Multiple of the fundamental frequency.

H band—Hellerscheibe; the region of the A band that is devoid of thin filaments.

heavy chain—A protein with a high molecular weight.

heavy meromyosin—The head-end fragment of the myosin molecule. This fragment can be further subdivided into subfragments 1 and 2 (S1 and S2).

hindlimb suspension—An experimental protocol in which the hindlimbs of a quadruped are lifted off the ground for a few weeks so that they can still move but cannot contact the ground or any support surface.

Hoffmann reflex—A response that is elicited artificially by electrical stimulation of a peripheral nerve and selective activation of the Group Ia (largest diameter) afferents. The afferent volley activates homonymous motor neurons and elicits an EMG and force response. The Hoffmann reflex, or H reflex, is used as a test of the level of excitability of the motor neuron pool.

hold-relax-agonist contract stretch—A stretch based on Proprioceptive Neuromuscular Facilitation. It involves the combination of hold-relax and agonist-contract stretches.

hold-relax stretch—A stretch based on Proprioceptive Neuromuscular Facilitation (PNF) that involves an initial maximum isometric contraction of the muscle to be stretched followed by relaxation and stretch of the muscle to the limit of the range of motion.

homonymous—An anatomical relation between sensory receptors and motor neurons that innervate the same muscle.

hyperplasia—An increase in muscle mass due to an increase in the number of muscle fibers.

hyperpolarization—An increase in the polarity of the membrane, characterized by the inside becoming more negative with respect to the outside.

hypertonus—An increase in muscle tone, such as occurs with spasticity and rigidity.

hypertrophy—An increase in muscle mass due to an increase in the cross-sectional area of the muscle fibers.

hypotonus—A decrease in muscle tone, such as occurs in patients with cerebellar lesions.

hysteresis loop—A history-dependent relationship in which the loading and unloading phases do not coincide. For example, the change in length (deformation) of a tissue when a force is released does not retrace the curve (the force-length relationship) obtained when the force was first applied.

Ia inhibitory interneuron—An interneuron that is excited by Group Ia afferents and by supraspinal input. This interneuron exerts an inhibitory effect on other neurons, especially motor neurons (e.g., reciprocal inhibition reflex).

I band—Isotropic or light band of skeletal muscle, named for the single refraction of light rays from a single source.

ideation—The generation of an idea.

impulse—The area under a force-time graph (N·s); the force-time integral. The application of an impulse changes the momentum of a system.

inertia—The resistance that an object offers to any changes in its motion.

inertial force—The force that an object can exert due to its motion.

inhibitory postsynaptic potential—A hyperpolarizing (inhibitory) synaptic potential that is elicited in the postsynaptic membrane (distal to the synapse).

innervation ratio—The number of muscle fibers innervated by a single motor neuron.

in parallel—Arranged side by side.

input resistant—The electrical resistance of a motor neuron to a current input that is injected by an intracellular microelectrode.

in series—Arranged end to end.

integration—A mathematical procedure for measuring the area under a curve, such as voltage-time or force-time relationship. This procedure is often applied to a rectified EMG and results in a measure of the area under the EMG-time signal. Sometimes, however, a procedure that smoothes the rectified EMG (filters out the high frequencies) is also called integration. This technique is more accurately referred to as *leaky integration*; however, the EMG signal is most frequently integrated by this procedure and described as integration.

integrin—Transmembrane protein that may serve to connect myofibrils to the extracellular matrix of connective tissue.

interference pattern—The unprocessed EMG that involves the overlapping of hundreds of muscle fiber action potentials. This is sometimes referred to as the *raw EMG*.

intermediate filaments—A component of the exosarcomeric cytoskeleton; arranged longitudinally along and transversely across sarcomeres.

interneuron—A neuron whose axon is confined to the limits of the spinal cord.

interposed nucleus—One of the deep subcortical nuclei of the cerebellum.

intersegmental dynamics—The inertial forces exerted by a moving body segment on its neighbors.

intraabdominal pressure—The pressure (Pa) inside the abdominal cavity. The intraabdominal pressure acts on the diaphragm and vertebral column to cause the trunk to extend.

intracellular—Inside the cell.

intradiscal pressure—The pressure inside the intervertebral discs.

intrafusal fiber—A miniature skeletal muscle fiber that forms part of the muscle spindle.

intramuscular pressure—The internal pressure in muscle that is associated with a muscle contraction.

intrathoracic pressure—The pressure (Pa) inside the thoracic cavity.

inverse dynamics—A dynamic analysis to determine the forces and torques acting on a system based on the kinematics of the motion.

inverse relation—For $y = k/x$, when x is positive and gets larger, y gets smaller.

irritability—The ability to respond to a stimulus.

isoform—A protein that is synthesized with a slightly different amino acid composition; also referred to as isoenzyme.

isokinetic—A movement in which the angular velocity of the displaced body segment is constant.

isometric—The mechanical condition in which the muscle torque is equal to the load torque and as a consequence whole-muscle length does not change.

isotonic—The condition in which a muscle contracts and does work against a constant load.

Jendrassik maneuver—The use of remote muscle activity to increase the excitability of a motor neuron pool and the size of the Hoffmann (H) reflex.

joint capsule—A tissue that encloses the articulating surfaces of a synovial joint and separates the joint cavity from surrounding tissues.

joint reaction force—The net force transmitted from one segment to another due to muscle, ligament, and bony contacts that are exerted across a joint.

joint receptor—A class of sensory receptors that sense joint-related events.

kilogram—The SI unit of measurement for mass.

kinematic—A description of motion in terms of position, velocity, and acceleration.

kinesiology—The study of movement.

kinesthesia—The ability of the system to use information derived from sensory receptors to determine the position of the limbs, to identify the agent (itself or something else) that causes it to move, to distinguish the senses of effort and heaviness, and to perceive the timing of movements.

kinetics—A description of motion that includes consideration of force as the cause of motion.

kinetic energy—The capacity of an object to perform work because of its motion.

lambda (λ) model—The concept that movements are controlled by variation in the threshold of the tonic stretch reflex.

laminar flow—Uniform flow of fluid (streamlines) around an object.

lateral sac—The enlargement of the sarcoplasmic reticulum that is adjacent to the transverse tubules; also referred to as the terminal cisternae.

lateral vestibular nucleus—Located between the pons and medulla; gives rise to the vestibulospinal tract.

law of acceleration—**F** = *m***a**

law of action-reaction—To every action, there is an equal and opposite reaction.

law of gravitation—All bodies attract one another with a force that is proportional to the product of their masses and inversely proportional to the square of the distance between them.

law of inertia—A force is required to stop, start, or alter motion.

lift—The component of the fluid resistance vector that acts perpendicular to the direction of the fluid flow.

light chain—A protein with a low molecular weight.

light meromyosin—The tail-end fragment of the myosin molecule.

limbic system—A set of forebrain structures that are interconnected with the hypothalamus and parts of the midbrain; sometimes described as the emotional motor system.

limb immobilization—An experimental protocol in which a limb is prevented from moving for several weeks (e.g., cast, splint).

linear—Characterized by a straight-line relationship.

linear region—The part of a stress-strain curve where the relationship is linear.

linked system—A system that is represented as a series of rigid links. In a biomechanical analysis, the human body is regarded as a linked system.

logarithm—The value (x) to which the number (b) must be raised to yield the specified number (n), e.g., if $b^x = n$, then $\log_b n = x$.

magnetic resonance imaging—A technique that involves the use of magnetic fields to determine the spatial localization of protons and that can be used to study the structure and function of the motor system. As with EMG, the intensity of one form (T2) of the magnetic resonance image is linearly related to the force of a muscle contraction.

magnetic stimulation—The use of a magnetic stimulator to generate a magnetic field that induces an electric field and elicits axonal action potentials.

magnitude—Size; amplitude.

mantissa—The part of a logarithm that is the logarithm of the digits in the number.

mass—A measure (kg) of the amount of matter in an object.

mass acceleration diagram—The graphic representation of the right-hand side of Newton's law of acceleration; shows the kinematic effects of the forces acting on the system.

maxima—The high points or peaks in a graph (e.g., position-time graph). The slope of a graph is zero at a maximum.

M band—Mittelscheibe; intrasarcomere connection of two sets of thick filaments.

medulla oblongata—The continuation of the brain into the spinal cord and the site of exit and entry for most of the cranial nerves into the brain.

Meissner corpuscle—A cutaneous mechanoreceptor that is sensitive to local, maintained pressure.

Merkel disk—A cutaneous mechanoreceptor that is sensitive to local vertical pressure.

mesencephalon—The rostral component of the brain stem; merges anteriorly into the thalamus and hypothalamus.

meter—The SI unit of measurement for length.

microfibril—An elongated bundle composed of five parallel rows of three-stranded collagen molecules arranged in series.

microtubule—A guiding structure involved in axonal transport.

miniature end-plate potential—Small synaptic potentials at the neuromuscular junction that occur in response to the spontaneous release of neurotransmitter.

minima—The low points or valleys in a graph (e.g., position-time graph). The slope of a graph is zero at a minimum.

modulus of elasticity—The slope of the elastic region of a stress-strain relationship; the ratio of stress to strain.

moment arm—The shortest distance (perpendicular) from the line of action of a force vector to an axis of rotation.

moment of force—The rotary effect of a force; torque.

moment of inertia—The resistance that an object offers to any change in its angular motion. It represents the distribution of the mass of the object about the axis of rotation. The symbol for moment of inertia is *I*, and the SI unit of measurement is $kg \cdot m^2$.

momentum—The quantity of motion possessed by an object; a vector quantity. The SI units of measurement are $kg \cdot m/s$ for linear momentum (**G**) and $kg \cdot m/s^2$ for angular momentum (**H**).

motion—A change in position (m) that occurs over an interval of time.

motion-dependent interactions—Interactive forces between body segments during human movement.

motor control—The control of movement.

motor neuron—A neuron whose axon connects directly to muscle fibers. Because it represents the final stage in the output from the nervous system and is the only means by which muscle can be activated, the motor neuron is referred to as the final common pathway.

motor neuron pool—The group of motor neurons that innervate a single muscle.

motor program—A stereotyped sequence of commands sent from the spinal cord to the muscles to elicit a specific behavior.

motor unit—The cell body and dendrites of a motor neuron, the multiple branches of its axon, and the muscle fibers that it innervates.

motor unit territory—The subvolume of muscle in which the muscle fibers belonging to a single motor unit are located.

muscle—A tissue that contains contractile cells capable of converting chemical energy into mechanical energy and that has the properties of irritability, conductivity, contractility, and a limited growth and regenerative capacity.

muscle architecture—The design of muscle that includes such factors as length, cross-sectional area, pennation, and the attachment points on the skeleton.

muscle cramp—A painful, involuntary shortening of muscle that appears to be triggered by peripheral stimuli.

muscle fatigue—A class of acute effects that impair motor performance.

muscle force—The force exerted by structural (passive) and active (crossbridge) elements of muscle.

muscle mechanics—The study of the mechanical properties of the force-generating units of muscle.

muscle soreness—See *delayed-onset muscle soreness*.

muscle spindle—An intramuscular sensory receptor that monitors unexpected changes in muscle length. It is arranged in parallel with skeletal muscle fibers.

muscle strain—A substantial strain of muscle that occurs as an acute and painful event and is immediately recognized as an injury. Muscle strains are also referred to as *pulls* and *tears*.

muscle strength—The magnitude of the torque exerted by a muscle in a single maximal isometric contraction of unrestricted duration.

muscle tone—The passive resistance of muscle to a change in its length.

muscle wisdom—The ability of muscle to reduce the discharge of its motor neurons to match the change in the biochemically mediated reduction in relaxation rate.

musculotendinous unit—The combination of muscle and associated connective tissue structures that are involved in transmitting the force exerted by muscle fibers to the skeleton.

M wave—An EMG and force response in muscle that is elicited by activation of the alpha axons with electrical stimulation. The M wave is used to test for the integrity of neuromuscular propagation.

myelin—A fatty, insulating sheath that, in peripheral nerves, is provided by Schwann cells.

myofibril—A unit within the muscle fiber that consists of the in-series arrangement of sarcomeres.

myofilament—The thick and thin filaments of a muscle fiber that contain the contractile apparatus.

myosin—The major protein in the thick filament of the muscle fiber; includes the crossbridge. There are several hundred myosin molecules in a single thick filament.

myosin ATPase—The enzyme that catalyzes the actomyosin reaction that is associated with the crossbridge cycle.

Na^+-K^+ pump—A membrane-bound protein that transports Na^+ from the intracellular to the extracellular fluid and transports K^+ in the reverse direction.

nebulin—A protein in the endosarcomeric cytoskeleton that may maintain the lattice array of actin.

negative feedback—A concept that describes the ability of a system to respond to a disturbance such that the response negates the disturbance.

negative work—The work done by the surroundings on a system. Energy is absorbed by the system from the surroundings during negative work. Muscle does negative work when the resultant muscle torque is less than the load torque.

neural strategy—Pattern of motor command and muscle activation.

neuroglia—One of two cell types in the nervous system. Neuroglia cells provide structural, metabolic, and protective support for neurons.

neuromechanical—The neurophysiological and biomechanical basis of an element, system, or discipline.

neuromuscular compartment—A volume of muscle supplied by a primary branch of the muscle nerve.

neuromuscular electrical stimulation—A clinical or experimental procedure in which electric shocks are used to artificially generate axonal action potentials and to elicit a muscle contraction.

neuromuscular junction—The synapse between the motor neuron and the muscle fiber; also known as the *motor end plate*.

neuromuscular propagation—The processes involved in the conversion of an axonal action potential into a sarcolemmal action potential.

neuron—One of two cell types in the nervous system; capable of generating and transmitting an electrical signal.

neurotransmitter—A chemical released in minute amounts from the terminals of nerve cells in response to the arrival of an action potential. The neurotransmitter diffuses across the gap between the nerve terminal and the target cell and activates or inhibits the innervated target cell.

neurotrophism—The sustaining influence that one biological element exerts on another.

nonlinear—Not characterized by a straight-line relationship.

nonresponders—Muscle fibers that do not respond (adapt) to certain types of physiological perturbations (e.g., weight training, immobilization, denervation).

normal component—A component that acts at a right angle to a surface.

nuclear bag fiber—An intrafusal fiber in which the nuclei cluster in a group.

nuclear chain fiber—An intrafusal fiber in which the nuclei are arranged end to end.

optimum—The most favorable condition for a criterion (e.g., minimum energy expenditure, maximum force, minimum stress).

orderly recruitment—The repeatable sequence of motor unit activation for a specific task.

orthogonal—Perpendicular; independent.

osteon—Basic structural unit of bone that consists of a series of concentric layers of mineralized matrix surrounding a central canal.

osteoporosis—A decline in the mass and strength of bone.

overlap of myofilaments—Interdigitation of the thick and thin filaments, which allows the crossbridges to cycle and to exert a force. The number of crossbridges that can be formed, and hence the force that can be exerted, depends on the amount of overlap.

overload principle—A training principle that states there is a threshold point that must be exceeded before an adaptive response will occur.

Pacinian corpuscle—Thickly encapsulated receptor with a parent axon of 8 to 12 μm; has a low threshold to mechanical stress and appears to detect acceleration of the joint.

parabola—A curve that is defined relative to a pair of pependicular lines (E and D) and a fixed point (F) on one of the lines (E). A parabola is the set of points (e.g., P) for which the distances FP and PO are equal and where PO is perpendicular to line D. The parabola is symmetric to line E, point V is the vertex, and point F is the focus. See figures on p. 362 for examples of parabolas.

parallel elastic element—One of the components in Hill-type models of muscle. It accounts for the elasticity in muscle that occurs in parallel with the contractile element.

passive component—The part of a length-force relationship of muscle due to the structural (non-contractile) properties of muscle and tendon.

paw-shake response—A rapid, alternating flexion-extension movement of the paw for the purpose of removing an irritant (e.g., tape stuck to the skin).

pennation—The angular deviation between the orientation of muscle fibers and the line of pull of the muscle.

perception—The process whereby afferent signals reach consciousness and contribute to sensation.

perimysium—The connective tissue matrix that collects bundles of muscle fibers into fascicles.

phasic—Intermittent.

phonomyography—A procedure that records the sounds associated with a muscle contraction. This technique has also been referred to as *sound myography* and *acoustic myography*.

piezoelectric—The generation of electric potentials due to pressure.

planar—Confined to a single plane.

plasma membrane—A lipid bilayer that is semipermeable and excitable.

point-slope equation of a line—$y - y_1 = m(x - x_1)$.

polynomial—A sum of terms that are powers of a variable.

pons—The intermediate section of the brain stem, which is between the medulla and the mesencephalon.

position—The location of an object relative to some reference.

positive work—The work done by a system on its surroundings. Energy flows from the system to the surroundings during positive work. Muscle does positive work when the resultant muscle torque exceeds the load torque.

postcontraction sensory discharge—The increase in neural activity, primarily due to increased muscle spindle discharge, in the dorsal roots after a muscle contraction.

posterior parietal cortex—A component of the sensorimotor cortex that is posterior to the central sulcus; Brodmann's areas 5 and 7.

postsynaptic—On the distal side of a synapse.

posttetanic potentiation—The increase in twitch force following a brief tetanus.

posture—a neuromechanical state that concerns the maintenance of equilibrium.

potential energy—The energy that a system possesses due to its location away from a more stable location. There are two forms of potential energy: gravitational and strain.

potentiation—An augmentation of a response (e.g., force, EMG) despite a constant input.

power—The rate of doing work; the rate of change in energy; the product of force and velocity.

power absorption—The flow of mechanical energy from the surroundings to the system. A system absorbs power when it does negative work.

power production—The flow of mechanical energy from the system to the surroundings. A system produces power when it does positive work.

preload effect—The increased force exerted by muscle at the start of the concentric phase during a stretch-shorten cycle.

premotor cortex—A component of the sensorimotor cortex that is anterior to the central sulcus; Brodmann's area 6.

pressure drag—The drag due to the nature of the fluid flow around the object. Pressure drag is greater for turbulent flow than for laminar flow.

presynaptic—On the proximal side of a synapse.

presynaptic inhibition—Hyperpolarization (inhibition by a synaptic potential) of an axon in the region just before it makes synaptic contact with a target cell.

presynaptic terminal—An enlargement of an axon that is involved in a synapse.

primary motor area—A component of the sensorimotor cortex that is anterior to the central sulcus; Brodmann's area 4.

primary somatosensory cortex—A component of the sensorimotor cortex that is posterior to the central sulcus: Brodmann's areas 1 to 3.

principle of transmissibility—The property that makes it possible to slide a vector along its line of action without altering the mechanical effect on the rigid body.

programming—The conversion of an idea by the brain into the proper strength and pattern of muscle activity necessary for a desired movement.

progressive resistance exercise—The prescription of exercise in which the load increases progressively from one set of repetitions to the next.

projected area—A term in the fluid resistance equation that accounts for the size (cross-sectional area) of the object moving through the fluid.

projectile—An object that is displaced so that it has a phase of nonsupport and the only forces it experiences are those due to gravity and air resistance.

propagation—The conduction of an action potential by active, regenerative processes that tend to preserve the quantity of the potential.

proprioceptor—A sensory receptor (e.g., muscle spindle, tendon organ) that can detect stimuli due to the actions that the system itself generates.

propriospinal—Interneurons that convey information between spinal segments.

psychophysics—The quantification of sensory experience; study of the relationship between a physical variable and its corresponding sensation.

putamen nucleus—One of the basal ganglia nuclei.

Pythagorean relationship—$a^2 = b^2 + c^2$.

quadratic—Second-degree polynomial.

qualitative—Describes the type or kind.

quantitative—Describes the quantity or how much.

quasistatic—A mechanical state in which the acceleration experienced by a system is small enough that it can be assumed to be zero.

radian—An angle that is represented as the quotient of a distance on the circumference of the circle relative to its radius. It is the dimensionless unit of measurement for angular position.

range of motion—The maximum angular displacement about a joint.

reaction time—The minimum time from the presentation of a stimulus to the onset of a response.

reciprocal inhibition reflex—A decrease in the excitability of motor neurons innervating an antagonist muscle due to the stretch of the agonist muscle and activation of its Group Ia afferents and the Ia inhibitory interneuron.

recruitment—The process of motor unit activation.

recruitment threshold—The force at which a motor unit is recruited (activated).

rectification—A process that involves eliminating or inverting the negative phase or phases of a signal. This process is frequently applied to the EMG interference pattern. Half-wave rectification refers to eliminating the negative phase, whereas full-wave rectification involves inverting the negative phases. Rectification may be accomplished numerically or electronically.

recurrent inhibition—A local reflex circuit whereby action potentials discharged by the motor neuron lead to activation, via axon collaterals, of the Renshaw cell and the subsequent generation of an inhibitory postsynaptic potential in the motor neuron.

red nucleus—A cluster of cells located in the mesencephalon that gives rise to the rubrospinal tract.

regulation—The process of maintaining a constant state by involving a system that is capable of accommodating disturbances.

regulator—A system that performs regulation.

reinnervation—A state in which a muscle or muscle fiber is reconnected to the nervous system by the development of a functional connection (synapse).

remodeling—The growth, reinforcement, and resorption experienced by living bone.

Renshaw cell—An interneuron in the spinal cord that elicits inhibitory postsynaptic potentials. The Renshaw cell receives input from collateral branches of motor neuron axons and from supraspinal sources.

repolarization—A return of membrane potential (polarity) to steady-state conditions.

residual analysis—A technique used in curve fitting to compare the difference between the filtered and unfiltered signals.

residual moment of force—The remaining term in an equation of motion when all of the other terms have been determined. This is the most common procedure for calculating the resultant muscle force.

resistance reflex—A negative feedback response.

resolution—The process of breaking a resultant vector down into several components.

responders—Muscle fibers that respond (adapt) to certain types of physiological perturbations (e.g., weight training, weightlessness, immobilization).

resting length of muscle—The muscle length at which the passive force begins.

resting membrane potential—The transmembrane voltage during steady-state conditions.

resultant muscle force—The net force exerted by a group of muscles about a joint.

reticular formation—A diffuse brain stem cluster of cells with concentrations in the pons and the medulla; gives rise to two reticulospinal tracts.

reticulospinal tract—Descending pathway from the reticular formation that terminates along the length of the spinal cord from the cervical to the lumbar segments.

reversibility principle—A training principle that states that training-induced adaptations are transient.

rheobase—A measure of neuron excitability that indicates the amount of current that must be injected into the neuron to generate an action potential.

right-hand-thumb rule—A technique for determining the direction of a vector product. For example, the calculation of torque involves the vector multiplication of moment arm and force, and the direction of the product (torque) is indicated by the right-hand-thumb rule.

rigidity—A condition that involves a bidirectional resistance to passive movement that is independent of the movement velocity and also involves no exaggerated tendon tap reflex.

rotation—A motion in which all parts of the system are not displaced by a similar amount.

rubrospinal tract—A descending pathway from the red nucleus to motor neurons in the spinal cord.

Ruffini ending—Two to six thinly encapsulated, globular corpuscles with a single myelinated parent axon that has a diameter of 5 to 9 µm. These mechanoreceptors are capable of signaling joint position and displacement, angular velocity, and intraarticular pressure.

running—A mode of human locomotion that includes a phase when neither foot is on the ground.

safety factor—The margin of safety before failure occurs.

sag—An unfused tetanus in which the force declines after the initial four to eight stimuli and then increases again.

saltatory conduction—Propagation of an action potential along a myelinated axon by the depolarization of only the axolemma that is exposed at each node of Ranvier.

sarcolemma—The excitable plasma membrane of a muscle fiber.

sarcomere—The component of a myofibril contained in the region from one Z band to the next. The sarcomere contains the contractile apparatus necessary to convert chemical energy into mechanical work.

sarcoplasm—The fluid enclosed within a muscle fiber by the sarcolemma.

sarcoplasmic reticulum—A hollow membranous system within the muscle fiber that bulges into lateral sacs in the vicinity of the transverse tubules.

scalar—A variable that is defined by magnitude only.

scalar product—The procedure used to multiply vectors to produce a scalar quantity (product).

secant—A trigonometric function equal to hypotenuse divided by the adjacent side.

segmental analysis—A biomechanical analysis that involves graphically separating body segments and drawing a free body diagram and equations of motion for each segment.

sense of effort—A perception based on the corollary discharge that indicates quantitative changes in the descending command.

sense of heaviness—The ability of the nervous system to integrate proprioceptive feedback and of the centrally generated motor command to estimate the weight of an object.

sensorimotor cortex—The regions of the cerebral cortex that are located immediately anterior and posterior to the central sulcus.

series elastic element—One of the components of Hill-type models of muscle. It accounts for the elasticity in muscle that occurs in series with the contractile element.

shear force—A side-to-side force.

shortening velocity—The rate at which a muscle shortens its length.

short-range stiffness—A mechanical property of muscle whereby the stiffness is high for the first few millimeters of stretch and then is reduced substantially.

sine—A trigonometric function equal to the opposite side divided by the hypotenuse.

size principle—An explanation for the orderly recruitment of motor units. The size principle states that motor units are recruited in the order of the size of the motor neuron, from the smallest to the largest.

sliding-filament theory—A concept describing the sliding of thick and thin myofilaments past one another during a contraction.

slope—Rate of change in one variable (dependent variable, or *y*-axis variable) relative to another variable (independent variable, or *x*-axis variable); ''rise over run.''

slope-intercept equation of a line—$y = mx + b$.

soma—The cell body of a neuron.

somatotopic—A body map in the brain.

somersault axis—An axis that passes through the human body from side to side.

space diagram—A schematic of the free body diagram that does not include the correct magnitudes of the force vectors.

spared-root preparation—An experimental model used to study sprouting in the CNS. In this model, all the dorsal roots innervating a hindlimb are transected except one. The dependent variables are the recovery of motor function and the development of collateral sprouts in the spinal cord.

spasticity—Pathologically induced state of heightened excitability of the stretch reflex.

specificity principle—A training principle that states that the exercise stress is specific in the kind of change that it induces.

specific tension—The force a muscle can exert per unit of cross-sectional area (N/cm^2).

speed—The magnitude of the velocity vector; how fast.

spinal transection—Separation of the supraspinal nervous system from the spinal system due to a transection of the spinal cord.

spinocerebellum—Comprises the vermis and intermediate hemispheres of the cerebellum; receives input from the spinal cord and sends output to the fastigial and interposed nuclei.

stability—A state of equilibrium to which the system returns after it has been perturbed.

staggered muscle fibers—Serial attachment of muscle fibers to span whole-muscle length.

staircase effect—The progressive increase in twitch force that occurs when a series of twitches are elicited; also known as *treppe*.

stance phase—The support phase of walking and running.

state-dependent response—A response that varies with the state of the system. Reflexes are described as state-dependent responses: The amplitude depends on the activity of the individual (e.g., walking vs. standing, support phase vs. nonsupport phase).

static analysis—Mechanical analysis in which the forces acting on the system are balanced and hence the system is not accelerating. The system is either stationary or moving at a constant velocity.

static stretch—A technique that attempts to increase the range of motion about a joint by sustaining a muscle stretch for 15 to 30 s.

step—A part of the human gait cycle from an event (e.g., takeoff) to the next appearance of the same event by the other foot. There are two steps in one complete gait cycle, or stride.

stiffness—The slope of a force-length graph; the change in force per unit change in length (N/mm). It is necessary to exert a large force to stretch a stiff spring even a few millimeters.

strain—The change in length relative to the initial length (expressed as a percentage).

strain energy—The potential energy (J) stored by a system when it is stretched from a resting position.

streamline—Schematized line of fluid flow around an object.

stress—Force applied per unit area (Pa).

stretch reflex—The response of a muscle to a sudden, unexpected increase in its length. The negative feedback response activates the stretched muscle to minimize the increase in its length.

stretch-shorten cycle—An eccentric-concentric sequence of muscle activation. This sequence enables muscle to perform more work during the concentric phase of the task.

stride—One complete cycle of human gait from an event (e.g., left foot takeoff) to the next appearance of the same event. A stride contains two steps.

stride rate—The frequency (Hz) at which strides are performed. The inverse of stride rate is the time it takes to complete a stride.

subfragment 1—The two globular heads of the myosin molecule. One globular head contains an ATP-binding site, and the other, an actin-binding site.

subfragment 2—The nonglobular head region of heavy meromyosin.

substantia nigra—One of the basal ganglia nuclei.

subthalamic nuclei—Some of the basal ganglia nuclei.

supplementary motor area—A component of the sensorimotor cortex that is anterior to the central sulcus; between areas 4 and 8 on the medial surface.

surface drag—The drag due to the friction between the boundary layer of the fluid and the object.

swing phase—The nonsupport phase of walking and running for each leg.

synapse—The structure by which a neuron transmits its signals to a target cell.

synaptic potential—A conducted (nonpropagated) excitatory or inhibitory electrical potential that is elicited in neurons and muscle fibers in response to a neurotransmitter. The amplitude of a synaptic potential is variable, depending on the excitability of the cell and the quantity of neurotransmitter released.

synaptogenesis—Formation of new synapses.

synchrony—A temporal association between the action potentials of motor units. Synchrony is caused by synaptic input that is common to the involved motor neurons.

synergy—A group of muscles that are constrained to act as a unit for a specific task; also referred to as a *muscle synergy* and a *functional synergy*.

synovial membrane—A vascular membrane that secretes synovial fluid into the joint cavity.

tangent—A trigonometric function equal to the opposite side divided by the adjacent side.

tangential component—A component that acts parallel to and along a surface.

task dependency—The dependence of an impairment process or of an effect on the details of the task. For example, the mechanisms that cause fatigue can vary and depend on the details of the task, such as intensity, duration, muscles involved, and motivation of the subject.

tectospinal tract—A descending pathway that arises from neurons in the superior colliculus and probably contributes to the head- and neck-orienting reactions to visual stimuli.

tendon tap reflex—A subset of the stretch reflex that involves the response of a muscle to the tap of its tendon. The afferents activated during the tendon tap are primarily the Group Ia afferents; a greater diversity of afferents are activated during the stretch reflex.

tensile force—A pulling force.

terminal velocity—The velocity of a falling object when the forces due to gravity and air resistance are equal in magnitude.

tetanus—A force response of muscle (single fiber, motor units, whole muscles) to a series of excitatory inputs (action potentials or electrical stimuli); represents a summation of twitch responses.

tetrodotoxin—A drug that blocks the propagation of action potentials by blocking Na^+ channels.

thalamus—A suprasegmental structure that integrates and distributes most of the sensory and motor information going to the cerebral cortex.

thixotropy—The property exhibited by various gels, such as muscle, of becoming fluid when shaken, stirred, or otherwise disturbed and of setting again when the gel is allowed to stand.

three-burst pattern of EMG—Sequence of agonist-antagonist muscle activity that is associated with a unidirectional movement to a target.

three-element model of muscle—a mechanical model of muscle that comprises three elements: parallel elastic, series elastic, and contractile elements.

titin—A protein in the endosarcomeric cytoskeleton that may be responsible for resting muscle elasticity.

toe region—The initial part of the stress-strain relationship.

tonic—Continuous, sustained.

tonic vibration reflex—An EMG and force response that is elicited in muscle by small-amplitude, high-frequency (50 to 150 Hz) vibration of the muscle. The vibration activates the muscle spindles, which leads to the reflex activation of the motor neurons.

torque—The rotary effect of a force; quantified as the product of force and moment arm (N·m).

trajectory—The position-time record of a projectile.

transduction—A process by which energy is converted from one form to another.

translation—A motion in which all parts of the system are displaced by a similar amount.

transmitter-gated—A property of some channels in an excitable membrane, especially those located close to a synapse. Transmitter-gated channels are opened by a neurotransmitter.

transverse tubule—Invaginations of the sarcolemma that facilitate a rapid communication between sarcolemmal events (action potentials) and myofilaments located in the interior of the muscle fiber.

treppe—The progressive increase in twitch force that occurs when a series of twitches are elicited. Also known as the *staircase effect*.

triad—One transverse tubule and the two adjacent lateral sacs of the sarcoplasmic reticulum.

trigonometry—A branch of mathematics dealing with the relations among the sides and angles of a triangle.

triphasic—Three phases.

triple helix of collagen—Three intertwined polypeptide chains that make up the collagen molecule.

tropomyosin—A thin-filament protein involved in regulating the interaction between actin and myosin.

troponin—A three-component molecule that forms part of the thin filament and is involved in regulating the interaction between actin and myosin.

turbulent flow—Nonuniform flow of fluid (streamlines) around an object.

twist axis—An axis that passes through the human body from head to toe.

twitch—The force response of muscle (single fiber, motor unit, or whole muscle) to a single excitatory input.

twitch superimposition—A technique that is used to test the maximality of a voluntary contraction of muscle. The technique involves superimposing one to three supramaximal electric shocks of the muscle nerve on a voluntary contraction to determine if the force increases. Also known as *twitch interpolation* and *twitch occlusion*.

Type FF motor unit—A fast-twitch, fatigable motor unit. This type of unit is characterized by a sag response and a decline in force during a standard fatigue test.

Type FG muscle fiber—A fast-twitch, glycolytic muscle fiber that exists in Type FF motor units.

Type FOG muscle fiber—A fast-twitch, oxidative glycolytic muscle fiber that exists in Type FR motor units.

Type FR motor unit—A fast-twitch, fatigue-resistant motor unit. This type of unit is characterized by a sag response and the absence of a decline in force during a standard fatigue test.

Type I muscle fiber—A slow-twitch muscle fiber, as defined by a low level of myosin ATPase.

Type II muscle fiber—A fast-twitch muscle fiber, as defined by a high level of myosin ATPase. Type II muscle fibers can be further separated into two groups (IIa and IIb) after preparation in baths with pHs of 4.3 (IIa) or 4.6 (IIb).

Type S motor unit—A slow-twitch motor unit. This type of unit is characterized by a zero sag response and the absence of a decline in force during a standard fatigue test.

Type SO muscle fiber—A slow-twitch, oxidative muscle fiber that exists in Type S motor units.

unfused tetanus—An irregular force response elicited by a low frequency of activation. The peaks of the individual twitch responses are evident.

unmasking—The recovery of function following an injury to the nervous system can be due to the ability of a dormant pathway to assume the function of the injured pathway. In this scheme, the injury has unmasked the previously hidden (functionally) pathway.

Valsalva maneuver—Voluntary pressurization of the abdominal cavity.

variable load—A training load that varies over a range of motion.

vector—Quantity that conveys both magnitude and direction.

velocity—Rate of change in position with respect to time (m/s); the derivative of position with respect to time; the slope of a position-time graph.

ventral root—The entry zone on the front side of the spinal cord. Efferent axons exit the spinal cord in the ventral root.

vertex—A point that is midway between the directrix and the focus of a parabola.

vestibulocerebellum—Occupies the flocculonodular lobe of the cerebellum; receives input from the vestibular nuclei and sends output back to these same nuclei.

vestibulospinal tract—A descending pathway from the lateral vestibular nucleus to ipsilateral motor neurons.

viscoelastic—A material that has both viscous and elastic properties.

viscosity—A measure of the shear stress that must be applied to a fluid to obtain a rate of deformation. Viscosity varies with temperature and depends on the cohesive forces between molecules and the momentum interchange between colliding molecules. The SI unit of measurement is $N \cdot s/cm^2$. Oil has a greater viscosity than water.

voltage-gated—A property of some channels in an excitable membrane. Voltage-gated channels are opened by a decrease in the electrical potential across the membrane.

walking—A mode of human locomotion in which at least one foot is always in contact with the ground.

wave resistance—The resistive effects of waves encountered by a swimmer. The resistance is probably due to the difference in the density of water and air.

weight—an expression of the amount of gravitational attraction between an object and Earth.

weighting coefficient—A coefficient used to modulate the amplitude of a variable and its contribution to a compound expression (e.g., the harmonic terms of a Fourier analysis).

weightlessness—A microgravity environment.

Wolff's law—All changes in the function of bone are attended by alterations in internal structure.

work—A scalar quantity that describes the extent to which a force can move an object in a specified direction. The symbol for work is U, and its SI unit of measurement is the joule (J).

Z band—Zwischenscheibe; intrasarcomere connection of the two sets of thin filaments in adjacent sarcomeres.

References

Adams, G.R., Duvoisin, M.R., & Dudley, G.A. (1992). Magnetic resonance imaging and electromyography as indexes of muscle function. *Journal of Applied Physiology*, **73**, 1578-1583.

Adams, G.R., Harris, R.T., Woodard, D., & Dudley, G.A. (1993). Mapping of electrical muscle stimulation using MRI. *Journal of Applied Physiology*, **74**, 532-537.

Adams, G.R., Hather, B.M., Baldwin, K.M., & Dudley, G.A. (1993). Skeletal muscle myosin heavy chain composition and resistance training. *Journal of Applied Physiology*, **74**, 911-915.

Alexander, R.M. (1981). Mechanics of skeleton and tendons. In V.B. Brooks (Ed.), *Handbook of physiology: Sec. 1. The nervous system Vol. 2. Motor control.* (Pt. 1, pp. 17-42). Bethesda, MD: American Physiological Society.

Alexander, R.M. (1984a). Optimal strengths for bones liable to fatigue and accidental fracture. *Journal of Theoretical Biology*, **109**, 621-636.

Alexander, R.M. (1984b). Walking and running. *American Scientist*, **72**, 348-354.

Alexander, R.M. (1991). Optimum timing of muscle activation for simple models of throwing. *Journal of Theoretical Biology*, **150**, 349-372.

Alexander, R.M., & Ker, R.F. (1990). The architecture of leg muscles. In J.M. Winters & S.L.-Y. Woo (Eds.), *Multiple muscle systems: Biomechanics and movement organization* (pp. 568-577). New York: Springer-Verlag.

Alexander, R.M., & Vernon, A. (1975). The dimensions of the knee and ankle muscles and the forces they exert. *Journal of Human Movement Studies*, **1**, 115-123.

Al-Falahe, N.A., Nagaoka, M., & Vallbo, A.B. (1990). Response profiles of human muscle afferents during active finger movements. *Brain*, **113**, 325-346.

Alford, E.K., Roy, R.R., Hodgson, J.A., & Edgerton, V.R. (1987). Electromyography of rat soleus, medial gastrocnemius, and tibialis anterior during hind limb suspension. *Experimental Neurology*, **96**, 635-649.

Allen, D.G., Lee, J.A., & Westerblad, H. (1989). Intracellular calcium and tension during fatigue in isolated single muscle fibres from *Xenopus Laevis*. *Journal of Physiology (London)*, **415**, 433-458.

Alshuaib, W.B., & Fahim, M.A. (1990). Effect of exercise on physiological age-related change at mouse neuromuscular junctions. *Neurobiology of Aging*, **11**, 555-561.

Alway, S.E., Grumbt, W.H., Stray-Gundersen, J., & Gonyea, W.J. (1992). Effects of resistance training on elbow flexors of highly competitive bodybuilders. *Journal of Applied Physiology*, **72**, 1512-1521.

Alway, S.E., MacDougall, J.D., & Sale, D.G. (1989). Contractile adaptations in the human triceps surae after isometric exercise. *Journal of Applied Physiology*, **66**, 2725-2732.

Alway, S.E., Stray-Gundersen, J., Grumbt, W.H., & Gonyea, W.J. (1990). Muscle cross-sectional area and torque in resistance-trained subjects. *European Journal of Applied Physiology*, **60**, 86-90.

Amar, J. (1920). *The human motor*. London: Routledge & Sons.

Amis, A.A., Dowson, D., & Wright, V. (1980). Elbow joint force predictions for some strenuous isometric actions. *Journal of Biomechanics*, **13**, 765-775.

An, K.N., Hui, F.C., Morrey, B.F., Linscheid, R.L., & Chao, E.Y. (1981). Muscles across the elbow joint: A biomechanical analysis. *Journal of Biomechanics*, **10**, 659-669.

An, K.N., Kwak, B.M., Chao, E.Y., & Morrey, B.F. (1984). Determination of muscle and joint forces: A new technique to solve the indeterminate problem. *Journal of Biomechanical Engineering*, **106**, 364-367.

Anderson, S.A., & Cohn, S.H. (1985). Bone demineralization during space flight. *Physiologist*, **28**, 212-217.

Andersson, G.B.J., Örtengren, R., & Nachemson, A. (1977). Intradiscal pressure, intra-abdominal pressure and myoelectric back muscle activity related to posture and loading. *Clinical Orthopedics Research*, **129**, 156-164.

Antonio, J., & Gonyea, W.J. (1993). Skeletal muscle fiber hyperplasia. *Medicine and Science in Sports and Exercise*, **25**, 1333-1345.

Aristotle. *Progression of Animals*. English translation by E.S. Forster (1968). Cambridge: Harvard University Press.

Armstrong, R.B. (1990). Initial events in exercise-induced muscular injury. *Medicine and Science in Sports and Exercise*, **22**, 429-435.

Armstrong, C.G., & Mow, V.C. (1980). Friction, lubrication and wear of synovial joints. In R. Owen, J. Goodfellow, & P. Bullough (Eds.), *Scientific foundations of orthopaedics and traumatology* (pp. 223-232). London: William Heinemann.

Armstrong, R.B. (1990). Initial events in exercise-induced muscular injury. *Medicine and Science in Sports and Exercise*, **22**, 429-435.

Arsenault, A.B., & Chapman, A.E. (1974). An electromyographic investigation of the individual recruitment of the quadriceps muscles during isometric contraction of the knee extensors in different patterns of movement. *Physiotherapy of Canada*, **26**, 253-261.

Asami, T., & Nolte, V. (1983). Analysis of powerful ball kicking. In H. Matsui & K. Kobayashi (Eds.), *Biomechanics VIII-B* (pp. 695-700). Champaign, IL: Human Kinetics.

Ashton-Miller, J.A., & Schultz, A.B. (1988). Biomechanics of the human spine and trunk. In K.B. Pandolf (Ed.), *Exercise and sport sciences reviews* (Vol. 16, pp. 169-204). New York: Macmillan.

Askew, M.J., & Mow, V.C. (1978). The biomechanical function of the collagen ultrastructure of articular cartilage. *Journal of Biomechanical Engineering*, **100**, 105-113.

Asmussen, E. (1956). Observations on experimental muscle soreness. *Acta Rheumatologica Scandinavica*, **2**, 109-116.

Asmussen, E. (1953). Positive and negative muscular work. *Acta Physiologica Scandinavica*, **28**, 365-382.

Atha, J. (1981). Strengthening muscle. In D.I. Miller (Ed.), *Exercise and sport sciences reviews* (Vol. 9, pp. 1-73). Philadelphia: Franklin Institute.

Atwater, A.E. (1970, April). *Overarm throwing patterns: A kinematographic analysis*. Paper presented at the national convention of the American Association for Health, Physical Education and Recreation, Seattle, WA.

Atwater, A.E. (1977, June). *Biomechanics of throwing: Correction of common misconceptions*. Paper presented at the joint meeting of the National College Physical Education Association for Men and the National Association for Physical Education of College Women, Orlando, FL.

Atwater, A.E. (1979). Biomechanics of overarm throwing movements and of throwing injuries. In R.S. Hutton & D.I. Miller (Eds.), *Exercise and sport sciences reviews* (Vol. 7, pp. 43-85). Philadelphia: Franklin Institute.

Babij, P., & Booth, F.W. (1988). Sculpturing new muscle phenotypes. *News in Physiological Sciences*, **3**, 100-102.

Bailey, A.J., Robins, S.P., & Balian, G. (1974). Biological significance of the intermolecular crosslinks of collagen. *Nature*, **251**, 105-109.

Baker, L.L., Bowman, B.R., & McNeal, D.R. (1988). Effects of waveform on comfort during neuromuscular electrical stimulation. *Clinical Orthopaedics and Related Research*, **233**, 75-85.

Baldissera, F., Hultborn, H., & Illert, M. (1981). Integration of spinal neuronal systems. In V.B. Brooks (Ed.), *Handbook of physiology: Sec. 1. The nervous system, Vol. 2. Motor control.* (Pt. I, pp. 509-595). Bethesda, MD: American Physiological Society.

Bárány, M. (1967). ATPase activity of myosin correlated with speed of muscle shortening. *Journal of General Physiology*, **50**, 197-218.

Baratta, R., Solomonow, M., Zhou, B.H., Letson, D., Chuinard, R., & D'Ambrosia, R. (1987). Muscular coactivation: The role of the antagonist musculature in maintaining knee stability. *American Journal of Sports Medicine*, **16**, 113-122.

Barclay, J.K. (1986). A delivery-independent blood flow effect on skeletal muscle fatigue. *Journal of Applied Physiology*, **61**, 1084-1090.

Barr, J.O., Nielsen, D.H., & Soderberg, G.L. (1986). Transcutaneous electrical nerve stimulation characteristics for altering pain perception. *Physical Therapy*, **66**, 1515-1521.

Bartee, H., & Dowell, L. (1982). A cinematographical analysis of twisting about the longitudinal axis when performers are free of support. *Journal of Human Movement Studies*, **8**, 41-54.

Bartelink, D.L. (1957). The role of abdominal pressure in relieving the pressure on the intervertebral discs. *Journal of Bone and Joint Surgery*, **39B**, 718-725.

Basmajian, J.V., & Latif, A. (1957). Integrated actions and functions of the chief flexors of the elbow. *Journal of Bone and Joint Surgery*, **39A**, 1106-1118.

Bassett, C.A.L. (1968). Biological significance of piezoelectricity. *Calcified Tissue Research*, **1**, 252-272.

Bates, B.T., Osternig, L.R., Mason, B., & James, S.L. (1978). Lower extremity function during the support phase of running. In E. Asmussen & K. Jørgensen (Eds.), *Biomechanics VIB* (pp. 30-39). Baltimore: University Park Press.

Bawa, P., & Calancie, B. (1983). Repetitive doublets in human flexor carpi radialis muscle. *Journal of Physiology (London)*, **339**, 123-132.

Behm, D.G., & Sale, D.G. (1993). Intended rather than actual movement velocity determines velocity-specific training response. *Journal of Applied Physiology*, **74**, 359-368.

Belanger, A.Y., & McComas, A.J. (1981). Extent of motor unit activation during effort. *Journal of Applied Physiology*, **51**, 1131-1135.

Belanger, A.Y., McComas, A.J., & Elder, G.B.C. (1983). Physiological properties of two antagonistic human muscle groups. *European Journal of Applied Physiology*, **51**, 381-393.

Belen'kii, V.Y., Gurfinkel, V.S., & Pal'tsev, Y.I. (1967). Elements of control of voluntary movements. *Biophysics*, **12**, 154-161.

Bell, J., Bolanowski, S., & Holmes, M.H. (1994). The structure and function of Pacinian corpuscles: A review. *Progress in Neurobiology*, **42**, 79-128.

Bellemare, F., & Garzaniti, N. (1988). Failure of neuromuscular propagation during human maximal voluntary contraction. *Journal of Applied Physiology*, **64**, 1084-1093.

Bennett, K.M.B., & Castiello, U. (1994). Reach to grasp: Changes with age. *Journal of Gerontology*, **49**, P1-P7.

Bergh, U., & Ekblom, B. (1979). Influence of muscle temperature on maximal muscle strength and power output in human skeletal muscles. *Acta Physiologica Scandinavica*, **107**, 33-37.

Bergmark, A. (1989). Stability of the lumbar spine: A study in mechanical engineering. *Acta Orthopaedica Scandinavica*, **60**(Suppl. 230), 1-54.

Bernstein, N. (1984). Biodynamics of locomotion. In H.T.A. Whiting (Ed.), *Human motor actions: Bernstein reassessed* (pp. 171-222). New York: Elsevier.

Bertolasi, L., De Grandis, D., Bongiovanni, L.G., Zanette, G.P., & Gasperini, M. (1993). The influence of muscular lengthening on cramps. *Annals of Neurology*, **33**, 176-180.

Beverly, M.C., Rider, T.A., Evans, M.J., & Smith, R. (1989). Local bone mineral response to brief exercise that stresses the skeleton. *British Medical Journal*, **299**, 233-235.

Bhagat, B., & Wheeler, N. (1973). Effect of amphetamine on the swimming endurance of rats. *Neuropharmacology*, **12**, 711-713.

Biewener, A.A. (1991). Musculoskeletal design in relation to body size. *Journal of Biomechanics*, **24**(Suppl. 1), 19-29.

Bigland, B., & Lippold, O.C.J. (1954). The relation between force, velocity and integrated electrical activity in human muscles. *Journal of Physiology (London)*, **123**, 214-224.

Bigland-Ritchie, B. (1981). EMG/force relations and fatigue of human voluntary contractions. In D.I. Miller (Ed.), *Exercise and sport sciences reviews* (Vol. 9, pp. 75-117). Philadelphia: Franklin Institute.

Bigland-Ritchie, B., Cafarelli, E., & Vøllestad, N.K. (1986). Fatigue of submaximal static contractions. *Acta Physiologica Scandinavica*, **128**(Suppl. 556), 137-148.

Bigland-Ritchie, B., Dawson, N.J., Johansson, R.S., & Lippold, O.C.J. (1986). Reflex origin for the slowing of motoneurone firing rates in fatigue of human voluntary contractions. *Journal of Physiology (London)*, **379**, 451-459.

Bigland-Ritchie, B., Furbush, F., & Woods, J.J. (1986). Fatigue of intermittent submaximal voluntary contractions: Central and peripheral factors. *Journal of Applied Physiology*, **61**, 421-429.

Bigland-Ritchie, B., Johansson, R., Lippold, O.C.J., Smith, S., & Woods, J.J. (1983). Changes in motoneurone firing rates during sustained maximal voluntary contractions. *Journal of Physiology (London)*, **340**, 335-346.

Bigland-Ritchie, B., Johansson, R., Lippold, O.C.J., & Woods, J.J. (1983). Contractile speed and EMG changes during fatigue of sustained maximal voluntary contractions. *Journal of Neurophysiology*, **50**, 313-324.

Bigland-Ritchie, B., Kukulka, C.G., Lippold, O.C.J., & Woods, J.J. (1982). The absence of neuromuscular transmission failure in sustained maximal voluntary contractions. *Journal of Physiology (London)*, **330**, 265-278.

Binder, M.D., Houk, J.C., Nichols, T.R., Rymer, W.Z., & Stuart, D.G. (1982). Properties and segmental actions of mammalian muscle receptors: An update. *Federation Proceedings*, **41**, 2907-2918.

Binder, M.D., Kroin, J.S., Moore, G.P., & Stuart, D.G. (1977). The response of Golgi tendon organs to single motor unit contractions. *Journal of Physiology (London)*, **271**, 337-349.

Binder, M.D., & Mendell, L.M. (Eds.) (1990). *The segmental motor system*. New York: Oxford University Press.

Binder-Macleod, S.A., & Guerin, T. (1990). Preservation of force output through progressive reduction of stimulation frequency in human quadriceps femoris muscle. *Physical Therapy*, **70**, 619-625.

Binkhorst, R.A., Hoofd, L., & Vissers, A.C.A. (1977). Temperature and force-velocity relationship of human muscles. *Journal of Applied Physiology*, **42**, 471-475.

Bishop, B. (1974). Vibratory stimulation: 1. Neurophysiology of motor responses evoked by vibratory stimulation. *Physical Therapy*, **54**, 1273-1282.

Bishop, B. (1975a). Vibratory stimulation: 2. Vibratory stimulation as an evaluation tool. *Physical Therapy*, **55**, 28-34.

Bishop, B. (1975b). Vibratory stimulation: 3. Possible applications of vibration in treatment of motor dysfunctions. *Physical Therapy*, **55**, 139-143.

Bishop, R.D., & Hay, J.G. (1979). Basketball: The mechanics of hanging in the air. *Medicine and Science in Sports*, **11**, 274-277.

Bizzi, E., & Abend, W. (1983). Posture control and trajectory formation in single- and multi-joint arm movements. In J.E. Desmedt (Ed.), *Motor control mechanisms in health and disease* (pp. 31-45). New York: Raven Press.

Blangé, T., Karemaker, J.M., & Kramer, A.E.J.L. (1972). Elasticity as an expression of cross-bridge activity in rat muscle. *Pflügers Archiv*, **336**, 277-288.

Blattner, S.E., & Noble, L. (1979). Relative effects of isokinetic and plyometric training on vertical jumping performance. *Research Quarterly*, **50**, 583-588.

Blinks, J.R., Rüdel, R., & Taylor, S.R. (1978). Calcium transients in isolated amphibian skeletal muscle fibres: Detection with aequorin. *Journal of Physiology (London)*, **277**, 291-323.

Blomstrand, E., Celsing, F., & Newsholme, E.A. (1988). Changes in plasma concentrations of aromatic and branched-chain amino acids during sustained exercise in man and their possible role in fatigue. *Acta Physiologica Scandinavica*, **133**, 115-121.

Bloom, W., & Fawcett, D.W. (1968). *A textbook of histology* (9th ed.). Philadelphia: Saunders.

Bobath, B. (1978). *Adult hemiplegia: Evaluation and treatment*. London: Heinemann Medical Books.

Bobbert, M.F., Schamhardt, H.C., & Nigg, B.J. (1991). Calculation of vertical ground reaction force estimates during running from positional data. *Journal of Biomechanics*, **24**, 1095-1105.

Bodine, S.C., Garfinkel, A., Roy, R.R., & Edgerton, V.R. (1988). Spatial distribution of motor unit fibers in the cat soleus and tibialis anterior muscles: Local interactions. *Journal of Neuroscience*, **8**, 2142-2152.

Bodine, S.C., Roy, R.R., Eldred, E., & Edgerton, V.R. (1987). Maximal force as a function of anatomical features of motor units in the cat tibialis anterior. *Journal of Neurophysiology*, **57**, 1730-1745.

Bongiovanni, L.G., & Hagbarth, K.-E. (1990). Tonic vibration reflexes elicited during fatigue from maximal voluntary contractions in man. *Journal of Physiology (London)*, **423**, 1-14.

Borelli, A. (1680). *De Motu Animalium*. Rome: Superiorum Permissu.

Bosco, C., & Komi, P.V. (1980). Influence of aging on the mechanical behavior of leg extensor muscles. *European Journal of Applied Physiology*, **45**, 209-219.

Botterman, B.R., Binder, M.D., & Stuart, D.G. (1978). Functional anatomy of the association between motor units and muscle receptors. *American Zoologist*, **18**, 135-152.

Botterman, B.R., Iwamoto, G.A., & Gonyea, W.J. (1986). Gradation of isometric tension by different activation rates in motor units of cat flexor carpi radialis muscle. *Journal of Neurophysiology*, **56**, 494-505.

Bouisset, S., & Zattara, M. (1990). Segmental movement as a perturbation to balance? Facts and concepts. In J.M. Winters & S.L.-Y. Woo (Eds.), *Multiple muscle systems: Biomechanics and movement organization* (pp. 498-506). New York: Springer-Verlag.

Bowditch, H.P. (1871). Ueber die Eigenthuemlichkeiten der Reizbarkeit, welch die Muskelfasern des Herzens zeigen. *Berichte Mathematische Physiologie*, **23**, 652-689.

Boulay, M.R., Lortie, G., Simoneau, J.A., Hamel, P., Leblanc, C., & Bouchard, C. (1985). Specificity of aerobic and anaerobic work capacities and powers. *International Journal of Sports Medicine*, **6**, 325-328.

Brancazio, P.J. (1984). *Sport science: Physical laws and optimum performance*. New York: Simon & Schuster.

Brand, R.A., Crowninshield, R.D., Wittstock, C.E., Pedersen, D.R., Clark, C.R., & van Krieken, F.M. (1982). A model of lower extremity muscular anatomy. *Journal of Biomechanical Engineering*, **104**, 304-310.

Brand, R.A., Pedersen, D.R., & Friederich, J.A. (1986). The sensitivity of muscle force predictions to changes in physiologic cross-sectional area. *Journal of Biomechanics*, **19**, 589-596.

Braune, W., & Fischer, O. (1889). Uber den Schwerpunkt des menschlichen Korpers mit Rucksicht auf die Ausrustung des deutschen Infanteristen. *Abhandlungen der Mathematisch Physischen Klasse der Saecksischen Akademie der Wissenschaften*, **26**, 561-672.

Bremner, F.D., Baker, J.R., & Stephens, J.A. (1991a). Correlation between the discharges of motor units recorded from the same and from different finger muscles in man. *Journal of Physiology (London)*, **432**, 355-380.

Bremner, F.D., Baker, J.R., & Stephens, J.A. (1991b). Effect of task on the degree of synchronization of intrinsic hand muscle motor units in man. *Journal of Neurophysiology*, **66**, 2072-2083.

Bremner, F.D., Baker, J.R., & Stephens, J.A. (1991c). Variation in the degree of synchrony exhibited by motor units lying in different finger muscles in man. *Journal of Physiology (London)*, **432**, 381-399.

Brighton, C.T. (1981). Current concept review: The treatment of non-unions with electricity. *Journal of Bone and Joint Surgery*, **63A**, 847-856.

Broberg, S., & Sahlin, K. (1989). Adenine nucleotide degradation in human skeletal muscle during prolonged exercise. *Journal of Applied Physiology*, **67**, 116-122.

Brooke, M.H., & Kaiser, K.K. (1974). The use and abuse of muscle histochemistry. *Annals of the New York Academy of Sciences*, **228**, 121-144.

Brooks, S.V., & Faulkner, J.A. (1991). Forces and powers of slow and fast skeletal muscle in mice during repeated contractions. *Journal of Physiology (London)*, **436**, 701-710.

Brooks, V.B. (1986). *The neural basis of motor control*. New York: Oxford University Press.

Brown, A.B., McCartney, N., & Sale, D.G. (1990). Positive adaptations to weight-lifting training in the elderly. *Journal of Applied Physiology*, **69**, 1725-1733.

Brown, G.L., & von Euler, U.S. (1938). The after effects of a tetanus on mammalian muscle. *Journal of Physiology (London)*, **93**, 39-60.

Buchanan, T.S., Rovai, G.P., & Rymer, W.Z. (1989). Strategies for muscle activation during isometric torque generation at the human elbow. *Journal of Neurophysiology*, **62**, 1201-1212.

Bulbulian, R., & Bowles, D.K. (1992). Effect of downhill running on motoneuron pool excitability. *Journal of Applied Physiology*, **73**, 968-973.

Burgeson, R.E., & Nimni, M.E. (1992). Collagen types. Molecular structure and tissue distribution. *Clinical Orthopaedics*, **282**, 250-272.

Burgess, P.R., Horch, K.W., & Tuckett, R.P. (1987). Mechanoreceptors. In G. Adelman (Ed.), *Encyclopedia of neuroscience* (Vol. 2, pp. 620-621). Boston: Birkhäuser.

Burke, D. (1988). Spasticity as an adaptation to pyramidal tract injury. *Advances in Neurology*, **47**, 401-422.

Burke, J.E., Kamen, G., & Koceja, D.M. (1989). Long-latency enhancement of quadriceps excitability from stimulation of skin afferents in young and old adults. *Journal of Gerontology*, **44**, M158-M163.

Burke, R.E. (1981). Motor units: Anatomy, physiology, and functional organization. In V.B. Brooks (Ed.), *Handbook of Physiology: Sec. 1. The nervous system; Vol. 2. Motor control* (Pt. 1, pp. 345-422), Bethesda, MD: American Physiological Society.

Burke, R.E. (1985). Integration of sensory information and motor commands in the spinal cord. In P.S.G. Stein (Organizer), *Motor control: From movement trajectories to neural mechanisms* (pp. 44-66). Class at the Society for Neuroscience: Washington, DC.

Burke, R.E., & Edgerton, V.R. (1975). Motor unit properties and selective involvement in movement. In J.H. Wilmore & J.F. Koegh (Eds.), *Exercise and sport sciences reviews* (Vol. 3, pp. 31-81). New York: Academic.

Burke, R.E., Levine, D.N., Tsairis, P., & Zajac, F.E. (1973). Physiological types and histochemical profiles in motor units of the cat gastrocnemius. *Journal of Physiology (London)*, **234**, 723-748.

Burke, R.E., Rudomin, P., & Zajac, F.E., III. (1970). Catch property in single mammalian motor units. *Science*, **168**, 122-124.

Burke, R.E., & Tsairis, P. (1973). Anatomy and innervation ratios in motor units of cat gastrocnemius. *Journal of Physiology (London)*, **234**, 749-765.

Burke, R.E., & Tsairis, P. (1974). The correlation of physiological properties with histochemical characteristics in single muscle units. *Annals of the New York Academy of Sciences*, **228**, 145-159.

Burstein, A.H., Reilly, D.T., & Martens, M. (1976). Aging bone tissue: Mechanical properties. *Journal of Bone and Joint Surgery*, **58A**, 82-86.

Butler, A.J., & Darling, W.G. (1990). Reflex changes accompanying isometric strength training of the contralateral limb. *Society for Neuroscience Abstracts*, **16**, 884.

Butler, D.L., Grood, E.S., Noyes, F.R., & Zernicke, R.F. (1978). Biomechanics of ligaments and tendons. In R.S. Hutton (Ed.), *Exercise and sport sciences reviews* (Vol. 6, pp. 125-181). Philadelphia: Franklin Institute.

Byrnes, W.C., Clarkson, P.M., White, J.S., Hsieh, S.S., Frykman, P.N., & Maughan, R.J. (1985). Delayed onset muscle soreness following repeated bouts of downhill running. *Journal of Applied Physiology*, **59**, 710-715.

Cafarelli, E. (1988). Force sensation in fresh and fatigued human skeletal muscle. In K.B. Pandolf (Ed.), *Exercise and sport sciences reviews* (Vol. 16, pp. 139-168). New York: Macmillan.

Caiozzo, V.J., Herrick, R.E., & Baldwin, K.M. (1991). Influence of hyperthyroidism on maximal shortening velocity and myosin isoform distribution in skeletal muscle. *American Journal of Physiology*, **261**, C285-C295.

Caiozzo, V.J., Herrick, R.E., & Baldwin, K.M. (1992). Response of slow and fast muscle to hypothyroidism: Maximal shortening velocity and myosin isoforms. *American Journal of Physiology*, **263**, C86-C94.

Calancie, B. (1991). Interlimb reflexes following cervical spinal cord injury in man. *Experimental Brain Research*, **85**, 458-469.

Calvert, T.W., & Chapman, A.E. (1977). Relationship between the surface EMG and force transients in muscle: Simulation and experimental studies. *Proceedings of the IEEE*, **65**, 682-689.

Campbell, M.J., McComas, A.J., & Petito, F. (1973). Physiological changes in ageing muscle. *Journal of Neurology, Neurosurgery, and Psychiatry*, **36**, 174-182.

Carolan, B., & Cafarelli, E. (1992). Adaptations in coactivation after isometric resistance training. *Journal of Applied Physiology*, **73**, 911-917.

Carter, D.R. (1984). Mechanical loading histories and cortical bone remodeling. *Calcified Tissue International*, **36**, S19-S24.

Carter, D.R., & Spengler, D.M. (1978). Mechanical properties and composition of cortical bone. *Clinical Orthopaedics and Related Research*, **135**, 192-217.

Cavagna, G.A. (1977). Storage and utilization of elastic energy in skeletal muscle. In R.S. Hutton (Ed.), *Exercise and sport sciences reviews* (Vol. 5, pp. 89-129). Santa Barbara, CA: Journal Publishing Affiliates.

Cavagna, G.A., & Citterio, G. (1974). Effect of stretching on the elastic characteristics of the contractile component of the frog striated muscle. *Journal of Physiology (London)*, **239**, 1-14.

Cavagna, G.A. Saibene, F.P., & Margaria, R. (1964). Mechanical work in running. *Journal of Applied Physiology*, **19**, 249-256.

Cavanagh, P.R. (Ed.) (1990). *Biomechanics of distance running*. Champaign, IL: Human Kinetics.

Cavanagh, P.R., & Ae, M. (1980). A technique for the display of pressure distributions beneath the foot. *Journal of Biomechanics*, **13**, 69-76.

Cavanagh, P.R., Andrew, G.C., Kram, R., Rodgers, M.M., Sanderson, D.J., & Henning, E.M. (1985). An approach to biomechanical profiling of elite distance runners. *International Journal of Sport Biomechanics*, **1**, 36-62.

Cavanagh, P.R., & Grieve, D.W. (1973). The graphical display of angular movement of the body. *British Journal of Sports Medicine*, **7**, 129-133.

Cavanagh, P.R., & Kram, R. (1989). Stride length in distance running: Velocity, body dimensions, and added mass effects. *Medicine and Science in Sports and Exercise*, **21**, 476-479.

Cavanagh, P.R., & Lafortune, M.A. (1980). Ground reaction forces in distance running. *Journal of Biomechanics*, **13**, 397-406.

Cavanagh, P.R., Pollock, M.L., & Landa, J. (1977). A biomechanical comparison of elite and good distance runners. *Annals of New York Academy of Sciences*, **301**, 328-345.

Chandler, R.F., Clauser, C.E., McConville, J.T., Reynolds, H.M., & Young, J.W. (1975). *Investigation of inertial properties of the human body* (AMRL-TR-74-137 as issued by the National Highway Traffic Safety Administration). Wright-Patterson Air Force Base, OH: Aerospace Medical Research Laboratories, Aerospace Medical Division. (NTIS No. AD-A016 485)

Chanaud, C.M., Pratt, C.A., & Loeb, G.E. (1987). A multiple-contact EMG recording array for mapping single muscle unit territories. *Journal of Neuroscience Methods*, **21**, 105-112.

Chao, E.Y.S. (1986). Biomechanics of the human gait, In G.W. Schmid-Schönbein, S.L.-Y. Woo, & B.W. Zweifach (Eds.), *Frontiers in biomechanics* (pp. 225-244). New York: Springer.

Chaouloff, F., Kennett, G.A., Serrurier, B., Merino, D., & Curson, G. (1986). Amino acid analysis demonstrates that increased plasma free tryptophan causes the increase of brain tryptophan during exercise in the rat. *Journal of Neurochemistry*, **46**, 1647-1650.

Chase, P.B., & Kushmerick, M.J. (1988). Effects of pH on contraction of rabbit fast muscle and slow skeletal muscle fibers. *Biophysical Journal*, **53**, 935-946.

Cheney, P.D. (1985). Role of cerebral cortex in voluntary movements: A review. *Physical Therapy*, **65**, 624-635.

Chi, M.M.-Y., Hintz, C.S., Coyle, E.F., Martin, W.H., Ivy, J.L., Nemeth, P.M., Holloszy, J.O., & Lowry, O.H. (1983). Effects of detraining on enzymes of energy metabolism in individual human muscle fibers. *American Journal of Physiology*, **244**, C276-C287.

Cholewicki, J., McGill, S.M., & Norman, R.W. (1991). Lumbar spine loads during the lifting of extremely heavy weights. *Medicine and Science in Sports and Exercise*, **23**, 1179-1186.

Christakos, C.N. (1982). A study of the muscle force waveform using a population stochastic model of skeletal muscle. *Biological Cybernetics*, **44**, 91-106.

Christakos, C.N., & Windhorst, U. (1986). Spindle gain increase during muscle unit fatigue. *Brain Research*, **365**, 388-392.

Claassen, H., Gerber, C., Hoppeler, H., Lüthi, J.-M., & Vock, P. (1989). Muscle filament spacing and short-term heavy-resistance exercise in humans. *Journal of Physiology (London)*, **409**, 491-495.

Clark, J.M., & Haynor, D.R. (1987). Anatomy of the abductor muscles of the hip as studied by computed tomography. *Journal of Bone and Joint Surgery*, **69A**, 1021-1031.

Clarke, T.E., Frederick, E.C., & Hamill, C.L. (1983). The effect of shoe design upon rearfoot control in running. *Medicine and Science in Sports and Exercise*, **15**, 376-381.

Clarkson, P.M., Cyrnes, W.C., McCarmick, K.M., Turcotte, L.P., & White, J.S. (1986). Muscle soreness and serum creatine kinase activity following isometric, eccentric, and concentric exercise. *International Journal of Sports Medicine*, **7**, 152-155.

Clarkson, P.M., & Tremblay, I. (1988). Exercise-induced muscle damage, repair, and adaptation in humans. *Journal of Applied Physiology*, **65**, 1-6.

Clauser, C.E., McConville, J.T., & Young, J.W. (1969). *Weight, volume and center of mass of segments of the human body* (AMRL-TR-69-70). Wright-Patterson Air Force Base, OH: Aerospace Medical Research Laboratories, Aerospace Medical Division.

Clement, D.B., Taunton, J.E., Smart, G.W., & McNicol, K.L. (1981). A survey of overuse running injuries. *Physician and Sports Medicine*, **9**, 47-58.

Coggan, A.R., & Coyle, E.F. (1987). Reversal of fatigue during prolonged exercise by carbohydrate infusion or ingestion. *Journal of Applied Physiology*, **63**, 2388-2395.

Cole, K.J. (1991). Grasp force control in older adults. *Journal of Motor Behavior*, **23**, 251-258.

Colleti, L.A., Edwards, J., Gordon, L., Shary, J., & Bell, N.H. (1989). The effects of muscle-building exercise on bone mineral density of the radius, spine, and hip in young men. *Calcified Tissue International*, **45**, 12-14.

Colliander, E.B., & Tesch, P.A. (1990). Effects of eccentric and concentric muscle actions in resistance training. *Acta Physiologica Scandinavica*, **140**, 31-39.

Collins, J.J., & O'Connor, J.J. (1991). Muscle-ligament interaction at the knee during walking. *Journal of Engineering in Medicine*, **205**, 11-18.

Condon, S.M., & Hutton, R.S. (1987). Soleus muscle electromyographic activity and ankle dorsiflexion range of motion during four stretching procedures. *Physical Therapy*, **67**, 24-30.

Cooke, P. (1985). A periodic cytoskeletal lattice in striated muscle. In J.W. Shay (Ed.), *Cell and muscle motility* (Vol. 6, pp. 287-313). New York: Plenum.

Cooke, R. (1990). Force generation in muscle. *Current Opinion in Cell Biology*, **2**, 62-66.

Cooke, R., Franks, K., Luciani, G.B., & Pate, E. (1988). The inhibition of rabbit skeletal muscle contraction by hydrogen ions and phosphate. *Journal of Physiology (London)*, **395**, 77-97.

Cope, T.C., Bodine, S.C., Fournier, M., & Edgerton, V.R. (1986). Soleus motor units in chronic spinal transected cat: Physiological and morphological alterations. *Journal of Neurophysiology*, **55**, 1201-1220.

Corcos, D.M., Gottlieb, G.L., & Agarwal, G.C. (1989). Organizing principles for single-joint movements: 2. A speed-sensitive strategy. *Journal of Neurophysiology*, **62**, 358-368.

Cordo, P.J., & Nashner, L.J. (1982). Properties of postural adjustments associated with rapid arm movements. *Journal of Neurophysiology*, **47**, 287-302.

Costill, D.L., & Hargreaves, M. (1992). Carbohydrate nutrition and fatigue. *Sports Medicine*, **13**, 86-92.

Côté, L., & Crutcher, M.D. (1991). The basal ganglia. In E.R. Kandel, J.H. Schwartz, & T.M. Jessell (Eds.), *Principles of neural science* (3rd ed., pp. 647-659). New York: Elsevier.

Cowin, S.C. (1983). The mechanical and stress adaptive properties of bone. *Annals of Biomedical Engineering*, **11**, 263-295.

Coyle, E.F., Coggan, A.R., Hemmert, M.K., & Ivy, J.L. (1986). Muscle glycogen utilization during prolonged strenuous exercise when fed carbohydrate. *Journal of Applied Physiology*, **61**, 165-172.

Crago, P.E., Houk, J.C., & Hasan, Z. (1976). Regulatory actions of human stretch reflex. *Journal of Neurophysiology*, **39**, 925-935.

Crenna, P., & Frigo, C. (1984). Evidence of phase-dependent nociceptive reflexes during locomotion in man. *Experimental Neurology*, **85**, 336-345.

Crenna, P., & Frigo, C. (1987). Excitability of the soleus H-reflex arc during walking and stepping in man. *Experimental Brain Research*, **66**, 49-60.

Cresswell, A.G. (1993). Responses of intra-abdominal pressure and abdominal muscle activity during dynamic trunk loading in man. *European Journal of Applied Physiology*, **66**, 315-320.

Cresswell, A.G., Grundström, H., & Thorstensson, A. (1992). Observations on intra-abdominal pressure and patterns of abdominal intra-muscular activity in man. *Acta Physiologica Scandinavica*, **144**, 409-418.

Crisco, J.J., III, & Panjabi, M.M. (1990). Postural biomechanical stability and gross muscular architecture in the spine. In J.M. Winters & S.L.-Y. Woo (Eds.), *Multiple muscle systems: Biomechanics and movement organization* (pp. 438-450). New York: Springer-Verlag.

Crone, C., & Nielsen, J. (1989). Methodological implications of the post activation depression of the soleus H-reflex in man. *Experimental Brain Research*, **78**, 28-32.

Crowninshield, R.D., & Brand, R.A. (1981). The prediction of forces in joint structures: Distribution of intersegmental resultants. In D.I. Miller (Ed.), *Exercise and sport science reviews* (Vol. 9, pp. 159-181). Philadelphia: Franklin Institute.

Cutts, A. (1988). The range of sarcomere lengths in the muscles of the human lower limb. *Journal of Anatomy*, **160**, 79-88.

Czéh, G., Gallego, R., Kudo, N., & Kuno, M. (1978). Evidence for the maintenance of motoneurone properties by muscle activity. *Journal of Physiology (London)*, **281**, 239-252.

Czerniecki, J.M., Gitter, A., & Munro, C. (1991). Joint moment and muscle power output characteristics of below knee amputees during running: The influence of energy storing prosthetic feet. *Journal of Biomechanics*, **24**, 63-75.

Dahl, H.A., Aaserud, R., & Jensen, J. (1992). Muscle hypertrophy after light and heavy resistance training. *Medicine and Science in Sports and Exercise*, **24**, S55.

Dahlkvist, N.J., Mayo, P., & Seedhom, B.B. (1982). Forces during squatting and rising from a deep squat. *Engineering in Medicine*, **11**, 69-76.

Dainis, A. (1981). A model for gymnastics vaulting. *Medicine and Science in Sports and Exercise*, **13**, 34-43.

Dalén, N., & Olsson, K.E. (1974). Bone mineral content and physical activity. *Acta Orthopaedica Scandinavica*, **45**, 170-174.

Daniels, J.T. (1985). A physiologist's view of running economy. *Medicine and Science in Sports and Exercise*, **17**, 332-338.

Darling, W.G., & Cole, K.J. (1990). Muscle activation patterns and kinetics of human index finger movements. *Journal of Neurophysiology*, **63**, 1098-1108.

Darling, W.G., & Hayes, K.C. (1983). Human servo responses to load disturbances in fatigued muscle. *Brain Research*, **267**, 345-351.

Datta, A.K., & Stephens, J.A. (1990). Synchronization of motor unit activity during voluntary contraction in man. *Journal of Physiology (London)*, **422**, 397-419.

Davies, C.T.M., Dooley, P., McDonagh, M.J.N., & White, M.J. (1985). Adaptation of mechanical properties of muscle to high force training in man. *Journal of Physiology (London)*, **365**, 277-284.

Davies, C.T.M., Rutherford, I.C., & Thomas, D.O. (1987). Electrically evoked contractions of the triceps surae during and following 21 days of voluntary leg immobilization. *European Journal of Applied Physiology*, **56**, 306-312.

Davies, C.T.M., & Young, K. (1983). Effect of temperature on the contractile properties and muscle power of triceps surae in humans. *Journal of Applied Physiology*, **55**, 191-195.

Davies, J., Parker, D.F., Rutherford, O.M., & Jones, D.A. (1988). Changes in strength and cross sectional area of elbow flexors as a result of isometric strength training. *European Journal of Applied Physiology*, **57**, 667-670.

Davis, B.L., & Cavanagh, P.R. (1993). Simulating reduced gravity: A review of biomechanical issues pertaining to human locomotion. *Aviation, Space, and Environmental Medicine*, **64**, 557-566.

Davis, B.L., & Vaughan, C.L. (1993). Phasic behavior of EMG signals during gait: Use of multivariate statistics. *Journal of Electromyography and Kinesiology*, **3**, 51-60.

Dawson, M.J., Gadian, D.G., & Wilkie, D.R. (1980). Mechanical relaxation rate and metabolism studied in fatiguing muscle by phosphorus nuclear magnetic resonance. *Journal of Physiology (London)*, **299**, 465-484.

Day, B.L., Marsden, C.D., Obeso, J.A., & Rothwell, J.C. (1984). Reciprocal inhibition between the muscles of the human forearm. *Journal of Physiology (London)*, **349**, 519-534.

de Haan, A., van Ingen Schenau, G.J., Ettema, G.J., Huijing, P.A., & Lodder, M.A.N. (1989). Efficiency of rat medial gastrocnemius muscle in contractions with and without an active prestretch. *Journal of Experimental Biology*, **141**, 327-341.

de Koning, F.L., Binkhorst, R.A., Vissers, A.C.A., & van't Hof, M.A. (1985). The force-velocity relationship of arm flexion in untrained males and females and arm-trained athletes. *European Journal of Applied Physiology*, **54**, 89-94.

de Koning, J.J., de Groot, G., & van Ingen Schenau, G.J. (1992). Ice friction during speed skating. *Journal of Biomechanics*, **25**, 565-571.

Delitto, A., Brown, M., Strube, M.J., Rose, S.J., & Lehman, R.C. (1989). Electrical stimulation of quadriceps femoris in an elite weight lifter: A single subject experiment. *International Journal of Sports Medicine*, **10**, 187-191.

Delitto, A., & Rose, S.J. (1986). Comparative comfort of three waveforms used in electrically eliciting quadriceps femoris muscle contractions. *Physical Therapy*, **66**, 1704-1707.

De Looze, M.P., Bussmann, J.B.J., Kingma, I., & Toussaint, H.M. (1992). Different methods to estimate total power and its components during lifting. *Journal of Biomechanics*, **25**, 1089-1095.

DeLorme, T.L. (1945). Restoration of muscle power by heavy-resistance exercises. *Journal of Bone and Joint Surgery*, **27**, 645-667.

DeLorme, T.L. (1946). Heavy resistance exercises. *Archives of Physical Medicine*, **27**, 607-630.

De Luca, C.J. (1984). Myoelectric manifestations of localized muscular fatigue in humans. *CRC Critical Reviews in Biomedical Engineering*, **11**, 251-279.

Dempster, W.T. (1955). *Space requirements of the seated operator* (WADC-TR-55-159). Wright-Patterson Air Force Base, OH: Aerospace Medical Research Laboratory. (NTIS No. AD-87892 as issued by the United States Air Force)

Dempster, W.T. (1961). Free-body diagrams as an approach to the mechanics of human posture and motion. In G.F. Evans (Ed.), *Biomechanical studies of the musculo-skeletal system* (pp. 81-135). Springfield, IL: Charles C Thomas.

Denny-Brown, D. (1949). Interpretation of the electromyogram. *Archives of Neurology and Psychiatry*, **61**, 99-128.

Desmedt, J.E., & Godaux, E. (1977). Fast motor units are not preferentially activated in rapid voluntary contractions in man. *Nature*, **267**, 717-719.

Devine, K.L., LeVeau, B.F., & Yack, H.J. (1981). Electromyographic activity recorded from an unexercised muscle during maximal isometric exercise of the contralateral agonists and antagonists. *Physical Therapy*, **61**, 898-903.

DeVita, P., & Skelly, W.A. (1992). Effect of landing stiffness on joint kinetics and energetics in the lower extremity. *Medicine and Science in Sports and Exercise*, **24**, 108-115.

Dick, R.W., & Cavanagh, P.R. (1987). An explanation of the upward drift in oxygen uptake during prolonged submaximal downhill running. *Medicine and Science in Sports and Exercise*, **19**, 310-317.

Dietz, V. (1978). Analysis of the electrical muscle activity during maximal contraction and the influence of ischaemia. *Journal of the Neurological Sciences*, **37**, 187-197.

Dietz, V. (1992). Human neuronal control of automatic functional movement: Interaction between central programs and afferent input. *Physiological Reviews*, **72**, 33-69.

Dietz, V., & Berger, W. (1982). Spatial coordination of bilateral leg muscle activity during balancing. *Experimental Brain Research*, **47**, 172-176.

Dimitrijevic, M.R., McKay, W.B., Sarjanovic, I., Sherwood, A.M., Svirtlih, L., & Vrbovà, G. (1992). Co-activation of ipsi- and contralateral muscle groups during contraction of ankle dorsiflexors. *Journal of the Neurological Sciences*, **109**, 49-55.

di Prampero, P.E. (1986). The energy cost of human locomotion on land and in water. *International Journal of Sports Medicine*, **7**, 55-72.

Doherty, T.J., Vandervoort, A.A., & Brown, W.F. (1993). Effects of aging on the motor unit: A brief review. *Canadian Journal of Applied Physiology*, **18**, 331-358.

Doherty, T.J., Vandervoort, A.A., Taylor, A.W., & Brown, W.F. (1993). Effects of motor unit losses on strength in older men and women. *Journal of Applied Physiology*, **69**, 2004-2011.

Dooley, P.C., Bach, T.M., & Luff, A.R. (1990). Effect of vertical jumping on the medial gastrocnemius and soleus muscles of rats. *Journal of Applied Physiology*, **69**, 2004-2011.

Dorlöchter, M., Irintchev, A., Brinkers, M., & Wernig, A. (1991). Effects of enhanced activity on synaptic transmission in mouse extensor digitorum longus muscle. *Journal of Physiology (London)*, **436**, 283-292.

Duchateau, J., & Hainaut, K. (1984). Isometric or dynamic training: Differential effects on mechanical properties of a human muscle. *Journal of Applied Physiology*, **56**, 296-301.

Duchateau, J., & Hainaut, K. (1986). Nonlinear summation of contractions in striated muscle: 1. Twitch potentiation in human muscle. *Journal of Muscle Research and Cell Motility*, **7**, 11-17.

Duchateau, J., & Hainaut, K. (1987). Electrical and mechanical changes in immobilized human muscle. *Journal of Applied Physiology*, **62**, 2168-2173.

Duchateau, J., & Hainaut, K. (1988). Training effects of submaximal electrostimulation in a human muscle. *Medicine and Science in Sports and Exercise*, **20**, 99-104.

Duchateau, J., & Hainaut, K. (1990). Effects of immobilization on contractile properties, recruitment and firing rates of human motor units. *Journal of Physiology (London)*, **422**, 55-65.

Duchateau, J., & Hainaut, K. (1991). Effects of immobilization on electromyogram power spectrum changes during fatigue. *European Journal of Applied Physiology*, **63**, 458-462.

Duchateau, J., & Hainaut, K. (1993). Behaviour of short and long latency reflexes in fatigued human muscles. *Journal of Physiology (London)*, **471**, 787-799.

Duchêne, J., & Goubel, F. (1993). Surface electromyogram during voluntary contraction: Processing tools and relation to physiological events. *Critical Reviews in Biomedical Engineering*, **21**, 313-397.

Dudley, G.A., & Djamil, R. (1985). Incompatibility of endurance- and strength-training modes of exercise. *Journal of Applied Physiology*, **59**, 1446-1451.

Dudley, G.A., Duvoisin, M.R., Adams, G.R., Meyer, R.A., Belew, A.H., & Buchanan, P. (1992). Adaptations to unilateral lower limb suspension in humans. *Aviation, Space, and Environmental Medicine*, **63**, 678-683.

Dudley, G.A., Tesch, P.A., Miller, B.J., & Buchanan, P. (1991). Importance of eccentric muscle actions in performance adaptations to resistance training. *Aviation, Space, and Environmental Medicine*, **62**, 543-550.

Dum, R.P., & Kennedy, T.T. (1980). Physiological and histochemical characteristics of motor units in cat tibialis anterior and extensor digitorum longus muscles. *Journal of Neurophysiology*, **43**, 1615-1630.

Dyck, P.J., Karnes, J., O'Brien, P.C., & Zimmerman, I.R. (1984). Detection thresholds of cutaneous sensation in humans. In P.J. Dyck, P.K. Thomas, E.H. Lambert, & R. Runge (Eds.), *Peripheral neuropathy* (Vol. 1, pp. 1103-1138). Philadelphia: Saunders.

Edgerton, V.R., Apor, P., & Roy, R.R. (1990). Specific tension of human elbow flexor muscles. *Acta Physiologica Hungarica*, **75**, 205-216.

Edgerton, V.R., & Roy, R.R. (1991). Regulation of skeletal muscle fiber size, shape and function. *Journal of Biomechanics*, **24**(Suppl. 1), 123-133.

Edgerton, V.R., Roy, R.R., Gregor, R.J., & Rugg, S. (1986). Morphological basis of skeletal muscle power output. In N.L. Jones, N. McCartney, & A.J. McComas (Eds.), *Human muscle power* (pp. 43-64). Champaign, IL: Human Kinetics.

Edgerton, V.R., Smith, J.L., & Simpson, D.R. (1975). Muscle fibre types of human leg muscles. *Histochemical Journal*, **7**, 259-266.

Edin, B.B. (1992). Quantitative analysis of static strain sensitivity in human mechanoreceptors from hairy skin. *Journal of Neurophysiology*, **67**, 1105-1113.

Edington, C.J., Frederick, E.C., & Cavanagh, P.R. (1990). Rearfoot motion in distance running. In P.R. Cavanagh (Ed.), *Biomechanics of distance running* (pp. 135-164). Champaign, IL: Human Kinetics.

Edman, K.A.P. (1992). Contractile performance of skeletal muscle fibres. In P.V. Komi (Ed.), *Strength and power in sport*, (pp. 96-114). Champaign, IL: Human Kinetics.

Edman, K.A.P., Elzinga, G., & Noble, M.I.M. (1978). Enhancement of mechanical performance by stretch during tetanic contractions of vertebral skeletal muscle fibres. *Journal of Physiology (London)*, **281**, 139-155.

Edman, K.A.P., & Lou, F. (1990). Changes in force and stiffness induced by fatigue and intracellular acidification in frog muscle fibres. *Journal of Physiology (London)*, **424**, 133-149.

Edström, L., & Kugelberg, E. (1968). Histochemical composition, distribution of fibres and fatiguability of single motor units. *Journal of Neurology, Neurosurgery, and Psychiatry*, **31**, 424-433.

Edwards, R.H.T., Hill, D.K., Jones, D.A., & Merton, P.A. (1977). Fatigue of long duration in human skeletal muscle after exercise. *Journal of Physiology (London)*, **272**, 769-778.

Eie, N., & Wehn, P. (1962). Measurements of the intra-abdominal pressure in relation to weight bearing of the lumbosacral spine. *Journal of Oslo City Hospitals*, **12**, 205-217.

Elftman, H. (1938). The measurement of the external force in walking. *Science*, **88**, 152-153.

Elftman, H. (1939). Forces and energy changes in the leg during walking. *American Journal of Physiology*, **125**, 339-356.

Eloranta, V., & Komi, P.V. (1980). Function of the quadriceps femoris muscle under maximal concentric and eccentric contractions. *Electromyography and Clinical Neurophysiology*, **20**, 159-174.

Elson, P. (1974, December). Strength increases by electrical stimulation. *Track Technique*, p. 1856.

Emerson, N.D., & Zahalak, G.I. (1981). Longitudinal electrode array for electromyography. *Medical & Biological Engineering & Computing*, **19**, 504-506.

Engelhorn, R. (1983). Agonist and antagonist muscle EMG activity pattern changes with skill acquisition. *Research Quarterly for Exercise and Sport*, **54**, 315-323.

Engin, A.E., & Chen, S.-M. (1986). Statistical data base for the biomechanical properties of the human shoulder complex: 1. Kinematics of the shoulder complex. *Journal of Biomechanical Engineering*, **108**, 215-221.

Engin, A.E., & Chen, S.-M. (1987). Kinematic and passive resistive properties of human elbow complex. *Journal of Biomechanical Engineering*, **109**, 318-323.

Engin, A.E., & Chen, S.-M. (1988). On the biomechanics of human hip complex in vivo: 1. Kinematics for determination of the maximal voluntary hip complex sinus. *Journal of Biomechanics*, **21**, 785-795.

English, A.W. (1984). An electromyographic analysis of compartments in cat lateral gastrocnemius during unrestrained locomotion. *Journal of Neurophysiology*, **52**, 114-125.

English, A.W., & Ledbetter, W.D. (1982). Anatomy and innervation patterns of cat lateral gastrocnemius and plantaris muscles. *American Journal of Anatomy*, **164**, 67-77.

Engstrom, C.M., Loeb, G.E., Reid, J.G., Forrest, W.J., & Avruch, L. (1991). Morphometry of the human thigh muscles: A comparison between anatomical sections and computer tomographic and magnetic resonance images. *Journal of Anatomy*, **176**, 139-156.

Enoka, R.M. (1979). The pull in Olympic weightlifting. *Medicine and Science in Sports*, **11**, 131-137.

Enoka, R.M. (1981). Muscular control of a learned movement. Ph.D. dissertation, University of Washington, Seattle.

Enoka, R.M. (1983). Muscular control of a learned movement: The speed control system hypothesis. *Experimental Brain Research*, **51**, 135-145.

Enoka, R.M. (1988a). Load- and skill-related changes in segmental contributions to a weightlifting movement. *Medicine and Science in Sports and Exercise*, **20**, 178-187.

Enoka, R.M. (1988b). Muscle strength and its development: New perspectives. *Sports Medicine*, **6**, 146-168.

Enoka, R.M., & Fuglevand, A.J. (1993). Neuromuscular basis of the maximum voluntary force capacity of muscle. In M.D. Grabiner (Ed.), *Current issues in biomechanics* (pp. 215-235). Champaign, IL: Human Kinetics.

Enoka, R.M., Hutton, R.S., & Eldred, E. (1980). Changes in excitability of tendon tap and Hoffmann reflexes following voluntary contractions. *Electroencephalography and Clinical Neurophysiology*, **48**, 664-672.

Enoka, R.M., Miller, D.I., & Burgess, E.M. (1982). Below-knee amputee running gait. *American Journal of Physical Medicine*, **61**, 66-84.

Enoka, R.M., Robinson, G.A., & Kossev, A.R. (1989). Task and fatigue effects on low-threshold motor units in human hand muscle. *Journal of Neurophysiology*, **62**, 1344-1359.

Enoka, R.M., & Stuart, D.G. (1984). Henneman's size principle: Current issues. *Trends in NeuroSciences*, **7**, 226-228.

Enoka, R.M., & Stuart, D.G. (1992). Neurobiology of muscle fatigue. *Journal of Applied Physiology*, **72**, 1631-1648.

Enoka, R.M., Trayanova, N., Laouris, Y., Bevan, L., Reinking, R.M., & Stuart, D.G. (1992). Fatigue-related changes in motor unit action potentials of adult cats. *Muscle & Nerve*, **14**, 138-150.

Ericson, M.O. (1988). Mechanical muscular power output and work during ergometer cycling at different work loads and speeds. *European Journal of Applied Physiology*, **57**, 382-387.

Etnyre, B.R., & Abraham, L.D. (1986). Gains in range of ankle dorsiflexion using three popular stretching techniques. *American Journal of Physical Medicine*, **65**, 189-196.

Etnyre, B.R., & Abraham, L.D. (1988). Antagonist muscle activity during stretching: A paradox re-assessed. *Medicine and Science in Sports and Exercise*, **20**, 285-289.

Evans, W.J., & Cannon, J.G. (1991). The metabolic effects of exercise-induced muscle damage. In J.O. Holloszy (Ed.), *Exercise and sport sciences reviews* (Vol. 19, pp. 99-125). Baltimore: Williams & Wilkins.

Evatt, M.L., Wolf, S.L., & Segal, R.L. (1989). Modification of human spinal stretch reflexes: Preliminary studies. *Neuroscience Letters*, **105**, 350-355.

Ewing, J.L., Wolfe, D.R., Rogers, M.A., Amundson, M.L., & Stull, G.A. (1990). Effects of velocity of isokinetic training on strength, power, and quadriceps muscle fibre characteristics. *European Journal of Applied Physiology*, **61**, 159-162.

Fahrer, H., Rentsch, H.U., Gerber, N.J., Beyeler, C., Hess, C.W., & Grünig, B. (1988). Knee effusion and reflex inhibition of quadriceps: A bar to effective training. *Journal of Bone and Joint Surgery*, **70B**, 635-638.

Fang, Z.-P., & Mortimer, J.T. (1991a). Alternate excitation of large and small axons with different stimulation waveforms: An application to muscle activation. *Medical and Biological Engineering and Computing*, **29**, 543-547.

Fang, Z.-P., & Mortimer, J.T. (1991b). A method to effect physiological recruitment order in electrically activated muscle. *IEEE Transactions on Biomedical Engineering*, **38**, 175-179.

Fang, Z.-P., & Mortimer, J.T. (1991c). Selective activation of small motor axons by quasitrapezoidal current pulses. *IEEE Transactions on Biomedical Engineering*, **38**, 168-174.

Feinstein, B., Lindegård, B., Nyman, E., & Wohlfart, G. (1955). Morphologic studies of motor units in normal human muscles. *Acta Anatomica*, **23**, 127-142.

Feldman, A.G. (1986). Once more on the equilibrium-point hypothesis (λ model) for motor control. *Journal of Motor Behavior*, **18**, 17-54.

Fellows, S.J., & Rack, P.M.H. (1987). Changes in the length of the human biceps brachii muscle during elbow movements. *Journal of Physiology (London)*, **383**, 405-412.

Feltner, M.E. (1989). Three-dimensional interactions in a two-segment kinetic chain: 2. Application to the throwing arm in baseball pitching. *International Journal of Sport Biomechanics*, **5**, 420-450.

Feltner, M., & Dapena, J. (1986). Dynamics of the shoulder and elbow joints of the throwing arm during a baseball pitch. *International Journal of Sport Biomechanics*, **2**, 235-259.

Feltner, M.E., & Dapena, J. (1989). Three-dimensional interactions in a two-segment kinetic chain: 1. General model. *International Journal of Sport Biomechanics*, **5**, 403-419.

Fenn, W.O. (1924). The relation between the work performed and the energy liberated in muscular contraction. *Journal of Physiology (London)*, **58**, 373-395.

Fenn, W.O. (1930). Work against gravity and work due to velocity changes in running. *American Journal of Physiology*, **93**, 433-462.

Fiatarone, M.A., Marks, E.C., Ryan, N.D., Meredith, C.N., Lipsitz, L.A., & Evans, W.J. (1990). High-intensity strength training in nonagenarians. *Journal of the American Medical Association*, **263**, 3029-3034.

Fick, R. (1904). *Handbuch der Anatomie des Menschen* (Vol. 2). Stuttgart, Germany: Gustav Fischer Verlag.

Fisher, M.A. (1992). H reflexes and F waves: Physiology and clinical indication. *Muscle & Nerve*, **15**, 1223-1233.

Fisher, M.J., Meyer, R.A., Adams, G.R., Foley, J.M., & Potchen, E.J. (1990). Direct relationship between proton T2 and exercise intensity in skeletal muscle MR images. *Investigative Radiology*, **25**, 480-485.

Fitts, R.H., Brimmer, C.J., Heywood-Cooksey, A. & Timmerman, R.J. (1989). Single muscle fiber enzyme shifts with hindlimb suspension and immobilization. *American Journal of Physiology*, **256**, C1082-C1091.

Fitts, R.H., McDonald, K.S., & Schulter, J.M. (1991). The determinants of skeletal muscle force and power: Their adaptability with changes in activity pattern. *Journal of Biomechanics*, **24**(Suppl. 1), 111-122.

Fitts, R.H., & Metzger, J.M. (1988). Mechanisms of muscular fatigue. In J.R. Poortmans (Ed.), *Principles of exercise biochemistry* (pp. 212-229). Basel, Switzerland: Karger.

Fitts, R.H., Metzger, J.M., Riley, D.A., & Unsworth, B.R. (1986). Models of disuse: A comparison of hindlimb suspension and immobilization. *Journal of Applied Physiology*, **60**, 1946-1953.

Fitzsimons, D.P., Herrick, R.E., & Baldwin, K.M. (1990). Isomyosin distribution in skeletal muscle: Effects of altered thyroid state. *Journal of Applied Physiology*, **69**, 321-327.

Flanders, M. (1991). Temporal patterns of muscle activation for arm movements in three-dimensional space. *Journal of Neuroscience*, **11**, 2680-2693.

Flanders, M., & Herrmann, U. (1992). Two components of muscle activation: Scaling with the speed of arm movement. *Journal of Neurophysiology*, **67**, 931-943.

Fleck, S.J., & Kraemer, W.J. (1987). *Designing resistance training programs*. Champaign, IL: Human Kinetics.

Fleckenstein, J.L., Canby, R.C., Parkey, R.W., & Peshock, R.M. (1988). Acute effects of exercise on MR imaging of skeletal muscle in normal volunteers. *American Journal of Roentgenology*, **151**, 231-237.

Fleckenstein, J.L., & Shellock, F.G. (1991). Exertional muscle injuries: Magnetic resonance imaging evaluation. *Topics in Magnetic Resonance Imaging*, **3**, 50-70.

Fleckenstein, J.L., Watumull, D., Bertocci, L.A., Parkey, R.W., & Peshock, R.M. (1992). Finger-specific flexor recruitment in humans: Depiction of exercise-enhanced MRI. *Journal of Applied Physiology*, **72**, 1974-1977.

Fletcher, J.G., & Lewis, H.E. (1960). Human power output: The mechanics of pole vaulting. *Ergonomics*, **3**, 30-34.

Foerhing, R.C., & Munson, J.B. (1990). Motoneuron and muscle-unit properties after long-term direct innervation of soleus muscle by medial gastrocnemius nerve in cat. *Journal of Neurophysiology*, **64**, 847-861.

Foehring, R.C., Sypert, G.W., & Munson, J.B. (1987a). Motor unit properties following cross-reinnervation of cat lateral gastrocnemius and soleus muscles with medial gastrocnemius nerve: 1. Influence of motoneurons on muscle. *Journal of Neurophysiology*, **57**, 1210-1226.

Foehring, R.C., Sypert, G.W., & Munson, J.B. (1987b). Motor unit properties following cross-reinnervation of cat lateral gastrocnemius and soleus muscles with medial gastrocnemius nerve: 2. Influence of muscle on motoneurons. *Journal of Neurophysiology*, **57**, 1227-1245.

Fournier, E., & Pierrot-Deseilligny, E. (1989). Changes in transmission in some reflex pathways during movement in humans. *News in Physiological Sciences*, **4**, 29-32.

Fournier, M., Roy, R.R., Perham, H., Simard, C.P., & Edgerton, V.R. (1983). Is limb immobilization a model of muscle disuse? *Experimental Neurology*, **80**, 147-156.

Frederick, E.C. (1986). Kinematically mediated effects of sport shoe design: A review. *Journal of Sports Sciences*, **4**, 169-184.

Freed, W.J., de Medinaceli, L., & Wyatt, R.J. (1985). Promoting functional plasticity in the damaged nervous system. *Science*, **227**, 1544-1552.

Fridén, J. (1984). Changes in human skeletal muscle induced by long term eccentric exercise. *Cell and Tissue Research*, **236**, 365-372.

Fridén, J., Kjörell, U., & Thornell, L.-E. (1984). Delayed muscle soreness and cytoskeletal alterations: An immunocytological study in man. *International Journal of Sports Medicine*, **5**, 15-18.

Fridén, J., & Lieber, R.L. (1992). Structural and mechanical basis of exercise-induced muscle injury. *Medicine and Science in Sports and Exercise*, **24**, 521-530.

Fridén, J., Seger, J., & Ekblom, B. (1988). Sublethal muscle fibre injuries after high-tension anaerobic exercise. *European Journal of Applied Physiology*, **57**, 360-368.

Fridén, J., Sfakianos, P.N., Hargens, A.R., & Akeson, W.H. (1988). Residual muscular swelling after repetitive eccentric contractions. *Journal of Orthopaedic Research*, **7**, 142-145.

Friederich, J.A., & Brand, R.A. (1990). Muscle fiber architecture in the human lower limb. *Journal of Biomechanics*, **23**, 91-95.

Frohlich, C. (1980, March). The physics of somersaulting and twisting. *Scientific American*, pp. 154-164.

Frolkis, V.V., Martynenko, O.A., & Zamostyan, V.P. (1976). Aging of the neuromuscular apparatus. *Gerontology*, **22**, 244-279.

Frolkis, V.V., Tanin, S.A., Marcinko, V.I., Kulchitsky, O.K., & Yasechko, A.V. (1985). Axoplasmic transport of substances in motoneuronal axons of the spinal cord in old age. *Mechanisms of Ageing and Development*, **29**, 19-28.

Frontera, W.R., Meredith, C.N., O'Reilly, K.P., Knuttgen, H.G., & Evans, W.J. (1988). Strength conditioning in older men: Skeletal muscle hypertrophy and improved function. *Journal of Applied Physiology*, **64**, 1038-1044.

Fu, F.H., Harner, C.D., Johnson, D.L., Miller, M.D., & Woo, S.L.-Y. (1993). Biomechanics of knee ligaments. *Journal of Bone and Joint Surgery*, **75-A**, 1716-1727.

Fuglevand, A.J., Winter, D.A., & Patla, A.E. (1993). Models of recruitment and rate coding organization in motor-unit pools. *Journal of Neurophysiology*, **70**, 2470-2488.

Fuglevand, A.J., Winter, D.A., Patla, A.E., & Stashuk, D. (1992). Detection of motor unit action potentials with surface electrodes: Influence of electrode size and spacing. *Biological Cybernetics*, **67**, 143-154.

Fuglevand, A.J., Zackowski, K.M., Huey, K.A., & Enoka, R.M. (1993). Impairment of neuromuscular propagation during human fatiguing contractions at submaximal forces. *Journal of Physiology (London)*, **460**, 549-572.

Fujiwara, M., & Basmajian, J.V. (1975). Electromyographic study of two-joint muscles. *American Journal of Physical Medicine*, **54**, 234-242.

Fukashiro, S., & Komi, P.V. (1987). Joint moment and mechanical power flow of the lower limb during vertical jump. *International Journal of Sports Medicine*, **8**, 15-21.

Fukunaga, T., Roy, R.R., Shellock, F.G., Hodgson, J.A., Day, M.K., Lee, P.L., Kwong-Fu, H., & Edgerton, V.R. (1991). Physiological cross-sectional area of human leg muscles based on magnetic resonance imaging. *Journal of Orthopaedic Research*, **10**, 926-934.

Funk, D.A., An, K.N., Morrey, B.F., & Daube, J.R. (1987). Electromyographic analysis of muscles across the elbow joint. *Journal of Orthopaedic Research*, **5**, 529-538.

Gaffney, F.A., Sjøgaard, G., & Saltin, B. (1990). Cardiovascular and metabolic responses to static contraction in man. *Acta Physiologica Scandinavica*, **138**, 249-258.

Galea, V., & Norman, R.W. (1985). Bone-on-bone forces at the ankle joint during a rapid dynamic movement. In D.A. Winter, R.W. Norman, R.P. Wells, K.C. Hayes, & A.E. Patla (Eds.), *Biomechanics IX-A* (pp. 71-76). Champaign, IL: Human Kinetics.

Galganski, M.E., Fuglevand, A.J., & Enoka, R.M. (1993). Reduced control of motor output in a human hand muscle of elderly subjects during submaximal contractions. *Journal of Neurophysiology*, **69**, 2108-2115.

Gandevia, S.C., McCloskey, D.I., & Burke, D. (1992). Kinaesthetic signals and muscle contraction. *Trends in Neurosciences*, **15**, 62-65.

Gandevia, S.C., & McKenzie, D.K. (1985). Activation of the human diaphragm during maximal static efforts. *Journal of Physiology (London)*, **367**, 45-56.

Gans, C. (1982). Fiber architecture and muscle function. In R.L. Terjung (Ed.), *Exercise and sport sciences reviews* (Vol. 10, pp. 160-207). Philadelphia: Franklin Institute.

Gardetto, P.R., Schluter, J.M., & Fitts, R.H. (1989). Contractile function of single muscle fibers after hindlimb suspension. *Journal of Applied Physiology*, **66**, 2739-2749.

Gardiner, P.F., & Lapointe, M.A. (1982). Daily in vivo neuromuscular stimulation effects on immobilized rat hindlimb muscles. *Journal of Applied Physiology*, **53**, 960-966.

Gareis, H., Solomonow, M., Baratta, R., Best, R., & D'Ambrosia, R. (1992). The isometric length-force models of nine different skeletal muscles. *Journal of Biomechanics*, **25**, 903-916.

Garfinkel, S., & Cafarelli, E. (1992). Relative changes in maximal force, EMG, and muscle cross-sectional area after isometric training. *Medicine and Science in Sports and Exercise*, **24**, 1220-1227.

Garhammer, J. (1980). Power production by Olympic weightlifters. *Medicine and Science in Sports and Exercise*, **12**, 54-60.

Garhammer, J. (1989). Weight lifting and training. In C.L. Vaughan (Ed.), *Biomechanics of sport* (pp. 169-211). Boca Raton, FL: CRC Press.

Garland, S.J. (1991). Role of small diameter afferents in reflex inhibition during human muscle fatigue. *Journal of Physiology (London)*, **435**, 547-558.

Garland, S.J., Enoka, R.M., Serrano, L.P., & Robinson, G.A. (in press). Behavior of motor units in human biceps brachii during fatigue. *Journal of Applied Physiology*.

Garner, S.H., Hicks, A.L., & McComas, A.J. (1989). Prolongation of twitch potentiating mechanism throughout muscle fatigue and recovery. *Experimental Neurology*, **103**, 277-281.

Garnett, R., & Stephens, J.A. (1981). Changes in the recruitment threshold of motor units produced by cutaneous stimulation in man. *Journal of Physiology (London)*, **311**, 463-473.

Garnett, R.A.F., O'Donovan, M.J., Stephens, J.A., & Taylor, A. (1978). Motor unit organization of human medial gastrocnemius. *Journal of Physiology (London)*, **287**, 33-43.

Garrett, W.E. (1990). Muscle strain injuries: Clinical and basic aspects. *Medicine and Science in Sports and Exercise*, **22**, 436-443.

Garrett, W.E., Califf, J.C., & Bassett, F.H. (1984). Histochemical correlates of hamstring injuries. *American Journal of Sports Medicine*, **12**, 98-103.

Garrett, W.E., Safran, M.R., Seaber, A.V., Glisson, R.R., & Ribbeck, B.M. (1987). Biomechanical comparison of stimulated and nonstimulated skeletal muscle pulled to failure. *American Journal of Sports Medicine*, **15**, 448-454.

Gebhard, J.S., Kabo, J.M., & Meals, R.A. (1993). Passive motion: The dose effects on joint stiffness, muscle mass, bone density, and regional swelling. *Journal of Bone and Joint Surgery*, **75-A**, 1636-1647.

Georgopoulos, A.P., Ashe, J., Smyrnis, N., & Taira, M. (1992). The motor cortex and the coding of force. *Science*, **256**, 1692-1695.

Gergely, J. (1974). Some aspects of the role of the sarcoplasmic reticulum and the tropomyosin-troponin system in the control of muscle contraction by calcium ions. *Circulation Research*, **34**(Suppl. 3), 74-82.

Gerilovsky, L., Tsvetinov, P., & Trenkova, G. (1989). Peripheral effects on the amplitude of monopolar and bipolar H-reflex potentials from the soleus muscle. *Experimental Brain Research*, **76**, 173-181.

Ghez, C. (1985). Introduction to the motor systems. In E.R. Kandel and J.H. Schwartz (Eds.), *Principles of Neural Science* (2nd ed., pp. 429-442). New York: Elsevier.

Ghez, C. (1991a). The cerebellum. In E.R. Kandel, J.H. Schwartz, & T.M. Jessell (Eds.), *Principles of neural science* (3rd ed., pp. 626-646). New York: Elsevier.

Ghez, C. (1991b). Voluntary movement. In E.R. Kandel, J.H. Schwartz, & T.M. Jessell (Eds.), *Principles of neural science* (3rd ed., pp. 609-625). New York: Elsevier.

Gibson, J.N.A., Halliday, D., Morrison, W.L., Stoward, P.J., Hornsby, G.A., Watt, P.W., Murdoch, G., & Rennie, M.J. (1987). Decrease in human quadriceps muscle protein turnover consequent upon leg immobilization. *Clinical Science*, **72**, 503-509.

Gielen, C.C.A.M., Ramaekers, L., & van Zuylen, E.J. (1988). Long-latency stretch reflexes as co-ordinated functional responses in man. *Journal of Physiology (London)*, **407**, 275-292.

Gielen, C.C.A.M., van Ingen Schenau, G.J., Tax, T., & Theeuwen, M. (1990). The activation of mono- and bi-articular muscles in multi-joint movements. In J.M. Winters & S.L.-Y. Woo (Eds.), *Multiple muscle systems: Biomechanics and movement organization* (pp. 302-311). New York: Springer-Verlag.

Godfrey, C.M., Jayawardena, A., & Welsh, P. (1986). Comparison of electro-stimulation and isometric exercise in strengthening the quadriceps muscle. *Physiotherapy (Canada)*, **31**, 265-267.

Goldberger, M., Murray, M., & Tessler, A. (1992). Sprouting and regeneration in the spinal cord: Their roles in recovery of function after spinal injury. In A. Gorio (Ed.), *Neuroregeneration* (pp. 241-264). New York: Raven.

Goldspink, G., Scutt, A., Loughna, P.T., Wells, D.J., Jaenicke, T., & Gerlach, G.F. (1992). Gene expression in skeletal muscle in response to stretch and force generation. *American Journal of Physiology*, **262**, R356-R363.

Gonyea, W.J., Sale, D.G., Gonyea, F.B., & Mikesky, A. (1986). Exercise induced increases in muscle fiber number. *European Journal of Applied Physiology*, **55**, 137-141.

Goodfellow, J., & O'Connor, J. (1978). The mechanics of the knee and prosthesis design. *Journal of Bone and Joint Surgery*, **60B**, 358-369.

Gordon, D.A., Enoka, R.M., Karst, G.M., & Stuart, D.G. (1990). Force development and relaxation in single motor units of adult cats during a standard fatigue test. *Journal of Physiology (London)*, **421**, 583-594.

Gordon, D.A., Enoka, R.M., & Stuart, D.G. (1990). Motor-unit force potentiation in adult cats during a standard fatigue test. *Journal of Physiology (London)*, **421**, 569-582.

Gordon, T., & Pattullo, M.C. (1993). Plasticity of muscle fiber and motor unit types. In J.O. Holloszy (Ed.), *Exercise and sport sciences reviews* (Vol. 21, pp. 339-362). Baltimore: Williams & Wilkins.

Gordon, T., Yang, J.F., Ayer, K., Stein, R.B., & Tyreman, N. (1993). Recovery potential of muscle after partial denervation: A comparison between rats and humans. *Brain Research Bulletin*, **30**, 477-482.

Goslow, G.E., Reinking, R.M., & Stuart, D.G. (1973). The cat step cycle: Hind limb joint angles and muscle lengths during unrestrained locomotion. *Journal of Morphology*, **141**, 1-41.

Gottlieb, G.L., Corcos, D.M., & Agarwal, G.C. (1989). Organizing principles for single-joint movements: 1. A speed-insensitive strategy. *Journal of Neurophysiology*, **62**, 342-357.

Gottlieb, G.L., Corcos, D.M., Agarwal, G.C., & Latash, M.L. (1990a). Organizing principles for single joint movements: 3. Speed-insensitive strategy as a default. *Journal of Neurophysiology*, **63**, 625-636.

Gottlieb, G.L., Corcos, D.J., Agarwal, G.C., & Latash, M.L. (1990b). Principles underlying single-joint movement strategies. In J.M. Winters & S.L.-Y. Woo (Eds.), *Multiple muscle systems: Biomechanics and movement organization* (pp. 236-250). New York: Springer-Verlag.

Gowland, C., deBruin, H., Basmajian, J.V., Plews, N., & Burcea, I. (1992). Agonist and antagonist activity during voluntary upper-limb movement in patients with stroke. *Physical Therapy*, **72**, 624-633.

Grabiner, M.D., & Jeziorowski, J.J. (1992). Isokinetic trunk extension discriminates uninjured subjects from subjects with previous low back pain. *Clinical Biomechanics*, **7**, 195-200.

Grabiner, M.D., Koh, T.J., & Andrish, J.T. (1992). Decreased excitation of vastus medialis oblique and vastus lateralis in patellofemoral pain. *European Journal of Experimental Musculoskeletal Research*, **1**, 33-39.

Grabiner, M.D., Koh, T.J., & Draganich, L.F. (1994). Neuromechanics of the patellofemoral joint. *Medicine and Science in Sports and Exercise*, **26**, 10-21.

Grange, R.W., & Houston, M.E. (1991). Simultaneous potentiation and fatigue in quadriceps after a 60-second maximal voluntary isometric contraction. *Journal of Applied Physiology*, **70**, 726-731.

Grange, R.W., Vandenboom, R., & Houston, M.E. (1993). Physiological significance of myosin phosphorylation in skeletal muscle. *Canadian Journal of Applied Sport Sciences*, **18**, 229-242.

Gransberg, L., & Knutsson, E. (1983). Determination of dynamic muscle strength in man with acceleration controlled isokinetic movements. *Acta Physiologica Scandinavica*, **119**, 317-320.

Gregor, R.J., Komi, P.V., & Järvinen, M. (1991). A comparison of the triceps surae and residual muscle moments at the ankle during cycling. *Journal of Biomechanics*, **24**, 287-297.

Gregor, R.J., Komi, P.V., & Järvinen, M. (1987). Achilles tendon forces during cycling. *International Journal of Sports Medicine*, **8**, 9-14.

Gregor, R.J., Roy, R.R., Whiting, W.C., Lovely, R.G., Hodgson, J.A., & Edgerton, V.R. (1988). Mechanical output of the cat soleus during treadmill locomotion: In vivo vs in situ characteristics. *Journal of Biomechanics*, **21**, 721-732.

Gregory, J.E. (1990). Relations between identified tendon organs and motor units in the medial gastrocnemius muscle of the cat. *Experimental Brain Research*, **81**, 602-608.

Gregory, J.E., Morgan, D.L., & Proske, U. (1986). Aftereffects in the responses of cat muscle spindles. *Journal of Neurophysiology*, **56**, 451-461.

Grieve, D.W. (1969, June). Stretching active muscles and leading with the hips. *Coaching Review*, pp. 3-4, 10.

Griffiths, R.I. (1991). Shortening of muscle fibres during stretch of the active cat medial gastrocnemius muscle: The role of tendon compliance. *Journal of Physiology (London)*, **436**, 219-236.

Grimby, G., & Saltin, B. (1983). The ageing muscle. *Clinical Physiology*, **3**, 209-218.

Grimby, L., Hannerz, J., & Hedman, B. (1981). The fatigue and voluntary discharge properties of single motor units in man. *Journal of Physiology (London)*, **316**, 545-554.

Grood, E.S., Suntay, W.J., Noyes, F.R., & Butler, D.L. (1984). Biomechanics of the knee-extension exercise. *Journal of Bone and Joint Surgery*, **66A**, 725-734.

Gross, A.C., Kyle, C.R., & Malewicki, D.J. (1983, December). The aerodynamics of human-powered land vehicles. *Scientific American*, pp. 142-145, 148-152.

Gross, T.S., & Bain, S.D. (1993). Skeletal adaptation to functional stimuli. In M.D. Grabiner (Ed.), *Current issues in biomechanics* (pp. 151-169). Champaign, IL: Human Kinetics.

Groves, B.K. (1989). Muscle differentiation and the origin of muscle fiber diversity. *CRC Critical Reviews in Neurobiology*, **4**, 201-234.

Guissard, N., Duchateau, J., & Hainaut, K. (1988). Muscle stretching and motoneuron excitability. *European Journal of Applied Physiology*, **58**, 47-52.

Gunning, P., & Hardeman, E. (1991). Multiple mechanisms regulate muscle fiber diversity. *FASEB Journal*, **5**, 3064-3070.

Gydikov, A., & Kosarov, D. (1973). Physiological characteristics of the tonic and phasic motor units in human muscles. In A.A. Gydikov, N.T. Tankov, & D.S. Kosarov (Eds.), *Motor control* (pp. 75-94). New York: Plenum.

Gydikov, A., & Kosarov, D. (1974). Some features of different motor units in human biceps brachii. *Pflügers Archiv*, **347**, 75-88.

Gydikov, A.A., Kossev, A.R., Kosarov, D.S., & Kostov, K.G. (1987). Investigations of single motor units firing during movements against elastic resistance. In B. Jonsson (Ed.), *Biomechanics X-A* (pp. 227-232). Champaign, IL: Human Kinetics.

Gydikov, A., Kossev, A., Radicheva, N., & Tankov, N. (1981). Interaction between reflexes and voluntary motor activity in man revealed by discharges of separate motor units. *Experimental Neurology*, **73**, 331-344.

Hagbarth, K.-E. (1962). Post-tetanic potentiation of myotatic reflexes in man. *Journal of Neurology, Neurosurgery, and Psychiatry*, **25**, 1-10.

Hagbarth, K.-E., Hagglund, J.V., Nordin, M., & Wallin, E.U. (1985). Thixotropic behaviour of human finger flexor muscles with accompanying changes in spindle and reflex responses to stretch. *Journal of Physiology (London)*, **368**, 323-342.

Hägg, G.M. (1992). Interpretation of EMG spectral alterations and alteration indexes at sustained contraction. *Journal of Applied Physiology*, **73**, 1211-1217.

Hainaut, K., & Duchateau, J. (1992). Neuromuscular electrical stimulation and voluntary exercise. *Sports Medicine*, **14**, 100-113.

Häkkinen, K. (1985). Research overview: Factors influencing trainability of muscular strength during short term and prolonged training. *National Strength and Conditioning Association Journal*, **7**(2), 32-37.

Häkkinen, K., Alén, M., & Komi, P.V. (1985). Changes in isometric force- and relaxation-time, electromyographic and muscle fiber characteristics of human skeletal muscle during strength training and detraining. *Acta Physiologica Scandinavica*, **125**, 573-585.

Häkkinen, K., & Keskinen, K.L. (1989). Muscle cross-sectional area and voluntary force production characteristics in elite strength-trained and endurance-trained athletes and sprinters. *European Journal of Applied Physiology*, **59**, 215-220.

Häkkinen, K., & Komi, P.V. (1986). Training-induced changes in neuromuscular performance under voluntary and reflex conditions. *European Journal of Applied Physiology*, **55**, 147-155.

Häkkinen, K., Komi, P.V., & Alén, M. (1985). Effect of explosive type strength training on isometric force- and relaxation-time, electromyographic and muscle fibre characteristics of leg extensor muscles. *Acta Physiologica Scandinavica*, **125**, 587-600.

Hales, J.P., & Gandevia, S.C. (1988). Assessment of maximal voluntary contraction with twitch interpolation: An instrument to measure twitch responses. *Journal of Neuroscience Methods*, **25**, 97-102.

Hall, S.J., & DePauw, K.P. (1982). A photogrammetrically based model for predicting total body mass centroid location. *Research Quarterly for Exercise and Sport*, **53**, 37-45.

Hallett, M. (1993). Physiology of basal ganglia disorders: An overview. *Canadian Journal of Neurological Sciences*, **20**, 177-183.

Hammond, P.H., Merton, P.A., & Sutton, G.G. (1956). Nervous gradation of muscular contraction. *British Medical Bulletin*, **12**, 214-218.

Hanavan, E.P. (1964). *A mathematical model of the human body* (AMRL-TR-64-102). Wright Patterson Air Force Base, OH: Aerospace Medical Research Laboratories. (NTIS No. AD-608463 as issued by the United States Air Force)

Hanavan, E.P. (1966). A personalized mathematical model of the human body. *Journal of Spacecraft and Rockets*, **3**, 446-448.

Hannaford, B., & Stark, L. (1985). Roles of the elements of the triphasic control signal. *Experimental Neurology*, **90**, 619-635.

Harman, E.A., Frykman, P.N., Clagett, E.R., & Kraemer, W.J. (1988). Intra-abdominal and intra-thoracic pressures during lifting and jumping. *Medicine and Science in Sports and Exercise*, **20**, 195-201.

Harman, E.A., Rosenstein, R.M., Frykman, P.N., & Nigro, G.A. (1989). Effects of a belt on intra-abdominal pressure during weight lifting. *Medicine and Science in Sports and Exercise*, **21**, 186-190.

Harre, D. (Ed.) (1982). *Principles of sports training: Introduction to the theory and methods of training*. East Berlin: Sportverlag.

Harrison, R.N., Lees, A., McCullagh, P.J.J., & Rowe, W.B. (1986). A bioengineering analysis of human muscle and joint forces in the lower limbs during running. *Journal of Sports Sciences*, **4**, 201-218.

Hasan, Z. (1986). Optimized movement trajectories and joint stiffness in unperturbed, inertially loaded movements. *Biological Cybernetics*, **53**, 373-382.

Hasan, Z. (1991). Biomechanics and the study of multijoint movements. In D.R. Humphrey & H.-J. Freund (Eds.), *Motor control: Concepts and issues* (pp. 75-84). Chichester, England: Wiley.

Hasan, Z., & Enoka, R.M. (1985). Isometric torque-angle relationship and movement-related activity of human elbow flexors: Implications for the equilibrium-point hypothesis. *Experimental Brain Research*, **59**, 441-450.

Hasan, Z., Enoka, R.M., & Stuart, D.G. (1985). The interface between biomechanics and neurophysiology in the study of movement: Some recent approaches. In R.L. Terjung (Ed.), *Exercise and sport sciences reviews* (Vol. 13, pp. 169-234). New York: Macmillan.

Hasan, Z., & Stuart, D.G. (1984). Mammalian muscle receptors. In R.A. Davidoff (Ed.), *Handbook of the spinal cord* (pp. 559-607). New York: Marcel Dekker.

Hasan, Z., & Stuart, D.G. (1988). Animal solutions to problems of movement control: The role of proprioceptors. *Annual Review of Neurosciences*, **11**, 199-223.

Hasselkus, B.R., & Shambes, G.M. (1975). Aging and postural sway in women. *Journal of Gerontology*, **30**, 661-667.

Hather, B.M., Adams, G.R., Tesch, P.A., & Dudley, G.A. (1992). Skeletal muscle responses to lower limb suspension in humans. *Journal of Applied Physiology*, **72**, 1493-1498.

Hather, B.M., Tesch, P.A., Buchanan, P., & Dudley, G.A. (1991). Influence of eccentric actions on skeletal muscle adaptations to resistance training. *Acta Physiologica Scandinavica*, **143**, 177-185.

Hatze, H. (1980). A mathematical model for the computational determination of parameter values of anthropomorphic segments. *Journal of Biomechanics*, **13**, 833-843.

Hatze, H. (1981a). Estimation of myodynamic parameter values from observations on isometrically contracting muscle groups. *European Journal of Applied Physiology*, **46**, 325-338.

Hatze, H. (1981b). *Myocybernetic control models of skeletal muscle: Characteristics and applications*. Pretoria: University of South Africa.

Hawkins, D. (1993). Ligament biomechanics. In M.D. Grabiner (Ed.), *Current Issues in Biomechanics* (pp. 123-150). Champaign, IL: Human Kinetics.

Hay, J.G. (1978). *The Biomechanics of Sports Techniques* (2nd ed.). Englewood Cliffs, NJ: Prentice Hall.

Hay, J.G. (1985). *The Biomechanics of Sports Techniques* (3rd ed.). Englewood Cliffs, NJ: Prentice-Hall.

Hay, J.G., Andrews, J.G., & Vaughan, C.L. (1980). The influence of external load on the joint torques exerted in a squat exercise. In J.M. Cooper & B. Haven (Eds.), *Proceedings of the Biomechanics Symposium* (pp. 286-293). Indiana University: Indiana State Board of Health.

Hay, J.G., Andrews, J.G., Vaughan, C.L., & Ueya, K. (1983). Load, speed and equipment effects in strength-training exercises. In H. Matsui & K. Kobayashi (Eds.), *Biomechanics VIII-B* (pp. 939-950). Champaign, IL: Human Kinetics.

Hay, J.G., Wilson, B.D., Dapena, J., & Woodworth, G.G. (1977). A computational technique to determine the angular momentum of the human body. *Journal of Biomechanics*, **10**, 269-277.

Hayes, K.C. (1982). Biomechanics of postural control. In R.L. Terjung (Ed.), *Exercise and sport sciences reviews* (Vol. 10, pp. 363-391). Philadelphia: Franklin Institute.

Heckathorne, C.W., & Childress, D.S. (1981). Relationships of the surface electromyogram to the force, length, velocity, and contraction rate of the cineplastic human biceps. *American Journal of Physical Medicine*, **60**, 1-19.

Heckman, C.J., & Binder, M.D. (1990). Neural mechanisms underlying the orderly recruitment of motoneurons. In M.D. Binder & L.M. Mendell (Eds.), *The segmental motor system* (pp. 182-204). New York: Oxford University Press.

Heckman, C.J., & Binder, M.D. (1991). Computer simulation of the steady-state input-output function of the cat medial gastrocnemius motoneuron pool. *Journal of Neurophysiology*, **65**, 952-967.

Henneman, E. (1957). Relation between size of neurons and their susceptibility to discharge. *Science*, **126**, 1345-1347.

Henneman, E. (1979). Functional organization of motoneuron pools: The size-principle. In H. Asanuma & V.J. Wilson (Eds.), *Integration in the nervous system* (pp. 13-25). Tokyo: Igaku-Shoin.

Henning, E.M., Cavanagh, P.R., Albert, H.T., & Macmillan, N.H. (1982). A piezoelectric method of measuring the vertical contact stress beneath the human foot. *Journal of Biomedical Engineering*, **4**, 213-222.

Henriksson-Larsén, K. (1985). Distribution, number and size of different types of fibres in whole cross-sections of female m tibialis anterior: An enzyme histochemical study. *Acta Physiologica Scandinavica*, **123**, 229-235.

Henriksson-Larsén, K., Fridén, J., & Wretling, M.-L. (1985). Distribution of fibre sizes in human skeletal muscle: An enzyme histochemical study in m tibialis anterior. *Acta Physiologica Scandinavica*, **123**, 171-177.

Herbert, M.E., Roy, R.R., & Edgerton, V.R. (1988). Influence of one week of hindlimb suspension and intermittent high load exercise on rat muscles. *Experimental Neurology*, **102**, 190-198.

Hermansen, L., Hultman, E., & Saltin, B. (1967). Muscle glycogen during prolonged severe exercise. *Acta Physiologica Scandinavica*, **71**, 129-139.

Hershler, C., & Milner, M. (1980a). Angle-angle diagrams in above-knee amputee and cerebral palsy gait. *American Journal of Physical Medicine*, **59**, 165-183.

Hershler, C., & Milner, M. (1980b). Angle-angle diagrams in the assessment of locomotion. *American Journal of Physical Medicine*, **59**, 109-125.

Hettinger, T. (1961). *Physiology of strength*. Springfield, IL: Charles C Thomas.

Heyes, M.P., Garnett, E.S., & Coates, G. (1985). Central dopaminergic activity influences rats' ability to exercise. *Life Sciences*, **36**, 671-677.

Heyes, M.P., Garnett, E.S., & Coates, G. (1988). Nigostriatal dopaminergic activity is increased during exhaustive exercise stress in rats. *Life Sciences*, **42**, 1537-1542.

Hicks, A.L., Cupido, C.M., Martin, J., & Dent, J. (1992). Muscle excitation in elderly adults: The effects of training. *Muscle & Nerve*, **15**, 87-93.

Hicks, A., Fenton, J., Garner, S., & McComas, A.J. (1989). M wave potentiation during and after muscle activity. *Journal of Applied Physiology*, **66**, 2606-2610.

Hicks, A., & McComas, A.J. (1989). Increased sodium pump activity following repetitive stimulation of rat soleus muscles. *Journal of Physiology (London)*, **414**, 337-349.

Higgins, S. (1985). Movement as an emergent form: Its structural limits. *Human Movement Science*, **4**, 119-148.

High, D.M., Howley, E.T., & Franks, B.D. (1989). The effects of static stretching and warm-up on prevention of delayed-onset muscle soreness. *Research Quarterly for Exercise and Sport*, **60**, 357-361.

Hill, A.V. (1928). The air-resistance to a runner. *Proceedings of the Royal Society of London, B*, **102**, 380-385.

Hill, A.V. (1938). The heat of shortening and the dynamic constraints of muscle. *Proceedings of the Royal Society of London, B*, **126**, 136-195.

Hinrichs, R.N. (1987). Upper extremity function in running: 2. Angular momentum considerations. *International Journal of Sport Biomechanics*, **3**, 242-263.

Hinrichs, R.N., Cavanagh, P.R., & Williams, K.R. (1987). Upper extremity function in running: 1. Center of mass and propulsion considerations. *International Journal of Sport Biomechanics*, **3**, 222-241.

Hof, A.L. (1984). EMG and muscle force: An introduction. *Human Movement Sciences*, **3**, 119-153.

Hof, A.L., Pronk, C.N.A., & van Best, J.A. (1987). Comparison between EMG to force processing and kinetic analysis for the calf muscle moment in walking and stepping. *Journal of Biomechanics*, **20**, 167-178.

Hof, A.L., & van den Berg, J.W. (1977). Linearity between the weighted sum of the EMGs of the human triceps surae and the total torque. *Journal of Biomechanics*, **10**, 529-539.

Hof, A.L., & van den Berg, J. (1981a). EMG to force processing: 1. An electrical analogue of the Hill muscle model. *Journal of Biomechanics*, **14**, 747-758.

Hof, A.L., & van den Berg, J. (1981b). EMG to force processing: 2. Estimation of parameters of the Hill muscle model for the human triceps surae by means of a calfergometer. *Journal of Biomechanics*, **14**, 759-770.

Hof, A.L., & van den Berg, J. (1981c). EMG to force processing: 3. Estimation of model parameters for the human triceps surae muscle and assessment of the accuracy by means of a torque plate. *Journal of Biomechanics*, 14, 771-785.

Hof, A.L., & van den Berg, J. (1981d). EMG to force processing: 4. Eccentric-concentric contractions on a spring-flywheel setup. *Journal of Biomechanics*, **14**, 787-792.

Högfors, C., Sigholm, G., & Herberts, P. (1987). Biomechanical model of the human shoulder: 1. Elements. *Journal of Biomechanics*, **20**, 157-166.

Hoh, J.F.Y. (1991). Myogenic regulation of mammalian skeletal muscle. *News in Physiological Sciences*, **6**, 1-6.

Hoh, J.F.Y., & Hughes, S. (1988). Myogenic and neurogenic regulation of myosin gene expression in cat jaw-closing muscles regenerating in fast and slow limb muscles. *Journal of Muscle Research and Cell Motility*, **9**, 57-72.

Hollerbach, J.M., & Flash, T. (1982). Dynamic interactions between limb segments during planar arm movement. *Biological Cybernetics*, **44**, 67-77.

Holst, E. von. (1954). Relations between the central nervous system and the peripheral organs. *British Journal of Animal Behaviour*, **2**, 89-94.

Holstege, G. (1992). The emotional motor system. *European Journal of Morphology*, **30**, 67-79.

Hopper, B.J. (1973). *The mechanics of human movement*. New York: Elsevier.

Horak, F.B., & Nashner, L.M. (1986). Central programming of postural movements: Adaptations to altered support-surface configurations. *Journal of Neurophysiology*, **55**, 1369-1381.

Horak, F.B., Shupert, C.L., & Mirka, A. (1989). Components of postural dyscontrol in the elderly: A review. *Neurobiology of Aging*, **10**, 727-738.

Hortobágyi, T., & Katch, F.I. (1990). Role of concentric force in limiting improvement in muscular strength. *Journal of Applied Physiology*, **68**, 650-658.

Houk, J.C. (1988). Control strategies in physiological systems. *Federation of American Societies for Experimental Biology Journal*, **2**, 97-107.

Houk, J.C., Keifer, J., & Barto, A.G. (1993). Distributed motor commands in the limb premotor network. *Trends in Neurosciences*, **16**, 27-33.

Houk, J.C., & Rymer, W.Z. (1981). Neural control of muscle length and tension. In V.B. Brooks, (Ed.), *Handbook of Physiology: Sec. 1. The nervous system: Vol. II. Motor control. Part 1* (pp. 257-324). Bethesda, MD: American Physiological Society.

Houston, M.E., Froese, E.A., Valeriote, S.P., Green, H.J., & Ranney, D.A. (1983). Muscle performance, morphology and metabolic capacity during strength training and detraining: A one leg model. *European Journal of Applied Physiology*, **51**, 25-35.

Houston, M.E., Norman, R.W., & Froese, E.A. (1988). Mechanical measures during maximal velocity knee extension exercise and their relation to fibre composition of the human vastus lateralis muscle. *European Journal of Applied Physiology*, **58**, 1-7.

Howard, J.D. (1987). *Central and peripheral factors underlying bilateral inhibition during maximal efforts*. Unpublished doctoral dissertation, University of Arizona, Tucson.

Howard, J.D., & Enoka, R.M. (1991). Maximum bilateral contractions are modified by neurally mediated interlimb effects. *Journal of Applied Physiology*, **70**, 306-316.

Howell, J.N., Chleboun, G., & Conatser, R. (1993). Muscle stiffness, strength loss, swelling and soreness following exercise-induced injury in humans. *Journal of Physiology (London)*, **464**, 183-196.

Hoy, M.G., Zajac, F.E., & Gordon, M.E. (1990). A musculoskeletal model of the human lower extremity: The effect of muscle, tendon, and moment arm on the moment-angle relationship of musculotendon actuators at the hip, knee, and ankle. *Journal of Biomechanics*, **23**, 157-169.

Hoy, M.G., & Zernicke, R.F. (1986). The role of intersegmental dynamics during rapid limb oscillation. *Journal of Biomechanics*, **19**, 867-879.

Hoy, M.G., Zernicke, R.J., & Smith, J.L. (1985). Contrasting roles of inertial and muscle moments at knee and ankle during paw-shake response. *Journal of Neurophysiology*, **54**, 1282-1295.

Hoyle, G. (1983). *Muscles and their neural control*. New York: Wiley.

Hubbard, M. (1980). Dynamics of the pole vault. *Journal of Biomechanics*, **13**, 965-976.

Hubley, C.L., & Wells, R.P. (1983). A work-energy approach to determine individual joint contributions to vertical jump performance. *European Journal of Applied Physiology*, **50**, 247-254.

Hufschmidt, A., & Mauritz, K.-H. (1985). Chronic transformation of muscle in spasticity: A peripheral contribution to increased tone. *Journal of Neurology, Neurosurgery, and Psychiatry*, **48**, 676-685.

Hugon, M. (1973). Methodology of the Hoffmann reflex in man. In J.E. Desmedt (Ed.), *New developments in electromyography and clinical neurophysiology* (Vol. 3, pp. 277-293). Basel, Switzerland: Karger.

Huijing, P.A. (1992a). Elastic potential of muscle. In P.V. Komi (Ed.), *Strength and power in sport* (pp. 151-168). Champaign, IL: Human Kinetics.

Huijing, P.A. (1992b). Mechanical muscle models. In P.V. Komi (Ed.), *Strength and power in sport* (pp. 130-150). Champaign, IL: Human Kinetics.

Huizar, P., Kuno, M., Judo, N., & Miyata, Y. (1978). Reaction of intact spinal motoneurones to partial denervation of the muscle. *Journal of Physiology (London)*, **265**, 175-193.

Hultborn, H., Lindström, S., & Wigström, H. (1979). On the function of recurrent inhibition in the spinal cord. *Experimental Brain Research*, **37**, 399-403.

Hultman, E., Bergström, M., Spriet, L.L., & Söderlund, K. (1990). Energy metabolism and fatigue. In A.W. Taylor, P.D. Gollnick, H.J. Green, C.D. Ianuzzo, E.G. Noble, G. Métivier, and J.R. Sutton (Eds.), *Biochemistry of exercise VII* (pp. 73-92). Champaign, IL: Human Kinetics.

Hultman, E., & Sjöholm, H. (1983). Electromyogram, force and relaxation time during and after continuous electrical stimulation of human skeletal muscle in situ. *Journal of Physiology (London)*, **339**, 33-40.

Hultman, E., Sjöholm, H., Jäderhaolm-Ek, I., & Krynicki, J. (1983). Evaluation of methods for electrical stimulation of human skeletal muscle in situ. *Pflügers Archiv*, **398**, 139-141.

Hutton, R.S. (1984). Acute plasticity in spinal segmental pathways with use: Implications for training. In M. Kumamoto (Ed.), *Neural and mechanical control of movement* (pp. 90-112). Kyoto: Yamaguchi Shoten.

Hutton, R.S., & Nelson, D.L. (1986). Stretch sensitivity of Golgi tendon organs in fatigued gastrocnemius muscle. *Medicine and Science in Sports and Exercise*, **18**, 69-74.

Hutton, R.S., Smith, J.L., & Eldred, E. (1973). Postcontraction sensory discharge from muscle and its source. *Journal of Neurophysiology*, **36**, 1090-1103.

Huxley, H.E. (1985). The crossbridge mechanism of muscular contraction and its implications. *Journal of Experimental Biology*, **115**, 17-30.

Ikai, M., & Fukunaga, T. (1968). Calculation of muscle strength per unit cross-sectional area of human muscle by means of ultrasonic measurement. *Internationale Zeitschrift für Angewandte Physiologie Einschliesslich Arbeitsphysiologie*, **26**, 26-32.

Imms, F.J., Hackett, A.J., Prestidge, S.P., & Fox, R.H. (1977). Voluntary isometric muscle strength of patients undergoing rehabilitation following fractures of the lower limb. *Rheumatology and Rehabilitation*, **16**, 162-171.

Imms, F.J., & MacDonald, I.C. (1978). Abnormalities of the gait occuring during recovery from fractures of the lower limb and their improvement during rehabilitation. *Scandinavian Journal of Rehabilitation Medicine*, **10**, 193-199.

Ingjer, F., & Strømme, S.B. (1979). Effects of active, passive or no warm-up on the physiological response to heavy exercise. *European Journal of Applied Physiology*, **40**, 273-282.

Inman, V.T., Ralston, H.J., Saunders, J.B. de C.M., Feinstein, B., & Wright, E.W. (1952). Relation of human electromyogram to muscular tension. *Electroencephalography and Clinical Neurophysiology*, **4**, 187-194.

Iversen, L.L. (1987). Neurotransmitters. In G. Adelman (Ed.), *Encyclopedia of neuroscience* (Vol. 2, pp. 856-861). Boston: Birkhäuser.

Jackson, M.J., Jones, D.A., & Edwards, R.H.T. (1984). Experimental skeletal muscle damage: The nature of the calcium-activated degenerative processes. *European Journal of Clinical Investigation*, **14**, 369-374.

Jami, L. (1992). Golgi tendon organs in mammalian skeletal muscle: Functional properties and central actions. *Physiological Reviews*, **72** 623-666.

Jami, L., Murthy, K.S.K., Petit, J., & Zytnicki, D. (1983). After-effects of repetitive stimulation at low frequency on fast-contracting motor units of cat muscle. *Journal of Physiology (London)*, **340**, 129-143.

Jarić, S., Gavrilović, P., & Ivančević, V. (1985). Effects of previous muscle contractions on cyclic movement dynamics. *European Journal of Applied Physiology*, **54**, 216-221.

Jasmin, B.J., Lavoie, P.-A., & Gardiner, P.F. (1987). Fast axonal transport of acetylcholine in rat sciatic motoneurons is enhanced following prolonged daily running, but not following swimming. *Neuroscience Letters*, **78**, 156-160.

Jeannerod, M. (1988). *The neural and behavioural organization of goal-directed movements*. Oxford: Clarendon Press.

Jiang, B., Roy, R.R., & Edgerton, V.R. (1990). Expression of a fast fiber enzyme profile in the cat soleus after spinalization. *Muscle & Nerve*, **13**, 1037-1049.

Johansson, H., Sjölander, P., & Sojka, P. (1991). Receptors in the knee joint ligaments and their role in the biomechanics of the joint. *CRC Critical Reviews in Biomedical Engineering*, **18**, 341-368.

Johansson, R.S., & Cole, K.J. (1992). Sensory-motor coordination during grasping and manipulative actions. *Current Opinion in Neurobiology*, **2**, 815-823.

Johansson, R.S., Häger, C., & Bäckström, L. (1992). Somatosensory control of precision grip during unpredictable pulling loads: 3. Impairments during digital anesthesia. *Experimental Brain Research*, **89**, 204-213.

Johansson, R.S., Häger, C., & Riso, R. (1992). Somatosensory control of precision grip during unpredictable pulling loads: 2. Changes in load force rate. *Experimental Brain Research*, **89**, 192-203.

Johansson, R.S., Riso, R., Häger, C., & Bäckström, L. (1992). Somatosensory control of precision grip during unpredictable pulling loads: 1. Changes in load force amplitude. *Experimental Brain Research*, **89**, 181-191.

Johansson, R.S., & Vallbo, Å.B. (1983). Tactile sensory coding in the glabrous skin of the human hand. *Trends in Neurosciences*, **6**, 27-32.

Johansson, R.S., & Westling, G. (1984). Roles of glabrous skin receptors and sensorimotor memory in automatic control of precision grip when lifting rougher or more slippery objects. *Experimental Brain Research*, **56**, 550-564.

Johansson, R.S., & Westling, G. (1987). Signals in tactile afferents from the fingers eliciting adaptive motor responses during precision grip. *Experimental Brain Research*, **66**, 141-154.

Johns, R.J., & Wright, V. (1952). Relative importance of various tissues in joint stiffness. *Journal of Applied Physiology*, **17**, 824-828.

Jones, D.A., Rutherford, O.M., & Parker, D.F. (1989). Physiological changes in skeletal muscle as a result of strength training. *Quarterly Journal of Experimental Physiology*, **74**, 233-256.

Jones, L.A., & Hunter, I.W. (1983). Effect of fatigue on force sensation. *Experimental Neurology*, **81**, 640-650.

Josephson, R.K. (1993). Contraction dynamics and power output of skeletal muscle. *Annual Reviews of Physiology*, **55**, 527-546.

Joyncr, M.J. (1991). Modeling: Optimal marathon performance on the basis of physiological factors. *Journal of Applied Physiology*, **70**, 683-687.

Kalaska, J.F., & Crammond, D.J. (1992). Cerebral cortical mechanisms of reaching movements. *Science*, **255**, 1517-1523.

Kamen, G., & DeLuca, C.J. (1989). Unusual motor unit firing behavior in older adults. *Brain Research*, **482**, 136-140.

Kanda, K., Burke, R.E., & Walmsley, B. (1977). Differential control of fast and slow twitch motor units in the decerebrate cat. *Experimental Brain Research*, **29**, 57-74.

Kanda, K., & Hashizume, K. (1989). Changes in properties of the medial gastrocnemius motor units in aging rats. *Journal of Neurophysiology*, **61**, 737-746.

Kanda, K., & Hashizume, K. (1992). Factors causing difference in force output among motor units in the rat medial gastrocnemius muscle. *Journal of Physiology (London)*, **448**, 677-695.

Kandel, E.R., & Jessell, T.M. (1991). Touch. In E.R. Kandel, J.H. Schwartz, & T.M. Jessell (Eds.), *Principles of neural science* (3rd ed., pp. 367-384). New York: Elsevier.

Kandel, E.R., Schwartz, J.H., & Jessell, T.M. (Eds.) (1991). *Principles of neural science*. New York: Elsevier.

Kandel, E.R., & Siegelbaum, S. (1985). Principles underlying electrical and chemical synaptic transmission. In E.R. Kandel & J.H. Schwartz (Eds.), *Principles of neural science* (3rd ed. pp. 89-107). New York: Elsevier.

Kane, T.R., & Scher, M.P. (1969). A dynamical explanation of the falling cat phenomenon. *International Journal of Solids and Structures*, **5**, 663-670.

Karst, G.M., & Hasan, Z. (1987). Antagonist muscle activity during human forearm movements under varying kinematic and loading conditions. *Experimental Brain Research*, **67**, 391-401.

Karst, G.M., & Hasan, Z. (1991a). Initiation rules for planar, two-joint arm movements: Agonist selection for movements throughout the work space. *Journal of Neurophysiology*, **66**, 1579-1593.

Karst, G.M., & Hasan, Z. (1991b). Timing and magnitude of electromyographic activity for two-joint arm movements in different directions. *Journal of Neurophysiology*, **66**, 1594-1604.

Katz, R., Mazzocchio, R., Pénicaud, A., & Rossi, A. (1993). Distribution of recurrent inhibition in the human upper limb. *Acta Physiologica Scandinavica*, **149**, 183-198.

Katz, R., Penicaud, A., & Rossi, A. (1991). Reciprocal Ia inhibition between elbow flexors and extensors in the human. *Journal of Physiology (London)*, **437**, 269-286.

Keen, D.A., Yue, G.H., & Enoka, R.M. (in press). Training-related enhancement in the control of motor output in elderly humans. *Journal of Applied Physiology*.

Keller, A., Arissian, K., & Asanuma, H. (1992). Synaptic proliferation in the motor cortex of adult cats after long-term thalamic stimulation. *Journal of Neurophysiology*, **68**, 295-308.

Kelly, J.P. (1985). Reactions of neurons to injury. In E.R. Kandel & J.H. Schwartz (Eds.), *Principles of neural science* (2nd ed., pp. 187-195). New York: Elsevier.

Kereshi, S., Manzano, G., & McComas, A.J. (1983). Impulse conduction velocities in human biceps brachii muscles. *Experimental Neurology*, **80**, 652-662.

Kernell, D. (1992). Organized variability in the neuromuscular system: A survey of task-related adaptations. *Archives Italiennes de Biologie*, **130**, 19-66.

Kernell, D., & Hultborn, H. (1990). Synaptic effects on recruitment gain: A mechanism of importance for the input-output relations of motoneurone pools? *Brain Research*, **507**, 176-179.

Kernell, D., & Monster, A.W. (1982a). Motoneurone properties and motor fatigue: An intracellular study of gastrocnemius motoneurones of the cat. *Experimental Brain Research*, **46**, 197-204.

Kernell, D., & Monster, A.W. (1982b). Time course and properties of late adaptation in spinal motoneurones of the cat. *Experimental Brain Research*, **46**, 191-196.

Keshner, E.A., & Allum, J.H.J. (1990). Muscle activation patterns coordinating postural stability from head to foot. In J.M. Winters & S.L.-Y. Woo (Eds.), *Multiple muscle systems: Biomechanics and movement organization* (pp. 481-497). New York: Springer-Verlag.

Kilgore, J.B., & Mobley, B.A. (1991). Additional force during stretch of single frog muscle fibres following tetanus. *Experimental Physiology*, **76**, 579-588.

Kirsch, R.F., & Rymer, W.Z. (1987). Neural compensation for muscular fatigue: Evidence for significant force regulation in man. *Journal of Neurophysiology*, **57**, 1893-1910.

Kirsch, R.F., & Rymer, W.Z. (1992). Neural compensation for fatigue-induced changes in muscle stiffness during perturbations of elbow angle in human. *Journal of Neurophysiology*, **68**, 449-470.

Kitai, T.A., & Sale, D.G. (1989). Specificity of joint angle in isometric training. *European Journal of Applied Physiology*, **58**, 744-748.

Knott, M., & Voss, D.E. (1968). *Proprioceptive neuromuscular facilitation: Patterns and techniques* (2nd ed.). New York: Hoeber Medical Division, Harper & Row.

Koceja, D.M., & Raglin, J. (1992). Changes in neuromuscular characteristics during periods of overtraining and tapering. *Medicine and Science in Sports and Exercise*, **24**(Suppl.), S80.

Koester, J. (1985a). Functional consequences of passive membrane properties of the neuron. In E.R. Kandel & J.H.

Schwartz (Eds.), *Principles of neural science* (2nd ed., pp. 66-74). New York: Elsevier.

Koester, J. (1985b). Resting membrane potential and action potential. In E.R. Kandel & J.H. Schwartz (Eds.), *Principles of neural science* (2nd ed., pp. 49-57). New York: Elsevier.

Koh, T.J., Grabiner, M.D., & Clough, C.A. (1993). Bilateral deficit is larger for step than for ramp isometric contractions. *Journal of Applied Physiology*, **74**, 1200-1205.

Komi, P.V. (1990). Relevance of in vivo force measurements to human biomechanics. *Journal of Biomechanics*, **23**(Suppl. 1), 23-34.

Komi, P.V. (1992). Stretch-shortening cycle. In P.V. Komi (Ed.), *Strength and power in sport*, (pp. 169-179). Champaign, IL: Human Kinetics.

Komi, P.V., & Bosco, C. (1978). Utilization of stored elastic energy in leg extensor muscles by men and women. *Medicine and Science in Sports*, **10**, 261-265.

Komi, P.V., & Buskirk, E.R. (1972). Effect of eccentric and concentric muscle conditioning on tension and electrical activity of human muscle. *Ergonomics*, **15**, 417-434.

Kornecki, S. (1992). Mechanism of muscular stabilization process in joints. *Journal of Biomechanics*, **25**, 235-245.

Koshland, G.F., Hasan, Z., & Gerilovsky, L. (1991). Activity of wrist muscles elicited during imposed or voluntary movements about the elbow joint. *Journal of Motor Behavior*, **23**, 91-100.

Koshland, G.F., & Smith, J.L. (1989a). Mutable and immutable features of paw-shake responses after hindlimb deafferentation in the cat. *Journal of Neurophysiology*, **62**, 162-173.

Koshland, G.F., & Smith, J.L. (198b). Paw-shake responses with joint immobilization: EMG changes with atypical feedback. *Experimental Brain Research*, **77**, 361-373.

Kots, J.M. (1971). Trenirovka mysecnoj sily metodom elektrostimuljacci. Soobscenie 1. Teoreticeskie predposylki. *Teoriya i Praktika Fizicheskoi Kultury*, **3**, 64-67.

Kovanen, V., Suominen, H., & Heikkinen, E. (1984a). Collagen of slow twitch and fast twitch muscle fibres in different types of rat skeletal muscle. *European Journal of Applied Physiology*, **52**, 235-242.

Kovanen, V., Suominen, H., & Heikkinen, E. (1984b). Mechanical properties of fast and slow skeletal muscle with special reference to collagen and endurance training. *European Journal of Applied Physiology*, **17**, 725-735.

Kranz, H., Williams, A.M., Cassell, J., Caddy, D.J., & Silberstein, R.B. (1983). Factors determining the frequency content of the electromyogram. *Journal of Applied Physiology*, **55**, 392-399.

Krarup, C. (1981). Enhancement and diminution of mechanical tension evoked by staircase and by tetanus in rat muscle. *Journal of Physiology (London)*, **311**, 355-372.

Krnjevic, K., & Miledi, R. (1958). Failure of neuromuscular propagation in rats. *Journal of Physiology (London)*, **140**, 440-461.

Krogh-Lund, C., & Jørgensen, K. (1992). Modification of myo-electric power spectrum in fatigue from 15% maximal voluntary contraction of human elbow flexor muscles, to limit of endurance: Reflection of conduction velocity variation and/or centrally mediated mechanisms? *European Journal of Applied Physiology*, **64**, 359-370.

Kugelberg, E., & Lindegren, B. (1979). Transmission and contraction fatigue of rat motor units in relation to succinate dehydrogenase activity of motor unit fibres. *Journal of Physiology (London)*, **288**, 285-300.

Kugelberg, E., & Thornell, L.-E. (1983). Contraction time, histochemical type, and terminal cisternae volume of rat motor units. *Muscle & Nerve*, **6**, 149-153.

Kukulka, C.G., & Clamann, H.P. (1981). Comparison of the recruitment and discharge properties of motor units in human brachial biceps and adductor pollicis during isometric contractions. *Brain Research*, **219**, 45-55.

Kulig, K., Andrews, J.G., & Hay, J.G. (1984). Human strength curves. In R.L. Terjung (Ed.), *Exercise and sport sciences reviews* (Vol. 12, pp. 417-466). New York: Macmillan.

Kuno, M. (1964). Mechanism of facilitation and depression of the excitatory synaptic potential in spinal motoneurons. *Journal of Physiology (London)*, **175**, 100-112.

Kuno, S.-Y., Katsuta, S., Akisada, M., Anno, I., & Matsumoto, K. (1990). Effect of strength training on the relationship between magnetic resonance relaxation time and muscle fibre composition. *European Journal of Applied Physiology*, **61**, 33-36.

Kuypers, H.G.J.M. (1985). The anatomical and functional organization of the motor system. In M. Swash & C. Kennard (Eds.), *Scientific basis of clinical neurology* (pp. 3-18). London: Churchill Livingstone.

Lacquaniti, F., & Soechting, J.F. (1986). EMG responses to load perturbations of the upper limb: Effect of dynamic coupling between shoulder and elbow motion. *Experimental Brain Research*, **61**, 482-496.

Ladin, Z., & Wu, G. (1991). Combining position and acceleration measurements for joint force estimation. *Journal of Biomechanics*, **24**, 1173-1187.

Laforest, S., St-Pierre, D.M.M., Cyr, J., & Guyton, D. (1990). Effects of age and regular exercise on muscle strength and endurance. *European Journal of Applied Physiology*, **60**, 104-111.

Lagasse, P.P. (1974). Muscle strength: Ipsilateral and contralateral effects of superimposed stretch. *Archives of Physical Medicine and Rehabilitation*, **55**, 305-310.

Lakie, M., & Robson, L.G. (1988a). Thixotropic changes in human muscle stiffness and the effects of fatigue. *Quarterly Journal of Experimental Physiology*, **73**, 487-500.

Lakie, M., & Robson, L.G. (1988b). Thixotropy: The effect of stretch size in relaxed frog muscle. *Quarterly Journal of Experimental Physiology*, **73**, 127-129.

Lakomy, H.K.A. (1987). Measurement of human power output in high intensity exercise. In B. Van Gheluwe & J. Atha (Eds.), *Current research in sport biomechanics* (pp. 46-57). Basel, Switzerland: Karger.

Lamarra, N. (1990). Variables, constants, and parameters: Clarifying the system structure. *Medicine and Science in Sports and Exercise*, **22**, 88-95.

Lance, J.W., Burke, D., & Andrews, C.J. (1973). The reflex effects of muscle vibration. In J.E. Desmedt (Ed.), *New developments in electromyography and clinical neurophysiology*, (Vol. 3, pp. 444-462). Basel, Switzerland: Karger.

Lander, J.E., Bates, B.T., & DeVita, P. (1986). Biomechanics of the squat exercise using a modified center of mass bar. *Medicine and Science in Sports and Exercise*, **18**, 468-478.

Lander, J.E., Hundley, J.R., & Simonton, R.L. (1992). The effectiveness of weight-belts during multiple repetitions of the squat exercise. *Medicine and Science in Sports and Exercise*, **24**, 603-609.

Lännergren, J., & Westerblad, H. (1989). Maximum tension and force-velocity properties of fatigued, single *Xenopus* muscle fibres studied by caffeine and high K^+. *Journal of Physiology (London)*, **409**, 473-490.

Lännergren, J., & Westerblad, H. (1991). Force decline due to fatigue and intracellular acidification in isolated fibres from mouse skeletal muscle. *Journal of Physiology (London)*, **434**, 307-322.

Lansing, R.W., & Banzett, R.B. (1993). What do fully paralyzed awake humans feel when they attempt to move? *Journal of Motor Behavior*, **25**, 309-313.

Lanyon, L.E., Hampson, W.G.J., Goodship, A.E., & Shah, J.S. (1975). Bone deformation recorded in vivo from strain gauges attached to the human tibial shaft. *Acta Orthopaedica Scandinavica*, **46**, 256-268.

Lanyon, L.E., & Rubin, C.T. (1984). Static vs dynamic loads as an influence on bone remodelling. *Journal of Biomechanics*, **17**, 897-905.

Laouris, Y., Kalli-Laouri, J., & Schwartze, P. (1990). The postnatal development of the air-righting reaction in albino rats: Quantitative analysis of normal development and the effect of preventing neck-torso and torso-pelvis rotations. *Behavioral Brain Research*, **37**, 37-44.

Larsson, L. (1983). Histochemical characteristics of human skeletal muscle during aging. *Acta Physiologica Scandinavica*, **117**, 469-471.

Larsson, L., Sjödin, B., & Karlsson, J. (1978). Histochemical and biochemical changes in human skeletal muscle with age in sedentary males, age 22-65 years. *Acta Physiologica Scandinavica*, **103**, 31-39.

Latash, M.L. (1993). *Control of human movement*. Champaign, IL: Human Kinetics.

Latash, M.L., & Gottlieb, G.L. (1991a). An equilibrium-point model for fast, single-joint movement: 1. Emergence of strategy-dependent EMG patterns. *Journal of Motor Behavior*, **23**, 163-177.

Latash, M.L., & Gottlieb, G.L. (1991b). An equilibrium-point model for fast, single-joint movement: 2. Similarity of single-joint isometric and isotonic descending commands. *Journal of Motor Behavior*, **23**, 179-191.

Laughman, R.K., Youdas, J.W., Garrett, T.R., & Chao, E.Y.S. (1983). Strength changes in the normal quadriceps femoris muscle as a result of electrical stimulation. *Physical Therapy*, **63**, 494-499.

Lawrence, D.G., & Kuypers, H.G.J.M. (1968). The functional organization of the motor system in the monkey: 2. The effects of lesions of the descending brain-stem pathways. *Brain*, **91**, 15-41.

Lawrence, J.H., & De Luca, C.J. (1983). Myoelectric signal versus force relationship in different human muscles. *Journal of Applied Physiology*, **54**, 1653-1659.

Lemon, R.N. (1993). Cortical control of the primate hand. *Experimental Physiology*, **78**, 263-301.

Lennmarken, C., Bergman, T., Larsson, J., & Larsson, L.-E. (1985). Skeletal muscle function in man: Force, relaxation rate, endurance and contraction-time dependence on sex and age. *Clinical Physiology*, **5**, 243-255.

Lephart, S.A. (1984). Measuring the inertial properties of cadaver segments. *Journal of Biomechanics*, **17**, 537-543.

Lesmes, G.R., Costill, D.L., Coyle, E.F., & Fink, W.J. (1978). Muscle strength and power changes during maximal isokinetic training. *Medicine and Science in Sports*, **10**, 266-269.

Levin, M.F., & Hui-Chan, C.W.Y. (1992). Relief of hemiparetic spasticity by TENS is associated with improvement in reflex and voluntary motor functions. *Electroencephalography and Clinical Neurophysiology*, **85**, 131-142.

Lexell, J., & Taylor, C.C. (1991). Variability in muscle fibre areas in whole human quadriceps muscle: Effects of increasing age. *Journal of Anatomy*, **174**, 239-249.

Li, C., & Atwater, A.E. (1984, June). *Temporal and kinematic analysis of arm motion in sprinters*, Paper presented at the Olympic Scientific Congress, Eugene, OR.

Liberson, W.T., Holmquest, H.J., Scot, D., & Dow, M. (1961). Functional electrotherapy: Stimulation of the peroneal nerve synchronized with the swing phase of the gait of hemiplegic patients. *Archives of Physical Medicine and Rehabilitation*, **42**, 101-105.

Lieber, R.L. (1992). *Skeletal muscle structure and function*. Baltimore: Williams & Wilkins.

Lieber, R.L., Fazeli, B.M., & Botte, M.J. (1990). Architecture of selected wrist flexor and extensor muscles. *Journal of Hand Surgery*, **15A**, 244-250.

Lieber, R.L., & Fridén, J. (1993). Muscle damage is not a function of muscle force but active muscle strain. *Journal of Applied Physiology*, **74**, 520-526.

Lieber, R.L., Fridén, J.O., Hargens, A.R., Danzig, L.A., & Gershuni, D.H. (1988). Differential response of the dog quadriceps muscle to external skeletal fixation of the knee. *Muscle & Nerve*, **11**, 193-201.

Lieber, R.L., & Kelly, M.J. (1991). Factors influencing quadriceps femoris muscle torque using transcutaneous neuromuscular electrical stimulation. *Physical Therapy*, **71**, 715-723.

Lindman, R., Eriksson, A., & Thornell, L.-E. (1990). Fiber type composition of the human male trapezius muscle: Enzyme-histochemical characteristics. *American Journal of Anatomy*, **189**, 236-244.

Lindman, R., Eriksson, A., & Thornell, L.-E. (1991). Fiber type composition of the human female trapezius muscle: Enzyme-histochemical characteristics. *American Journal of Anatomy*, **190**, 385-392.

Lindström, L., Magnusson, R., & Petersén, I. (1970). Muscular fatigue and action potential conduction velocity changes studied with frequency analysis of EMG signals. *Electromyography*, **4**, 341-356.

Lloyd, A.R., Gandevia, S.C., & Hales, J.P. (1991). Muscle performance, voluntary activation, twitch properties and perceived effort in normal subjects and patients with chronic fatigue syndrome. *Brain*, **114**, 85-98.

Lloyd, D.P.C. (1949). Post-tetanic potentiation of response in monosynaptic reflex pathways of the spinal cord. *Journal of General Physiology*, **33**, 147-170.

Loeb, G.E., & Gans, C. (1986). *Electromyography for experimentalists*. Chicago: University of Chicago Press.

Lombardi, V., & Piazzesi, G. (1990). The contractile response during steady lengthening of stimulated frog muscle fibres. *Journal of Physiology (London)*, **431**, 141-171.

Lømo, T., & Waerhaug, O. (1985). Motor endplates in fast and slow muscles of the rat: What determines their differences? *Journal of Physiology (Paris)*, **80**, 290-297.

Lotz, B.P., Dunne, J.W., & Daube, J.R. (1989). Preferential activation of muscle fibers with peripheral magnetic stimulation of the limb. *Muscle & Nerve*, **12**, 636-639.

Lovely, R.G., Gregor, R.J., Roy, R.R., & Edgerton, V.R. (1990). Weight-bearing hindlimb stepping in treadmill-exercised adult spinal cats. *Brain Research*, **514**, 206-218.

Lowrie, M.B., & Vrbová, G. (1992). Dependence of postnatal motoneurones on their targets: Review and hypothesis. *Trends in Neurosciences*, **15**, 80-84.

Lucas, S.M., Ruff, R.L., & Binder, M.D. (1987). Specific tension measurements in single soleus and medial gastrocnemius muscle fibers of the cat. *Experimental Neurology*, **95**, 142-154.

Luethi, S.M., & Stacoff, A. (1987). The influence of the shoe on foot mechanics in running. In B. van Gheluwe & J. Atha (Eds.), *Current research in sports biomechanics* (pp. 72-85). New York: Karger.

Luhtanen, P. (1988). Kinematics and kinetics of serve in volleyball at different age levels. In G. de Groot, A.P. Hollander, P.A. Huijing, & G.J. van Ingen Schenau (Eds.), *Biomechanics XI-B* (pp. 815-819). Amsterdam: Free University Press.

Luhtanen, P., & Komi, P.V. (1978). Mechanical factors influencing running speed. In E. Asmussen & K. Jørgensen (Eds.), *Biomechanics VI-B* (pp. 23-29). Baltimore: University Park Press.

Luhtanen, P., & Komi, P.V. (1980). Force-, power-, and elasticity-velocity relationships in walking, running, and jumping. *European Journal of Applied Physiology*, **44**, 279-289.

Lusby, L.A., & Atwater, A.E. (1983). Speed-related position-time profiles of arm motion in trained women distance runners. *Medicine and Science in Sports and Exercise*, **15**, 171.

Lüscher, H.-R., Ruenzel, P., & Henneman, E. (1979). How the size of motoneurones determines their susceptibility to discharge. *Nature*, **282**, 859-861.

MacDougall, J.D., Sale, D.G., Alway, S.E., & Sutton, J.R. (1984). Muscle fiber number in biceps brachii in bodybuilders and control subjects. *Journal of Applied Physiology*, **57**, 1399-1403.

Macefield, G., Hagbarth, K.-E., Gorman, R., Gandevia, S.C., & Burke, D. (1991). Decline in spindle support to α-motoneurones during sustained voluntary contractions. *Journal of Physiology (London)*, **440**, 497-512.

Madsen, N., & McLaughlin, T. (1984). Kinematic factors influencing performance and injury risk in the bench press exercise. *Medicine and Science in Sports and Exercise*, **16**, 376-381.

Magladery, J.W., & McDougal, D.B. (1950). Electrophysiological studies of nerve and reflex activity in normal man: 1. Identification of certain reflexes in the electromyogram and the conduction velocity of peripheral nerve fibres. *Bulletin of Johns Hopkins Hopsital*, **86**, 265-290.

Magnus, R. (1922). Wie sich die fallende Katze in de Luft umdrecht. *Archives Neerlandaises de Physiologie de l'Homme et des Animaux*, **7**, 218-222.

Mancini, D.M., Coyle, E., Coggan, A., Beltz, J., Ferraro, N., Montain, S., & Wilson, J.R. (1989). Contribution of intrinsic skeletal muscle changes to 31P NMR skeletal muscle metabolic abnormalities in patients with chronic heart failure. *Circulation*, **80**, 1338-1346.

Mann, R.V. (1981). A kinetic analysis of sprinting. *Medicine and Science in Sports and Exercise*, **13**, 325-328.

Manter, J.T. (1938). The dynamics of quadrupedal walking. *Journal of Experimental Biology*, **15**, 522-540.

Marey, E.-J. (1874). *Animal mechanism: A treatise on terrestial and aerial locomotion*. New York: Appleton & Co.

Marey, E.-J. (1894). Des mouvements que certains animaux executent pour retomber sur leurs pieds, lorsqu'ils sont precipites d'un lieu eleve. *Academie des Sciences*, **119**, 714-718.

Marino, A.A. (1984). Electrical stimulation in orthopaedics: Past, present and future. *Journal of Bioelectricity*, **3**, 235-244.

Markhede, G., & Grimby, G. (1980). Measurements of strength of hip joint muscles. *Scandinavian Journal of Rehabilitation Medicine*, **12**, 169-174.

Marras, W.S., & Mirka, G.A. (1992). A comprehensive evaluation of trunk response to asymmetric trunk motion. *Spine*, **17**, 318-326.

Marras, W.S., & Sommerich, C.M. (1991a). A three-dimensional motion model of loads on the lumbar spine: I Model structure. *Human Factors*, **33**, 123-137.

Marras, W.S., & Sommerich, C.M. (1991b). A three-dimensional motion model of loads on the lumbar spine: II. Model validation. *Human Factors*, **33**, 139-149.

Marras, W.S., Joynt, R.L., & King, A.I. (1985). The force-velocity relation and intra-abdominal pressure during lifting activities. *Ergonomics*, **28**, 603-613.

Marsden, C.D., Meadows, J.C., & Merton, P.A. (1983). "Muscular wisdom" that minimized fatigue during prolonged effort in man: Peak rates of motoneuron discharge and slowing of discharge during fatigue. In J.E. Desmedt (Ed.), *Motor control mechanisms in health and disease* (pp. 169-211). New York: Raven Press.

Marsden, C.D., Merton, P.A., & Morton, H.B. (1976). Servo action in the human thumb. *Journal of Physiology (London)*, **257**, 1-44.

Marsden, C.D., Merton, P.A., & Morton, H.B. (1983). Rapid postural reactions to mechanical displacement of the hand in man. In J.E. Desmedt (Ed.), *Motor control mechanisms in health and disease* (pp. 645-659). New York: Raven Press.

Marsh, E., Sale, D., McComas, A.J., & Quinlan, J. (1981). The influence of joint position on ankle dorsiflexion in humans. *Journal of Applied Physiology*, **51**, 160-167.

Martin, J.H., & Jessell, T.M. (1991). Anatomy of the somatic sensory system. In E.R. Kandel, J.H. Schwartz, & T.M. Jessell (Eds.), *Principles of neural science* (3rd ed., pp. 353-366). New York: Elsevier.

Martin, J.P. (1977). A short essay on posture and movement. *Journal of Neurology, Neurosurgery, and Psychiatry*, **40**, 25-29.

Martin, P.E., Mungiole, M., Marzke, M.W., & Longhill, J.M. (1989). The use of magnetic resonance imaging for measuring segment inertial properties. *Journal of Biomechanics*, **22**, 367-376.

Maton, B. (1991). Central nervous system changes in fatigue induced by local work. In G. Atlan, L. Beliveau, and P. Bouissou (Eds.), *Muscle fatigue: Biochemical and physiological aspects*, (pp. 207-221). Paris: Masson.

Maton, B., Petitjean, M., & Cnockaert, J.C. (1990). Phonomyogram and electromyogram relationships with isometric force reinvestigated in man. *European Journal of Applied Physiology*, **60**, 194-201.

Matthews, P.B.C. (1972). *Mammalian muscle receptors and their central actions*. Baltimore: Williams & Wilkins.

Matthews, P.B.C. (1987). Muscle sense. In G. Adelman (Ed.), *Encyclopedia of neuroscience* (Vol. 2, pp. 720-721). Boston: Birkhäuser.

Matthews, P.B.C. (1990). The knee jerk: Still an enigma? *Canadian Journal of Physiology and Pharmacology*, **68**, 347-354.

Matthews, P.B.C. (1991). The human stretch reflex and the motor cortex. *Trends in Neurosciences*, **14**, 87-91.

Matyas, T.A., Galea, M.P., & Spicer, S.D. (1986). Facilitation of the maximum voluntary contraction in hemiplegia by concomitant cutaneous stimulation. *American Journal of Physical Medicine*, **65**, 125-134.

Maughan, R.J., Nimmo, M.A., & Harmon, M. (1985). The relationship between myosin ATP-ase activity and isometric endurance in untrained male subjects. *European Journal of Applied Physiology*, **54**, 291-296.

Mayer, R.F., Burke, R.E., Toop, J., Hodgson, J.A., Kanda, K., & Walmsley, B. (1981). The effect of long-term immobilization on the motor unit population of the cat medial gastrocnemius muscle. *Neuroscience*, **6**, 725-739.

McCafferty, W.B., & Horvath, S.M. (1977). Specificity of exercise and specificity of training: A subcellular review. *Research Quarterly*, **48**, 358-371.

McCartney, N., Heigenhauser, G.J.F., & Jones, N.L. (1983). Power output and fatigue of human muscle in maximal cycling exercise. *Journal of Applied Physiology*, **55**, 218-224.

McCartney, N., Moroz, D., Garner, S.H., & McComas, A.J. (1988). The effects of strength training in patients with selected neuromuscular disorders. *Medicine and Science in Sports and Exercise*, **20**, 362-368.

McCloskey, D.I. (1987). Kinesthesia, kinesthetic perception. In G. Adelman (Ed.), *Encyclopedia of neuroscience* (Vol. 1, pp. 548-551). Boston: Birkhäuser.

McCloskey, D.I., Ebeling, P., & Goodwin, G.M. (1974). Estimation of weights and tensions and apparent involvement of a "sense of effort." *Experimental Neurology*, **42**, 220-232.

McCloskey, M. (1983, April), Intuitive physics. *Scientific American*, pp. 122-130.

McComas, A.J. (1977). *Neuromuscular function and disorders*. Boston: Butterworths.

McComas, A.J. (1991). Motor unit estimation: Methods, results, and present status. *Muscle & Nerve*, **14**, 585-597.

McComas, A.J., Sica, R.E., Upton, A.R.M., & Aguilera, G.C. (1973). Functional changes in motoneurons of hemiparetic patients. *Journal of Neurology, Neurosurgery, and Psychiatry*, **36**, 183-193.

McCully, K.K., & Faulkner, J.A. (1985). Injury to skeletal muscle fibers of mice following lengthening contractions. *Journal of Applied Physiology*, **59**, 119-126.

McDonagh, J.C., Binder, M.D., Reinking, R.M., & Stuart, D.G. (1980). A commentary on muscle unit properties in cat hindlimb muscles. *Journal of Morphology*, **166**, 217-230.

McDonagh, M.J.N., & Davies, C.T.M. (1984). Adaptive response of mammalian skeletal muscle to exercise with high loads. *European Journal of Applied Physiology*, **52**, 139-155.

McDonagh, M.J.N., Hayward, C.M., & Davies, C.T.M. (1983). Isometric training in human elbow flexor muscles: The effects on voluntary and electrically evoked forces. *Journal of Bone and Joint Surgery*, **65B**, 355-358.

McGill, S.M., & Norman, R.W. (1993). Low back biomechanics in industry: The prevention of injury through safer lifting. In M.D. Grabiner (Ed.), *Current issues in biomechanics* (pp. 69-120). Champaign, IL: Human Kinetics.

McLaughlin, T.M., Dillman, C.J., & Lardner, T.J. (1977). A kinematic model of performance in the parallel squat by champion powerlifters. *Medicine and Science in Sports*, **9**, 128-133.

McMahon, T.A. (1984). *Muscles, reflexes, and locomotion*. Princeton, NJ: Princeton University Press.

McNitt-Gray, J.L. (1991). Kinematics and impulse characteristics of drop landings from three heights. *International Journal of Sport Biomechanics*, **7**, 201-224.

Melvill-Jones, G., & Watt, D.G.D. (1971). Muscular control of landing from unexpected falls in man. *Journal of Physiology (London)*, **219**, 729-737.

Meriam, J.L., & Kraige, L.G. (1987). *Engineering mechanics*. New York: Wiley.

Merletti, R., Knaflitz, M., & De Luca, C.J. (1990). Myoelectric manifestations of fatigue in voluntary and electrically elicited contractions. *Journal of Applied Physiology*, **69**, 1810-1820.

Metzger, J.M., & Fitts, R.H. (1986). Fatigue from high- and low-frequency muscle stimulation: Role of sarcolemma action potentials. *Experimental Neurology*, **93**, 320-333.

Miller, C., & Thépaut-Mathieu, C. (1993). Strength training by electrostimulation conditions for efficacy. *International Journal of Sports Medicine*, **14**, 20-28.

Miller, D.I. (1976). A biomechanical analysis of the contribution of the trunk to standing vertical jump take-off. In J. Broekhoff (Ed.), *Physical education, sports and the sciences* (pp. 355-374). Eugene, OR: Microform.

Miller, D.I. (1978). Biomechanics of running: What should the future hold? *Canadian Journal of Applied Sport Sciences*, **3**, 229-236.

Miller, D.I. (1979). Modelling in biomechanics: An overview. *Medicine and Science in Sports*, **11**, 115-122.

Miller, D.I. (1980). Body segment contributions to sport skill performance: Two contrasting approaches. *Research Quarterly for Exercise and Sport*, **51**, 219-233.

Miller, D.I. (1990). Ground reaction forces in distance running. In P.R. Cavanagh (Ed.), *Biomechanics of distance running* (pp. 203-224). Champaign, IL: Human Kinetics.

Miller, D.I., Enoka, R.M., McCulloch, R.G., Burgess, E.M., Hutton, R.S., & Frankel, V.H. (1979). *Biomechanical analysis of lower extremity amputee extra-ambulatory activities* (Contract No. V5244P-1540/VA). New York: Veterans Administration.

Miller, D.I., Henning, E., Pizzimenti, M.A., Jones, I.C., & Nelson, R.C. (1989). Kinetic and kinematic characteristics of 10-m platform performances of elite divers: 1. Back takeoffs. *International Journal of Sport Biomechanics*, **5**, 60-88.

Miller, D.I., & Morrison, W.E. (1975). Prediction of segmental parameters using the Hanavan human body model. *Medicine and Science in Sports*, **7**, 207-212.

Miller, D.I., & Munro, C.F. (1984). Body segment contributions to height achieved during the flight of a springboard dive. *Medicine and Science in Sports and Exercise*, **16**, 234-242.

Miller, D.I., & Munro, C.F. (1985). Greg Louganis' springboard takeoff: 2. Linear and angular momentum considerations. *International Journal of Sport Biomechanics*, **1**, 288-307.

Miller, D.I., & Nelson, R.C. (1973). *Biomechanics of sport*. London: Kimpton.

Miller, R.G., Giannini, D., Milner-Brown, H.S., Layzer, R.B., Koretsky, A.P., Hooper, D., & Weiner, M.W. (1987). Effects of fatiguing exercise on high-energy phosphates, force, and EMG: Evidence for three phases of recovery. *Muscle & Nerve*, **10**, 810-821.

Milliron, M.J., & Cavanagh, P.R. (1990). Sagittal plane kinematics of the lower extremity during distance running. In P.R. Cavanagh (Ed.), *Biomechanics of distance running* (pp. 65-105). Champaign, IL: Human Kinetics.

Milner-Brown, H.S., & Miller, R.G. (1986). Muscle membrane excitation and impulse propagation velocity are reduced during muscle fatigue. *Muscle & Nerve*, **9**, 367-374.

Milner-Brown, H.S., & Miller, R.G. (1989). Increased muscular fatigue in patients with neurogenic muscle weakness: Quantification and pathophysiology. *Archives of Physical Medicine and Rehabilitation*, **70**, 361-366.

Milner-Brown, H.S., & Stein, R.B. (1975). The relation between the surface electromyogram and muscular force. *Journal of Physiology (London)*, **246**, 549-569.

Milner-Brown, H.S., Stein, R.B., & Lee, R.G. (1975). Synchronization of human motor units: Possible roles of exercise and supraspinal reflexes. *Electroencephalography and Clinical Neurophysiology*, **38**, 245-254.

Milner-Brown, H.S., Stein, R.B., & Yemm, R. (1973). The contractile properties of human motor units during voluntary isometric contractions. *Journal of Physiology (London)*, **228**, 285-306.

Monster, A.W., & Chan, H.C. (1977). Isometric force production by motor units of extensor digitorum communs muscle in man. *Journal of Neurophysiology*, **40**, 1432-1443.

Monster, A.W., Chan, N.C., & O'Connor, D. (1978). Activity patterns of human skeletal muscle: Relation to muscle fiber type composition. *Science*, **200**, 314-317.

Moore, G.P., Perkel, D.H., & Segundo, J.P. (1966). Statistical analysis and functional interpretation of neuronal spike data. *Annual Review of Physiology*, **28**, 493-522.

Moore, M.A., & Hutton, R.S. (1980). Electromyographic investigation of muscle stretching techniques. *Medicine and Science in Sports and Exercise*, **12**, 322-329.

Moore, M.A., & Kukulka, C.G. (1991). Depression of Hoffman reflexes following voluntary contraction and implications for proprioceptive neuromuscular facilitation therapy. *Physical Therapy*, **71**, 321-333.

Moreno-Aranda, J., & Seireg, A. (1981a). Electrical parameters for over-the-skin muscle stimulation. *Journal of Biomechanics*, **14**, 579-585.

Moreno-Aranda, J., & Seireg, A. (1981b). Force response to electrical stimulation of canine skeletal muscles. *Journal of Biomechanics*, **14**, 595-599.

Moreno-Aranda, J., & Seireg, A. (1981c). Investigation of over-the-skin electrical stimulation parameters for different normal muscles and subjects. *Journal of Biomechanics*, **14**, 587-593.

Morey, E.R. (1979). Spaceflight and bone turnover: Correlation with a new rat model of weightlessness. *Bioscience*, **29**, 168-172.

Morgan, D.L. (1990). New insights into the behavior of muscle during active lengthening. *Biophysical Journal*, **57**, 209-221.

Moritani, T. (1992). Time course of adaptations during strength and power training. In P.V. Komi (Ed.), *Strength and power in sport* (pp. 266-278). Champaign, IL: Human Kinetics.

Morris, A.F. (1974). Myotatic reflex effects on bilateral reciprocal leg strength. *American Corrective Therapy Journal*, **28**, 24-29.

Morris, J.M., Lucas, D.B., & Bressler, B. (1961). Role of the trunk in stability of the spine. *Journal of Bone and Joint Surgery*, **43A**, 327-351.

Mow, V.C., Proctor, C.S., & Kelly, M.A. (1989). Biomechanics of articular cartilage. In M. Nordin & V.H. Frankel (Eds.), *Basic biomechanics of the musculoskeletal system* (pp. 31-58). Philadelphia: Lea & Febiger.

Munro, C.F., Miller, D.I., & Fuglevand, A.J. (1987). Ground reaction forces in running: A reexamination. *Journal of Biomechanics*, **20**, 147-155.

Munson, J.B., Foehring, R.C., Lofton, S.A., Zengel, J.E., & Sypert, G.W. (1986). Plasticity of medial gastrocnemius motor units following cordotomy in the cat. *Journal of Neurophysiology*, **55**, 619-633.

Nachemson, A.L., Andersson, G.B.J., & Schultz, A.B. (1986). Valsalva maneuver biomechanics. *Spine*, **11**, 476-479.

Nakazawa, K., Kawakami, Y., Fukunaga, T., Yano, H., & Miyashita, M. (1993). Differences in activation patterns in elbow flexor muscles during isometric, concentric and eccentric contractions. *European Journal of Applied Physiology*, **66**, 214-220.

Nardone, A., Romanò, C., & Schieppati, M. (1989). Selective recruitment of high-threshold human motor units during voluntary isotonic lengthening of active muscles. *Journal of Physiology (London)*, **409**, 451-471.

Narici, M.V., Bordini, M., & Ceretelli, P. (1991). Effect of aging on human adductor pollicis muscle function. *Journal of Applied Physiology*, **71**, 1277-1281.

Narici, M.V., Roi, G.S., & Landoni, L. (1988). Force of knee extensor and flexor muscles and cross-sectional area determined by nuclear magnetic resonance imaging. *European Journal of Applied Physiology*, **57**, 39-44.

Narici, M.V., Roi, G.S., Landoni, L., Minetti, A.E., & Ceretelli, P. (1989). Changes in force, cross-sectional area and neural activation during strength training and detraining of the human quadriceps. *European Journal of Applied Physiology*, **59**, 310-319.

Nashner, L.M. (1971). A model describing vestibular detection of body sway motion. *Acta Otolaryngology*, **72**, 429-436.

Nashner, L.M. (1972). Vestibular postural control model. *Kybernetik*, **10**, 106-110.

Nashner, L.M. (1976). Adapting reflexes controlling the human posture. *Experimental Brain Research*, **26**, 59-72.

Nashner, L.M. (1977). Fixed patterns of rapid postural responses among leg muscles during stance. *Experimental Brain Research*, **30**, 13-24.

Nashner, L.M. (1982). Adaptation of human movement to altered environments. *Trends in Neurosciences*, **5**, 358-361.

Nashner, L.M., Woollacott, M.H., & Tuma, G. (1979). Organization of rapid responses to postural and locomotor-like perturbation of standing man. *Experimental Brain Research*, **36**, 463-476.

Németh, G., Ekholm, J., Arborelius, U.P., Harms-Ringdahl, K., & Schuldt, K. (1983). Influence of knee flexion on isometric hip extensor strength. *Scandinavian Journal of Rehabilitation Medicine*, **15**, 97-101.

Németh, G., & Ohlsén, H. (1986). Moment arm lengths of trunk muscles to the lumbosacral joint obtained in vivo with computed tomography. *Spine*, **11**, 158-160.

Nemeth, P.M., & Pette, D. (1981). Succinate dehydrogenase activity in fibres classified by myosin ATPase in three hind limb muscles of the rat. *Journal of Physiology (London)*, **320**, 73-80.

Nemeth, P.M., Solanki, L., Gordon, D.A., Hamm, T.M., Reinking, R.M., & Stuart, D.G. (1986). Uniformity of metabolic enzymes within individual motor units. *Journal of Neuroscience*, **6**, 892-898.

Newham, D.J. (1988). The consequences of eccentric contractions and their relationship to delayed onset muscle pain. *European Journal of Applied Physiology*, **57**, 353-359.

Newham, D.J., Jones, D.A., & Clarkson, P.M. (1987). Repeated high-force eccentric exercise: Effects on muscle pain and damage. *Journal of Applied Physiology*, **63**, 1381-1386.

Newham, D.J., Jones, D.A., & Edwards, R.H.T. (1983). Large delayed plasma creatine kinase changes after stepping exercise. *Muscle & Nerve*, **6**, 380-385.

Nicks, D.K., Beneke, W.M., Key, R.M., & Timson, B.F. (1989). Muscle fibre size and number following immobili[illegible]on atrophy. *Journal of Anatomy*, **163**, 1-5.

Niels[illegible], J., Crone, C., & Hultborn, H. (1993). H-reflexes are smaller in dancers from The Royal Danish Ballet than in well-trained athletes. *European Journal of Applied Physiology*, **66**, 116-121.

Nielsen, J., Kagamihara, Y., Crone, C., & Hultborn, H. (1992). Central facilitation of Ia inhibition during tonic ankle dorsiflexion revealed after blockade of peripheral feedback. *Experimental Brain Research*, **88**, 651-656.

Nigg, B.M. (Ed.) (1986). *Biomechanics of running shoes*. Champaign, IL: Human Kinetics.

Nilsson, J., & Thorstensson, A. (1987). Adaptability in frequency and amplitude of leg movements during human locomotion at different speeds. *Acta Physiologica Scandinavica*, **129**, 107-114.

Nilsson, J., & Thorstensson, A. (1989). Ground reaction forces at different speeds of human walking and running. *Acta Physiologica Scandinavica*, **136**, 217-228.

Nissinen, M. (1978). *Kinematic and kinetic analysis of the hurdle and support phases of a running forward somersault*. Unpublished master's thesis, University of Washington, Seattle.

Nissinen, M., Preiss, R., & Brüggemann, P. (1985). Simulation of human airborne movements on the horizontal bar. In D.A. Winter, R.W. Norman, R.P. Wells, K.C. Hayes, & A.E. Patla (Eds.), *Biomechanics IX-B* (pp. 373-376). Champaign, IL: Human Kinetics.

Nordin, M., & Frankel, V.H. (1989). Biomechanics of bone. In M. Nordin & V.H. Frankel (Eds.), *Basic biomechanics of the musculoskeletal system* (pp. 3-29). Philadelphia: Lea & Febiger.

Nordstrom M.A., Enoka, R.M., Callister, R.J., Reinking, R.M., & Stuart, D.G. (1994). *Effects of six weeks of limb immobilization on the cat tibialis posterior: 1. Motor units.* Manuscript submitted for publication.

Nordstrom, M.A., Fuglevand, A.J., & Enoka, R.M. (1992). Estimating the strength of common input to motoneurons from the cross-correlogram. *Journal of Physiology (London)*, **453**, 547-574.

Nordstrom, M.A., & Miles, T.S. (1990). Fatigue of single motor units in human masseter. *Journal of Applied Physiology*, **68**, 26-34.

Noyes, F.R. (1977). Functional properties of knee ligaments and alterations induced by immobilization. *Clinical Orthopaedics and Related Research*, **123**, 210-242.

Noyes, F.R., Butler, D.L., Grood, E.S., Zernicke, R.F., & Hefzy, M.S. (1984). Biomechanical analysis of human ligament grafts used in knee-ligament repairs and reconstructions. *Journal of Bone and Joint Surgery*, **66A**, 344-352.

Nygaard, E., Houston, M.E., Suzuki, Y., Jørgensen, K., & Saltin, B. (1983). Morphology of the brachial biceps muscle and elbow flexion in man. *Acta Physiologica Scandinavica*, **117**, 287-292.

Ochs, S. (1987). Axoplasmic transport. In G. Adelman (Ed.), *Encyclopedia of neuroscience* (Vol. I, pp. 105-108). Boston: Birkhäuser.

O'Hara, T.E., & Goshgarian, H.G. (1991). Quantitative assessment of phrenic nerve functional recovery mediated by the crossed phrenic reflex at various time intervals after spinal cord injury. *Experimental Neurology*, **111**, 244-250.

Ohtsuki, T. (1983). Decrease in human voluntary isometric arm strength induced by simultaneous bilateral exertion. *Behavioural Brain Research*, **7**, 165-178.

Oishi, Y., Ishihara, A., & Katsuta, S. (1992). Muscle fibre number following hindlimb immobilization. *Acta Physiologica Scandinavica*, **146**, 281-282.

Orizio, C., Perini, R., Diemont, B., Figini, M.M., & Veicsteinas, A. (1990). Spectral analysis of muscular sound during isometric contraction of biceps brachii. *Journal of Applied Physiology*, **68**, 508-512.

Orizio, C., Perini, R., & Veicsteinas, A. (1989). Changes in muscular sound during sustained isometric contraction up to exhaustion. *Journal of Applied Physiology*, **66**, 1593-1598.

Oster, G. (1984, March). Muscle sounds. *Scientific American*, pp. 108-115.

Osternig, L.R. (1986). Isokinetic dynamometry: Implications for muscle testing and rehabilitation. In K.B. Pandolf (Ed.), *Exercise and sport sciences reviews* (Vol. 14, pp. 45-80). New York: Macmillan.

Otten, E. (1988). Concepts and models of functional architecture in skeletal muscle. In K.B. Pandolf (Ed.), *Exercise and sport sciences reviews* (Vol. 16, pp. 89-137). New York: Macmillan.

Ounjian, M., Roy, R.R., Eldred, E., Garfinkel, A., Payne, J.R., Armstrong, A., Toga, A.W., & Edgerton, V.R. (1991). Physiological and developmental implications of motor unit anatomy. *Journal of Neurobiology*, **22**, 547-559.

Pachter, B.R., & Eberstein, A. (1986). The effect of limb immobilization and stretch on the fine structure of the neuromuscular junction in rat muscle. *Experimental Neurology*, **92**, 13-19.

Palay, S.L., & Chan-Palay, V. (1987). Neuron. In G. Adelman (Ed.), *Encyclopedia of neuroscience* (Vol. 2, pp. 812-815). Boston: Birkhäuser.

Panin, N., Lindenauer, H.J., Weiss, A.A., & Ebel, A. (1961). Electromyographic evaluation of the cross exercise effect. *Archives of Physical Medicine and Rehabilitation*, **42**, 47-53.

Parker, P.A., Körner, L., & Kadefors, R. (1984). Estimation of muscle force from intramuscular pressure. *Medical and Biological Engineering and Computing*, **22**, 453-457.

Parkkola, R., Alanen, A., Kalimo, H., Lillsunde, I., Komu, M., & Kormano, M. (1993). MR relaxation times and fiber type predominance of the psoas and multifidus muscle. *Acta Radiologica*, **34**, 16-19.

Patla, A.E. (1991). Visual control of human locomotion. In A.E. Patla (Ed.), *Adaptability of Human Gait* (pp. 55-97). North Holland: Elsevier Science Publishers B.V.

Pawson, P.A., & Grinnell, A.D. (1990). Physiological differences between strong and weak frog neuromuscular junctions: A study involving tetanic and posttetanic potentiation. *Journal of Neuroscience*, **10**, 1769-1778.

Pearson, K. (1993). Common principles of motor control in vertebrates and invertebrates. *Annual Reviews in Neuroscience*, **16**, 265-297.

Pepe, F.A., & Drucker, B. (1979). The myosin filament IV: Myosin content. *Journal of Molecular Biology*, **130**, 379-393.

Perry, J., & Bekey, G.A. (1981). EMG-force relationships in skeletal muscle. *CRC Critical Reviews in Biomedical Engineering*, **7**, 1-22.

Person, R.S. (1958). An electromyographic investigation on co-ordination of the activity of antagonist muscles in man during development of a motor habit. *Pavlov Journal of Higher Nervous Activity*, **8**, 13-23.

Person, R.S. (1963). Problems in the interpretation of electromyograms: 1. Comparison of electromyograms on recording with skin and needle electrodes. *Biophysics*, **8**, 89-97.

Person, R.S. (1974). Rhythmic activity of a group of human motoneurones during voluntary contraction of a muscle. *Electroencephalography and Clinical Neurophysiology*, **36**, 585-595.

Person, R.S., & Kudina, L.P. (1972). Discharge frequency and discharge pattern of human motor units during voluntary contraction of muscle. *Electroencephalography and Clinical Neurophysiology*, **32**, 471-483.

Peters, S.E. (1989). Structure and function in vertebrate skeletal muscle. *American Zoologist*, **29**, 221-234.

Peterson, S.R., Bell, G.J., Bagnall, K.M., & Quinney, H.A. (1991). The effects of concentric resistance training on eccentric peak torque and muscle cross-sectional area. *Journal of Orthopaedic and Sports Physical Therapy*, **13**, 132-137.

Petrofsky, J.S., & Hendershot, D.M. (1984). The interrelationship between blood pressure, intramuscular pressure, and isometric endurance in fast and slow twitch skeletal muscle in the cat. *European Journal of Applied Physiology*, **53**, 106-111.

Petrofsky, J.S., & Stacy, R. (1992). The effect of training on endurance and the cardiovascualr responses of individuals with paraplegia during dynamic exercise induced by functional electrical stimulation. *European Journal of Applied Physiology*, **64**, 487-492.

Pette, D., & Vrbová, G. (1992). Adaptation of mammalian skeletal muscle to chronic electrical stimulation. *Reviews of Physiology, Biochemistry, and Pharmacology*, **120**, 115-202.

Phillips, C.G. (1986). *Movements of the hand.* Liverpool: Liverpool University Press.

Phillips, S.J., & Roberts, E.M. (1980). Muscular and nonmuscular moments of force in the swing limb of Masters runners. In J.M. Cooper & B. Haven (Eds.), *Proceedings of the Biomechanics Symposium* (pp. 256-274). Bloomington, IN: Indiana State Board of Health.

Phillips, S.J., Roberts, E.M., & Huang, T.C. (1983). Quantification of intersegmental reactions during rapid swing motion. *Journal of Biomechanics*, **16**, 411-418.

Pitman, M.I., & Peterson, L. (1989). Biomechanics of skeletal muscle. In M. Nordin & V.H. Frankel (Eds.), *Basic biomechanics of the musculoskeletal system* (pp. 89-111). Philadelphia: Lea & Febiger.

Pollack, G.H. (1983). The cross-bridge theory. *Physiological Reviews*, **63**, 411-417.

Powers, R.K., Robinson, F.R., Konodi, M.A., & Binder, M.D. (1992). Effective synaptic current can be estimated from measurements of neuronal discharge. *Journal of Neurophysiology*, **68**, 964-968.

Proske, V., Morgan, D.L., & Gregory, J.E. (1993). Thixotropy in skeletal muscle spindles: A review. *Progress in Neurobiology*, **41**, 705-721.

Psek, J.A., & Cafarelli, E. (1993). Behavior of coactive muscles during fatigue. *Journal of Applied Physiology*, **74**, 170-175.

Putnam, C.A. (1983). Interaction between segments during a kicking motion. In H. Matsui and K. Kobayashi (Eds.), *Biomechanics VIII-B* (pp. 688-694). Champaign, IL: Human Kinetics.

Putnam, C.A. (1991). A segment interaction analysis of proximal-to-distal sequential segment motion patterns. *Medicine and Science in Sports and Exercise*, **23**, 130-144.

Putnam, C.A. (1993). Sequential motions of body segments in striking and throwing skills: Descriptions and explanations. *Journal of Biomechanics*, **26**, 125-135.

Rab, G.T., Chao, E.Y.S., & Stauffer, R.N. (1977). Muscle force analysis of the lumbar spine. *Orthopedic Clinics of North America*, **8**, 193-199.

Rack, P.M.H., & Westbury, D.R. (1969). The effects of length and stimulus rate on tension in the isometric cat soleus muscle. *Journal of Physiology (London)*, **204**, 443-460.

Rack, P.M.H., & Westbury, D.R. (1974). The short range stiffness of active mammalian muscle and its effect on mechanical properties. *Journal of Physiology (London)*, **240**, 331-350.

Rafuse, V.F., Gordon, T., & Orozco, R. (1992). Proportional enlargement of motor units after partial denervation of cat triceps surae muscles. *Journal of Neurophysiology*, **68**, 1261-1276.

Rall, W. (1987). Neuron, cable properties. In G. Adleman (Ed.), *Encyclopedia of neuroscience* (Vol. 2, pp. 816-820). Boston: Birkhäuser.

Ralston, H.J., Inman, V.T., Strait, L.A., & Shaffrath, M.D. (1947). Mechanics of human isolated voluntary muscle. *American Journal of Physiology*, **151**, 612-620.

Ranatunga, K.W., Sharpe, B., & Turnbull, B. (1987). Contractions of a human skeletal muscle at different temperatures. *Journal of Physiology (London)*, **390**, 383-395.

Rankin, L.L., Enoka, R.M., Volz, K.A., & Stuart, D.G. (1988). Coexistence of twitch potentiation and tetanic force decline in rat hindlimb muscle. *Journal of Applied Physiology*, **65**, 2687-2695.

Riley, D.A., Slocum, G.R., Bain, J.L.W., Sedlak, F.R., Sowa, T.E., & Mellender, J.W. (1990). Rat hindlimb unloading: Soleus histochemistry, ultrastructure, and electromyography. *Journal of Applied Physiology*, **69**, 58-66.

Ríos, E., Ma, J., & González, A. (1991). The mechanical hypothesis of excitation-contraction (EC) coupling in skeletal muscle. *Journal of Muscle Research and Cell Motility*, **12**, 127-135.

Ríos, E., & Pizarró, G. (1988). Voltage sensors and calcium channels of excitation-contraction coupling. *News in Physiological Sciences*, **3**, 223-227.

Robinson, G.A., Enoka, R.M., & Stuart, D.G. (1991). Immobilization-induced changes in motor unit force and fatigability in the cat. *Muscle & Nerve*, **14**, 563-573.

Rodgers, M.M. (1993). Biomechanics of the foot during locomotion. In M.D. Grabiner (Ed.), *Current issues in biomechanics* (pp. 33-52). Champaign, IL: Human Kinetics.

Roesler, H. (1987). The history of some fundamental concepts in bone biomechanics. *Journal of Biomechanics*, **20**, 1025-1034.

Rogers, M.W. (1991). Motor control problems in Parkinson's disease. In M.J. Lister (Ed.), *Contemporary management of motor control problems* (pp. 195-208). Alexandria, VA: Foundation for Physical Therapy.

Rogers, M.W., Kukulka, C.G., & Soderberg, G.L. (1992). Age-related changes in postural responses preceding rapid self-paced and reaction time arm movements. *Journal of Gerontology*, **47**, M159-M165.

Rogers, M.W., & Pai, Y.-C. (1990). Dynamic transitions in stance support accompanying leg flexion movements in man. *Experimental Brain Research*, **81**, 398-402.

Roland, P.E., Larsen, B., Lassen, N.A., & Skinhøj, E. (1980). Supplementary motor area and other cortical areas in organization of voluntary movements in man. *Journal of Neurophysiology*, **43**, 118-136.

Roman, W.J., Fleckenstein, J., Stray-Gundersen, J., Alway, S.E., Peshock, R., & Gonyea, W.J. (1993). Adaptations in the elbow flexors of elderly males after heavy-resistance training. *Journal of Applied Physiology*, **74**, 750-754.

Rome, L.C., Funke, R.P., Alexander, R.M., Lutz, G., Aldridge, H., Scott, F., & Freadman, M. (1988). Why animals have different muscle fibre types. *Nature*, **335**, 824-827.

Rosenheimer, J.L., & Smith, D.O. (1990). Age-related increase in soluble and cell surface-associated neurite-outgrowth factors from rat muscle. *Brain Research*, **509**, 309-320.

Rossignol, S., Julien, C., Gauthier, L., & Lund, J.P. (1981). State dependent responses during movements. In A. Taylor & A. Prochazka (Eds.), *Muscle receptors and movement* (pp. 389-402). London: Macmillan.

Rothwell, J.C. (1987). *Control of human voluntary movement*. Kent, UK: Croom Helm.

Roy, R.R., Baldwin, K.M., & Edgerton, V.R. (1991). The plasticity of skeletal muscle: Effects of neuromuscular activity. In J.O. Holloszy (Ed.), *Exercise and sport sciences reviews* (Vol. 19, pp. 269-312). Baltimore: Williams & Wilkins.

Roy, R.R., & Edgerton, V.R. (1992). Skeletal muscle architecture and performance. In P.V. Komi (Ed.), *Strength and power in sport* (pp. 115-129). Champaign, IL: Human Kinetics.

Roy, R.R., Sacks, R.D., Baldwin, K.M., Short, M., & Edgerton, V.R. (1984). Interrelationships of contraction time, Vmax and myosin ATPase after spinal transection. *Journal of Applied Physiology*, **56**, 1594-1601.

Rube, N., & Secher, N.H. (1990). Effect of training on central factors in fatigue following two- and one-leg static exercise in man. *Acta Physiologica Scandinavica*, **141**, 87-95.

Rüegg, J.C. (1983). Muscle. In R.F. Schmidt & F. Thews (Eds.), *Human physiology* (pp. 32-50). Berlin: Springer-Verlag.

Rugg, S.G., Gregor, R.J., Mandelbaum, B.R., & Chiu, L. (1990). In vivo moment arm calculations at the ankle using magnetic resonance imaging (MRI). *Journal of Biomechanics*, **23**, 495-501.

Rutherford, O.M., & Jones, D.A. (1986). The role of learning and coordination in strength training. *European Journal of Applied Physiology*, **55**, 100-105.

Rydell, N. (1965). Forces at the hip joint: Pt. 2. Intravital measurements. In R.M. Kenedi (Ed.), *Biomechanics and related bioengineering topics* (pp. 351-357). Oxford: Pergamon Press.

Sadamoto, T., Bonde-Petersen, F., & Suzuki, Y. (1983). Skeletal muscle tension, flow, pressure, and EMG during sustained isometric contractions in humans. *European Journal of Applied Physiology*, **51**, 395-408.

Sahrmann, S.A., & Norton, B.J. (1977). The relationship of voluntary movement to spasticity in the upper motor neuron syndrome. *Annals of Neurology*, **2**, 460-464.

Saito, M., Kobayashi, K., Miyashita, M., & Hoshikawa, T. (1974). Temporal patterns in running. In R.C. Nelson & C.A. Morehouse (Eds.), *Biomechanics IV* (pp. 106-111). Baltimore: University Park Press.

Sale, D.G., (1988). Neural adaptation to resistance training. *Medicine and Science in Sports and Exercise*, **20**, S135-S145.

Sale, D.G., & MacDougall, J.D. (1981). Specificity in strength training: A review for the coach and athlete. *Canadian Journal of Applied Sport Sciences*, **6**, 87-92.

Sale, D.G., & MacDougall, J.D. (1984). Isokinetic strength in weight-trainers. *European Journal of Applied Physiology*, **53**, 128-132.

Sale, D.G., MacDougall, J.D., Jacobs, I., & Barner, S. (1990). Interaction between concurrent strength and endurance training. *Journal of Applied Physiology*, **68**, 260-270.

Sale, D.G., MacDougall, J.D., Upton, A.R.M., & McComas, A.J. (1983). Effect of strength training upon motoneuron excitability in man. *Medicine and Science in Sports and Exercise*, **125**, 57-62.

Sale, D.G., Quinlan, J., Marsh, E., McComas, A.J., & Belanger, A.Y. (1982). Influence of joint position on ankle plantarflexion in humans. *Journal of Applied Physiology*, **52**, 1636-1642.

Saltin, B., & Gollnick, P.D. (1983). Skeletal muscle adaptability: Significance for metabolism and performance. In L.D. Peachey (Ed.), *Handbook of physiology: Sec. 10. Skeletal muscle* (pp. 555-631). Bethesda, MD: American Physiological Society.

Sanderson, D.J., & Henning, E.M. (1992, August). *In-shoe pressure distribution in cycling and running shoes during steady-state cycling*. Paper presented at the 16th Annual Conference of the American Society of Biomechanics, Chicago, IL.

Sanes, J.N. (1987). Proprioceptive afferent information and movement control. In G. Adelman (Ed.), *Encyclopedia of neuroscience* (Vol. II, pp. 982-984). Boston: Birkhäuser.

Sanes, J.N., & Evarts, E.V. (1984). Motor psychophysics. *Human Neurobiology*, **2**, 217-225.

Sapega, A.A., Quedenfeld, T.C., Moyer, R.A., & Butler, R.A. (1981). Biophysical factors in range-of-motion exercise. *Physician and Sports Medicine*, **9**(12), 57-65.

Sargeant, A.J., & Boreham, A. (1982). Measurement of maximal short-term (anaerobic) power output during cycling. In J. Borms, M. Hebbelinck, & A. Venerando (Eds.), *Women and sport* (pp. 119-124). Basel, Switzerland: Karger.

Savolainen, S. (1989). Theoretical drag analysis of a skier in the downhill speed race. *International Journal of Sport Biomechanics*, **5**, 26-39, 1989.

Schantz, P., Fox, E.R., Norgren, P., & Tydén, A. (1981). The relationship between mean muscle fibre area and the muscle cross-sectional area of the thigh in subjects with large differences in thigh girth. *Acta Physiologica Scandinavica*, **113**, 537-539.

Schantz, P.G., Moritani, T., Karlson, E., Johansson, E., & Lundh, A. (1989). Maximal voluntary force of bilateral and unilateral leg extension. *Acta Physiologica Scandinavica*, **136**, 185-192.

Schiff, J.M. (1858). *Lehrbuch der Physiologie des Menschen: 1. Muskelund Nervenphysiologie*. Lahr: Schauenburg.

Schleihauf, R.E., Jr. (1979). A hydrodynamic analysis of swimming propulsion. In J. Terauds & E.W. Bedingfield (Eds.), *Swimming III* (pp. 70-109). Baltimore: University Park Press.

Schmidt, R.F., & Thews, G. (Eds). (1983). *Human physiology*. Berlin: Springer-Verlag.

Schneider, K., & Zernicke, R.F. (1992). Mass, center of mass, and moment of inertia estimates for infant limb segments. *Journal of Biomechanics*, **25**, 145-148.

Schultz, A.B. (1992). Mobility impairment in the elderly: Challenges for biomechanics research. *Journal of Biomechanics*, **25**, 519-528.

Schwartz, J.H. (1985). Synthesis and distribution of neuronal protein. In E.R. Kandel and J.H. Schwartz (Eds.), *Principles of Neural Science* (2nd ed., pp. 37-48). New York: Elsevier.

Scudder, G.N. (1980). Torque curves produced at the knee during isometric and isokinetic exercises. *Archives of Physical Medicine and Rehabilitation*, **61**, 68-72.

Seals, D.R., Washburn, R.A., Hanson, P.G., Painter, P.L., & Nagle, F.J. (1983). Increased cardiovascular response to static contraction of larger muscle groups. *Journal of Applied Physiology*, **54**, 434-437.

Secher, N.H., Rube, N., & Elers, J. (1988). Strength of two- and one-leg extension in man. *Acta Physiologica Scandinavica*, **134**, 333-339.

Segal, R.L. (1992). Neuromuscular compartments in the human biceps brachii muscle. *Neuroscience Letters*, **140**, 98-102.

Selkowitz, D.M. (1985). Improvement in isometric strength of the quadriceps femoris muscle after training with electrical stimulation. *Physical Therapy*, **65**, 186-196.

Seyffarth, H. (1940). The behavior of motor units in voluntary contractions. *Avhandlinger Utgitt Norske Videnskap-Akad Oslo. I. Matematisk-Naturvidenskapelig Klasse*, **4**, 1-63.

Shadwick, R.E. (1990). Elastic energy storage in tendons: Mechanical differences related to function and age. *Journal of Applied Physiology*, **68**, 1033-1040.

Shanebrook, J.R., & Jaszczak, R.D. (1976). Aerodynamic drag analysis of runners. *Medicine and Science in Sports*, **8**, 43-45.

Sheetz, M.P., Steuer, E.R., & Schroer, T.A. (1989). The mechanism and regulation of fast axonal transport. *Trends in Neurosciences*, **12**, 474-478.

Shellock, F.G. (1986). Physiological, psychological, and injury prevention aspects of warm-up. *National Strength and Conditioning Association Journal*, **8**, 24-27.

Shellock, F.G., Fukunaga, T., Mink, J.H., & Edgerton, V.R. (1991a). Acute effects of exercise on MR imaging of skeletal muscle: Concentric and eccentric actions. *American Journal of Roentgenology*, **156**, 765-768.

Shellock, F.G., Fukunaga, T., Mink, J.H., & Edgerton, V.R. (1991b). Exertional muscle injury: Evaluation of concentric versus eccentric actions with serial MRI imaging. *Radiology*, **179**, 659-664.

Shellock, F.G., & Prentice, W.E. (1985). Warming-up and stretching for improved physical performance and prevention of sports-related injuries. *Sports Medicine*, **2**, 267-278.

Shepherd, G. (1988). *Neurobiology*. New York: Oxford.

Sherrington, C.S. (1931). Quantitative management of contraction in lowest level co-ordination. *Brain*, **54**, 1-28.

Shindo, M., Harayama, H., Kondo, K., Yanagisawa, N., & Tanaka, R. (1984). Changes in reciprocal inhibition during voluntary contraction in man. *Experimental Brain Research*, **53**, 400-408.

Shorten, M.R. (1987). Muscle elasticity and human performance. In B. Van Gheluwe & J. Atha (Eds.), *Current research in sport biomechanics* (pp. 1-18). Basel, Switzerland: Karger.

Shumskii, V.V., Merten, A.A., & Dzenis, V.V. (1978). Effect of the type of physical stress on the state of the tibial bones of highly trained athletes as measured by ultrasound techniques. *Mekhanika Polimerov*, **5**, 884-888.

Simard, C.P., Spector, S.A., & Edgerton, V.R. (1982). Contractile properties of rat hind limb muscles immobilized at different lengths. *Experimental Neurology*, **77**, 467-482.

Simoneau, J.-A., & Bouchard, C. (1989). Human variation in skeletal muscle fiber-type proportion and enzyme activities. *American Journal of Physiology*, **257**, E567-E572.

Singh, S., & Katz, J.L. (1989). Electromechanical properties of bone: A review. *Journal of Bioelectricity*, **7**, 219-238.

Sjøgaard, G., Kiens, B., Jørgensen, K., & Saltin, B. (1986). Intramuscular pressure, EMG and blood flow during low-level prolonged static contraction in man. *Acta Physiologica Scandinavica*, **128**, 475-484.

Sjøgaard, G., Savard, G., & Juel, C. (1988). Muscle blood flow during isometric activity and its relation to muscle fatigue. *European Journal of Applied Physiology*, **57**, 327-335.

Sjöström, M., Ängquist, K.-A., Bylund, A.-C., Fridé, J., Gustavsson, L., & Scherstén, T. (1982). Morphometric analyses of human muscle fiber types. *Muscle & Nerve*, **5**, 538-553.

Sjöström, M., Kidman, S., Larsén, K.H., & Ängquist, K.-A., (1982). Z- and M-band appearance in different histochemically defined types of human skeletal muscle fibers. *Journal of Histochemistry and Cytochemistry*, **30**, 1-11.

Sjöström, M., Lexell, J., Eriksson, A., & Taylor, C.C. (1991). Evidence of fibre hyperplasia in human skeletal muscles from healthy young men? *European Journal of Applied Physiology*, **62**, 301-304.

Smith, F. (1982). Dynamic variable resistance and the Universal system. *National Strength & Conditioning Association Journal*, **4**(4), 14-19.

Smith, J.L., Betts, B., Edgerton, V.R., & Zernicke, R.F. (1980). Rapid ankle extension during paw shakes: Selective recruitment of fast ankle extensors. *Journal of Neurophysiology*, **43**, 612-620.

Smith, J.L., & Zernicke, R.F. (1987). Predictions for neural control based limb dynamics. *Trends in Neurosciences*, **10**, 123-128.

Snow-Harter, C., & Marcus, R. (1991). Exercise, bone mineral density, and osteoporosis. In J.O. Holloszy (Ed.), *Exercise and sport sciences reviews* (Vol. 19, pp. 351-388). Baltimore: Williams & Wilkins.

Snyder-Mackler, L., Binder-Macleod, S.A., & Williams, P.R. (1993). Fatigability of human quadriceps femoris muscle following anterior cruciate ligament reconstruction. *Medicine and Science in Sports and Exercise*, **25**, 783-789.

Snyder-Mackler, L., Ladin, Z., Schepsis, A.A., & Young, J.C. (1991). Electrical stimulation of the thigh muscles after reconstruction of the anterior cruciate ligament. *Journal of Bone and Joint Surgery*, **73A**, 1025-1036.

Somjen, G.G. (1987). Glial cells, functions, In G. Adelman (Ed.), *Encyclopedia of neuroscience* (Vol. 1, pp. 465-466). Boston: Birkhäuser.

Spector, S.A., Simard, C.P., Fournier, M., Sternlicht, E., & Edgerton, V.R. (1982). Architectural alterations of rat hind-limb skeletal muscles immobilized at different lengths. *Experimental Neurology*, **76**, 94-110.

Sperry, R.W. (1950). Neural basis of the spontaneous optokinetic response produced by visual neural inversion. *Journal of Comparative and Physiological Psychology*, **43**, 482-489.

Spira, M.E., Yarom, Y., & Parnas, I. (1976). Modulation of spike frequency by regions of special axonal geometry and by synaptic inputs. *Journal of Neurophysiology*, **39**, 882-899.

Spoor, C.W., van Leeuwen, J.L., Meskers, C.G.M., Titulaer, A.F., & Huson, A. (1990). Estimation of instantaneous moment arms of lower-leg muscles. *Journal of Biomechanics*, **23**, 1247-1259.

Spring, E., Savolainen, S., Erkkilä, J., Hämäläinen, T., & Pihkala, P. (1988). Drag area of a cross-country skier. *International Journal of Sport Biomechanics*, **4**, 103-113.

Stacoff, A., & Kaelin, X. (1983). Pronation and sportshoe design. In B.M. Nigg & B.A. Kerr (Eds.), *Biomechanical aspects of sport shoes and playing surfaces* (pp. 143-151). Calgary, AB: University of Calgary.

Stainsby, W.N., Brechue, W.F., O'Drobinak, D.M., & Barclay, J.K. (1990). Effects of ischemic and hypoxic hypoxia on $\dot{V}O_2$ and lactic acid output during tetanic contractions. *Journal of Applied Physiology*, **68**, 574-579.

Stålberg, E. (1980). Marcro EMG: A new recording technique. *Journal of Neurology, Neurosurgery, and Psychiatry*, **43**, 475-482.

Stålberg, E., & Fawcett, P.R.W. (1982). Macro EMG in healthy subjects of different ages. *Journal of Neurology, Neurosurgery, and Psychiatry*, **45**, 870-878.

Stanish, W.D., & Hubley-Kozey, C.L. (1984). Separating fact from fiction about common sports activity: Can stretching prevent athletic injuries? *Journal of Musculoskeletal Medicine*, **1**, 25-32.

Stauber, W.T. (1989). Eccentric actions of muscles: Physiology, injury, and adaptation. In K.B. Pandolf (Ed.), *Exercise and sport sciences reviews* (Vol. 17, pp. 157-185). Baltimore: Williams & Wilkins.

Stein, R.B., & Capaday, C. (1988). The modulation of human reflexes during functional motor tasks. *Trends in Neurosciences*, **11**, 328-332.

Stein, R.B., Gordon, T., Jefferson, J., Sharfenberger, A., Yang, J.F., Tötösy de Zepetnek, J., & Belanger, Y. (1992). Optimal stimulation of paralzyed muscle after human spinal cord injury. *Journal of Applied Physiology*, **72**, 1393-1400.

Stein, R.B., & Yang, J.F. (1990). Methods for estimating the number of motor units in human muscles. *Annals of Neurology*, **28**, 487-495.

Stevenson, J.M. (1985). The impact force of entry in diving from a ten-meter tower. In D.A. Winter, R.W. Norman, R.P. Wells, K.C. Hayes, & A.E. Patla (Eds.), *Biomechanics IX-B* (pp. 106-111). Champaign, IL: Human Kinetics.

Stokes, I.A.F., Moffroid, M.S., Rush, S., & Haugh, L.D. (1988). Comparison of acoustic and electrical signals from erectores spinae muscles. *Muscle & Nerve*, **11**, 331-336.

Stokes, M., & Young, A. (1984). The contribution of reflex inhibition to arthrogenous muscle weakness. *Clinical Science*, **67**, 7-14.

Stokes, M.J., & Cooper, R.G. (1992). Muscle sounds during voluntary and stimulated contractions of the human adductor pollicis muscle. *Journal of Applied Physiology*, **72**, 1908-1913.

Stroup, F., & Bushnell, D.L. (1970). Rotation, translation, and trajectory in diving. *Research Quarterly*, **40**, 812-817.

Stuart, D.G. (1987a). Muscle receptors, mammalian. In G. Adelman (Ed.), *Encyclopedia of neuroscience* (Vol. 2, pp. 716-718). Boston: Birkhäuser.

Stuart, D.G. (1987b). Muscle receptors, mammalian, spinal actions. In G. Adelman (Ed.), *Encyclopedia of neuroscience* (Vol. 2, pp. 718-719). Boston: Birkhäuser.

Stuart, D.G., & Enoka, R.M. (1983). Motoneurons, motor units, and the size principle. In R.N. Rosenberg (Ed.), *The clinical neurosciences* (pp. V:471-V:517). New York: Churchill Livingstone.

Stucke, H., Baudzus, W., & Baumann, W. (1984). On friction characteristics of playing surfaces. In E.C. Frederick (Ed.), *Sport shoes and playing surfaces* (pp. 87-97). Champaign, IL: Human Kinetics.

Sullivan, M.J., Green, H.J., & Cobb, F.R. (1990). Skeletal muscle biochemistry and histology in ambulatory patients with long term heart failure. *Circulation*, **81**, 518-527.

Suzuki, S., & Hutton, R.S. (1976). Postcontractile motoneuronal discharge produced by muscle afferent activation. *Medicine and Science in Sports*, **8**, 258-264.

Suzuki, S., & Sugi, H. (1983). Extensibility of the myofilaments in vertebrate skeletal muscle as revealed by stretching rigor muscle fibers. *Journal of General Physiology*, **81**, 531-546.

Sweeney, H.L., Bowman, B.F., & Stull, J.T. (1993). Myosin light chain phosphorylation in vertebrate striated muscle: Regulation and function. *American Journal of Physiology*, **264**, C1085-C1095.

Sweeney, H.L., & Stull, J.T. (1990). Alteration of cross-bridge kinetics by myosin light chain phosphorylation in rabbit skeletal muscle: Implications for regulation of actin-myosin interaction. *Proceedings of the National Academy of Sciences*, **87**, 414-418.

Tamaki, T., Uchiyama, S., & Nakano, S. (1992). A weight-lifting exercise model for inducing hypertrophy in the hindlimb muscles of rats. *Medicine and Science in Sports and Exercise*, **24**, 881-886.

Tang, A., & Rymer, W.Z. (1981). Abnormal force: EMG relations in paretic limbs of hemiparetic human subjects. *Journal of Neurology, Neurosurgery, and Psychiatry*, **44**, 690-698.

Tax, A.A.M., Denier van der Gon, J.J., & Erkelens, C.J. (1990). Differences in coordination of elbow flexor muscles in force tasks and in movement tasks. *Experimental Brain Research*, **81**, 567-572.

Taylor, D.C., Dalton, J., Seaber, A.V., & Garrett, W.E. (1990). The viscoelastic properties of muscle-tendon units. *American Journal of Sports Medicine*, **18**, 300-309.

ter Haar Romeny, B.M., Denier van der Gon, J.J., & Gielen, C.C.A.M. (1982). Changes in recruitment order of motor units in the human biceps brachii muscle. *Experimental Neurology*, **78**, 360-368.

Tesch, P.A., Dudley, G.A., Duvoisin, M.R., Hather, B.M., & Harris, R.T. (1990). Force and EMG signal patterns during repeated bouts of concentric and eccentric muscle actions. *Acta Physiologica Scandinavica*, **138**, 263-271.

Tesch, P.A., & Karlsson, J. (1985). Muscle fiber types and size in trained and untrained muscles of elite athletes. *Journal of Applied Physiology*, **59**, 1716-1720.

Thomas, C.K., Bigland-Ritchie, B., & Johansson, R.S. (1991). Force-frequency relationships of human thenar motor units. *Journal of Neurophysiology*, **65**, 1509-1516.

Thomas, C.K., Johansson, R.S., & Bigland-Ritchie, B. (1991). Attempts to physiologically classify human thenar motor units. *Journal of Neurophysiology*, **65**, 1501-1508.

Thomas, C.K., Stein, R.B., Gordon, T., Lee, R.G., & Elleker, M.G. (1987). Patterns of reinnervation and motor unit recruitment in human hand muscles after complete ulnar and median nerve section and resuture. *Journal of Neurology, Neurosurgery, and Psychiatry*, **50**, 259-268.

Thomas, C.K., Woods, J.J., & Bigland-Ritchie, B. (1989). Impulse propagation and muscle activation in long maximal voluntary contractions. *Journal of Applied Physiology*, **67**, 1835-1842.

Thomason, D.B., & Booth, F.W. (1990). Atrophy of the soleus muscle by hindlimb unweighting. *Journal of Applied Physiology*, **68**, 1-12.

Thomason, D.B., Herrick, R.E., & Baldwin, K.M. (1987). Activity influences on soleus muscle myosin during rodent hindlimb suspension. *Journal of Applied Physiology*, **63**, 138-144.

Thornell, L.-E., & Price, M.G. (1991). The cytoskeleton in muscle cells in relation to function. *Biochemical Society Transactions*, **19**, 1116-1120.

Thorstensson, A. (1977). Observations on strength training and detraining. *Acta Physiologica Scandinavica*, **100**, 491-493.

Thusneyapan, S., & Zahalak, G.I. (1989). A practical electrode-array myoprocessor for surface electromyography. *IEEE Transactions on Biomedical Engineering*, **36**, 295-299.

Tidball, J.G. (1991). Force transmission across muscle cell membranes. *Journal of Biomechanics*, **24**(Suppl. 1), 43-52.

Tötösy de Zepetnek, J.E., Zung, H.V., Erdebil,, S., & Gordon, T. (1992). Innervation ratio is an important determinant of force in normal and reinnervated rat tibialis anterior muscles. *Journal of Neurophysiology*, **67**, 1385-1403.

Toussaint, H.M., van Baar, C.E., van Langen, P.P., de Looze, M.P., & van Dieën, J.H. (1992). Coordination of the leg muscles in backlift and leglift. *Journal of Biomechanics*, **25**, 1279-1289.

Townend, M.S. (1984). *Mathematics in sport*. New York: Halstead.

Trimble, M.H., & Enoka, R.M. (1991). Mechanisms underlying the training effects associated with neuromuscular electrical stimulation. *Physical Therapy*, **71**, 273-282.

Trotter, J.A. (1990). Interfiber tension transmission in series-fibered muscles of the cat hindlimb. *Journal of Morphology*, **206**, 351-361.

Trotter, J.A. (1993). Functional morphology of force transmission in skeletal muscle. *Acta Anatomica*, **146**, 205-222.

Trotter, J.A., & Purslow, P.P. (1992). Functional morphology of the endomysium in series fibered muscles. *Journal of Morphology*, **212**, 109-122.

Tsika, R.W., Herrick, R.E., & Baldwin, K.M. (1987). Subunit composition of rodent isomyosins and their distribution in hindlimb skeletal muscles. *Journal of Applied Physiology*, **63**, 2101-2110.

Tsujimoto, T., & Kuno, M. (1988). Calcitonin gene-related peptide prevents disuse-induced sprouting of rat motor nerve terminals. *Journal of Neuroscience*, **8**, 3951-3957.

Tucker, V.A. (1975). Aerodynamics and energetics of vertebrate fliers. In T.Y.T. Wu, C.J. Brokaw, & C. Brennan (Eds.), *Swimming and flying in nature* (pp. 845-867). New York: Plemum.

Ulfhake, B., & Kellerth, J.-O. (1984). Electrophysiological and morphological measurements in cat gastrocnemius and soleus α-motoneurones. *Brain Research*, **307**, 167-179.

Upton, A.R.M., McComas, A.J., & Sica, R.E.P. (1971). Potentiation of ''late'' responses evoked in muscles during effort. *Journal of Neurology, Neurosurgery, and Psychiatry*, **34**, 699-711.

Valiant, G.A. (1990). Transmission and attenuation of heel-strike acceleration. In P.R. Cavanagh, (Ed.), *Biomechanics of Distance Running* (pp. 225-247). Champaign, IL: Human Kinetics.

Vallbo, Å.B., & Wessberg, J. (1993). Organization of motor output in slow finger movements in man. *Journal of Physiology (London)*, **469**, 673-691.

Vandenburgh, H.H. (1992). Mechanical forces and their second messengers in stimulating cell growth in vitro. *American Journal of Physiology*, **262**, R350-R355.

Vandenburgh, H.H., Hatfaludy, S., Karlisch, P., & Shansky, J. (1991). Mechanically induced alterations in cultered skeletal muscle growth. *Journal of Biomechanics*, **24**(Supplement), 91-99.

Vandenburgh, H.H., Hatfaludy, S., Sohar, I., & Shansky, J. (1990). Stretch-induced prostaglandins and protein turnover in cultured skeletal msucle. *American Journal of Physiology*, **259**, C232-C240.

Vander, A.J., Sherman, J.H., & Luciano, D.S. (1990). *Human physiology. The mechanisms of body function*. New York: McGraw-Hill.

van der Helm, F.C.T., & Veenbaas, R. (1991). Modelling the mechanical effect of muscles with large attachment sites: Application to the shoulder mechanism. *Journal of Biomechanics*, **24**, 1151-1163.

van der Vaart, A.J.M., Savelberg, H.H.C.M., de Groot, G., Hollander, A.P., Toussaint, H.M., & van Ingen Schenau, G.J. (1987). An estimation of drag in front crawl swimming. *Journal of Biomechanics*, **20**, 543-546.

Vandervoort, A.A., & Hayes, K.C. (1989). Plantarflexor muscle function in young and elderly women. *European Journal of Applied Physiology*, **58**, 389-394.

Vandervoort, A.A., Quinlan, J., & McComas, A.J. (1983). Twitch potentiation after voluntary contraction. *Experimental Neurology*, **81**, 141-152.

Vandervoort, A.A., Sale, D.G., & Moroz, J. (1984). Comparison of motor unit activation during unilateral and bilateral leg extension. *Journal of Applied Physiology*, **56**, 46-51.

van Ingen Schenau, G.J. (1984). An alternative view of the concept of utilisation of elastic energy in human movement. *Human Movement Science*, **3**, 301-336.

van Ingen Schenau, G.J., Bobbert, M.F., Huijing, P.A., & Woittiez, R.D. (1985). The instantaneous torque-angular velocity relation in plantar flexion during jumping. *Medicine and Science in Sports and Exercise*, **17**, 422-426.

van Ingen Schenau, G.J., Bobbert, M.F., & van Soest, A.J. (1990). The unique action of bi-articular muscles in leg extensions. In J.M. Winters & S.L.-Y. Woo (eds.), *Multiple muscle systems: Biomechanics and movement organization* (pp. 639-652). New York: Springer-Verlag.

van Ingen Schenau, G.J., & Cavanagh, P.R. (1990). Power equations in endurance sports. *Journal of Biomechanics*, **9**, 865-881.

van Ingen Schenau, G.J., van Woensel, W.W.L.M., Boots, P.J.M., Snackers, R.W., & de Groot, G. (1990). Determination and interpretation of mechanical power in human movement: Application to ergometer cycling. *European Journal of Applied Physiology*, **61**, 11-19.

van Mameren, H., & Drukker, J. (1979). Attachment and composition of skeletal muscles in relation of their function. *Journal of Biomechanics*, **12**, 859-867.

van Zuylen, E.J., Gielen, C.C.A.M., & Denier van der Gon, J.J. (1988). Coordination and inhomogeneous activation of human arm muscles during isometric torques. *Journal of Neurophysiology*, **60**, 1523-1548.

Varon, S.S., & Somjen, G.G. (1979). Neuron-glia interactions. *Neuroscience Research Progress Bulletin*, **17**, 1-239.

Vaughan, C.L. (1980). A kinetic analysis of basic trampoline stunts. *Journal of Human Movement Studies*, **6**, 236-251.

Vaughan, C.L. (1984). Biomechanics of running gait. *CRC Critical Reviews in Biomedical Engineering*, **12**, 1-48.

Verillo, R.T. (1979). Change in vibrotactile thresholds as a function of age. *Sensory Processes*, **3**, 49-59.

Videman, T. (1987). Connective tissue and immobilization: Key factors in musculoskeletal degeneration. *Clinical Orthopaedics and Related Research*, **221**, 26-32.

Vrbová, G., & Wareham, A.C. (1976). Effects of nerve activity on the postsynaptic membrane of skeletal muscle. *Brain Research*, **118**, 371-382.

Wachholder, K., & Altenburger, H. (1926). Beiträge zur Physiologie der willkürlichen Bewegung: X. Mitteilung, Einzelbewegungen. *Pflügers Archiv für die Physiologie*, **214**, 642-661.

Wallin, D., Ekblom, B., Grahn, R., & Nordenborg, T. (1985). Improvement of muscle flexibility: A comparison between two techniques. *American Journal of Sports Medicine*, **13**, 263-268.

Walmsley, B., Hodgson, J.A., & Burke, R.E. (1978). Forces produced by medial gastrocnemius and soleus muscles during locomotion in freely moving cats. *Journal of Neurophysiology*, **41**, 1203-1216.

Warabi, T., Noda, H., & Kato, T. (1986). Effect of aging on sensorimotor functions of eye and hand movement. *Experimental Neurology*, **92**, 686-697.

Ward-Smith, A.J. (1983). The influence of aerodynamic and biomechanical factors on long jump performance. *Journal of Biomechanics*, **16**, 263-268.

Ward-Smith, A.J. (1984). Air resistance and its influence of adverse and favourable winds on sprinting. *Journal of Biomechanics*, **17**, 339-347.

Ward-Smith, A.J. (1985). A mathematical analysis of the influence of adverse and favourable winds on sprinting. *Journal of Biomechanics*, **18**, 351-357.

Warren, G.L., Hayes, D.A., Lowe, D.A., & Armstrong, R.B. (1993). Mechanical factors in the initiation of eccentric contraction-induced injury in rat soleus muscle. *Journal of Physiology (London)*, **464**, 457-475.

Warren, G.L., Hayes, D.A., Lowe, D.A., Prior, B.M., & Armstrong, R.B. (1993). Materials fatigue initiates eccentric contraction-induced injury in rat soleus muscle. *Journal of Physiology (London)*, **464**, 477-489.

Waterman-Storer, C.M. (1991). The cytoskeleton of skeletal muscle: Is it affected by exercise? A brief review. *Medicine and Science in Sports and Exercise*, **23**, 1249-1249.

Watkins, J. (1983). *An introduction to the mechanics of human movement*. Boston: MTP Press Ltd.

Webb, M.R., Hibberd, M.G., Golman, Y.E., & Trentham, D.R. (1986). Oxygen exchange between P_i in the medium and water during ATP hydrolysis mediated by skinned fibers from rabbit skeletal muscle. *Journal of Biological Chemistry*, **261**, 15557-15564.

Weeks, O.I. (1989). Vertebrate skeletal muscle: Power source for locomotion. *BioScience*, **39**, 791-799.

Weis-Fogh, T., & Alexander, R.M. (1977). The sustained power output from striated muscle. In T.J. Pedley (Ed.). *Scale effects in animal locomotion* (pp. 511-525). London: Academic.

West, S.P., Roy, R.R., & Edgerton, V.R. (1986). Fiber type and fiber size of cat ankle, knee and hip extensors and flexors following low thoracic spinal cord transection at an early age. *Experimental Neurology*, **91**, 174-182.

Westerblad, H., & Lännergren, J. (1990). Recovery of fatigued Xenopus muscle fibres is markedly affected by the extracellular tonicity. *Journal of Muscle Research and Cell Motility*, **11**, 147-153.

Westing, S.H., Cresswell, A.G., & Thorstensson, A. (1991). Muscle activation during maximal voluntary eccentric and concentric knee extension. *European Journal of Applied Physiology*, **62**, 104-108.

Westing, S.H., Seger, J.Y., & Thorstensson, A. (1990). Effects of electrical stimulation on eccentric and concentric torque-velocity relationships during knee extension in man. *Acta Physiologica Scandinavica*, **140**, 17-22.

Westling, G., Johansson, R.S., Thomas, C.K., & Bigland-Ritchie, B. (1990). Measurement of contractile and electrical properties of single human thenar motor units in response to intraneural motor axon stimulation. *Journal of Neurophysiology*, **64**, 1331-1338.

Whalen, R.G. (1985). Myosin isoenzymes as molecular markers for muscle physiology. *Journal of Experimental Biology*, **115**, 43-53.

Whalen, R.T., Carter, D.R., & Steele, C.R. (1988). Influence of physical activity on the regulation of bone density. *Journal of Biomechanics*, **21**, 825-837.

Wickiewicz, T.L., Roy, R.R., Powell, P.L., & Edgerton, V.R. (1983). Muscle architecture of the human lower limb. *Clinical Orthopaedics and Related Research*, **179**, 275-283.

Wickiewicz, T.L., Roy, R.R., Powell, P.L., Perrine, J.J., & Edgerton, V.R. (1984). Muscle architecture and force-velocity relationships in humans. *Journal of Applied Physiology*, **57**, 435-443.

Wiegner, A.W. (1987). Mechanism of thixotropic behaviour at relaxed joints in the rat. *Journal of Applied Physiology*, **62**, 1615-1621.

Wiktorsson-Möller, M., Öberg, B., Ekstrand, J., & Gillquist, J. (1983). Effects of warming up, massage, and stretching on range of motion and muscle strength in the lower extremity. *American Journal of Sports Medicine*, **11**, 249-252.

Wilkie, D.R. (1960). Man as a source of mechanical power. *Ergonomics*, **3**, 1-8.

Williams, K.R. (1985). Biomechanics of running. In R.L. Terjung (Ed.), *Exercise and sport sciences reviews* (Vol. 13, pp. 389 441). New York: Macmillan.

Williams, M., & Lissner, H.R. (1962). *Biomechanics of human motion*. Philadelphia: Saunders.

Williams, P.L., & Warwick, R. (Eds.) (1980). *Gray's anatomy* (36th ed.). Edinburgh: Churchill Livingstone.

Williams, R.A., Morrissey, M.C., & Brewster, C.E. (1986). The effect of electrical stimulation on quadriceps strength and thigh circumference in menisectomy patients. *Journal of Orthopaedic and Sports Physical Therapy*, **8**, 143-146.

Wilson, B.D. (1977). Toppling techniques in diving. *Research Quarterly*, **48**, 800-804.

Windhorst, U. (1988). *How brain-like is the spinal cord?* Berlin: Springer-Verlag.

Windhorst, U., Christakos, C.N., Koehler, W., Hamm, T.M., Enoka, R.M., & Stuart, D.G. (1986). Amplitude reduction of motor unit twitches during repetitive activation is accompanied by relative increase of hyperpolarizing membrane potential trajectories in homonymous α-motoneurones. *Brain Research*, **398**, 181-184.

Windhorst, U., Hamm, T.M., & Stuart, D.G. (1989). On the function of muscle and reflex partitioning. *Behavioral and Brain Sciences*, **12**, 629-682.

Winiarski, A.M., Roy, R.R., Alford, E.K., Chiang, P.C., & Edgerton, V.R. (1987). Mechanical properties of rat skeletal muscle after hindlimb suspension. *Experimental Neurology*, **96**, 650-660.

Winter, D.A. (1983). Moments of force and mechanical power in jogging. *Journal of Biomechanics*, **16**, 91-97.

Winter, D.A. (1990). *Biomechanics and motor control of human movement*. (2nd ed.). New York: John Wiley & Sons.

Winter, D.A., Patla, A.E., & Frank, J.S. (1990). Assessment of balance control in humans. *Medical Progress Through Technology*, **16**, 31-51.

Winter, D.A., Sidwall, H.G., & Hobson, D.A. (1974). Measurement and reduction of noise in kinematics of locomotion. *Journal of Biomechanics*, **7**, 157-159.

Winter, D.A., Wells, R.P., & Orr, G.W. (1981). Errors in the use of isokinetic dynamometers. *European Journal of Applied Physiology*, **46**, 397-408.

Winters, J.M. (1990). Hill-based muscle models: A systems engineering perspective. In J.M. Winters & S.L.-Y. Woo (Eds.), *Multiple muscle systems: Biomechanics and movement organization* (pp. 69-93). New York: Springer-Verlag.

Winters, J.M., & Woo, S.L.-Y. (Eds.) (1990). *Multiple muscle systems: Biomechanics and movement organization*. New York: Springer-Verlag.

Woittiez, R.D., Huijing, P.A., & Rozendal, R.H. (1983). Influence of muscle architecture on the length-force diagram of mammalian muscle. *Pflügers Archiv*, **399**, 275-279.

Wolpaw, J.R., & Carp, J.S. (1990). Memory traces in spinal cord. *Trends in Neurosciences*, **13**, 137-142.

Wolpaw, J.R., Herchenroder, P.A., & Carp, J.S. (1993). Operant conditioning of the primate H-reflex: Factors affecting the magnitude of the change. *Experimental Brain Research*, **97**, 31-39.

Wong, T.S., & Booth, F.W. (1990a). Protein metabolism in rat gastrocnemius muscle after stimulated chronic concentric exercise. *Journal of Applied Physiology*, **69**, 1709-1717.

Wong, T.S., & Booth, F.W. (1990b). Protein metabolism in rat tibialis anterior muscle after stimulated chronic eccentric exercise. *Journal of Applied Physiology*, **69**, 1718-1724.

Wood, G.A. (1982). Data smoothing and differentiation procedures in biomechanics. In R.L. Terjung (Ed.), *Exercise and sport sciences reviews* (Vol. 10, pp. 308-362). Philadelphia: Franklin Institute.

Wu, G., & Ladin, Z. (1993). The kinematometer: An integrated kinematic sensor for kinesiologic measurements. *Journal of Biomechanical Engineering*, **115**, 53-62.

Yamaguchi, G.T., Sawa, A.G.U., Moran, D.W., Fessler, M.J., & Winters, J.M. (1990). A survey of human musculotendon actuator parameters. In J.M. Winters & S.L.-Y. Woo (Eds.), *Multiple muscle systems: Biomechanics and movement organization* (pp. 717-773). New York: Springer-Verlag.

Yamashita, N. (1988). EMG activities in mono- and bi-articular thigh muscles in combined hip and knee extension. *European Journal of Applied Physiology*, **58**, 274-277.

Yang, J.F., & Stein, R.B. (1990). Phase-dependent reflex reversal in human leg muscles during walking. *Journal of Neurophysiology*, **63**, 1109-1117.

Yarasheski, K.E., Campbell, J.A., Smith, K., Rennie, M.J., Holloszy, J.O., & Bier, D.M. (1992). Effect of growth hormone and resistance exercise on muscle growth in young men. *American Journal of Physiology*, **262**, E261-E267.

Young, A., Stokes, M., & Iles, J.F. (1987). Effects of joint pathology on muscle. *Clinical Orthopaedics and Related Research*, **219**, 21-27.

Young, J.L., & Mayer, R.F. (1981). Physiological properties and classification of single motor units activated by intramuscular microstimulation in the first dorsal interosseus muscle in man. In J.E. Desmedt (Ed.), *Motor unit types, recruitment and plasticity in health and disease* (pp. 17-25). Basel, Switzerland: Karger.

Young, K., McDonagh, M.J.N., & Davies, C.T.M. (1985). The effects of two forms of isometric training on the mechanical properties of the triceps surae in man. *Pflügers Archiv*, **405**, 384-388.

Yue, G., Alexander, A.L., Laidlaw, D.H., Gmitro, A.F., Unger, E.C., & Enoka, R.M. (1994). Sensitivity of muscle proton spin-spin relaxation time as an index of muscle activation. *Journal of Applied Physiology*, **77**, a84-a92..

Yue, G., & Cole, K.J. (1992). Strength increases from the motor program: A comparison of training with maximal voluntary and imagined muscle contractions. *Journal of Neurophysiology*, **67**, 1114-1123.

Yue, G., Fuglevand, A.J., Nordstrom, M.A., & Enoka, R.M. (1992). Motor unit synchronization assessed from surface EMG increases with discharge rate of reference motor unit. *Society for Neuroscience Abstracts*, **18**, 1406.

Zahalak, G.I. (1990). Modeling muscle mechanics (and energetics). In J.M. Winters & S.L.-Y. Woo (Eds.), *Multiple muscle systems: Biomechanics and movement organization* (pp. 1-23). New York: Springer-Verlg.

Zajac, F.E. (1993). Muscle coordination of movement: A perspective. *Journal of Biomechanics*, **26**, 109-124.

Zajac, F.E., & Gordon, M.E. (1989). Determining muscle's force and action in multi-articular movement. In K.B. Pandolf (Ed.), *Exercise and sport sciences reviews* (Vol. 17, pp. 187-230). Baltimore: Williams & Wilkins.

Zatsiorsky, V., & Seluyanov, V. (1983). The mass and inertia characteristics of the main segments of the human body. In H. Matsui & K. Kobayashi (Eds.), *Biomechanics VIII-B* (pp. 1152-1159). Champaign, IL: Human Kinetics.

Zemková, H., Teisinger, J., Almon, R.R., Vejsada, R., Hník, P., & Vyskocil, F. (1990). Immobilization atrophy and membrane properties in rat skeletal muscle fibres. *Pflügers Archiv*, **416**, 126-129.

Zernicke, R.F., Vailis, A.C., & Salem, G.J. (1990). Biomechanical response of bone to weightlessness. In K.B. Pandolf (Ed.), *Exercise and sport sciences reviews* (Vol. 18, pp. 167-192). Baltimore: Williams & Wilkins.

Zimny, M.L., & Wink, C.S. (1991). Neuroreceptors in the tissues of the knee joint. *Journal of Electromyography and Kinesiology*, **3**, 148-157.

Zinovieff, A.N. (1951). Heavy-resistance exercise: The Oxford Technique. *British Journal of Physical Medicine*, **14**, 159-162.

Zuurbier, C.J., & Huijing, P.A. (1992). Influence of muscle geometry on shortening speed of fibre, aponeurosis and muscle. *Journal of Biomechanics*, **25**, 1017-1026.

Zytnicki, D., Lafleur, J., Horcholle-Bossavit, G., Lamy, F., & Jami, L. (1990). Reduction of Ib autogenetic inhibition in motoneurons during contractions of an ankle extensor muscle in the cat. *Journal of Neurophysiology*, **64**, 1380-1389.

Index

Subject

Author